D1091060

Basic statistics

Basic statistics

DICK A. LEABO
Professor of Statistics
University of Michigan

Fifth Edition 1976

RICHARD D. IRWIN, INC. Homewood, Illinois 60430
Irwin-Dorsey International Arundel, Sussex BN18 9AB
Irwin-Dorsey Limited Georgetown, Ontario L7G 4B3

Fifth Edition

First Printing, April 1976

ISBN 0-256-01835-9
Library of Congress Catalog Card No. 75-35101
Printed in the United States of America

To
Artis and Tom
with gratefulness

and to the I.R.S.
à contre-coeur

Preface

THE BASIC OBJECTIVES of the fifth edition of this text remain virtually the same as those of the four previous editions. Namely, the goal is to present a modern introduction to statistical methods and data analysis for students in business administration and economics whether they are undergraduates or graduates. Further objectives are to help students develop their ability to apply critical judgment in the analysis of business and economic data and to serve as a discipline through emphasis on thinking logically and analytically.

Certain organizational and content changes were made in the fourth edition, and the fifth edition builds upon these changes and modifies and refines the text material. In this latest revision the chapter on sampling theory has been expanded to include an explicit application of the multistage cluster sample. Chapter 5 has been changed so that much of the material previously contained in the appendix is now included in the chapter. The chapter on probability has been expanded to include an illustration of joint probability.

Chapters 8 and 9 have been revised substantially. Much of the material on sampling distributions has been moved to Chapter 9. Also, Chapter 9 has been enlarged to include discussion and applications of the Poisson and exponential distributions. These changes give Chapter 8 a better opportunity to introduce the student to the general theoretical distributions first before discussing in Chapter 9 the more specialized sampling distributions. In addition, some of the sampling concepts relating to sample variation and the central limit theorem previously contained in Chapter 8 are now a vital introductory part of Chapter 9.

Chapter 12 has been revised to reflect the changes being made and contemplated in the Consumers Price Index. Chapters 13 and 14 dealing with time series analysis have been reorganized significantly. Chapter 13 now contains all of the material, both linear and nonlinear, on trends. Chapter 14 now concentrates on seasonal and cyclical variations. Also, the material on exponential smoothing has been expanded to include more examples and more discussion of this important topic.

Chapters 15 and 16 have been improved by the addition and discussion of more computer printout of correlation and regression. Another significant addition to Chapter 16 is the major expansion of the section on the use of dummy variables in regression. An example is developed to help clarify the application of this tool. In addition, the discussion of the analysis of variance tables as a part of the computer output has been strengthened and explained better.

Chapter 17 has been changed to make use of a previous example from Chapter 4 in demonstrating the application of chi-square analysis. This helps coordinate the text in a better way and assists the student to see the relationship of several statistical techniques. Two appendix sections have been eliminated. They are the presentation of research results and the review of some elementary math. It was felt that in order to reduce the size of the fifth edition that these topics can be handled by reference to separate texts dealing with these problems. However, because of the expansion of the text material on the use of the Durbin-Watson statistic, Appendix L was a necessary addition. Also, because of the additions of the material on the Poisson and exponential distributions to Chapter 9, Appendixes H and I are vital additions to the fifth edition.

An important addition in the fourth edition has been the use of the *concept* of flowcharting where appropriate to illustrate the solution to a problem in order to help the student think in a logical and analytical process. Increased emphasis on the role of high-speed digital computing equipment in business and economic research also is a vital part of this edition. Chapter 1 discusses the impact of the computer on data analysis and the way this tool has forced the statistician to think more clearly and precisely about the methods and procedures he plans to use. The computer's value to the research is vividly demonstrated in Chapter 16 where the use of stepwise regression is employed to help develop the model. While the text can be used without employing any computing equipment at all (or perhaps only a desk calculator), computer printouts have been used in some parts as pedagogical tools.

Many of the problems are new and others have been improved by clarification. Problem 16.19 provides the basis for a term paper that can be used to apply many of the techniques learned or those being developed. This project might even be expanded to coordinate it with work in other classes to make statistical analysis more meaningful to the nonquantitative student. Hopefully, he or she can be made to appreciate the interrelationship of statistics and other academic work.

Some instructors may wish to utilize selected articles from the Related Readings provided at the end of each chapter as a means of exposing the student to more advanced work. Others may wish to use these readings as a way of adapting the text to a two-term course. Some instructors may find that by using Chapters 1–10 for the first term and Chapters 11–18 in the second term, their needs are met. Still others might wish to use selected chapters for a one-term course. Backgrounds of students entering a first course in statistical methods usually vary broadly, and in some cases the instructor can correctly assume previous exposure to the descriptive topics contained in Part II.

An understanding of elementary algebra is a prerequisite to achieving an appreciation of the basis of analysis. As someone involved for more than 20 years in the teaching of beginning students of statistics, one vital fact strikes me. That is, the *academic classification* of the student is not the important prerequisite, but rather the level of the mathematical preparation of the individual is the one that should separate the classes. A first-year graduate student with modest training in elementary mathematics cannot effectively handle a rigorous course in statistics any better than an undergraduate with the same preparation. Whereas in some disciplines academic maturity seems to be an important prerequisite for success, in the area of statistics the mathematical preparation is *the* dividing line. This text was written with the idea that elementary algebra is the only mathematical prerequisite needed and that this is the student's *initial* exposure to the subject of statistics, whether he or she be an undergraduate or graduate student.

While I assume full responsibility for any errors of omission or commission, the author and publisher would sincerely appreciate receiving notice of any annoying typographical errors.

March 1976 Dick A. Leabo

Acknowledgments

I AM INDEBTED to Sir Ronald A. Fisher, F.R.S., Cambridge, and Oliver and Boyd Ltd., Edinburgh, and to the Hafner Publishing Company for permission to reprint as Appendixes A, K, and J of this text, Tables III, V–A, and V–B respectively from their book, *Statistical Methods for Research Workers.* Acknowledgment is also made to Sir Ronald A. Fisher, F.R.S., Cambridge, and to Frank Yates, F.R.S., Rothamsted, and to Oliver and Boyd Ltd., Edinburgh, for permission to use as Appendix D of this text, Table III from their book, *Statistical Tables for Biological, Agricultural and Medical Research.* Appreciation is also due Frederick E. Croxton and Dudley H. Cowden for permission to reproduce Chart 26.1 on page 703 of the second edition of *Applied General Statistics.* Appreciation is expressed to the McGraw-Hill Book Co. for permission to reproduce part of Table I of Burington and May, *Handbook of Probability and Statistics* as Appendix E. A word of acknowledgment also is due Harper & Row, Publishers, for their permission to use Tables 1 and 2 of Taro Yamane, *Statistics, An Introductory Analysis* as Tables C.2 and C.3 respectively. Acknowledgment is made to William A. Spurr and Charles P. Bonini and to Richard D. Irwin, Inc. publishers for their permission to reproduce Table H Poisson Distribution and Table I Values of e^{-X} from their text *Statistical Analysis for Business Decisions,* pp. 696–97 and 700. I am grateful also for the permission of the Board of Trustees of *Biometrika* for the opportunity to reproduce as Appendix Table L, The Durbin-Watson Statistic, the table from Vol. 41 (1951), pp. 173 and 175. In other instances of indebtedness, specific acknowledgment is made in the text.

Previous editions benefited from the thoughtful criticism of the late George R. Davies of the University of Iowa, Anthony Constantino, Jerry Alig, Richard Duvall, William S. Peters, James E. Holstein, Lee Cobb, David Walker, and Mildred Massey.

Special acknowledgment is recognized for Frank J. Katusak formerly of the University of Scranton (currently with the National Realtors Association) and to Donald W. Satterfield of Memphis State University for their excellent detailed suggestions and comments that were extremely helpful in the production of the fifth edition of this text. Their recommendations for reorganization of several chapters and the expansion of some topics were appropriate and should make this edition a more useful and readable text. In addition, I wish to express my appreciation to Dean Floyd A. Bond and the University of Michigan for enabling me to complete this project.

D. A. L.

Contents

Part I
The nature of statistical analysis

Part VII
Nonparametric statistics: Data analysis with less restrictive assumptions

Part VIII
Elementary decision making using prior information

Part I

The nature of statistical analysis

Statistical thinking will one day be as necessary for efficient citizenship as the ability to read and write.

H. G. Wells

Are Statistics trite? On the contrary, they are The Poetry of Science.

F. Emerson Andrews

1

Role of statistics in research[1]

Morally, a researcher who uses his professional competence for anything except a disinterested search for truth is guilty of a kind of treachery.

Bertrand Russell

WHAT IS statistics? Just what does the average person associate with the word "statistics"? Perhaps it is the batting average of his or her favorite baseball player; maybe it is the July 4 traffic fatalities; or possibly it is the trite quotation: "There are lies, damned lies, and statistics." Or as a Paris banker stated: "Statistics are like miniskirts. They cover up the essentials but give you ideas." To most people the word means a mass of numerical figures.

In reality, statistics is also an area of mathematics concerned with much more sophisticated techniques than the mere tabulating of sets of numbers. *Mathematics Dictionary* (Van Nostrand, 1966) defines statistics as:

> Statistics, *n*. 1. Methods of obtaining and analyzing quantitative data. The following aspects are applicable only in reference to some phase of the experimental logic of quantitatively measured, variable, multiple phenomena: (*a*) inference from samples to population by means of probability (commonly called statistical inference); (*b*) characterizing and summarizing a given set of data without direct reference to inference (called descriptive statistics); (*c*) methods of obtaining samples for statistical inference (called sampling statistics). 2. Sometimes used in all

[1] For a concise and lucid explanation of the scope of statistics, see Donald G. Watts, ed., *The Future of Statistics* (New York: Academic Press, 1968); and John W. Tukey, "The Technical Tools of Statistics," *American Statistician*, April 1965, pp. 23–28.

three of above senses. 3. A set of quantitatively measured data (obsolete, technically). 4. Plural of statistic.[2]

In the statistical literature the word *statistic* generally is taken to mean an estimate of a population characteristic (parameter). For example, the arithmetic mean of a random sample might be considered the expected value of the true population. Let us bring some of these concepts into sharper focus.

Descriptive statistics: Data analysis without probability

This area concerns itself primarily with providing a useful, clear, and informative *description* of a mass of numerical data. This is done by considering such topics as the collection and processing of original data, tabular and graphic presentation, sources of data, frequency distributions, measures of central position (averages), and measures of dispersion or scatter.[3] In the past, traditional elementary statistics courses were largely directed to the discussion of these techniques.

Inductive statistics: Data analysis with probability

However, if statisticians are restricted to descriptive methods, they are deprived of the use of the powerful tool of the scientific method.[4] Science is based on inductive methods, using combined observations.[5] Inductive reasoning usually is defined as reasoning from the partial to the general. Therefore, inductive statistics, which is concerned with making

---

[2] Professor Oscar Kempthorne prefers the following crisp and short definition: "Statistics deals with the collection and interpretation of data." Professor H. O. Hartley, another well-known authority in the field, seems to be in agreement with the notion that statistics sometimes has been classified as that area of applied mathematics concerned with stochastic (random) phenomena. Hartley sees the contribution of the applied statistician as: (*a*) his ability to understand the problem posed by his fellow scientist and (*b*) the mathematical formulation and solution of that problem. See Watts, *Future of Statistics*, pp. 104–5 and 115–19.

[3] See Chapters 3 through 6.

[4] See "Scientific Method," *Encyclopaedia Britannica*, 14th ed., vol. 20, pp. 127–33. Also see Vernon Clover and Howard Balsley, *Business Research Methods* (Columbus: Grid, Inc., 1974), chap. 2, pp. 15–34.

[5] In discussing experimental science and the role of scientific method in research, Conant states:

"(1) A problem is recognized and an objective formulated;
(2) all the relevant information is collected;
(3) a working hypothesis is formulated;
(4) deductions from the hypothesis are drawn;
(5) the deductions are tested by actual trial;
(6) depending on the outcome, the working hypothesis is accepted, modified, or discarded."

See James B. Conant, *Science and Common Sense*, rev. ed. (New Haven, Conn.: Yale University Press, 1960), p. 50.

estimates, predictions, and generalizations, or reaching decisions based on sample observations or experiments, is a useful tool of analysis of business and economic problems.

Modern statistical methods are based on mathematical probability theory. Probability theory enables us to use partial information (sample data) to produce generalizations relative to the large group from which the sample was properly drawn (population or universe[6]). Statistical theory allows us to make estimates of population parameters (e.g., the arithmetic mean, or standard deviation) based on sample statistics. It also enables the statistician to reach decisions in such a manner that the accuracy of the conclusion may be judged.[7]

By way of contrast, data collected on the deaths per thousand of people by age groups in the United States over a given period may be presented in tabular and graphic form in a summary fashion. *In a descriptive sense, interpretations would not go beyond the data or phenomena at hand.* The inductive statistician would arrange such data in the form of a mortality table. This information would be used to *estimate future deaths,* that is, the probability of death at various ages. Insurance companies base their premiums, indeed their existence, on such inferences. They are not concerned with *who* will die but with an estimate of *how many* will die in each age bracket during the coming year. Descriptive statistics can tell us what we have seen or measured; however, it is powerless to put the information to work for us as a means of prediction or estimation.

More advanced fields of quantitative methods in business and economic research, which would normally be considered to be segments of inductive statistics, include such topics as operations research, linear programming, queueing theory, game theory, quality control, stochastic processes, and others.[8] More and more, management is turning to mathematics as a tool of decision making.[9]

---

[6] If we were to perform an experiment over and over infinitely, the unlimited set of measurements which would evolve might be referred to as the population or universe. Or one might define the population as the total set of items (actual or theoretical) described by some characteristic of the items. Therefore, the totality of *potential* measurements of a product of fixed length is a population; or the totality of golf clubs mass-produced under prescribed conditions is the population of that defined set. If the population is infinitely large, it is called an *infinite production.*

[7] This is not to imply that the inductive problem always is solely one of dealing with sampling error. For example, what is the sampling error of such indicators as the Wholesale Price Index or gross national product? The example which follows in the text also is divorced from the sampling error concept.

[8] Students wishing to become acquainted with these topics may want to consult any one of a number of excellent introductory texts in these areas.

[9] See D. F. Mulvihill, ed., *Guide to the Quantitative Age,* Readings from *Fortune* (New York: Holt, Rinehart & Winston, Inc., 1966); "The New Computerized Age," *Saturday Review,* July 23, 1966; or A. J. Alwan and D. G. Parisi, *Quantitative Methods for Decision Making* (Morristown, N.J.: General Learning Press, 1974).

Significance of inductive reasoning

Why must we rely on the uncertain inferences involved with sampling? That is, why do we assume this risk of making an error by using partial information to draw conclusions about the population? Why not take a census or complete count? We rely on sample data because (1) the events which usually are of concern to us may be too numerous or too inaccessible for a census; (2) we may subject the product to a test which is destructive, that is, the electronic tube manufacturer cannot burn out the entire production run to estimate average life; or (3) in some cases a sample may be more accurate than a census. In addition, in market research the element of timeliness coupled with (1) above makes sampling a necessity.

Almost always, the statistician is interested in the sample from the point of view of learning something about the characteristics of the universe. Because of the significance of sampling and inductive reasoning, the statistician usually begins a study by asking some pertinent questions concerning samples. Some of these questions are discussed in the remainder of this chapter as a method of introducing the student to the field of statistics and to some of the terminology employed. *It is imperative that the student of elementary statistics begin to learn the language and symbols as soon as possible.*

IMPORTANT QUESTIONS ABOUT SAMPLES

Some questions concerning samples and sampling theory which seem significant are: (1) How do we design a sample in order that the information collected is as revealing and dependable as possible? (2) How do we describe the sample data concisely and clearly and still be informative? (3) On the basis of the sample evidence, what generalizations or conclusions can we make relative to the population? (4) What reliability can be associated with these inferences? Answering these four questions easily can occupy the time of the student and instructor in any first course in statistics.

Design of sample

Question one is the basis for volumes of material on the construction of a sample design.[10] Suffice it to say here that the most acceptable method,

---

[10] Excellent texts in this area include Morris H. Hansen, William N. Hurwitz, and William G. Madow, *Sample Survey Methods and Theory* (New York: John Wiley & Sons, Inc., 1953), vol. I: *Methods and Applications,* and vol. II: *Theory;* W. Edwards Deming, *Sample Design in Business Research* (New York: John Wiley & Sons, Inc., 1960); Leslie Kish, *Survey Sampling* (New York: John Wiley & Sons, Inc., 1965);

which permits answering question four, is a probability or random sample. *Defined, a simple random sample is one in which each item in the universe has an equal opportunity of being selected in the sample.* Prejudices or personal factors of choice do not enter into a probability sample. Remember, in most cases the primary objective of a sample is to be as representative of the universe as possible. That is, the sample should be an approximate small-scale replica of the population from which it came. If some items are more likely to be selected than others, the sample is biased. Any bias[11] in a sample may lead to invalid inferences or conclusions. It seems obvious that the most common values in a population are also the items most likely to be selected by a random method. This is true because there are more items with the common value, each with an equal probability of being selected. Chapter 3 is devoted to an elementary discussion of sampling theory and several general types of samples utilized in collecting business and economic data.

Describing the sample

Question two fairly well covers the subject matter contained in Chapters 3–6 and Chapters 12, 13, and 14. That is, how do we summarize the mass of data in order that it is as useful and informative as possible? We may want to organize the data in tabular form, or we may want to use some of the more vivid visual techniques of graphic presentation. Tables, charts, pie diagrams, pictorial graphs, and statistical maps are all ways of summarizing the original data. These devices are useful in highlighting essential characteristics of the data. It may be desirable to estimate a measure of central position, that is, the arithmetic mean, median, or mode. However, an average in and of itself does not tell the whole story. It is necessary to supplement the average with a measure of dispersion or scatter, that is, the standard deviation, the quartile deviation, or maybe just the crude range. We use such measures to summarize data because it is not feasible to describe each individual item.

This means we must group similar measurements together. When

Taro Yamane, *Elementary Sampling Theory* (Englewood Cliffs, N.J.: Prentice-Hall, Inc., 1967); J. B. Lansing and J. N. Morgan, *Economic Survey Methods* (Ann Arbor: Institute for Social Research, University of Michigan, 1971); and D. P. Warwick and C. A. Lininger, *The Sample Survey: Theory and Practice* (New York: McGraw-Hill Book Co., 1975).

[11] Bias is sometimes loosely defined as an error running in one direction. Kahn and Cannell state it as follows: "When errors of measurement result in a systematic piling up of inaccuracies in a single direction—either a consistent overestimation or a consistent underestimation—we refer to the measurement as biased." See Robert L. Kahn and Charles F. Cannell, *The Dynamics of Interviewing* (New York: John Wiley & Sons, Inc., rev. ed., 1967), p. 169. See also Warwick and Lininger, *The Sample Survey*, pp. 43, 75, and 79.

this is done in tabular form, a frequency distribution results. When these data are presented graphically, a frequency distribution curve evolves. In elementary statistics the two most useful theoretical frequency curves are the normal or Gaussian probability curve and the binomial distribution. Both of these are discussed in Chapters 8 and 9. Other theoretical frequency curves which prove helpful in more advanced work are the Student t distribution, the Poisson distribution, and the chi-square distribution.

In attempting to answer question two, the statistician sometimes throws some light, usually rather dim, on question three. However, if we stop here, we limit the function of the analyst.

Inferences from the sample

Only after we move to questions three and four do we reach the heart of modern statistical methods. Given the sample statistics, what conclusions can be reached concerning the nature of the universe? This is the concern of question three and Chapters 9, 10, 11, and 14 through 17. In attempting to shed some light on the problem, we make use of the concept of probability as related to a frequency distribution. Here, we take up such topics as point estimates, confidence intervals, confidence coefficients, testing an hypothesis, power curves, correlation and regression, nonparametric statistics, and Bayesian inference.

Reliability of inferences

Once some conclusions have been reached, it seems reasonable to pose question four: What reliability can be associated with these inferences? Questions three and four are so closely related that Chapters 9, 10, and 11 cover both. Reliability may be considered in terms of probability as indicated by the confidence coefficient, whereas the precision of our estimate is reflected in the width of the confidence interval. The size of the confidence interval depends upon the standard deviation of the sampling distribution, commonly called the standard error. Therefore statisticians usually express their answers in two parts: part one indicates the nature of their best estimate (the confidence interval); part two indicates the degree of reliability we may place in this estimate (confidence coefficient).

COMPENDIUM

Science makes use of two main forms of the logical thought process. *Deductive reasoning*, moving from general axioms and deducing from these assumptions a useful array of implied propositions, is definite and absolute. Sometimes, deductive reasoning is described as going from the general to the specific. *Specific* inferences follow inescapably from the general assumptions or axioms. Much of mathematics relies heavily on this form of logical thinking. *Inductive reasoning* proceeds in the opposite direction from deduction and deals with *uncertain* inferences. Behaving inductively, one begins with the facts of experience and infers general conclusions. Unlike deductive logic, the facts of experience that provide the impetus for inductive reasoning do not unrelentingly lead to categorical conclusions. Instead, they lead to *judgments* concerning the credibility of these conclusions.

Statistics and inductive reasoning

Statistical inference is the art which deals with uncertain inferences. It uses sample evidence to discover information about the larger group (population) from which the sample was selected. Aside from trivial exceptions, the events and phenomena that the economists or business analysts are interested in are too numerous, too extensive, or too inaccessible to permit a complete observation or census. Consequently, sampling theory is of great significance to assist in the rational decision-making process. Just as surely as the physicist uses statistical inference to draw conclusions about molecular speed, so does the economist use statistical inference in making estimates of unemployment on the basis of sample evidence.

The whole area of statistical inference, whether we are concerned with interval estimation problems, tests of hypotheses, or correlation, leans extremely heavily upon the theory of probability. Therefore, Chapters 7 through 11 provide some background to the concept of a probability statement and its interpretation. Although inductive reasoning depends on probability theory for its foundation, it also supplements it. In effect, on the basis of known population parameters, probability theory tells us what characteristics the sample is likely to have. Statistical inference really does the reverse because it utilizes the sample evidence to *estimate* the population characteristics. *We might say that probability theory deduces from the known characteristics of the population the probable characteristics of the sample, whereas statistical inference infers the population contents from the sample observations.*

Schools of thought in terms of probability interpretation

The beginning statistics student should recognize that at the mathematical level there is virtually no disagreement about the foundations of the theory of probability. A firm underpinning in set theory was laid in 1933 by A. Kolmogorov, a Russian mathematician. At the interpretative level, there currently are two extreme views. *The classical school maintains that probability is applicable only to events that can be repeated frequently under the same conditions.* Followers of this theory prefer to be called *objectivists,* and they are content to discuss probabilities in terms of coin tossing or the mass production of a product or similar events that may be repeated over and over. They view probability as a *long-run relative frequency concept,* for example, the probability of a defective mass-produced product being the long-run ratio of the number of defectives to the total number produced. Objectivists will not attempt to make a probability interpretation for unique events. Therefore a large class of problems is categorized by the classical school as not appropriate for a probability interpretation simply because there is no long-run relative frequency in sight. In addition, the classical school likes to make probability applications only for events that can be repeated over and over, and usually refuses to fully utilize prior information.

The second group is the *Bayesian* school,[12] sometimes called the subjectivists or personalists. This increasing legion of followers regards probability as a measure of personal belief in a stated proposition. It assumes that various reasonable, rational individuals may differ in their *degrees of belief,* perhaps, even when offered the same evidence. Consequently, their personal probabilities may vary for the same proposition. Generally, the Bayesian statistician makes use of some additional theory, too.

In this basic text, as in most elementary treatments, the emphasis generally will be on the classical interpretations and applications of probability. This is done deliberately. The author feels that it is impossible properly to understand the concept of probability, and to make valid inferences, without a thorough grounding in this approach. Students are urged to pursue a second course in statistics where time may permit a rigorous treatment of the Bayesian approach.

IMPACT OF THE COMPUTER ON DATA ANALYSIS

The computer revolution in management is in full swing. Its impact upon business and managers has become obvious. For example, in 1955 there were only 10 or 15 computers installed in industry worth approximately $30 million. Ten years later the total had reached almost 31,000

[12] See Robert Schlaifer, *Probability and Statistics for Business Decisions* (New York: McGraw-Hill Book Co., 1959); Robert Schlaifer, *Introduction to Statistics for Business Decisions* (New York: McGraw-Hill Book Co., 1961); Harry V. Roberts, "Bayesian Statistics in Marketing," *Journal of Marketing*, vol. 28, no. 1 (January 1963), pp. 1–4; Paul E. Green, "Bayesian Decision Theory in Pricing Strategy," *Journal of Marketing*, vol. 28, no. 1 (January 1963), pp. 5–14; Leonard J. Savage, "Bayesian Statistics," in Robert E. Machol and Paul Gray, eds., *Recent Developments in Information and Decision Processes* (New York: Macmillan Co., 1962), pp. 161–94; and Howard Raiffa, "Bayesian Decision Theory," in *ibid.*, pp. 92–101.

computers with a combined value of $7.8 billion. Estimates for 1975 *alone* were over 55,000 computers in use at that time with a worth of more than $10 billion. The worldwide population of only U.S.-made computers estimated in 1975 was more than 225,000, of which 150,000 were located in the United States. Clearly, in the immediate future no business firm or research organization will escape the impact of computers.

The impact of the computer upon the statistician or business analyst is equally startling. The *logic* of the computer is impeccable; therefore, to use it in problem solving statisticians are forced to think clearly and precisely about the methods and procedures they plan to use. The computer will not work for the analyst until it is clear in his or her own mind precisely what he or she wants done. From this point of view, the computer has had an influence on the statistician's thinking of how to approach and solve a problem. Basically, the computer, like the old desk calculator, accepts input data, processes the information, and outputs results. The big difference, in terms of business analysts, is that the computer makes it practical to use some of the more high-powered statistical techniques available to them, to process large volumes of data, and to provide the results in short order. However, it should be pointed out that sometimes the computer becomes impractical for solving small problems and the so-called conventional "shortcut" methods of computation remain the better approach. One should also recognize that there has been a subrevolution in the development of desk calculators and that current technology provides rapid solutions to certain problems. It is not always economical and efficient to program a problem for solving by a high-speed digital computer. However, in the area of correlation and regression analyses in particular, the advantages of the computer become immediately obvious even to the beginning student of statistics. Correlation problems that might take 40–50 hours to solve "by hand" easily can be handled by the computer in a matter of several minutes. It should be pointed out that a certain investment of capital (in the form of effort and time) is required to (*a*) develop or adapt the necessary computer software and (*b*) prepare the data for computer consumption. Frequently, these two requirements consume much time and effort; but if the program is reused or already available, the savings are significant.

Where feasible in this text, the computations are done in detail to illustrate to the beginning student the basic procedures in order that he or she may better appreciate the computer output. In other cases the problem initially has been reduced for us by the use of an IBM 360 Model 67. As a pedagogical device, an algorithm or method of solution will be presented where appropriate for some of the illustrations. This graphic tool for presenting a logical approach to a problem is called *flowcharting*. It is not the purpose of this text to teach flowcharting but rather our aim is to begin to help the student *to think in logical terms in an analytical way.*[13] The emphasis then is on the statistical method with computer assistance when required. Flowcharting is a graphic method of formulating solutions to problems in such a manner that they may be readily prepared for computer assistance within the restrictions such help imposes.

That the computer has had an impact upon the student in terms of the student's classroom instruction and research is evident in the type of computation facilities the typical university has available. At the University of Michigan a pair of IBM 360 Model 67 computers has been in use operating in multiprocessing, multiprogramming mode. Students have access to time-sharing to the extent that some 100–125 remote terminals are in use at present. On-line batch processing of programs is permitted in most systems of this type. Other "hardware" supporting the Model 67s usually include card-read-punch units, high-speed line printers, disk files, magnetic tape drives, data cells, and numerous teletypewriters. More experienced computer users have available a CalComp 780/763 continuous plotter. Ideally, students should have a one-semester (three-hour) course in computers and computer programming prior to, or concurrent with, a first course in statistics. Individuals who deprive themselves of this assistance will be limited in their application of statistical methods in any data analysis their work or research requires them to perform.

ROLE OF STATISTICS IN RESEARCH

There are probably as many definitions of research as there are researchers. However, most definitions would include the objective of the discovery of new facts and their *correct interpretation.* Research also aims at modifying, revising, or verifying accepted theories or conclusions in the light of new information. To most academicians, the term research implies the continued search for knowledge or the practical applications of scientific conclusions.

Scientific reasoning essentially is composed of two elements: (*a*) *observation,* by which is meant the collection of new facts through various methods; and (*b*) *logical reasoning,* by which the meaning of these facts is ascertained. The methods of statistics contribute greatly to these two elements. For example, experimental and survey investigations invariably lead to the use of statistical techniques. When properly used, statistics is an integral compound of the scientific method of research and allows for more efficient, critical, and exhaustive inquiry. Because statistics is a valuable tool for most research, the techniques sometimes get overworked and at times provide a weak justification for poorly designed or badly

[13] Students interested in developing their flowcharting skills are referred to Thomas J. Schriber, *Fundamentals of Flowcharting* (New York: John Wiley & Sons, Inc., 1969).

executed research. At times, what this author refers to as Gresham's "law of statistics" seems to operate.[14] That is, insufficient attention is paid to the basic assumptions required for the valid application of certain statistical methods and the "bad statistics drives out the good." Unfortunately, there is a wide gap between statistical theory and the application in business of some of the conclusions of good research. Perhaps this gap is narrowing, but because of it statistics sometimes is referred to as a science and at other times as an art. Theoretical statisticians look upon their work as a science because the methods are logical and systematic with broad applications. The applied statistician views his or her work as an art because successful application of the methods depends on the skill, special experience, and imagination of the individual. *The interpretative function of data analysis is one of the most important contributions of the applied statistician.* It is with this concept in mind that the scope and approach of this text is directed.

OBJECTIVES OF TEXT

Important objectives of this text are (*a*) to develop the student's ability to apply critical judgment in the analysis of business and economic data and (*b*) to use the methods of statistics as a discipline through emphasis on thinking logically and analytically. Basically, this text is concerned with developing techniques and tools to shed some light on the previously mentioned four questions. By focusing attention on statistical inference and these questions concerning samples, *it is not implied that all research is based on partial information.* Frequently, data already published are employed, and in such instances a knowledge of sources is mandatory. Chapter 2 deals with the art of locating published information for business and economic research projects. However, some of the problems of the collection of original data are discussed in Chapter 3. In addition to the subjects mentioned in connection with the four questions, index numbers, time series, simple, multiple, and partial correlation and regression analysis, analysis of variance, and chi-square analysis are discussed in Chapters 12–17. Concluding the text of each chapter is a list of symbols used in the chapter.

PROBLEMS

1.1. What is research? Should academic research complement instruction? In what way?

---

[14] Economists refer to Gresham's theorem which briefly stated holds that less valuable money drives the good out of circulation. The law is named after a British financier, Sir Thomas Gresham, who originated the concept.

1.2. What is the meaning of "data analysis without probability"? What is the meaning of "data analysis with probability"?
1.3. Distinguish between inductive and deductive reasoning.
1.4. What is a random sample?
1.5. In what general way are probability theory and statistical inference related?
1.6. In terms of a probability interpretation, identify and distinguish between the two major schools of thought.
1.7. Has the computer changed the statistician's approach to problem solving? In what way?
1.8. What is an algorithm? How does it relate to a flowchart?

RELATED READINGS

ACKOFF, R. L. "Management Misinformation Systems." *Management Science,* vol. 14, no. 4 (December 1967), pp. B147–56.

AMERICAN STATISTICAL ASSOCIATION. *Careers in Statistics.* Washington, D.C.: ASA, 806 15th Street, N.W., 20005, undated.

BERNSTEIN, LEONARD A. *Statistics for the Executive.* New York: Hawthorn Books, Inc., 1970.

CHOU, VICTOR. "The Use of Gaming Models in Company Planning." *Business Economics,* vol. 1 (Winter 1965–66), pp. 62–64.

DEDERICK, ROBERT G. "National Association of Business Economists and the Business Forecaster." *Business Economics,* vol. 10, no. 1 (January 1975), pp. 7–11.

EVANS, M. K. "A Study of Industry Investment Decisions." *Review of Economics and Statistics,* vol. 49 (May 1967), pp. 151–64.

FRANCIS, I.; HEIBERGER, R. M.; and VELLEMAN, P. F. "Criteria and Considerations in the Evaluation of Statistical Program Packages." *The American Statistician,* vol. 29, no. 1 (February 1975), pp. 52–56.

GODFREY, JAMES T., and SPIVEY, W. ALLEN. *Models for Cash Flow Estimation in Capital Budgeting.* Working Paper No. 17. Ann Arbor: Division of Research, University of Michigan, 1970.

GRAYSON, LESLIE E. "The Economist in Business." *Business Economics,* vol. 4 (May 1969), pp. 53–56.

HERTZ, D. B. *New Power for Management.* New York: McGraw-Hill Book Co., 1969.

KAHLE, T. V. "Mathematical Techniques in Corporate Planning." *Business Economics,* vol. 1 (Winter 1965–66), pp. 58–61.

MILLER, DAVID W., and STARR, MARTIN K. *Executive Decisions and Operations Research.* 2d ed. Englewood Cliffs, N.J.: Prentice-Hall, Inc., 1969.

NAIMAN, ARNOLD; ROSENFELD, ROBERT; and ZIRKEL, GENE. *Understanding Statistics.* New York: McGraw-Hill Book Co., 1972.

REICHARD, ROBERT S. *The Numbers Game: Uses and Abuses of Managerial Statistics.* New York: McGraw-Hill, Inc., 1972.

SALTZMAN, SIDNEY. "An Econometric Model of a Firm." *Review of Economics and Statistics.* vol. 49 (August 1967), pp. 332–41.

SELVIN, HANAN C. "Data-Dredging Procedures in Survey Analysis." *The American Statistician,* June 1966, pp. 20–23.

SPIVEY, W. ALLEN. *Optimization in Complex Management Systems.* Working Paper No. 18. Ann Arbor: Division of Research, University of Michigan, 1970.

STOKES, CHARLES J. "Linear Programming and All That." *Business Economics,* vol. 3 (Fall 1967), pp. 48–54.

TUKEY, JOHN W. "Is Statistics a Computing Science?" *The Future of Statistics.* New York Academic Press, 1968, pp. 19–38.

WHITE, JOHN R. "The Use of Input-Output Economics in Corporate Planning." *Business Economics,* vol. 4 (September 1969), pp. 19–38.

2

Basic sources of data

If we could first know where we are and whither we are tending, we could better judge what to do and how to do it.

Abraham Lincoln

ALL SERIOUS research projects begin by defining the problem and stating clearly the purpose of the investigation. Once this is accomplished, the researcher is in a position to plan the data collection. The problems involved in the collection of original data are discussed in Chapter 3. At this point, our objective is to suggest the various types of available *published* information and to describe the nature of some of the manifold public and private sources.

Definitions

Generally, business and economic data are classified as being either (1) primary or secondary, (2) internal or external.

Primary sources. All data are considered primary to the investigator or organization that collects, presents, and interprets such information. Usually, the original data are published in some form by the collecting agency. For example, the Bureau of Labor Statistics (BLS), U.S. Department of Labor, collects, combines, presents, and interprets the basic data comprising the Consumer Price Index (CPI). The BLS also publishes the CPI as primary data in its *Monthly Labor Review*. The same is true for the Index of Industrial Production as prepared and released by the Federal Reserve Board in the *Federal Reserve Bulletin*. The Business Failure Index compiled by Dun & Bradstreet, Inc., as presented in *Dun's Review & Modern Industry*, is another example of primary data.

A primary source is preferred to a secondary source because it usually contains information on (1) how the data are collected, (2) when and where the data are gathered, (3) some possible uses of the data, and (4) the limitations.

Secondary sources. The Consumer Price Index is presented in such diverse periodicals as *Business Week*[1] *Economic Indicators,*[2] *Survey of Current Business,*[3] and in most daily and weekly newspapers. As given in such sources, the CPI would be considered to be presented by a secondary source. Secondary data are, in effect, *republished* information. Such data generally are characterized by their absence of explanations.

Internal data. Basic data that come from within a firm and relate to the organization's operation are called internal. Generally, the accounting and sales departments of a firm are the most fruitful fountains of internal information. Internal data of a selected nature may be found in *Automotive Industries*[4] and *Moody's Manual of Investments, Domestic and Foreign.*[5]

External data. Information drawn from outside the firm or organization, such as industry data when used by a given firm, is external data. Data relating to the economy as a whole also would be so classified. Usually, a firm or organization will utilize external data as bench marks or averages to make comparisons of its own operations.

Public sources of data

Federal government. Undoubtedly, the federal government is by far the largest single publisher of business and economic information. Such agencies as the U.S. Department of Labor and the U.S. Department of Commerce, with their multitude of bureaus, are excellent sources of bench-mark data. Much of the information collected by these agencies and the Bureau of the Census is too expensive to gather on a private basis. Yet these data are vital in making business and government decisions with social ramifications.[6] Naturally, only a very minor part of the volume

---

[1] Published by McGraw-Hill Book Co., New York.

[2] Published by the U.S. Government Printing Office, Washington, D.C. 20402.

[3] Published by the Office of Business Economics, U.S. Department of Commerce.

[4] Published by Chilton Co., Philadelphia, Pennsylvania.

[5] Published by Moody's Investors Service, New York, New York.

[6] Mr. Arthur C. Nielsen, Jr., president of one of the nation's largest private research organizations, stated the following:

> "The Census exists for the good of all . . . for the good of business, whose progress and profit depend on responsiveness to the constantly changing identities of markets.
> For the good of government—which first must organize itself, and then proceed to deal with some of the most pressing social problems in the history of our republic.
> For the good of the man on the street who simply wants to live in an area possessing adequate business services, adequate schools and a host of adequate other things . . . the adequacy of which is determined by the census."

From "The U.S. Census: Mirror of Our Times," an address to the American Management Association Special Census Seminar, New York City, June 23, 1969. Published by A. C. Nielsen Company, 2101 Howard Street, Chicago, Illinois 60645.

of releases by the federal government can be mentioned here. However, it should be noted that the various censuses, for example, business, manufacturers, and population, are frequently extremely useful in the preliminary stages of many research projects.

The Office of Business Economics of the U.S. Department of Commerce publishes the *Survey of Current Business* monthly. This bulletin contains three sections. The first part deals with "The Business Situation" and includes a discussion of business activity for the month; it usually ventures a modest forecast regarding business conditions. In addition, this lead article attempts to pinpoint the underlying factors affecting economic trends. The second section deals with what are termed "Special Articles" devoted to business subjects of broad interest or studies of particular industries. This material generally is treated in rather factual form, utilizing tabular and graphic presentation. The third section of this monthly is strictly statistical in nature. That is, data are reported in tabular form, for example, indexes of prices, construction, and employment; or data relating to finance, hours, earnings, consumer credit, and many other useful economic indicators are provided. Weekly supplements keep the monthly data up to date and are included in the subscription price. More recently, this agency began publishing a new document, *Commerce Today.* This new periodical gives timely biweekly reports of new programs, achievements, and services of all of the Department of Commerce's 20 agencies.

The December 1970 issue of the *Survey* presents some supplementary historical statistics for the national income and product accounts. Current issues may be used to keep the series up to date. Included in this supplement are the following statistical series showing percent change from the preceding period: (1) GNP in current prices, (2) GNP in constant prices, (3) GNP implicit price deflator, (4) gross private product in current prices, (5) gross private product in constant prices, (6) gross private product implicit price deflator, (7) the implicit price deflator for final sales of goods and services, and (8) the personal saving rate. Annual data are presented beginning with 1929, and quarterly data, seasonally adjusted, beginning with 1947. Regular publication of the saving rate in the national income and product accounts were first published in the July 1968 *Survey* for annual data and in the February 1969 issue for quarterly data. Regular publication of the other series mentioned was initiated with the July 1970 issue of the *Survey.*

Biennial "Supplements" to the *Survey,* called *Business Statistics,* combine under one cover the material contained in the statistical section. Another useful publication is the periodic supplement entitled *National Income.* This volume contains a comprehensive description of the nation's economic accounts. In addition to providing an historical review of the basic trends in the economy since 1929, the 1954 edition also contains (1) a detailed description of the conceptual framework of the national income accounts; (2) a discussion of the sources and methods of estimating national income data; (3) gross national product in constant dollars from 1929 to 1953; and (4) a large statistical section providing annual, quarterly, and monthly data. Later issues keep the data current.

U.S. Income and Output may be considered a supplement to the *Survey of Current Business* and introduces an enlarged set of national income statistics. It features a comprehensive account of postwar economic developments. Also, the development of the national income measures is traced and presented with data sources and procedures employed in preparing the estimates.

Periodic "Supplements" consolidate the monthly information into separate volumes. For example, *Personal Income by States since 1929* presents a comprehensive account of the widely used state personal income series. Data are presented by state, by source, and by industry from 1929 to 1955. This supplement contains an analysis of geographic income changes as well as being the best source of procedures and definitions used in preparing the estimates. Usually, the August or September issue of the *Survey of Current Business* presents comparable data for the previous year and any revisions, thereby keeping the supplement up to date. Another excellent supplement to the *Survey of Current Business* is *The National Income and Product Accounts of the United States, 1929–1965.*

The Board of Governors of the Federal Reserve System publishes the *Federal Reserve Bulletin* monthly. As indicated above, the *Bulletin* is the primary source for the Index of Industrial Production, but this publication also is probably the best *single* source of financial information. Marketing data such as department store sales and inventories, consumer credit, collections, and accounts receivable are provided, most of them by Federal Reserve districts.

Two publications by the U.S. Department of Commerce that are useful sources of data for business are the *U.S. Industrial Outlook* (published annually) prepared by the Office of Domestic and International Business Administration, and *Social Indicators, 1973.* The latter is prepared by the Statistical Policy Division of the Office of Management and the Budget but published by the Social and Economics Statistics Administration of Commerce. The most recent (1975) issue of *Outlook* contains estimates and projections for 200 major industries to 1980. *Social Indicators* contains statistics on life expectancy, disability, medical care, safety of life and property from crime, education, employment opportunities, quality of employment, levels of income, distribution of income, expenditure of income, housing, leisure and recreation, and population.

The Bureau of Labor Statistics (BLS) publishes the *Monthly Labor Review,* which has two major sections. Section one contains articles dealing with labor-management problems, and the other part is entitled "Current Labor Statistics." This, of course, is the primary source of the

Consumer Price Index, but it also provides data on labor turnover, hours, earnings, and employment trends in selected industries and areas. *Patterns of U.S. Economic Growth,* Bulletin 1672, is a publication of the BLS that presents projections of employment by industry for 1980, based on projections of the labor force, potential GNP, and industry output and output per man-hour. Each of the elements in the sequence of projections is discussed in considerable detail. These 1980 projections are part of a coordinated program of the BLS in the field of manpower projections. A summary of this document is presented in *The U.S. Economy in 1980,* Bulletin 1673. A related publication is *Seasonality and Manpower in Construction,* BLS Bulletin 1642. *Employment and Earnings Statistics for States and Areas,* a biennial publication of the BLS provides historical data on a geographical basis. Included are annual average data on industry employment, hours, and earnings for each of the 50 states and 146 major areas. This publication groups together all information available for a single state or area. Current state and area statistics are published monthly by appropriate state agencies and in the Bureau of Labor Statistics monthly report, *Employment and Earnings.* Comparable historical national industry data are available in *Earnings Statistics for the United States.* The *Handbook of Labor Statistics 1973* recently was published by the BLS and is a useful document providing historical series that bear on labor economics and labor institutions.

Another excellent source of aggregate economic data is *Economic Indicators,* prepared for the Joint Economic Committee by the Council of Economic Advisers. Data on the nation's output, income and spending, employment, wages, production activity, prices, currency, credit and security markets, and federal finance are given in tabular and graphic form by months or quarters and by years, usually since 1939. Recently, the Bureau of the Census released *Long Term Economic Growth, 1860–1970,* which is a good source of aggregate data and growth trends of the economy.

A valuable document which summarizes the economic status of the nation is the *Economic Report of the President.*[7] This annual statement presents an economic review of the previous year, discusses current international developments, and usually outlines the administration's legislative proposals for consideration by the new Congress. This publication is a useful guide to assist management in attempting to assess the role and possible impact of the federal government upon business activity. The statistical appendix also proves helpful because of the number of series reported. Broad subject matters covered, which are broken down in great detail, include national income or expenditure, employment and wages, production and business activity, prices, money supply, credit and finance,

[7] Published by the U.S. Government Printing Office, Washington, D.C. 20402.

government finance, corporate profits and finance, agriculture, and international transactions.

Business Conditions Digest, published monthly by the Bureau of the Census, brings together the many economic time series found most useful by business economists. Many of the indicators contained in this report emphasize the cyclical indicators approach to the analysis of business conditions based to a large extent on the list of leading, roughly coincident, and lagging time series maintained by the National Bureau of Economic Research (NBER). The NBER indicators are discussed in greater detail in Chapter 13.

The federal government publishes annually a counterpart of the National Industrial Conference Board's *Economic Almanac.* It is called the *Statistical Abstract of the United States* and appears about two months after the close of the calendar year. The *Abstract* is a good assist for historical series and is also of some use in locating original sources of data by referring to the source note of the appropriate table.

Although there are many other publications of the federal government which may be of interest to a specific industry,[8] the above-mentioned sources are intended only as an introduction to some of the more widely used documents. The student attempting to locate specific data should consult such references as *United States Government Publications, Monthly Catalog;*[9] *List of Available Publications of the Department of Commerce;* and *Census Publications, Catalog and Subject Guide.* However, finding the appropriate government document is an art! Most library card catalogs classify them by issuing department, which in many cases is no help at all! A description of the agencies and suggestion of areas in which they do research can be found in *The United States Government Organization Manual.* Once the proper issuing department is located, the card catalog may be used in the same manner as an author's index.

Another useful document is the monthly *Statistical Reporter* prepared by the Office of Statistical Standards. It is prepared primarily for the interchange of information among governmental employees but also reports on current developments in many areas.

State Government. Although no two states have exactly the same list of sources of useful data, there is a degree of similarity which allows a general discussion. The sources cited below are available in most states.

The *tax commissions* in many states publish data on retail sales or gross receipts taxes. These data generally are broken down by source

---

[8] For example, *Construction Review* and *Construction Statistics, 1950–1964,* a supplement to the *Review* (U.S. Government Printing Office, Washington, D.C. 20402).

[9] The December number of each year contains a cumulative index by subject matter. All documents are serially arranged in the monthly issues for the year, and each subject reference shows a serial number.

and by kind of business, sometimes even by counties or cities. Many state laws require such organizations to publish annual reports; and in some instances, quarterly reports are made. In many cases the commissions are willing to supply tabulations on almost any type of tax collected. Such information has considerable application in the areas of economics, marketing, and market research.

Another excellent source of data at the state level is the *employment security commissions.* These agencies have a wealth of information on such topics as number of unemployment compensation claims, employment, wages, and hours, all of which are usually given by industry and frequently by county. Special tabulations are readily obtained because much of the data is kept on punch cards or tapes. In addition, such state offices publish regular reports as well as a monthly *labor review.*

Data on freight movements and tariffs by commodities and by points of origin and termination are found in the annual reports and occasional pamphlets of the various state *commerce commissions.* Data of this type can be effectively utilized in an analysis of the economy of a state or region. The flow of commodities produced, as well as the sources of manufacturing raw materials and other products consumed but not produced within a region, can be fairly accurately measured.

In areas where agriculture is an important segment of the state's economy, the state *department of agriculture* may provide data on crops, livestock production, and many times on inventories. Most of this information is found in monthly and annual reports of these departments.

Most states have agencies that also publish data on housing, population, vital statistics, and many other economic subjects.

At least 45 states have a *bureau of business and economic research* as an integral department of a state university or private institution, for example, Houston, Syracuse, and Temple. Such agencies usually are engaged in broad economic research and are helpful in locating other sources of data.

Private sources of data

An accurate roll of private publications is practically infinite; therefore, it is impossible to complete the subject here. A few of these private publications, however, may be listed as examples, with no intention of providing an exhaustive list.

The *National Industrial Conference Board* (NICB) publishes widely used magazines, pamphlets, and research monographs. The *Business Management Record,* a monthly, is the private counterpart of the *Survey of Current Business.* This magazine includes such data as the NICB Consumers' Price Index, inventory changes, wages, production, distribution, commodity prices, and other economic series. The *Record* also contains brief articles on current economic topics. The NICB also publishes the *Economic Almanac.* This private agency carries on research projects and publishes the results of such efforts in *Studies in Personnel Policy and Management Memoranda.*

Dun & Bradstreet, Inc., New York, publishes *Dun's Review & Modern Industry.* Probably the best-known economic barometer of this organization is its data on business failures by industries which are combined into a "Failure Index." The *Review* also contains a section on the "Trend of Business" and articles written by staff members and outside contributors.

The F. W. Dodge Corporation is a leading source of data for the construction industry. Data which this firm collects include number of building projects and valuation and floor space by residential, nonresidential, public buildings, and public works.

For financial data, *Standard & Poor's Trade and Securities Statistics,* used with the monthly supplements, is one of the best private sources available. The same organization publishes a *Bond Guide* on a semimonthly basis which contains information on a large number of corporate, municipal, and real estate bonds and stocks. This booklet contains such data as 12-year high and low prices, 5-year earnings, and bid-and-asked prices for many securities.

Moody's Investors' Service, New York, publishes in the finance area. Three of its sources are *Moody's Dividend Record,* a semiweekly record of dividends and dates of payment; *Moody's Bond Record,* a semimonthly which provides quotations on several thousand corporate, government, and municipal bonds; and *Moody's Manual of Investments, Domestic and Foreign,* an annual publication. The latter contains five-year comparisons of income and balance sheets for large companies and two-year comparisons for smaller firms.

The *American Statistics Index,* a useful reference tool on the publications of the federal government, was initiated in 1973 by the Congressional Information Service (CIS) of Washington, D.C. This is sort of a central catalog of data available in government agencies. It indexes the output of over 100 agencies, congressional committees, and statistics-producing programs.

Other sources

It is possible to assemble a comprehensive bibliography of books on most subjects by first using the current issues of the *Publishers' Weekly* and then going back to the last monthly issue of the *Cumulative Book Index.*[10] Finally, the researcher would proceed from there to the earlier issues of the latter index. One drawback to the *Publishers' Weekly* is that

[10] Formerly known as the *United States Catalogue.*

it lists only by authors; however, the advantage of beginning here is that this source may well contain references that have not yet reached the library shelf or card catalog.

The *Cumulative Book Index* is available in most libraries; however, because librarians use it more frequently than researchers, one usually must request permission to see it. This source contains a complete list of all books in print in the English language. This index is most helpful for locating information on fringe areas and for determining whether or not a book exists on the subject.

All research projects require an awareness of current articles on the subject under analysis. Fortunately, we need not check every periodical in our quest for information. Although there are many indexes of articles, the two most commonly consulted are *Readers' Guide to Periodical Literature,* 1900 to date, and the *Business Periodicals Index.* Both of these indexes list articles by author and by subject, patterned after those found in most library card catalogs. The *Applied Science and Technology Index* is a subject index to at least 225 periodicals in such fields as aerodynamics, automation, ceramics, chemistry, construction, electricity, food technology, and telecommunications. *The Engineering Index* "indexes and annotates selectively, on the basis of engineering significance, the available current technical periodicals received by and permanently housed in the Engineering Societies Library," in its description. Two indexes to newspapers that might be useful are *The New York Times Index* and *The Wall Street Journal Index.* The former carries more items than any other newspaper index. The latter is an important index, too. It is compiled from the final Eastern Edition of the *Journal* with the first section indexing corporate or business news and the second containing general news. One will usually discover that the supply of material on almost any topic is not only abundant but generally distributed through a dozen or more journals. Consequently, one of these research aids is a "must."

Some professional journals which are useful for describing the current methodology and for reporting on significant research in the field include the *American Economic Review,* the *Journal of the American Statistical Association,* the *Journal of Marketing,* the *Review of Economics and Statistics, Journal of Marketing Research, The Bell Journal of Economics and Management Science, Business Economics,* and the *Journal of Finance.* In addition, the *International Journal of Abstracts: Statistical Theory and Method* provides summaries of all articles published in the *Annals of Mathematical Statistics, Biometrika,* the *Journal of the Royal Statistical Society,* the *Bulletin of Mathematical Statistics,* and the *Annals, Institute of Statistical Mathematics.* The *International Journal of Abstracts on Statistical Methods in Industry* is a similar but complementary publication.

One final word: It is recommended that the researcher also consult the following publications:

Almanac of Business and Industrial Financial Ratios. 1975 ed. Englewood Cliffs, N.J.: Prentice-Hall, Inc., 1975. (Published annually.)

COMAN, E. T. *Sources of Business Information.* Rev. ed. New York: Prentice-Hall, Inc. 1964.

HASS, J. E.; MITCHELL, E. J.; and STONE, B. K. *Financing the Energy Industry.* Cambridge, Mass.: Ballinger Publishing Co., 1974.

JOHNSON, H. W., and MCFARLAND, S. W. *How to Use a Business Library.* Cincinnati: South-Western Publishing Co., 1964.

MORTON, J. E. "A Student's Guide to American Federal Government Statistics." *Journal of Economic Literature,* vol. 10, no. 2 (June 1972), pp. 371–97.

NATIONAL REFERRAL CENTER FOR SCIENCE AND TECHNOLOGY. *A Directory of Information Resources in the United States.* Social Sciences, Library of Congress, October 1967. Washington, D.C.: U.S. Government Printing Office.

SILK, LEONARD S., and CURLEY, M. LOUISE. *A Primer on Business Forecasting, with a Guide to Sources of Business Data.* New York: Random House, 1970.

TAX FOUNDATION, INC. *Facts and Figures on Government Finance.* 18th biennial ed. New York, 1975.

U.S. DEPARTMENT OF COMMERCE, BUREAU OF THE CENSUS. *County and City Data Book.* A Statistical Abstract Supplement. Washington, D.C.: U.S. Government Printing Office.

———. *County and City Data Book, 1972.*

WASSERMAN, PAUL; ALLEN, ELEANOR; KRUZAS, ANTHONY; and GEORGI, CHARLOTTE. *Statistics Sources.* 4th ed. Detroit: Gale Research Co., 1971.

3

Introduction to sampling theory

The most difficult portion of any inquiry is its initiation.

F. S. C. Northrop

THE COLLECTION of data by organizations, bureaus, and agencies such as those discussed in Chapter 2 is a formidable problem. Sometimes the data are collected by censuses; but more commonly, the organization collecting data uses some kind of sampling method. The problems of obtaining interviews, constructing questions, providing for tabulation, and editing data are present in both methods. In general, these problems, other than tabulation, are best resolved by experts in the field and have no place in a discussion of elementary statistics. Primarily, this chapter is devoted to gathering data by sampling, even though many of the same problems discussed here are present when a census is taken.

SAMPLING CONCEPTS

A definition of sampling

Sampling implies the selection of a few items from a given group to be investigated in such a way as to secure data on the basis of which reasonable conclusions can be drawn regarding the entire mass. Thus a few drops of cream tested for butterfat content represent a sample selected to reveal the butterfat content of a large container of cream. On the same principle, a small number of price changes sometimes is used to represent change in a price level, or unemployment is estimated on the basis of sampling of a relatively few households. Public opinion polls use sample evidence to gain an understanding of the group's attitude. Quality control experts inspect a few items as they come off the production line in order to insure good products. Indeed, the sampling process is a large part of our everyday life. For example, the person who takes the first cautious sip of a hot cup of coffee is "sampling" in order to "estimate" the temperature.

It is not intended to imply that *all* samples are small, that is, less than 30. Indeed, some designs of national probability samples frequently contain several thousand observations. However, even with a sample of this size it is "relatively small" given the number of items in the population. In the past, a well-known public opinion poll interviewed less than 2,000 voting-age individuals to estimate how 70 million people might vote in a presidential election. The size of the sample relative to the population is important but it is not the most vital thing. What is important is the *method* of selection of the observations to be included in the sample.

Nature of a population or universe

The group which is sampled is referred to as the *population* or *universe*. Obviously, this population may be made up of people, prices, production records, interest rates, wages, inventories, accounts receivable, or any similar quantitative measure. The problem being investigated defines the population. In many instances, for theoretical considerations, a statistical universe is regarded as infinite rather than finite as the above populations. *In this text, unless otherwise specified, the population is considered to be infinite.*

One very real problem encountered in sampling has to do with the definition of the universe. Sometimes, it may be easy to define the universe, for example, all eligible voters in a given forthcoming election; whereas in other cases, it may be a real task. Once the population is defined, then we must be sure that we draw the sample from the universe as we have defined it. Here, we encounter the concepts of the *target population* and *sampled population.* We hope these are the same! In effect, the target population is the one we have carefully defined and the one we *intend* to sample. The analyst must always bear in mind that although we may apply our statistical inference procedures to draw conclusions about the sampled population, we cannot assume that these statements validly apply to the target population unless the two universes are identical.

For example, the political polls sometimes imply inferences concerning a given election. These are based on a sample of eligible voters, that is, the sampled population. However, the implication frequently is made that these conclusions will apply equally to the target population, that is, those eligible voters who *actually* vote. The latter may be a much smaller universe and even unlike the sampled population in some important respects.

An excellent illustration relates to the much cited "sample" of the now defunct *Literary Digest*. During the Franklin D. Roosevelt–Alfred Landon presidential campaign, the *Digest* wished to forecast the outcome of this election. Reports indicated that approximately 2,000,000 opinions were collected, primarily through mail (postcard) techniques using *Literary Digest* mailing lists. The target population was clear: those eligible voters in the 1936 presidential election. The actual sampled population turned out to be something else. Obviously, there existed a heavy Republican bias in the *Digest* "sample." Equally obvious is the fact that the sample did not lack size!

The *Literary Digest* boldly announced that the election would result in a landslide victory! So far so good. Their only mistake was that they predicted this overwhelming victory for Landon! At least they were half right.

Part of the difficulty was in the method of sample selection. The *Literary Digest* apparently relied heavily upon its readers. Also, this group must have been more conservative politically than the population in general. Perhaps, the *Digest* conclusions were appropriate for the *sampled* population; however, they did not reflect the views of the *target* population. Unfortunately, their prediction related to the latter population.

Principle of statistical regularity

All sampling practice is dependent upon the phenomenon described by the principle of *statistical regularity*. This principle is stated thus: *A sample properly drawn from a population will within certain variable limits reveal the characteristics of that population, even though the number of items in the sample is small as compared with the number of items in the population.* There is a key word in the above statement. The word "properly" is the center of the problem of how to sample so that the principle functions. The discussion of methods of sampling which follows is designed to evaluate the various methods in the light of being *properly* drawn.

Stability of mass data

The *principle of statistical regularity* defined above springs from a condition ordinarily referred to as *the stability of mass data*. A group of items frequently reveals certain stable characteristics, and statistical method is concerned with the measurement of these stable characteristics. For example, the highly useful business of life insurance is dependent upon estimating how many people in a given age group will die each year. Experience tables of insurance companies cannot show *who* probably will die, but they do show *how many* probably will die. Some people live to be over 100 years old; some die at age of 1 year, 2 years, 40 years, 60 years, and at various other ages. The life-spans are *erratic*. Yet a probable given number in several thousand will die at a particular age. This is a somewhat *stable* characteristic.

Some food prices advance, others decline, and some remain unchanged during a given period of time. Yet, taken as a group, such food prices will reveal a stable characteristic to rise, decline, or remain unchanged during such a period.

Multiplicity of causes

Erratic items reveal stable characteristics when they are subject to a multiplicity of causation. For example, people die at all ages, but the tendency for a given number of people to die at age 35, or at some other age, comes about because their life-span is affected by a host of causes. Some of the factors determining how long a man will live are diet, medical services available and used, kind of employment, recreational activities, climate, clothing, housing, ability to make adjustments to changes, temperament, physical attributes inherited and acquired, and many others. Some of the factors affecting prices are interest rates, rents, weather, competition, government policy, and transportation charges. Other examples of the multiplicity of causation can be given, but these two should be sufficient to illustrate the phenomenon.

REASONS FOR SAMPLING

There are various reasons why the sampling technique is used as a means of gathering statistical data. Two of the most important of these are (1) that it may be the only possible method and (2) that it is the quick and economical method.

Speaking of the role of modern methods of sampling, W. Edwards Deming, one of the leading authorities, wrote the following in the Prefatory Note to Jean Namias' *Handbook of Selected Sample Surveys in the Federal Government* (St. John's University Press, 1968):

> All statistical work has improved through application of statistical theory, complete censuses as well as smaller studies. A complete study or a complete census no longer need be presented as a glob of figures with unknown reliability. Through use of sampling in conjunction with a complete study, flaws in execution may be detected and corrected before it is too late. Variances between interviewers are as important for a complete study as for a sample, and may be measured in either one by appropriate statistical design. Imperfections such as the proportion of nonresponse by area, by sex, by age-group, under-coverage or over-coverage, and other operational imperfections in a complete census as well as in a sample can now be evaluated by intensive studies of a small sample of units. The user of data

derived from a complete census may now have the same protection as the user of data from a sample less than 100 per cent.

The only possible method

In the field of production a test of quality of the product produced often changes the form of the product so that it must be reprocessed or scrapped. Where this is the case, there is no alternative to sampling for measuring quality. A test for butterfat content of cream spoils the cream, so the test is made of a sample. A test of the drug content of a capsule or pill requires an analytical breakdown and destroys its form, so drug manufacturers test their products by analyzing the contents of a bottle or vial *properly* selected from the produced lot (population).

The quick and economical method

If the only method of obtaining information about a population consisted of observing each item in that universe, the timeliness of the information might well be past when the findings became available. By comparison with the observation of all items, taking a sample from the population and inferring the characteristics of the universe from that sample is a brief process. Sampling is the economical and quick method. A sample is by nature less costly than the observation of all items. Sampling is a boon to individuals or organizations concerned with estimating within limits certain characteristics of a population.

PROBABILITY SAMPLES

Simple random samples

The term "simple random sample" of n items means a sample selected from a population in such a way that each possible combination of n units has the same chance or probability of being selected.

Simple random sampling is not used widely in practice. However, it is important in sampling theory, it is simple, and it is the *basis* for other types of sampling used.

The discussion can be undertaken by setting up a simple, hypothetical population of eight households and assuming that the objective is to estimate the average (mean) monthly expenditure for rent from a sample. This population is given in Table 3.1.

Selection of a simple random sample. The selection of a simple random sample may proceed in various ways. We might obtain eight identical chips, each having one of the eight letters A, B, C, . . . , H on it, with no two chips having the same letter. Then the chips could be placed in a container and mixed thoroughly and two chips drawn from the container.

TABLE 3.1
Rents of a hypothetical population of eight households

Household	Rent (dollars)
A	160
B	158
C	174
D	192
E	162
F	148
G	170
H	148
Total rent (Σ)	1,312
Average rent ($\bar{X}$)	164

Suppose chips C and F are drawn. This combination CF has been obtained by a simple random sample. If the process is repeated again and again and a count is made of the number of times each combination occurs, the proportion of times CF occurs will be the same as that for any other combination. Table 3.2 lists all the possible combinations. The

TABLE 3.2
All possible combination samples of two drawn from the population in Table 3.1

Households in Sample	Sample Average	Households in Sample	Sample Average
AB	159	CE	168
AC	167	CF	161
AD	176	CG	172
AE	161	CH	161
AF	154	DE	177
AG	165	DF	170
AH	154	DG	181
BC	166	DH	170
BD	175	EF	155
BE	160	EG	161
BF	153	EH	155
BG	164	FG	159
BH	153	FH	148
CD	183	GH	159

$\Sigma = 4{,}587$; $\bar{X} = 4{,}587 \div 28 = 164$

number for all possible combinations without replacement is

$$C_n^N = \binom{N}{n} = \frac{N!}{n!(N-n)!} \tag{3.1}$$

where N is the number in the population and n is the sample size. For the above example:

$$C_2^8 = \frac{8!}{2!6!} = \frac{8\cdot7\cdot6\cdot5\cdot4\cdot3\cdot2\cdot1}{2\cdot1\cdot6\cdot5\cdot4\cdot3\cdot2\cdot1} = \frac{8\cdot7}{2\cdot1} = 28$$

Table 3.3 lists the frequency and probability of the sample means. All these possible combinations have the same chance of being drawn in a simple random sample. Therefore, whichever combination occurs in a drawing of two items actually selected as a sample is called a simple random sample of two.

TABLE 3.3
Frequency array and probability of occurrence of sample means in Table 3.2

Sample Mean $\bar{X}$	Frequency of Occurrence f	Probability of Occurrence $P(\bar{X})$
148	1	.0357
153	2	.0714
154	2	.0714
155	2	.0714
159	3	.1071
160	1	.0357
161	4	.1432
164	1	.0357
165	1	.0357
166	1	.0357
167	1	.0357
168	1	.0357
170	2	.0714
172	1	.0357
175	1	.0357
176	1	.0357
177	1	.0357
181	1	.0357
183	1	.0357
	$\Sigma = 28$	$\Sigma = 1.0000$

If the population is of any considerable size, the task of selecting a simple random sample in the way just described would indeed be difficult. A simpler method is to make use of a random numbers table such as Table 3.4. the genesis of a random numbers table dates back to the 1920s and some of the work of the well-known English statisticians, L. C. H. Tippett and Karl Pearson. In a letter to the editor of *The American Statistician* and published in the February 1965 issue, Mr. Tippett recalled his early research in this area.

TABLE 3.4
Random numbers

24269	71529	60311	42678	53565	21785
97965	88302	49739	39808	07541	47458
54444	74412	78626	94614	20483	40405
44893	10408	66692	00582	49022	71209
90120	46759	44071	97702	07772	85561
78435	64937	13173	44053	49368	11657
37836	02985	86716	91691	06312	91339
04345	53424	92581	01748	60943	99396
32192	16218	12470	85736	92329	57649
75583	16136	01016	60555	87936	28977
13895	88190	72682	37301	54480	45990
22502	00244	21443	92003	88940	76668
63680	16016	01176	16543	00307	39014
63266	30432	80582	99837	98932	81232
23896	64672	13177	07362	63574	76447
34030	85301	77567	94770	18555	26667
50259	54066	79457	64238	32350	12025
73959	98525	40455	12836	21529	80217
46874	90391	03577	22995	12058	10875
60883	26629	31370	94050	06651	54127
17994	85149	14924	88158	78920	42824
53119	37559	70312	24240	98041	71484
58660	70360	90580	24069	81105	51594
86861	91113	24781	68990	36222	13986
08289	80504	40902	37981	71643	28091
29490	88977	95889	43584	55452	38101
47724	15243	92613	14338	54716	39641
24432	24335	67667	36292	23417	84054
56411	61105	24700	62004	36372	47468
24472	19087	76479	57326	44302	43321
64281	25807	10404	94083	41468	35439
66847	06170	97583	85774	94550	20094
72461	60808	68224	34108	41615	06343
21032	80940	48139	07201	50273	72535
95362	19516	51247	53969	41396	47749
93241	49712	99236	49554	61826	23554
92348	58275	89535	49678	70495	69321
57411	89514	10252	64146	33230	92674
50395	15472	27799	41794	91050	46347
78736	12120	50241	90670	67011	51924

The circumstances that led to the development of the random sampling numbers that were published under my name in 1927 [Tippett's, *Random Numbers*] are briefly described in Karl Pearson's foreword to the publication. I repeat here the substance of that description and, from a somewhat dim recollection, add some colour.

As a research student working under K.P.'s direction, I had calculated some tables of the mean range and some associated distributions, and sought to test and confirm the results by making a sampling experiment. I constructed a normal population of, I think, 10,000 individuals grouped in sub-ranges of one-tenth of a standard deviation by writing the values on 10,000 tickets, and attempted to sample this by taking one ticket at a time, replacing it and shuffling the lot, taking another ticket, and so on. The

> work was laborious to a discouraging degree, and the resulting sampling distributions did not agree with theory. We did not discard the theory (I felt slightly uncomfortable about this) but decided that the sampling technique was wrong, and K.P. suggested the idea of random numbers. I cast around for a source and, following a suggestion of K.P., happened upon U.K. census returns, in which the areas of parishes seemed to be a random set of numbers. In order to avoid possible biases due to such factors as rounding, I discarded the first two and the last two digits of each area, and copied down all the remaining digits, more or less in the order in which they appeared in the returns. I still faintly remember during the hot summer days in the library of the Statistics Department of University College, London, slogging my way through this work. Pedestrian but useful!

Using electronic computers to generate a random process, tables of random numbers today are created by a process of generating the digits "0" through "9" one after the other in such a way that the *order* of the digits defies any method of predicting a series of numbers by referring to the table.

To obtain a simple random sample of two households by using Table 3.4, we could first number the households in Table 3.1 from 1 through 8 in the order in which they are listed. Then starting at any point in the random numbers table, obtain two random numbers between 1 and 8, inclusive. For example, starting at the lower right corner of Table 3.4, these numbers are 4 and 7, and they give the combination DG. This procedure insures that every possible combination has an equal chance or probability of being selected. The random numbers table accomplishes the same thing that thorough mixing in a container accomplishes. Alternatively, computer programs may be used to randomly select sample items.

Thus, it can be seen that the probability that an item or individual will be drawn in a simple random sample of any size turns out to be simply n/N, where n is the number drawn in a sample and N is the number of items or individuals in the population. The simple random sample, therefore, is a probability design, and the probability of any item being drawn is known. Certainly, this method is sharply different from one in which an interviewer stands in a downtown area and picks the first n people who come along. Simple random sampling requires positive action to insure that each population item has the same chance of being drawn in the sample.

The sample estimate. The rent of the households in the sample is an estimate of the average rent of the population. If we drew a sample of two and happened to get B and D, our estimate of the average rent would be $175. When all possible samples of two are considered, as in Table 3.2, the average of the 28 sample means is $164. This is exactly the average of the population in Table 3.1. *Thus, it can be seen that on the average,* *the sample estimate is exactly equal to the population average that is being estimated. This will be true for sample means based on simple random sampling for any sample size from any population.* Therefore the sample average is said to be an *unbiased* estimate of the population average.[1]

Simple stratified random samples

Knowledge of the population to be sampled may be used to divide the population into groups so that the items in each group are more nearly alike than are the items of the population as a whole. If a sample is drawn from each group by simple random sampling and the samples are combined for all groups, we have insured the proper representation from each group and thus have a probability sample. *The division of the population into similar groups and the taking of a random sample from each group is called stratified random sampling.*

Just what should be the basis for division of the population into groups or strata depends upon the kind of investigation to be undertaken. If the problem is to investigate attitude toward a proposed school bond issue, the population may well be divided into families with school-age children and those who have no school-age children. On the other hand, if the problem is to estimate the average income in an area, the population may well be divided into occupational groups. Thus, it is seen that the division of the population into strata requires both a knowledge of the population and a knowledge of the effect of a certain characteristic upon the estimate to be made.

Table 3.5 shows the hypothetical population in Table 3.1 in stratified form. The assumption is that the size of rent paid depends in part upon

TABLE 3.5
Households in Table 3.1 stratified according to location

Area 1		Area 2	
Household	Rent (dollars)	Household	Rent (dollars)
A	160	C	174
B	158	D	192
F	148	E	162
H	148	G	170

[1] An unbiased estimator is one whose expected value (namely, the mean of its sampling distribution) equals the parameter (population characteristic) that it is intended to estimate. Chapter 8 discusses the concept of a sampling distribution, and Chapter 9 deals with the problems of estimation.

TABLE 3.6
All possible combination samples of two drawn from the stratified population in Table 3.5

Households in Sample	Sample Average	Households in Sample	Sample Average
AC	167	FC	161
AD	176	FD	170
AE	161	FE	155
AG	165	FG	159
BC	166	HC	161
BD	175	HD	170
BE	160	HE	155
BG	164	HG	159

$$\Sigma = 2{,}624; \bar{X} = \frac{2{,}624}{16} = 164$$

the area in which the rental unit is located. Thus, Table 3.5 shows how the population might be divided into two strata so that the rents in each stratum are more nearly similar. If we take random samples of one from each stratum and combine the samples, all possible combinations are as shown in Table 3.6. Again, the average of the sample averages is exactly equal to the population average; thus, any combination is said to result in an unbiased estimate of the population average. Table 3.7 lists the frequency and probability of the sample means.

The stratified random sample properly carried out will be an improvement upon the simple random sample. This can be shown simply by noting that in Table 3.2 the smallest sample average is 148 and the largest aver-

TABLE 3.7
Frequency array and probability of occurrence of sample means in Table 3.6

Sample Mean $\bar{X}$	Frequency of Occurrence f	Probability of Occurrence $P(\bar{X})$
155	2	.1250
159	2	.1250
160	1	.0625
161	3	.1875
164	1	.0625
165	1	.0625
166	1	.0625
167	1	.0625
170	2	.1250
175	1	.0625
176	1	.0625
	$\Sigma = 16$	$\Sigma = 1.0000$

age is 183; therefore the combinations result in averages ranging from 148 through 183, or 35. In Table 3.6, on the other hand, the smallest sample average is 155 and the largest average is 176; the combinations result in averages ranging from 155 through 176, or 21. The smaller the range of all possible sample averages, the more reliable the results for a given sample size. Thus the statement that "a sample properly stratified will be more reliable than a simple random sample of the same size" is a useful one.

Cluster samples

This type of sample design is widely used in economic and market analysis as an efficient method of collecting original data. It should be noted that the systematic sample from a list and the so-called "area sample" are basically cluster types. These two sample designs will be discussed below.

With the simple random and stratified sampling plans, it was discovered that each observation is chosen individually. In cluster sampling, groups of items are chosen at random. To illustrate, assume that we use our abbreviated hypothetical population of Table 3.1, except that we arrange the households arbitrarily into four clusters:

Cluster	Elements
1	A, B
2	C, D
3	E, F
4	G, H

Assume that it is desired to select a probability sample of four items from this new universe. One way, of course, would be to choose a simple random sample of four items; or we might look upon the clusters as strata and choose one from each. For our purposes, further assume that neither of these alternatives is feasible. Then, another way would be to select two clusters from the four and enumerate their items. Although we have oversimplified our universe, this sampling plan would yield a probability sample because every element would have a known chance of being selected, namely, $\frac{2}{4}$ or $\frac{1}{2}$. Obviously, every possible combination of population elements would *not* have the same chance of being chosen. This follows because the selection of one item in a cluster automatically includes all other items in the cluster. That is, it would be impossible to obtain some sample combinations, for example, A, B, C, and E.

Actually, this method amounts to sampling from a population of clusters rather than selection from a universe of individual elements. In effect, our original universe of eight items has been redefined as a population of four clusters, and a random sample of two clusters is then chosen. Such a scheme is widely used in sampling human populations where a city may be divided into blocks and then the sample blocks chosen at random and every unit in the selected blocks comprising the sample.

Health interview survey.[2] The Public Health Service of HEW weekly conducts a survey by interviewing a probability sample of civilian and noninstitutional households to obtain information on acute illnesses, accidental injuries, chronic conditions, and many other health topics. This survey incorporates many of the features of a cluster sample. Actually, this sample is a *multistage* probability design which permits a continuous sampling of the population of the United States. The *first stage* of this design is the drawing of 357 from about 1,900 geographically defined primary sampling units (PSUs) into which the United States has been divided. A PSU might be a county, a group of contiguous counties, or a standard metropolitan statistical area. In the interest of brevity, but with no loss in understanding, we might combine the remaining stages and view them as an *ultimate stage*. Within these PSUs ultimate stage units (called segments) are defined in such a manner that each segment contains an expected six households. A segment consists of a *cluster* of neighboring households or addresses. The Public Health Service uses two general types of segments in this sampling plan: (1) area segments which are defined geographically, and (2) segments which are defined from a list of addresses from the Decennial Census and Survey of Construction. Then, each week a random sample of approximately 90 households is selected. These 90 segments contain about 800 households, and members of these households are then interviewed concerning the factors related to health.

Because the household members interviewed each week are a representative sample of the entire population of the United States, successive samples from later weeks may be combined into larger samples. This feature makes it possible to avoid the problem of low frequencies (of rare diseases for example) because the design permits both continuous measurement of high incidence of certain health problems, and through the combined larger samples more detailed analysis of the less common diseases and smaller categories. The sample size for the nation for a 12-month period (which ends in June) recently included about 134,000 persons from 42,000 households living in approximately 4,700 segments. The data are collected by an experienced field staff of the Bureau of Census in accordance with these standards provided by the National Center for Health Statistics. The Center tabulates, analyzes, and publishes the results of the surveys.

---

[2] See *Health Survey Procedure, Concepts, Questionnaire Development, and Definitions in the Health Interview Survey,* Public Health Service Publication No. 1,000, Series 1–No. 2 (May 1964), Department of Health, Education, and Welfare, Washington, D.C.

The estimating procedures used result in each statistic produced by the survey is based on two stages of ratio estimation. The first of these stages is based on the control factor of the ratio of the 1960 total decennial population count to the count on the first-stage sample of PSUs in 1960. These factors are now applied to 24 color-residence classes. The second stage results in computing ratios of sample-produced estimates of the population to the official estimates of the Bureau of Census in about 60 age-sex color classes. The net effect of this ratio estimating procedures is to make the Health Interview Survey more closely representative of the U.S. population by age, color, sex, and residence. This helps to reduce sampling variance. However, because these estimates are based upon sample evidence they will differ to some degree from what would have been obtained if a complete census had been taken using the same schedules, procedures, questionnaire, and so forth. The results, also are subject to measurement error which is not a condition limited to sampling. Sampling and measurement errors are published with the results of the survey as a means of inferring reliability.

Clearly, this type of sampling plan has its merit; however, one may ask: Is the cluster method as efficient as random sampling? It should be noted that in analyzing sample effectiveness, we encounter both statistical and economic efficiency, the former relating to the sampling error, whereas economic efficiency relates to cost per completed observation. At the moment, let us concern ourselves with contrasting the statistical efficiency of two types of samples studied so far. Let us compare cluster sampling and random sampling. One method will be considered superior if it has a smaller sampling error (standard error) for the same size sample.

It should be obvious that the relative statistical efficiency of the two plans rests on the degree of similarity among items in each cluster. For example, the greater the similarity of the items in a cluster, the *less* efficient will be the cluster sample. On the other hand, the larger the dissimilarity among the elements in a cluster, the *more* efficient will be the cluster sample. Therefore, where the variation within the clusters is a minimum, the simple random sample will be more efficient, that is, have a lower standard error. On the other hand, where the clusters have a maximum variation (maximum dissimilarity of items within each cluster), the standard error of the means of the clusters will be a minimum and the cluster sample will be the more efficient. To verify this intuitively, assume that every item in cluster 1 has the same value, every item in cluster 2 has equal values, and so forth, so that we have maximum similarity of items within each cluster, or a minimum dissimilarity. This kind of cluster grouping will yield a larger standard error than would the simple random sample. The reason lies in the fact that no matter which two clusters are selected, we shall have information on only two of the four different values. The other items merely repeat the information. One can easily see that if we were to construct a universe of clusters with maximum variation, the cluster scheme can be more efficient than the random sample.

This intuitive approach emphasizes the fact that the statistical efficiency of cluster sampling depends on how the clusters are designed, that is, the degree to which each cluster can be made to include all the values in the universe. In theory, this is the ideal cluster. In the real world, however, clusters frequently are relatively homogeneous. For example, if a city block is the cluster, the households in each tend to have similar economic or other characteristics. This implies that in practice, cluster sampling is frequently less efficient from a statistical point of view; however, when the dollar cost per completed observation is considered, the reverse may be true.

Figure 3.1 contrasts the three sampling plans mentioned above on the basis of statistical efficiency. One can see that for the same size sample in the real world, we *generally* may expect a larger standard error for a cluster type than for either of the other two samples.

Figure 3.2 contrasts the three plans in terms of their economic efficiency, that is, dollar cost per completed observation. Here, we find the cluster sampling generally having the lowest cost per observation, and therein lies the key to its utilization in social research. That is, in terms of the relative net efficiency—statistical efficiency versus economic efficiency—cluster sampling frequently rates as a very effective plan and one that is feasible in many situations.

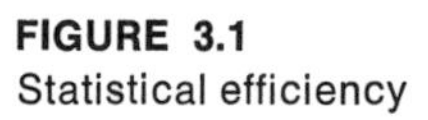

FIGURE 3.1
Statistical efficiency

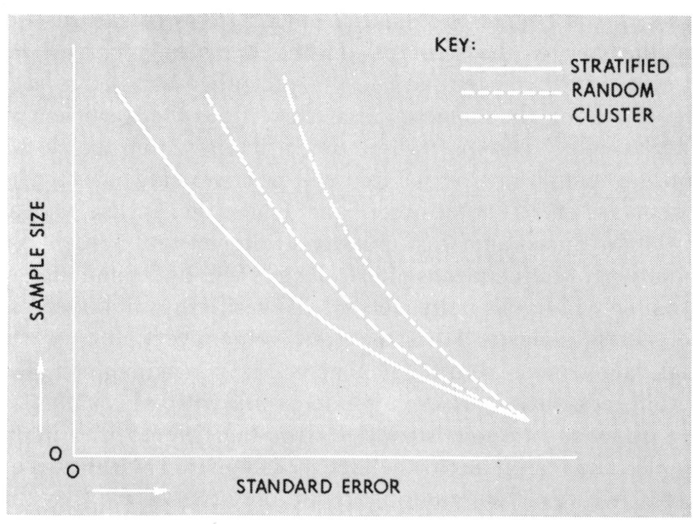

FIGURE 3.2
Economic efficiency

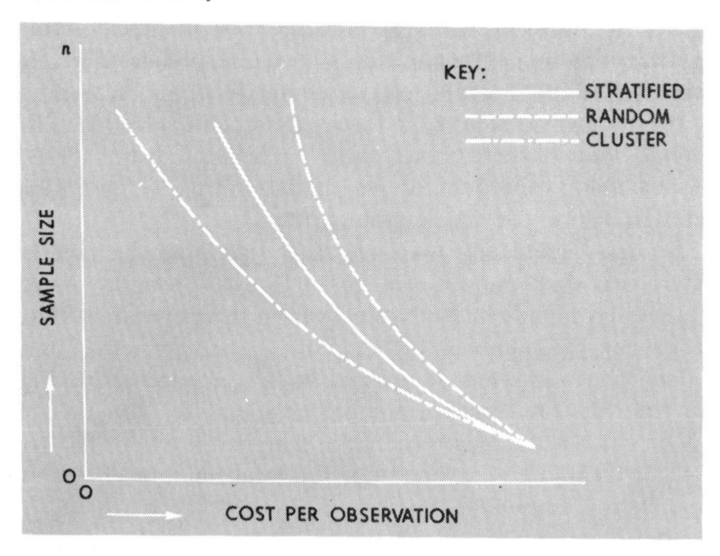

Systematic samples

There are times when the use of a random numbers table such as Table 3.4 is not practical. If, for example, the population listed were 10,000 in size, the numbering of the list would be laborious and 9 of 10 random numbers would be discarded. In general, reliable samples can be chosen by selecting the first numbered item at random and then taking every qth item. A sample of 200 from a population of 10,000 is a 1/50 sample. Thus a number between 1 and 50 selected by starting in the upper left corner of Table 3.4 would be 24. The sample then can be selected by taking the 24th, 74th, 124th, 174th item, and so on, until 200 items have been selected.

Although one can generally assume that a systematic sample will be a sound method of selecting items, the method does have one main limitation. It results in biased measures if the variate to be estimated is subjected to periodic fluctuations. For example, the selection of the number 6 from the interval 1–7 and then taking every seventh day to estimate average daily sales would result in the same day of the week being selected each time.

Area samples

Frequently, lists are not available. This certainly is true for most consumer attitude studies. If a list is available, many times it is badly out of date. To overcome this problem, a special type of cluster design, called area sampling, is utilized. The basic principles of an area sample are easy to comprehend; however, their application to a complex problem requires considerable ingenuity and a great deal of experience.[3]

To illustrate the fundamentals, let us assume that our research task is to estimate the number of unemployed workers in households located within the corporate limits of the city of Detroit. No accurate list of these households is available; consequently we might approach the problem of sample design by using an area sampling plan.

First, from a current city map, we may determine the number (N) of the city blocks that comprise the city of Detroit. Second, we may then randomly select a sample of n city blocks from this population. Third, we could determine the number of unemployed workers (using Bureau of Census definitions) in every household in the sample blocks. Four, we could multiply the total sample number of unemployed workers by the ratio of N/n which is the reciprocal of the sampling fraction. The resulting figure will be an unbiased estimate of the total number of unemployed workers in Detroit households.

Obviously, the above sampling procedure is a *probability sample* because each Detroit household has a known chance (n/N) of being included in the sample. You note, however, that the universe (or population) of all Detroit households was not sampled directly. Instead a universe of areas (city blocks) was defined and then a sample of these areas was chosen at random. In our example, *every* household within the sample blocks was contacted to estimate the number of unemployed workers.

The logic behind this simple one-stage area sampling plan is that Detroit may be defined precisely in terms of a particular area. It is also possible to subdivide this universe into a number of blocks such that each area is clearly delineated in terms of location and size. The total number of blocks (N) in the universe easily can be determined. Because all of these items are identified and their total number known, it is a simple assignment to draw a probability sample from this population. We may note that the original population (all of the households in Detroit) for which no accurate list existed was restructured into a universe of areas (blocks) for which a listing (city map) is available. This permits the selection of a probability sample because the cluster of households associated with each block has a known chance of being chosen. This proba-

[3] The Survey Research Center at The University of Michigan has demonstrated a real skill for designing samples of this type for use in their research. For example, see J. B. Lansing and J. N. Morgan, *Economic Survey Methods* (Ann Arbor: Institute for Social Research, University of Michigan), 1971. Also see D. P. Warwick and C. A. Lininger, *The Sample Survey: Theory and Practice* (New York: McGraw-Hill Book Co., 1975), chap. 5, pp. 111–25.

bility is equal to the chance that the block within which each cluster of households exists will be selected.

One distinguishing feature of this particular one-stage area sample is that *every* household in the selected blocks is contacted to estimate the number of unemployed workers. Theoretically, this may result in a sample that is less efficient from a statistical point of view. This is true because the households in these area clusters (each block may be looked upon as a cluster of households) are probably more similar in economic and social characteristics than would be the same number of households chosen at random from the *entire* universe. This intracluster similarity results in a larger sampling error (standard error of the mean) than would be the case of a simple random sample. The economic efficiency—cost per sampled observation—may more than offset this loss of statistical efficiency in some surveys.

The above example could be made into a two-stage area sample by merely randomly sampling the households within the first-stage units selected. That is, the random selection of the blocks would be the first stage; the random selection of households within the chosen blocks would then be the second stage. Many area samples are multistage. For example, stage one might be the random selection of a state; stage two would be the random selection of the county within this state; stage three would be the random selection of the household within this county; stage four might be the random selection of the person within the household.

Let us illustrate these fundamentals by using an artificial example. Suppose that the nicely symmetrical city of Bayesian Heights contains 2,000 blocks with each block having 10 households. Assume also that our study indicates that we need 400 households in our sample. Because we have 20,000 households this amounts to an overall sampling rate for households of 400/20,000, or 2 out of 100, or 1/50. Using this information, we could select our households as follows:

1. Sample the blocks at the rate of 1 in 5;
2. Then subsample the households on the selected blocks at a rate of 1 in 10.

This procedure would yield a sample of 400 households because we would first select at random 400 blocks from the universe of 20,000 blocks. Then we would randomly choose one household from each of the 400 blocks.

One check available is that the product of the block sampling rate (1/5) and the intrablock subsampling rate (1/10) must equal the over-sampling rate of 1/50. Obviously, there are several ways to select such a sample. For instance, we might have chosen blocks at the rate of 1 in 20 and then subsampled the households at the rate of 4 in 10. The check would be: $(1/20)(4/10) = 4/200$, or 1/50.

From a statistical point of view which sampling scheme would be more efficient? Clearly, the first plan would provide more information because of the fact of intrablock similarity. Economic efficiency, however, might dictate the second plan because of a reduction in travel and interviewing costs. The cost per completed interview under the second plan probably would be lower; however, this must be weighed against the probable higher standard error that this scheme would generate.

The reader should recognize that many of the variations actually used by social science researchers have been ignored here for simplicity. Some area samples also make use of stratification; some are three- or four-stage; still other area samples might make use of sampling units with varying probabilities (our examples used equal probabilities). In addition, the whole problem of estimation of parameters from sample evidence is not discussed here. This latter topic is taken up in Chapter 6; the other items are left for specific courses in experimental design or survey sampling.

Sequential sampling

Perhaps, a brief statement about sequential sampling is in order here. Normally, most people think of this technique as being associated with quality control type of inspection. However, its uses are not limited to lot-by-lot sampling. Basically, sequential sampling involves small size samples. Indeed, in some cases the observations might be drawn one by one or group by group (cluster) in order, and the results of the sampling at any stage decide whether or not sampling is to continue. The sample size is not fixed before sampling; rather it depends on the actual results and might vary from one sample to the next. In sequential sampling the process terminates according to some predetermined rules which are decided upon by the degree of precision required. In more practical terms these rules are usually determined in terms of the economic consequences of making a wrong decision. It has been statistically shown that in certain situations sequential sampling involving a given sample size provides more accurate, more efficient, and more reliable results than one large sample of the same size.

A NONPROBABILITY SAMPLE

The so-called "quota-controlled" sampling method has been used in marketing research. It is essentially a stratified sample of the nonrandom type. The population is divided into two or more groups according to a characteristic that is thought to have an effect upon the variate to be investigated, that is, estimated. Quotas are then set up, for example, by telling an interviewer to take a specified number from each group. The interviewer may take any item or individual that is a member of the population group, and naturally will take those that are conveniently available. If restrictions are placed on the convenience of the interviewer, this may possibly reduce biases. However, there is no way to determine bias except by a sample properly drawn, and just how effective the restrictions are is never known.

If it is important that reliable results be obtained, then a method for which the error can be controlled should be used. On the other hand, if conditions are such that fairly rough estimates are all that is required, then a nonprobability sample may be justified. Thus, we cannot say that the *quota method* should never be used. But when used, its results must be looked upon as only a rough estimate that cannot be tested for reliability.

THE PROBLEM OF SAMPLE SIZE

A discussion of sampling inevitability raises the question as to how large a sample should be. Courses in sampling are designed to provide the answer to this question, and the answer cannot be given here. However, some general ideas of sample size required can be obtained from a *commonsense* discussion.

If the variate—rent, price, income, or some such factor—does not have a wide range in the population, the sample can be relatively small and still give a good picture of the population. For example, a very small sample would give a reliable average for the price of choice T-bone steak in an area because the range of prices charged in all stores is not great.

If, on the other hand, the range of the variate in the population is great, then a relatively large sample is needed to estimate the population average within reasonable limits. For example, incomes vary widely, so a large sample is needed to estimate average income. Chapters 7 and 8 attempt to quantify the terms "relatively small" and "relatively large."

Of course, if the sample is to be taken for the purpose of measuring a *rare* or *unique* condition in the population, then a large sample is necessary. For example, if only very few children are allergic to a particular drug, in order to estimate the percentage allergic the sample by necessity must be large, for a small sample may not include even one allergic child.

The principle of decreasing variation

It is often said that the *reliability* of a simple random sample increases as the square root of the number (n) in the sample increases. This is the *principle of decreasing variation.* Thus a sample of $n = 100$ is twice as reliable as a sample of $n = 25$ because the square root of the former is twice the square root of the latter. This principle is used in arriving at a crude approximation of proper sample size in Chapters 9, 10, and 11.

SUMMARY OF SAMPLING CONCEPTS

Sampling is an indispensable way of gathering business and economic data. It is important, therefore, that samples be designed so that the characteristics of the sample are a reasonably good picture of the characteristics in the population. The basis of all probability samples is the simple random sample, and all other kinds of probability samples are a variation of the simple random sample. Nonrandom samples do not provide measures which can be tested for reliability, and the result of such a sample must be looked upon as being at best a rough estimate of the population measure. Just what the size of a sample should be is a question that is considered in Chapters 9, 10, and 11. The following section briefly describes how data are gathered from individuals, households, firms, or whatever the sampling unit may be.

DATA COLLECTION

Most surveys are concerned with human populations. A number of methods have been used to gather data from the so-called "respondents." Interviewers are often used, and ordinarily these interviewers call on the respondents and ask the questions person to person. Questionnaires frequently are used, and normally these reach the respondents by mail. Occasionally, an attempt is made to gather data by telephone interview, and sometimes the panel method is used for obtaining information. A brief examination of each of these will reveal their advantages, disadvantages, and general worthwhileness.

Personal interviews

All other things being equal, the personal interview is the most desirable method for gathering statistical information from human populations. The schedule of questions is in the possession of the interviewer, and he or she can explain to the respondent the meaning of any particularly difficult question. Further, the interviewer can use probing in an attempt to induce the respondent to answer the questions. Third, the interviewer can arrange to call on the respondent at a time which is convenient for the respondent and thus avoid nonresponse due to not-at-homes. Thus the interview method makes possible obtaining answers to difficult questions and recording such answers so that they are in the desired form.

The disadvantages of the interview method arise from all other things *not* being equal. It is expensive to train, supervise, and transport interviewers; and further, it is difficult to hire interviewers who have a personal makeup fitted to the task of meeting and persuading people. Thus, although the interview method is the most reliable method of gathering data, it is an expensive and difficult one.

Mail questionnaires: Self-enumeration

Questionnaires sent through the mails can be used for wide areas of coverage relatively cheaply. They can be answered at the convenience of the respondent without prior arrangement, and thus costliness connected with the interview method is largely eliminated. Thus, at first glance, this appears to be an excellent way in which to gather data.

Questions on questionnaires generally are simpler than those asked by interviewers. This must be the condition if the respondent is to interpret the questions. This in itself restricts the use to which the questionnaire can be put. Further, this method of gathering information has been used so much that people are weary of receiving questionnaires; as a result, many intended respondents fail to return the questionnaire to the sender. Often the questionnaire is returned because the respondent has a special prejudice or pride, and the questionnaire appeals to that prejudice or pride. For example, who is likely to return a questionnaire which asks: "Do you favor lower interest rates on FHA loans?" Will it be the person who already has a home loan, the person who owns a home clear of mortgage, or the person who is in the market for a home? Obviously, persons wanting to obtain FHA loans will be the ones *most likely* to return the questionnaire. Thus the nonresponse is likely to be high, and the returns probably will not be representative of the group or population sampled.

One further example: Suppose that 500 questionnaires are mailed to households selected at random. The survey has to do with homeownership. Homeowners probably are much more likely to return the questionnaire than are renters. If 120 return the questionnaire, the results might be something like those in the first two columns of Table 3.8. In the last two columns of Table 3.8 is shown what the condition could be if 500 returned the questionnaire but the additional 380 were all renters. Thus, it is clear

TABLE 3.8
Possible questionnaire results

	Possible Results of 120 Questionnaires Returned in an Intended Sample of 500 Households		Possible True Condition if All 500 Had Returned the Questionnaire	
	Number	Percent	Number	Percent
Homeowners	84	70	84	16.8
Renters	36	30	416	83.2

that we cannot rely on a partial return of a questionnaire to give a reasonably good picture of the population from which the sample came. We shall never be able to determine the amount or direction of bias unless there is a check on nonrespondents.

Potential nonresponse encountered in a questionnaire survey has brought about a concerted effort by those using it to obtain returns from all who receive the questionnaire. These efforts take the form of follow-up letters, telephone calls, monetary reward, and a host of other techniques. Some research organizations have used the mail questionnaire method of gathering statistical data with a high degree of success, receiving returns from 95 percent or more of the intended respondents. Obviously, we cannot quarrel with the results of such surveys. Nevertheless, potential nonresponse is a serious shortcoming of mailed questionnaires. The check on nonresponse may involve more costly procedures than the interview method. The decision whether to use the questionnaire method, therefore, depends upon a host of factors, for each survey presents special problems.

Telephone interviews

A considerable amount of sheer nonsense has been promoted to indicate that data can be quickly and successfully gathered by telephone interview. Nothing could be further from the true condition. Such a method may be satisfactory for getting crude estimates of TV audience size, but it is not of value in gathering business and economic data. Busy signals, party lines, no answer, and a host of other impediments ordinarily work toward the sample's not being completed, and thus the nonresponse is high and the bias unknown. Perhaps, the only possible valid use of telephone interviews might be as economical follow-ups to a previously completed personal interview. This might be done to clarify some point or to obtain additional information.

Panels

Panel technique means that data are collected repeatedly from the same people. This method makes possible measuring change in rents, expenditures for food, or any other economic variable. It is a continuous process, which is one of its chief advantages in terms of market research. Ordinarily, there is some compensation for participants, and if one or more panelists drop out of the sample, statistical devices have been formulated for handling the problem.

Public opinion polling

Public opinion polling by the politicians for many offices, both national and local, has become big business since the 1950s. Following the momentary setback to polling suffered by the shock of the 1948 presidential election, this tool for keeping track of the voters' pulse has been much used and misused. In 1948 when some of the polls predicated flatly that Dewey would defeat Truman, the public (and the politicians) cooled to the "findings" of these surveys. Following that disaster, which was brought about by the pollsters' insistence that the decision was clear in August and thereby quit sampling, the confidence of the public had been regained. By quitting sampling in August and September the polls missed the impact of Truman's "whistle-stop" hit'em-hard campaign of October. The population had changed, but their conclusions were based on the sampled population not the target universe! In addition, some voters apparently changed their minds in the voting booths–this makes it difficult to predict even with sampling continuing into October. For 1972 (as well as previous presidential elections since 1948) the polls did an excellent job of measuring public opinion and were usually within 1 percent or less of the final vote distribution.

Different methods of gathering the data and different ways of asking essentially the same questions can bring about noticeable differences in surveys. For example, in mid-1975 both the Gallup Poll and the Harris Survey sampled the public on the topic of how well President Ford was doing his job. Primarily because of the way the questions were asked, the Gallup Poll suggested that the pro-Ford group was the larger while the Harris Survey said the anti-Ford persons were in the majority. The problem stems from the fact that asking for opinions about a subject does not pigeonhole individuals with the same "preciseness" as an election. Clearly, both Gallup and Harris (as well as many other firms doing the same or similar sampling) reach representative samples of the eligible voters, they have slightly different ways of wording questions, and this affects the way people respond. The reader may try to answer the questions of both just to make the point. The Gallup question was: "Do you approve or disapprove of the way [President] Ford is handling his job as president?" In August 1975 Gallup reported that 45 percent said they approved, 37 percent said they disapproved, and 18 percent had no opinion. This large latter category is another key to the differences between the two polls.

The Harris Survey question was: "How would you rate the job President Ford is doing–excellent, pretty good, only fair, or poor?" Harris reported in August 1975 that 60 percent gave "negative ratings," only fair or poor. Thirty-eight percent said the president was doing excellent or pretty good. In the Harris poll only 2 percent held no opinion.

While the questions were asked by both polls during the first week of August 1975 of a "representative sample of the population" the conclusions were almost the opposite. Why? Again, it seems that the differences stem largely from the way the questions were asked rather than from any other major methods of sampling. The way the questions were worded apparently affected the percentage of "no opinion" voters in the two surveys. Therein lies a clue to the basic objective of this text: consumers of statistics should not blindly accept the findings of surveys but rather should look behind the data for subtle differences. Surveys on public opinion tend to mask a lot of "noise," misinformation, and voter apathy behind the apparent preciseness of the numbers reported in the press. The way in which questions are asked can screen out this ambiguity. For example, generally it appears that the respondent can pigeonhole himself or herself easier with the Harris questions than with the Gallup. If the voter is not too well informed about national and international affairs (and it is estimated that 54 percent of the voters do not know the names of their Congress representative!), then the respondent is likely to answer the Harris question in terms of his or her own personal situation. If things have not been going "right" for the voter, the response of "only fair" might seem generous to the respondent. In trying to respond to the Gallup question though, the same respondent might not wish to take a stance on this sharp difference and therefore list "no opinion." Consequently, nine times as many in the Gallup Poll as in the Harris Survey took refuge in the "no opinion" category. (Some writers have referred to the "no opinion" voters as "public opinion dropouts.")

It is interesting to note that when the *trends* of the two polls are compared for many months, the "lines" for President Ford's performance have become remarkably parallel. Both polls show that Ford started his term in office with high public approval, lost it quickly after the Nixon pardon, climbed halfway back in the Spring of 1975, and then went into another decline in the summer of the same year. In this writer's judgment, no matter how the questions were worded these two trends probably would have appeared no matter which question the "representative sample" was asked. However, a case can be made that the Gallup question is better in the sense that it forces the less informed voter to stand aside. On the other hand, a case may be made that the Harris question provides every voter with a question he or she can answer! No matter what your stance is, it appears that public opinion polling is here to stay for a long time.

SUMMARY

Most surveys are concerned with people, and thus the problem of gathering statistical information for business and economic purposes must of necessity consider the human element. Sometimes a survey is carried out solely by the use of interviewers. This is ordinarily the most reliable way of achieving a representative sample. For wide areas of coverage the mail questionnaire may be used, with an interview of nonrespondents. However, the probability of obtaining a representative sample by questionnaire alone is slight. Telephone interviewing is generally as likely to result in considerable nonresponse as is the mail questionnaire and thus cannot be successfully used to collect business and economic data. Panels have a specialized use for continuous gathering of information and are a reliable way of gathering data of limited use.

SYMBOLS

Σ, summation
$\bar{X}$, average of X series
N, number of items in the population
n, number of items in sample
$C = \binom{N}{n} = \frac{N!}{n!(N-n)!}$, the number of possible combinations of samples of size n
P, probability

PROBLEMS

3.1. *Required:*

a. Table showing all possible samples of size 2, using multiplicity of choice (sampling with replacement).

b. From information at the right, what is the range of the sample means?

X'
1
2
3
4
5

3.2. *Required* (assume items not replaced):

a. Show that the sample mean is an unbiased estimate of the population mean, using combinations of size 2.

b. From information at right, what is the range of the sample mean?

	X'
A	1
B	2
C	3
D	4
E	5
F	6

3.3. *Required:*

Assume that the data in Problem 3.2 are broken into two strata, as shown below.

a. Show that the sample mean is an unbiased estimate of the population mean, using combinations of size 2.

b. What is the range of the sample mean?

Stratum 1		Stratum 2	
	X'_1		X'_2
A	1	D	4
B	2	E	5
C	3	F	6

3.4. "The size of the sample is no criterion of its precision, nor of its accuracy, nor of its usefulness." Defend or refute this statement.

3.5. Describe succinctly but distinctly the following types of sample designs. That is, what are their essential and identifying characteristics?
a. Simple random sample.
b. Stratified random sample.
c. Cluster sample.
d. Quota sample.

3.6. Compare and contrast a target population with a sampled population. Illustrate your response with an example of a human population.

3.7. Distinguish between economic efficiency and statistical efficiency as these terms are used in sampling theory.

3.8. How might we define the target population if—
a. We wish to estimate the proportion of tenant-occupied dwellings in a given city?
b. We wish to estimate the number of unemployed in a given city?
c. The equipment of a community senior citizens' recreation center will be influenced by our survey of the use of leisure time by residents in the vicinity?
d. It is desired to know the direction and amount of change in the prices of food in our city from one month to the next?

3.9. In the accompanying table why do you think that a random sample of one thousand households was selected? Why do you suppose a census of automobile ownership was not taken? Would a census have provided more reliable data? Would the concepts of economic and statistical efficiency have any significance in this kind of situation? Would you think a market analyst for one of the automobile manufacturers might be interested in such information? Why? Would this same person be interested in the table of Problem 4.7? Why?

3.10. A state senator from the 18th District in Michigan sent out "about 10,000" questionnaires to constituents. Thirty-two questions were asked and "about

Automobile ownership, by income size of household, 1975

Households by Income-Size Groups	Number of Households (thousands)	Automobiles per Household				
		0	1 or More	1	2	3 or More
All households	52,888	22%	78%	57%	19%	2%
Less than $2,000	9,983	54	46	40	5	1
$ 2,000–$ 3,999	9,780	29	71	61	9	1
4,000– 5,999	11,547	15	85	68	15	2
6,000– 7,999	9,241	8	92	66	23	2
8,000– 9,999	5,330	6	94	58	32	3
10,000– 14,999	4,861	4	96	50	40	6
15,000– 24,999	1,537	5	95	34	50	11
25,000 and over	609	2	98	28	57	12

2,000 returned completed questionnaires." No follow-up was made of the nonrespondents.

a. Comment on some of the following widely publicized findings in terms of sampling theory:

(1) "The poll shows that 75 percent of the residents of the 18th District oppose lowering the voting age to 18 years."

(2) "If a state income tax is enacted, 46 percent of the constituents of the 18th District would favor a graduated income tax."

(3) "More than one half (51 percent) of the residents of the 18th District are against lengthening the term of house members to four years."

b. How might the state senator have improved the "survey"?

3.11. Following the 1964 presidential election, the Bureau of Census sampled the voters to determine who did and who did not vote. According to the tabulations released by the Bureau an estimated 69 percent of the voting-age population said yes, indeed, they did go to the polls on November 3. Assume that the sample was representative of the population of voting-age citizens, the estimated total vote was 76.6 million. Official election counts indicated that only 70.6 million votes were cast by about 62.3 percent of those eligible to vote.

a. What might account for the discrepancies between the sample results and the actual vote?

b. What problems of survey sampling does this example illustrate?

3.12. What is the target population of the survey mentioned in Problem 3.11? Define it carefully.

3.13. A congressional committee has criticized the Gallup Poll (and other similar surveys) on the grounds that the interviewers' returns are "doctored" prior to publication. For example, in 1944 Gallup subtracted 2 percentage points from Roosevelt's strength in preliminary forecasts because he believed that many FDR partisans would fail to vote. In 1948, Gallup subtracted 1.8 percentage points from his final estimate of the "majority" for Thomas E. Dewey and added 1.1 points to Harry S. Truman's strength. In 1960, Gallup made similar adjustments to the raw data of the Kennedy-Nixon preliminary polls. In his defense, Dr. George Gallup said: "Of course we make corrections. This is the real science of polling. People tend to confuse a simple straw vote with a scientific sampling poll such as we do. The raw figures from our interviews would be an absolute distortion of the truth without corrections." The special committee of the Social Science Research Council commenting on the problem said the adjustment of returns is ". . . desirable procedure when performed by proper methods." In your judgment, why should such adjustments or "corrections" be necessary?

3.14. Some social critics of public opinion polls claim that an election outcome can be changed by the widespread publicity given the results of these preliminary surveys, especially when interpreted as a forecast. That is, by the mere fact that some people interpret the reported results as a "prediction," some argue that the election outcome might be influenced. These social critics argue that such polls should be used only to gather public opinion on a national issue and not be used to forecast elections. Comment.

3.15. "Princeton, N.J., October 31, 1948. Gov. Thomas E. Dewey will win the presidential election with a substantial majority of electoral votes." Gallup Poll.

"September 9, 1948. Gov. Dewey is amost as good as elected . . . my whole inclination is to predict the election of Dewey by a heavy margin and devote my time and efforts to other things." Elmo Roper of the Roper Poll.

Actually, Roper did not present any more figures during the course of the campaign and said crisply at the end, "Dewey is in." Since that time the pollsters have made many refinements and improvements in their sampling procedures and in their reporting procedures; however, what do you think went wrong with the surveys?

3.16. Assume that a carton of 144 bottles of pills is chosen at random from a group of such cartons and that a number of bottles is selected at random from each chosen carton. Now assume that from each chosen bottle two pills are selected. Explain what type of sample this might be.

3.17. On what basis is the statement made that a properly designed stratified random sample will tend to be more accurate than a simple random sample?

RELATED READINGS

Barnard, G. A. "Summary Remarks," pp. 696–741. *New Developments in Survey Sampling.* New York: John Wiley & Sons, Inc., 1969.

Cochran, W. G. *Sampling Techniques.* 2d. ed. New York: John Wiley & Sons, Inc., 1963.

Cyert, Richard M., and Trueblood, Robert M. *Sampling Techniques in Accounting.* Englewood Cliffs, N.J.: Prentice-Hall, Inc., 1957.

Deming, W. Edwards. *Sample Design in Business Research.* New York: John Wiley & Sons, Inc., 1960.

Godambe, V. P. "Foundations of Survey Sampling." *The American Statistician,* vol. 24, no. 1 (February 1970), pp. 33–38.

Hansen, Morris H.; Hurwitz, William N.; and Madow, William G. *Sample Survey Methods and Theory.* vol. I: *Methods and Application;* vol. II: *Theory.* New York: John Wiley & Sons, Inc., 1953.

Hess, Irene; Riedel, Donald C.; and Fitzpatrick, Thomas B. *Probability Sampling of Hospitals and Patients.* Ann Arbor: Graduate School of Business Administration, University of Michigan, 1961.

Hyman, Herbert. *Survey Design and Analysis.* Glencoe, Ill.: Free Press, Inc., 1955.

Johnson, Palmer O., and Rao, Munamarty. *Modern Sampling Methods: Theory, Experimentation, Application.* Minneapolis: University of Minnesota Press, 1959.

Kahn, Robert L., and Cannell, Charles F. *The Dynamics of Interviewing* New York: John Wiley & Sons, Inc., 1967.

KISH, LESLIE. *Survey Sampling.* New York: John Wiley & Sons, Inc., 1965.

LANSING, JOHN B. *An Investigation of Response Error.* Urbana, Ill.: Bureau of Business and Economic Research, University of Illinois, 1961.

MONROE, JOHN, and FINKNER, A. L. *Handbook of Area Sampling.* Philadelphia: Chilton Co., 1959.

MORGAN, JAMES, and SONQUIST, J. A. "Problems in the Analysis of Survey Data and a Proposal." *Journal of the American Statistical Association,* June 1963, pp. 415–34.

NAMIAS, JEAN. *Handbook of Selected Sample Surveys in the Federal Government* (with Annotated Bibliography). New York: St. John's University Press, 1969.

PEARSON, KARL. "The Fundamental Problem of Practical Statistics." *Biometrika,* vol. 13, pp. 1–16.

RAIMI, RALPH A. "The Peculiar Distribution of First Digits." *Scientific American,* December 1969, p. 109.

ROSENBERG, MORRIS. *The Logic of Survey Analysis.* New York: Basic Books, Inc., 1968.

SOLNIM, MORRIS JAMES. "Sampling in a Nutshell." *Journal of the American Statistical Association,* vol. 52, no. 278 (June 1957), pp. 143–61.

U.S. DEPARTMENT OF LABOR, BUREAU OF LABOR STATISTICS. *BLS Handbook of Methods for Surveys and Studies,* Bulletin No. 1458, October 1966.

WARWICK, DONALD P., and LININGER, CHARLES A. *The Sample Survey: Theory and Practice.* New York: McGraw-Hill Book Co., 1975.

YAMANE, TARO. *Elementary Sampling Theory.* Englewood Cliffs, N.J.: Prentice-Hall, Inc., 1967.

ZEMACH, RITA. "Social Surveys and the Responsibilities of Statisticians." *The American Statistician,* December 1969, pp. 46–48.

Part II

Fundamental tools of descriptive statistics

I keep six honest serving-men
(They taught me all I knew)
Their names are What and Why and When
And How and Where and Who.

Rudyard Kipling

4

Empirical frequency distributions: Nature and construction

If a man can group his ideas, he is a good writer.

Robert Louis Stevenson

THE TERM *frequency distribution* is used to denote the classification of numerical data into convenient groupings. A very simple illustration is the statement that in a certain city 30 percent of the families received incomes under $5,000, 50 percent received incomes of $5,000 to—but not including—$10,000, and 20 percent received incomes of $10,000 or more. Expressed in dollar units, and excluding negative incomes, the classes are $0–$4,999, inclusive; $5,000–$9,999, inclusive; and $10,000 and above—a so-called "open" class. And the frequencies, respectively, are 30, 50, and 20 percent. Or more likely, the actual numbers would be given as 6,000, 10,000, and 4,000, respectively.

This matter of classifying or grouping data seems at first glance to be very simple in principle. Just send out enumerators to ask pertinent questions, and later have the answers recorded on punched tabulating cards which are run through machines. Then we have the answers either recorded as "raw" data—read from the cards in haphazard order—or neatly bunched by appropriate classes in frequency distribution.

But even aside from the fact that some of the enumerated answers may be inaccurate or false and that some families may have been overlooked, the problem of classification is by no means as simple as it seems. In fact, classification of data is a complex initial step in the development of science.[1]

[1] See Norman Campbell, *What Is Science?* (New York: Dover Publications, Inc., 1963), chap. 3.

DISCRETE AND CONTINUOUS DATA

It is customary to approach the problem of classification by dividing data into two categories: on the one hand, the discrete or discontinuous; and on the other hand, the continuous. The former category is assumed to include all data that by their very nature form distinct units, such as people, cities, residences, or books. These are discrete units, that is, separate, distinct, though we may have difficulty in defining each class exactly.

On the other hand, we have substances that, though measured by arbitrary units, may merge into each other, such as acres, water, or oil. To be sure, we find land fenced as separate farms, water delimited as lakes and ponds, and oil sold in quart containers which are counted as discrete units. Yet we could vary the latter units by very small graduations, making them practically continuous. But we could also break down discrete units like residences into number of rooms or square feet of floor space, and books could be counted by chapters, or pages, or words. Finally, all matter could be thought of in terms of molecules, atoms, or potential energy.

Thus, we could split hairs over the distinction between discrete and continuous until we were involved in the disputes of the Greek philosophers over the nature of the universe, or the arguments of the atomic scientists over the quantum theory. But since neither the philosophers nor the physicists have yet a clear-cut answer, we had better retreat from such abstruse distinction.[2] We should recognize the distinction in its most obvious forms, of course. But we shall find it more practical to turn attention to the units of counting and measurement that are commonly utilized in business.

UNITS OF BUSINESS DATA

When we examine the units of value and quantity in which most business data are expressed, the first thing we note is that they are all, in effect, discrete to a certain limit of accuracy—a limit that depends largely on convenience and common sense. For example, when total capital investments are listed by states, the value unit might very well be a thousand dollars. And when population by countries is tabulated, the figures might plausibly be rounded to millions. Very large units, however, may unduly lose accuracy, but very small units clutter up a tabulation and fail to give a clear-cut impression. The limit of accuracy, whether to the nearest thousand dollars or the nearest cent, whether in tons or ounces or intermediate units, all depends upon the research problem at hand, and is a matter of common sense, not of a logical or mathematical rule.

---

[2] For a lucid exposition of the principle involved, see Edna E. Kramer, *The Main Stream of Mathematics* (New York: Oxford University Press, 1952), chap. 13.

Not only are continuous units like dollars treated as discrete but distinctly discrete units are treated as continuous. For example, the United States census of 1970 reported 3.5 persons as the average-sized household or, in effect, the family.[3] Statistically, the discrete unit, a person, is treated as if intermediate fractional units are meaningful, as indeed they are. The statistic 3.5 is understood as indicating a central point in the distribution of households of various discrete sizes. Compared from census to census, such an average gives a much clearer picture of the trend than it would if logically rounded to the nearest unit.

DEFINITION OF TERMS

Before a real problem in classification is presented, it may be well to define terms by means of another hypothetical miniature case. Suppose there is at hand a set of raw data listing the number of employees per shop in a limited district of small shops. Inspection shows that they range from 1 to 12 employees to a shop. It is decided by considerations to be discussed later to classify them in four groups. Stated inclusively, the classes may be indicated by lower (L_1) and upper (L_2) limits of the range of each class, thus:

L_1		L_2
1	to	3, inclusive
4	to	6, inclusive
7	to	9, inclusive
10	to	12, inclusive

An inspection of these classes shows that they precisely include all the data. Shops having one, two, or three employees fall in the first class; those with four, five, or six employees fall in the second class; and so on. The limit of accuracy is indicated by the gaps between the classes. Between L_2 of one class and L_1 of the next class, there is a gap of one unit. In this case the unit is determined by the nature of the research, though the scale would change if workers were entered in terms of years of experience or tested efficiency.

Nevertheless, from the standpoint of computations, the data are treated as if they were continuous. The average number employed might come out to 4.32, or some other decimal number. Other averages discussed later, such as the median, might also have decimal values.

Because a discrete series becomes continuous, in effect, when employed in computations, mathematicians generally advise that the data should be treated as if each unit were expressed as a unit space rather than

---

[3] For successive censuses from 1900 to 1970, inclusive, the figures run 4.7, 4.4, 4.2, 4.0, 3.7, 3.4, 3.6, and 3.5. Though reported as persons per household, in effect they represent the average-sized family. In 1975 it was estimated that the figure is less than 3.

by a point, that is, as if one were represented by a space on the graphic scale from .5 to 1.5; two, from 1.5 to 2.5, and so on, putting the discrete item at the midpoint of this space.[4] Hence, as arranged for computation, the class limits above should have been written, or at least mentally extended, thus:

L_1		L_2
.5	to	3.5
3.5	to	6.5
6.5	to	9.5
9.5	to	12.5

It will be seen that the midpoint of each class has not been changed; it is still the average of L_1 and L_2. But the class range, or *interval* (i), which for the discrete classification first given seemed to be $L_2 - L_1 = (3 - 1) = (6 - 4) \ldots = 2$, now becomes

$$
\begin{aligned}
i = L_2 - L_1 &= 3.5 - .5 = 3 \\
&= 6.5 - 3.5 = 3 \\
&= 9.5 - 6.5 = 3 \\
&= 12.5 - 9.5 = 3
\end{aligned}
$$

which is the correct interval when the data are thought of as continuous for the purpose of computation. Of course, the intervals are not necessarily equal; the higher classes where the frequencies are small are often broadened, but i is still computed as indicated. However, if the classification is left in the original discrete form, the continuous interval is indicated as the difference between a given L_1 and the next L_1.

To summarize, it may be said that class limits are preferably set at first as discrete, in effect, to whatever limit of accuracy seems appropriate to the problem at hand, and are made *inclusive,* so that both discrete L_1 and L_2 represent items actually present in the data. Then, for purposes of computation, they are either actually rewritten or mentally assumed to be broadened to the continuous form, thus making the upper limit of one class identical with the lower limit of the next class.

It should be kept in mind that computations based upon a frequency distribution are almost always less accurate than the same computations based upon the raw data, even though theoretically they may suggest the nature of the universe from which they are drawn. When raw data are available, computations are usually made directly from them. The chief reason for considering computation from a frequency distribution is that

[4] An exception occurs with classes such as value or quantity units of 0–2, 3–5, and so on. Negative values being proscribed, the first class will be shortened 0–2.5, with a midpoint 1.25 and an interval of 2.5. Other classes of changing intervals, however, will be readily handled. Thus the classes 0–2, 3–5, 6–12 would be made continuous as 0 to 2.5, 2.5 to 5.5, 5.5 to 12.5, with midpoints of 1.25, 4.0, 9.0, and intervals (i) of 2.5, 3.0, and 7.0, respectively.

so often data are presented in that form, without reference to the raw data. Also, if many observations are made, it is more convenient to summarize the data in a frequency distribution. But it cannot be too often repeated that the assumptions made in such computations are only approximations that may sometimes lead to appreciable error.[5]

A REALISTIC PROBLEM

The discussion of frequency distributions thus far has dealt chiefly with definitions and theoretical considerations. It may be well, therefore, to turn to an actual problem in the classification of business data. For this purpose the weekly wages (including some piecework and overtime) of 200 factory workers have been selected. They represent a typical distribution expressed in dollars and cents. The figures have been slightly modified in the interest of convenience of computation. It will not be necessary to record them here in full; they were copied as miscellaneous items selected as a stratified random sample from different classes of workers. In order to indicate the nature of the sample, it should be sufficient to list a few items at the beginning and at the end of the list, as follows: $66.54, $78.12, $61.91, $119.06, $52.60, . . . , $72.15, $51.23, $84.81, $101.13, $73.74.

It is now necessary to decide on the number of classes and the class limits. As has been noted, this is largely an arbitrary decision. The purpose is chiefly to present the data in such a way as to indicate the nature of the distribution. Obviously, this requires several classes, but not so many that there will be insufficient items in each class.

The frequency array

As a preliminary step, it is advisable to arrange the raw data in the form of an array, in order to determine whether they scatter normally or "bunch" in irregular groups. If, for example, they are more frequent at even 5 or 10 dollars, or some other point on the scale, it would be well to arrange the classes so that these items would fall near the middle of the class, so that the midpoint fairly represents them. In Table 4.1 the data are all arrayed at dollar values, thus making, in a sense, a preliminary fre-

[5] Sometimes, apparent errors in recording and classifying data conveniently offset. Suppose, for instance, that ages of adults as of the *last* birthday are entered in classes with limits of 20 to, but not including, 22, and so on. Only ages 20 and 21 will be entered in this class. But the midpoint will be 21, which is half a year greater than the average of 20 and 21. But half a year added to the recorded ages of 20 and 21 gives practically the true ages in years and fractions of years, approximated as of the nearest birthday. Hence the bias involved in recording ages as of the last birthday is here offset by the bias of class limits. But in general, it is safer to avoid bias by collecting data as of the nearest unit and selecting unbiased class limits.

TABLE 4.1
Frequency array of weekly wages of 200 workers of Corporation A, June, 1976 (X is the wage; f is the number of workers)

X	f	X	f	X	f	X	f
45	1	64	9	83	5	99	2
46	1	66	3	84	5	100	2
47	1	67	2	85	4	101	2
48	1	68	5	86	5	104	1
50	3	69	10	87	4	105	1
51	3	70	12	90	4	107	1
53	2	71	12	91	3	108	1
54	2	72	5	92	2	110	1
55	4	74	5	93	3	113	1
56	3	75	8	94	1	114	1
58	5	78	7	95	1	119	1
59	6	80	7	96	2	121	1
60	6	81	6	97	1	$\Sigma f=$	200
62	7	82	8	98	1		

quency distribution. For convenience the "nearest dollar" was used; that is, the data were rounded. This was done so that they could be readily pictured, as in Figure 4.1. This speeds the plotting, but it involves a danger of error, for it hides a possible tendency of the data to cluster at dollar values. However, an inspection of the data shows that the cents that have been eliminated scatter widely, so the frequency array plotted as of the nearest dollar will suffice to describe the overall scatter. Since there is no apparent tendency for the data to "bunch" consistently, the choice of class limits may be considered merely from the standpoint of appropriateness.

FIGURE 4.1
Distribution of weekly wages of 200 unskilled workers of Corporation A, June, 1976

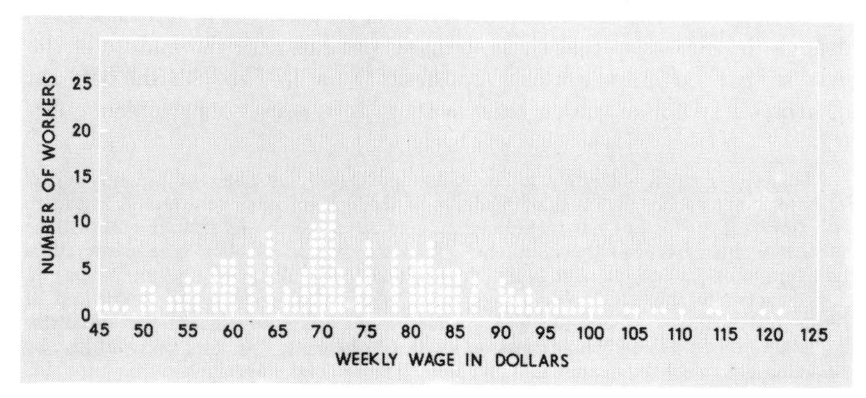

Number of classes

It has already been noted that the choice of the number of classes must be a compromise. We must choose between literal accuracy and the ability to generalize. At least three classes are required to give any idea of the nature of the distribution. And the maximum number of classes is limited by n—the number of items to be classified. Frequently, a rough estimate of the number of classes is made by use of the expression

$$K = 1 + (3.222 \log n) \tag{4.1}$$

where K is the estimated number of classes and n is the number of items to be classified, in this case 200. The estimated number of classes for the 200 wage items using formula (4.1) is

$$\begin{aligned} K &= 1 + (3.222 \log 200) \\ &= 1 + (3.222 \cdot 2.3010) \\ &= 8.41 \end{aligned}$$

About eight classes seem appropriate, and Table 4.2 has been constructed on this assumption. However, this is only a rough approximation. The satisfactory number of classes may be more or less than eight.

This formula is based on the number of items and classes in symmetrical binomials. Actually, this is the so-called "Sturge's rule" which merely is nothing more than an approximation of the number of classes that might be desirable for valid analyses. In reality, determining the optimum number of classes is both an art and a science. The researcher must use good judgment in compiling the frequency distribution. Knowledge of the scope and limitations of the data is a necessary prerequisite

TABLE 4.2
Frequency distribution of average weekly wages of unskilled workers of Corporation A, 1976

Wage L_1	Wage L_2	Number of Workers f
\$ 45.00–	\$ 54.99	14
55.00–	64.99	40
65.00–	74.99	54 (modal class)
75.00–	84.99	46
85.00–	94.99	26
95.00–	104.99	12
105.00–	114.99	6
115.00–	124.99	2
		$n = \Sigma f = 200$

to good data reduction and analyses. The suggestion to use Sturge's rule may not *always* be justified; it is merely one way of beginning the analysis.

The class interval

The class interval has already been discussed from a theoretical point of view. However, the practical problem at hand requires establishing a class interval for the frequency distribution used to classify the 200 wages. A simple method of approximating the interval is to divide the range of the raw data by the chosen number of classes. The smallest wage was $45.12, and the largest wage was $121.32; then the approximate class interval (i_a) can be determined as

$$i_a = \frac{\$121.32 - \$45.12}{8} = \$9.525$$

Obviously, such a class interval as $9.525 would be illogical as well as inconvenient. It is only a guide, which suggests a class interval of $10. Such an interval as 10 is not only convenient but also provides for the complete inclusion of all the raw data.

Class limits

The lower limit of the first class is set at a convenient number reasonably close to the lowest item in the data. This limit is $45. Since the class interval has been set at $10, the discrete limits of all the classes are now determined as shown in Table 4.2, namely, $45.00–$54.99, $55.00–$64.99, and so on. As has been explained, the theoretical continuous classes have overlapping limits such as $44.995–$54.995, with a midpoint at the average of these limits, but this half-cent change is relatively insignificant and is disregarded. Hence, in computing and charting, L_1 is taken as $45, $55, and so on, and the midpoints as $50, $60, and so on.

The basis of classification thus determined proves satisfactory from the standpoint of the frequencies, which rise from the first frequency to a fairly well-defined apex, indicating a *mode* (class or point of greatest frequencies), and then descend somewhat regularly to the last frequency. But if the class interval had been only $5, Table 4.3 would have been the result. This table has three modal classes; it is trimodal. But with the class interval enlarged to $10, the trimodal condition disappears. However, data may be bimodal or even trimodal by nature. That is, a bimodal distribution, for example, may show that noncomparable data have been put in one tabulation, as may be illustrated by a distribution of both men and women classified by height.

TABLE 4.3
Frequency distribution of average weekly wages of unskilled workers of Corporation A, 1976

Wage L_1	Wage L_2	Number of Workers f
$ 45.00–	$ 49.99	4
50.00–	54.99	10
55.00–	59.99	18
60.00–	64.99	22 (modal class)
65.00–	69.99	20
70.00–	74.99	34 (modal class)
75.00–	79.99	15
80.00–	84.99	31 (modal class)
85.00–	89.99	13
90.00–	94.99	13
95.00–	99.99	7
100.00–	104.99	5
105.00–	109.99	3
110.00–	114.99	3
115.00–	119.99	1
120.00–	124.99	1
		$n = \Sigma f = 200$

A basic assumption

To repeat what has been said before, computations from frequency distributions are at best only estimates. Therefore, ideally, all computations should be made from raw data. However, it is frequently inconvenient to do this even when the raw data are available. In any case, calculations made from a frequency distribution are customarily based upon the convenient but unrealistic assumption that the items are evenly distributed through each class, and that the midpoint is thus the average of the items in a class. This assumption may be checked by reference to the items which made up the first class of Table 4.2, and which were: $45.12, $45.78, $46.84, $47.52, $50.02, $50.12, $50.40, $50.90, $51.10, $51.30, $52.60, $53.30, $54.10, and $54.48. The sum of these 14 items is $703.58, and the average is $50.26. The midpoint of the first class is $50, which is slightly less than the average of the 14 items. But in computing averages, such a difference tends to be offset in the frequency distribution as a whole because in those classes above the modal class the midpoint ordinarily is larger than the average of the items in the class.

CHARTING THE FREQUENCY DISTRIBUTION

The advantages of graphic presentation may be realized by plotting a frequency distribution in a form that pictures the pattern of the data. For this purpose the histogram and frequency polygon are the two types of charts most commonly used.

Histogram

The histogram is plotted on ordinary cross-section arithmetic paper. It is composed of contiguous, vertical rectangles representing the frequencies. This type of graph, based upon the data of Table 4.2, appears

FIGURE 4.2
Histogram of weekly wages of 200 unskilled workers of Corporation A, June, 1976

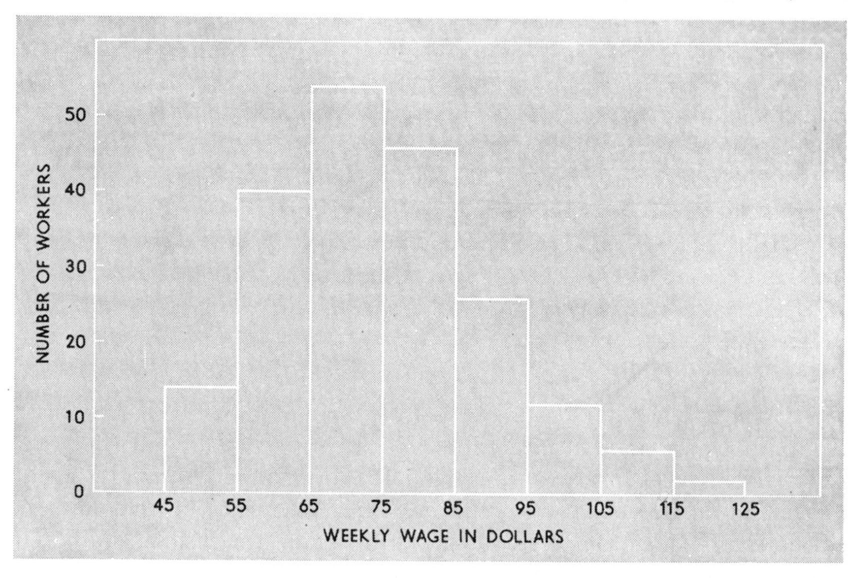

in Figure 4.2. It will be seen that the width of each vertical rectangle is the class interval rounded for convenience to $10, the first beginning with $45. The height of each rectangle, or column, on the Y scale is the frequency.

The classes here have the same interval, and therefore the columns have the same width. When the intervals vary, however, the columns require a compensating adjustment. For example, suppose that the last two classes had been combined as one class, making the limits $105.00–$124.99 and the frequency eight. In order to maintain the same area as before, the height of this column would have to be four and the average of the original frequencies, six and two. The base having been doubled, the combined frequency, eight, is cut in half. This readjustment is plotted in Figure 4.3.

The histogram is essentially an *area* diagram, though the customary method of plotting the frequencies on the Y axis violates this principle.

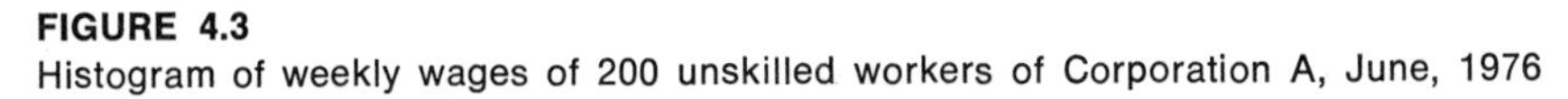

FIGURE 4.3
Histogram of weekly wages of 200 unskilled workers of Corporation A, June, 1976

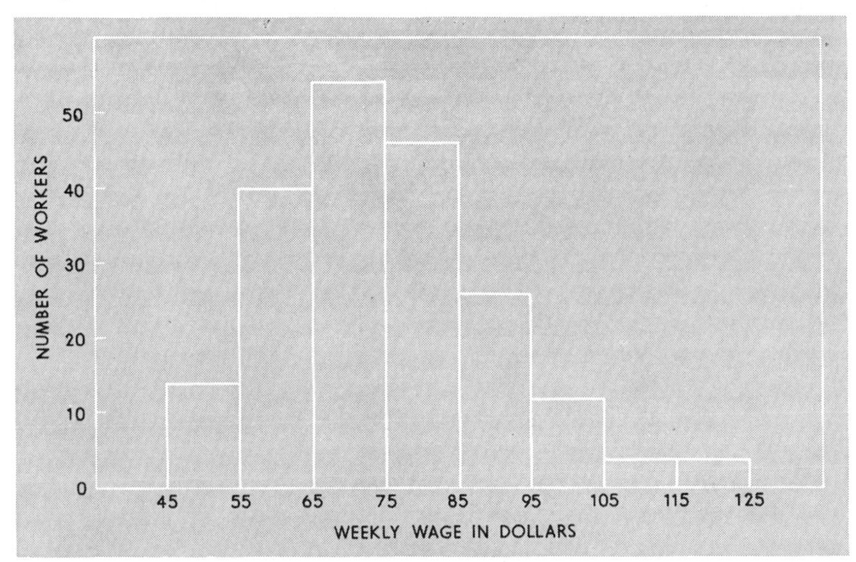

Theoretically, each frequency should be plotted at the graphic height f/i if it is to represent the area correctly. When the class intervals vary throughout the distribution, this is the only adjustment that can be made. But usually, there are at most only two or three classes that vary from the common standard with respect to the class interval; in such a case, frequencies in the standard classes are usually plotted on the Y axis, and the frequencies of the class with an irregular i are adjusted by dividing them by the ratio of the adjusted i to the standard i, as illustrated in Figure 4.3. The graphic height of f in the open-end class, however, can only be roughly estimated, preferably by the use of probability paper, as discussed later.

Frequency polygon

The data of Table 4.2 can be plotted as an arithmetic line chart which, because of the several angles it forms, is called a frequency polygon. The principles used in constructing a simple line chart are adhered to in drawing this polygon. Each Y value is a frequency, and the X is the midpoint of its class.

In order to retain the same area under the polygon as that of the histogram, the line is ordinarily extended to the midpoint of classes at the extremes having zero frequencies, as is shown in Figure 4.4. However,

FIGURE 4.4
Frequency polygon and histogram of weekly wages of 200 unskilled workers of Corporation A, June, 1976

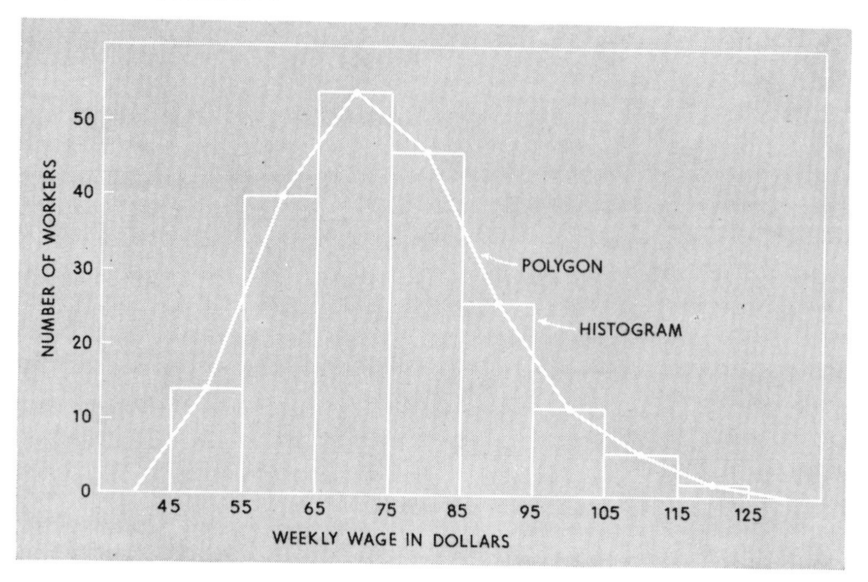

this is strictly logical only if the data represent a sample.[6] Again, if the class interval is not identical for every class, an adjustment such as that used for Figure 4.3 may be made, though the area may then be distorted.

Normal curve

No discussion of frequency distribution charts can be undertaken without mention of the so-called "normal" curve. This is illustrated by the continuous bell-shaped curve shown in Figure 4.5. As discussed in previous chapters, it is the foundation of modern statistical theory. However, it can be noted here that numerous measurements of the same variable often present a high degree of regularity, and the scatter of data when suitably grouped tends to follow this so-called "law of the normal curve," either on the ordinary arithmetic scale or on some other scale adapted to the type of distribution.

It is sufficient here to note that a normal distribution of four items in classes of equal intervals would approximate frequencies in ratios of 1:2:1, and 16 items would tend to distribute in ratios of 1:4:6:4:1. Frequencies are typically large in the central classes and fall away toward the extremes.

---

[6] This may well cause difficulty if the lower limit of the smallest class is zero. To extend the line into quadrant II in such a case would ordinarily be unrealistic.

FIGURE 4.5
The normal curve of distribution

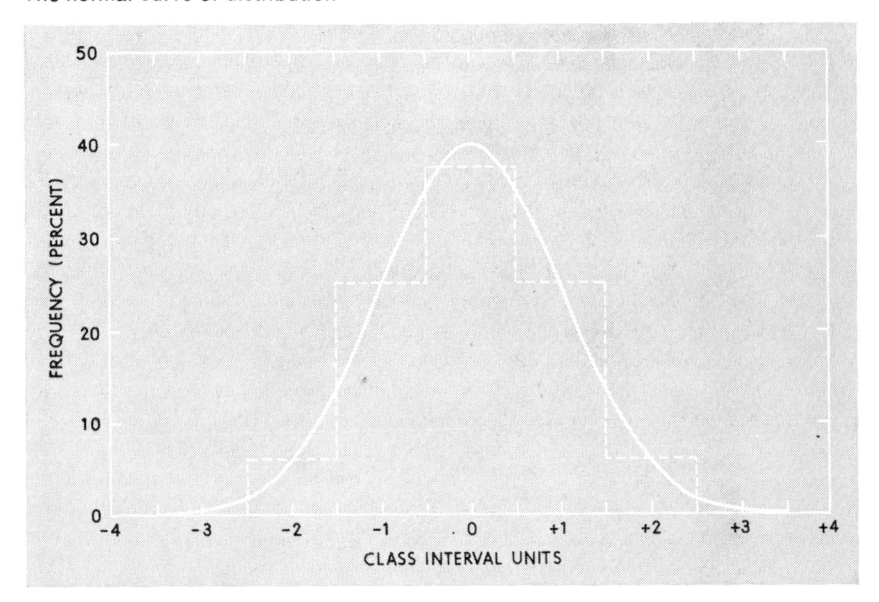

The normal curve is plotted in Figure 4.5 against a background histogram of frequencies having the ratios just noted, namely, 1:4:6:4:1, with a centered X scale of unit intervals. These are *standard deviation* units. In terms of these units and of the Y scale as plotted, the normal curve has unit area.

A comparison of the frequency polygon of 200 wages, Figure 4.4, with Figure 4.5 shows that the polygon is pulled out to the right. As will appear in later discussions, this characteristic is called skewness—positive when, as here, it extends to the right, and negative when it extends to the left.[7] As is usual with business data, it does not conform to the theoretical normal curve on an arithmetic scale, though it may on some other scale. The further it deviates from such an ideal curve, the more complex will be the analysis.

Smoothed histogram

In Figure 4.6 the histogram of Figure 4.4 has been smoothed to a line of continuous relative frequencies in order to approximate a little more closely the universe from which the sample of 200 wages was drawn. It

[7] This is not to say that the data have been improperly classified. Figure 4.1 shows that the data were pulled out to the right, toward the higher values (positive skewness), when expressed merely as an array at dollar intervals.

FIGURE 4.6
Smoothed histogram of weekly wages of 200 unskilled workers of Corporation A, June, 1976

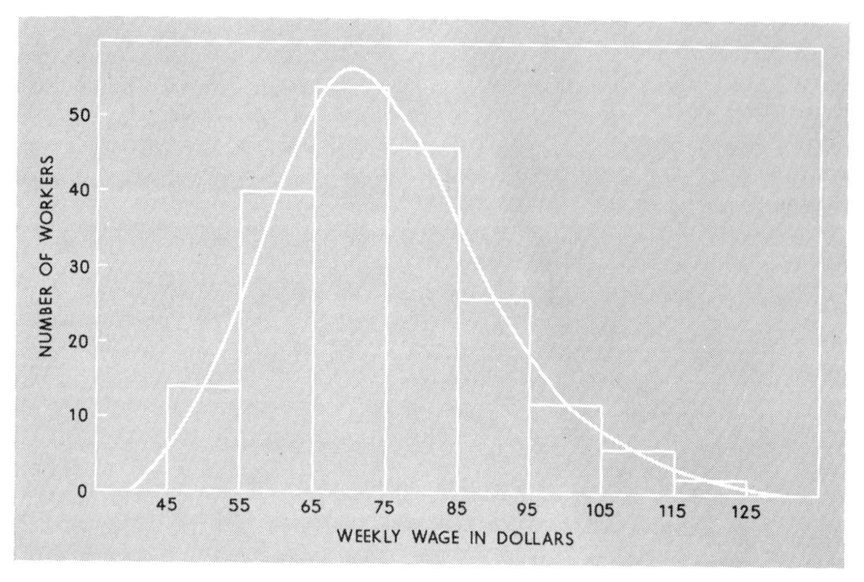

will be seen that the smoothing has removed the irregularities occasioned by the wide grouping, while retaining approximately the same area. It has not, however, removed the skewness, that is, the extension of the frequencies to the right, which appears abnormal in comparison with the theoretical normal curve. Possibly the distribution will approach the normal more closely if plotted on another base, such as the logarithmic scale.

OTHER DISTRIBUTIONS

Thus far, only the simple frequency distribution has been discussed. In it the frequency of each class is shown separately, and the total frequencies are the sum of those in individual classes. However, for reasons discussed below, it is often desirable—in fact, it is often necessary—to change the form of the frequency distribution from a simple distribution to a cumulative one.

Cumulative frequency distribution

For some purposes the most convenient form of a frequency distribution is one that accumulates or subtotals the frequencies consecutively with reference to the lower or the upper limits of the classes, assumed to

TABLE 4.4
Cumulative frequency table (data from Table 4.2)

Wage L_1	Wage L_2	Number of Workers f	Less than the Upper Limit*	More than the Lower Limit*
$ 45.00–	$ 54.99	14	14	200
55.00–	64.99	40	54	186
65.00–	74.99	54	108	146
75.00–	84.99	46	154	92
85.00–	94.99	26	180	46
95.00–	104.99	12	192	20
105.00–	114.99	6	198	8
115.00–	124.99	2	200	2
		$\Sigma f = 200$		

* The cumulatives are exact in terms of the *continuous* class limits: $44,995–$54,995 and so on, which are theoretically assumed and closely approximated by the stated limits.

be taken at the continuous levels they approximate, namely, $45 to $55, and so on. Table 4.4 shows the cumulative on both these bases. In the "Less than the Upper Limit" column, cumulative frequencies are totaled from the beginning of the f column. That is, 14 workers made less than $55, 54 workers made less than $65, 108 made less than $75, and so on. In the "More than the Lower Limit" column, the reverse procedure is used, and frequencies are totaled from the bottom of the f column. That is, two workers made $115 or more, eight made $105 or more, and so on. The table clearly shows the number of workers making less than or more than the wage at any continuous class limit.

Cumulative frequency chart, or ogive

The cumulative frequency distribution may be plotted as an arithmetic line graph, as is illustrated in Figure 4.7. This chart shows the cumulatives on both a "more than" and a "less than" basis. Such a double chart is convenient for ready reference. If, for example, management wants to know approximately how many of the 200 workers made less than $66, the chart will answer the question. The procedure consists of locating $66 on the X axis and reading up to the "less than" curve, which shows the number of workers at that point to be 60. And the number theoretically receiving more than $66 is read on the other line as 140 workers.

From what has been said above, it is evident that the frequency distribution may be broken at any point by use of the cumulative frequency chart. Thus, such a chart is a useful tool for reclassifying data that are available only in frequency distribution form. For example, a frequency

FIGURE 4.7
Ogive, or cumulative frequency chart

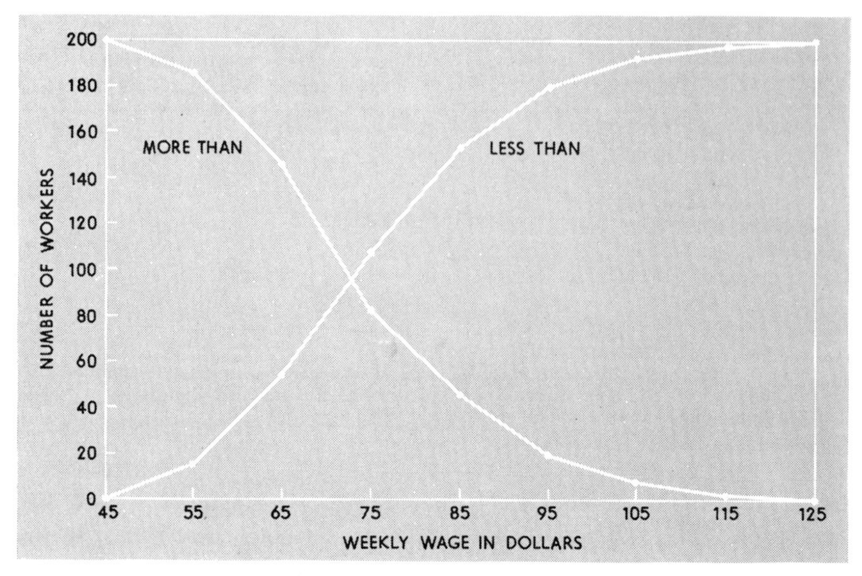

Source: Data of Table 4.4.

distribution of population by age groups may be available in classes that are not suitable for a given piece of research. The data may be reclassified by reading the cumulatives on the chart at the desired limits and subtracting to find the specific frequencies. Or, as will appear later, the cumulatives at the required limits may be mathematically interpolated.

Normal probability paper

An inspection of Figure 4.7 will immediately suggest that the lines connecting any two plotted points would be curved if more data and more classes were available. The straight lines express the convenient but slightly erroneous assumption that the items in any class tend to be spaced at equal distances apart. Fortunately, a form of graph paper has been invented that implies a more realistic assumption. This is the so-called "normal probability" paper, which is so constructed that the ogive cumulating the theoretical normal probability curve will appear as a straight line.

The first step in the use of probability paper is the reduction of the frequencies to percentages of their total, that is, they are rewritten (unless they already total 100) as $100f/n$. This is done in Table 4.5. There may be a slight loss of accuracy in so doing, in that the percentage fre-

TABLE 4.5
Cumulative percentage table (data from Table 4.2)

Wage L_1	L_2	Number of Workers f	Percent of Workers $f\%$	Less than the Upper Limit*	More than the Lower Limit*
$ 45.00–	$ 54.99	14	7	7	100
55.00–	64.99	40	20	27	93
65.00–	74.99	54	27	54	73
75.00–	84.99	46	23	77	46
85.00–	94.99	26	13	90	23
95.00–	104.99	12	6	96	10
105.00–	114.99	6	3	99	4
115.00–	124.99	2	1	100	1
		$\Sigma f = 200$	$\Sigma f\% = 100$		

* These cumulatives are exact only in terms of the *continuous* class limits: $44,995–$54,995, and so on, which are theoretically assumed and closely approximated by the stated limits.

quencies are seldom whole numbers, as here; but for graphic purposes, this is negligible.

The ogive of percentage frequencies is presented in Figure 4.8. This does not reproduce all the details of the original form as sold by stationers, but it does indicate the usual "more than" Y scale, on the left hand, and the "less than" scale, on the right. The former scale takes the place of the "more than" cumulative of Figure 4.7, and only the "less than" cumulative is drawn–the one that is commonly made the basis of interpolations and other computations. The percentage scales could be adapted, of course, to other cumulative totals than 100, but it is usually more convenient to adapt the data to the percentage scales as drawn.

Figure 4.8–the probability chart–can be used for the purposes discussed in connection with Figure 4.7, with more accurate results. That is, it can be used to estimate the percentage of workers receiving a wage both below and above a specified sum, or the percentage of workers between any given limit. And it can be used to reclassify the data at any chosen limits. When it is appropriate to assume the possibility of population frequencies beyond the limits of the data, these frequencies may be estimated by extending the ogive tentatively at the two ends, as here indicated by dotted lines. But the ogive is not completed at the extremes, since the 0 percent and 100 percent limits are at infinite distances beyond the chart.

If management wishes a careful estimate, for example, of the maximum wage received by the lowest fifth of the workers in the total population of which the given 200 workers are a sample, it could be read from the chart at approximately $62.50. Similarly, other interpolations may be graphically estimated.

FIGURE 4.8
Ogive, or cumulative percentage chart, on "probability paper"

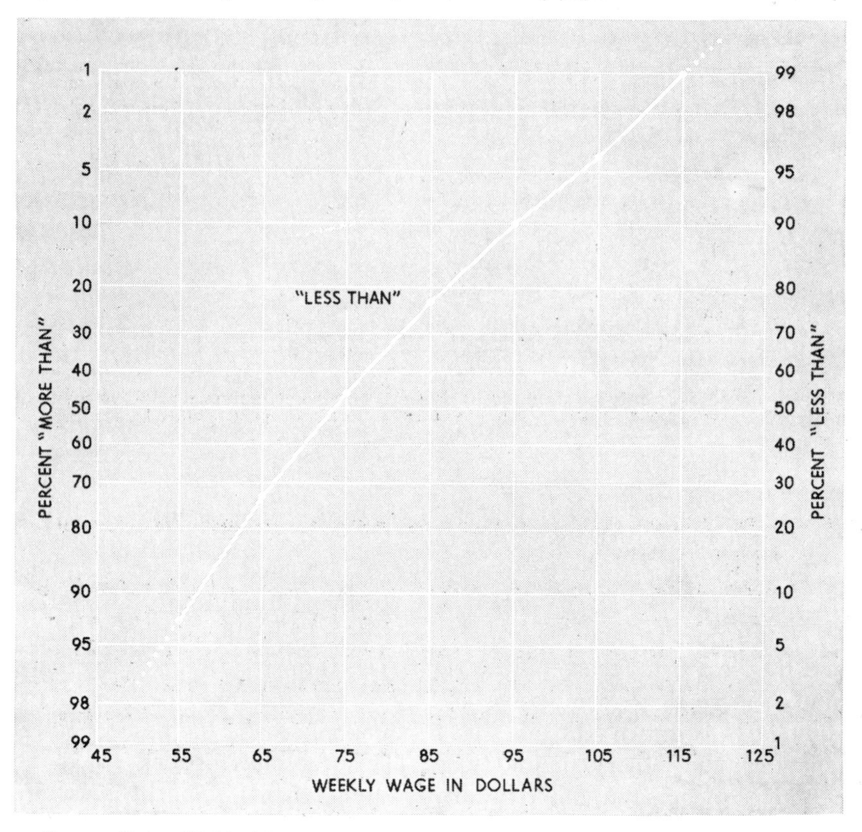

Source: Data of Table 4.5.

Suppose, further, that management wishes to know how many workers made between \$66 and \$86, inclusive. This information can be given approximately by reading Figure 4.8 on the ordinates $X = 66$ and $X = 86$ and reading the interpolations 30 percent and 80 percent, respectively. Hence, 50 percent $(80 - 30 = 50)$ of the workers, or 100 workers, in the sample made between \$66 and \$86, inclusive. The estimate may be made more precise by interpolation on a logarithmic chart, but the one just given is probably close enough.

The probability chart is perhaps the most underrated graphic device. There are a great many uses to which it can be put beyond those discussed here. For example, a straight-line ogive, either on an arithmetic or on a logarithmic base (X) scale, drawn so as to approximate the plotted frequencies, indicates or at least suggests the theoretical distribu-

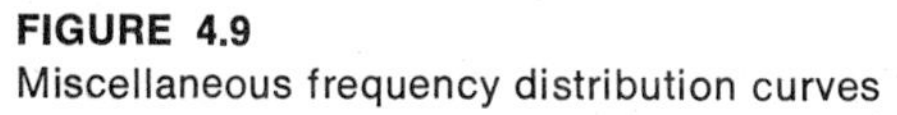

FIGURE 4.9
Miscellaneous frequency distribution curves

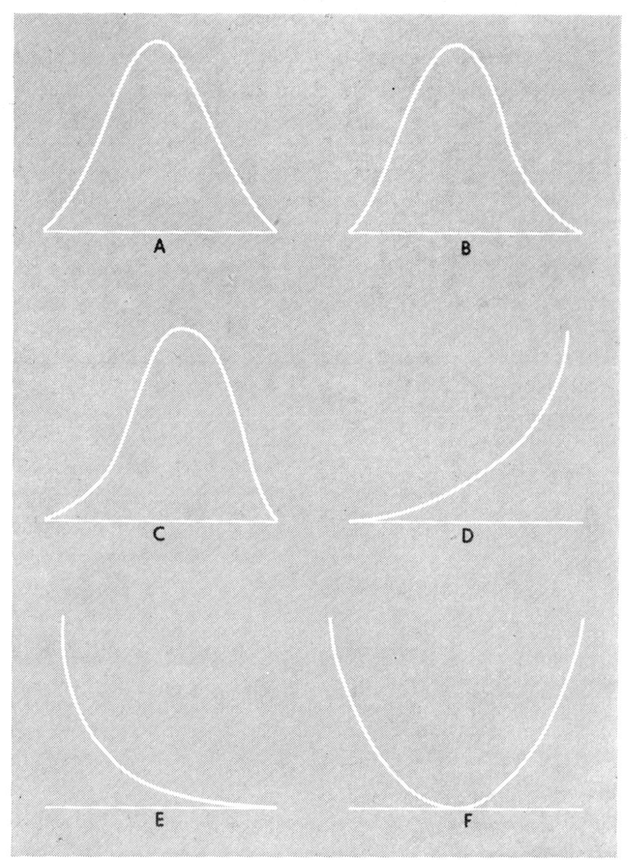

tion from which the sample is drawn. In practice, other uses will be discovered.

Miscellaneous distributions

Many varieties of frequency distributions may be found in the statistics of business, but the ones discussed here exemplify the most important. Obviously, an elementary treatise does not provide space for a discussion of all possible types. It should, however, be noted that a frequency distribution, such as Table 4.2, when charted, appears approximately like curve A in Figure 4.9, positively skewed (pulled out) to the right like curve B, or negatively skewed to the left like curve C.

Then, too, a frequency polygon may look like curve D, which shows that the frequencies are in an ascending order. This is sometimes called a **J** curve. Accident rates by age groups sometimes take this form. Curve E is one resulting from frequencies being in a descending order. Number of cars parked in business districts by length of time parked sometimes takes this form, which is called a *reverse* **J**. Curve F is a **U** curve which shows high frequencies at the ends and low frequencies in the central classes. Such a curve would result from a frequency percentage table of the unemployed by age groups. In none of the three cases, curves D, E, or F, is it feasible to compute a measure of central tendency, that is, an average. Such extremely irregular distributions belong to a class quite apart from those that have been tentatively analyzed in this chapter.

SYMBOLS

L_1, lower limit of class
L_2, upper limit of class
i, class interval
i_a, approximate class interval
X, a variable
f, frequency of occurrence
Σf, sum of the frequencies
n, number of items; sample size
K, estimated number of classes

PROBLEMS

4.1. What type of population model would best represent the following populations? Describe as symmetrical, **U**-shaped, or skewed to the left, and so on. Sketch the smooth curve you have selected; and in each case, explain the reasons for your choice.

a. The monthly rents of three-room apartments in a college community (use your city).
b. The diameters of a mass-produced product.
c. Scores on the Graduate Management Admissions Test (GMAT) examinations.
d. Ages of employed persons 14 years and older.
e. The ages of readers of a space-age comic book.

4.2. Would you think it might be reasonable to conclude that a frequency distribution of *Fortune* magazine subscribers in 1976, classified by 1976 family income, would be symmetrical? If so, why? If not, why not?

4.3. Given the following hypothetical information:

Average weekly earnings of production workers on manufacturing payrolls in 50 states, 1976

$ 89	$122	$127	$120	$120	$ 79	$ 88	$105	$119	$129
88	111	123	111	116	82	96	108	93	121
99	117	145	105	112	86	79	118	118	129
105	111	121	110	91	96	78	111	119	165
92	132	116	107	114	104	113	113	133	99

a. Array the above data as a first step in selecting class interval and class limits.
b. Present in a frequency table the distribution you believe to be an acceptable portrayal of the gross average weekly earnings.
c. Present this distribution as a frequency polygon and as a histogram on one chart.
d. Present your distribution of (*b*) as a cumulative frequency distribution in acceptable tabular form.
e. Present the table of (*d*) as a cumulative frequency table and chart. Interpret your results.

4.4. Using the following hypothetical data:

Average weekly hours of production workers on manufacturing payrolls in 50 states, 1976

41.5	40.2	41.7	40.7	41.0	42.0	41.7	42.0	40.1	39.6
41.1	41.3	43.3	42.3	41.2	41.2	41.5	40.6	41.5	40.8
43.0	40.8	42.0	45.6	41.6	42.5	41.4	40.7	40.6	42.3
40.7	42.4	41.5	43.1	40.5	40.9	42.5	38.5	40.4	40.4
43.2	41.9	41.2	42.9	41.4	40.8	41.7	41.3	39.7	40.4

a. As a first step in selecting class interval and class limits, array the above data.
b. Present in acceptable tabular form the frequency distribution you believe best portrays the average weekly hours.
c. Present your distribution in acceptable graphic form as a frequency polygon superimposed on a histogram.
d. Present your distribution as a cumulative frequency table and chart. Interpret your results.

4.5. Given the following hypothetical information:

Average hourly earnings of production workers on manufacturing payrolls in 50 states, 1976

$2.15	$2.82	$3.02	$2.91	$2.79	$2.82	$2.55	$2.67	$2.94	$3.28
2.14	2.77	2.95	2.72	2.84	1.92	2.16	2.51	2.87	3.25
2.30	2.84	3.35	2.48	2.72	1.96	2.31	2.57	2.32	3.05
2.57	2.72	2.87	2.42	2.95	2.09	1.90	2.91	2.85	3.16
2.27	3.10	2.80	2.47	2.18	2.26	1.89	2.73	2.94	3.90

a. As a first step in selecting class interval and class limits, array the above data.
b. Present in acceptable tabular form the frequency distribution you believe best portrays the gross average hourly earnings.
c. Present your distribution in acceptable graphic form as a frequency polygon superimposed on a histogram.
d. Present your distribution as a cumulative frequency table and chart. Interpret your results.

4.6. Given the following hypothetical data:

Consumption taxes in relation to family personal income class, 1976

Income	Taxes
Under $2,000	8.8%
$2,000– 3,999	7.2
4,000– 5,999	6.8
6,000– 7,999	6.6
8,000– 9,999	6.4
10,000– 14,999	5.9
15,000 and over	3.9

a. Is this a frequency distribution? How can you tell?
b. Comment on the class intervals and class limits.

4.7. Given the following:

Age distribution of the automobile stock (percent distribution)

Age of Automobile	1941	1947	1952	1957	1962	1972*	1975*
Under 4 years	38	13	46	40	36	38	37
4 to under 8 years	38	30	20	39	36	37	41
8 to under 12 years	15	37	11	16	20	16	15
12 years and over	9	20	23	4	8	9	7
All registrations	100	100	100	100	100	100	100
Median age (years)	4·5	8.5	4.0	4.5	5.5	5.4	5.6
Mean age (years)	5.0	8.0	6.0	5.0	5.5	5.4	5.5

* Estimated by author.
Source: *Survey of Current Business,* September 1963, p. 19.

a. Present these seven distributions in graphic form for the purpose of comparison.
b. What conclusions are warranted from the above table and your chart(s)? Write a brief paragraph summarizing your conclusions.
c. Are these discrete or continuous distributions?

4.8. Use the information in the table which appeared in Problem 4.7.
a. Are these frequency distributions? How can you tell?
b. Comment on the class intervals and class limits.
c. How do these distributions differ from your distributions of Problems 4.3, 4.4., and 4.5?
d. Write a brief paragraph based on the above table. Would you think a similar table based upon a sample of one thousand households in Russia would lead you to the same general conclusions? If not, why not?

4.9. Construct a cumulative frequency table and chart from the following:

All New York Stock Exchange listed common stocks

Number of Consecutive Years Dividends Paid	Number of Stocks (annual record)
100 or more	16
75 to 99	34
50 to 74	149
25 to 49	492
20 to 24	123
10 to 19	118
Less than 10	358

Source: New York Stock Exchange, *1970 Fact Book,* p. 23.

4.10. Construct a cumulative frequency chart and table from the following:

Annual common stock price changes, New York Stock Exchange, 1969

	Number of Issues	
Price Change	Up	Down
50% and over	12	144
40 to 50	13	153
30 to 40	18	178
20 to 30	36	251
10 to 20	53	166
Less than 10%	69	106
Totals	201	998

Source: New York Stock Exchange, *1970 Fact Book,* p. 18.

4.11. "With respect to a **U**-shaped distribution, the computed average has no meaning in such a situation." Comment on this statement.

4.12. Assume that undergraduate engineering majors start at higher earnings than statisticians and that the highest earnings of the two are equal. Further assume that in the middle-income ranges of the two groups, statisticians characteristically earn more than engineers. Under these conditions, approximate the "less than" ogives for both professions on the same graph. Label your chart carefully.

4.13. Distinguish between the frequency polygon and the ogive.

RELATED READINGS

Alder, Henry L., and Gillespie, Edward B. *Introduction to Probability and Statistics,* chap. 2. San Francisco: W. H. Freeman & Co., 1960.

BLALOCK, HUBERT M., JR. *Social Statistics,* chap. 5. New York: McGraw-Hill Book Co., 1960.

CHURCHMAN, C. W., and ROTTOOSH, P. *Measurement, Definitions, and Theories.* New York: John Wiley & Sons, Inc., 1959.

COOMBS, C. H. *A Theory of Data.* New York: John Wiley & Sons, Inc., 1964.

EHRENBERG, A. S. C. *Data Reduction.* New York: John Wiley & Sons, Inc., 1975.

KU, HARRY H. *Precision Measurement and Calibration.* Washington, D.C.: U.S. Government Printing Office, 1969.

MORGENSTERN, OSKAR. *On the Accuracy of Economic Observations.* 2d ed. Princeton: Princeton University Press, 1963.

PFANGEL, J. *Theory of Measurement.* New York: John Wiley & Sons, Inc., 1968.

WHITLA, D. K., ed. *Handbook of Measurement and Assessment in Behavioral Sciences.* Reading, Mass.: Addison-Wesley Publishing Co., Inc., 1968.

5

Measures of central position

If you are out to describe the truth, leave elegance to the tailor.

Albert Einstein

CHAPTER 4 indicated that a frequency distribution is nothing more than a grouping of data into an organized form for the purpose of summarization. Even with the use of high-speed computers it is not always convenient and desirable to describe each and every item. In the grouping process, a certain amount of detail is sacrificed; however, often more important aspects of the data may be emphasized. In order to analyze a distribution thoroughly one must know something about (*a*) the measures of central position; (*b*) the scatter of the items around their average, that is, the dispersion; and (*c*) the degree to which the distribution is or is not pulled to one direction or the other, that is, a measure of skewness.

This chapter will help further to reduce the data to an understandable form by calculating or locating the various averages. Chapter 6 will provide even more assistance by introducing us to the concepts of variation and skewness. The averages alone do not reveal enough about the distribution; one needs to know how the items are dispersed about the measure of central position. Also, the standard deviation, a measure of this scatter, is one of the most important concepts in statistics; its square, called the variance, has been referred to as the "cornerstone" of statistical analysis.

Prior to selecting a specific average to represent a distribution of values, the analyst should keep several factors in mind. The precise function that the average is intended to serve must be considered first. What do you want to have the average do? If the distribution is symmetrical and single-peaked (normal), then it probably will make little difference whether you use the mean, median, or mode. If the distribution is not symmetrical, you would want to know which way does it tail off. Should extreme values or simply the number of observations be permitted to influence the average? Is the average to be used in further algebraic operations? Are the data to be averaged original observations, ratios, averages, or rates? Will the average need to be tested for reliability? Answers to these and other questions usually dictate which average is chosen.

After statistical data have been gathered and arranged so that items can be compared as to size or magnitude, it is often desirable—in fact, it is frequently necessary—to compute their *average*. Thus, in the case of the wages of 200 workers discussed in preceding pages, the question arises as to what is the average wage of these workers. The average is sometimes described as a number which is typical of the whole group. As will appear later, however, it is not always typical of the group, and may only express the *summary* character of data.

The averages most commonly used are the *arithmetic mean* (usually called the *mean* or common average), the *median*, the *mode*, the *geometric mean*, and the *harmonic mean*.

ARITHMETIC MEAN

The *arithmetic mean* is the most widely used of all averages. It is the average which is most easily computed and understood, for it is calculated as the sum of the items divided by the number of items. Thus the mean of 8, 10, and 15 is 11, which is $(8 + 10 + 15)/3 = 33/3$. The symbol for the mean is $\bar{X}$ (read X bar). The formula for the mean of a series of numbers is

$$\bar{X} = \frac{\Sigma X}{n} \tag{5.1}$$

where X represents items to be averaged and n the number of items.

Arithmetic mean of grouped data

When the mean of a frequency distribution is computed, the original data ordinarily are not available. But if the ungrouped data are available, the mean is most accurately calculated by totaling the items and dividing by the number of items. If, on the other hand, only the frequency distribution is available, the mean must be computed from it.

When the arithmetic mean of a frequency distribution is computed, the midpoint of each class is taken as representative of the items in the class. The midpoint is multiplied by the class frequency, and the resulting product is the total for the class. These products are summed and divided by the sum of the frequencies, n. See Formula (5.2).

In Table 5.1 the approximate midpoints are given as 50, 60, . . . , 120.

TABLE 5.1
The arithmetic mean, direct method (grouped data as in Table 4.2)

L_1	L_2	X	f	fX
$ 45.00–	$ 54.99	50	14	700
55.00–	64.99	60	40	2,400
65.00–	74.99	70	54	3,780
75.00–	84.99	80	46	3,680
85.00–	94.99	90	26	2,340
95.00–	104.99	100	12	1,200
105.00–	114.99	110	6	660
115.00–	124.99	120	2	240
			$n = \Sigma f = 200$	$\Sigma fX = 15{,}000$

$$\bar{X} = \frac{\Sigma fX}{n} = \frac{15{,}000}{200} = \$75.00 \qquad (5.2)$$

The data are discrete to the penny, so the midpoints are accurate to a half penny. This inaccuracy usually is disregarded, but it should be recognized as existing. Not only are the midpoints not quite accurate due to the discrete form of the frequency table but also it is almost always true that none of the midpoints is actually the average of the items in a class. As was stated in Chapter 4, there is a tendency for these latter errors to offset each other, insofar as the mean is concerned.

The subtraction of a constant

The subtraction of a constant from each of a set of numbers to be averaged changes the average of the set by the value of the constant. This is simply demonstrated in Table 5.2, where X is variable and $\bar{X}_0$ is a constant, 25.

TABLE 5.2

X	$d = X - \bar{X}_0$
10	−15
20	−5
30	5
40	15
50	25
$\Sigma X = 150$	$\Sigma d = 25$
$\bar{X} = 30$	$\bar{d} = 5$

The division of each of a set of numbers to be averaged by a constant divides the average of the set by the value of the constant. This is shown in Table 5.3, where X is variable and i is a constant, 5. Thus

$$\bar{X} = \bar{X}_0 + i\left(\frac{\Sigma d_i}{n}\right) = 25 + 5\left(\frac{5}{5}\right) = 30 \qquad (5.3)$$

TABLE 5.3

X	$d = X - \bar{X}_0$	$d_i = (X - \bar{X}_0)/i$
10	−15	−3
20	−5	−1
30	5	1
40	15	3
50	25	5
$\Sigma X = 150$	$\Sigma d = 25$	$\Sigma d_i = 5$
$\bar{X} = 30$	$\bar{d} = 5$	$\bar{d}_i = 1$

Usually, the constant subtracted is referred to as the assumed mean. Here the assumed mean, $\bar{X}_0$ is 25. Table 5.4 illustrates the so-called "assumed mean" method, which is nothing more than use of the first axiom.

The method of subtracting a constant or assumed mean from ungrouped items, as in Table 5.4, has little advantage over the direct method, but it is often advantageous when applied to a frequency distribution when the use of an electronic computer is not feasible.

Arithmetic mean of grouped data—shortcuts

The assumed mean method is useful when computing the mean of a frequency distribution. Table 5.5 shows how the technique may be used. The midpoint of the first class is selected as the assumed mean. However, any other midpoint or other convenient figure can be used. The d column is the original variable, X, minus a constant; the assumed mean, $\bar{X}_0 = 50$. In the second class, for example, the midpoint, X, is 60, and $d = (X - \bar{X}_0 = 60 - 50 = 10$. The d's for the other classes are determined in the

TABLE 5.4
Methods of computing the mean

I Direct Method		II Assumed Mean, $\bar{X}_0 = 10$		III Assumed Mean, $\bar{X}_0 = 20$	
X	$x = (X - \bar{X})$	X	$d = (X - \bar{X}_0)$	X	$d = (X - \bar{X}_0)$
10	−6	$\bar{X}_0 = 10$	0	10	−10
12	−4	12	12	12	−8
15	−1	15	5	15	−5
21	+5	21	11	21	+1
22	+6	22	12	22	+2
$\Sigma X = 80$	$\Sigma x = 0$		$\Sigma d = 30$		$\Sigma d = -20$
$\bar{X} = \frac{80}{5} = 16$		$\bar{X} = \bar{X}_0 + \frac{\Sigma d}{n}$		$\bar{X} = \bar{X}_0 + \frac{\Sigma d}{n}$	
		$= 10 + \frac{30}{5} = 16$		$= 20 + \frac{-20}{5} = 16$	

TABLE 5.5
The arithmetic mean, assumed mean method—assumed mean ($\bar{X}_o$) = 50 (grouped data as in Table 4.2)

L_1	L_2	f	d	fd
$ 45.00–	$ 54.99	14	0	0
55.00–	64.99	40	10	400
65.00–	74.99	54	20	1,080
75.00–	84.99	46	30	1,380
85.00–	94.99	26	40	1,040
95.00–	104.99	12	50	600
105.00–	114.99	6	60	360
115.00–	124.99	2	70	140
		$n = \Sigma f = 200$		$\Sigma fd = 5{,}000$

$$\bar{X} = \bar{X}_0 + \frac{\Sigma fd}{n} = 50.00 + \frac{5{,}000}{200} = 50.00 + 25.00 = \$75.00$$

same manner, whether for like or variable class intervals. Next the d's are multiplied by their respective frequencies and the products entered in the fd column. For example, in the second class, there are 40 items, the midpoint of which minus $\bar{X}_0$ is 10, so the fd for the class is 400. $\Sigma fd = 5{,}000$ represents the aggregate of the variable d, and

$$\bar{d} = \frac{\Sigma fd}{n} = \frac{5{,}000}{200} = 25$$

is the average of the variable d. Hence the sample mean of X is

$$\bar{X} = \bar{X}_0 + \frac{\Sigma fd}{n} = 50 + 25 = 75 \tag{5.4}$$

TABLE 5.6
The arithmetic mean, assumed mean with deviations expressed in terms of class intervals (grouped data as in Table 4.2)

L_1	L_2	f	d_i	fd_i
$ 45.00–	$ 54.99	14	0	0
55.00–	64.99	40	1	40
65.00–	74.99	54	2	108
75.00–	84.99	46	3	138
85.00–	94.99	26	4	104
95.00–	104.99	12	5	60
105.00–	114.99	6	6	36
115.00–	124.99	2	7	14
		$n = \Sigma f = 200$		$\Sigma fd_i = 500$

$$\bar{X} = \bar{X}_0 + i\left(\frac{\Sigma fd_i}{n}\right) = 50.00 + 10\left(\frac{500}{200}\right) = 50.00 + 25.00 = \$75.00$$

The classes of Table 5.6 have the same interval, 10. When the class interval is constant, as here, the computation of $\bar{X}$ can be further simplified by dividing the d column by 10. That is, d is divided by the class interval, i, and written as d_i; then $\frac{\Sigma f d_i}{n}$ gives the same result as before (see Table 5.5). This method is especially useful when the class interval is large and when there are many classes and large frequencies. In Table 5.6 the assumed mean, $\bar{X}_0$ is 50, but it could be 60, 70, 80, or any other midpoint. The use of 50 as $\bar{X}_0$, however, gives all positive d's, and this helps to avoid possible errors of computation.

An evaluation of the arithmetic mean

The arithmetic mean is the most widely used of all averages. When it is possible to do so, it should be used to represent a group of items, except in the case of very irregular data where its use would be misleading.

If a frequency distribution having an open-end class on either or both ends is the only available data, it is impossible to compute $\bar{X}$ except by estimating the missing end intervals. Also, extreme items render the mean useless. For example, the $\bar{X}$ of 18, 22, 24, 26, and 100 is 38. This average is larger than four of the five items and is not typical of them. Obviously, such an average, which gives weight to extreme values, can be misleading. Therefore, it is useful to think of the mean as a computed value and probably a value which does not actually exist.

The mean, when properly used, has many advantages over the other averages discussed here. It may be treated algebraically, and thus a series of means may be averaged. For example, $\bar{X} = \Sigma X/n$; $n\bar{X} = \Sigma X$; $n = \Sigma X/\bar{X}$. The arithmetic mean has two important mathematical properties that prove helpful in calculating measures of dispersion and in some later work. First, the sum of the deviations of the items from the mean is equal to zero. That is, $\Sigma x = 0$, where $x = X - \bar{X}$. Second, Σx^2 = a minimum. The algebraic proof of this is simple: A deviation, x, is found as $x = (X - \bar{X})$.

$$\Sigma x = \Sigma(X - \bar{X}) = \Sigma X - n\bar{X}$$

Since

$$\bar{X} = \Sigma X/n, n\bar{X} = \Sigma X$$

hence,

$$\Sigma x = 0$$

For an illustration, see Table 5.4, Part I.

Further, the mean is simple to compute and easily understood. In addition to these attributes, it can be tested for reliability when it represents a sample. This latter advantage is discussed in Chapter 9.

MEDIAN

The median may be described as that value which divides an array of items so that an equal number are on either side of it. For example, the median of the numbers used on the opposite page, namely, 18, 22, 24, 26, and 100, is 24. The number of the median item is $(n + 1)/2$, or the third item. There are two items below 24 and two items above it. If a sixth item is added so that the items are 18, 22, 24, 26, 100, and 120, the median is located as $(n + 1)/2 = 3.5$, or between the third and fourth item. Thus, it is seen that the value of the median may not be the value of an existing item. It should be carefully noted, however, that if the array is counted by *spaces* between items, the median item is the space numbered $n/2$. Thus, for the case last mentioned, $n/2 = 3$, or the third *space*, which is interpreted as midway between the two adjacent items, or 25. In grouped data, $n/2$ is used, since a continuous scale is measured by spaces.

Median of grouped data

The median of a frequency distribution may be estimated by counting halfway through the frequencies and interpolating the corresponding point in the classified items considered as continuous. In Table 5.7 the class in which $n/2$, or 100, lies is located by reference to the "less than

TABLE 5.7
The median (grouped data as in Table 4.2)

L_1	L_2	f	Σf_1	Σf_2
$ 45.00–	$ 54.99	14	0	14
55.00–	64.99	40	14	54
65.00–	74.99	54	54 = F	108 (median class)
75.00–	84.99	46	108	154
85.00–	94.99	26	154	180
95.00–	104.99	12	180	192
105.00–	114.99	6	192	198
115.00–	124.99	2	198	200
		$n = \Sigma f = 200$		

$$Md = L_1 + i\left(\frac{n/2 - F}{f_{Md}}\right) = 65.00 + 10\left(\frac{100 - 54}{54}\right) = 65.00 + 10\left(\frac{46}{54}\right) = \$73.52$$

the lower limit" (Σf_1) and "less than the upper limit" (Σf_2) columns.[1] Thus, in the class \$65.00–\$74.99, there are 54 items less than \$65 and 108 items less than \$74.99 (literally, \$74.995). Therefore the halfway mark is in this class. Forty-six items in this class are less than the median, and since there are 54 items in the class, the median is determined as 65 + 46/54 of the interval, 10, or $65 + 10(46/54) = 65 + 8.52 = 73.52$. The formula describing this interpolation, as indicated in Table 5.7, is

$$Md = L_1 + i\left(\frac{n/2 - F}{f_{Md}}\right) \tag{5.5}$$

where L_1 is the lower limit of the median class, i is the class interval, F is the number of items less than L_1 of the median class, and f_{Md} is the frequency of the median class.

Perhaps the process of determining the median can be shown by more detailed description of the method. First, there are 200 items in the distribution shown in Table 5.7. Second, the median is the interpolated value, on either side of which fall one half of these items. Third, 54 items fall below the value \$65, so 46 more items are needed to reach the space numbered $n/2$. Therefore, only 46 of the 54 items in this class are less than the median. The 54 items of the class are assumed to be equally distributed through the class, that is, through the interval (i) of \$10. For this reason, 46/54 of 10 determines the amount, \$8.52, to be added to \$65 to locate the median value. The primary difficulty encountered in explaining Table 5.7 stems from the fact that the frequency of the median class is 54 and the sum of the frequencies down to that class is also 54, which may at first glance lead to confusion.

Median of discrete data

As was stated in Chapter 4, a discrete series becomes continuous when employed in computations. As has been seen, when the class limits are almost continuous, as in Table 5.7, correction is unnecessary. But sometimes, appreciable error will result if the limits are not written as continuous, or at least treated as such. This is illustrated in Table 5.8, where the limit of accuracy is one worker and the class limits remain discrete. The interpolation of the median follows the rule already explained, except that the lower limit of the median class, written as 4–6, is entered in the formula as if the classes had been extended to the form 0.5–3.5, 3.5–6.5, etc. This

[1] As was noted in Chapter 4, there is a negligible error in computing on the basis of class limits that may be actual items. The continuous limits here are \$44.995–\$54.995, and so on, the limit of accuracy being 1 cent. If this limit is large, the class limits should be more accurately stated. This point is discussed more fully in Chapter 6.

TABLE 5.8
The median of discrete data (data from Chapter 4)

L_1 L_2	f	Σf_1	Σf_2
1– 3	8	0	8
4– 6	16	8	24 (median class)
7– 9	15		
10–12	5		
	$n = 44$		

$$Md = L_1 + i\left(\frac{n/2 - F}{f_{Md}}\right) = 3.5 + 3\left(\frac{22 - 8}{16}\right) = 3.5 + 3\left(\frac{14}{16}\right) = 3.5 + \frac{42}{16} = 6.1$$

makes the median 6.1 instead of 6.6, as it would have been if the correction had been neglected.

An evaluation of the median

The values of the arithmetic mean and median have now been determined. The mean of the 200 wages exceeds the median because the distribution is pulled off to the right; that is, it has positive skewness. Had the wage distribution been symmetrical, the mean and the median would have been identical values.

The mean of a distribution pulled off to the left (negative skewness) is always less than the median. The median is a positional measure of central position and is not affected by the size of extreme values, whereas the mean is affected by these extremes. Therefore, the median is a location along a continuous scale instead of a calculated value such as the arithmetic mean. Under conditions of marked skewness the median may be preferable to the mean. Then, too, the median can be determined for the open-end distributions that are often encountered. And it also can be estimated whether the class intervals are equal or unequal.

Although the median has definite advantages, it also has certain disadvantages. One of these is that it cannot be handled algebraically, as can the mean. Then, too, the median cannot be computed as exactly as can the mean, and its reliability cannot be so readily estimated. These drawbacks will be given consideration later.

MODE

The mode of a distribution is the value that has the greatest concentration of frequencies. Thus the modal class in Table 5.9 is the class with the greatest frequency. In smooth distributions with equal class intervals,

TABLE 5.9
Approximation of the mode (grouped data, same as Table 4.2)

L_1 — L_2	f	d_1	d_2
$ 45.00–$ 54.99	14		
55.00– 64.99	40		
65.00– 74.99	54	14	8 (modal class)
75.00– 84.99	46		
85.00– 94.99	26		
95.00– 104.99	12		
105.00– 114.99	6		
115.00– 124.99	2		

$$M_0 = L_1 + i\left(\frac{d_1}{d_1 + d_2}\right)$$

$$d_1 = 54 - 40 = 14,\ d_2 = 54 - 46 = 8$$

$$M_0 = 65 + 10\left(\frac{14}{14 + 8}\right) = 65 + \left(\frac{140}{22}\right)$$

$$= 65 + 6.36 = \$71.36$$

this class is, of course, easily determined by inspection; but the mode itself—one value—usually is not readily found.

The formula commonly employed in interpolating a mode is valid only as applied to a comparatively smooth distribution. It is

$$M_0 = L_1 + i\left(\frac{d_1}{d_1 + d_2}\right) \tag{5.6}$$

where L_1 is the lower limit of the modal class, d_1 is the difference between the frequency of the modal class and the preceding class, and d_2 is the difference between the frequency of the modal class and the succeeding class. This formula is derived by the use of calculus and assumes a parabola through the midpoints of the three classes in question and is reasonably in line with the general contour of the distribution. Hence, as a rule, the mode is at best a rough estimate. It may be pertinent to more advanced statistical analysis, but it is not an important average as such. However, the modal class or the concept of the mode is often important in the field of business, for it represents the typical group or value. For example, the modal income group represents the concentration of a given class of potential consumers.

Evaluation of the mode

The mode, like the median, is a positional average and is not affected by extreme values. In Table 5.9 the mode is less than the median of the same distribution, which is in turn less than the mean.[2] This is typical of a distribution pulled off to the right, and the opposite condition prevails when the distribution is pulled off to the left. When the distribution is normal, the mean, median, and mode are identical. Like the median, the mode is not directly influenced by extreme items or by open-end distributions.

The mode, although possessing certain desirable characteristics as an average, has several shortcomings. As has been stated before, ordinarily the mode is only a rough estimate. In addition, it cannot be treated algebraically or tested for reliability.

GEOMETRIC MEAN

Although the arithmetic mean, median, and mode are the most widely used and simplest averages, under certain conditions other more complex averages are required. Of these, the two most important are the geometric mean and the harmonic mean. The geometric mean (G) is defined as the nth root of the product of n factors.[3] Thus the geometric mean of a room 9 feet by 16 feet is 12 feet, that is,

$$G = \sqrt{9 \cdot 16} = \sqrt{144} = 12$$

And G of a room 9 feet by 12 feet by 16 feet is 12 feet, that is,

$$G = \sqrt[3]{9 \cdot 12 \cdot 16} = \sqrt[3]{1{,}728} = 12$$

When the items are looked upon as being factors, the geometric mean is the proper average, for it can be substituted for each item without changing the product. Thus the average dimension of a room 4 feet by 9 feet is 6 feet, not 6.5 feet; a room with an average dimension of 6.5 feet has 42.25 square feet in it, not 36 square feet. With variable items, excluding zero, the geometric mean is always smaller than the arithmetic mean. If all items are identical, $G = \bar{X}$.

A simple way of taking the root of a number is to divide its logarithm by the root index and look up the antilogarithm of the quotient. Further, it is not necessary to compute the product of the factors. The sum of the

[2] In a theoretical distribution, moderately skewed, the distance from Mo to Md is twice that from Md to $\bar{X}$. For example, they might be 70, 74, and 76, respectively. But this is only roughly approximated in actual distributions, as in the data of Table 5.9, where $Mo = 71.36$, $Md = 73.52$, and $\bar{X} = 75.00$. The differences are $Md - Mo = 2.16$, and $\bar{X} - Md = 1.48$–a ratio of 1.46 instead of 2.

[3] Mathematically, the even root (n, even) of a product is positive or negative; but in practice, geometric means are confined to positive numbers greater than zero, having positive roots.

logarithms of the factors is the logarithm of the product. Thus the geometric mean of 2, 4, and 8 is found as

$$\begin{aligned} \log 2 &= 0.3010 \\ \log 4 &= 0.6020 \\ \log 8 &= 0.9030 \\ \text{Sum of logs} &= 1.8060 \end{aligned}$$

$$\log G = \text{Mean of logs} = \frac{1.8060}{3} = 0.6020$$

$$G = \text{Antilog } 0.6020 = 4$$

Geometric mean of a frequency distribution

In addition, the geometric mean is used as the average of a frequency distribution, which is or tends to be logarithmic. A hypothetical example, using abbreviated data, can be used to illustrate a case where the geometric mean is the appropriate average of a series. Table 5.10 is a comparison of the arithmetic mean and the geometric mean of a series which is logarithmic. The variable (X) is logarithmic, for it is a geometric progression. The frequencies of 1:4:6:4:1 make it symmetrical; and obviously, 8 is an appropriate average of the series. However, the arithmetic mean is 10.125, and 8 is determined as the average only by use of the geometric mean. The logarithms of the items are symmetrically distributed, and the average logarithm is 0.9030. It should be recalled at this point that the geometric mean of a series is the *n*th root of the product of the series.

It would be a unique circumstance, indeed, for a distribution to take

TABLE 5.10

A comparison of the arithmetic mean with the geometric mean of a logarithmic distribution

I Arithmetic Mean				II Geometric Mean		
X	f	fX	X	$\log X$	f	$f(\log X)$
2	1	2	2	0.3010	1	0.3010
4	4	16	4	0.6020	4	2.4080
8	6	48	8	0.9030	6	5.4180
16	4	64	16	1.2040	4	4.8160
32	1	32	32	1.5050	1	1.5050
	$\Sigma fX =$	162			$\Sigma f(\log X) =$	14.4480

$$\bar{X} = \frac{162}{16} = 10.125 \qquad \log G = \frac{14.4480}{16} = 0.9030$$

$$G = \text{antilog } 0.9030 = 8$$

TABLE 5.11
The geometric mean

L_1	L_2	X	$\log X$	f	$f(\log X)$
$ 45.00–	$ 54.99	50	1.6990	14	23.7860
55.00–	64.99	60	1.7782	40	71.1280
65.00–	74.99	70	1.8451	54	99.6354
75.00–	84.99	80	1.9031	46	87.5426
85.00–	94.99	90	1.9542	26	50.8092
95.00–	104.99	100	2.0000	12	24.0000
105.00–	114.99	110	2.0414	6	12.2484
115.00–	124.99	120	2.0792	2	4.1584
				$n = \Sigma f = 200$	$\Sigma f(\log X) = 373.3080$

$$\log G = \frac{\Sigma f(\log X)}{n} = \frac{373.3080}{200} = 1.866540$$

$$G = \text{antilog } 1.866540 = \$73.54$$

precisely the form of that in Table 5.10; but distributions such as those of incomes, prices, and wages *tend* to take this form.

The geometric mean of the 200 wages is computed in Table 5.11 and is found to be $73.54, which is $1.46 less than the arithmetic mean of this distribution, but very close to the median.

Averaging ratios

When the mean of ratios is desired, the geometric mean is the correct average to use. However, one must recognize that this is what is wanted. There are conditions under which the ratio of means is desired, and under such conditions the arithmetic mean should be used. Ordinarily, the common error made, however, is in computing a ratio of averages, when what was intended was an average of ratios. This can be demonstrated by a highly simplified case (Table 5.12) where the average ratio of prices is desired. When 1970 is the base, 1971 shows a price ratio above 1970; and when 1971 is the base, 1970 shows a price ratio above 1971. This is not possible. The geometric mean of the ratios (Table 5.13), on the other hand, shows no relative difference in prices.

TABLE 5.12

Commodity	1970 Price	1971 Price	Ratio, 1970 Price to 1971 Price	Ratio, 1971 Price to 1970 Price
Milk (quart)	$.20	$.40	50	200
Eggs (dozen)	.40	.20	200	50
Sum			250	250
Mean			125	125

TABLE 5.13

Commodity	Log of Ratio, 1970 Price to 1971 Price	Log of Ratio, 1971 Price to 1970 Price
Milk	1.6990	2.3010
Eggs	2.3010	1.6990
Sum	4.0000	4.0000
Log G	2.0000	2.0000
G	100	100

Average percentage rates of change

When an average percentage rate of change is wanted, the geometric mean is the only correct average to use. The classic example is that of average rate of population change. An example using first the mean and then the geometric mean is illustrated in Table 5.14. Application of 112 percent each year involves beginning with 10,000 and multiplying by 112 for three successive years and recognizing it as a percentage, thus using two decimals. Such a process yields a population figure for 1976 of 14,049, which does not agree with the actual population for that year.

The geometric mean (Table 5.15), on the other hand, yields the correct population figure.

Evaluation of the geometric mean

The geometric mean, like the mean, is a computed measure of central tendency. It is the correct average of percentage rate of increase. Further, it is the appropriate average of distributions which tend to be logarithmic in form. These uses of the geometric mean make it an important measure of tendency. A geometric mean of a sample can be tested for reliability and is on a par with the arithmetic mean in this respect.

The primary disadvantages of the geometric mean are that it is not as

TABLE 5.14

Year	Population	Percent of Previous Period	112 Applied
1973	10,000		10,000
1974	11,000	110	11,200
1975	13,200	120	12,544
1976	13,992	106	14,049
Sum		336	
Mean		112	

TABLE 5.15

Year	Percent of Previous Period	Log Percent	111.85 Applied
1973			10,000
1974	110	2.0414	11,185
1975	120	2.0792	12,510
1976	106	2.0253	13,992
Sum		6.1459	
Log G		2.0486	
G		111.85	

simple as the other measures of central tendency and is more difficult to compute. In general, its use is limited because of these two factors; however, the geometric mean is useful in index number construction.

HARMONIC MEAN

The harmonic mean is useful when data are given in terms of rates. Sometimes, it is convenient to give data in miles per hour, units purchased per dollar, or units produced per hour. For example, suppose that an individual drove from city A to city B, which is 120 miles away. The speed driven was 40 miles per hour on the trip to B and 30 miles per hour on the return trip to A. What was the average speed? At first glance, one might say 35 miles per hour. This is not the correct average speed. Three hours were driven at 40 miles per hour and 4 hours at 30 miles per hour. Therefore, seven hours were spent in the round trip covering 240 miles, and the average is $240/7 = 34.3$ miles per hour. Of course, the 34.3 is the arithmetic mean. The harmonic mean will give the same results as the arithmetic mean when the latter is computed by the use of weights such as three hours and four hours used above.

The harmonic mean H is the reciprocal of the arithmetic mean of the reciprocals of the values. Since these values ordinarily are rates, the formula for the harmonic mean is frequently written as

$$H = \frac{1}{\dfrac{1/r_1 + 1/r_2 + \cdots + 1/r_n}{n}} = \frac{1}{\dfrac{\Sigma 1/r}{n}} \tag{5.7}$$

or more conveniently for purposes of computation:

$$H = \frac{n}{1/r_1 + 1/r_2 + \cdots + 1/r_n} \tag{5.8}$$

Thus, for the illustration of the round trip from city A to city B:

$$H = \frac{2}{1/30 + 1/40} = \frac{2}{4/120 + 3/120} = \frac{2}{7/120} = \frac{240}{7} = 34.3$$

The harmonic mean is a very special average and is not often required. It does give the greatest weight to the smallest numbers, and thus is smaller than either the mean or the geometric mean. Usually, a weighted mean is preferred over the harmonic mean.

The brief description of various averages in this chapter and the advantages and disadvantages of each is not intended to be complete. However, if students are well acquainted with the material presented here, they will have the ability to make a reasonably sound decision when faced with the problem of averaging statistical data.

SYMBOLS

X, a variable

$\bar{X} = \frac{\Sigma X}{n}$, the arithmetic mean of a variable

$x = X - \bar{X}$, deviation of a variable from the mean

$\bar{X}_0$, an assumed constant

$d = X - \bar{X}_0$, a variable after a constant has been subtracted

$\bar{d} = \frac{\Sigma d}{n}$, the arithmetic mean of d

i, a constant, ordinarily the class interval of a frequency distribution

$d_i = (X - \bar{X}_0)/i$, a variable, after a constant $\bar{X}_0$ has been subtracted, divided by a constant i (the class interval)

$Md = L_1 + i\left(\frac{n/2 - F}{f_{Md}}\right)$, the estimated median of a frequency distribution

$F = \Sigma f_1$, sum of the frequencies down to the median class

f_{Md}, the frequency of the median class

$M_0 = L_1 + i\left(\frac{d_1}{d_1 + d_2}\right)$, the estimated mode of a frequency distribution

d_1, the difference between the modal frequency and the frequency of the preceding class

d_2, the difference between the frequency of the modal class and the frequency of the succeeding class

$G = \text{antilog}\,\frac{\Sigma f \log X}{n}$, the geometric mean of a frequency distribution

$H = \frac{n}{1/r_1 + 1/r_2 + \cdots + 1/r_n}$, the harmonic mean

PROBLEMS

Problems 5.1 through 5.5 are made up of highly simplified frequency distributions. Work all parts of each problem.

5.1. *Required:*
The arithmetic mean, using f_1, f_2, and f_3.

X	f_1	f_2	f_3
1	1	1	1
2	1	2	4
3	1	4	3
4	1	2	1
5	1	1	1

5.2. *Required:*
The median, using f_1, f_2, f_3.

5.3. *Required:*
The mode, using f_2 and f_3.

5.4. *Required:*
The mean, median, and mode.

L_1	L_2	f
10	20	1
20	30	1
30	40	3
40	50	4
50	60	1

5.5. *Required:*
The geometric mean, using f_1 and f_2.

X	log X	f_1	f_2
1	0.0000	1	1
2	0.3010	1	2
3	0.4771	1	4
4	0.6020	1	2
5	0.6990	1	1

5.6. If A can complete a job in 30 minutes, B in 40 minutes, C in 25 minutes, and D in 30 minutes, what is the average that should be used as a guide for incentive payments?

5.7. Comment on the statement that the mean, median, and mode become identical when the whole population, and not just a sample, is examined.

5.8. Given the following:

Accredited members of the American Association of Collegiate Schools of Business

Period of Founding	Number of Schools Founded
1900 and earlier	6
1901–1905	1
1906–10	7
1911–15	11
1916–20	23
1921–25	14
1926–30	10
1931–35	3
1936–40	4
1941–45	4
1946–50	8
1957 and later	59
	150

a. Is this a frequency distribution? How can you tell?
b. If it is not a frequency distribution, can it be converted into one? If so, present it as such, in acceptable tabular form.
c. Is your distribution discrete or continuous?
d. What measure of central position best describes your new frequency distribution? Compute this measure and interpret it.
e. Are there any other measures of central position which you can estimate? If any, compute these measures. Why did you prefer the measure computed in (*d*)?
f. Are there any measures you cannot determine from this distribution? Why not?

5.9. Given the data from Problem 4.3 in Chapter 4 as you presented them in part (*b*), determine and interpret the following:
a. The mean.
b. The median.

5.10. Given the frequency distribution illustrated in the random sample of sales slips shown at the right, determine the following, showing work in a neat, orderly manner, and *interpret* your answers:
a. The mean, using $\bar{X}_0 = \$15$.
b. The median.
c. The mode.

Daily Sales				Frequency
L_1		L_2	X	f
\$10	up to	\$20....	\$15	10
20	up to	30....	25	30
30	up to	40....	35	40
40	up to	50....	45	15
50	up to	60....	55	5
			$n =$	100

5.11. The weights (in ounces) of a number of mass-produced parts taken from a given production run are as follows: 16.1, 15.9, 16.3, 16.2, 16.0, 16.1, 16.0, 15.9, 15.8, 16.0, 16.1, 16.0, 15.9, 16.1, 16.0, and 16.0. Compute the following and *interpret* each measure:
a. Arithmetic mean.
b. Median.
c. Mode.

5.12. Refer to your distribution of Problem 4.4, Chapter 4, and determine the following:
a. Arithmetic mean.
b. Median.
c. Mode.
Which measure best describes the frequency distribution? Why?

5.13. The "less than" cumulative frequency distribution below may be assumed to represent a normal distribution of statistics exam scores. As accurately as you can from the chart answer the following:
a. Determine the median and interpret this measure.
b. Determine the arithmetic mean and interpret this measure.

5.14. Refer to your distribution of Problem 4.5, Chapter 4, and determine the following:

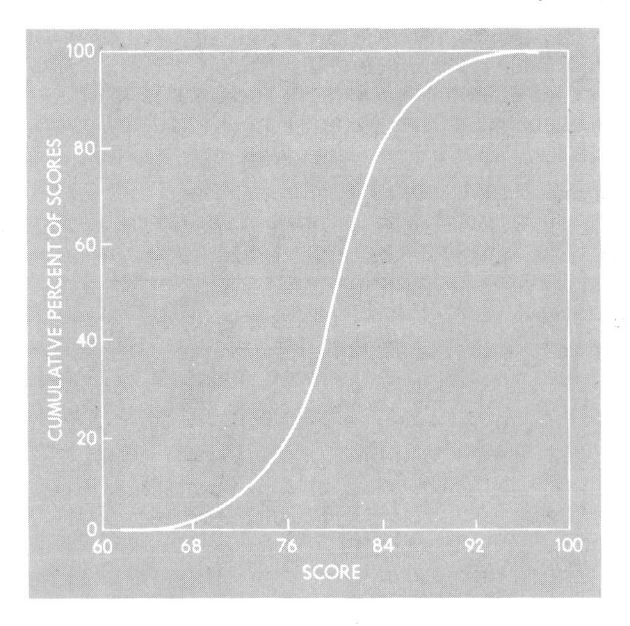

a. Arithmetic mean.
b. Median.
c. Mode.

Which measure best describes the frequency distribution? Why?

5.15. Two professors teach the same statistics course at the university. In section 1 Professor Sigma gives a final grade on the basis of the arithmetic mean of a student's population of weekly quiz grades. In section 2 Professor Beta grades on the basis of the median of a student's population of weekly quiz grades. Assume that teaching abilities, degree of difficulty of quizzes, and any other pertinent factors are constant between the two sections. Answer the following:

a. If you had your choice of with whom to take the course, which professor would you select?
b. Why? Explain your choice carefully.

5.16. In each of the following situations, indicate whether the description applies to the mean, median, or both:

a. Can be determined from a frequency distribution with open-end class intervals.
b. Can be determined from a frequency distribution with unequal class intervals.
c. The value of every item in the frequency distribution is taken into consideration in the determination of this measure.
d. The values of the extreme items in a frequency distribution do not influence the average.

e. It is closer to the concentration of the distribution with a single peak and moderate skewness to the right.

5.17. Given the following distribution, determine the most appropriate average for each year and interpret this measure.

Common stock yields, by group, 1965, 1967, 1969, and 1975 (based on cash payments during the year and price at end of year for NYSE dividend-paying stocks only)

Yield Group	Number of Issues			
	1965	1967	1969	1975p
Less than 2%	185	285	256	321
2 to 2.9%	290	215	193	238
3 to 3.9%	341	221	156	149
4 to 4.9%	192	208	187	158
5 to 5.9%	70	140	141	92
6 to 6.9%	17	32	111	87
7 to 7.9%	11	10	46	53
8% and over	5	5	31	41
Total	1,111	1,116	1,121	1,139

p = preliminary estimate of the author.
Source: New York Stock Exchange, *1970 Fact Book,* p. 23.

5.18. Given the following information, determine the mean, median, and mode for each year. What descriptive comments can you make about the total population in the United States on the basis of this information?

Total population, by age, 1960 and 1980 (000 omitted)

Age Group	1960	1980
16 to 19 years	10,673	16,940
20 to 24 years	11,100	20,997
25 to 34 years	22,952	36,997
35 to 44 years	24,226	25,376
45 to 54 years	20,586	22,147
55 to 64 years	15,634	21,032

Source: U.S. Department of Commerce, Bureau of the Census, Bulletin 1673, *The U.S. Economy in 1980,* p. 39. 1980 data based on Series C population projections.

5.19. Assume that we have a machine producing a certain type of auto part. Machine A produces one part in five minutes; machine B produces one part in ten minutes; and machine C produces one part in six minutes. If we receive an order for 6,000 parts how much time do we have to run the three machines?

5.20. Assume that we are putting a fixed dollar amount, say $500, each quarter into a mutual stock fund. If we wish to determine the average cost per unit over a year, which average would be most appropriate to compute?

5.21. Assume the following sales data:

Nassau Press, Ltd. gross revenue (in millions of dollars)

Type of Revenue	1975	1976	Percent Gain
Private golf courses	400	440	10
Public golf courses	100	102	2
Other	50	60	20
Total	550	602	9.5

a. What is the average percentage gain in revenue by type? Which average did you use and why?

b. The company sales manager reported the average sales gain by type using the arithmetic mean. Was this an accurate description of the gain in revenue by type? Why or why not?

5.22. Let us consider the revenue of the Nassau Press, Ltd. company for private golf courses only. Given the data below respond to the questions:

Nassau Press, Ltd. gross revenue (millions of dollars) from private golf course sales, 1972–76

Gross Sales	1972	1973	1974	1975	1976
Private golf courses	225	300	350	400	440
Percent annual gain	—	33%	17%	14%	10%

a. What is the average growth rate of Nassau Press, Ltd. revenue from private golf courses from 1972 to 1976? Which average did you use? Why? What is the meaning of *average growth rate*?

b. The company sales manager reported an average rate of growth for these years to be 18.5 percent. Which average was used? Is this the correct one to use? Why or why not?

c. The company assistant sales manager decided that it would be more descriptive to report the average rate of growth to be $(440/225)/4 =$ 24 percent. Is this the correct way to report the sales gain? If not why not?

5.23. Assume that the earnings per share of the Eagle Manufacturing Company were as follows:

1973	$5.50
1974	5.60
1975	5.75
1976	6.00

a. What is the average rate of growth in earnings per share since 1973? Which average did you use and why?
b. Why is the arithmetic mean inappropriate for this situation?

5.24. Assume that a machine purchased at an original cost of $100,000 with an expected life of 5 years is to be depreciated at a uniform rate of diminishing book value. At the end of the 5 years it is expected that the machine has a salvage value of $2,000.
a. Determine the declining constant rate that would reduce the book value to the $2,000 level at the end of the 5th year.
b. What is the depreciation charge for each year?
c. Which average did you use in your computations? Why did you consider this one to be the most appropriate?

5.25. A firm quadrupled its sales in one year and then doubled the sales in the next. The former company statistician reported that "on the average the sales tripled."
a. What average did the company statistician use? Was he or she correct in so doing?
b. Assume that the actual sales rose from $1 billion to $4 billion and then up to $8 billion. Using the company statistician's approach, what would be his or her estimate of sales for the last year?
c. What is the actual average growth rate using the data of part (*b*)?

5.26. What is the relationship of the geometric mean to the simple compound interest rate?

5.27. Why does the geometric mean work for problem 5.25 when the arithmetic mean does not?

5.28. What are some of the many limitations of the geometric mean that restrict its use?

5.29. Salesperson A travels from one customer to the next by auto at 30 miles per hour (mph) and then returning to the original point of departure at 60 mph.
a. What was A's average speed during the trip?
b. A reported to the sales manager that 45 mph were averaged during this round trip. Is this the correct figure? Why or why not?

5.30. Assume that an auto replacement part can be sold on the West Coast at the rate of $5,000 per 1,000 parts. On the East Coast that same part can be sold through similar channels of distribution at the rate of $5,000 per 500 parts.
a. What is the average revenue per part?
b. What average did you select and why?

5.31. "When all the ratios being averaged have the same value in the denominator, the arithmetic mean should be used. When all the ratios being averaged have the same value in the numerator, the harmonic mean must get the nod."

a. Defend or refute this statement.

b. Give an illustration to demonstrate these "rules of thumb."

RELATED READINGS

CHISWICK, B. R., and CHISWICK, S. J. *Statistics and Econometrics: A Problem-Solving Text,* chap. 1. Baltimore: University Park Press, 1975.

DANIEL, W. W., and TERRELL, J. C. *Business Statistics,* chap. 2. Boston: Houghton Mifflin Co., 1975.

HAMBURG, MORRIS. *Basic Statistics,* chap. 3. New York: Harcourt Brace Jovanovich, Inc., 1974.

HUNTSBERGER, D. V.; BILLINGSLEY, P.; and CROFT, D. JAMES. *Statistical Inference for Management and Economics,* chap. 3. Boston: Allyn & Bacon, Inc., 1975.

NETER, JOHN; WASSERMAN, W.; and WHITMORE, G. A. *Fundamental Statistics for Business and Economics.* Abridged 4th ed. Boston: Allyn & Bacon, Inc., 1973.

6

Measures of dispersion

Who can number the sands of the sea, the drops of rain, or the days of eternity?

Unknown

THE SELECTION of a mean or average that is taken as representative of a group of items implies a measurable degree of variability. Some of the items are smaller and some are larger than the point of central tendency, though a majority usually cluster about it. Hence the question arises: How representative is the mean? In other words, how much, numerically, do the items spread around the mean?

It should be observed, first, that this tendency of like items to scatter about a central point is one of the most common features of the world about us. Even the stars differ in magnitude, and so do the neutrons, protons, and electrons as described by the physicist. So do living creatures, particularly man. The census classifies a population as varying in respect to age, race, sex, occupation, marital status, and other characteristics. For other purposes, individuals may be classified by height, weight, income, or IQ. No two items or individuals are exactly alike.

MEANING OF DISPERSION

Sometimes, classification does not lend itself to exact numerical measurement. Citizens may be classified with respect to their attitude toward immigrants as (1) unfriendly, (2) indifferent, and (3) friendly. Or a person's attitude toward a radio commerical may be described as (1) very hostile, (2) hostile, (3) indifferent, (4) favorable, or (5) very favorable. Or people may be designated as good or bad, rich or poor, native or foreign-born. Such twofold classifications are called dichotomies; and when they are measurable, they give rise to so-called "binomial" distributions, which will be examined later.

From the standpoint of theory, the way in which like items scatter is best described by the laws of chance, or probability. If we toss four coins at once, how many heads are likely to occur? As will be discussed later, a little experience and observation will show that two heads are more likely to occur than all four heads or no heads. Or if we toss two dice, we are more likely to throw a total of 7 points than of the extremes, 2 and 12. Similarly, for items that vary considerably, moderate values occur more often than extreme values. There are more common people than imbeciles or geniuses. As was noted earlier, this tendency of items to cluster about a mode is generalized in the form of the normal curve (see page 70).

RANGE

A small sample of the number of residences in certain suburban city blocks yields the numbers 5, 9, 7, 3, and 6; or, listed as an array, 3, 5, 6, 7, and 9. The arithmetic mean is 6. How shall the scatter of the items about the mean be measured?

The simplest measure is the range (R). The items vary from 3 to 9, hence the range is $9 - 3$, or 6. This figure expresses the extreme variability or scatter of the items. But this measure is seldom used because it is subject to the chance of erratic extreme items and fails also to take account of the nature of the scatter within the range. In a normal distribution, it is sometimes useful to estimate the standard deviation from the range by taking one sixth of R. This is based on the assumption that the mean plus and minus three standard deviations include almost the entire range of the data.

AVERAGE DEVIATION

When an informal measure of dispersion is required—one that is not to be used in complex mathematical problems—a mere average of the absolute deviations from the mean is often sufficient. An *absolute* deviation is one that is taken without regard to the sign. It is symbolized by x, or by d if not taken from the mean. It may be illustrated by the use of an ungrouped small sample of the number of residences in suburban blocks previously cited, namely,

$$X = 3, 5, 6, 7, 9$$

which, centered as deviations from the mean, are

$$x = -3, -1, 0, 1, 3$$

If these deviations are averaged merely as deviations without regard to the direction (sign), they give the following result, called the average deviation:

$$\text{A.D.} = \frac{\Sigma|x|}{n} = \frac{8}{5} = 1.6 \tag{6.1}$$

and for comparative purposes the coefficient of variation may be written as

$$V = \frac{\text{A.D.}}{\bar{X}} = \frac{1.6}{6} = .267 = 26.7 \text{ percent} \tag{6.2}$$

It will be noted that the average deviation is smaller than the standard deviation, which is found to be 2. (See page 111.) Some such disparity is to be expected, since squaring gives exceptional weight to large deviations. On the basis of the theoretical normal distribution, A.D. is approximately $.8\sigma$, which happens to be true of the case at hand. More precisely, the ratio is $\sqrt{2/\pi}$, or .7979. In the case of samples the actual ratio may vary widely from the theoretical. For example, in the case of two items the standard deviation and the average deviation are alike, both being half the difference between the two items.

The direct method of computing the average deviation is obvious. For grouped data, it follows the pattern of part II, Table 6.3, where the absolute sum of the negative deviations equals the sum of the positive deviations, each being 1,220, making the algebraic sum 0. This is always the case when the deviations are taken from the mean. The absolute sum is therefore 2(1,220), or 2,440. And the average deviation is

$$\text{A.D.} = \frac{\Sigma|x|}{n} = \frac{2{,}440}{200} = 12.2$$

with a coefficient of variation,

$$V = \frac{\text{A.D.}}{\bar{X}} = \frac{12.2}{75} = .163 = 16.3 \text{ percent}$$

Average deviation from median or other bases

Average deviation is sometimes computed from the median on the theoretical ground that it is then a minimum or, more precisely, is as small as or smaller than the same measure based on any other origin. The direct method of computation follows that already described. Deviations are written as $X - Md$ instead of $X - \bar{X}$, and summed without regard to sign, taking account of frequencies in the case of grouped data. The absolute deviations thus obtained are averaged. Shortcuts described above, however, must be modified, because the summed deviations of X's greater than the median do not necessarily balance the absolute sum of the smaller.

It may be said that the average deviation computed from the mean or other required origin has recently been gaining in favor except for those cases requiring mathematical elaboration. The disadvantage of the standard deviation is that, in effect, extreme items are weighted by the process of squaring. Hence a chance extreme item, either very large or very small, may distort this measure of dispersion. On the other hand, the average deviation merely sums each deviation rather than its square.

STANDARD DEVIATION

There are several possible measures of dispersion such as the average deviation (an average of the absolute differences of items from the mean) and the quartile deviation (based on the range of the central half of the items). These will be considered later. First, attention should be directed to the mathematical measure most commonly used, namely, the *standard deviation.* As measured in a sample, this is usually denoted by s, or as estimated for a universe[1] by $\hat{\sigma}$. It will be shown (Chapter 9) that in estimating the characteristics of a universe from a sample, the mean may be taken as such an estimate, but in computing an estimate of the standard deviation, $n - 1$ must be used instead of n.

Because of the wide use of the standard deviation as the measure of dispersion, several different methods of computing it have been devised. These include shortcuts, particularly those applied to data grouped in classes. But it is well to begin a discussion of the standard deviation by applying it to simple data, using the most direct method.

Applied to the sample of number of homes in suburban blocks quoted above, we have

$$X = 3, 5, 6, 7, 9; \bar{X} = \frac{\Sigma X}{n} = 6$$

The deviations (x) of each of the items from the mean (i.e., $x = X - \bar{X}$) are

$$x = -3, -1, 0, +1, +3$$

which are described as the data *centered.* If these deviations are averaged, they obviously give zero, unless the signs are neglected, as was done in finding the average deviation. But the omission of a sign is likely to get us into mathematical difficulties if the process is utilized in more

[1] The symbol σ, strictly speaking, indicates the actual standard deviation of the universe, whereas a modification of it, such as $\hat{\sigma}$, is often used to indicate a population estimate. The symbol s is restricted to the sample. But in the literature of statistics, usage with respect to these symbols varies considerably.

complex procedures such as correlation. So the standard deviation averages the squares of the deviations, obtaining a result that is called the *variance* of the sample (s^2). Then, to compensate for the squaring of the items, the square root is taken. The variance may be defined as

$$s^2 = \frac{\Sigma(X - \bar{X})^2}{n} = \frac{\Sigma x^2}{n} \tag{6.3}$$

and the standard deviation is

$$s = \sqrt{\frac{\Sigma x^2}{n}} \tag{6.4}$$

though it is hardly necessary to write the formula for s, since it is implied in the formula for the variance.[2]

Applied to the data quoted above, we have

$$x = -3, -1, 0, 1, 3;\ \Sigma x = 0$$
$$x^2 = 9, 1, 0, 1, 9;\ \Sigma x^2 = 20$$
$$s^2 = \frac{\Sigma x^2}{n} = \frac{20}{5} = 4 \text{ (the variance)}$$

and

$$s = \pm\sqrt{4} = \pm 2$$

thus indicating a scatter of the items above (+) and below (−) the mean. Hence, the variability of the sample in question may be described by writing

$$\bar{X} \pm s = 6 \pm 2$$

and the *relative* variability, written as a fraction or a percentage, may be measured by the coefficient of variation (V), thus:

$$V = \frac{s}{\bar{X}} = \frac{2}{6} = .33 = 33 \text{ percent} \tag{6.5}$$

It should be noted that s is expressed in the same units as X; for example, s, above, is two houses, and fractions are interpreted in terms of degree of scatter. The name of the unit, however, is usually omitted. But the coefficient of variation is an abstract number, expressing only a relationship, and is comparable from one sample to another expressed in different units. Thus, by V, a comparison could be made between the variability of IQ's and of wages, or between wage dispersions under different incentive plans.

---

[2] Mathematicians sometimes call the standard deviation the root-mean-square deviation. See James and James, *Mathematics Dictionary* (Princeton, N.J.: D. Van Nostrand Co., Inc., 1966), p. 111.

Effects of addends and factors

The methods of correcting a sum of squares considered later depend on a very simple, axiomatic proposition, namely, that the variability of a set of numbers is not changed if a given number is added to or subtracted from each item. Thus the standard deviation of the items $X = 2, 8, 10, 16$ is exactly the same as that of $X = 102, 108, 110, 116$. The process described as taking deviations from a guessed or assumed mean is merely a subtraction of a constant from each item (i.e., the addition of a negative constant). Hence, in finding a standard deviation, one may freely add or subtract any convenient constant and then apply a correction which in principle is the same as that which would be applied if no so-called "deviations" had been taken.

It should be noted, however, that adding a constant (positive or negative) changes the *relative* standard deviation, that is, the coefficient of variation, $s/\bar{X}$. For example, the coefficient of variation (V) of 2, 8, 10, and 16 is 5/9 or .556; but of 102, 108, 110, 116, it is 5/109, or .046. Relatively speaking, the former variability is much the larger.

But suppose we multiply a list of items by a constant factor. How would this affect the standard deviation? Apparently, it would multiply s because the deviations from the mean would each be multiplied. For an illustration, the items of Table 6.1 may be multiplied by 10, making them

$$X = 30, 50, 60, 70, 90; \Sigma X = 300, \bar{X} = 60$$

TABLE 6.1
Standard deviation computations (using Formulas [6.7] or [6.9])

I: $\bar{X}_0 = 7$			II: $\bar{X}_0 = 0$		III: $\bar{X}_0 = 0$	
X	d	d^2	X	X^2	X	X^2
3	−4	16	3	9	3	9
5	−2	4	5	25	5	25
6	−1	1	6	36	6	36
$\bar{X}_0 = 7$	0	0	7	49	7	49
9	2	4	9	81	9	81
$\bar{X} = 6$	−5	25	30	200	5)30	5)200
	$C = \bar{d}\Sigma d$			$C = \bar{X}\Sigma X$	$\bar{X} = 6$	40
$\bar{d} = -1, C = 5$			$\bar{X} = 6, C = 180$		$\bar{X}^2 = 36 = C$	
$\Sigma x^2 = 20$			$\Sigma x^2 = 20$		$s^2 = 4$	
$s^2 = 4$			$s^2 = 4$		$s = \pm 2$	
$s = \pm 2$			$s = \pm 2$			

$\bar{X} = \bar{X}_0 + \bar{d} = 7 - 1 = 6$
$\Sigma x^2 = \Sigma d^2 - \bar{d}\Sigma d$ or $\Sigma X^2 - \bar{X}\Sigma X$; $s^2 = \Sigma x^2/n$
$\hat{\sigma}^2 = \Sigma x^2/(n - 1) = 20/4 = 5$; $\hat{\sigma} = \sqrt{5} = 2.236$
$\hat{\sigma}^2 = s^2 n/n - 1) = 4(5/4) = 5$; $\hat{\sigma} = \sqrt{5} = 2.236$

The deviations from the mean now are

$$x = -30, -10, 0, 10, 30; \Sigma x^2 = 2{,}000 \quad \text{and} \quad s \text{ is } \sqrt{2{,}000/5} = 20$$

which is 10 times s of the original items.

Similarly, it may be shown that if the items are divided by 10 (i.e., multiplied by .1), the standard deviation will likewise be divided. But multiplication does not change the *relative* scatter of the items. V remains the same, since both the standard deviation and the mean are multiplied. Thus, for the above multiplied items, using Formula (6.5),

$$V = \frac{s}{\bar{X}} = \frac{20}{60} = .33, \quad \text{or} \quad 33 \text{ percent}$$

which is the same coefficient of variation as was obtained for the original data.

These two rules which apply to the usual methods of measuring variability will be found very useful both for understanding the nature of the measurements and for devising shortcuts. To summarize, they are:

1. Adding a constant (positive or negative) to each of a series of items does not change s but it does change V.
2. Multiplying each of a series of items by a constant (greater or less than unity, positive or negative) likewise multiplies s but does not change V. To validate this statement completely, however, it must be recalled that the sign of s is plus or minus, though it is usually written without a sign prefixed.

It will be recalled that if the X's are changed by an addend ($\bar{X}_0$) or a factor ($1/k$), the mean is similarly changed and a correction is called for:

$$\bar{X} = \bar{X}_0 + \bar{d}; \bar{X} = k\bar{d}$$

Standard deviation shortcuts

The shortcuts utilized in finding the standard deviation of large numbers of items, particularly as grouped in classes, may be made clear by first applying them to small numbers such as the sample utilized above. The first shortcut involves the principle that if the deviations are taken from some other origin than the mean, the resulting sum of squares will be abnormally increased and will require a correction. In other words, the variance (and its square root, s) is a minimum when computed from the mean. However, the process of correction when the origin is not the mean but is a "guessed mean" or other arbitrary origin remains to be seen.

The logic of the shortcut rule is readily seen if the deviations of $X = 3$, 5, 6, 7, 9 from $\bar{X} = 6$ are expressed so that they are positive, thus:

$\bar{X} - X$	$X - \bar{X}$
$6 - 3$	$7 - 6$
$6 - 5$	$9 - 6$
$12 - 8 = 4$	$16 - 12 = 4$

In each column there is an $\bar{X}$ for each item, and each column gives half the sum of the absolute deviations. Hence, only one column need be computed. The rule is essentially the same for a frequency distribution. With a little ingenuity, it may be modified to suit i units, or deviations from other origins than the mean.

It should be noted, however, that in the case at hand the use of an arbitrary origin ($\bar{X}_0$) does not simplify the computation. The mean itself happens to be a round number, as are also the deviations. It is when the mean is an inconvenient fraction or mixed number that an assumed mean is really advantageous. But to illustrate the principle, three different computations of variability are presented in Table 6.1. The first, part I, assumes an origin of 7 and takes the deviations as $X - 7$. These deviations, d, are squared; and the summed squares are found to be 25, which must be too large, since the origin is not the mean. In other words, the deviations are not centered.

The required corrections may be algebraically derived; but with the simple numbers at hand, it may also be experimentally discovered. Obviously, it is a function of the average of the noncentered deviations ($\bar{d}$) which indicates how far the assumed origin, $\bar{X}_0$, differs from the true origin, $\bar{X}$. As has been shown, $\bar{X} = \bar{X}_0 + \bar{d}$.

From a previous computation, it is known that the true Σx^2 is 20, so the correction to be subtracted is 5, which, it may be noted, is the product of the sum of the noncentered deviations (-5) times its mean (-1). Is this correction, namely, $\bar{d}\Sigma d$, the theoretical correction (C) to be generally applied? Algebraic analysis shows that it is.

Experimentation may help to justify this correction. For example, suppose the origin is taken as 4. Then the deviations are

$$d\text{: } -1, 1, 2, 3, 5;\ \Sigma d = 10;\ \bar{d} = 2$$

But Σd^2 is 40, with the correction $\bar{d}\Sigma d$. Hence, $\Sigma x^2 = 40 - (2)(10) = 20$, which was previously found to be the true total.

In parts II and III of Table 6.1, deviations are nominally taken from an origin of zero, which simply means that d is X and is so labeled. In part II the correction applied to ΣX^2 is $\bar{X}\Sigma X$; hence, $\Sigma x^2 = \Sigma X^2 - \bar{X}\Sigma X = 200 - 6(30) = 20$, as before.

In part III of Table 6.1 a minor modification in the correction process is made. The sum of the X squares, as well as the sum of the X's, is divided by n, that is, by 5. Obviously, the correction must be similarly reduced: instead of $\bar{X}\Sigma X$, it must be $\bar{X}\Sigma X/n$, or $\bar{X}^2$; and this correction, 6^2, applied to $\Sigma X^2/n$ gives the variance directly without indicating Σx^2.

The two correction methods may be compared thus: In part II,

$$\Sigma x^2 = \Sigma X^2 - \bar{X}\,\Sigma X$$

and in part III,

$$s^2 = \frac{\Sigma X^2}{n} - \bar{X}^2 \qquad (6.6)$$

It will be seen that the latter equation is merely the former divided through by n.

These two methods, which begin with X rather than with an actual set of deviations, are the ones most used when the data are presented as a series or array of ungrouped items. It will be seen that each is well adapted to machine calculations since in the process of cumulative squaring, ΣX will appear on one dial and ΣX^2 on another. When these totals have been found, either method of correction, II or III, may be used. The method indicated in part I will later be found to be very convenient as applied to grouped data, particularly if the mean is not a round number.

It will be noted that at the foot of Table 6.1 an estimate of the standard deviation of the population from which the sample (X) was drawn is obtained by the formula

$$\hat{\sigma}^2 = \frac{\Sigma x^2}{n-1} \qquad (6.7)$$

and

$$\hat{\sigma} = \sqrt{\frac{\Sigma x^2}{n-1}} \qquad (6.8)$$

Or the same result may be obtained by correcting s^2, thus:

$$\hat{\sigma}^2 = s^2 \cdot \frac{n}{n-1} \qquad (6.9)$$

and

$$\hat{\sigma} = s\sqrt{\frac{n}{n-1}} \qquad (6.10)$$

As will be shown later, this correction makes up for the fact that small samples—usually failing to draw extremely large or small items—tend to underestimate the variance of the whole population. Dividing by $n - 1$ instead of n appropriately increases the variance and standard deviation. But if n is large, this correction makes little difference, and often is disregarded. The first method of Table 6.1 is the one readily adapted to machine calculation and therefore may be regarded as the most important. It parallels part II of Table 6.1; and with slight change in the last steps, it may be made to parallel part III.

Standard deviation, grouped data

The methods employed in finding the standard deviation, as just discussed, may be readily applied to grouped data. The data of weekly wages classified in previous chapters are employed to illustrate the two methods of computation shown in Table 6.2.

It will be seen that the only thing that distinguishes the computation of grouped data is that the totaling of the X's and their squares requires the use of the factor f. The correction formula previously used was

$$\Sigma x^2 = \Sigma X^2 - \bar{X}\Sigma X$$

which may now be written

$$\Sigma f x^2 = \Sigma f X^2 - \bar{X}\Sigma f X$$

TABLE 6.2
Standard deviation, grouped data—using Formulas 6.9 and 6.10 (wage distribution; see Table 4.2)

Data		I		II		
X	f	fX	fX^2	d_i	fd_i	fd_i^2
50	14	700	35,000	0	0	0
60	40	2,400	144,000	1	40	40
70	54	3,780	264,600	2	108	216
80	46	3,680	294,400	3	138	414
90	26	2,340	210,600	4	104	416
100	12	1,200	120,000	5	60	300
110	6	660	72,600	6	36	216
120	2	240	28,800	7	14	98
	$n = 200$	15,000	1,170,000		500	1,700

Part I: $\bar{X} = 75$, $C = 1,125,000$; $\Sigma x^2 = 45,000$; $s^2 = 225$; $s = 15$

Part II: $\bar{d}_i = 3.5$, $C = 1,250$; $\Sigma x_i^2 = 450$; $s_i^2 = 2.25$; $s_i = 1.5$

Part II:
$\bar{X} = \bar{X}_0 + i\bar{d} = 50 + 10(2.5) = 75$
$s = is_i = 10(1.5) = 15$, $V = 15/75 = 20\%$
$\hat{\sigma}^2 = s^2 n/(n-1) = 225(200/199) = 226.13$, $\hat{\sigma} = 15.038$

or

$$\hat{\sigma} = s\sqrt{n(n-1)} = 15\sqrt{(200)(199)} = 15(1.00251) = 15.038$$

It is evident, however, that the former expression and others like it are more general than the latter. The symbol ΣX, for example, means add *all* the X's, which implies taking each X as many times as it occurs, that is, times in each class. Hence, in general mathematical formulas, f's are not written, though to the beginner they serve as a convenient reminder.

In laboratory or office work, where calculating machines are nearly always available, the products fX and fX^2 are not written in detail, multiplication by f being cumulative. If necessary, each X^2 is read from a table, such as *Barlow's Tables*,[3] and entered on the machine times f. But sometimes, it is more convenient to write the column fX and obtain ΣfX^2 as a cumulative total of each fX times X. Similar machine calculations may be utilized in other shortcut methods. An alternative correction formula, such as

$$s^2 = \frac{\Sigma fX^2}{n} - \bar{X}^2 = 5{,}850 - 5{,}625 = 225$$

may be used, which is a special case of the general formula previously written without the factor f.

Class interval units

Part II of Table 6.2 illustrates a shortcut method in which deviations (d_i) are written in class interval (i) units. These deviations are here taken with the base $\bar{X}_0$ as 50, though the calculation may be reduced a little more if $\bar{X}_0$ is selected nearer the middle of the array. But in any case the correction process for both the mean and the standard deviation is complicated by the fact that computed s will be reduced by the factor $1/i$, while at the same time the arbitrary origin must be taken into account.

The mean of the fd_i column (2.5) must be corrected by the factor i before a correction for the arbitrary origin $\bar{X}_0 = 50$ can be made, since d_i and X are not expressed in comparable units. But when $\bar{d}_i$ has been corrected to read $\bar{d} = i\bar{d}_i = 10(2.5) = 25$, then it may be added algebraically to $\bar{X}_0$ to obtain the true average, $\bar{X} = 50 + 25 = 75$. The corrected standard deviation is easily obtained merely by multiplying s_i by i.

Table 6.3, part I, illustrates a common method of computing the standard deviation of grouped data. In principle, it resembles part I of the preceding example, but in this case the constant subtracted from the variable X is the midpoint of a class (80). It corresponds to part I of Table 6.1. The correction formula, as before, expressed in general terms, is

$$\Sigma x^2 = \Sigma d^2 - \bar{d}\Sigma d$$

[3] New York: Spon and Chamberlain.

or, with frequencies specifically indicated,

$$\Sigma fx^2 = \Sigma fd^2 - \bar{d}\Sigma fd$$

If it is recalled that the d column is merely a modified X column which has the same variance as the original X's, then all the cases thus far considered, including the one using i units, may be brought under the one general method:

$$\Sigma x^2 = \Sigma X^2 - \bar{X}\Sigma X;\ s^2 = \Sigma x^2/n$$

Part II of Table 6.3 illustrates the computation of the standard deviation by means of deviations (x) taken from the mean, in which case x^2 is obtained directly without a correction. This method is not commonly used, however, since the mean is not usually an even number and the computation is complicated by decimal fractions.

THE PROBLEM OF ESTIMATION

The use of a sample for estimating the characteristics of a population involves the so-called "problem of inference." This problem, in turn, depends upon a revision of n in terms of what are known as *degrees of freedom,* which now may be given consideration.

The term "degrees of freedom" was first applied in a very technical sense to indicate the divisor used in computing a standard deviation, and other like statistics, when the objective is an estimated measure of the population or universe, instead of the sample at hand. For example, the variance of a random sample is computed as (see Formula [6.3])

TABLE 6.3
Standard deviation, grouped data (wage distribution; see Table 4.2)

	Data		I			II		
	X	f	d	fd	fd^2	x	fx	fx^2
	50	14	−30	−420	12,600	−25	−350	8,750
	60	40	−20	−800	16,000	−15	−600	9,000
	70	54	−10	−540	5,400	−5	−270	1,350
$\bar{X}_0 =$	80	46	0	0	0	5	230	1,150
	90	26	10	260	2,600	15	390	5,850
	100	12	20	240	4,800	25	300	7,500
	110	6	30	180	5,400	35	210	7,350
	120	2	40	80	3,200	45	90	4,050
	$n =$	200		−1,000	50,000		−1,220	$45{,}000 = \Sigma x^2$
				$\bar{d} = -5,\ C =$	5,000		+1,220	$225 = s^2$
				$\Sigma x^2 =$	45,000			$15 = s$

Variance: $s^2 = 225$
Standard deviation: $s = 15$
$\bar{X} = \bar{X}_0 + \bar{d} = 80 - 5 = 75$

$$s^2 = \frac{\Sigma(X - \bar{X})^2}{n} = \frac{\Sigma x^2}{n}$$

If the sample is very large, the result may be considered a fair estimate of the variance of the universe; but if the sample is small, the estimate will probably be too low. In drawing a small sample, one is more likely to draw items near the mode than at the extremes. Only as the sample is expanded to high values of n will the extremely large and small items give an adequate extension to the population variance.

In a later chapter, it will be demonstrated that the formula for estimating the variance of the universe is

$$\hat{\sigma}^2 = \frac{\Sigma(X - \bar{X})^2}{n - 1} = \frac{\Sigma x^2}{n - 1} \tag{6.11}$$

where the change in the denominator from n to $n - 1$ considerably increases the variance and standard deviation when the sample is small, but makes little difference when the sample is large. In this case, it is said that one degree of freedom is lost.

It is not possible at this stage of the work to review the reasoning on which the theory of degrees of freedom rests. It may be possible, however, to suggest certain rationalizations which are helpful in determining degrees of freedom in a particular case. One of these rationalizations runs as follows.

For our purposes here, let us define *degrees of freedom* as the number of items in a sample that are effective in estimating a specified characteristic of a population. Suppose we begin drawing a sample with a view to estimating the mean of the universe. We draw one item. Obviously, this item provides an estimate, though a crude one; it is more likely to be at the mode than at any other position, and therefore, in a limited sense, is effective. Then we draw a second item, assumed to be different. This also is effective, since the average of the two items drawn is a better estimate than the first item alone. Other items included in the sample further improve the estimate. Hence, in this case, there are no degrees of freedom lost because every item in the sample is effective, and the estimate of the population average is $\Sigma X/n$, just as it was for the sample.

Next, suppose we are sampling to determine the variance and standard deviation of a certain universe. We draw a single item. This item, however, in itself gives not the slightest clue to the dispersion. Considered alone, it is not effective. But a second item, placed with the first, and assumed to differ from it, is effective. A variance can be computed from a sample of two items. Hence, it is said that one degree of freedom is lost, presumably because the first sample item drawn is not effective. The divisor, therefore, is $n - 1$ rather than n. In effect, counting the number of items in a sample drawn for the purpose of estimating the variance and standard deviation of the universe begins with the second item.

The use of the formula for estimating the variance of the universe may be illustrated by applying it to a small sample used earlier in this chapter, as follows (see Table 6.1):

$$X = 3, 5, 6, 7, 9; \bar{X} = 6, \Sigma x^2 = 20$$
$$\hat{\sigma}^2 = \frac{\Sigma x^2}{(n-1)} = \frac{20}{4} = 5$$
$$\hat{\sigma} = \sqrt{5} = 2.236$$

Though $\Sigma x^2/(n-1)$ is an unbiased estimate of the variance, its square root is not precisely an unbiased estimate of the standard deviation. The slight error involved may be likened to that which would occur if the geometric mean were taken as the best estimate of the arithmetic mean of two numbers. However, the error involved in estimating the standard deviation of the population from the estimated variance is so small that for practical purposes it may be ignored.

Hence, characteristics of the universe as estimated from the sample are

$$\bar{X} = 6; \hat{\sigma} = 2.236; V = 37 \text{ percent}$$

When the variance and standard deviation of the sample have been computed, they may readily be adjusted to express estimates of the universe:

$$\hat{\sigma}^2 = s^2 \cdot \frac{n}{n-1} = 4 \cdot \frac{5}{4} = 5$$

or from Formula (6.10),

$$\hat{\sigma} = s\sqrt{\frac{n}{n-1}} = 2\sqrt{1.25} = 2(1.118) = 2.236$$

In this case, the sample being small, the estimated population variance and standard deviation are considerably increased. But when the sample is large, as in Table 6.2 (page 115), the estimate is not appreciably increased. In that case,

$$\bar{X} = 75, s^2 = 225, s = 15, V = 20 \text{ percent}$$

and

$$\hat{\sigma}^2 = 225\left(\frac{200}{199}\right) = 225(1.005025) = 226.13$$
$$\hat{\sigma} = \sqrt{226.13} = 15.038; V = 20.05 \text{ percent}$$

It is evident that estimates based on small samples are not, in general, as reliable as those based on large samples. In a later chapter, consideration will be given to the problem of reliability.

QUARTILE DEVIATION

It was noted earlier that the range (largest item less the smallest) is not a good measure of dispersion because of the irregularity of items in most samples. The chance occurrence of an extremely large or small item may make the range unrepresentative. Only when the data are mathematically arranged—as in the case of a binomial—is the range, or some function of it, a satisfactory measure of the dispersion.

If the irregular scattering of extreme items could be lopped off, the limited range thus obtained might be very useful. Consequently, there has been developed the quartile deviation, which is based upon the quartile range—the range of the middle half of the arrayed items. That is, one quarter of the items at the lower end of the array and one quarter at the upper end are disregarded in measuring the quartile range. This description must be modified in practice because interpolation will be necessary, but it serves to introduce the process.

The computation of the quartile range and deviation is illustrated in Table 6.4. The problem is set up in greater detail than is necessary in practice; in fact, the work could be done rather easily with columns L_1, f, and Σ_1. However, the i column, for example, will be found very useful when

TABLE 6.4
Quartile deviation, grouped data (wage distribution; see Table 4.2)

L_1 L_2	i	f	Σ_1	Σ_2	
\$ 45–\$ 54.99	10	14	0	14	
55– 64.99	10	40	14(50)	54	$Q_1 = \frac{50-14}{40}(10) + 55 = 64.00$
65– 74.99	10	54	54(100)	108	
75– 84.99	10	46	108(150)	154	
85– 94.99	10	26	154	180	$Q_3 = \frac{150-108}{46}(10) + 75 = 84.13$
95– 104.99	10	12	180	192	
105– 114.99	10	6	192	198	
115– 124.99	10	2	198	200	
	$\Sigma f =$	200			

$$QD = \frac{Q_3 - Q_1}{2} = \frac{84.13 - 64.00}{2} = 10.065$$

$$\text{Coefficient } QD = \frac{Q_3 - Q_1}{Q_3 + Q_1} = \frac{20.12}{148.13} = .136 = 13.6\%$$

$$Sk_Q = \frac{Q_3 + Q_1 - 2Q_2}{Q_3 - Q_1} = \frac{1.09}{20.13} = .054 = 5.4\%$$

Note: The outside limits of the distribution are $Q_0 = 45$ and $Q_4 = 125$. Each L_2 is usually taken as the next L_1 (54.99 as 55, etc.) though the theoretical limits are 44.995, 54.995 and so on. The median (Q_2) was previously computed 73.52.

the class intervals vary and when computation is reversed, as in finding the position that a given wage occupies in the whole array.

The formulas for the quartiles can be written

$$Q_1 = L_1 + i\left(\frac{n/4 - F}{f}\right) \tag{6.12}$$

and

$$Q_3 = L_1 + i\left(\frac{3n/4 - F}{f}\right) \tag{6.13}$$

The process of finding the first and third quartiles parallels that of finding the median, as described in Table 6.5. The median—the second quartile—was found as the wage at the middle of the array, that is, at the point indicated by $.5n$. Similarly, the first quartile is the wage at the position in the array indicated by $.25n$, and the third quartile at $.75n$. If we generalize the fraction by writing it as F, we obtain the following equation for the magnitude at any given position in the array considered as a percentage, P; that is, the first quartile is the 25th percentile, the third quartile the 75th, and so on,

$$P = \frac{Fn - \Sigma_1}{f}(i) + L_1 \tag{6.14}$$

where the symbols refer to the class (row) indicated by Fn.

The equation (6.14)

$$P = \frac{Fn - \Sigma_1}{f}(i) + L_1$$

may be algebraically transformed to

$$Fn = \frac{P - L_1}{i}(f) + \Sigma_1$$

For example, a wage of $P = \$80$ is located in the array as

$$Fn = \frac{80 - 75}{10}(46) + 108 = 131$$

which means that a wage of \$80 is theoretically number 131 in an array of 200, or

$$F = 131/n = .655 = 65.5 \text{ percent}$$

along the array.

To illustrate: The first quartile of Table 6.4 is found by first taking Fn as $.25(200) = 50$. This, in effect, counts off the first 50 items in the array, ending somewhere in the second class, which begins with a cumulative total of 14 smaller items and ends with a total of 54. If Fn happens to coincide with Σ_1 or Σ_2, the corresponding L_1 or L_2 is, of course, the required percentile.

As in the case of the median, the assumption is made that the wage array is evenly distributed through the class and that a wage may be interpolated, as on a continuous scale, even though it is not a sum actually paid. In terms of the formula the quartile wage is located at nine tenths (36/40, where 36 is obtained as 50 − 14) of the interval \$10, above the lower limit, \$55. This gives the result $Q_1 = \$64$. That is, as theoretically interpolated, a quarter of the workers in the distribution under discussion receive a weekly wage at or below this limit.

The third quartile is similarly found, beginning with $.75n = 150$, which falls in the fourth class, between the summations 108 and 154. The quartile as interpolated is \$84.13, implying that a quarter of the workers receive a wage at or above this limit.

Dispersion is usually measured as a certain distance out from the center of the array, as in the case of both the average deviation and the standard deviation. Hence the quartile range, $Q_3 - Q_1$, is divided by two to obtain the quartile deviation. In a perfectly normal distribution, this is equivalent to taking the distance from the median out to either the first or the third quartile. But few samples are normal, so that the central point from which the quartile deviation is measured is by implication the midpoint of the quartile range, $Q = (Q_3 + Q_1)/2$, which will be identical with the median only occasionally.

In the problem at hand the quartile deviation therefore is

$$QD = \frac{Q_3 - Q_1}{2} = \frac{84.13 - 64.00}{2} = 10.065 \qquad (6.15)$$

and the midpoint from which it is measured is the average of the two quartiles, or

$$\bar{Q} = \frac{Q_3 + Q_1}{2} = \frac{84.13 + 64.00}{2} = 74.065 \qquad (6.16)$$

Hence the coefficient of variation is

$$V = \frac{QD}{\bar{Q}} = \frac{Q_3 - Q_1}{Q_3 + Q_1} = .136 = 13.6 \text{ percent} \qquad (6.17)$$

It will be seen that substituting given algebraic values in the ratio $QD/\bar{Q}$, results in the cancellation of the denominator, 2, of each term.

The quartile analysis of the data of Table 6.4 may be summarized by listing the quartiles in order, including the lowest and highest limits (in dollars), Q_0 and Q_4:

$Q_0 = 45.00$ $\quad Q_1 = 64.00$ $\quad Q_2 = 73.52$ $\quad Q_3 = 84.13$ $\quad Q_4 = 125.00$

and the two measures of dispersion $QD = \$10.065$ and $V = 13.6$ percent.

It was noted above that the first quartile is described as the 25th percentile, the median as the 50th, and the third quartile as the 75th. Similarly, the array may be divided into quintiles, deciles, or other divisions. Obviously, the quintiles could be computed by the same formula as the quartiles, merely by writing an appropriate value of F. Thus the first quintile is the 20th percentile; the first decile, the 10th percentile; and so on.

THEORETICAL CONSIDERATIONS

Interpolation in grouped data really involves more complexities than have been considered here. Theory may be suggested by Figure 6.1, in which the frequencies of Table 6.4 have been plotted cumulatively and diagonals drawn through them to form a so-called *ogive,* which may be drawn independently merely by plotting Σ_1 against L_1 and (overlapping) Σ_2 against L_2. To be exact, the classes should be corrected to read \$44.995 to \$54.995, \$54.995 to \$64.995, and so on. As noted above, this correction for the limit of accuracy is sometimes large enough to be significant. When it is made, the items in each class, as represented on the diagonal, are each placed at the center of f equal subdivisions.

For fairly large distributions with somewhat regular classes the method of estimating quartiles, quintiles, and other measures, as already described, and illustrated in Figure 6.1, is sufficiently accurate. But theory might demand that the ogive should be curved within each class so as to make a smooth, continuous curve. In fact, artists sometimes correct the plotted ogive in this way and estimate the quartiles graphically. The effect is usually to decrease the quartile deviation a little. The frequencies may be reduced to percentages (making $n = 100$), and probability paper may be used for interpolation (see Chapter 4).

SKEWNESS

A good deal of statistical theory is based upon the assumption that data may generally be expected to form distributions approximating the normal curve. In elementary form, this is illustrated when it is stated that the average deviation approximates .7979 times the standard deviation, and that the quartile deviation is .6745 times that measure, or that 68.3 percent of a distribution (area) lies between -1σ and $+1\sigma$. In more advanced statistical theory the normal curve is even more in evidence.[4]

The departure of sample data from normal theory may be due to

[4] See *The Kelley Statistical Tables* (Cambridge: Harvard University Press, 1948), preface to 1938 edition.

FIGURE 6.1. Cumulative frequencies (blocks) of weekly wages of 200 workers distributed equally through their respective classes (diagonals), with interpolation of quartiles (horizontal broken lines to intersect diagonals and perpendiculars to *Q*'s on *X* scale) (for data, see Table 6.4)

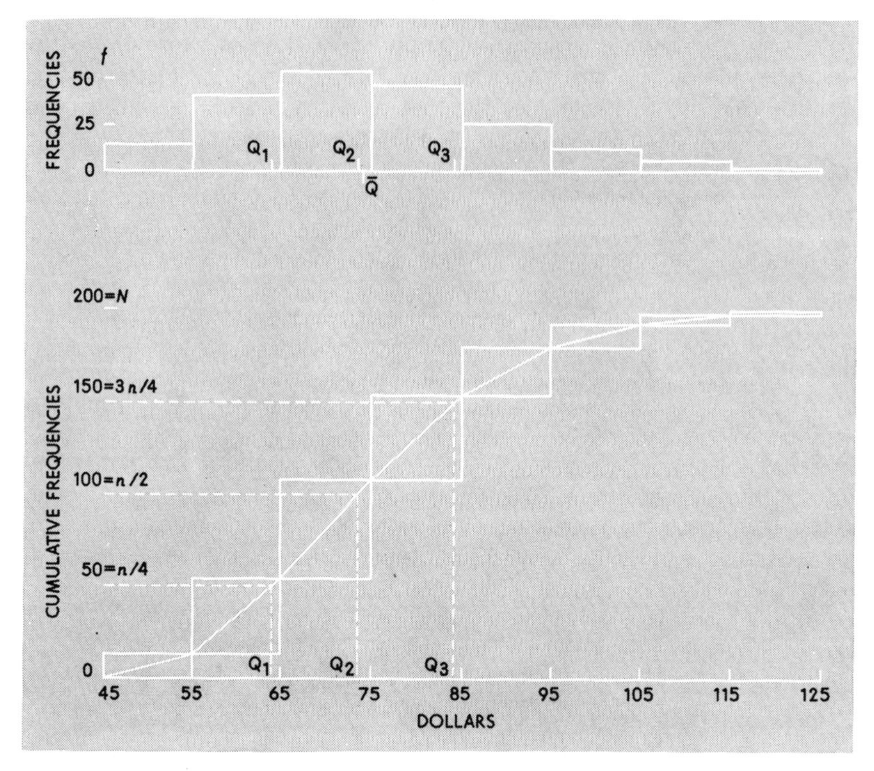

many causes, such as inadequate samples or mere chance. Or more likely, it may be merely a condition of *skewness*, a term which is generally used to imply that the distribution, when plotted, is seen to extend farther from the modal frequency on one side than on the other (see "Frequencies" in Figure 6.1). The skewness is said to be positive when frequencies extend from the mode farther to the right, and negative when they extend farther to the left. Positive skewness is more common than negative, particularly in data such as production and prices, which have a lower limit of zero. In any case, it is desirable to have a measure of skewness so as to be on guard against the consequence of an assumption of normality.

There are several mathematical measures of skewness available, but only the two that appear to be the most convenient and useful in practice are discussed here. The first is based on the fact that the mode is not appreciably affected by skewness, whereas the mean is. Hence, $\bar{X} - M_0$ indicates skewness, positive or negative. To make it comparable from one distribution to another, it is reduced, in effect, to a relative measure by giving it the form

$$Sk = \frac{\bar{X} - M_0}{s} \tag{6.18}$$

as measured from the sample. The chief objection to this measure is that the mode can seldom be exactly located.[5]

A second measure, and probably the most useful, is one that is based upon the quartiles. This measure, like the quartile deviation, may be applied to any data that can be satisfactorily grouped, even though open-end classes are present. And of course, skewness is chiefly a function of grouped data.

The quartile measure (Sk_Q) depends upon the fact that normally Q_3 and Q_1 are equidistant from the median (Q_2). That is, normally, $Q_3 - Q_2 = Q_2 - Q_1$—an equation that may be transposed to the form $Q_3 + Q_1 - 2Q_2 = 0$. But skewness disturbs this equality in that one quartile is farther from the median than the other. Hence the expression $Q_3 + Q_1 - 2Q_2$ will reveal skewness by a positive value if the skewness is positive; otherwise, by a negative value. To make the measure a comparable one, not dependent on the relative size of the X scale, the quartile range $Q_3 - Q_1$ is chosen as a denominator. The measure then is

$$Sk_Q = \frac{Q_3 + Q_1 - 2Q_2}{Q_3 - Q_1} \tag{6.19}$$

which will be zero when the distribution is symmetrical, and will approach the limits, -1 and $+1$, for various degrees of skewness.

It has already been noted that when applied to grouped data, measures of dispersion are only estimates. Certain minor corrections for these estimates are available. They relate chiefly to the error arising from taking the midpoint of a class (X as average of L_1 and L_2) as typical of the class, whereas usually the typical X is somewhat closer to the mode.

It is interesting to note that some current research in investment theory is centering on the impact of skewness on common stock portfolio selection. In the past the primary concern was on the variance (of the rate of return) as a measure of risk. Now some researchers are attempting to determine the extent to which a consideration of skewness contributes to a better understanding of the concepts of riskiness and any possible affect on the theory of diversification. It is asserted by some financial analysts

[5] Skewness may be measured without the mode, using

$$Sk = \frac{3(\bar{X} - M_d)}{s}$$

that investors should be more interested in a positively skewed distribution of expected returns, given the same average rate of return.

COMPARISON OF THE LOCATION OF THE THREE MAJOR AVERAGES IN A DISTRIBUTION

When a frequency distribution is absolutely symmetrical, the mean, median, and the mode are the same values. That is, the value that occurs most frequently and the value that divides the data in half are identical with the computed mean. However, if we were to add a few high values to this same distribution, then it would be pulled to the right and have positive skewness. Because the mean is a computed value and is influenced by each and every value, the additional high values would tend to increase the mean (i.e., move to the right on the continuous scale). The median which is a positional measure is responsive only to the number of items in the distribution. It would shift *slightly* to the right but only because of the number of items added, not because of their magnitude. The mode probably would remain exactly where it was previously. In summary, in a positively skewed distribution, the mean is greater than the median and the median is larger than the mode.

On the other hand, assume instead of adding some high values to our symmetrical distribution we included some low values. Then our new series would be pulled to the left and have negative skewness. In such a situation the mean is the smallest of the three measures and the mode is the largest.

SUMMARY

The various measures of dispersion and the accompanying measure of central tendency, with other computations, may be summarized as follows:

Standard deviation: $s = \sqrt{\Sigma x^2/n}$

Estimated universe standard deviation, $\hat{\sigma} = \sqrt{\Sigma x^2/(n-1)}$
Estimated coefficient of variation, $\hat{\sigma}/\bar{X} = \hat{V}$
Based on arithmetic mean
Normally, 68 percent of items in range $\bar{X} \pm \hat{\sigma}$
Most commonly used measure
Particularly useful in complex computations
Somewhat overstresses extreme items

Average deviation: A.D. $= \Sigma|x|/n$

Coefficient of variation, A.D.$/\bar{X}$
Based on arithmetic mean (sometimes on M_d)
Normally, A.D. $= .7979\sigma$

A convenient, informal measure
Not adapted to complex computations
Unit weight to all items

Quartile deviation: $(Q_3 - Q_1)/2$

Coefficient of variation $(Q_3 - Q_1)/(Q_3 + Q_1)$
Particularly useful in open-end class distributions
Normally includes half of sample items in $Q_3 - Q_1$
Not adapted to complex computations
Understresses extreme items
Percentiles (P) interpolated at fraction of n, that is, at Fn

SYMBOLS

$s^2 = \frac{\Sigma x^2}{n}$, the variance of the sample

$s = \sqrt{\frac{\Sigma x^2}{n}}$, the standard deviation of the sample

$V = \frac{s}{\bar{X}}$, the coefficient of variation of the sample

$\hat{V} = \frac{\hat{\sigma}}{\bar{X}}$, the estimated coefficient of variation of the population

$\bar{X}_0$, a constant called the assumed mean or "guessed mean" subtracted or added to a variable

$d = X - \bar{X}_0$, a variable after a constant has been subtracted

$\bar{d}$, the mean of the variable d

$d_i = (X - \bar{X}_0)/i$, a variable, after a constant has been subtracted, divided by a constant i (the class interval)

$\bar{d}_i$, mean of variable d_i

$\hat{\sigma} = s\sqrt{\frac{n}{n-1}} = \sqrt{\frac{\Sigma x^2}{n-1}}$, the estimate of the standard deviation of the population

$\sigma = \sqrt{\frac{(X' - \mu)^2}{N}}$, the standard deviation of the population

$\text{A.D.} = \frac{\Sigma|x|}{n}$, the average deviation from the mean for the sample

OR

$\text{A.D.} = \frac{\Sigma|(X - Md)|}{n}$, the average deviation from the median

$Q_1 = L_1 + i\left(\frac{n/4 - F}{f}\right)$, quartile one

$$Q_3 = L_1 + i\left(\frac{3n/4 - F}{f}\right),$$ quartile three

$$QD = (Q_3 - Q_1)/2,$$ quartile deviation

$$V = \frac{Q_3 - Q_1}{Q_3 + Q_1},$$ coefficient of variation of the middle half of a distribution

$$Sk_Q = \frac{Q_3 + Q_1 - 2Q_2}{Q_3 - Q_1},$$ a measure of skewness or asymmetry

PROBLEMS

6.1. Compute the standard deviation for the problems (*a*) and (*b*) at the right.
Is $s_2 = s_1$? Is $V_1 = V_2$? Why?

a) X_1	b) X_2
1	10
2	20
3	30
4	40
5	50

6.2. The variable X in the table at the right represents the number of barbers in ten cities in a state:

X	f
20	1
30	2
50	3
60	2
80	2
	$n = 10$

Required:

a. Compute the standard deviation by the formula

$$s = \sqrt{\frac{\Sigma fx^2}{n}}$$

using $\Sigma fx^2 = \Sigma fX^2 - (\Sigma fX)^2/n$.

b. Compute the standard deviation by the formula

$$s = i\sqrt{\frac{\Sigma fd_i^2}{n} - \left(\frac{\Sigma fd_i}{n}\right)^2}$$

using $\bar{X}_0 = 20$, $i = 10$.

6.3. This problem and Problem 6.4 use the frequency distribution of five different product values purchased by various consumers shown in the table at right below.

Required:

a. The average deviation, using f_1, f_2, and f_3.

b. What is the range of X in this problem?

X	f_1	f_2	f_3
1.00	1	1	1
2.00	1	2	4
3.00	1	4	3
4.00	1	2	1
5.00	1	1	1

6.4. *Required:*
The quartile deviation, using f_2 and f_3 of table in Problem 6.3.

6.5. The distribution of family income among consumer units in any given year forms an open-end percentage frequency distribution. In 1965 and 1975 the distributions were:

L_1	L_2	$f\%$ 1965	$f\%$ 1975
0 and under	2,000	16	14
2,000 and under	4,000	25	22
4,000 and under	6,000	26	25
6,000 and under	8,000	16	17
8,000 and under	10,000	7	9
10,000 and under	15,000	6	8
15,000 and over		4	5

Required:
Determine quartiles for each distribution. Is the 1975 distribution more uniform than the 1965 distribution?

6.6. What is the standard deviation of the set of numbers shown at the right? (Show all of your work.)

X
100
110
120
130
140
150

6.7. Subtract a constant, $k = 100$, from each of the values in the set of numbers in Problem 6.6. What is the standard deviation of the $(X - k)$ set?

6.8. Divide the $(X - k)$ of Problem 6.7 by 10. What is the standard deviation of the $(X - k)/10$ set?

6.9. If the standard deviation of the $(X - k)/10$ set is multiplied by 10, does it equal the standard deviation determined in Problem 6.6? Why, or why not?

6.10. Label the following distributions with respect to the location of the mean, median, and mode. Indicate the direction of the skewness.

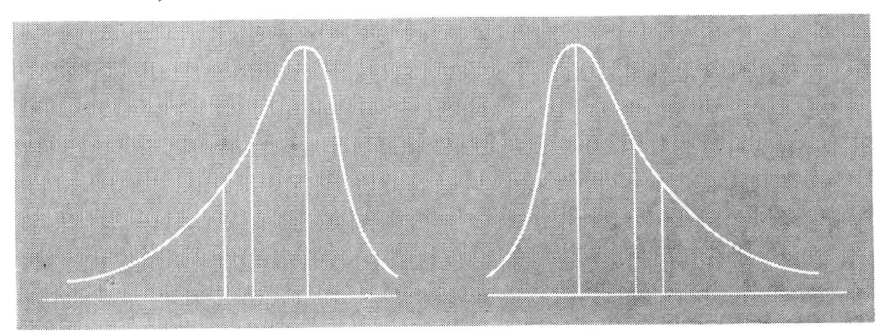

6.11. In 1974 the mean monthly employment in the Gauss Company was 25,000; the standard deviation was 1,000. In 1975 the mean monthly employment was 50,000; the standard deviation was 1,500.

a. How would you measure employment stability in such a situation?

b. According to your measure, in which year was employment more stable?

6.12. Given the following data: 3, 5, 6, 8, 11, 13, 14, and 17. Find the following:

a. Mean.
b. Median.
c. Range.
d. Standard deviation.
e. Variance.
f. Coefficient of variation.

6.13. Assume a markedly right-skewed frequency distribution with a standard deviation of ten units. Assume that you withdraw an item equal to the mean and recalculate the standard deviation. What will be the effect upon this measure of dispersion? Why will the standard deviation be affected in this manner?

6.14. What is meant by the term "relative dispersion," and how would this concept be used?

6.15. Refer to Problem 4.3, Chapter 4; determine the following from your frequency distribution and interpret your results:

a. Variance.
b. Standard deviation.
c. Quartiles.
d. Quartile deviation.
e. A measure of skewness.

6.16. Refer to Problem 4.4, Chapter 4; determine the following from your frequency distribution and interpret your results:

a. Variance.
b. Standard deviation.
c. Quartiles.
d. Quartile deviation.
e. A measure of skewness.

Which distribution has less variation (i.e., refer to Problems 4.3 and 4.4 of Chapter 4)?

6.17. Refer to Problem 4.5, Chapter 4; determine the following from your frequency distribution and interpret your results:

a. Variance.
b. Standard deviation.
c. Quartiles.
d. Quartile deviation.
e. A measure of skewness.

Refer to Problems 4.3, 4.4, and 4.5, Chapter 4, and indicate which distribution has the least variation.

6.18. The following appeared on the editorial page of a Detroit newspaper:

> Parents complain from time to time that they can't follow school marking systems, and hence can't tell whether their child's report card means the child is a slow reader or a fast runner.
>
> Those who think they're mystified may realize they're lucky after all on reading the rules for getting on the honor roll at the high school in Scottsdale, Arizona. They go:
>
> > A student whose average exhibited performance in all credited subjects in relation to the performances of all other students falls

> at a level which places him or her on the normal curve of probability at a point falling on the plus side of the mean and between the second and third standard deviations will have made the honor roll first class.

The editors went on to suggest that it would be easier just to put on the honor roll any youngster who can explain what that paragraph means.

Required:

a. How would you explain this paragraph to a layman? Use a diagram to help illustrate the requirements.

b. Is there anything wrong with the requirements aside from the technical jargon?

6.19. Refer to Problem 5.14, Chapter 5.

a. What are the range limits?

b. Estimate the standard deviation.

c. What is the third quartile value?

d. What is the first quartile value.

e. What is the quartile deviation?

6.20.

Estimates of the male labor force, by age, 1975 (in millions)

Age L_1 — L_2	Midpoint of Class X	Male Labor Force f
15 and under 25	20	13
25 and under 35	30	12
35 and under 45	40	11
45 and under 55	50	11
55 and under 65	60	8
65 and under 75	70	2
		$n = 57$

Calculate and interpret the following measures:

a. Mean.

b. Median.

c. Mode.

d. First quartile value.

e. Third quartile value.

f. Quartile deviation.

g. Standard deviation.

h. A measure of skewness using the quartile values.

6.21. In a distribution of a large number of customers handled per sales station per day in a major department store, the mean number of customers handled is 35 and the median is 57.

Required:

a. Is this distribution positively or negatively skewed? Why?

b. Is the mode probably less than 35? Greater than 57? Why?

6.22. Comment on the following statements:

a. The standard deviation of a normal distribution is greater than that of a positively skewed distribution.

b. The midpoint of the first and third quartile cutoff points of a normal distribution is at the median value.

c. If distribution *A* has greater relative dispersion than distribution *B*, and $\bar{X}_a < \bar{X}_b$, then the standard deviation of *A* must be greater than that of *B*.

6.23. Comment on the following statements:

a. The percentage of items within the interquartile range equals the percentage of items above the median in a normal distribution.

b. Assume two distributions have the same mean and the same sum of squared deviations of their values around their respective means. Then if the first distribution includes more items than the second, the former has less relative dispersion.

6.24. In comparing the dispersion in the distribution of incomes for the United States with that of the distribution of intelligence test scores of the population in the United States, what measure should be used? Why? Which average is likely to be representative of each distribution? Why?

6.25. Given the frequency distribution shown at right:

a. How were the measurements in the table made?

b. What is the precise midpoint of the class 2.15–2.24?

c. Using the shortcut formulas, determine the mean and standard deviation of the distribution.

Class Limit (percentage) L_1 L_2	Frequency f
1.75–1.84	1
1.85–1.94	6
1.95–2.04	8
2.05–2.14	24
2.15–2.24	26
2.25–2.34	32
2.35–2.44	23
2.45–2.54	9
2.55–2.64	6
2.65–2.74	3
	$n = 138$

6.26. Comment on the following:

a. The arithmetic mean computed from an array of items will always be the same as the one determined from a frequency distribution of the same series.

b. The sum of the squared deviations of the items in a series taken from their mean is less than the same measure taken from any other value.

c. The most appropriate measure of dispersion to use in evaluating the variation of an open-end distribution is the variance.

d. The appropriateness of a mean as a reasonable measure of central position requires that the measure be based on a frequency distribution with small variation.

6.27. In a given community, the distribution of the population by age had a mean of 35.7 years and a standard deviation of 14.3 years. The distribution of family income for the same city had a mean of $6,742 and a standard deviation of $3,371. Which distribution contains the greater variability?

6.28. The Savon Company, Inc. divided one of its "classes" of sales trainees into two groups equal in number. While the sales trainees in each group were approximately comparable in background, the groups were given different types of training programs. The sales manager was interested in making an evaluation of the effectiveness of the two programs, analyzed the sales of trainees, and at the end of the first six months obtained the following results:

	Amount of Sales per Trainee	
	Program S_1	Program S_2
Mean	$45,897	$40,352
Median	38,678	43,651

Required:

a. Assume that the differences in the averages reflect differences in training programs (not differences in individuals) and that the next six months is likely to be similar, which training program will lead to the larger volume of total sales at the end of the first year? Why?

b. Assuming that each of the frequency distributions (for the first six months) is unimodal, how can you account for the different *direction* of skewness for the two programs? Explain.

6.29. The Ace Manufacturing Company is producing two similar parts by a patented process. The engineers have indicated that one of the dimensions is vital and specified that this measurement should be .450 ± .003 inches for one of the items and .765 ± .003 inches for the other part. Would the quality control experts be interested in the standard deviation as a measure of absolute variation or would a measure of relative variation be more meaningful in controlling the production process? Why?

6.30. The mean daily output of the above-mentioned parts in department A of the Ace Manufacturing Company was 1,224.3 units per operator and a standard deviation of 75.8 units. The company held some special classes of instruction for the less efficient workers. After the training sessions the average (mean) daily output per operator was 1,452.7 units and the standard deviation was lowered to 62.1 units. On the basis of this information respond to the following:

a. What changes in the frequency distribution of daily outputs per operator probably took place? Explain.

b. Which frequency distribution of outputs per operator, that is, before and after the training sessions, contained less variability? Which measure of variation did you use and why?

6.31. If the average height of males in the United States is 68 inches and the standard deviation is 1.75 inches, and the corresponding data for females are 64 and 1.40 inches, which distribution shows greater variability?

6.32. The average daily closing price of the common stock of the consulting firm of L & R, Inc. was $50 during the past month. The standard deviation was $20. The average daily closing price on the related stock warrants was $5 with a standard deviation of $1. Which one showed the greatest stability?

6.33. Assume that potential employees in a given firm must take an "employment aptitude test" that is determined on the basis of (*a*) abstract thinking and (*b*) manual dexterity tests. Given the following information respond to the question below:

	Test Scores	
	Abstract Thinking	Manual Dexterity
Employee	150	100
Company average	130	85
Company standard deviation	10	7.5

a. What is the prospective employee's standing relative to the company averages in both of these test areas?

b. Basically, what is involved here is the standardizing of the two related scores. Why does this method facilitate comparisons?

RELATED READINGS

HAMBURG, MORRIS *Basic Statistics*, chap. 3. New York: Harcourt, Brace, Jovanovich, Inc., 1974.

HUNTSBERGER, D. V.; BILLINGSLEY, PATRICK; and CROFT, D. J. *Statistical Inference: for Management and Economics*, chap. 3. Boston: Allyn & Bacon, Inc., 1975.

NETER, JOHN; WASSERMAN, WILLIAM; and WHITMORE, G. A. *Fundamental Statistics for Business and Economics*, chap. 4. Abridged 4th ed. Boston: Allyn & Bacon, Inc., 1973.

Part III

Probability, the basis for statistical inference

Far better an approximate answer to the *right* question, which is often vague, than an *exact* answer to a wrong question, which can always be made precise.

John W. Tukey

7

Introduction to probability theory

All business proceeds on beliefs or judgments of probabilities and not on certainties.

Charles W. Elliot

THE STATISTICAL METHODS that have been discussed so far all belong to the field of descriptive statistics. They can be reasonably described as being safe, inasmuch as the only errors to which we expose ourselves might be mistakes in arithmetic, selection of the wrong formula, perhaps misreading of a tabular or graphic value, or some other error due to careless habits. Although this constitutes a sufficient risk for most of us, we now consider an entirely different kind of risk, namely, errors which cannot be eliminated through more careful procedures. This new kind of error appears precisely at the very moment we use partial information, for example, a sample, and derive general conclusions or make decisions by statistical inference.[1] Professor J. Newman has referred to this as the use of inductive behavior. Naturally, it is not necessary for us to expose ourselves to this new kind of error; but if we restrict ourselves to the descriptive methods we have found useful thus far, we shall, in effect, be burying our heads in the sand. In other words, we would be depriving ourselves of perhaps the most powerful source of knowledge we have ever had, that is, the scientific method of statistical inference.

Science is based on the inductive procedures which are described in most logic books as "reasoning from the partial to the general." Scientific knowledge consists of generalizations which are made after observation and experimentation. We might escape this new kind of error mentioned above if we were to restrain ourselves from making statistical inferences (generalizations or conclusions), but we would only be able to report

[1] See George A. W. Boehm, "The Science of Being Almost Certain," *Fortune,* vol. 69 (February 1964), pp. 104–7, 142–48.

what we have seen and what we have measured. Never would we be able to put this information fully to work for us. Consequently, if we want to learn from experience and use our knowledge of the past in making estimations, we must face the risk which is intrinsic in the scientific method and inductive statistics.

INDUCTIVE STATISTICS AND PROBABILITY

One might say that the primary task of inductive statistics consists of the development of methods which enable the analyst to calculate and subsequently minimize the gamble which automatically is assumed in every problem of estimation or prediction. Therefore, we shall attempt to evaluate these risks because it is the use of the "calculated risk" which distinguishes the scientist from the fortuneteller.

The next few pages will be concerned with a consideration of the concept of probability. This is done because it is virtually impossible to understand intelligently the factors of chance which are the foundation of inductive statistics without some appreciation of the notion of probability. That is, we do not move immediately to "practical problems of the real world" because pedagogical experience suggests that an introduction to the oversimplified games of chance makes it easier to understand some *basic* principles of probability than if these same concepts are discussed at the outset in the context of an applied example.

The whole area of statistical inference relies heavily upon probability theory; but in addition, it also supplements it. When we collect data, we might use statistical theory to assist us in choosing among appropriate alternative mathematical models. For example, in a given city, let us consider selecting a random sample of family units to estimate the fraction of unemployment. For a given percentage unemployed in the city, probability theory indicates what the percentage of unemployment in the sample is likely to be. Here, we have used probability to deduce from the known characteristic of the population the probable statistic of the sample. Inductive statistics does the exact opposite: that is, it utilizes the sample evidence to estimate the fraction unemployed in the universe (the city). In other words, statistical reasoning infers the content of the population from the characteristics of the sample evidence. *In general terms, we might say that probability theory deduces from an appropriate mathematical model the properties of a physical process, whereas statistical reasoning infers the characteristics of the model from the sample data.*

INTERPRETATIONS OF PROBABILITY

There is very little disagreement about the foundations of probability, and, therefore, the foundation of modern statistics, or about the mathematical consequences. The foundation of probability was laid in 1933 by the Russian probabilist A. Kolmogorov in his theory of sets, as was indicated in Chapter 1. However, at the level of application and interpretation, two rather extreme views currently exist.

First of all, there is the *classical or objectivist position,* which presently seems to be the more popular. This position holds that the probability interpretation should be made only for events that can be repeated under much the same conditions. As a result, the classical school is perfectly content to talk about probabilities with respect to the tossing of a coin or a pair of dice or the manufacture of some mass-produced product. For example, statisticians in this school can readily think of many electronic tubes being mass-produced and of the probability of an acceptable tube as the long-run relative frequency of the number of good tubes to the total number produced. It is at the interpretation of a probability with respect to unique events that the classical statistician draws the line. For example, the classical statistician would not consider talking about the probability that East and West Berlin would unite to become a single city in the next two years. As a result, a large number of problems is set aside by the classical school as not being appropriate for the application of a probability interpretation. This view is held because no long-run ratio can be seen. That is, the objectivist prefers to make interpretations only from events that may be repeated over and over again under much the same conditions, and prefers not to introduce other kinds of evidence into his or her statistical inferences.

The second school of thought is sometimes called the *personalistic, Bayesian, or subjective school of probability.* This school regards probability as a measure of personal belief in a particular event or proposition —for example, the proposition that it will snow tomorrow. Subjectivists believe that different "reasonable" people may differ in their degrees of belief even though they reach this conclusion by examining the same evidence (perhaps some economists or business analysts may have this characteristic). As a result, the personal probabilities of these people may differ for the same event. This is referred to as subjective probability. It is also worth noting that the subjectivists will make a probability interpretation or application to all of the problems the classical statistician studies and perhaps to many more. For example, in principle at least, the subjective school would take the Berlin question in stride. The personalists also make a more effective use of Bayes theorem than the classical statisticians and as a consequence are referred to as Bayesians.[2] However, when

[2] See Harry V. Roberts, "Bayesian Statistics in Marketing," *Journal of Marketing,* vol. 27, no. 1 (January 1963), pp. 1–4; Paul E. Green, "Bayesian Decision Theory in Pricing Strategy," *Journal of Marketing,* vol. 27, no. 1 (January 1963), pp. 5–14; and Leonard J. Savage, "Bayesian Statistics," in Robert E. Machol and Paul Gray, ed., *Recent Developments in Information and Decision Processes* (New York: Macmillan Co., 1962), pp. 161–94.

the amount of data studied is large, both of these schools will usually obtain similar answers.

Bayes theorem is as follows:

$$P(A_i|B) = \frac{P(A_i)P(B|A_i)}{P(A_1)P(B|A_1) + \cdots P(A_n)P(B|A_n)} \tag{7.1}$$

which actually is nothing more than the mathematical definition of conditional probability written out to indicate how calculations are made. Bayes theorem is used to revise probabilities in the light of additional sample evidence; these revised probabilities are termed *posterior probabilities.* (See page 142 and Chapter 18.)

Although the final word is never said on matters relating to various new schools of thought, it may well be that the distinction between probability as a long-run relative frequency and probability as a subjective measure of the degree of belief is one that students may wish to reexamine from time to time as the issues become clearer.

Basically, we encounter three types of problems in the study of probability: (1) the problem of definition and interpretation; (2) the problem of calculation of probabilities based upon known values; and (3) the problem of obtaining numerical probabilities or the problems of statistical inference. Those issues associated with statistical inference will be taken up in detail in Chapters 9, 10, and 11. The first two problems will be discussed below.

THE MEANING OF PROBABILITY AND PROBABILITY INTERPRETATIONS

One very serious obstacle to our study of probability is the multiplicity of meanings assigned the word. For example, the following words are frequently used as synonyms for the layman's interpretation of probability: "possible," "probable," "probably," "likely," "chance," and "odds."[3] Although it is well to remember that all definitions are arbitrary, the statistician's use of the word in the *classical* interpretation relates specifically to the important scientific concept of the relative frequency—to be more exact, the limit of a relative frequency or the relative frequency in the long run.

In many texts, the standard definition of probability runs something as follows: "If an event can lead to the occurrence of n equally likely results, of which s are denoted as success, the probability of a success is given by

[3] Bennett Cerf told the story about the barber: "The shears snipped merrily away and the barber's dog lay close beside his master's chair, his eyes riveted on the customer in the chair. The customer remarked: 'That dog of yours seems mighty fond of watching you cut hair.' 'T'aint that,' chuckled the barber. 'He knows the chances are about two to one that before I've finished, I'll snip off a bit of your ear.' "

the ratio of s divided by n." It is clear that this is not really a definition of probability and cannot be accepted as such, because it explains probability in terms of events which are equally likely without indicating what "equally likely" means. Normally, if two events are said to be equally likely, we usually imply that they are equally probable, that is, they have the same probability; as a result, in this definition, then, we are using the word we are trying to define. It should be noted that in this text we shall concentrate on the classical interpretation of a probability for the simple reason that before we can appreciate the subjectivist's interpretations, it is necessary fully to understand the objectivist's point of view. Therefore, let us look at a few artificial examples of a probability in terms of a relative frequency.

If on the average, 7 out of 10 points after a touchdown are successful, one might say that the relative frequency of successful extra-point attempts is 7 over 10, or .70. This assumes that the *population* characteristics is .70 and that the universe is not changing over time. The population involved here is all *past* attempts at the point after touchdown and all *future* attempts covered by the generalization. It would be valid to say that the probability that any *random* attempt is successful is .70; it would be absurd to say that the probability that the *next* attempt will be successful is .70. The next attempt will either be successful or it will not. *Remember that a probability interpretation relates to the population and not to a specific event.* Or if out of every 100 students entering a university, 67 eventually graduate, we might say that the relative frequency with which students graduate from this university is .67. Or to look at an imaginary (theoretical) distribution for the moment, we might consider flipping an honest coin, where we would expect the head to appear with a relative frequency of .5. Obviously, this does not mean that if we flip a coin 30 or 40 times, we must always get 50 percent heads and 50 percent tails. We cannot expect precisely an equal number of heads and tails every time; but if we were to flip a coin a large number of times, we would come fairly close to the 50 percent heads and 50 percent tails. *From our point of view, then, it is this relative frequency in the long run which we define as a probability of the occurrence of a particular event.*

Estimation, prediction, and the testing of hypotheses

Probability interpretations are used to judge the reliability of results in problems of estimation, prediction, and testing of hypotheses. For example, let us consider the problem of the political pollster attempting to predict the outcome of an election. If we ask a random sample of the eligible voting public how they might vote, this would provide us with some information about their voting plans. That is, we would have some sample evidence. Then, if on the basis of this sample evidence and with some pretty fancy statistical methods, we estimate that candidate K will receive from 55 to 58 percent of the votes, we cannot be absolutely certain that our estimate is correct. That is, a very real question may be posed concerning the reliability of this generalization, and it leaves us open to the question as to how sure we really are that candidate K will receive from 55 to 58 percent of the total vote cast during this election. Perhaps, upon analyzing the data and the methods we have used, we might say that we are 95 percent sure of our estimate. This means that we have assigned our generalization a probability of .95 of being correct. Obviously, we are interested in one particular election; and as a result the public is interested in one particular estimation. Consequently, our estimate could not possibly be meaningfully interpreted that candidate K would receive from 55 to 58 percent of the votes about 95 percent of the time if he ran for office a large number of times. (This is an important concept and will be discussed in Chapter 9). *What we really mean when we assign a given probability to our response is that we are using a method of estimation which we expect to be successful about 95 percent of the time.* The point is, statisticians discuss the accuracy of their statements by giving the success ratio of the *methods* employed. In precisely the same manner, we shall express the goodness of statistical decisions which are based on sample information in terms of the success ratio of the techniques utilized. (This will be done in Chapter 10.)

It should be noted that experimental data are used to verify the theorems of probability and really involve two basic ideas: (1) the random selection of the individual items and (2) "the law of large numbers." The latter, of course, is the average person's term for the central-limit theorem which, briefly stated, indicates that the sample mean, for example, tends to be closer and closer to the population mean (μ) as n increases.

To repeat, from this page forward the probabilities we shall assign to our results of estimations, tests of hypotheses, and so forth, will always refer to the goodness of the methods we have employed. That is, they will stand for the percentage of the time we might anticipate that these methods will provide us the correct results if they (the methods) are used a large number of times.

Probability and probability distributions

As indicated previously, probability theory represents the foundation of modern statistics and sampling theory. When a person makes a probability statement with respect to the occurrence of a particular event, the estimate may be based (1) on an intuitive feeling; (2) on an analysis of the "laws of chance" operating on the problem, that is, a priori probability; or (3) perhaps upon a systematic analysis of past experience (empirical probability).

Although one probably should not forget Kipling's famous quotation, "A woman's guess is more accurate than a man's certainty," modern statistics is based primarily on a priori probability and empirical probability; thus, we may ignore (1) above. In addition, the Bayesian statistician makes use of the posterior probability distribution. The latter probability distribution is utilized in Bayesian statistics and recognizes the role played by Bayes theorem in indicating how a particular a priori probability distribution when combined with sample evidence leads to a posterior probability distribution for the unknown parameter. (See footnote 2.)

A priori and empirical probability

Perhaps it would be worthwhile to make a clearer distinction between the a priori and the empirical probabilities. First of all, a priori probability is normally encountered in problems dealing with games of chance—for example, dice and cards. It is worthy of note that a consideration of the simple a priori problems led to the development of some important theoretical concepts which are applicable to more complex problems. Normally, we think of a priori probability as being deductive in nature, that is, from cause to effect and based on theory instead of the evidence of experience or experimentation. Probability derived from past experience is called empirical probability and is used in many practical statistical problems, the classic example being the preparation of insurance mortality tables which, obviously, are based on past experience. Actually, in the analysis of most practical business problems, these two kinds of probability concepts are combined.

A priori probability may be determined rather easily whenever complete information is available on the various ways an event may occur. If we designate a particular outcome as success and the total number of possible outcomes, each equally likely, is known, then in a large number of trials the ratio of the number of possible successes to the total number of outcomes is the probability of success with a long-run relative frequency. For example, what is the probability of randomly drawing any one of the four kings from a deck of 52 playing cards? There are 52 possible outcomes, each of which is equally likely. Four of these outcomes are kings, and the drawing of any one of these would represent success. Consequently, the probability of drawing a king is $4/52$, or $1/13$. If the cards are replaced after each draw, the probability of drawing a king on the first draw and also the second draw is the product of $4/52 \times 4/52$.

Another example may be in order. A die has six sides, and each of these sides is presumed to be equally likely to turn up on any one throw. (It might be noted that some wags have defined a die as a "piece of equipment used in fleecing poor little lambs.") The probability of throwing a six is $1/6$. Many people would indicate that the odds are $1/5$. They make this mistake by relating the possible favorable outcomes to the possible unfavorable ones. Although it is true that the odds are five to one against, it must also be remembered that there are six possible outcomes, only one of which represents success.

Probability in statistical work

Usually, probability is expressed as a decimal fraction from zero to one. The number of ways a particular outcome may occur is the numerator of the fraction, and the total number of possible outcomes is the denominator. The reasons the limits of probability are from zero to one are: (1) the proportion of successes can never be negative, and (2) the proportion of successes can never exceed 100 percent. In our work, we shall usually express probability as a decimal fraction such as $P = .20$, rather than 1/5. The probability that an event will occur (success) is designated by the symbol s. If the event is certain not to occur, then $P(E) = 0$. If the event is certain to occur, then $P(E) = 1$.

What do we mean when we say that the probability of drawing a spade from a deck of 52 cards is 1/4, or .25? Precisely what does this statement imply? We mean that our best possible estimate, that is, our a priori expectation, is that out of four random selections (each from a full deck), one card will be a spade. Certainly, we do not mean that one should expect to obtain exactly one spade in *every* draw of four cards from a full deck. As many a gambler has found in the short run, numerous cards *may* be drawn in succession before obtaining a spade. Our expectation is confirmed more and more fully as the number of trials increases, that is, as n increases. It is certainly much more reasonable to assume that about one fourth of 10,000 draws will actually turn out to be spades than that about one fourth of 20 draws will be spades. The point illustrated here is that we find from experimentation that the tendency is for the ratio of spades to the total cards drawn to approach one fourth more and more closely as n increases; again, we are talking about the relative frequency in the long run.

Probability is readily understood and, perhaps, conveniently handled if it is treated as a fraction or a proportion. If we divide a population, either a real or an imaginary one, into $s + f$ elements, of which s represents success and f represents failure, then the probability of success is $P(E) = s \div (s + f)$. Again, we may refer to the tossing of a fair coin and recall the theoretical distribution which would be generated by repetitively tossing the coin. Approximately 50 percent of the outcomes would be heads and 50 percent tails. Therefore, we may say that the probability of heads is equal to .5. Clearly, this type of reasoning is not necessarily confined to games of chance and can be applied to any kind of population. It is also abundantly clear that the numerical value of the probability cannot always be so precisely determined as with coins.

However, a very important aspect of the definition of probability should be clear. *That is, any probability interpretation relates to the population and not to any specific individual or event in it.* For example, we might speak of the probability of snow on any *random* day in a given area, but we cannot relate that probability to a particular day. Quite possibly, all of the snow is concentrated in a short period, and it would be ridiculous to be talking in terms of the probability of snow on any *given* day. *The point is, a probability interpretation is the exclusive possession of the universe.*

PROBABILITY, RANDOMNESS, AND PREDICTION

The concept of randomness needs clarification for the simple reason that any definition of probability becomes meaningful only with reference to the random selection of items from a population. The concept of randomness seems to be taken for granted by most analysts who assume that everyone knows what the word means but relatively few can actually define the term! In this sense, randomness is not unlike the term entropy in engineering. Entropy is a theoretical measure of energy which all engineers are expected to understand yet few exacting definitions of the word are readily available.

The word random was first introduced here with respect to the selection of a simple random sample. To recall, a sample is considered to be random if every item in the population (or that each possible combination of *n* units) has the same chance of being selected. Please note that probability has crept into our definition in the use of the word "chance." Perhaps, this is an indication that randomness might be explained in terms of the concept of probability.

The 1969 draft lottery was labeled "statistically unfair" with indications that it would result in twice as many men born in December being drafted as those born in January. Dr. Fred T. Haddock of the University of Michigan[4] concluded that the draft lottery was "not random" after working on the problem with the aid of some mathematicians. Dr. Haddock stated:

> The draft lottery is definitely not random. Inspection of the lottery results clearly shows a systematically increasing number of men being drafted as their birth date falls later in the year. The odds against this

[4] As reported in the December 6, 1969, *Ann Arbor News*. For a discussion of the coefficient of correlation see Chapter 15. Indelibly inscribed in the author's own mind are his similar protests about the nonrandomness of the draft during World War II. For the record, this author had precisely the same degree of "success" in filing his complaint as did Dr. Haddock!

> trend resulting from random selection are over 100,000 to 1. For example, twice as many men with December birth dates will be drafted compared to those having January birthdays.
>
> This can be seen by plotting the average monthly draft number from January through December. The plot gives a nearly linear decrease in average draft number (increasing draft risk) with the date of birth.
>
> It is as if the capsules containing the birth dates were placed in the glass bowl in monthly order with January on the bottom and December on top and mixed too little for a random mixture to be obtained.
>
> The monthly average draft numbers from January to December are approximately: 201, 203, 226, 204, 208, 196, 182, 173, 157, 182, 149, and 122. Note that the first six months all have averages above the overall average of 183.5, and that the last six months averages are all below the overall average. The coefficient of correlation between the order number of the lottery drawing and the order of the birth date from January 1, is −0.222, with a standard deviation of 0.052. If the drawings were random the coefficient would be very near 0. The chance of the coefficient being this far from 0 is less than 1 in 100,000.

Dr. Haddock went on to recommend that the men born in November and December with draft card numbers below 184 should be given a new deal by having their 47 birth dates redrawn from a new lottery which would give them order numbers to be multiplied by 366 divided by 47 and then interlaced with the remaining present numbers. Do you think this would have made the lottery random?

Randomness: Intuitive understanding

As indicated in Chapter 3, tables of random numbers are constructed by the use of a computer generating digits "0" through "9" one after the other in such a manner that it is impossible to uncover any method of predicting some numbers in the table simply by referring to the table. To help us in gaining an intuitive understanding of the concept of randomness, let us develop a brief table of random numbers without the aid of a computer. Assume that we have ten tags numbered with the digits 0, 1, 2, 3, 4, 5, 6, 7, 8, and 9. We then follow this procedure: (1) place all ten tags in a container; (2) thoroughly mix the tags; (3) withdraw one tag and record the number; (4) return the tag to the container; (5) thoroughly mix the tags; (6) withdraw another tag and record the number; and (7) repeat this process over and over again. This procedure was repeated for 100 trials with the following results:

31994	43622	91386	08105	71712
28619	13484	77114	37978	44968
42125	87439	59748	06364	33591
58418	91369	27151	89024	93573

The results produced in this fashion are called *random numbers* or random digits. These numbers form what is called a *random variable,* and the procedure generating this sequence is called a *random process.* The population is one in which each number (item) has an equal probability of being selected, namely, .10. *The concept of randomness must be produced by a procedure which in itself is random.* It should be noted that if the above procedure were to be repeated again, a different sequence undoubtedly would be generated from the same population. Perhaps it is worthwhile to remind the reader that randomness is not synonymous with haphazard. To achieve randomness requires an extremely careful design and appropriate planning. A test for randomness frequently is the study of the sequence the process has produced, for example, Dr. Haddock's report.

If a variable can assume any of a number of values, each of which has a given probability, it is called a random variable. The variable can be either discrete or continuous. A *discrete variable has a finite* number of values whose probabilities are merely their relative frequencies. A *continuous variable* dictates that we refer to the probability that its value will lie winthin an *interval.* While the sequence of most random variables is irregular, when arranged in a frequency distribution form they will have certain distinct attributes. The implication here is that of the concept of the stability of mass data and that large numbers generate regularity and stability. This, of course, makes possible the prediction of mass behavior in terms of a probability interpretation. (See Chapter 3, pages 26–29.)

Assume that we did not know the actual digits on the tags in our example. From the process of 100 random selections we obtained the number 3 twelve times. The relative frequency associated with the number 3 in our experiment is 12/100, or .12. With this empirical evidence we might wish to predict the outcomes of the appearance of the number 3 in future selections. To illustrate, if we call the number 3 every time before a tag is withdrawn, we would *expect* to be right 12 percent of the time. However, the reliability of our prediction might be in question because we based our relative frequency (probability) on only 100 trials. *Had we made 100,000 trials, the relative frequency would have approached its mathematical (a priori) probability which is .10 or 10 percent of the time.* In effect, we discover that the concept of randomness operates with the so-called law of large numbers. The more observations we have on which to base our relative frequency, the more confidence we have in making our predictions. Estimates of deaths by ages (mortality tables) made by life insurance companies are based on the relative frequencies of many observations.

It would seem obvious that while randomness makes prediction of mass behavior practical, the outcome of a single event is irregular and, therefore, has no probability interpretation associated with it. As the insurance executive might say: "We do not know *who* will die in a given age group, but we do know about how *many* will die."

Joint, marginal, and conditional probabilities

Before examining in detail the relationships among probabilities, a few more definitions might be appropriate. Perhaps, for the concepts involved here an illustration might be the best method of achieving some understanding of the terms. Assume that a random sample of 1,000 of the eligible voters in a given city are interviewed to determine their opinions on several sensitive issues. One question dealt with the possible ban on smoking in public places as proposed by the state legislature. The results of this survey on this specific question are given in Table 7.1. Assume that

TABLE 7.1
Opinions of 1,000 voters on the smoking ban question (percent of total)

Response	Male (M)	Female (F)	Total
Favors smoking ban (Y)	.20	.40	.60
Does not favor smoking ban (N)	.30	.10	.40
Total	.50	.50	1.00

we were now to select a voter at random from those 1,000 respondents. Then:

1. The *simple probability* of choosing a male is $P(M) = .50$. The symbol $P(M)$ is used to denote the probability of an event E. The event "not-E," could be written $\sim E$. Therefore, the simple probability of selecting a female could have been denoted $P(\sim M) = .50$. (In Table 7.1 we used F instead of $\sim M$.)

2. The probability of selecting a male (M) voter who also is a person in favor of the ban on smoking in public places (Y) is referred to the *joint probability* of drawing a respondent with two (or more) specific characteristics. In our example, the $P(M, Y) = .20$ and the $P(F, Y) = .40$. The probability of a female voter who does not favor the ban may be written $P(F, N) = .10$.

3. The sum of the joint probabilities is equal to the *marginal probabilities*. This is nothing more than a special case of the simple probabilities. For example, the probability of drawing a male respondent is $P(M) = P(M, F) + P(M, N) = .20 + .30 = .50$.

4. Assume that we know that a voter drawn at random from our 1,000 respondents is a male. What is the probability that this person also favors

the ban on smoking? We refer to this concept as the *conditional probability* and symbolize it as $P(M, Y) = .20$. Where the symbol $M|Y$ is read "male given he favors the ban." Because 50 percent of the voters in the sample are male, and 20 percent favor the ban, $P(M|Y) = .20/.50 = .40$. Therefore, the general mathematical rule for defining conditional probability is:

Conditional probability of M given Y:

$$P(M|Y) = \frac{P(M, Y)}{P(M)} = \frac{\text{joint probability of } M \text{and } Y}{\text{marginal probability of } M} \tag{7.2}$$

and we may then determine the probability of a female respondent who favors the smoking ban:

$$P(F|Y) = \frac{P(F, Y)}{P(F)} = \frac{.40}{.50} = .80$$

Consider $P(Y, M)$, that is, the probability that a voter respondent who favors the ban is a male.

$$P(Y, M) = \frac{P(Y, M)}{P(Y)} = \frac{.20}{.60} = .33$$

which we note is not equal to the probability of $P(M, Y) = .40$.

5. Another useful concept is that of *statistical independence.* In our example, the knowledge that a respondent selected at random from our 1,000 voters was a female gave us some indication about her position on the proposed legislation (given Table 7.1). Statistical independence implies that our knowledge of one event (drawing a female) is of no value in forecasting the other event (favors ban). However, in this illustration, sex and stance on the issue of banning smoking in public places are *not* independent. According to our sample females are more likely to favor the ban on smoking in public places. This concept of statistical independence will be taken up in more detail in a section below.

Independence

The numbers produced by our experiment of 100 random draws are *independent* as well. *In statistics it is said that an event is independent if the probability of its outcome does not influence and is not influenced by the outcome of other possible events.* Independence in our example was provided by the fact that each tag was replaced prior to the selection of the next one. That is, the probability of each outcome at each selection remains the same, namely, .10. Another term for this replacement of the items is multiplicity of choice which was discussed in Chapter 3.

Referring to the results of our 100 trials, we see that the random sequence has produced some clustering. This should not surprise the reader nor is it cause for alarm and suspect. In fact, quite the opposite is true! If we had taken only ten independent random draws from our container in the hope of producing the uniform sequence of 0, 1, 2, 3, 4, 5, 6, 7, 8, and 9, the probability of success would be $(10)^{10}$ or virtually impossible. *Generally, if a sequence has too few runs or irregularities, one might wish to question the process of randomness.*

Sometimes the amateur gambler is misled by a misunderstanding of this property of clustering. It is not uncommon for the neophyte to favor (place a bet) the occurrence of the outcomes that have not turned up as frequently as some others in the past. What this person ignores is that *the probability of the occurrence of an independent random event is not affected by the previous outcomes.* Dice, for example, do not have a memory of what has come up in the past. In a crap game a 7 has occurred five times in a row. What is the probability that a 7 will appear again in the next random toss of a pair of dice? The answer: still 1/6. (See page 153.) One might argue that before the game begins it is highly unlikely to expect five 7s in a row; however, once five tosses are made, the probability of the next random toss is independent of the first five. The probability of a 7 will neither increase or decrease just because of this clustering that has taken place previously. At least, theoretically there is absolutely no reason to expect that there will be an excess of some other numbers to offset the excess of 7s. *In the long run, when many more tosses are made, the relative frequency of a 7 will approach the a priori probability of 1/6.* The hope that springs eternally in the heart of the gambler is based upon short-run considerations!

Summary

Several points seem worthy of repetition:

1. Probability is a unique possession of the population and is not associated with any particular outcome of an event.
2. Randomness makes it possible to predict mass behavior, but the irregularities of a random process deem individual events unpredictable.
3. Large numbers tend to generate stability and regularity; therefore, prediction of mass behavior is possible.
4. The law of large numbers is applicable only to random processes in which the probability of each possible outcome remains constant.
5. In the "real world" such probabilities are frequently unknown.

For example, assume your friend wishes to engage you in a game of matching coins. After ten tosses of this coin you discover that all ten trials turned up tails. Originally you thought the coin might be honest and perfect; however, you may wish to reconsider on the basis of the "sample" evidence. That is, on the basis of a sample of ten observations you have made the inference that the coin was not a fair one and you thereby decline the "opportunity" to play the game.

RELATIONSHIPS AMONG PROBABILITIES

There are five basic rules for handling probabilities upon which probability theory rests. They are (1) the rule of complementation, (2) the special rule of addition, (3) the special rule of multiplication, (4) the general rule of multiplication, and (5) the general rule of addition. We shall now take up each of these independently in more detail.

Rule of complementation: Complementary events

The rule of complementation states that the maximum limit of a probability equals one. This stems directly from the fact that probability represents the proportion of success in the population. In fact, the proportion of success cannot be negative. It should be noted that the probability of success plus the probability of failure must equal one. This, of course, makes the assumption that every item in the population is either a success or a failure. In terms of symbols, this relationship may be written as follows:

$$P(A) + P(B) = 1, \quad \text{or} \quad P(A) = 1 - P(\mathrm{B}) \tag{7.3}$$

Special rule of addition: Mutually exclusive events

In situations where the possibilities of several outcomes exclude the likelihood of some other outcomes, the probability of occurrence of either one or another of these several possible outcomes is equal to the sum of the probabilities of the several possible outcomes. This follows directly from the definition of probability. In terms of symbols, this axiom may be stated as follows:

$$P(A \text{ or } B) = P(A) + P(B) \tag{7.4}$$

For example, the probability of tossing a tail with an honest coin equals .5. The probability of tossing a head with this same coin is also .5, but the probability of "either a tail or a head" is equal to $.5 + .5$, or 1, which is the old proposition of "heads I win, tails you lose." Or let us consider the case of an honest die which contains six sides with each face numbered from 1 to 6. In this example the probability of 1/6 is associated with each number; and if we were to throw an honest die and denote success as having "either a 3 or a 4" turn up, the probability of success is $1/6 + 1/6 = 1/3$. *A limitation to this axiom is that it applies only to events that are mutually exclusive.* Events that are mutually exclusive are such that no more than one of them can occur at the same time. It should be noted that the term "mutually exclusive" does *not* mean that one outcome has nothing to do with the other. For example, in selecting a card from a full deck of playing cards, "either a queen or a jack" would be an example of a mutually exclusive event. However, if the proposition were stated "either a queen or a heart," it would refer to events that are not mutually exclusive. It is obvious that we may achieve "success" in the selection of one card, the queen of hearts.

The diagrams in Figure 7.1 emphasize the distinction between events that are mutually exclusive and those that are not. The shaded area indicates points that are common to both and might be comparable to the event "queen of hearts" described above.

FIGURE 7.1

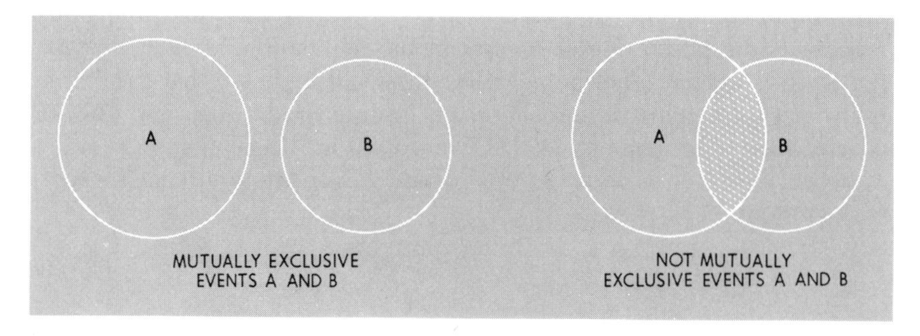

Figure 7.2 graphically explains the special rule of addition of probabilities. Assume that we have a population of size N and it is composed of subpopulations of n_1 whose items have characteristic 1; n_2 whose items have characteristic 2; and n_3 whose items have characteristic 3. The relative area that these three subpopulations occupy within population N is equal to the probability that an item selected at random from N has one of the characteristics 1, 2, or 3. Notice there is no overlap in the smaller circles; therefore the events are mutually exclusive.

Special rule of multiplication: Independent events

If the situation requires the occurrence of at least two *independent* events in order to achieve "success," *the probability of the occurrence of obtaining a combination of several independent outcomes either simultaneously or in succession is equal to the product of the probabilities of the individual independent events.* In terms of symbols, this may be stated as follows:

$$P(A \text{ and } B) = P(A) \times P(B) \tag{7.5}$$

FIGURE 7.2

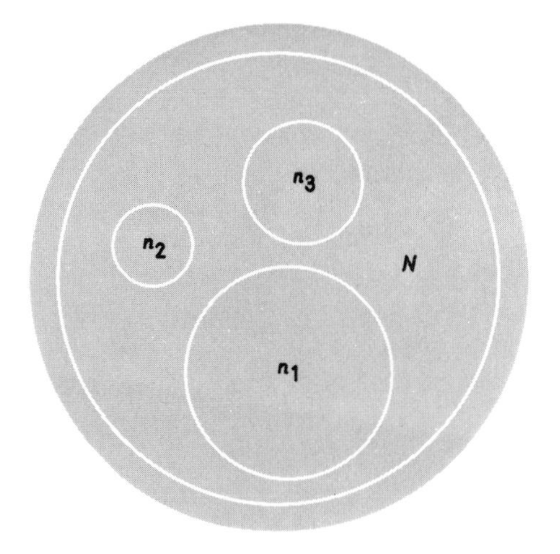

Again, this stems directly from our definition. For example, the probability of throwing two dice and obtaining a 3 and a 4 is 2/36. That is, the probability of obtaining either of these two alternative ways of getting a 3 and a 4 is 2/36. The probability of obtaining a 3 and also a 4 simultaneously is 1/6 × 1/6. We might also obtain a 4 on the first die and a 3 on the second die; therefore the probability is

$$P(3 \text{ and } 4) = (1/6 \times 1/6) + (1/6 \times 1/6) = 2/36$$

Referring to Table 7.2, we may see that the special rule of addition and the special rule of multiplication may be combined to determine the probabilities from the theoretical distribution of the 36 possible outcomes of a throw of a pair of honest dice. For example, the probability of obtaining a 7 is as follows:

$$P(7) = (1/6 \times 1/6) + (1/6 \times 1/6) + (1/6 \times 1/6) + (1/6 \times 1/6) + (1/6 \times 1/6) + (1/6 \times 1/6) = 6/36 \text{ or } 1/6$$

TABLE 7.2
Thirty-six possible outcomes of a throw of a pair of honest dice

Blue Die	Maize Die 1	2	3	4	5	6
1	1, 1	2, 1	3, 1	4, 1	5, 1	6, 1
2	1, 2	2, 2	3, 2	4, 2	5, 2	6, 2
3	1, 3	2, 3	3, 3	4, 3	5, 3	6, 3
4	1, 4	2, 4	3, 4	4, 4	5, 4	6, 4
5	1, 5	2, 5	3, 5	4, 5	5, 5	6, 5
6	1, 6	2, 6	3, 6	4, 6	5, 6	6, 6

FIGURE 7.3

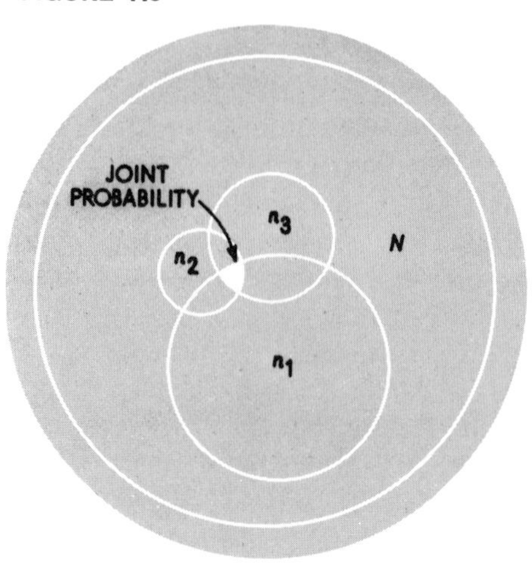

The individual probabilities are 1/6 for any one number, and the probability included in each of the parentheses represents the probability of obtaining the following numbers, respectively: 5, 2; or 4, 3; or 1, 6; or 2, 5; or 3, 4; or 6, 1. The probability of obtaining a 9 is equal to 1/9, which may be determined in a similar manner. The student should check this to insure his understanding of the computations involved.

Figure 7.3 illustrates the multiplication law of independent events. Assume that N is a population similar to the one shown in Figure 7.2. The shaded area illustrates the joint or conditional probability that an item selected at random has all three characteristics. In this case the events are not mutually exclusive. An item might be selected that has all three characteristics.

General rule of multiplication: Conditional probability

A significant concept of probability is that of a conditional probability. *The probability that an event B will take place provided that event A has taken place, is taking place, or will take place for sure is called the conditional probability of B given A.* To illustrate, if A stands for an individual being a physician and if B stands for the probability that this person's income is over \$20,000, then the probability of B given A is the probability that a physician has an income over \$20,000. In terms of symbols, this may be written $P(B|A)$. It also follows that the probability that a person with an income over \$20,000 is a physician can be written as follows: $P(A|B)$. We say that if two events are mutually exclusive, they are necessarily dependent; and in terms of conditional probabilities, we now add that if A and B are *mutually exclusive* then the probability of B given A and the probability of A given B are both equal to zero.[5] All we need to recall is the example of the toss of an honest coin. That is, we cannot have a head given that a tail has appeared.

Utilizing the concept of conditional probabilities, we can now formulate a more general rule for the probability that two events, A and B, will both occur simultaneously. For example, we may write

$$P(A \text{ and } B) = P(A) \times P(B|A), \quad \text{or} \quad P(B \text{ and } A) = P(B) \times P(A|B) \tag{7.6}$$

We should note that these formulas really represent the same thing because we are merely interchanging A and B on the left-hand side of the equation. "A and B" is the same as "B and A." We should also recognize that the general rule of multiplication is no longer restricted to events that are independent. This is a particular advantage because it is easier to determine examples of events that are dependent than events that are independent. However, if the events are independent, then the following hold true:

$$P(B|A) = P(B) \quad \text{and} \quad P(A|B) = P(A) \tag{7.7}$$

The reason for this is as follows:

$$P(B|A) = [P(B) \times P(A)] \div P(A) = P(B)$$

and

$$P(A|B) = [P(A) \times P(B)] \div P(B) = P(A)$$

Therefore the general equation reduces to $P(A \text{ and } B) = P(A) \times P(B)$, which, of course, is our special rule of multiplication.

Use of conditional probability. Assume that A stands for a student's passing an examination in college algebra and B stands for a student's passing an examination in statistics. Then we may write $P(B)$ equals the probability of passing the statistics examination and $P(B|A)$ equals the probability of passing the examination in statistics provided the student passes in college algebra. Usually, a student who is good in one of these subjects also is likely to be a good student in the other. Therefore the two events are dependent, and $P(B|A)$ is in fact greater than $P(B)$. For example, if we assume the following information is given: $P(A) = .90$; $P(B) = .95$; $P(B|A) = .96$, then, substituting in the general multiplication formula, we have the following: $P(A \text{ and } B) = P(A) \cdot P(B|A) = .90 \times .96 = .86$, which is the probability that a student will pass both examinations.

[5] If either A or B are the null event (a null set is one that is empty—has no items) then $P(A \text{ and } B) = P(A) \times P(B)$ and we have independence but A and B are mutually exclusive. This is a special case and should not confuse the student.

General rule of addition

Assume that $P(A)$ is equal to the probability that it will sleet in Ann Arbor, Michigan, on a random day and that the $P(B)$ is equal to the probability that it will snow. If we were to substitute into our special rule of addition, we might have some difficulty because we would be using the formula that applies only to mutually exclusive events and would be making the mistake of double counting. To illustrate, if $P(A) = .85$ and $P(B) = .20$, by our special rule of addition, $P(A \text{ or } B) = .85 + .20 = 1.05$! This is beyond certainty! Here, by using the rule which relates to mutually exclusive events, we count twice all the days on which there is or was both sleet and snow. In order to compensate for this double counting, it will be necessary to subtract the days that were counted twice or the proportion of days that were counted twice. In this case, we would refer to the general rule of addition, which is for events that are not mutually exclusive. In terms of symbols, this rule is as follows:

$$P(A \text{ or } B) = P(A) + P(B) - P(A \text{ and } B) \tag{7.8}$$

In our example, assume we knew that there is both sleet and snow in Ann Arbor, Michigan about 15 percent of the time. Then, using the general rule of addition, we would have the following result: $P(A \text{ or } B) = .85 + .20 - .15 = .90$, which is the probability that it will either sleet or snow, given certain atmospheric conditions. Going back to our example of college algebra and statistics, substituting in our general formula, we would have the following: $P(A \text{ or } B) = .90 + .95 - .86 = .99$, which indicates the probability that a student would pass at least one of the two examinations.

It is important to note that the "or" in our statement $P(A \text{ or } B)$ is what is referred to as the "inclusive or." That is, $P(A \text{ or } B)$ stands for the probability that A, B, or both will occur. What we are saying is that *at least* one of the two events will occur. It should be clear that if A and B are mutually exclusive events, then $P(A \text{ and } B) = 0$ because the two events cannot both happen at the same time, and the general rule of addition reduces to our special rule of addition.

These five simple rules or relationships among probabilities stem directly from the reasoning stated previously concerning a priori and empirical probability. It is hoped the student will find that a careful study of these ideas and axioms will contribute a great deal to an appreciation of the concept and later the interpretations of a probability.

PERMUTATIONS AND COMBINATIONS

To calculate probabilities, one must be able to count efficiently. Generally, writing out all possible outcomes is both tedious and inefficient because we usually are interested only in knowing the total number of ways an event can occur. Permutations and combinations really are shortcut methods of counting. In addition, we may see their application to probability calculations, and they may help us to understand the formula for the binomial distribution.

Definition: The factorial symbol

Let us first consider a basic definition and a special notation which is helpful. Assume that we call n a positive integer (i.e., a whole number) and read $n!$ as "n factorial." This means $n!$ is the product of all positive integers from one through n. Then:

$$n! = (n)\ldots(4)(3)(2)(1), \qquad \text{or} \qquad n! = n(n-1)\ldots(3)(2)(1)$$

Furthermore, we define $0! = 1$. To illustrate, $4! = (4)(3)(2)(1) = 24$. Also, $5! = (5)(4)(3)(2)(1) = 120$. Because factorials are products, the regular rules of cancellation can be used to simplify computations when dealing with quotients of factorial terms. For example:

$$\frac{6!3!}{4!} = \frac{(6\cdot5\cdot4\cdot3\cdot2\cdot1)(3\cdot2\cdot1)}{(4\cdot3\cdot2\cdot1)} = 180$$

To improve our grasp of the factorial notation, we should note that a positive integer can be written as a difference between two positive integers. Assume we seek $(12 - 7)!$. It is permissible either to perform the subtraction inside the parentheses:

$$(12-7)! = 5! = (5)(4)(3)(2)(1) = 120$$

or to write in more laborious fashion:

$$\begin{aligned}(12-7)! &= 12-7\cdot[(12-7)-1]\cdot[(12-7)-2]\cdot[(12-7)-3]\\ &\qquad\cdot[(12-7)-4]\\ &= 5\cdot(\;-1)\cdot(5-2)\cdot(5-3)\cdot(5-4)\\ &= 5\cdot4\cdot3\cdot2\cdot1\\ &= 5!\end{aligned}$$

Then we may generalize this by replacing n for 12 and x for 7, as follows:

$$(n-x)! = (n-x)[(n-x)-1]\cdot[(n-x)-2]\cdot[(n-x)-3], \quad \text{and so on}$$

$$(n-x)! = (n-x)(n-x-1)\ldots(2)(1)$$

Basic principle

If an event can occur in E_1 different ways, and after it happens a second event can occur in E_2 different ways, then the number of ways in which *the sequence of two events can occur is* $E_1 \cdot E_2$. To generalize, then, all events that can happen in the order stated are the product of $(E_1)(E_2)(E_3)(E_4) \ldots (E_n)$. To illustrate, assume that the board of directors of a corporation is filling four positions at the officer level, that is, president, executive vice president, vice president, and secretary. In how many ways can these positions be filled by making selections from a group of six candidates? Let E_1 = selection of president; E_2 = selection of executive vice president; E_3 = selection of vice president, and E_4 = selection of secretary. Then we may demonstrate the principle by use of a modified tree diagram as shown in Figure 7.4.

That is, first the office of president (E_1) can be filled by selection from any one of the group of six candidates. When E_1 is filled, there are five candidates for the office of executive vice president (E_2). When E_1 and E_2 are filled, there are four candidates for the office of vice president

FIGURE 7.4

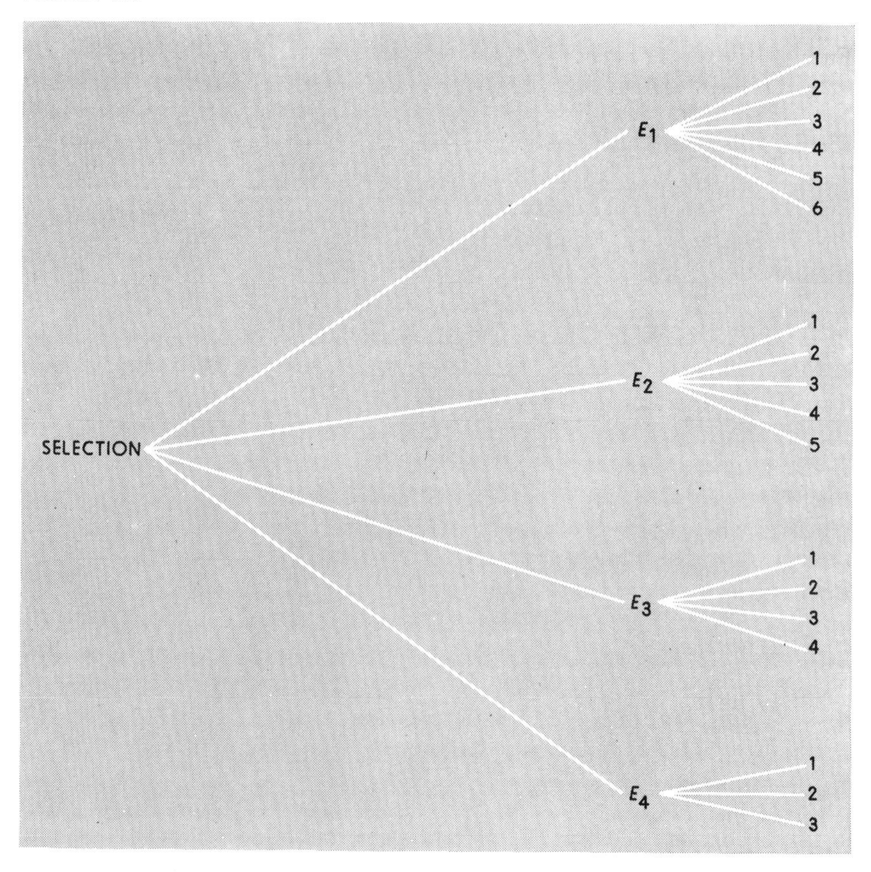

(E_3). When E_1, E_2, and E_3 selections have occurred, there are three candidates for the position of secretary (E_4). Therefore, making use of our basic principle, the office of president can be filled from one of six persons; the offices of president and executive vice president can be filled in $6 \times 5 = 30$ ways; the offices of president, executive vice president, and vice president can be filled in $30 \times 4 = 120$ ways; and the offices of president, executive vice president, vice president, and secretary can be filled in $120 \times 3 = 360$ ways. Or we could write $6 \times 5 \times 4 \times 3 = 360$ ways.

To emphasize the need for an efficient method of counting, consider how many ways five different economics texts, six different mathematics texts, and seven different statistics texts can be arranged on a book shelf *if* those in each subject are placed together?

The five economic books can be arranged in $(5)(4)(3)(2)(1) = 120$ ways.

The six mathematics books can be arranged in $(6)(5)(4)(3)(2)(1) = 720$ ways.

The seven statistics books can be arranged in $(7)(6)(5)(4)(3)(2)(1) = 5{,}040$ ways.

Therefore, for a *given* subject order, for example, economics, mathematics, and statistics, there are $(120)(720)(5{,}040) = 43{,}545{,}600$ ways of arranging all of the texts in a row in order that all books in the same subject are grouped together! But wait a minute; we have not completed the task! The three subject matter areas may be ordered in $(3)(2)(1) = 6$ ways. Therefore, there are $(43{,}545{,}600)(6) = 261{,}273{,}600$ ways of arranging the books in a row such that all the books on the same subject are grouped together. No wonder librarians are so busy!

We are now ready to consider counting techniques that are concerned with the mathematical properties or alternative ways in which objects or events may be ordered or combined. We shall consider two different but related counting methods—permutations and combinations.

Permutations

A permutation is an arrangement of a set of objects in which there is a first, a second, and a third order through n. Note that the order of arrangement is important in permutations. The arrangements of a, b, c, d and a, d, b, c are *different* permutations. If we have a set of four letters, in how many ways can four letters be ordered? The answer can be determined by applying our basic principle. There are four different choices for the first letter; once that choice is made, there are three different choices for the second letter, and so on. Therefore, there are $(4)(3)(2)(1) = 24$ permutations. This may be verified by writing out the arrangements as follows:

a, b, c, d	b, a, c, d	c, a, b, d	d, a, b, c
a, b, d, c	b, a, d, c	c, a, d, b	d, a, c, b
a, d, b, c	b, c, d, a	c, b, a, d	d, c, a, b
a, d, c, b	b, c, a, d	c, b, d, a	d, c, b, a
a, c, b, d	b, d, c, a	c, d, b, a	d, b, a, c
a, c, d, b	b, d, a, c	c, d, a, b	d, b, c, a

By now, it should be abundantly clear that it is easy to generalize that the number of n objects, taken n things at a time, is $n!$ Let P^n_n be the number of permutations of n objects or events taken n things at a time. This also may be written

$$P^n_n = n! \tag{7.9}$$

The number of permutations of P^n_x of n different objects or events, taken x at a time, is

$$P^n_x = n(n-1)(n-2)\ldots(n-x+1) \tag{7.10}$$

This latter formula covers the situations where we may wish to consider all possible arrangements of a certain number of the different objects or events in a set. That is, we can justify this formula as follows:

The first selection can be made in n ways;
The second selection can be made in $n - 1$ ways;
The third selection can be made in $n - 2$ ways; and so on until
The x selection can be made in $n - (x - 1)$ or $n - x + 1$ ways.

At this point, it is desirable to derive a second formula which is equivalent to the above expression but frequently is more useful and is helpful in understanding combinations. Given

$$P^n_x = n(n-1)(n-2)\ldots(n-x+1)$$

we can multiply both sides of the expression by $(n - x)!$. Recall that $(n - x)!$ is merely a positive integer; therefore, multiplying both sides of the above expression by $(n - x)!$ is no more complicated than multiplying both sides by a positive integer, say 10. We obtain

$$(n-x)!\,P^n_x = n(n-1)(n-2)\ldots(n-x+1)(n-x)! \tag{7.11}$$

and we note that

$$(n-x)! = (n-x)(n-x+1)\ldots(2)(1).$$

Therefore,

$$(n-x)!\,P^n_x = n(n-1)(n-2)\ldots(n-x+1)(n-x)(n-x-1)\ldots(2)(1)$$

But the right side of this expression can be written

$$(n-x)!\,P^n_x = n!$$

Again dividing both sides of the equation by the positive integer $(n - x)!$, we obtain an *alternative expression* for the permutations of n objects or events taken x at a time:

$$P_x^n = \frac{n!}{(n-x)!} \tag{7.12}$$

To illustrate, in how many ways can six persons be assigned offices if there are ten rooms available?

$$P_6^{10} = \frac{10!}{(10-6)!} = \frac{10!}{4!} = 151{,}200$$

As indicated above, a permutation requires that the objects or events assume a definite order. Frequently, order is of no significance. For example, according to our understanding of permutations, the ordered set of A, A, A, K, K, Q, Q, J, J, J, 10, 10, 10 is different from the ordered set of A, K, K, A, A, Q, Q, J, J, J, 10, 10, 10. However, to the bridge player, assuming the suits are the same in each case, they represent the same hand (a pretty good one at that!). The bridge player is interested only in what the cards are, rather than how they are ordered. For this and similar situations, we need to define another concept, that is, combinations.

Combinations

A combination is a set of objects (or events) taken from a set of n objects, disregarding order. All possible combinations of n objects taken x at a time are denoted by C_x^n where x is equal to or less than n. Therefore, although the four letters (a, b, c, d) gave rise to 24 permutations of four letters each, there is only *one combination* of four letters taken four at a time. This is because we no longer require an order, and the combination of a, b, c, d is the same as a, c, b, d, and so on.

Further to illustrate the point, and to emphasize the distinction between a permutation and a combination, how many different combinations taken three at a time can we make from the set of a, b, c, d? The different combinations are

a, b, c	a, b, d
a, c, d	b, c, d

Now, from each combination, we can form 3! different permutations of the set of four letters taken three at a time. That is, from the combination a, b, c the following permutations can be made:

a, b, c	b, a, c	c, a, b
a, c, b	b, c, a	c, b, a

Consequently, there are four combinations taken three at a time, whereas there are $(4)(6) = 24$ permutations of the four letters taken three at a time.

The formula to indicate the number of combinations of n different things taken x at a time is

$$C_x^n = \frac{n!}{x!(n-x)!} \tag{7.13}$$

In effect, what this formula does is to "knock out" the permutations which form the same combination—that is, we eliminate double counting. All we really have done is to divide the formula for permutations by $x!$.

APPENDIX: FINITE SAMPLE SPACES[6]

An excellent approach to the understanding of the concept of probability is the notion of the *sample space*. Mathematicians look upon the sample space as a *set* of elements such that any trial of the experiment produces an event which corresponds to one element of the set. The concept of a set is basic to all mathematics, yet it is a simple idea with intuitive counterparts in our daily life. For example, the mathematician defines the word "set" to denote "any well-delineated collection of objects, things, or symbols." Any object contained in the set is called an element or member. The students of a class in statistics constitute a set; each person is an element in the set. The individuals of a family represent a set; each person is a member or element. Other examples of a set might be the varsity football team at a university, the debating team, the faculty of a school, the 50 states in the nation, the members of the United Nations, and many other illustrations we can think of.

If we toss an honest coin, there are two possible outcomes (events) which we can label {H, T}, where H = heads, and T = tails. Or if we toss an honest die, there exist six possible outcomes which may be indicated by the numbers on the faces, that is, 1, 2, 3, 4, 5, and 6. These listings of all possible outcomes of our experiment also are sample spaces.

A box contains a large number of marbles, exactly alike in every respect except that some are maize (m), some blue (b), and some gold (g). Assume that we select a random sample of two marbles from a box (without replacement). What is the sample space for this experiment?

---

[6] Some instructors may prefer to discuss elementary probability theory from the point of view of finite sample spaces. This section may be substituted for or assigned in addition to the material on pages 151–56. In addition, most instructors using this method would want to supplement this Appendix with reading assignments on the theory of sets and provide a more rigorous approach to probability in terms of finite sample spaces.

Again, it is merely a listing of all possible outcomes for the items selected. Giving the first marble chosen first, and the second one next, here is our sample space:

m, m	m, b	m, g
b, m	b, b	b, g
g, m	g, b	g, g

Recall the example used in Table 7.2 (see page 153), where we listed all the 36 possible outcomes of a throw of a pair of honest dice. Table 7.2 is a *set* of all possible outcomes and is a systematic way of listing all possible events in the finite sample space for our experiment of throwing a pair of honest dice. We may also refer to this table as the *universal set,* whereas the ordered pairs (1, 1; 1, 2; etc.) are elements of this sample space. Sometimes the elements are called *sample points* or simply *points.*

Outcomes or events and probability assignments

As implied above, a point, or even a set of points, of the sample space is called an *event* or outcome. We can assume that each point in our sample space of Table 7.2 has the same probability of occurring, namely, one in 36, or $1/36$. (To simplify matters, let us assume that the first die is maize-colored and the other is blue.) Then the probability of obtaining a maize 3 and a blue 4 is $1/36$. If the event we are interested in is a subset of the universal set or sample space, we assign its probability in the following manner:

1. Count the number of points in it to get s; the number of successful outcomes; and
2. Take s/n as the probability of the event, where n is the number of points in the sample space. In Table 7.2, $n = 36$.

To illustrate, assume that we are interested in the outcome "the sum of the dots on the two dice is seven." This event contains the set of points $\{(1, 6), (2, 5), (3, 4), (4, 3), (5, 2), \text{and } (6, 1)\}$. Therefore, $s = 6$ and $n = 36$, so $P(7) = 6/36 = 1/6$.

> Note: In the real world, we may encounter serious problems by assigning equal probabilities to a point. In fact, we may be absolutely wrong! In our dice example, we would be intuitively safe; however, we may not always be in a position to assign probabilities. For example, in our marble problem, we would need extensive observation or prior knowledge of the composition of the contents in order to assign reasonable probabilities.

Returning to Table 7.2, and using the concepts just developed, we can answer some interesting probability questions merely by *counting.*

What is the probability of the sum of eight appearing? The event contains five points: $\{(2, 6), (3, 5), (4, 4), (5, 3), \text{and a } (6, 2)\}$. Therefore, $P(8) = 5/36$. What is the probability that the maize die will turn up a 4? Using Table 7.2 to count, we discover that there are six points with the maize 4 (first die). The answer to our question, then, is $P(4m) = 6/36 = 1/6$. It is reassuring to note that this result agrees with our original assignment of probabilities. We should expect that $P(4) = 1/6$ with a single die, just as intuitively indicated earlier.

Probabilities and sets

The following discussion will make more use of the language of sets. An event or outcome (E) is a set of points in the finite sample space (S). Generally, the probability of that event, $P(E)$, is the *sum* of the probabilities of the points. Although probabilities of the points need not be equal, to simplify our discussion the examples that follow will continue to use equal probabilities.

Our initial task is to identify the set of points involved with the event—no matter how the outcome is stated. Because we are limiting ourselves to the finite sample space of n points, the probability assigned to each point is $1/n$. The probability of an event may be determined by counting the number of points in the event and multiplying that figure by $1/n$.

Let us consider some rules for handling probabilities in terms of sets of points in a sample space.

Rule of complementation: Complementary events

The rule of complementation states that the maximum limit of a probability equals one, or 100 percent. The event A and the event which consists of all the other points of the same sample space are *complementary events*. It follows that the sum of the events A (success) and B (not success) must equal one; and together, they comprise the whole sample space. Therefore,

$$P(A \cup B) = 1 \tag{7a.1}$$

which is read $P(A \text{ union } B) = 1$ or $P(A \text{ or } B) = 1$. $A \cup B$ is the set of all elements which belong either to A or to B *or* to both A and B. Because A and B are *mutually exclusive* events,

$$P(A \cup B) = P(A) + P(B) \tag{7a.2}$$

so we may also write

$$P(A) = 1 - P(B) \tag{7a.3}$$

or

$$P(B) = 1 - P(A) \tag{7a.4}$$

For example, assume that we are interested in the event $m + b \neq 6$; what is the probability we would assign this outcome? Let A equal the sum of the dots that total six and B equal the sum of the dots that do *not* total six. Then the set of points contained in A is $\{(1, 5), (2, 4), (3, 3), (4, 2),$ and $(5, 1)\}$. Therefore, from Formula (7a.4),

$$P(B) = 1 - P(A)$$
$$P(B) = 1 - 5/36$$
$$P(B) = 31/36$$

and from Formula (7a.2),

$$P(A \cup B) = P(A) + P(B)$$
$$P(A \cup B) = 5/36 + 31/36$$
$$P(A \cup B) = 1$$

Special rule of addition: Mutually exclusive events

There are situations where the possibilities of several outcomes *exclude* the likelihood of some other outcomes. It also follows from our definition that the probability of occurrence of either one or another of such several possibilities is equal to the *sum* of the probabilities of the several individual possible events. What is the probability that the throw of two dice yields a total of five or eight? There are four sample points where the total is five—$\{(1, 4), (2, 3), (3, 2),$ and $(4, 1)\}$. There are five sample points where the total is eight—$\{(2, 6), (3, 5), (4, 4), (5, 3),$ and $6, 2)\}$. These sets have no common points; therefore the probability of a total of five or eight is $9/36$, or $1/4$. In set language, we can call A the set of points that total five; let B equal the set of points for which the sum is eight. Then we have from Formula (7a.2)

$$P(A \cup B) = P(A) + P(B)$$
$$P(A \cup B) = 4/36 + 5/36$$
$$P(A \cup B) = 9/36, \text{ or } 1/4$$

The limitation of this axiom is that it applies only to *mutually exclusive* events, that is, *outcomes such that no more than one of them can occur at the same time*. We cannot obtain both sums five and eight at the same time; if our throw yields one sum, the other is excluded. Therefore, for mutually exclusive events, $P(A \cap B) = 0$; which might be read $P(A$ intersection $B) = 0$. $A \cap B$ is the subset of elements which belong to both A and B. To illustrate further, if in drawing a card from a regular bridge deck, we denote success as "being an ace or a king," it is an example of a mutually exclusive event because both cannot occur at once. However, if we say "either an ace or a spade," we refer to events that are *not* mutually exclusive; drawing the ace of spades satisfies both conditions.

Special rule of multiplication: Independent events

Where we are dealing with compound events, that is, where the problem requires the occurrence of *at least two independent* outcomes, the probability is equal to the product of the individual events. What is the probability of obtaining a 3 with one die and also a 4 with the other die? Let A be the set of points satisfying our first condition and B be the set of points satisfying the latter. *Our interest is the number of points they have in common.* The set of points in A is $\{(1, 3), (2, 3), (3, 3), (4, 3), (5, 3),$ and $(6, 3)\}$. The set of points in B is $\{(1, 4), (2, 4), (3, 4), (4, 4), (5, 4),$ and $(6, 4)\}$. The common pairs are $(4, 3)$ and $(3, 4)$, and so the probability is

$$P(4,3) = P(4) \cdot P(3) = (1/6)(1/6)$$

and

$$P(3,4) = P(3) \cdot P(4) = (1/6)(1/6)$$

Therefore,

$$P(A \cap B) = P(A) \cdot P(B) \tag{7a.5}$$

Or combining the special rule of addition with the special rule of multiplication,

$$P(A \cap B) = (1/6 \cdot 1/6) + (1/6 \cdot 1/6)$$
$$P(A \cap B) = 2/36, \quad \text{or} \quad 1/18$$

This combination of the two special rules is necessary because we can obtain a 4 on the maize die and a 3 on the blue die. We also can yield a 3 on the maize and a 4 on the blue; therefore the products of the individual probabilities must be added to obtain the correct answer.

Unfortunately, this formula is not applicable generally because the events must be *independent.* We can be reasonably sure that the roll of the maize die is independent of the roll of the blue one. However, what is the probability that the sum of our points is both 7 and 11? By the special rule relating to mutually exclusive events (clearly, both cannot occur at once), the answer is $P(A \cap B) = 0$. That is, mutually exclusive events are not independent, so we shall need a general rule to cover such situations.

General rule of multiplication: Conditional probability

Let us proceed to this general rule and the useful statistical concept of conditional probability. The probability that event B will take place provided event A has taken place, is taking place, or will take place for sure is called the conditional probability of B given A. In terms of symbols, it is written $P(B|A)$. To illustrate, let us consider again the roll of two dice. Given that the sum of the dots on both dice is ≤ 6, what is the probability that the maize die (first one) is equal to three? Obviously, some throws of the dice will result in a sum ≤ 6; others will not. We are concerned with those sets of points ≤ 6; we ignore all other possible points. The set of points in the sample space that interests us is {(1, 1), (1, 2), (1, 3), (1, 4), (1, 5), (2, 1), (2, 2), (2, 3), (2, 4), (3, 1), (3, 2), (3, 3), (4, 1), (4, 2), and (5, 1)}. The probability of the maize die (first one) being equal to three is called the conditional probability that $m = 3$ given $m + b \leq 6$. The concept of conditional probability may be more meaningful if we consider the sets in our original sample space of Table 7.2. Given $m + b \leq 6$ and $m = 3$, let

$$\begin{aligned} B = \{&(1,1), (1,2), (1,3), (1,4), (1,5), (2,1), (2,2), (2,3), \\ &(2,4), (3,1), (3,2), (3,3), (4,1), (4,2), \text{and } (5,1)\} \\ A = \{&(3,1), (3,2), (3,3), (3,4), (3,5), \text{and } (3,6)\} \end{aligned}$$

However, we are interested only in the events that are in both A and B. We label this set "$A \cap B$" or "$B \cap A$." Then

$$(A \cap B) \text{ or } (B \cap A) = \{(3,1), (3,2), \text{and } (3,3)\}$$

and in our original sample space of Table 7.2, these outcomes have the following probabilities:

$$P(B) = 15/36; P(A) = 6/36; P(A \cap B) = 3/36$$

Let us consider the following as a *new* sample space: {(1, 1), (1, 2), (1, 3), (1, 4), (1, 5), (2, 1), (2, 2), (2, 3), (2, 4), (3, 1), (3, 2), (3, 3), (4, 1), (4, 2), and (5, 1)}. Also, we assign probabilities to the points as though they were the only possible ones, that is, $1/15$. The event $m = 3$ consists of (3, 1), (3, 2), and (3, 3), and

$$P(m = 3|m + b \leq 6) = 1/15 + 1/15 + 1/15 = 3/15$$

This is called the conditional probability of $m = 3$ given $m + b \leq 6$. Then,

$$P(A|B) = 3/15$$

and

$$\begin{aligned} P(A \cap B) &= P(B) \cdot P(A|B) \\ &= (15/36)(3/15) \\ &= 1/12 \end{aligned} \tag{7a.6}$$

and

$$P(A|B) = [P(A) \cdot P(B)] \div P(B) = P(A) \tag{7a.7}$$

Therefore the general rule reduces to the special rule of multiplication of independent events.

We note also that

$$P(A \cap B) = P(A) \cdot P(B|A) \tag{7a.8}$$

or

$$P(B \cap A) = P(B) \cdot P(A|B) \tag{7a.9}$$

which are really the same thing, since "A and B" is no different from "B and A." One should also remember that this general rule of multiplication is no longer restricted to independent events. But *if* the events *are* independent, then

$$P(B|A) = P(B) \tag{7a.10}$$

and

$$P(A|B) = P(A) \tag{7a.11}$$

because

$$P(B|A) = [P(B) \cdot P(A)] \div P(A) = P(B) \tag{7a.12}$$

General rule of addition: Events need not be mutually exclusive

What is the probability that $m \leqq 2$ or $b \leqq 3$? For the event $m \leqq 2$ the maize die must show either a 1 or a 2; for the event $b \leqq 3$ the blue die must turn up either a 1, a 2, or a 3. The set of events A corresponding to the outcome $m \leqq 2$ consists of the 12 points in the first two columns of Table 7.2. Also, the set of events B corresponding to $b \leqq 3$ consists of the first three rows of Table 7.2. However, it would be misleading to add these two $(12 + 18)$ because six points are in both sets and we should avoid this double counting. Consequently, for the events $m \leqq 2$ or $b \leqq 3$, the count of different points is $18 + 12 - 6 = 24$, and the desired probability is $^{24}/_{36}$, or $^{2}/_{3}$.

In our calculations, there were 12 points in A and 18 points in B, with six points in both A and B. If we had divided by 36, the total number of points in the sample space, we would have had

$$^{12}/_{36} + ^{18}/_{36} - ^{6}/_{36} = ^{24}/_{36}, \quad \text{or} \quad ^{2}/_{3}$$

The general formula for this kind of a situation is

$$P(A \cup B) = P(A) + P(B) - P(A \cap B) \tag{7a.13}$$

and

$$P(A) = ^{12}/_{36}; P(B) = ^{18}/_{36}; P(A \cap B) = ^{6}/_{36}$$

so

$$P(A \cup B) = {}^{12}\!/_{36} + {}^{18}\!/_{36} - {}^{6}\!/_{36} = {}^{24}\!/_{36}, \quad \text{or} \quad {}^{2}\!/_{3}$$

Of course, if we had used our special rule of addition, we would have overstated the probability by including points that are in both sets A and B.

SYMBOLS

$P(E) = \frac{s}{s+f}$, probability of success
$P(A) + P(B) = 1$, rule of complementation
$P(A \text{ or } B) = P(A) + P(B)$, special rule of addition
$P(A \text{ and } B) = P(A) \times P(B)$, special rule of multiplication
$P(A \text{ and } B) = P(A) \times P(B|A)$ } Conditional probability, general rule of multiplication
$P(B \text{ and } A) = P(B) \times P(A|B)$
$P(A \text{ or } B) = P(A) + P(B) - P(A \text{ and } B)$, general rule of addition
$n!$, read n factorial, $n! = (n) \cdots (3)(2)(1)$
$P_x^n = n(n-1)(n-2) \cdots (n-x+1)$, the number of permutations of n objects taken x at a time
$P_x^n \; \frac{n!}{(n-x)!}$, alternative expression for the number of permutations of n objects taken x at a time
$C_x^n = \frac{n!}{x!(n-x)!}$, number of combinations of n different things taken x at a time
(E), an event or outcome
(S), sample space
$\{ \; \}$, set
$\cup$, union of sets
$\cap$, intersection of sets

PROBLEMS

7.1. In the two-dice experiment, what is the probability of a sum of nine? Twelve? Sum less than seven? Greater than nine? Show how you can verify your responses by using the rules of Chapter 7.

7.2. In the two-dice problem, what is the probability that the maize die is less than five and the blue die is greater than four?

7.3. If A and B are mutually exclusive events, $P(A) = .30$, and $P(B) = .50$, determine each of the following probabilities:
a. $P(A|B)$.
b. $P(B|A)$.
c. $P(A \text{ or } B)$.
d. $P(A \text{ and } B)$.
e. $P(\text{neither } A \text{ nor } B)$.

7.4. Three business executives have finished their luncheon and agree to toss a coin to see who picks up the check. When the three coins are tossed, what is the probability that all three turn up heads? All three turn up tails? Two heads and one tail? Two tails and one head? Interpret your probabilities.

7.5. If two persons are chosen at random from five job applicants and all choices have equal probabilities, what is the probability that two specified individuals will be selected? That these two specified persons will not be chosen? That neither of them will be chosen?

7.6. On the basis of standard mortality tables, it was expected that in a corporation the probability that A, the chairman of the board, would die within the next 15 years was .10, and that B, the president, would die within the next 15 years was .05.

Required:

a. What is the probability that both A and B will die within the next 15 years?

b. That neither A nor B will die within the next 15 years?

c. That A will die and B will live during the next 15 years?

d. Would this information be of any interest to the board of directors? Why?

e. If one year later B was discovered to have a bad heart, would this alter your answer to (*a*), (*b*), or (*c*)? If so, why? If not, why not?

7.7. If Q stands for a female MBA candidate being in the Management Science Program and H stands for her being on the honor roll, write each of the following probabilities in symbolic form:

a. The probability that a female candidate in the Management Science Program is on the honor roll.

b. The probability that a member of the honor roll is a female candidate in the Management Science Program.

c. The probability that a female MBA candidate in the Management Science Program is not on the honor roll.

7.8. In a laboratory the probability of success of an experiment is known to be .95. A similar experiment has been performed successfully in this laboratory 95 times.

Required:

a. Does this mean that the next experiment will be a failure? Next five?

b. Does the concept of the long-run relative frequency interpretation of a probability have any meaning here?

7.9. A corporation has 20 plants. The president wishes to place five new top-management executives in these plants, but does not want any more than one new executive to each plant. In how many different ways can these executives be assigned to the plants?

7.10. A campus restaurant near a large state university features two luncheon meals: (*a*) a student special and (*b*) a salad bowl. Ninety-five percent of the male customers order the student special, and the rest order the salad bowl. The women customers prefer the salad bowl (and its lower number of calories). Eighty-five percent order the salad, and the remainder order the student special. Eighty percent of the restaurant's patrons are males. If you were the manager, what ratio of student specials to salads would you normally prepare for?

7.11. Given the information of Problem 7.10, what is the probability that a customer ordering a student special is a female? What is the probability that a customer ordering a salad is a female?

7.12. Four vice presidents of a large corporation are being considered for the offices of president and executive vice president. In how many different ways can these posts be filled?

7.13. A classic problem in probability is the so-called "birthday problem." Assume that there are five people in a room. What is the probability that at least two of these people have the same birthday? (That is, their birthday is on the same day and month of the year.) What is the smallest value of n to insure that the probability is .50 or better that at least two of the people in the room have the same birthday? (Ignore the extra day from "leap year.")

7.14. Assume that the task is to select at random two persons from ten in a room and that all choices are "equally likely." What is the probability that *two specified* persons will both be chosen? What is the probability that these two specified persons will *not* be chosen? What is the probability that neither of these two persons will be chosen?

7.15. The manufacturer of typewriters uses an identification code for a particular model consisting of two letters (except O and I) and three numbers.

Required:

a. How many different codes are possible if the two letters always appear first? (Assume replacement of letters.)

b. How many different codes are possible if the two letters may appear in any position in the code number?

7.16. Assume that the Alpha Company is considering an expensive site for the location of a new plant. It has been determined from prior research that the probability that the site is acceptable for both finished-product market and supply-market considerations is .60. Later it is determined that the site is, in fact, acceptable from the supply-market viewpoint, too. Given the probability of this latter event is .80, what is the probability now that the site is acceptable for the finished-product market, too?

7.17. In the past weather conditions were reported as "scattered showers this afternoon" or a "chance of rain tonight." In most areas the current method of forecasting is similar to the following: "Precipitation probability 30 percent this afternoon, 50 percent tonight, and less than 5 percent tomorrow."

Required:

a. From a statistical point of view, how would you interpret the latter forecast?

b. In what way does the current method of reporting a forecast improve your understanding of what type of weather to expect?

7.18. A questionnaire is submitted to a random sample of voters, 30 percent of whom are Republicans, 35 percent are Democrats, 25 percent are Independents, and 10 percent are Others. Eighty percent of the Republicans, 10 percent of the Democrats, 45 percent of the Independents, and 5 percent of the Others are in favor of a given public policy. If a voter is selected at random from this population, what is the probability that the voter will be in favor of this public policy?

7.19. Fifteen young men and one attractive young lady attend a statistics lecture. The young lady sits in the center of row 1. How many different pairs of men might she sit next to if one sits on each side of her, the seats are not preassigned, and she has made no prior commitments?

7.20. If the number of permutations of n objects taken 5 at a time equals 24,000, how many different combinations of the n objects taken 5 at a time could there be? (Assume that all n objects are different.)

7.21. Cash, a typical MBA student who is thoroughly enthralled with statistics, wishes to take two more statistics courses following the successful completion of the first course. If four statistics courses are available to Cash, how many different options may be selected ignoring sequencing and assuming prerequisites are no problem?

7.22. An adaptation of the Insurance Commissioner's Annuity Table for 1949—Male is reproduced below:

Age	Deaths per 1,000	Life Expectancy (years)	Probability of Death
18.	.58	56.17	.00058
19.	.60	55.20	.00060
20.	.62	54.23	.00062
21.	.65	53.27	.00065
22.	.67	52.30	.00067

Source: *1970 Life Insurance Fact Book,* New York Institute of Life Insurance, p. 118.

Assume that two males are selected at random: A, who is 20 years old and B, who is 21 years old. Answer the following questions using the data of the above table:

a. What is the probability that A will live and that B will die within the next year? That A will die and B will live? That both A and B will live? That both A and B will die? What rule or rules did you use in computing these probabilities?

b. What is the probability that either A or B but not both will live for one year? That A or B or both will live for one year? What rule or rules did you use for computing these probabilities?

c. What is the probability that at least one of three males selected at random who are 20, 21, and 22 years old will live for one or more years?

d. Assume that an individual male aged 19 is known to have a serious heart disease. What is the probability that he will die within the next year?

e. What assumptions are necessary for the valid construction and application of the probabilities associated with such a table?

7.23. One fair six-sided die is randomly rolled four times.

Required:

a. What is the probability that the value of the die shown on the top side will be less than three for at least three of the tosses?

b. What is the probability that the value of the die shown on the top side will be more than two for at least three of the tosses?

7.24. If automobile license plates are to contain three alphabetic letters (always in the first three positions) followed by three digits, how many possible different license plates can there be? (Assume there are no blanks and letters are replaced.) If the vowels are not used, how many different license plates can there be?

7.25. A public opinion poll reports that the probability candidate N will win an election is .62, while the probability of N's losing the election is .45. Would you accept these findings? Why or why not?

7.26. A young lady is given the opportunity to receive any 5 of 20 magazines free if she will pay the postage. If the young lady accepts the offer, how many different options does she have?

7.27. A firm has eight secretaries available to be assigned to four executives. The firm wishes to assign each executive one secretary, and the others will be assigned to a general secretarial pool. How many different assignments of the eight secretaries might be made?

7.28. If the "expected number" of 5s is to be 40, how many dice must be randomly tossed?

7.29. In a bridge game, assume that a card is randomly dealt from the deck of 52 cards, what is the probability that it will be an honor (ace, king, queen, jack, or ten) or a spade?

7.30. What rule or axiom for handling probabilities did you use in responding to Problem 7.29?

7.31. Assume that a card is randomly selected from a deck of 52 playing cards.

Required:

a. What is the probability that this card is either a heart or the queen of spades?

b. What rule or rules for handling probabilities did you use in responding to the question above?

7.32. Assume that we draw two cards at random successively from a pack of 52 playing cards with the first card being replaced before the second one is selected.

Required:

a. What is the probability that the first card is a "heart" and the second one "not a king"?

b. What rule or rules did you use?

7.33. A large city has a morning and an evening newspaper. A recent probability survey revealed that 60 percent of the residents read the morning newspaper, 35 percent read the evening paper, and 20 percent read both papers. Answer the following:

a. What is the probability that a family selected at random does not read either paper? Is this the same as the percent of families that do not read either paper? Explain.

b. What rule is useful in helping you respond to the above? Why?

7.34. Precisely and concisely define the following terms:

a. Random variable.
b. Permutation.
c. Combination.
d. Mutually exclusive events.
e. A priori probability.
f. Empirical probability.
g. Independent event.

RELATED READINGS

ALEXANDER, TOM. "Helping the Executive to Make up His Mind." *Fortune,* April 1962.

AYER, A. J. "Chance." *Scientific American,* vol. 213, no. 4 (October 1965), pp. 44–54.

CARVER, HARRY C. *Probabilities in Bridge Hands.* Ann Arbor: University Club, University of Michigan, March 5, 1963.

CRAMER, HAROLD. *The Elements of Probability Theory.* New York: John Wiley & Sons, Inc., 1955.

DRAKE, ALVIN W. *Fundamentals of Applied Probability Theory.* New York: McGraw-Hill Book Co., 1967.

DRAPER, NORMAN, and LAWRENCE, WILLARD E. *Probability: An Introductory Course.* Chicago: Markham Publishing Co., 1970.

DWASS, MEYER. *First Steps in Probability.* New York: McGraw-Hill Book Co., 1967.

FELLER, WILLIAM. *An Introduction to Probability Theory and Its Applications,* vol. I. 3d ed. New York: John Wiley & Sons, Inc., 1968.

FREUND, JOHN E. *Mathematical Statistics,* chaps. 1–8. 2d ed. Englewood Cliffs, N.J.: Prentice-Hall, Inc., 1971.

GORDON, CHARLES K., JR. "On the Probability of Past vs. Future Events." *The American Statistician,* vol. 18 (April 1965), pp. 20–22.

GRIDGEMAN, N. T. "Probability and Sex." *The American Statistician,* vol. 22 (June 1968), pp. 29–30.

HADLEY, G. *Introduction to Probability and Statistical Decision Theory.* San Francisco: Holden-Day, Inc., 1967.

HOEL, PAUL G. *Introduction to Mathematical Statistics.* 4th ed. New York: John Wiley & Sons, Inc., 1971.

HOGG, ROBERT V., and CRAIG, ALLEN T. *Introduction to Mathematical Statistics.* 3d ed. New York: Macmillan Co., 1970.

MENDENHALL, WILLIAM, and SCHEAFFER, R. L. *Mathematical Statistics with Applications,* chaps. 2–5. North Scituate, Mass.: Duxbury Press, 1973.

MEYER, PAUL L. *Introductory Probability and Statistical Applications.* Reading, Mass.: Addison-Wesley Publishing Co., Inc., 1965.

MOOD, ALEXANDER M.; GRAYBILL, FRANKLIN A.; and BOES, D. C. *Introduction to the Theory of Statistics,* chaps. 1–5. 3d ed. New York: McGraw-Hill Book Co., 1974.

MOSTELLER, FREDERICK; ROURKE, ROBERT E. K.; and THOMAS, GEORGE B., JR.

Probability with Statistical Applications. 2d ed. Reading, Mass.: Addison-Wesley Publishing Co., Inc., 1970.

NEYMAN, J. *First Course in Probability and Statistics.* New York: Henry Holt & Co., Inc., 1953.

PARZEN, EMANUEL. *Modern Probability Theory and Its Applications.* New York: John Wiley & Sons, Inc., 1960.

PORTER, RICHARD C. "Extra Point Strategy in Football." *The American Statistician,* vol. 21 (December 1967), pp. 14–15.

SPRINGER, H. H. *et al. Statistical Inference,* vol. III, Mathematics for Management Series, chap. 2. Homewood, Ill.: Richard D. Irwin, Inc., 1966.

WADSWORTH, G. P., and BRYAN, J. G. *Applications, Probability and Random Variables.* 2d ed. New York: McGraw-Hill Book Co., 1974.

WILKS, S. S. *Mathematical Statistics.* New York: John Wiley & Sons, Inc., 1962.

WOLF, FRANK L. *Elements of Probability and Statistics.* 2d. ed. New York: McGraw-Hill Book Co., 1974.

8

Theoretical frequency distributions: Probability distributions

Though this be madness, yet there is method in it.
William Shakespeare

Chapters 4, 5, and 6 indicated how *actual* frequency distributions may be constructed and used as a convenient way of summarizing the various observed data obtained from either a complete census or a random sample. This chapter discusses some *theoretical frequency distributions or probability distributions* that might be used as models of how outcomes (or observations) are *expected* to vary. As such models, probability distributions provide us with the necessary statistical basis for making estimations of parameters or to help us in the testing of an hypothesis about a population being sampled. That is, these theoretical distributions are vital in making statistical inferences about population characteristics based on some sample evidence or statistic.

If we were to present the distribution of a universe, which either is taken from the real world or is a theoretical population, according to the various classifications into which it has been divided and stating the probability of each classification, we would have a probability distribution. For example, Table 8.1 represents the simplest probability distribution of the theoretical outcomes of many tosses of a fair coin. If we were to toss a fair coin four times, the probability of obtaining four heads in these four tosses might be said to be 1/16; that is, $P(4 \text{ heads}) = (1/2)^4 = 1/16 = .0625$. The other possible sample compositions for this probability distribution are given in Table 8.2. If many such random samples were taken, that is, if the procedure of tossing a fair coin four times were repeated over and over, all possible values of this statistic (number of heads) varying from 0 to 4 would be obtained. Later, when

TABLE 8.1
Probability distribution of the outcomes of many tosses of a fair coin

Outcome	Probability
Heads	.5
Tails	.5
Total	1.0

TABLE 8.2
Probability distribution of four tosses of a fair coin ($P(H) = .5$)

Number of Heads (sample composition)	Probability
0	.0625
1	.2500
2	.3750
3	.2500
4	.0625
	1.0000

discussing the binomial, we shall review how the probabilities in Table 8.2 were computed. Because it is the probability distribution of a statistic taken from a random process, we normally refer to such tables as a *sampling distribution*. Since these sampling distributions are in reality relative frequency distributions, we may also present them as in Table 8.3. For example, if we were to repeat our experiment (four tosses of a fair coin) one hundred times, we might expect the theoretical frequencies given in Table 8.3. These are our a priori expectations, that is, the theoretical frequencies we might expect if we were to perform this experiment many times.

THEORETICAL DISTRIBUTIONS TO APPROXIMATE ACTUAL DISTRIBUTIONS

Frequency distributions generally are represented by a smooth curve rather than by a histogram or a frequency polygon. This is particularly true in situations where there are a large number of values to be depicted and where the values are ungrouped. Also, in many problems, it is desirable for purposes of further analysis not only to approximate the original discrete distribution by a smooth curve but also to find a mathematical formula which represents the smooth curve. Obviously, such a

TABLE 8.3
Theoretical distribution of the outcomes of four tosses of a fair coin in 100 trials

Number of Heads	Theoretical Frequency
0	6
1	25
2	38
3	25
4	6
	100

formula can be found, for there are mathematical formulas which will approximate the distribution of any shape. Just as obviously, they get increasingly complex as the shape of the curve becomes more complex or as the degree of approximation is reduced. And although there are many theoretical mathematical curves which may be fitted to a distribution, there are a few basic and useful ones which approximate a large variety of "real-life" situations. Among these distributions are: (1) the binomial, (2) the normal, (3) the Poisson, (4) the exponential, (5) the Student t, (6) the F, and (7) the chi-square. In this chapter we will discuss the first four theoretical distributions. Later, in Chapter 9 we shall cover the Student t distribution; in Chapter 15 the F distribution will be discussed; and in Chapter 17 we shall take up the chi-square distribution.

Binomial as a probability distribution

The probability distribution of the outcomes of four tosses of a fair coin is a binomial distribution, and its probability may be determined by the use of the familiar binomial expression $(p + q)^n$. The binomial distribution is of great importance in elementary statistics because it enables us to determine the probability that a sample of, say, n observations will contain any number of "successes" from zero to n, given the probability of success in the universe. To illustrate, if in a given suburb of a large city, 80 percent of the suburb's consumers shop in the city's central shopping district at least once a week, what is the probability that a random sample of 20 consumers will yield less than 15 consumers who so shop? Less than 10? None? Exactly 15? Questions of this kind may be readily answered by a consideration of the binomial distribution. Alternatively, we might use tables showing binomial probabilities and binomial coefficients,[1] or we might even use the normal distribution as an approximation of the binomial in such a situation.

The binomial distribution is applicable in situations in which an event can occur in one of two ways, for example, (1) heads-tails on a coin tossed, (2) accept-reject on an inspection operation, (3) yes-no on a voters' decision, or (4) whether or not a home possesses a color television set. When the number of trials is specified and the probability of success is known, a binomial distribution can indicate the theoretical relative frequency of each of all possible outcomes. To illustrate, what is the probability of obtaining two heads in four tosses of a single coin? The answer, as indicated in Table 8.2, is .375. It can also be obtained from the histogram of the binomial distribution shown in Figure 8.1. Table 8.4 indicates

---

[1] See Harry G. Romig, *50–100 Binomial Tables* (New York: John Wiley & Sons, Inc., 1947).

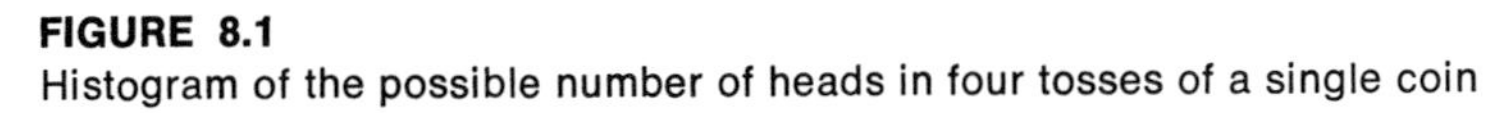

FIGURE 8.1
Histogram of the possible number of heads in four tosses of a single coin

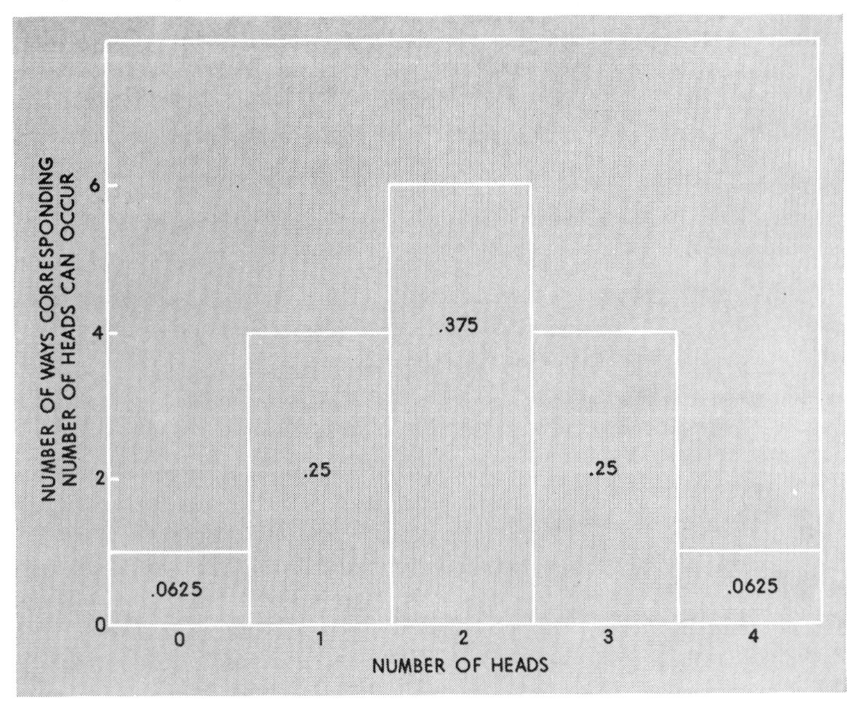

the 16 possible ways the four tosses of our single coin can occur, with the four losses listed from left to right for each possible outcome.

If we were to analyze the theoretical distribution of four tosses of a single coin, we would find that there are five possible outcomes; that is, we could receive zero heads, one head, two heads, three heads, or four heads. The total number of ways the five outcomes can occur is 16; and from Tables 8.2 and 8.4, we discover that the outcome of two heads is

TABLE 8.4
Sixteen possible outcomes of four tosses of a single coin

0 Heads	1 Head	2 Heads	3 Heads	4 Heads
T T T T	T T T H	T T H H	T H H H	H H H H
	T T H T	T H T H	H T H H	
	T H T T	T H H T	H H T H	
	H T T T	H T T H	H H H T	
		H T H T		
		H H T T		

seen to occur with a probability of .375, or in six different ways (therefore the probability of obtaining two heads is 6/16, or .375). We may also look at Figure 8.1 and think of the total area under the histogram as being equal to 16 and the area under the bar for two heads as equal to 6.

Also, the binomial distribution can be derived from a consideration of the special rule of addition and the special rule of multiplication for handling probabilities. If a single coin is tossed, there are two possible outcomes of one head and one tail; and if the probability of obtaining a head is symbolized by p and the probability of obtaining a tail is equal to q, we might write $p = q = .5$. The outcomes of a single coin toss could be expressed as follows:

Possible outcomes: H + T = all possible outcomes
In terms of symbols: $p + q = 1$
Probability in numbers: $.5 + .5 = 1$

Returning to our four tosses of a single coin, we may use the general mathematical formula for the binomial distribution and express it as follows: $(p + q)^4$. When expanded, this formula then gives the previously determined probabilities if we substitute .5 for each of the p's and q's. Let us expand this general formula, however, substituting the symbol H for p to indicate heads and the symbol T for q to indicate tails. Then our formula would be $(H + T)^4$. Our expansion would be as follows: $H^4 + 4H^3T + 6H^2T^2 + 4HT^3 + T^4$. Each of these terms specifies the number of "successes" resulting from any number of binomial trials where H is equal to the probability of a head (or "success") and T is equal to the probability of a tail (or "failure"). We should also note that in general, the binomial will not be symmetrical as we have it here, where $H = T = .5$.

It should be obvious from the above that if the number of trials increases, that is, n becomes large, the enumerations of all possible outcomes in terms of Table 8.4 would become quite involved, whereas the corresponding algebraic expressions would remain relatively convenient to write out. In our problem the desired information is the probability that a given number of heads will result from a specified number (four) of coin tosses. The exponent of H of each of the terms in our expansion of $(H + T)^4$ gives the number of heads associated with the probability specified by the corresponding term of the equation. Therefore, H^4 indicates the probability of all four tosses coming up heads. The second term, $4H^3T$, equals the probability of the mixed outcome of three heads and one tail. The third term, $6H^2T^2$, indicates the mixed outcome of two heads and two tails; that is, there are six possible outcomes which provide the two heads and two tails out of a total of 16 possible outcomes. The fourth term is the mixed outcome of one head and three tails, and the final term indicates the probability of all four tosses coming up tails.

The terms of the binomial expansion for our formula $(H + T)^4$ are derived as follows. The number of terms in our binomial expansion always equals $n + 1$; therefore, in our case, we shall have five terms. The second consideration is that when added together, the exponents of H and T for any single term always sum to n. This is true because the two exponents are always x for H and $n - x$ for T. The sum of these two is n. The exponents of H decrease from n by 1, and the exponents of T increase by 1. For example, the first term has H^4. The second term has a binomial coefficient of 4, and the H is raised to the third power, that is, it is decreased 1, and T is increased by 1. For the third term, we have further decreased the H exponent and increased the exponent of T. The coefficients for the $n + 1$ terms are always symmetrical and ascending to the middle of the series and then descending. For example, our binomial coefficients are 1, 4, 6, 4, and 1. We may determine the coefficient of each term from the previous term by the following procedure. First, consider the terms to be numbered consecutively beginning with 1. For any term, multiply its binomial coefficient by the exponent of H and divide by the number of the term to obtain the coefficient of the next term. For example, the binomial coefficient of the first term is always 1 and therefore is never written opposite the H. But for the second term the binomial coefficient is found by multiplying the exponent of H (i.e., 4) times the coefficient (1) which equals 4, and then dividing by the number of the term (1) giving the binomial coefficient for the second term. The coefficient for the third term may be found by multiplying 3×4 and dividing by 2, the number of the second term, and obtaining the result 6. The binomial coefficient for the fourth term would be found by taking 2×6 divided by 3, and the coefficient for the final term would be found by taking the exponent 1 of H times the coefficient 4 of the fourth term and dividing by 4, which equals 1. Normally, a table of binomial coefficients and binomial probabilities is provided in most texts and can be very conveniently used.

General case

We might proceed from the above results to the more general case where we have n independent events, with the probability of success being denoted as p, and where we are attempting to determine the probability of x successes and $n - x$ failures. Obviously, x must be smaller than or equal to n. In this case the probability of x consecutive successes followed by $n - x$ consecutive failures is p^x multiplied by q^{n-x} or $p^x q^{n-x}$. However, we must recognize that x successes followed by $n - x$ failures is only one of the ways in which n events can yield x successes and $n - x$ failures. As a matter of fact, there are $\frac{n!}{x!(n-x)!}$ different orders in which these successes and failures might be arranged. Each has the same probability of $p^x q^{n-x}$ because they all consist of the same number of successes and failures. Our expression *n!* is read "*n* factorial" and this is the product of $n \ldots 4 \times 3 \times 2 \times 1$. More generally, we might say that n factorial equals $n(n-1) \ldots (4)(3)(2)(1)$ because there are $\frac{n!}{x!(n-x)!}$ different orders in which x successes may occur in n trials; that is, from n trials, we may have $n!$ arrangements of outcomes if each outcome is distinguishable from all others. However, x of these outcomes ("successes") cannot be distinguished from each other; and likewise, $(n-x)$ of these results ("failures") cannot be distingiushed from one another. This reduces the total number of arrangements by a factor of $x!(n-x)!$, since the x "successes," if distinguishable, could be arranged in $x!$ ways. Thus the number of possible arrangements of the results of the n trials in which there are x "successes" and $(n-x)$ "failures" is

$$\frac{n!}{x!(n-x)!} \tag{8.1}$$

Notice that this is the same form as the formula for the number of combinations of n things chosen x at a time, C_x^n. The probability of x successes in n events is the sum of the probability of all these different orders, that is, $\frac{n!}{x!(n-x)!}$ times the probability of one of these orders. And a general formula for the binomial probability is as follows:

$$P_x^n = P(x \text{ successes in } n \text{ events}) = \frac{n!}{x!(n-x)!} \cdot p^x q^{n-x} \tag{8.2}$$

In our example of the computation of the probabilities associated with the four tosses of a single coin, we might use this general formula and determine the probabilities as follows, where $p^0 = 1$ and $q = 1 - p$:

$$P(0 \text{ heads}) = \frac{4!}{0!(4-0)!}(.5)^4 = .0625; \quad \text{where } 0! = 1 \text{ by definition}$$

$$P(1 \text{ head}) = \frac{4!}{1!(4-1)!}(.5)(.5)^3 = .2500$$

$$P(2 \text{ heads}) = \frac{4!}{2!(4-2)!}(.5)^2(.5)^2 = .3750$$

$$P(3 \text{ heads}) = \frac{4!}{3!(4-3)!}(.5)^3(.5) = .2500$$

$$P(4 \text{ heads}) = \frac{4!}{4!(4-4)!}(.5)^4 = .0625$$

Summing these probabilities, we should get 1.0000.

Arithmetic mean and standard deviation of the binomial distribution

Like most other distributions, either real or theoretical, the binomial distribution has a measure of central position and a measure of dispersion. These parameters may be computed from the theoretical formulas which are specifically applicable to the binomial distribution. For the mean the formula is equal to $\mu = np$, and for the standard deviation the formula is $\sigma = \sqrt{npq}$. In our example of the four tosses of a single coin, the computations would be as follows:

$$\mu = (4)(.5) = 2$$
$$\sigma = \sqrt{(4)(.5)(.5)} = 1$$

Since the arithmetic mean is 2, we may expect, on the average, two heads in samples of four from this universe.

Application of the binomial to the problem of the consumers and the downtown shopping center

Recall that we raised some questions at the beginning of the discussion of the binomial concerning the shopping habits of consumers in a given suburban area of a large city. We indicated that 80 percent of the consumers shop in the downtown center at least once a week, and we took a sample of 20 consumers, selected at random. The following questions were asked: What is the probability that our sample of 20 will contain less than 15 consumers who shop in the downtown area at least once a week? Less than 10? None? Exactly 15? Normally, one might use tables showing binomial probabilities and binomial coefficients, but we could also compute these probabilities by using our general formula. For example, take the question concerning the probability of getting precisely 0 consumers out of 20 who shop in the downtown district. We could determine this probability as follows using Formula (8.2):

$$P(0 \text{ successes in 20 events}) = \frac{20!}{0!(20-0)!} \times (.20)^{20} \simeq .000$$

In other words, it is unlikely that given a universe that has the characteristics of $p = .80$ and $q = .20$, we shall select none out of 20 that have the characteristic. What is the probability of selecting precisely 15 out of 20 that have the characteristic? This may be determined as follows:

$$P(15 \text{ successes out of 20 events}) = \frac{20!}{15!(20-15)!} \times (.80)^{15}(.20)^{5} = .1746$$

The other questions are left for the student's practice in determining the binomial probabilities.

Normal distribution and the limit of the binomial

The binomial is a discrete distribution where the values of x are discrete integral numbers, and a histogram or frequency polygon of the binomial generally has a broken-line character. In addition, there is no interpolation between points on the frequency polygon because for the discrete case, there are no intermediate points. However, as the number of trials (n) increases, the number of possible outcomes also increases, and the frequency polygon approaches a smooth distribution curve. The shape of this smooth distribution is suggested by Figures 8.2, 8.3, 8.4, and 8.5, where $p = q = .5$ and $n = 2$, 4, 10, and 20, respectively. The distribution which the binomial resembles under such circumstances is called the normal probability distribution, and it takes the shape of a bell. This distribution is also known as the Gaussian probability distribution and was first derived by De Moivre in 1733. Approximately one hundred years later, this distribution was derived independently by Gauss and La Place. The normal distribution was derived from the standardized form of the binomial, using the variable in the form of

$$\frac{X - \mu}{\sigma} = \frac{X - np}{\sqrt{npq}} \tag{8.3}$$

and permitting n to increase without limit. Although the mathematics of the derivation is not particularly useful or interesting to us here, the conclusions are of extreme significance.

The derivation made no assumption about the equality of p and q. Thus, it can be shown that even though p and q are not equal and the binomial distribution therefore is not symmetrical, as n increases the binomial approaches as its limit the normal probability distribution

FIGURE 8.2

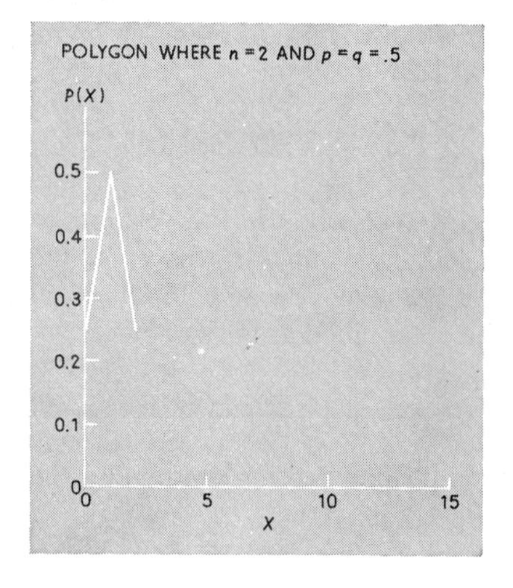

FIGURE 8.3

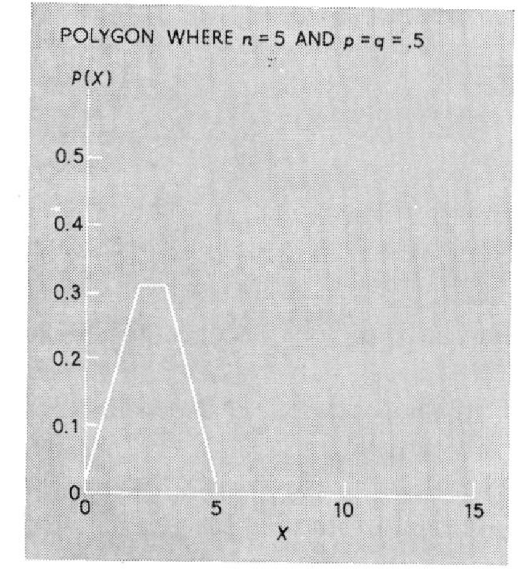

FIGURE 8.4

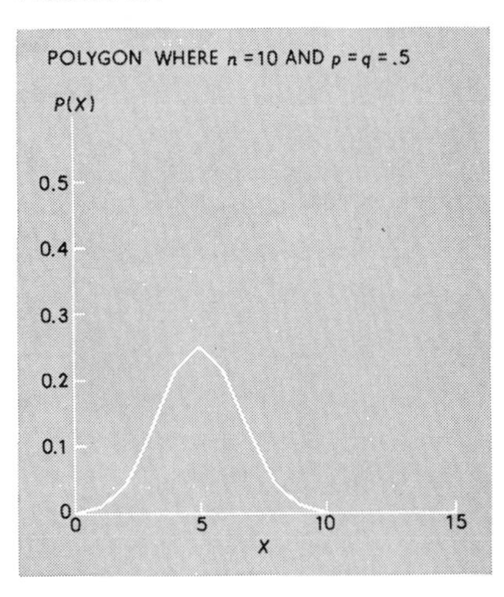

FIGURE 8.5

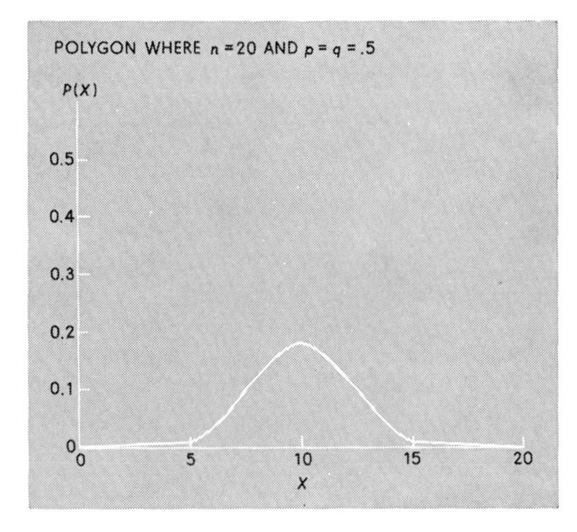

which *is* symmetrical. Figures 8.6, 8.7, and 8.8 illustrate this fact graphically.

NORMAL FREQUENCY FUNCTION

A variate normally distributed takes all values from minus infinity to plus infinity, and the frequency at each size of the variate is given by a mathematical frequency function as follows:

$$Y = \frac{1}{\sigma\sqrt{2\pi}} e^{-1/2\left(\frac{x-\mu}{\sigma}\right)^2} \tag{8.4}$$

FIGURE 8.6

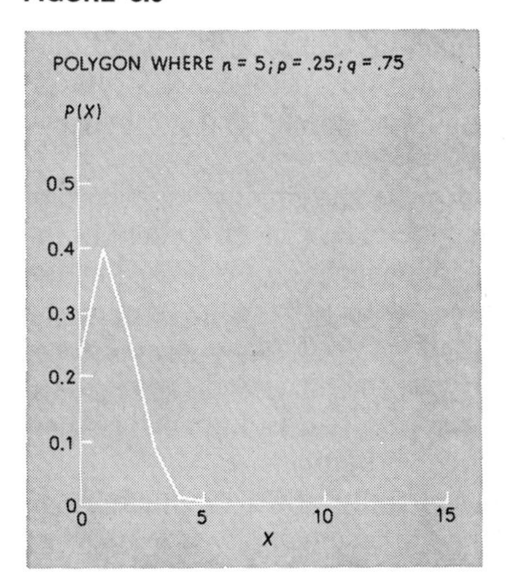

FIGURE 8.7

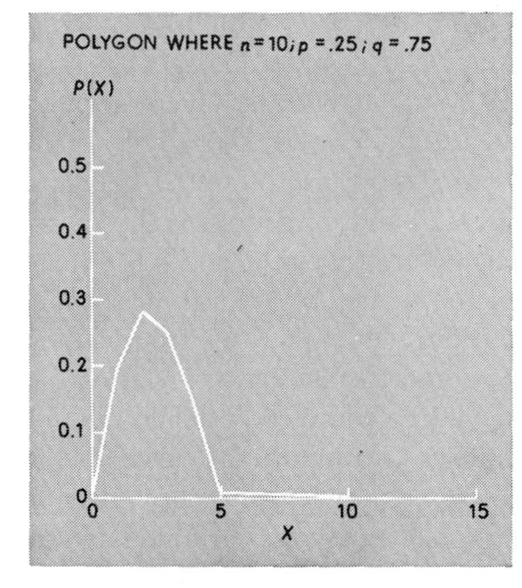

FIGURE 8.8

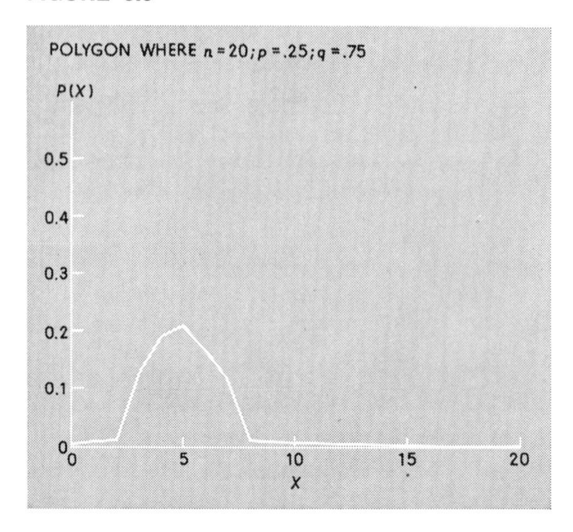

This formula says that the logarithm of the frequency at any distance, x, from the center of the distribution is less than the logarithm of the frequency at the center of the distribution by a quantity proportional to x^2. Therefore the normal curve is symmetrical and takes the shape of a bell. (See Figure 8.9.) Table C.1 in Appendix C clearly shows that if $z = 1.96$, the area included between 0 and x is .4750. Recall that if the total area under the normal curve is 1; therefore an area of .4750 represents 47.5 percent of the items. Any other value of

$$\frac{-x}{\sigma} = -z = \frac{(0 - x)}{\sigma} \tag{8.5}$$

can be read from the table. Therefore, if a variate is normally distributed with a mean $\mu = 100$ and a standard deviation $\sigma = 10$, there are only about 2.5 chances in 100 that an item will exceed $\mu + 1.96\sigma = 100 + 19.60 = 119.60$. Also, there are only 2.5 chances in 100 that an item will be less than $\mu - 1.96\sigma = 100 - 19.60 = 80.40$.

It should also be noted that in the formula, π is equal to the constant 3.14159 and, of course, is the ratio of the circumference of a circle to the diameter of a circle, and e is equal to 2.7182, which is, of course, the base of the system of natural logarithms. Therefore, in order to determine the various values of Y or f from values of X, all we need to know is the mean (μ) and the standard deviation (σ). That is, the mean and the standard deviation completely define the normal probability distribution. This permits us to compute the curve for the entire distribution.

The normal probability distribution generally arises in situations in which a *measured* variable has a tendency to assume a given fixed mean value and in which the variable differs from the mean because of chance alone. The normal curve is symmetrical and tapers off in both directions asymptotically towards the base line. The reasons are: (1) purely chance or random variations are equally likely to occur in either direction from the mean, and (2) large random variations are less likely to occur than are small variations.

Probability from frequency distribution curves in general

As indicated previously, frequency distributions are sometimes represented by a frequency polygon, and this is particularly true in cases where there are a large number of values to be presented and where the values are ungrouped. In such situations the horizontal scale is considered as a continuous scale along which the value of an individual item changes continuously rather than in discrete steps. The vertical axis, which for a histogram indicates the actual frequency of occurrence of the items falling within the various classes, becomes a scale which can give only a *relative measure of frequency*. Figure 8.9 illustrates this point; the Y axis indicates the probability of X or the relative frequency, and the X scale indicates a continuous scale for a measured variable. The

FIGURE 8.9
Areas under a curve

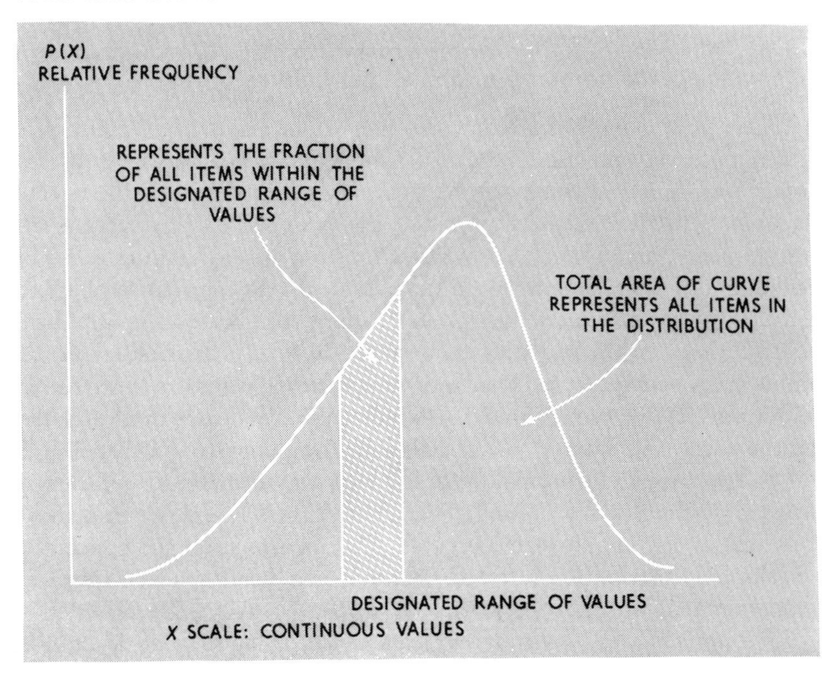

vertical scale represents relative frequency only, rather than absolute frequency, because there are no specific group intervals to relate to. The total area under the curve represents all of the items in the original data. Consequently, the percentage of the items falling within any arbitrary vertical strip of a frequency curve is equal to the ratio of the area of that strip to the total area of the curve. Also, by definition the probability is one, or certainty that any item in the distribution will be included somewhere under the probability curve. In statistical and probability theory the total area under a distribution curve is always defined as being equal to one. It follows, then, that if a single item is to be selected at random from a distribution, the probability that it will have a value which lies somewhere between the lower and upper values of a given interval of values is equal to the ratio of the area of the vertical strip bounded by that interval to the total area of the curve. To illustrate, the probability that an item in a symmetrical distribution has a value equal to or greater than the mean is .5. Also, the semi-interquartile range of any distribution includes 25 percent of the total area of the curve, and the probability equals .25 that any single item falls within this range.

It should be noted that as determined from a smooth distribution curve, the probability that an event will occur with precisely one given value is equal to zero. This is so because a unique value on a continuous scale is represented by a straight vertical line which has no area, rather than by an interval which has a finite width. This has particular significance when we discuss point estimates in Chapter 9.

Importance of the normal probability distribution

Although it may be said that the normal probability distribution is seldom observed precisely in social science phenomena, it is of importance in statistical theory primarily for two reasons. First, the normal probability distribution is the limit of the binomial as this sample size becomes large; and as such, it serves as a useful approximation to the binomial under that specific condition. Second, the normal probability distribution has a remarkable property derived from the central-limit theorem that certain statistics tend to be normally distributed as the sample size becomes large. That is, if samples of a large-size n are drawn from a universe or population that is not normally distributed, nonetheless the successive sample means will form by themselves a distribution that is approximately normal. We say that this distribution of sample means is asymptotically normal. That is, it approaches closer and closer to normal as the sample size increases. This, of course, is the concept of the sampling distribution referred to previously, which will be taken up in more detail in Chapter 9.

What are some examples of the occurrence of the normal probability distribution? The distribution of the actual diameters of a mass-produced product which is being machined to a specified dimension would tend to be normally distributed about this specified diameter—the mean of the distribution. This is true if the variations from the mean are due only to chance, which is the case unless there exists some consistent accountable bias in the production process. The distribution of many close measurements made on any given part would normally be affected by chance errors which, of course, would result in a normal probability distribution. The heights of all the males in the United States tend to be normally distributed about an average height of 5 feet 9 inches or 5 feet 10 inches. There are very few extremely tall people, and there are very few extremely short people. Most of them tend to be clustered about the average height. Many economic, biological, and physical measurements tend to be approximated by the normal probability distribution; however, frequently when a distribution is not known precisely but the variations appear to be affected by chance alone, or when it is the result of many independent factors which are approximately equal in magnitude, the distribution is assumed to be a normal one. Later, when we discuss the tests of hypotheses, we discover that this assumption may be checked by comparing the actual distribution or the real-world situation with the theoretical normal distribution which has the same mean and the same standard deviation as the actual.

What does the normal theoretical probability distribution indicate? First, it indicates that the expected value for the variable is the one in the center of the distribution, that is, the mean or the typical value. Second, it shows that small deviations in either direction from this expected value occur much more often than large deviations. Third, the normal theoretical probability distribution shows that plus and minus deviations or variations exactly offset each other, and that the curve is symmetrical. This latter fact is because the deviations are due to chance alone. As implied previously, the normal probability distribution is a basic one in sampling theory because samples are normally picked at random and the values of the items in a sample are subject to random or chance variations. In sampling theory, it is necessary to know the probability that a given item will fall within a specified range of values of a normal probability distribution. Generally, the standard deviation is used to specify these ranges, and an analysis of the normal curve reveals the following relationships:

1. Within the range of the mean plus and minus one standard deviation, approximately 68.26 percent of all of the items will be included.
2. Within the mean plus and minus 1.645 standard deviations, approximately 90 percent of the items fall.
3. The mean plus and minus 1.96 standard deviations includes 95 percent of the items in the distribution.

4. The mean plus and minus 2.58 standard deviations includes 99 percent of the items.
5. Within the range of the mean plus and minus three standard deviations, virtually all of the items in the distribution are included, or about 99.73 percent.

The reason this is not 100 percent is because the normal distribution is a theoretical distribution which allows the values of x to be between $-\infty$ and $+\infty$. However, the probability which it gives to values of x which are much more than three standard deviations away from the mean is so small that in most practical applications it can be ignored. The above relationships are useful and are frequently worthwhile to memorize in order that we need not refer to Table C.1 in Appendix C every time. That is, given these relationships, it is possible to make numerical statements about the probability or expectation that an item drawn at random from a normal distribution will have a value lying within a specified interval. Actually, we shall make additional uses of the normal probability curve in our problems of estimation and tests of hypotheses in Chapters 9, 10, and 11.

THE USE OF A STANDARDIZED SCALE

Because the normal probability distribution is a good approximation of the binomial distribution in many situations, it is necessary to provide some drill in determining probabilities from a normal curve. The question which might be asked at this time is: How does one convert any distribution into a standard form in order to compare it with the theoretical normal curve? The answer would be: *Any distribution is changed into its standardized form by means of a z scale; that is, the X scale is the scale in which the measurements were originally obtained, and a z scale is merely the device which reduces all possible normal distributions to a common denominator. A distribution which has a zero mean and a standard deviation of one is said to be in its standardized form.* For example, $\frac{x}{\sigma} = z = \frac{X - \mu}{\sigma}$, and it may be approximated by $z = \frac{X - \bar{X}}{\hat{\sigma}}$. (See Formula [8.3].)

As indicated earlier, the mean and the standard deviation completely define a normal curve, and from these two measures the percentage of the distribution falling within any specified range of x values may be determined. Consequently, this technique based upon the table of the areas of the normal curve can be used to construct a normal probability distribution to contrast with a sample distribution or to represent a population distribution. The latter relates to the problems of estimation, and the former relates to the problems of a test of hypotheses. Incidentally, the table is called the table of areas because it gives the percentage of the area under the curve which lies between the vertical line erected at two points along the horizontal axis. That is, the table of the area of the normal curve indicates the probability that an item chosen at random from a population will deviate not more than a stated number of standard deviations in any specified direction. Because the normal distribution is a continuous one, we talk of the probability that a random variable lies between two x values rather than that it has a particular value, as we do in the discrete binomial distribution.

Drill in the use of the table of the areas under the normal curve

In the section that follows, we shall make use of Table C.1 of Appendix C to provide some practice in using a table of the areas and to give some idea of how we might use the area in problems involving probability interpretations. To begin, let us determine some of the commonly used relationships.

Let us assume that we have a distribution with a population mean of 200 units and a population standard deviation of 10 units. (See Figure 8.10.) That is,

$$\mu = 200, \qquad \text{and} \qquad \sigma = 10.$$

What is the area between 190 and 210? First, we determine the area between 200 and 210 using Formula (8.3):

$$z = \frac{X - \mu}{\sigma}$$

$$z = \frac{210 - 200}{10}$$

$$z = +1.0 \qquad \begin{matrix} \text{Area} \\ .34134 \end{matrix}$$

FIGURE 8.10

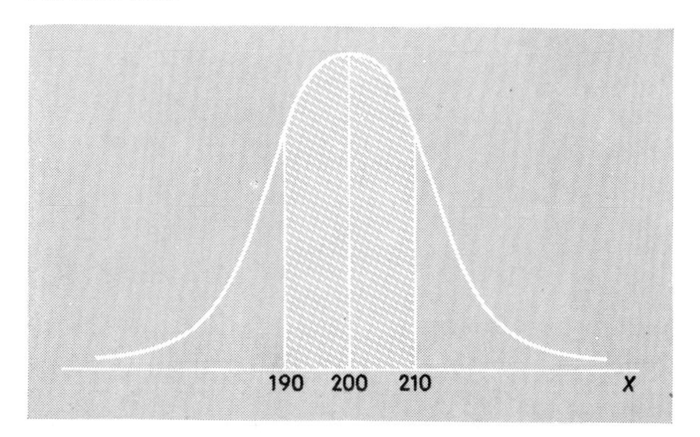

If we look in Table C.1, we discover that the area associated with this z value, or interval between 200 and 210, is equal to .34134. What is the area between 190 and 210? First, we determine the z value again:

$$z = \frac{190 - 200}{10}$$

$$z = -1.0 \qquad \begin{matrix} \text{Area} \\ .34134 \end{matrix}$$

Of course, since the normal curve is symmetrical, the area again is .34134. To determine the area between 190 and 210, then, all we need do is add the two areas to obtain .68268. Clearly, we would need to do the computations for only one side of the curve.

What is the probability that an item selected at random from this distribution will fall between 183.55 and 216.45? (See Figure 8.11.) Again, we use our standardizing formula:

$$z = \frac{216.45 - 200}{10} = +1.645 \qquad \begin{matrix} \text{Area} \\ .45000 \end{matrix}$$

$$z = \frac{183.55 - 200}{10} = -1.645 \qquad \begin{matrix} .45000 \\ \hline .90000 \end{matrix}$$

That is, adding the two areas, as in the previous example, the answer is .90. This is the basis for an earlier statement that the mean plus and minus 1.645 standard deviations includes 90 percent of the items in a normal distribution.

What is the probability that an item selected at random from this distribution will fall between 180.4 and 219.6? (See Figure 8.12.) Again, we determine z:

$$z = \frac{219.6 - 200}{10} = +1.96 \qquad \begin{matrix} \text{Area} \\ .47500 \end{matrix}$$

FIGURE 8.11

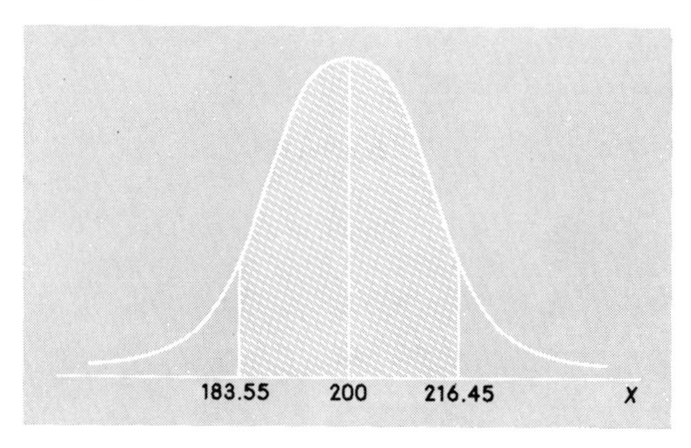

FIGURE 8.12

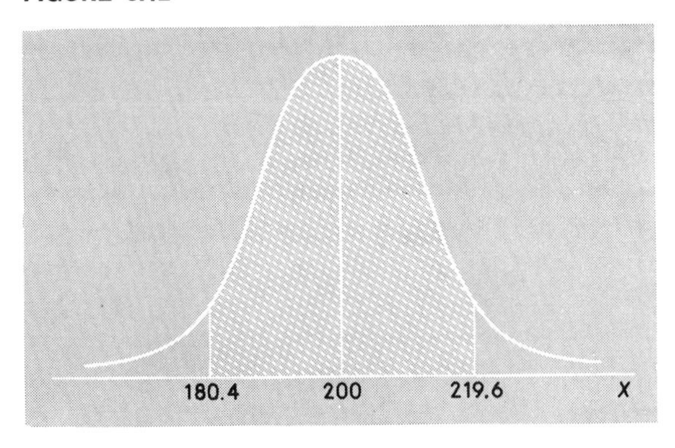

$$z = \frac{180.4 - 200}{10} = -1.96 \qquad \begin{array}{r} \text{Area} \\ .47500 \\ \hline .95000 \end{array}$$

Therefore the probability is .95 that an item selected at random will lie between 180.4 and 219.6. This also illustrates the basis for our statement that the mean plus and minus 1.96 standard deviations includes 95 percent of the area.

What is the area between 174.2 and 225.8? (See Figure 8.13.) Again, we determine the areas and add them as follows:

$$z = \frac{225.8 - 200}{10} = +2.58 \qquad \begin{array}{r} \text{Area} \\ .49506 \end{array}$$

$$z = \frac{174.2 - 200}{10} = -2.58 \qquad \begin{array}{r} .49506 \\ \hline .99012 \end{array}$$

What is the area between 170 and 230? The answer is .99730 and will be left for the student to verify.

FIGURE 8.13

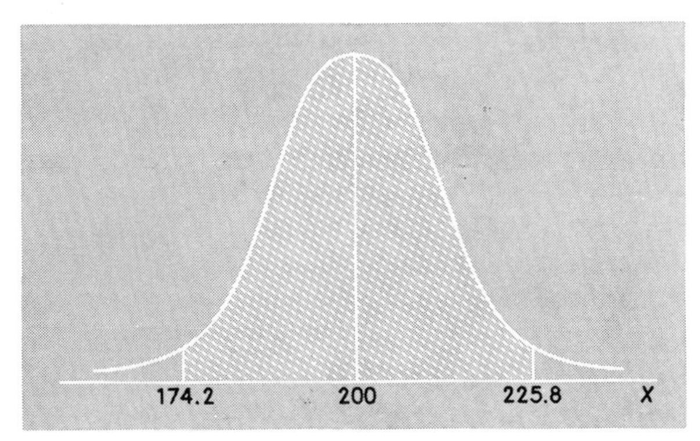

FIGURE 8.14

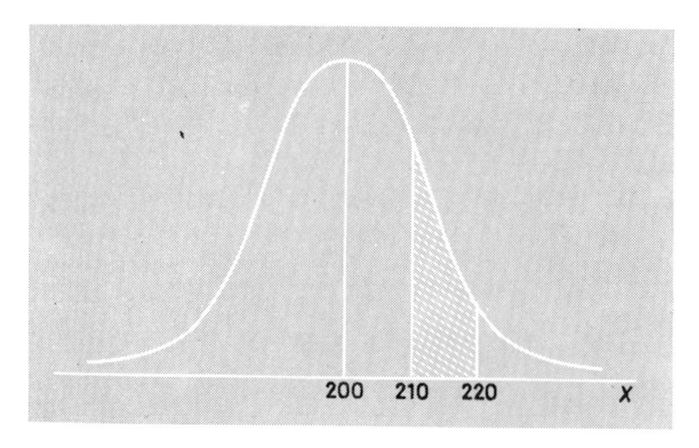

What is the area between 210 and 220? This value cannot be determined directly and involves the subtraction of areas. (See Figure 8.14.) First, we must find the area between 200 and 220, and from that we subtract the area between 200 and 210 as follows:

$$z = \frac{220 - 200}{10} = +2.0 \qquad \begin{array}{r} \text{Area} \\ .47725 \end{array}$$

$$z = \frac{210 - 200}{10} = +1.0 \qquad \begin{array}{r} .34134 \\ \hline .13591 \end{array}$$

The answer is .13591. That is, the probability of a randomly selected item from this universe having a value between 210 and 220 units is about .14. Although it should be obvious, the student is cautioned not to subtract the z values directly; we must work with the areas.

What is the probability of a randomly selected item from this universe having a value $\leq$ 215 units? Greater than 215? (See Figure 8.15.) Here,

FIGURE 8.15

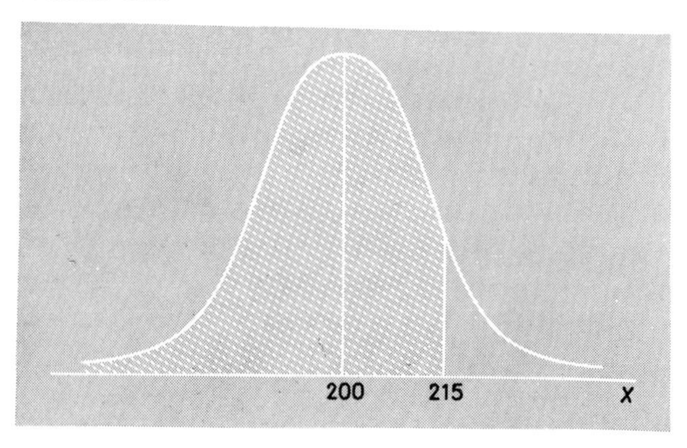

we can determine the probability directly. First, we find the area between 200 and 215:

$$z = \frac{215 - 200}{10} = +1.5 \qquad \begin{array}{c}\text{Area}\\ .43319\end{array}$$

And we know that the area to the left of 200 equals .5 by definition; adding these two, we obtain $.5 + .43319 = .93319$ as the probability of an item being ≤ 215 units.

The probability that an item is > 215 units is found by subtracting the area between 200 and 215, that is, .43319, from .5. The answer is .06681.

For drill purposes, let us take a slightly different approach. Given the *symmetrical* area around the mean as .92160, what are the values of X? (See Figure 8.16.) Because we assumed that the areas about the mean are equal, we need to divide .92160 by 2:

$$\frac{.92160}{2} = .46080$$

Reading in Table C.1, we find that the z value associated with an area of .46080 is ± 1.76. Then,

$$+1.76 = \frac{X - 200}{10}$$

and

$$(1.76)(10) + 200 = X$$

so

$$X = 217.6$$

Also,

$$-1.76 = \frac{X - 200}{10}$$

$$(-1.76)(10) + 200 = X$$

$$X = 182.4$$

APPROXIMATION OF THE BINOMIAL DISTRIBUTION WITH THE NORMAL DISTRIBUTION

As indicated above, it can be derived mathematically; but it is obvious graphically that the normal probability distribution can provide a reasonable approximation of the binomial for large values of n. Generally, the normal approximation is satisfactory for most real-world problems when the value of n is greater than 30 and the values of p and q are not less

FIGURE 8.16

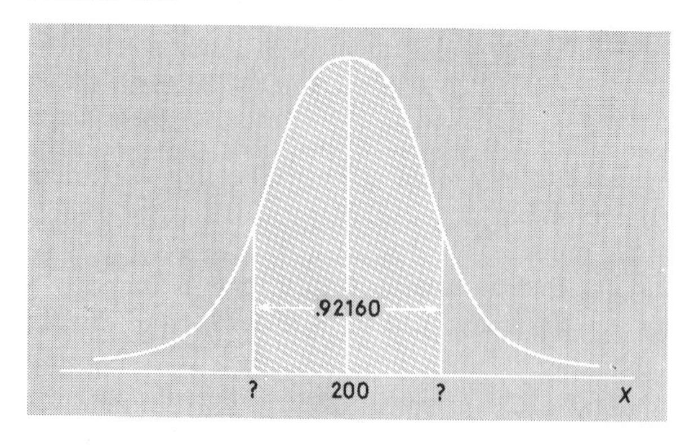

than .1 and not greater than .9. Some writers suggest that the approximation is appropriate if np and $nq > 25$ or $npq > 10$. This fact that the normal probability distribution is a reasonable approximation of the binomial is of great practical importance in reducing the amount of computations necessary to handle problems where the binomial underlies the situation.[2] This is true because the normal distribution is easier to compute mathematically and the tables for the normal curve are shorter yet more complete. If in an example we knew the value of p and q for a large-size n, we could estimate the probability that any specified number or numbers from zero to n would have the attribute, that is, the characteristic, indicated by p. We could do this by expanding the usual binomial formula of $(p + q)^n$ and substituting our values for p and q in the appropriate terms of the expansion. Or we might substitute in the formula

$$P^n_x = \frac{n!}{x!(n-x)!} \cdot p^x q^{n-x}$$

to determine the values. Alternatively, we could approximate the binomial distribution by the normal probability distribution with the mean value of $np = \mu$ and a standard deviation $\sigma = \sqrt{npq}$. Then we would refer to the table of the normal distribution and determine the probabilities; or we might settle for less refined answers and utilize the more commonly encountered relationships given above.

Recall the problem of the shopping habits of the consumers given on page 183. Let us assume the following: $n = 100$, $p = 80$ percent, $q = 20$ percent, where n is the number of consumers interviewed, p is the per-

[2] This is based on the central-limit theorem. If n is large but π is so small that $n\pi \leq 5$, then a better approximation than the normal distribution is the Poisson distribution. The latter is discussed below. Students interested in reading more on this subject should consult one of the related readings of Chapter 7.

centage of them who shop in the central shopping district (CSD), and q represents those who do not have this attribute. Then, using the normal curve to approximate this binomial situation, we can answer such questions as: What is the probability that between 70 and 75 consumers in samples of 100 "shop in the CSD at least once a week?" (See Figure 8.17.)

$$\mu = np \qquad \sigma = \sqrt{npq}$$

$$\mu = (100)(.80) = 80 \qquad \sigma = \sqrt{(100)(.80)(.20)}$$

$$\sigma = 4$$

$$z = \frac{X - \mu}{\sigma}$$

	Area
$z = \dfrac{70 - 80}{4} = -2.5$	.49379
$z = \dfrac{75 - 80}{4} = -1.25$	.39435

Therefore, .49379 − .39435 = .09944, or the probability is about .10; or 10 percent of the time, between 70 and 75 out of 100 consumers will shop in the CSD at least once a week.

What is the probability that between 72 and 90 will so shop? (See Figure 8.18.)

	Area
$z = \dfrac{72 - 80}{4} = -2.0$	.47725
$z = \dfrac{90 - 80}{4} = 2.5$	.49379
	.97104

Therefore, .47725 + .49379 = .97104; or about 97 percent of the time, we would expect between 72 and 90 shoppers among 100 consumers who so shop in the CSD.

FIGURE 8.17

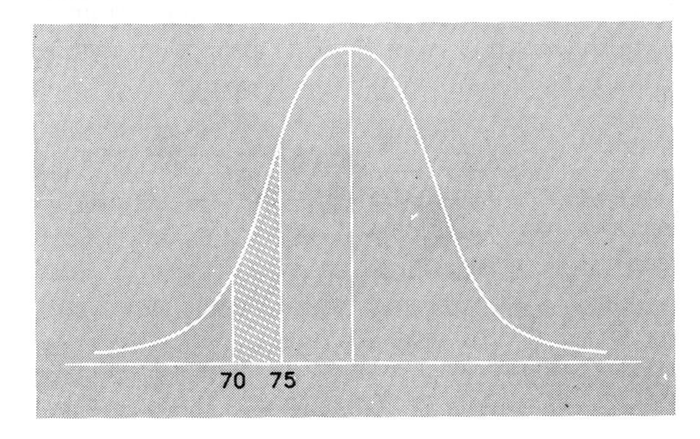

FIGURE 8.18

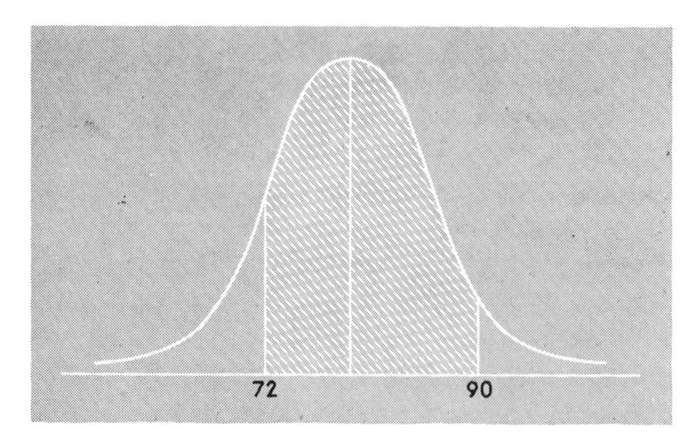

What is the probability that 95 or more so shop? (See Figure 8.19.)

$$z = \frac{95 - 80}{4} = 3.75 \qquad \begin{array}{c}\text{Area} \\ .49991\end{array}$$

$$.50000 - .49991 = .00009$$

Therefore the probability is extremely small that we would expect 95 or more shoppers among 100 consumers who so shop.[3]

THE POISSON DISTRIBUTION

Undoubtedly, as indicated previously, the binomial distribution is the most generally applied theoretical frequency distribution of a discrete random variable. However, there are other useful probability distributions widely used in management science and operations research. These theoretical distributions include the multinomial, exponential, hypergeometric, and the Poisson. We will now discuss the latter distribution from two points of view: (1) as a probability distribution in its own right, and (2) as an approximation to the binomial under certain conditions.

The Poisson distribution was named after Siméon Denis Poisson (1781–1840) the French mathematician who wrote in the area of applied

[3] When a statistic is discontinuous (discrete) but its distribution function is represented as an approximation by a continuous distribution, the probability levels usually can be more accurately determined by using slightly corrected values of the statistic. These values are then said to be corrected for continuity. In the above example a better approximation might be:

$$z = \frac{94.5 - 80}{4} = 3.625;\ .50000 - .499855 = .000145$$

This theoretical correction is ignored in later calculations in this chapter.

FIGURE 8.19

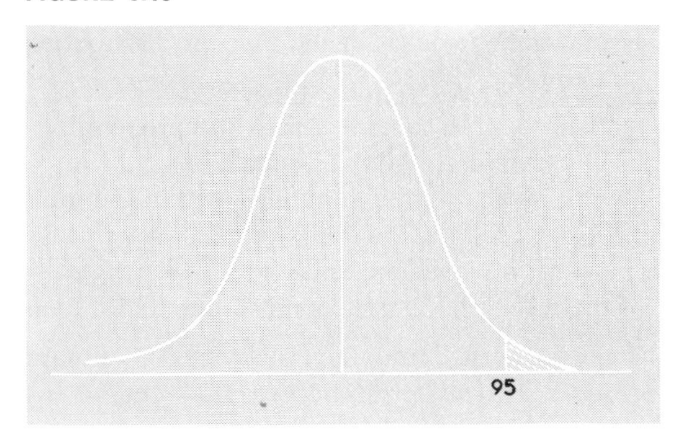

probability as well as in other areas of math. The Poisson distribution relates to discrete random variables that frequently are denoted x where x can take on the values of 0, 1, 2, 3 . . . , and so on ad infinitum. Mathematically, the domain of this theoretical distribution may be described as a probability function consisting of all nonnegative integers. The mathematical form of the Poisson distribution is

$$P(X) = \frac{\mu^x e^{-\mu}}{X!} \tag{8.6}$$

where μ is the parameter called the *mean or variance,* which are equal in the Poisson. The standard deviation is $= \sqrt{\mu}$. The probability of exactly x outcomes in the Poisson distribution may be calculated by substituting in the above formula. The constant $e = 2.71828\ldots$, the base of the natural logs. The Poisson often appears where one is observing events which are very improbable but which occur occasionally because so many trials occur. For example, traffic accidents, traffic deaths, demands for service, number of phone calls coming through a given switchboard, number of airplane arrivals at a given airport, arrivals of ships at docks, arrival of passenger cars at toll stations, number of persons waiting in line for a service, and so on, are situations where the Poisson probability distribution might be useful in estimating the probability of occurrence of these events. There is a common pattern here that the reader should notice. All of these outcomes might be described in terms of a discrete random variable that assumes values of 0, 1, 2 . . . , and so on, and may be viewed as a stochastic process that produces random events in a continuous manner. It should be noted here, but discussed later in a section below, that if $n^{\to\infty}$ and $p^{\to 0}$ *in the binomial distribution* in such a fashion that $np = \mu$, the binomial becomes the Poisson distribution.

Poisson as a probability distribution

The basic assumptions underlying the Poisson distribution are similar to those of the binomial:

1. Large number of possible outcomes with the probability of an occurrence of a specific outcome being very small;
2. The random variable X must be an integer within the unit of measurement;
3. The events must be independent; and
4. The value of the mean (μ) must remain stable, that is, constant.

Because the Poisson distribution is widely used in situations where a waiting line or queue is likely to develop we shall choose an example using this application. The reader should keep in mind that another important user of this probability distribution is the person in quality control where the number of defects is small relative to the total observations. Let us assume that we wish to estimate incoming phone calls to a particular switchboard during the hours of 9 A.M. to 12:00 noon each day of the working week. If we view our unit of measurement as a minute of time there would be no incoming calls during most minutes. Sometimes only one call will come in during a minute while at other times two, three, four, or even more calls might "arrive" at the same minute. This makes the Poisson more appropriate than the binomial in such a stochastic process. The reason for limiting the time during the hours of 9 A.M. to 12:00 noon is to avoid violating the stability assumption of the mean. Calls in the afternoon may be more or less which would cause the mean to be unstable or vary. The probability that a unit of time (one minute) will contain no calls may be determined using formula (8.6):

$$P(X = 0 \mid \mu = 2) = \frac{(2^0)(2.71828)^{-2}}{0!}$$
$$= .1353$$

if on the average two calls per minute usually came into this switchboard during these hours. The probabilities of one, two, and three calls coming in are:

$$P(X = 1 \mid \mu = 2) = \frac{(2^1)(2.71828)^{-1}}{1!}$$
$$= .2707$$

$$P(X = 2 \mid \mu = 2) = \frac{(2^2)(2.71828)^{-2}}{2!}$$
$$= .2707$$

$$P(X = 3 \mid \mu = 2) = \frac{(2^3)(2.71828)^{-3}}{3!}$$
$$= .1804$$

The probability of three or less calls coming in per minute to this switchboard is equal to the sum of the probabilities:

$$P(X \leq 3 \mid \mu = 2) = .1353 + .2707 + .2707 + .1804$$
$$= .8571$$

Whereas the probability of four or more calls coming in per minute are:

$$P(X \geq 4 \mid \mu = 2) = 1 - P(X \leq 3 \mid \mu = 2)$$
$$= 1 - .8571$$
$$= .1429$$

If one did not wish to calculate the probabilities (and they become tedious quickly) Appendix H may be used. This table shows individual probabilities of the Poisson distribution for selected values of μ from .001 to 10. If $\mu > 10$ then the normal distribution may be used as an approximation with a mean equal to μ and $\sigma = \sqrt{\mu}$. Remember that the Poisson is a discrete distribution and to allow for this we must add or subtract a factor of .5 to X depending on the circumstances. To find the probability of r or less successes, add .5 to the value of X in calculating the standard normal deviate z. If you wish to determine the probability of r or more success then one must subtract .5 from the value of X in calculating z. (See footnote 3, page 198.)

Poisson as an approximation to the binomial

Specifically, if $n \to \infty$ and $p \to 0$ in the *binomial* distribution in such a fashion that np remains constant and equal to μ, then the binomial becomes the Poisson. Therefore, the Poisson distribution becomes a good approximation of the binomial under these conditions. This is useful because the computations for the binomial become wearisome when n is large. Mathematically, we can say that the Poisson is the limiting distribution to the binomial as n approaches infinity and p becomes small. A good general rule of thumb to follow is that if

$$n \geq 10 \text{ and } p \leq .01 \textit{ or } \text{if } n \geq 20 \text{ and } p \leq .03$$

or if

$$n \geq 50 \text{ and } p \leq .05 \textit{ or } \text{if } n \geq 100 \text{ and } p \leq .08$$

then the Poisson is a reasonable approximation to the binomial. The reader is cautioned that these give a moderate degree of accuracy and for a finer precision the sample size n must be increased to the point of becoming large.

To approximate a binomial under these conditions we merely set up $np = \mu$ and look up the values in Appendix H. In this way the Poisson gives the probability of x successes in n observations with $p = P$ (success on a single trial). In effect, x, n, and p are interpreted in exactly the same way as in the binomial. To illustrate assume that we sample 2,000 items of which .002 have the characteristic p where p might represent the fraction defective or some other item of interest to the researcher. Here $n = 2{,}000$, $p = .002$, and $np = \mu = 4$. We can estimate the probability of various defectives (those items with the attribute) in our sample by using the Poisson table of Appendix H just as we did before.

THE EXPONENTIAL DISTRIBUTION

The exponential distribution is extensively used in queuing theory to describe the interarrival times. In this sense the exponential distribution bears a unique relationship to the Poisson. In our example of the telephone calls coming into the switchboard, the Poisson distribution could be used to depict the number of calls per unit of measurement (minute). The exponential distribution could be used to describe the value of the interarrival times, for example, the time lapse between successive calls. Other uses of the exponential distribution might be to describe the distribution of the average life of a product, the time between breakdowns of a certain piece of equipment, or the interarrival times of customers at a specific department and the time required to service such customers.

The exponential has the shape of a reverse J curve as shown in Chapter 4. The probability function of the exponential is:

$$f(t) = \lambda e^{-\lambda t} \tag{8.7}$$

where t is a random variable representing the time between successive arrivals (e.g., time between successive calls, or between successive customers); λ (lambda) is the average *rate* of arrivals (the same as μ in the Poisson distribution). The reciprocal $1/\lambda$ is the average time between arrivals (calls), and e again is the constant 2.71828 the base of the natural logs. While the mean of the exponential is $1/\lambda$, the variance is easily determined as $1/\lambda^2$. Lambda is the only parameter of this distribution, and as such it defines the whole curve. Both t and λ must be positive values.

An example of the application of the exponential may be found in a continuation of the switchboard problem. Assume that the time *between* calls is an average of .5 minutes. What is the probability of an interval of less than 20 seconds (.33 minutes) between a given phone call and the next one? If the mean of .5 minutes $= 1/\lambda$ then $\lambda = 2$. Therefore,

$$\begin{aligned} P(t < .33) &= 1 - P(t > .33) \\ &= 1 - e^{-(2)(.33)} = 1 - e^{-.2} \\ &= 1 - .819 \\ &= .181 \end{aligned}$$

Note that the .819 was obtained from Appendix I; the .181 indicates that the probability of an interarrival time of the phone calls of less than 20 seconds (.33 minutes) is about 1 in 5.5 chances.

SYMBOLS

$(p+q)^n$, binomial expression raised to nth power

$n!$, read n factorial, $n! = n(n-1)$

$P^n_x = \frac{n!}{(n-x)!}$, alternative expression for the number of permutations of n objects taken x at a time

$C^n_x = \frac{n!}{x!(n-x)!}$, number of combinations of n different things taken x at a time

$P(x;n) = \frac{n!}{x!(n-x)!} \cdot p^x q^{n-x}$, general formula for binomial probabilities

$\mu = np$, arithmetic mean of a binomial distribution

$\sigma = \sqrt{npq}$, standard deviation of a binomial distribution

$z = \frac{X-\mu}{\sigma}$, formula for standardizing a distribution, population standard deviation known

$t = \frac{X-\mu}{\hat{\sigma}}$, formula for standardizing a distribution, population standard deviation unknown

PROBLEMS

8.1 Write out in acceptable tabular form the total number of possible outcomes of a single toss of six coins. Label them T H H H H T, and so on. On the basis of your table, answer the following:

a. P(0 heads).
b. P(1 head).
c. P(2 heads).
d. P(3 heads).
e. P(4 heads).
f. P(5 heads).
g. P(6 heads).
h. Present the probabilities in acceptable form as a histogram and a frequency polygon, using one chart.

8.2. Referring to Problem 8.1, what is the binomial expression that refers to this set of events? Expand the binomial and determine the probabilities associated with each term. Are these the same probabilities as computed in Problem 8.1? Is there another method of computing these probabilities? If so, how?

8.3. Given the following data: $p = .60$; $q = .40$; $n = 4$. Assume that we take at random 1,000 samples of size n from an infinite population, and answer the following:

a. Present in acceptable tabular form the probabilities and theoretical frequencies for each of the five possible sample combinations.

b. Draw the frequency polygon of the probabilities and the histogram of the sample frequencies.

8.4. The Sigma Manufacturing Corporation is producing a large lot of a mass-produced product which is known to be 10 percent defective. Compute the probability distribution, and complete the table below for a number of defectives in one hundred samples of size 4.

Probability distribution of number of defectives in 100 samples of size 4 from a lot 10 percent defective

Number of Defectives X	$P(X)$	Theoretical Frequencies f
0		
1		
2		
3		
4		

a. What is the probability of selecting:
(1) Precisely two defectives?
(2) Two or more defectives?
(3) Two or less defectives?

b. What is the arithmetic mean and standard deviation of this binomial probability distribution? Interpret these measures.

8.5. For each of the following, draw a normal curve and illustrate your answer by crosshatching the appropriate area.

a. Determine the normal curve areas which lie–
(1) To the right of $z = +1.72$.
(2) To the left of $z = -1.02$.
(3) Between $z = 0$ and $z = -2.06$.
(4) Between $z = +1.80$ and $z = +1.92$.
(5) Between $z = -1.43$ and $z = +1.62$.
(6) To the left of $z = +2.30$.

b. Determine z if–
(1) The normal curve area between zero and z is .44630.
(2) The normal curve area between $+z$ and $-z$ is .9216.
(3) The normal curve area to the left of $-z$ is .2946.
(4) The normal curve area to the right of $+z$ is .0233.
(5) The normal curve area between $0 + z$ and $0 - z$ is .9861 (assume symmetrical limits about $z = 0$).

8.6. Assume that the population average of examination scores for the Graduate Management Admission Test (GMAT) is 500 and the population standard deviation is 100. What is the probability of a student scoring 700 or above? Less than 300? Between 550 and 600?

8.7. An instructor in statistics has been requested by a student to "grade on the basis of the normal curve." The instructor, being a generous soul by nature, as most statisticians are, has suggested that 5 percent be given A's; 20 percent B's; 50 percent C's; 20 percent D's; and 5 percent F's.

a. If the population average is 75 and the standard deviation is 10, what are the numerical grades which divide the 5 letter grades?

b. What assumptions are necessary for the instructor to make a valid application of grading on the normal probability distribution?

c. If you were in this course, would you want the instructor to use the normal curve to determine the final grades? Why, or why not?

8.8. Suppose we have 3,280 accounts receivable in a large furniture store, distributed normally with a mean and a standard deviation of $281 and $35, respectively. Answer the following:

a. How many of the accounts will lie above the point $316?

b. Above what point will 25 percent of the accounts lie?

c. The probability is 90 out of 100 that a single account selected at random will lie above what point?

d. Ten percent of the accounts will lie more than what distance from the mean?

8.9. The heights of a number of female students are normally distributed with a mean of 5 feet 4 inches. If 5 percent are 5 feet 10 inches or more tall, what is the standard deviation of this distribution? What is the probability of selecting at random from this population a girl who is between 6 feet 2 inches and 6 feet 4 inches in height?

8.10. In a normally distributed universe the population mean is 28, and the population standard deviation is 6. Determine:

a. The area below 24.

b. The area between 22 and 34.

c. The area between 16 and 40.

d. The area between 13 and 43.

e. The area above 39.

f. The point that has 90 percent of the area below it.

g. The two points containing the middle 80 percent of the area.

8.11. A large department store has found that generally about 10 percent of the items sold are returned for refunds. What is the probability that of the next four items sold, one will be returned? Two? Three? All four? None?

8.12. Ten percent of the students score over 700 on a college admissions test. What is the probability that of the next 10 students tested, no more than 2 would be in the 700-or-over group?

8.13. A production process manufacturing inexpensive ball-point pens has a defective rate of .10. These ball-point pens are sold to a local bank for use by its customers. If the bank has just placed six of these pens on a counter, what is the probability that—

a. At least two of the pens would be defective?

b. Exactly two would be defective?

c. That all pens would be perfect?

8.14. Given the following information with respect to a binomial distribution, determine the mean and standard deviation in the first three cases.

a. $p = q = .5$; $n = 100$.
b. $p = q = .5$; $n = 500$.
c. $p = 6$; $q = .4$; $n = 30$.
d. What is the meaning of your measures in part (*a*)?

8.15. A poll is taken among 100 faculty members in a given university on the question of affiliation with a union. The results of the poll are tabulated below (assume that the sample represents the population):

Response	Assistant Professor	Associate Professor	Professor	Totals
Favor affiliation	50	8	2	60
Oppose affiliation	20	6	4	30
No opinion	8	0	2	10
Total	78	14	8	100

Required:

a. What is the probability that a faculty member selected at random at this university will oppose union affiliation?
b. What is the probability that a faculty member selected at random at this university will be an assistant professor?
c. What is the probability that a faculty member selected at random at this university will be an associate professor who favors union affiliation?
d. What is the probability that a faculty member selected at random at this university will be an assistant professor who favors union affiliation?

8.16. From a recent previous survey it is known that 90 percent of the faculty members in a given university are in favor of the establishment of a faculty luncheon club. If a random sample of four faculty members is selected from this population, what would be the probabilities of 0, 1, 2, 3, and 4 "yes" responses in the sample?

8.17. If 100 samples of $n = 4$ are taken from the population in Problem 8.16, how would the results be expected to distribute among the number of "yes" responses ranging from 0 to 4? Present your data in an acceptable tabular form.

8.18. Given a population with a mean of 100 and a standard deviation of 20:
a. What is the probability that a random sample of 100 items from this population will yield a sample mean of less than 98? Greater than 106? 106 or less?
b. Would your response to the above questions be the same if n were 1,000 instead of 100 given the fact that both are "large" samples?

8.19. Assume that for practical purposes a production process is considered to be infinite and that the sampling of a single item does not change the probabilities for the remaining items. That is, the events are really independent. Further assume that this process produces a low-cost product,

70 percent of which are classified as grade A and 30 percent of which are rated grade B.

Required:

a. Determine the probability distribution, and complete the table at right for a number of grade B items in 100 samples where each sample is of size 3.

Probability distribution of number of grade B items in 100 samples of size 3 from a production process where 30 percent are rated grade B

No. of Grade B X	$P(X)$	Theoretical Frequencies f
0		
1		
2		
3		

b. What is the probability of sampling at random exactly two grade B items ($n = 3$)?

c. What is the probability of sampling at random two or more grade B items ($n = 3$)?

d. What is the probability of sampling at random two or less grade B items ($n = 3$)?

e. What is the mean and standard deviation of this probability distribution?

8.20. The distribution of scores on a nationwide aptitude test for graduate study tends to be normally distributed with the following parameters: (1) the mean equals 500 units and the (2) standard deviation equals 60 units.

Required:

a. If a school wishes to admit only those students who score in the top 20 percent of this distribution, what would the cutoff test score be?

b. What proportion of the students taking this aptitude test score 440 or above?

c. What proportion score 600 or above?

d. What proportion score 700 or above?

e. What proportion score between 600 and 700?

8.21. An internal auditor wants to develop a decision rule to be used in the evaluation of accounts payable of the LeBeaux-Nassau Corporation, Ltd. It is estimated that there are 10,000 of these accounts. Generally accepted auditing principles suggest that if there are errors in only 1 percent of the accounts payable the auditor will consider this satisfactory and will be willing to certify their accuracy. However, if the errors are 5 percent or more, then the auditor will insist upon a complete census of the accounts before certification. Because of the size of the universe of accounts payable, the decision is made to randomly sample 50 accounts and verify their accuracy. The decision rule agreed upon by the auditing staff is that if only 1 of the sample of 50 accounts contains an error then the internal auditor will certify the entire population as being accurate. If 3 or more of the sampled accounts contain errors then a much further investigation will be done to determine the accuracy of the entire 10,000.

a. On the basis of the sample evidence, the internal auditor will be making a responsible statement. Therefore, he or she wishes to know before sampling what is the probability of certifying the accounts as being accurate if indeed 1 percent contain errors. What is the probability that the auditor will decide that further investigation is necessary because the accounts contain too many errors if the true proportion of errors in the 10,000 is 1 percent?

b. On the other hand, the auditor is interested in knowing what is the probability of recommending a complete census of the accounts payable if indeed the true proportion of errors in the population of accounts is 5 percent. If the true proportion is indeed 5 percent what is the probability that the auditor will certify the accounts as being accurate?

8.22. Assume that the time between arrivals of customers passing an important bridge toll station is known to be described by an exponential distribution with a mean time of .40 minutes. What is the probability of a gap of 6 seconds between a given customer arrival and the next one? Note that $P(t) = .10$ which relatively is an infinitesimal number. Also, if the mean waiting time between customers is .40, then $\lambda = 1/.40 = 2.5$.

8.23. Assume that the T. Watson Electronics Corporation manufactures a part whose known life can be described by an exponential distribution with an average life of .25 days.

a. What is the probability that a given part randomly selected from the production run will have a life in excess of .25 days?

b. Assume that we have selected a part at random and it has been used for .25 days. What is the probability of an *additional* .50 days life of this part?

8.24. A manufacturer of electronic air cleaners is debating whether or not to expand its market by setting up a branch sales and service outlet in a large metropolitan area. The area contains an estimated 100,000 single-family dwelling units. In the past, this quality air cleaner has appealed largely to homeowners in the $35,000 and up price range. Given the current interest in pollution control and the environment, the sales manager wishes to estimate the sales potential of this new market area. The only information that he has available is the following from a random sample of houses (single-family dwelling units) sold during the past year: $N = 100{,}000$; $n = 2{,}000$; $\bar{X} = \$30{,}000$; $s = \$5{,}000$. (Assume that because n is so large the sample standard deviation (s) is a good approximation of the population variation without further correction.) Answer the following questions:

a. By use of the Table of the Areas of the Normal Distribution, estimate the number of single-family dwelling units in this metropolitan area which are in the price bracket of $35,000 and up.

b. If a smaller and less expensive unit were produced which would be appropriate capacitywise and pricewise for houses in the $25,000 to $35,000 price range, by what *percentage* would the potential market be increased? That is above what you estimated in part (*a*).

c. Estimate the number of houses that are in the $35,000 to $45,000 price range.

d. What is the potential sales for the homes $45,000 and over?

RELATED READINGS

BENSON, PURNELL H. "Fitting and Analyzing Distribution Curves of Consumer Choices." *Journal of Advertising Research,* vol. 5, no. 1 (March 1965), pp. 28–34.

CHEW, VICTOR. "Some Useful Alternatives to the Normal Distribution." *The American Statistician,* vol. 22 (June 1968), pp. 22–24.

CURETON, EDWARD E. "Z, *t*, *F* or Beta." *The American Statistician,* vol. 18 (April 1964), pp. 26 ff.

NETER, JOHN, and WASSERMAN, WILLIAM. *Applied Linear Statistical Models.* Homewood, Ill.: Richard D. Irwin, Inc., 1974.

ROSS, SHELDON M. *Applied Probability Models with Optimization Applications.* San Francisco: Holden-Day, Inc., 1970.

SHONICK, W. "Stochastic Model for Occupancy Related Random Variables in General-Acute Hospitals." *Journal of American Statistical Association,* vol. 65, no. 332 (December 1970), pp. 1474–99.

Part IV

Basic concepts of statistical inference – I

. . . the great difficulty in the social sciences . . . of applying scientific method is that we have not yet established an agreed standard for the disproof of an hypothesis.

Joan Robinson

9

Sampling distributions and estimation of parameters

A woman's guess is more accurate than a man's certainty.

Rudyard Kipling

THERE ARE TWO major types of sampling problems involving statistical inference. First, there is the problem of estimating a parameter on the basis of the sample evidence. In this section, we shall consider the problem of estimation of both μ and π under differing conditions. The other major type of problem involving statistical inference is the one commonly referred to as testing of hypotheses or statistical decision making. This latter kind of problem involves testing for the significance of the differences between samples or between a sample and a universe. This second type will be taken up in Chapters 10 and 11. Both of these types of statistical inference problems deal with the question of drawing valid conclusions concerning the population characteristic (e.g., μ or π) based on the information from a sample ($\bar{X}$ or p). The major reasons for estimating the population values from partial information relate to cost and timeliness. Usually it is just too expensive to take a census every time. In addition, even if the budget can afford a 100 percent sample, a census normally takes a much longer time to execute properly. Many finite populations are too large to enumerate each and every item. And, by definition an infinite universe precludes a complete count.

Before moving directly into problems of estimation and tests of hypothesis we need to have some additional background. Specifically, we need to know something about the theoretical sampling distributions and their measures of dispersion, namely, the standard error. The central-limit theorem is the basis for this aspect of statistical methodology and will be discussed below. That is, in order to be able to validly make estimates of population characteristics based on random sample observations the statistician must have a solid foundation in these concepts. Let us now turn to these important ideas as they relate to statistical inference.

SAMPLING DISTRIBUTIONS

If we were to draw a number of random samples from a population and calculate a mean for each sample, the distribution of these sample means would be a normal one. This would be true even if the population or universe itself were not a normal one. There are two basic reasons for this phenomenon which characterize the central-limit theorem: (1) The variations among the statistics of different samples are caused only by a random process when probability samples are drawn. (2) As the number of statistics increases, the average of the many statistics approaches the parameter of the universe from which the samples were drawn. This is partly an averaging process, which accounts for a smaller variation in this latter distribution.

To illustrate, assume that a simple random sample is taken of the families of the residents of the city of Ann Arbor, Michigan. The purpose is to estimate the average family income in this given city. It is quite likely that the universe depicting family incomes in this city is skewed to the right. If we take a random sample and determine the average family income, we shall have a statistic, that is, the mean income. Merely for purposes of illustration, let us further assume that we take a second sample. The mean income of the second sample will probably differ from the mean of the first, as will also the mean of a third sample and any subsequent ones we may select. This simple illustration relates to the whole concept of the pattern of sampling variation. If the various means (statistics) themselves are arrayed in a relative frequency distribution, this particular distribution will tend to be a normal one, and the mean of this theoretical distribution will be equal to the mean income for the city (the parameter μ). This fact is true despite the knowledge that the distribution of individual incomes in the city is quite likely to be decidedly skewed to the right. It should be noted that the scatter of values, that is, the dispersion, in the distribution of sample means (sampling distribution) is much less than the scatter of individual items either in the universe or in a single sample largely because of the averaging process.

Pausing to summarize momentarily, then, three basic frequency distributions have been mentioned. They include:

1. *The universe or population.* This is the total group from which the sample is selected; the frequency distribution of a universe may be approximated by
2. *The random sample.* The random sample may be thought of as a

small-scale replica of the universe with characteristics (e.g., the mean, median, or variance) approximately the same as those of the population. If many simple random samples are selected from a given universe, the frequency distribution of the means of all these samples would be a

3. *Normal distribution of sample means or the sampling distribution of sample means.* This theoretical distribution is sometimes called the normal probability distribution of sample means and tends to take the shape of a normal probability curve, even though the population may be skewed. As noted above, it has less dispersion than either the universe or the individual samples. These distributions are illustrated graphically in Figure 9.1.

THE STANDARD DEVIATION OF THE SAMPLING DISTRIBUTION

It will be recalled that the standard deviation indicates the scatter or dispersion of the individual items in a distribution about their arithmetic mean. The sampling distribution of means also has a standard deviation of its own which is somewhat different from both the standard deviation of the universe and the standard deviation of an individual sample. In order to distinguish it from the others, this standard deviation of the sampling distribution is called the *standard error* or sometimes the standard deviation of the probability distribution. The formula for calculating the standard error differs depending on whether the problem deals with the enumeration of the presence or absence of a specified attribute (that is, situations in which the binomial case is applicable) or with measured data, that is, measurements of values of items along a continuous numerical scale. Attribute data, of course, are normally thought of as being one of two classes, whereas measurement data really are an attribute with an infinite number of values.

SAMPLING DISTRIBUTION OF SAMPLE MEANS

In problems involving *measurement data,* the standard error of the mean is required. The formula for determining the true standard error of the mean is as follows:

$$\sigma_{\bar{x}} = \frac{\sigma_x}{\sqrt{n}} \tag{9.1}$$

In cases where the population standard deviation (σ_x) is unknown, we must substitute our best estimate for this parameter. In such cases the estimated population standard deviation ($\hat{\sigma}_x$) is used in the above for-

FIGURE 9.1
Three basic distributions

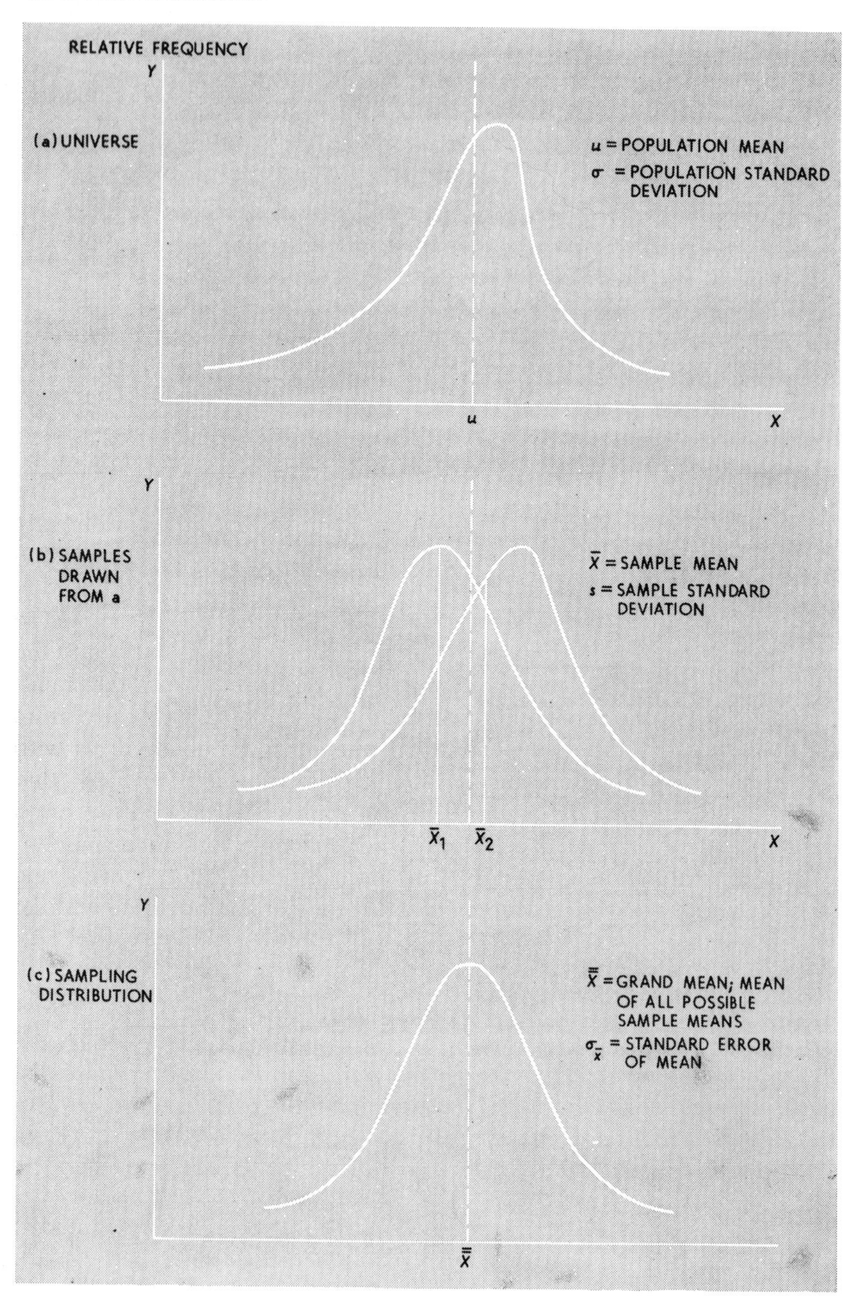

mula; therefore, our standard error also becomes an estimate. The formula for the estimated standard error of the mean is as follows:

$$\hat{\sigma}_{\bar{x}} = \frac{\hat{\sigma}_x}{\sqrt{n}} \tag{9.2}$$

The estimated standard error of the mean may also be found by the following formula:

$$\hat{\sigma}_{\bar{x}} = \frac{s_x}{\sqrt{n-1}} \tag{9.3}$$

We recall that the estimated standard deviation of the population is found as follows:

$$\hat{\sigma}_x = \sqrt{\frac{\Sigma f(X - \bar{X})^2}{n-1}}, \quad \text{or} \quad \hat{\sigma}_x = s_x\sqrt{\frac{n}{n-1}} \tag{9.4}$$

It should be noted that in estimating the standard error of the percentage, even though we use the sample proportion in our formula, we do not use $n - 1$ in the denominator. This is true because there is no standard deviation of the universe for attribute data and no degree of freedom really is lost here.

Central-limit theorem

For measurement data, we have indicated that the measure of dispersion of the theoretical sampling distribution is the standard error of the mean; and *from the central-limit theorem, we learn that if we select a large number of simple random samples from any population and determine the mean for each sample, the distribution of these sample means will tend to be described by the normal probability distribution with a mean μ and variance σ^2/n.* We also found that this is true even if the population itself is not normal because the variations among the means of the many samples are caused by chance fluctuations when the samples are drawn at random; and as the number of simple random sample means increases, the mean of the many sample means, that is, the grand mean ($\bar{X}$), approaches the mean of the universe (μ) from which the samples were drawn. Just as the population has its own standard deviation, so does the individual sample have a standard deviation; and the theoretical mathematical model, called our sampling distribution, also has its own measure of dispersion, in this latter case the standard error of the mean. Although each of these units of dispersion, in effect, measures the same thing in concept, they are slightly different. Whereas the population standard deviation is equal to the square root of the mean of the squared deviations of the items taken from the population mean (μ), the sample standard deviation is really the square root of the mean of the squared deviations of the sample items taken from their own mean $\bar{X}$. The deviations involved in the theoretical sampling distribution are the deviations of the individual sample means from the grand mean. That is, just as the standard deviation of the universe helps us to locate an individual population item and just as the sample standard deviation helps us locate an individual sample item, in our problems of estimation and tests of hypotheses, the standard error of the mean helps us to locate the individual sample mean with respect to the population mean or the mean of all of the means.

Problems requiring the use of the standard error of the mean, of course, are those concerned with the value of some variable such as weight, height, cost, length, or some other measurable unit rather than with a proportion value, as in the binomial case, which calls for the use of the standard error of a proportion.

Suppose a population trait is normally distributed. Then items repeatedly drawn from that population in random fashion will evidence a normal distribution of that trait. Thus, samples of $n = 1$ repeatedly drawn from a normally distributed population show a normal distribution. It is also true that samples of size $n = 2, 3, 4$, or any other size will be normally distributed.

If a population is symmetrical, means of samples properly drawn from it will be symmetrically distributed. For example, a simple illustrative case is a population consisting of only 16 items, as in Table 9.1. Table 9.2 shows all the sample sums of samples of size $n = 2$, and Figure 9.2 pictures the distribution of the sample means.

It is not necessary to make the detailed computation shown in Table 9.3. All one needs to know in order to determine the standard error of the mean (the standard deviation of all possible sample means around the

TABLE 9.1

A symmetrical population of $n = 16$ ($a = 2$, $b = 4$, $c = 4$, $d = 4$, $e = 4$, $f = 6$, $g = 6$, $h = 6$, $i = 6$, $j = 6$, $k = 6$, $l = 8$, $m = 8$, $n = 8$, $o = 8$, $p = 10$)

Population Item X'	Number of Items f	$X' - \mu$	$(X' - \mu)^2$	$f(X' - \mu)^2$
2	1	−4	16	16
4	4	−2	4	16
6	6	0	0	0
8	4	2	4	16
10	1	4	16	16
	$N = 16$			$\Sigma f(X' - \mu)^2 = 64$

$$\sigma_x^2 = \frac{\Sigma f(X' - \mu)^2}{N} = \frac{64}{16} = 4$$

TABLE 9.2
Combination sample sums of $n = 2$ taken from a population $N = 16$

	a 2	*b* 4	*c* 4	*d* 4	*e* 4	*f* 6	*g* 6	*h* 6	*i* 6	*j* 6	*k* 6	*l* 8	*m* 8	*n* 8	*o* 8	*p* 10
a 2	x	6	6	6	6	8	8	8	8	8	8	10	10	10	10	12
b 4	x	x	8	8	8	10	10	10	10	10	10	12	12	12	12	14
c 4	x	x	x	8	8	10	10	10	10	10	10	12	12	12	12	14
d 4	x	x	x	x	8	10	10	10	10	10	10	12	12	12	12	14
e 4	x	x	x	x	x	10	10	10	10	10	10	12	12	12	12	14
f 6	x	x	x	x	x	x	12	12	12	12	12	14	14	14	14	16
g 6	x	x	x	x	x	x	x	12	12	12	12	14	14	14	14	16
h 6	x	x	x	x	x	x	x	x	12	12	12	14	14	14	14	16
i 6	x	x	x	x	x	x	x	x	x	12	12	14	14	14	14	16
j 6	x	x	x	x	x	x	x	x	x	x	12	14	14	14	14	16
k 6	x	x	x	x	x	x	x	x	x	x	x	14	14	14	14	16
l 8	x	x	x	x	x	x	x	x	x	x	x	x	16	16	16	18
m 8	x	x	x	x	x	x	x	x	x	x	x	x	x	16	16	18
n 8	x	x	x	x	x	x	x	x	x	x	x	x	x	x	16	18
o 8	x	x	x	x	x	x	x	x	x	x	x	x	x	x	x	18
p 10	x	x	x	x	x	x	x	x	x	x	x	x	x	x	x	x

TABLE 9.3
Distribution of sample means of all possible combinations of two taken from the population of Table 9.1 as shown in Table 9.2

Sample Mean X	Number of Means f	$(\mu - \bar{X})$	$(\mu - \bar{X})^2$	$f(\mu - \bar{X})^2$
3	4	−3	9	36
4	12	−2	4	48
5	28	−1	1	28
$\bar{X} = \mu = 6$	32	0	0	0
7	28	1	1	28
8	12	2	4	48
9	4	3	9	36
Number of means =	120		$\Sigma f(\mu - \bar{X})^2 =$	224

$$\sigma_{\bar{X}}^2 = \frac{224}{120} = 1.86$$

$$\sigma_{\bar{X}} = \sqrt{1.86} = 1.36$$

FIGURE 9.2
Distribution of symmetrical population and sampling distribution of means of samples of size $n = 2$

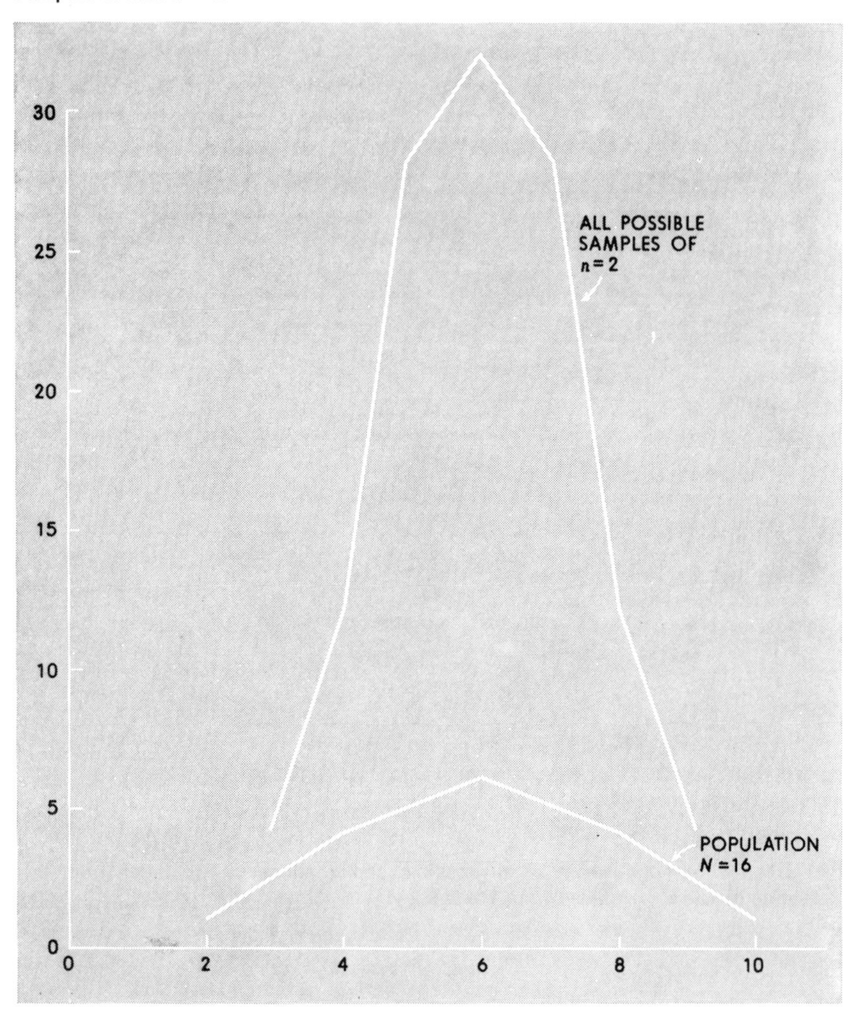

mean of the population) is the standard deviation (σ_x) of the population.
In Table 9.1, $\sigma_x = 2$. The precise formula for the standard error of the mean ($\sigma_{\bar{x}}$) is

$$\sigma_{\bar{x}} = \sqrt{\frac{\sigma^2}{n} \cdot \frac{N}{N-1}\left(1 - \frac{n}{N}\right)} \tag{9.5}$$

σ^2 = population variance
n = number in the sample
N = number in the population

$$\sigma_{\bar{x}} = \sqrt{\frac{4}{2} \cdot \frac{16}{15}\left(1 - \frac{2}{16}\right)} = \sqrt{1.86} = 1.36$$

When the population to be sampled is large, say several hundred, the formula for all practical purposes becomes the same as (9.1):

$$\sigma_{\bar{x}} = \sqrt{\frac{\sigma^2}{n}} = \frac{\sigma}{\sqrt{n}} \qquad (9.6)$$

and the distribution of sample means virtually is a normal distribution if the sample is large, say, $n = 30$. Figure 9.6 pictures the normal distribution of sample means when n is large. (See Table 9.3.)

Sample standard deviation as an estimate of σ

The sample standard deviation is not an unbiased estimate of the population standard deviation. On the average, the sample standard deviation is smaller than σ. If, however, all possible sample standard deviations of a given sample size are computed as (same as Formula [9.4]):

$$\hat{\sigma} = \sqrt{\frac{\Sigma x^2}{n - 1}}$$

on the average, the sample $\hat{\sigma}$ is equal to σ. And an unbiased estimate of σ^2 is

$$\hat{\sigma}^2 = \frac{\Sigma x^2}{n - 1}$$

Probability and the sampling distribution of means

Assume a population with $\mu = 200$ units and $\sigma_x = 20$ units. What is the probability that $\bar{X}$ is between 190 and 210 if $n = 4$? (See Figure 9.3.) Then, using Formula (9.1):

$$\sigma_{\bar{x}} = \frac{20}{\sqrt{4}}$$

$$\sigma_{\bar{x}} = 10 \text{ units}$$

and

	Area
$z = \dfrac{190 - 200}{10} = -1.0$	.34134
$z = \dfrac{210 - 200}{10} = +1.0$	.34134
	.68268

and adding the two areas, the probability is .68268 that $\bar{X}$ is between 190 and 210.

FIGURE 9.3

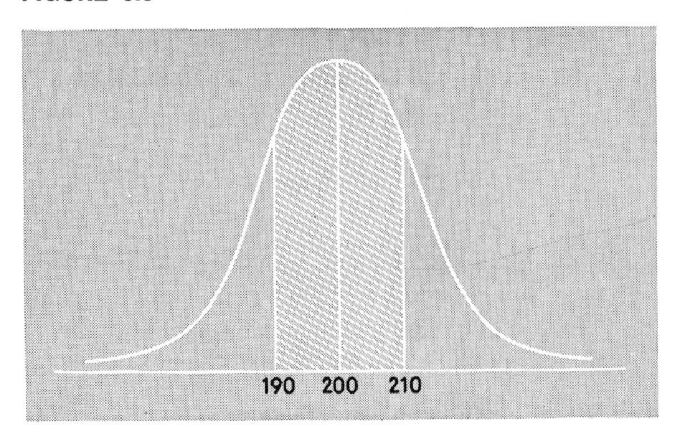

Now, let us assume that $n = 25$. What is the probability that $\bar{X}$ is between 200 and 204? (See Figure 9.4.) Then,

$$\sigma_{\bar{x}} = \frac{20}{\sqrt{25}}$$

$$\sigma_{\bar{x}} = 4 \text{ units}$$

and

$$z = \frac{204 - 200}{4} = +1.0 \qquad \begin{array}{c}\text{Area}\\ .34134\end{array}$$

Therefore the probability is .34134 that $\bar{X}$ lies between 200 and 204.

Finally, let us assume that $n = 100$. What is the probability that $\bar{X}$ is between 198 and 200? (See Figure 9.5.) Then,

$$\sigma_{\bar{x}} = \frac{20}{\sqrt{100}}$$

$$\sigma_{\bar{x}} = 2 \text{ units}$$

FIGURE 9.4

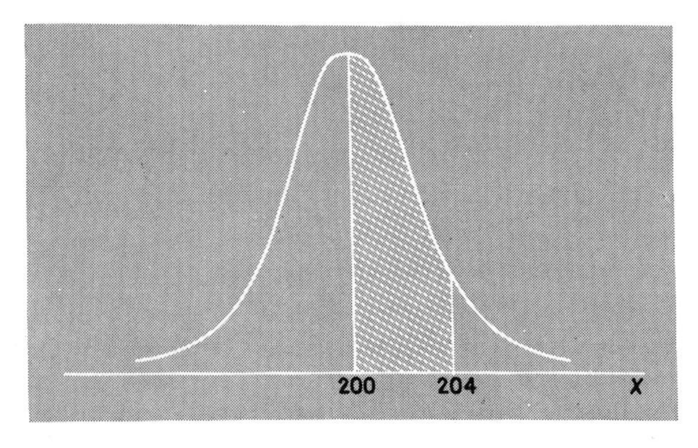

FIGURE 9.5

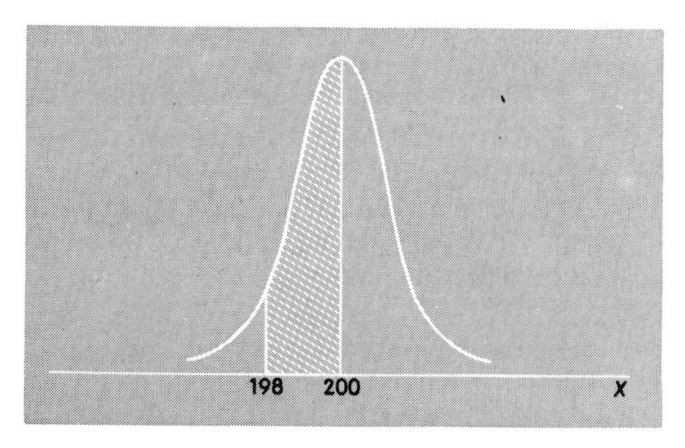

and

$$z = \frac{198 - 200}{2} = -1.0 \qquad \begin{array}{c}\text{Area} \\ .34134\end{array}$$

Therefore the probability is .34134 that $\bar{X}$ lies between 198 and 200. It should be clear that the shape of the sampling distribution depends upon the population standard deviation and the sample size. In all three examples the distance from X to μ was either a plus or a minus one standard deviation. The dispersion in the sampling distribution as measured by $\sigma_{\bar{x}}$ is reduced as n increases, since σ_x was held constant at 20 units.

Substitution of $\hat{\sigma}$ for σ

If the population is large, the true standard deviation of all possible sample means will be (see Formula [9.6])

$$\sigma_{\bar{x}} = \sqrt{\frac{\sigma^2}{n}}$$

and its unbiased estimate $\hat{\sigma}_{\bar{x}}$ is as given in Formula (9.3)

$$\hat{\sigma}_{\bar{x}} = \frac{\hat{\sigma}}{\sqrt{n}}, \qquad \text{or} \qquad \hat{\sigma}_{\bar{x}} = \sqrt{\frac{\frac{\Sigma x^2}{n-1}}{n}} = \sqrt{\frac{\Sigma x^2}{n(n-1)}} = \frac{s}{\sqrt{n-1}}$$

Further, when the sample is large—for example, 30 or more—the distribution of sample means will approximate a normal curve. This is true even though the population from which the sample was drawn is not normally distributed. Figure 9.6 pictures the population mean, μ, in the center of

FIGURE 9.6
Distribution of sample means when n is large

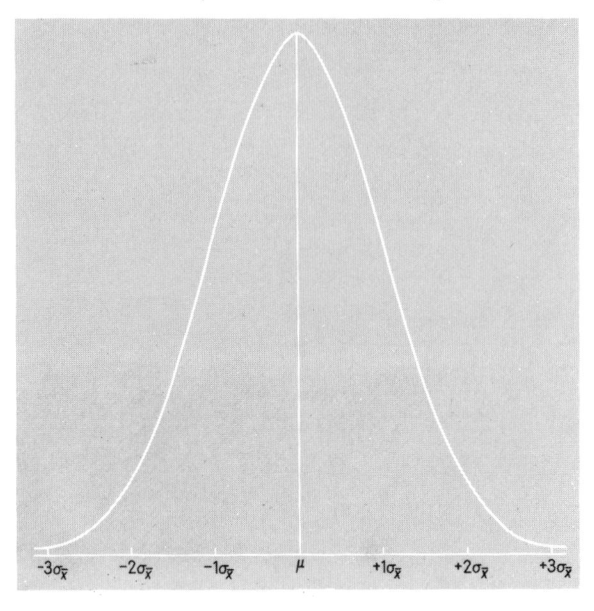

the distribution of sample means and shows the areas included in the interval $\mu - 2\sigma_{\bar{x}}$ and $\mu + 2\sigma_{\bar{x}}$. This area is approximately 95 percent of the total area under the curve. Thus the mean of the population plus and minus two standard errors of the mean includes 95.45 percent of all the possible sample means of a given sample size if the size of the sample, n, is 30 or more.

Because the standard deviation of the universe (σ_x) is usually unknown, we must use the standard deviation of the sample (s_x) as a rough approximation to our parameter. If the sample size is large (generally considered 30 or over), the error introduced by this approximation is rather small; or we can better estimate the standard deviation of the population by one of the formulas given previously. For example, if we use Formula (9.4):

$$\hat{\sigma}_x = \sqrt{\frac{\Sigma f(X - \bar{X})^2}{n - 1}}$$

we divide by $n - 1$ because the use of the sample standard deviation instead of the population standard deviation leads to an underestimation of the value of the standard error. The reason is that the sample standard deviation is likely to be less than the population standard deviation as the sample size decreases. This is partly true because we tend to miss the extreme values which, of course, are important in the determination of a standard deviation. As will become evident later, this understatement would lead to an inference about the parameter which would look more reliable than it merits; therefore, when working with small samples, and particularly when the population standard deviation is unknown, this tendency can be corrected with a better approximation to the universe standard deviation by using $n - 1$ in our denominator, as indicated in the above formula. This correction is, of course, the distinguishing feature of the so-called "*t* distribution" often used in such problems.

Further to illustrate the point, we know that $\Sigma(X - \bar{X})^2$ is as small or smaller than $\Sigma(X - k)^2$ where k is any other number; then, $\Sigma(X - \bar{X})^2 < \Sigma(X - \mu)^2$. If we knew μ, we would use it and not lose any degree of freedom; indeed, we would not be drawing a sample in the first place. However, when μ is unknown, we have to estimate μ with $\bar{X}$, which means that the numerator of the standard deviation is going to be smaller than if we had used μ. Therefore, we decrease the denominator by one, since we have estimated μ with $\bar{X}$.

SAMPLING DISTRIBUTION OF A PERCENTAGE

Where we are dealing with *attribute data,* that is, where the binomial distribution is appropriate, the measure of dispersion of our theoretical mathematical model is called the standard error of a percentage. Although the scale for measurement data on the x axis is a continuous scale, for attribute data the scale goes from 0 to 1 or from 0 to 100 percent; and although the universe for attribute data may have a parameter π, that is, the population proportion, there is no measure of dispersion in the universe. However, if we were to take many samples of size n from this given population and determine the sample percentages, and then array the sample percentages in a relative frequency distribution, these sample percentages would tend to take the shape of a normal distribution centered about π (it should be noted that the x scale here would be an infinite scale).

The formula for the true standard error of a proportion is

$$\sigma_p = \sqrt{\frac{\pi(100 - \pi)}{n}} \tag{9.7}$$

and the estimated standard error of a proportion may be found by substituting the sample percentage for π, as follows:

$$\hat{\sigma}_p = \sqrt{\frac{p(100 - p)}{n}} \tag{9.8}$$

The mean and standard deviation which were used above in the discussion of the reliability of quantitative sample data are also used in the discussion of qualitative data. Qualitative data are gathered in such a way as to reveal a particular characteristic—for example, a brand preference by cigarette smokers, or a preference for a particular political candidate. In other words, the data are tabulated as either having a particular attribute or having some other characteristics. Thus a survey of cigarette brand preference is tabulated only as $X = 0$ or $X = 1$, the 1 being a preference for brand A cigarette, and the 0 being a preference for any other brand or no preference. Thus, if in a group of ten people, six prefer brand A, two prefer brand B, one prefers brand C, and one has no preference, the tabulation is

X'	f
0	4
1	6

If the ten people constitute a universe, the variance is $\pi(1 - \pi)$, where π is the proportion that prefer brand A[1]. Here, $\pi = .60$ and $1 - \pi = .40$, and the variance is $\sigma^2 = .60 \cdot .40 = .24$. This is shown in Table 9.4, where X' is the population characteristic and f is the number having that characteristic.

TABLE 9.4*

Computation of σ^2 for qualitative data

X'	f	fX'	$f(X')^2$
0	4	0	0.0
1	6	6	6.0
		$\Sigma fX' = 6$	$\Sigma f(X')^2 = 6.0$
		$\bar{X}' = .60$	$\bar{X}'\Sigma fX' = 3.6$
			$\Sigma f(x')^2 = 2.4$

$$\sigma^2 = \frac{2.4}{10} = .24 = \pi(1 - \pi) = .60 \cdot .40 = .24$$

* The generalization is:

0	$1 - \pi$	0	0.0
0	π	π	π
		$\Sigma fX' = \pi$	$\Sigma f(X')^2 = \pi$
		$\bar{X}' = \pi$	$\bar{X}'\Sigma fX' = \pi^2$
			$\Sigma f(x')^2 = \pi - \pi^2 = \pi(1 - \pi)$

[1] Technically, the hypergeometric distribution is applicable here; however, to minimize the computations in the problem, it is assumed that the binomial is appropriate. Generally, if the population is finite and we sample without replacement, the exact distribution for the number of successes is given by the hypergeometric. The binomial will give *exact* probabilities only if the sample is drawn with replacement, or if the population is infinite. However, if the population is large relative to the sample size, the error will be negligible. Correction factors are available, but for most practical problems the uncorrected results will be adequate.

TABLE 9.5
All possible sums of the combinations of two taken from population of Table 9.4

X	0	0	0	0	1	1	1	1	1	1
0	X	0	0	0	1	1	1	1	1	1
0	X	X	0	0	1	1	1	1	1	1
0	X	X	X	0	1	1	1	1	1	1
0	X	X	X	X	1	1	1	1	1	1
1	X	X	X	X	X	2	2	2	2	2
1	X	X	X	X	X	X	2	2	2	2
1	X	X	X	X	X	X	X	2	2	2
1	X	X	X	X	X	X	X	X	2	2
1	X	X	X	X	X	X	X	X	X	2
1	X	X	X	X	X	X	X	X	X	X

The sample distribution of the proportion is shown in Table 9.5, Table 9.6, and Figure 9.7. Again, as in the sampling distribution of means, the variance of the sample proportions of $n = 2$ is

$$\sigma_p^2 = \frac{\sigma^2}{n} \cdot \frac{N}{N-1}\left(1 - \frac{n}{N}\right) = \frac{.24}{2} \cdot \frac{10}{9}\left(1 - \frac{2}{10}\right) = .106 \tag{9.9}$$

However, when the number in the population is large, say several hundred,

$$\sigma_p^2 = \frac{\sigma^2}{n} = \frac{\pi(100 - \pi)}{n} \tag{9.10}$$

and when the number in the sample is large, the distribution of the sample proportion p is approximately a normal curve, even though π is not equal to .50.

TABLE 9.6
Distribution of all possible sample proportions of combinations of two shown in Table 9.5

p	f	fp	fp^2
.0	6	.0	.0
.5	24	12.0	6.0
1.0	15	15.0	15.0
	$K = 45$	$\Sigma fp = 27.0$	$\Sigma fp^2 = 21.0$
		$\bar{p} = .6$	$\bar{p} \cdot \Sigma fp = 16.2$
			$(p - p^2) = 4.8$

$$\sigma_p^2 = \frac{4.8}{45} = .106$$

FIGURE 9.7
Distribution of sample proportions shown in Table 9.6

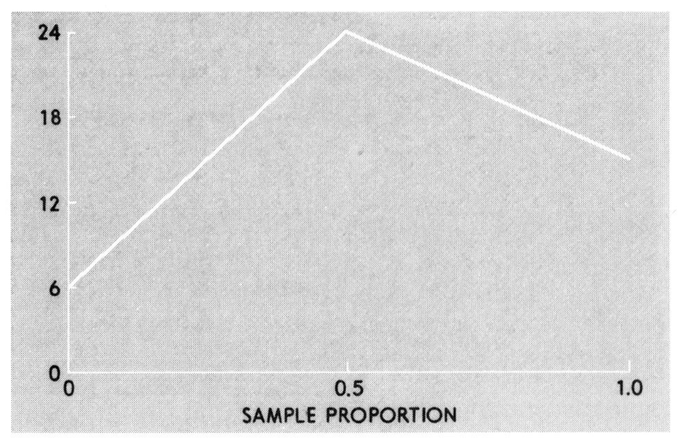

Probability and the sampling distribution of percentages

Assume a population with $\pi = 50$ percent. What is the probability that p is between 45 and 55 percent if $n = 100$? (See Figure 9.8.) Then using Formula (9.7):

$$\sigma_p^2 = \frac{(50)(50)}{100}$$

$$\sigma_p = 5 \text{ percent}$$

and

$$z = \frac{45 - 50}{5} = -1.0 \qquad \begin{array}{r} \text{Area} \\ .34134 \end{array}$$

$$z = \frac{55 - 50}{5} = +1.0 \qquad \begin{array}{r} .34134 \\ \hline .68268 \end{array}$$

and adding the two areas, the probability is .68268 that p is between 45 and 55 percent.

FIGURE 9.8

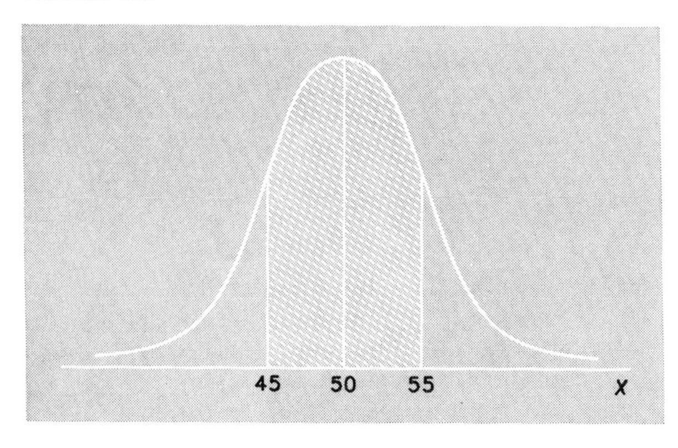

FIGURE 9.9

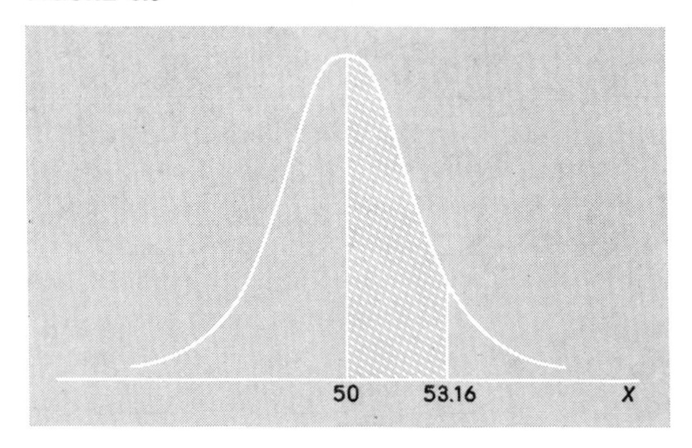

Now, let us assume that $n = 250$. What is the probability that p is between 50 and 53.16? (See Figure 9.9.) Then

$$\sigma_p^2 = \frac{(50)(50)}{250}$$
$$\sigma_p = 3.16 \text{ percent}$$

and

$$z = \frac{53.16 - 59}{3.16} = +1.0 \qquad \begin{matrix}\text{Area} \\ .34134\end{matrix}$$

Therefore the probability is .34134 that p lies between 50 and 53.16 percent.

Finally, let us assume that $n = 500$. What is the probability that p is between 47.76 and 50 percent? (See Figure 9.10.) Then

FIGURE 9.10

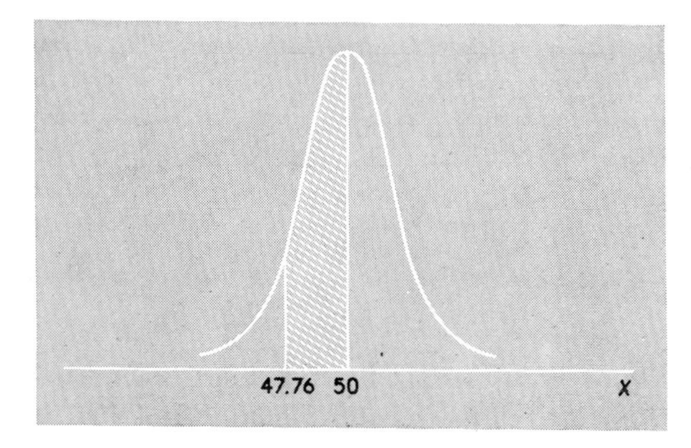

$$\sigma_p^2 = \frac{(50)(50)}{500}$$

$$\sigma_p = 2.24 \text{ percent}$$

and

$$z = \frac{47.76 - 50}{2.24} = -1.0 \qquad \begin{matrix}\text{Area} \\ .34134\end{matrix}$$

Therefore the probability that p is between 47.76 and 50 percent is .34134.

Again, all three examples illustrate the impact on the dispersion of the sampling distribution as n increases with π held constant in all three cases.

The sample variance pq as an estimate of the population variance π $(100 - \pi)$

Here again, as in the case of quantitative data, the sample variance $p(100 - p)$, or as it is more commonly written, pq, is not an unbiased estimate of σ^2; here, $\pi(100 - \pi)$. But $pq \cdot \frac{n}{n-1}$ will on the average be equal to $\pi(100 - \pi)$. Therefore the standard deviation of all possible sample percentages around π is estimated by Formula (9.8)

$$\hat{\sigma}_p = \sqrt{\frac{p(100 - p)}{n}}$$

This can most readily be shown by use of a simple table showing samples taken by the multiplicity-of-choice or replacement method. Using the population

X'	f
0	4
1	6

the distribution of all possible samples is as shown in Table 9.7.

TABLE 9.7

X	0	0	0	0	1	1	1	1	1	1
0	0, 0	0, 0	0, 0	0, 0	0, 1	0, 1	0, 1	0, 1	0, 1	0, 1
0	0, 0	0, 0	0, 0	0, 0	0, 1	0, 1	0, 1	0, 1	0, 1	0, 1
0	0, 0	0, 0	0, 0	0, 0	0, 1	0, 1	0, 1	0, 1	0, 1	0, 1
0	0, 0	0, 0	0, 0	0, 0	0, 1	0, 1	0, 1	0, 1	0, 1	0, 1
1	0, 1	0, 1	0, 1	0, 1	1, 1	1, 1	1, 1	1, 1	1, 1	1, 1
1	0, 1	0, 1	0, 1	0, 1	1, 1	1, 1	1, 1	1, 1	1, 1	1, 1
1	0, 1	0, 1	0, 1	0, 1	1, 1	1, 1	1, 1	1, 1	1, 1	1, 1
1	0, 1	0, 1	0, 1	0, 1	1, 1	1, 1	1, 1	1, 1	1, 1	1, 1
1	0, 1	0, 1	0, 1	0, 1	1, 1	1, 1	1, 1	1, 1	1, 1	1, 1
1	0, 1	0, 1	0, 1	0, 1	1, 1	1, 1	1, 1	1, 1	1, 1	1, 1

TABLE 9.8

pq	f	$f(pq)$
.00	52	0
.25	48	12
	$k = 100$	$\Sigma f(pq) = 12$
		Average $pq = 12/100 = .12$

$pq \cdot n/(n-1) = .12 \cdot 2/(2-1) = .24 = \sigma^2$ (see Table 9.4)

A sample of 0, 0 means $p = 0$, $q = 1$; a sample of 0, 1 means $p = .5$, $q = .5$; a sample of 1, 1 means $p = 1$, $q = 0$. Therefore the distribution of all possible pq's is as shown in Table 9.8.

Ordinarily, it is more convenient to change the sample proportion to a percentage and work with this rather than to work with the proportion. This merely involves shifting the decimal point two places to the right. The arithmetic is simpler, and the term "percentage" is in somewhat wider use. The results are precisely the same, the difference being only in the arithmetic. Therefore the following summary is appropriate:

π = population percentage

p = sample percentage

$\hat{\sigma}_p = \sqrt{\dfrac{pq}{n}}$ = estimated standard error of a sample percentage if n is several hundred

When the number in the sample, n, is large, the sampling distribution of a percentage approximates a normal curve like that of Figure 9.11.

INFERENCES FROM SAMPLES

Statistical inference procedures allow us to make some statement concerning a population without observing or measuring all of the items. Estimation problems dealing with population values include both *point estimates* and *interval estimates*. The sample mean is our "best estimate" of the parameter mu or the "true" population average. However, this point estimate is a single value and might well be "wrong," that is, not precisely the actual mean. In most market research projects—for example, estimating family income or expenditures—a high degree of precision may not be required or justified. But, a certain amount of reliability is expected in these cases. A *confidence interval* always includes some probability statement (confidence coefficient) with respect to the degree of reliability one may place in the estimate of the parameter. For example, a confidence coefficient of .95 theoretically means that if one were

FIGURE 9.11
Distribution of all possible sample percentages when n is large

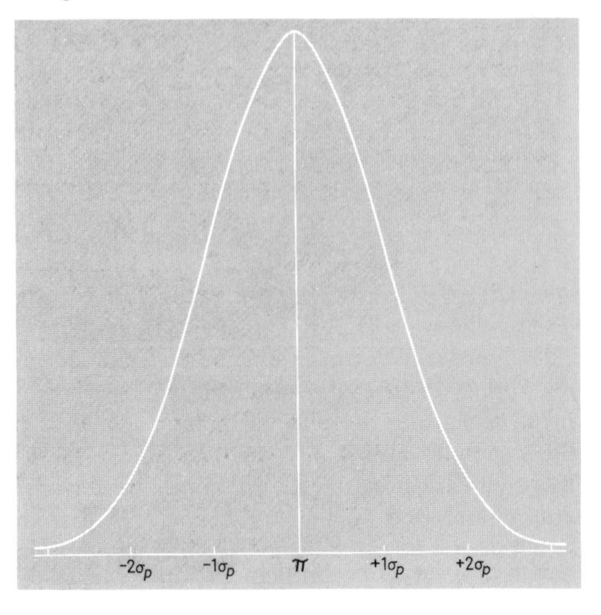

to select all possible samples of size n from a given population and then calculate the mean for each, and using proper estimating procedures (including a correctly drawn sample) to construct confidence intervals for each mean, 95 percent of these intervals would include the true mean. Note that the probability interpretation relates to the statistical methods used, not to a specific interval. A population value is either in or out of a *specific* interval; it is nonsense to talk of 95 percent probability of a value being within a *certain* confidence interval! Later, however, we discover that to be practical (and not to construct all possible intervals) we do behave inductively in our interpretation of a given interval estimate; we act as though the specific interval is one of the 95 percent.

We shall discuss estimation procedures under differing circumstances. First, merely to illustrate the techniques and to better understand a confidence interval, we will use an unreal case of measurement data where the population variance is known and the sample size is large enough to permit us to use the z distribution. The lack of reality stems from the assumption of known variance. Obviously, if one knew the variance, the standard deviation is easily determined; and, more importantly, to calculate the variance one needs to know the value of the mean! Therefore, it is unrealistic to be estimating the mean under these conditions but it does provide the basis for further understanding the theory in more practical situations. Next we relax the requirement of the known standard deviation. Eventually, the small sample case with unknown variance is developed in detail where the student t distribution is appropriately used. Confidence intervals also are constructed for attribute data with large samples. No new theory is required for this latter case except the formula for the measurement of the sampling error changes.

Estimation of μ when σ is known (n is large)

This is the situation where we are attempting to estimate the parameter μ using the true standard error of the mean. The probability distribution that approximates our sampling distribution is the normal curve because n is large and we know the population standard deviation. Although this knowledge of the population variation may be somewhat unrealistic, there are cases where this information can be achieved based upon past experience.

For example, a study made by the staff officers of an Air Force wing indicated that the number of planes in operating condition on any one day may be looked upon as a value assumed by a random variable having a normal distribution with a population standard deviation of 200. If a random sample of 50 days indicated that there were on the average 2,000 planes in operating condition, construct a .95 confidence interval for the true daily average number of planes which this wing has in operating condition. Clearly, this is a problem of estimation of the population parameter where the sample size is considered large and the population standard deviation is known. Our estimate of the parameter might be as follows. First, we would determine the true standard error of the mean, using the Formula (9.1):

$$\sigma_{\bar{x}} = \frac{\sigma_x}{\sqrt{n}} = \frac{200}{\sqrt{50}}$$

which is approximately 28 planes. Therefore, our 95 percent confidence interval estimate would be $\bar{X} \pm z\sigma_{\bar{x}}$, or $2{,}000 \pm 1.96(28) = 1{,}945 - 2{,}055$ planes. Our *point estimate* of the parameter, of course, would be the sample mean, that is, 2,000 planes. The degree of reliability one may place in this point estimate is, of course, zero because the population mean is estimated from a sample by a single number; and in terms of the areas of the normal curve, one unique value has no area and therefore no probability. The reliability of a point estimate also is in question because it can be expected to contain an error due to sampling. We might interpret our confidence interval for the number of planes in operation as follows. We do not know what the true mean of the population is, but our best guess is that it is somewhere between 1,945 and 2,055 planes. The

FIGURE 9.12
Possible confidence intervals

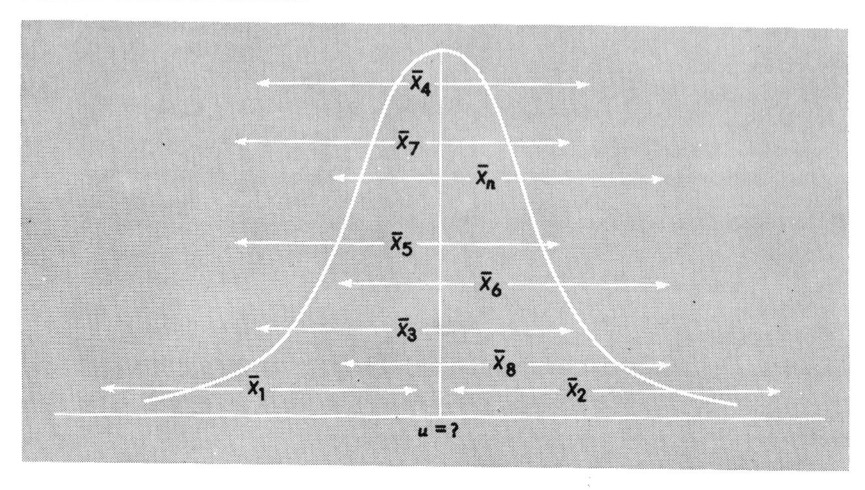

degree of reliability one may place in this statement is .95. That is, we act as if this statement is correct because a *procedure* has been used which would lead us to the correct statement 95 percent of the time. *If we were to take may samples of size 50 from this population and determine the average number of planes in operating condition in each case and construct confidence intervals in this same way using the same methods, 95 percent of these intervals would include the true mean.* (See Figure 9.12.)

It should be noted that we *cannot* say that the *true* mean is within ± 55 planes from the sample mean. We *can* say that the *sample* mean is within ± 55 planes of the true mean with a probability of .95. The reason this is true is that the sample mean varies, whereas the population mean remains a fixed constant even though it may be unknown.

What if we indicated that we would need to have a higher level of confidence, that is, use a .99 confidence interval? What impact would this have upon the estimate of μ? Given a confidence coefficient of .99, the standard normal deviate, $z = 2.58$. Therefore, our interval estimate would be $\bar{X} \pm 2.58\sigma_{\bar{x}}$, or $2{,}000 \pm 2.58(28)$, or $1{,}928 - 2{,}072$. It should be noted, of course, that the second interval estimate is wider than the first and therefore involves less risk of making an incorrect statement. It also is a less precise statement. However, one generalization may be made from this comparison. *For any given size sample the confidence interval can be narrowed only by increasing the risk of an incorrect statement, or the risk of an incorrect statement can be reduced only by widening the confidence interval.* Therefore the decision as to which level of confidence we choose to decide upon must be a compromise between the desirability

FIGURE 9.13
Flowcharting symbols used in graphic algorithm

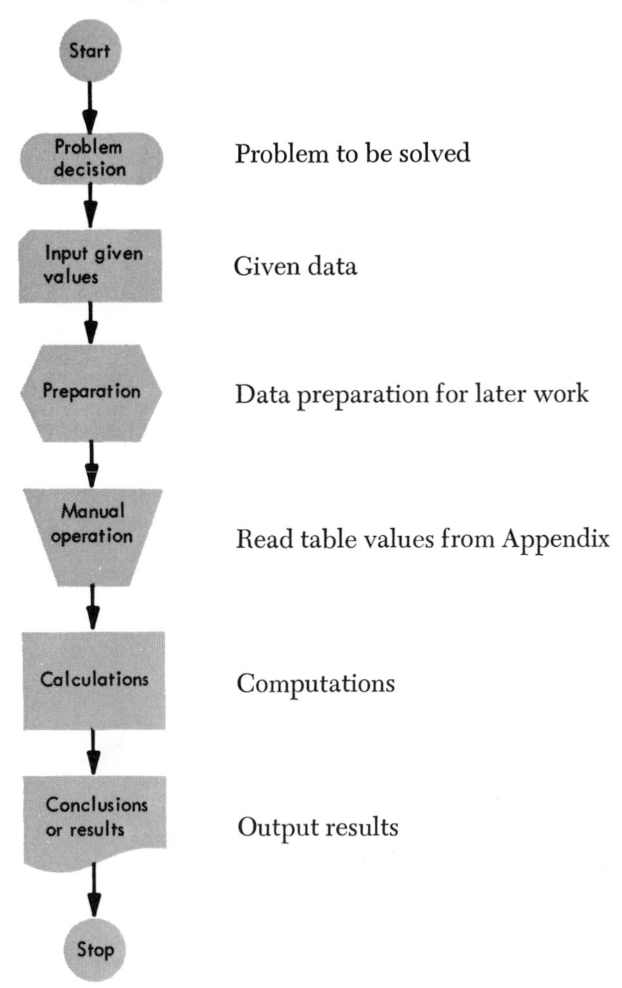

for precision (i.e., the width of the estimate) and the desirability for reliability. As someone has said: "The surer we want to be, the less we have to be sure about."

Figure 9.13 illustrates the typical flowcharting symbols to be used in this and following chapters where it is deemed appropriate to assist the student. As used here, the flowchart is nothing more than a graphic algorithm or visual presentation of procedures used in solving a particular problem. A more normal use of flowcharting techniques is as a guide to the development of a computer program. Perhaps, these charts will help

FIGURE 9.14

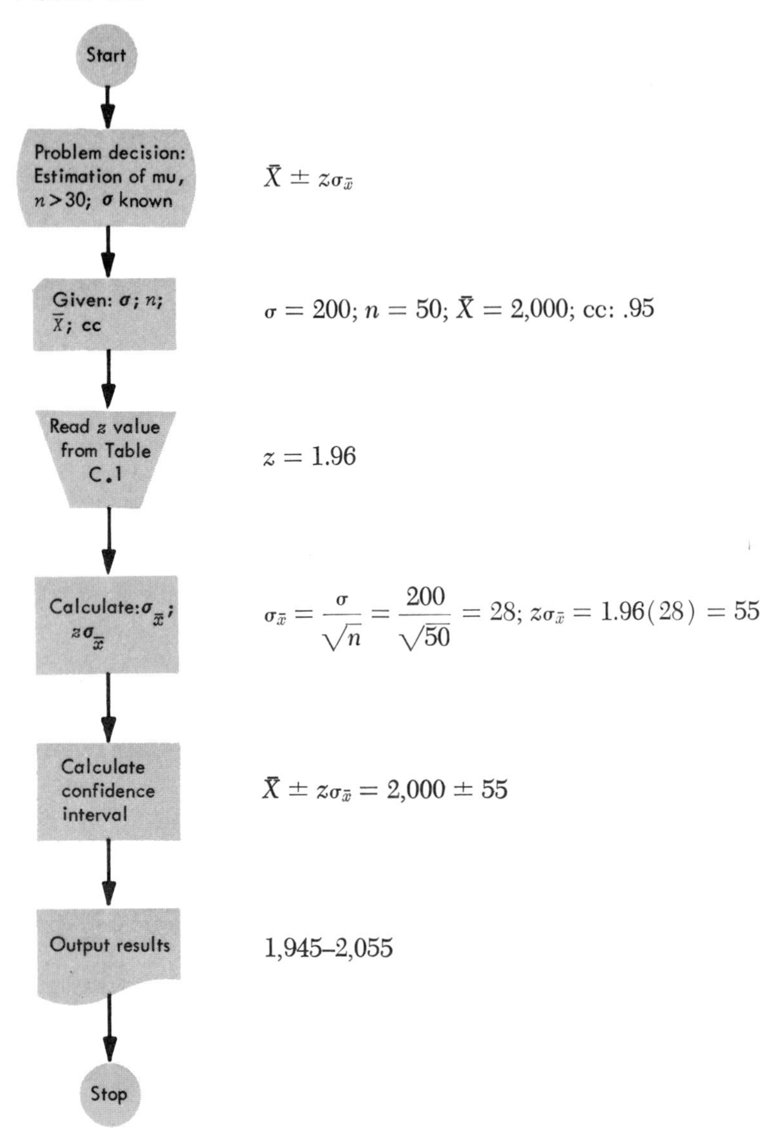

us to follow the instructions for an orderly handling of estimation and later more complicated problems. In all cases, the problem is worked in the text in the usual manner and the flowchart of the same solution is presented. Each individually shaped box in Figure 9.13 represents some procedure, activity, or instruction. *The flowcharts in this text are employed as a problem-structuring device and as a pedagogical tool to achieve a better understanding of the statistical procedures to be followed.*

Figure 9.14 graphically presents the solution to the Air Force problem of estimating the number of planes in operating condition on any one given day. The chart should be helpful in visualizing the procedures for the estimation of mu when our conclusions are based on a large sample and the population standard deviation is known. By following the steps in Figure 9.11, a logical solution to the problem evolves. The student now should be in a position to *structure* problems of estimation for measurement data for either large or small samples.

Estimation of μ when σ is unknown (n is large)

Here the procedure is virtually the same as in the above case except that we do not know the population standard deviation and therefore must substitute some estimate for it in our formula for the standard error of the mean. Even though the standard deviation is unknown, the z distribution is appropriate because n is large. Let us assume that a credit manager of a large department store wishes to evaluate the effectiveness of its credits and collections policy by estimating the average delinquent charge account. The credit manager does this by selecting a random sample of 100 delinquent charge accounts and calculates that the mean is \$58.14 with a sample standard deviation of \$15.30 (assume that this sample standard deviation has been corrected for the loss of one degree of freedom and therefore represents an estimate of the population standard deviation). Construct a .95 confidence interval for the actual average size of the delinquent charge accounts at this particular department store.

In this case, since we do not know the standard deviation of the population, we are using an estimate, that is, the standard deviation of the sample, corrected for the degrees of freedom lost in substituting $\bar{X}$ for μ. We now use Formula (9.2):

$$\hat{\sigma}_{\bar{x}} = \frac{\hat{\sigma}_x}{\sqrt{n}}$$

For our problem, the figures would be

$$\hat{\sigma}_{\bar{x}} = \frac{\$15.30}{\sqrt{100}} = \$1.53$$

This value, then, is the estimated standard error of the theoretical sampling distribution of all possible samples of size 100 from this given universe. Our 95 percent confidence estimate would be \$58.14 $\pm$ 1.96 (\$1.53), or \$55.14 to \$61.14. Again, our point estimate would be \$58.14. If we decided to give up some of the probability of being right and settled

FIGURE 9.15

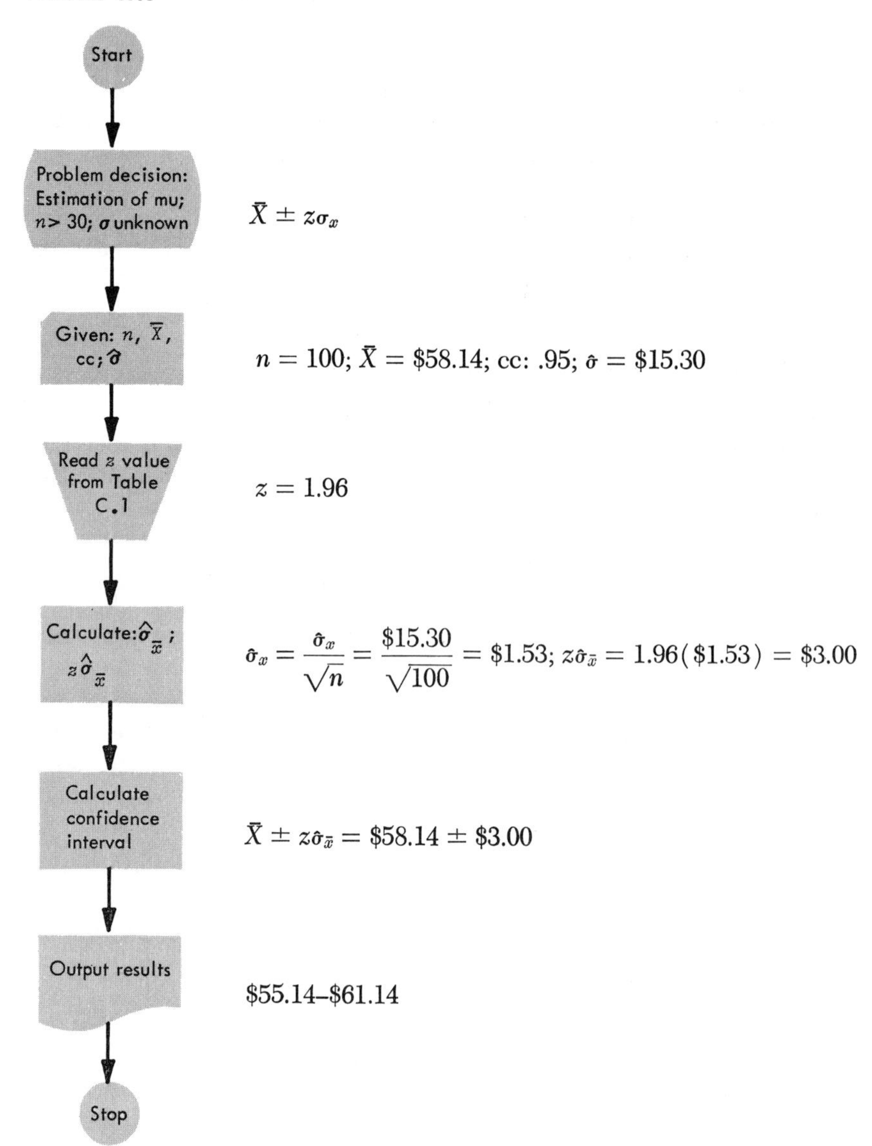

on a confidence coefficient of .90, our interval estimate would be \$58.14 ± 1.645 (\$1.53), or \$55.62 to \$60.66. In other words, in giving up some of the probability of being right, we have narrowed our interval estimate of the parameter. If we had insisted upon keeping the same confidence coefficient of .95 and also insisted on reducing our sampling error, that is, making a narrower confidence interval, we would have needed to increase our sample size. This fact will become clearer in sections that follow. (See Figure 9.15. Note: The use of z instead of t is justified even though σ is unknown because the two values happen to coincide when n is large. See below for discussion of use of t.)

Estimation of π when n is large

Let us now examine the case for attribute data where the task is to estimate the parameter π based upon the sample percentage p. Assume that a magazine publisher desires to know what proportion of the magazine's subscribers read the editorials. The publisher has an independent research agency conduct a survey utilizing a probability sample from the list of subscribers. It is found that of 200 subscribers, 120 read the editrials. Between what limits can the magazine publisher expect the true proportion to lie? Use a .90 confidence coefficient.

Obviously, here we do not know π, since this is our task—estimating this parameter. We must utilize the formula for the estimated standard error of a proportion, or Formula (9.8):

$$\hat{\sigma}_p = \sqrt{\frac{p(100 - p)}{n}}$$

For our example,

$$\hat{\sigma}_p = \sqrt{\frac{(60)(40)}{200}} = 3.46$$

where

$$p = \frac{120}{200} \times 100 = 60\%$$

The 90 percent confidence interval estimate, then, is equal to 60 ± 1.645 $(3.46) = 54$ to 66 percent. Again, our point estimate of the parameter π would be 60 percent with a zero degree of reliability. A better estimate might be that we do not know what the true proportion of subscribers who read the editorials really is; but *if we were to take many samples of size 200 from this universe and compute their sample proportions, and thereby construct many intervals using the same methods, 90 percent of these intervals would include the true proportion.* In other words, we act as though the above interval is one of the 90 percent. (See Figure 9.16.)

Finite correction

It should be noted that in case of either measurement or attribute data, if the sample size *exceeds 5 percent* of the population, we should use what

FIGURE 9.16

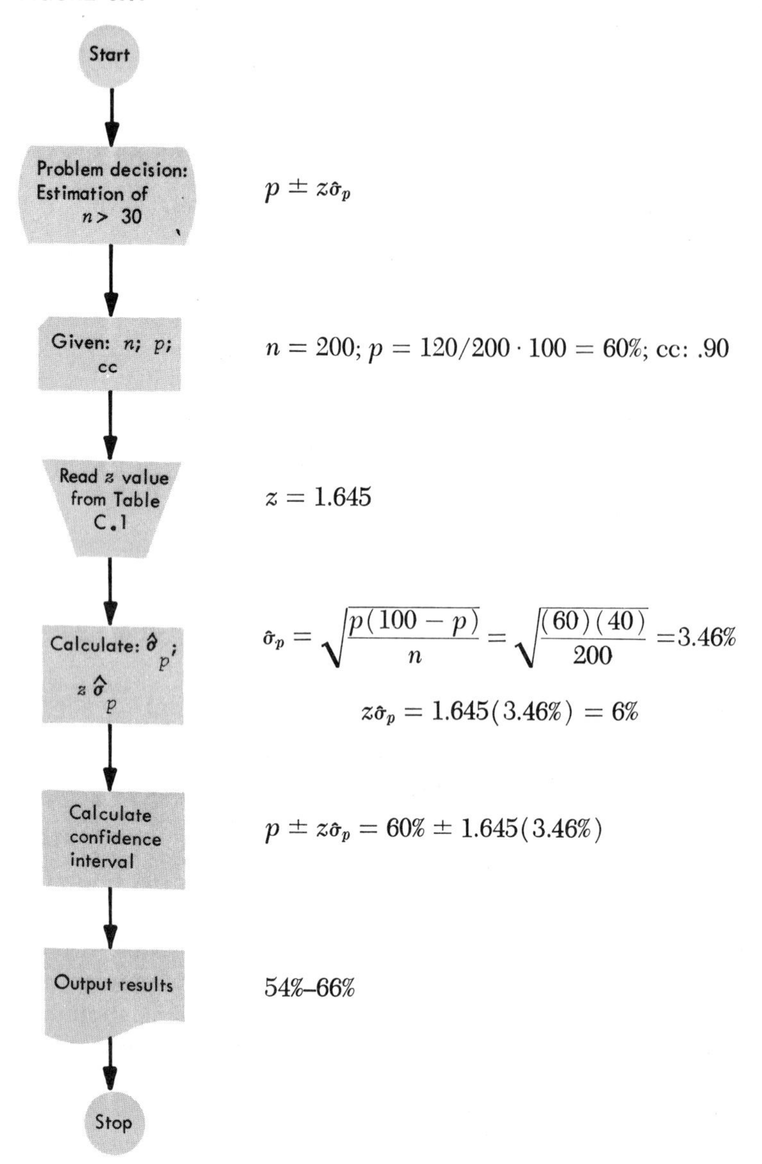

is referred to as the finite correction factor; that is, we should multiply the appropriate standard error formula by

$$\sqrt{\frac{N-n}{N-1}}, \quad \text{finite correction} \tag{9.11}$$

Obviously, where the sample size is small relative to the population, this correction factor will make a minor increase in the standard error. However, where the sample size exceeds 5 percent of the population, this correction factor would tend to reduce the size of the standard error.

For example, in the above illustration let us now assume that the publisher has 3,500 subscribers to a particular magazine. Then, our sample of 200 would exceed 5 percent of the total population and we should make use of the finite correction factor to avoid a slight overstatement of the sampling error. The finite correction would be:

$$\sqrt{\frac{3{,}500 - 200}{3{,}500 - 1}} = .97$$

Multiplying the estimated standard error of the percentage by this figure we obtain:

$$(3.46 \text{ percent})(.97) = 3.36 \text{ percent}$$

And, the new confidence interval would be:

$$60 \text{ percent} \pm 1.645(3.36 \text{ percent}) = 54.5\text{–}65.5 \text{ percent}$$

If the data are rounded according to the even-digits rule, the interval is 54 percent–66 percent, the same as the previously computed values. In other cases the use of the finite correction factor can become important; however, in analyzing much of the data of the social sciences, this additional precision may not be warranted.

STUDENT *t* DISTRIBUTION

Sometimes, it is necessary to work with less than 30 observations or with sample sizes that are considered small or, more importantly, *where the population standard deviation is unknown.* Previously, we indicated that means of random samples taken from a population, whether normal or not, approach a normal probability distribution centered around their mean. This is particularly true as the size of the sample becomes large. When the sample size is small, and especially when the population standard deviation is unknown, the probability distribution of the sample means is no longer closely approximated by the normal distribution. W. S. Gosset, who published under the pen name of Student, demonstrated that in such cases another theoretical distribution becomes applicable. This is the so-called "Student *t* distribution" and is found in Table D.1 in Appendix D. It is reported that Gosset was a statistician for a brewery, and that the management did not want him to publish his scholarly theoretical work under his real name and bring shame to his employer! Consequently, he selected the pen name of Student. The Student *t* distribution is symmetrical, but it is also flatter than the normal probability distribution. However, it does approach the normal curve as

FIGURE 9.17
Normal distribution

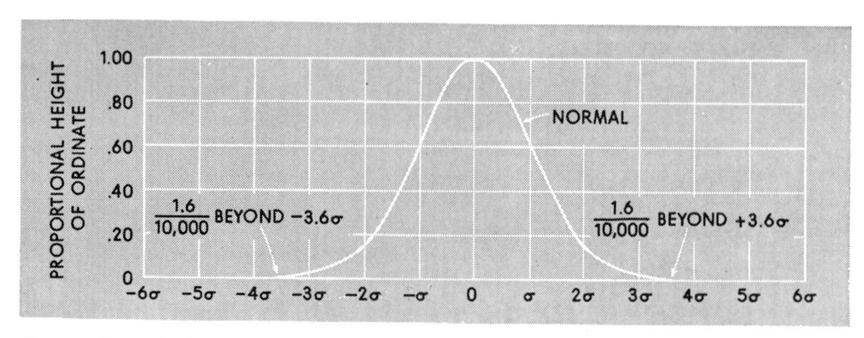

Comparison of the normal distribution and the *t* distribution when $n = 20$

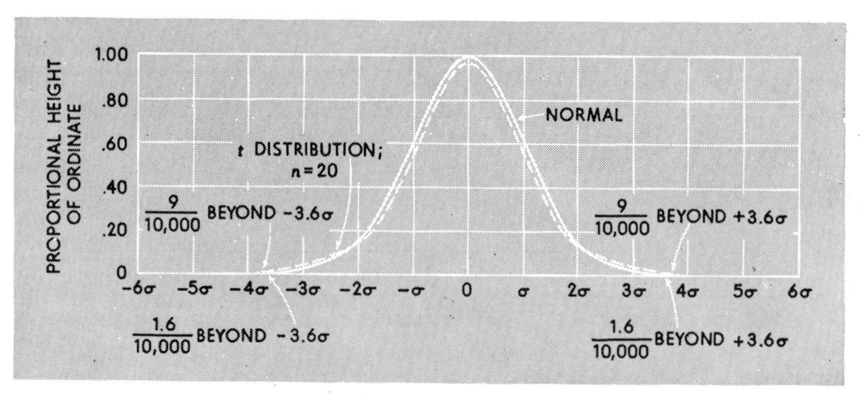

n becomes large; and as Table D.1 indicaes, when n becomes large, that is, greater than 120, the t values and z values coincide. (See Figure 9.17.) Because the t distribution is symmetrical, we need calculate only the areas for one half of the curve. Also, like the normal distribution, only the positive values are indicated in Table D.1. However, even this is quite an undertaking because t assumes different values, depending on the size of the sample. The point which distinguishes this theoretical distribution from the normal one is that the Student t distribution reflects the fact that the standard deviation of the sample mean was estimated. The t distribution is not a single distribution, as is the normal curve; rather, it is a family of distributions, one for each number of the degrees of freedom remaining.

If a variable X is normally distributed, then the statistic

$$z = \frac{x}{\sigma_x} = \frac{X - \mu}{\sigma_x} \tag{9.12}$$

has a normal distribution. But the statistic

$$t = \frac{x}{\hat{\sigma}_x} = \frac{X - \mu}{\hat{\sigma}_x} \tag{9.13}$$

has a t distribution, the sample standard deviation, s, being the square root of an unbiased estimate of σ_x^2. That is,

$$s = \sqrt{s^2} \quad \text{and} \quad \hat{\sigma}^2 = \Sigma(X - \bar{X})^2 \div (n - 1)$$

However, even though the variable X is not normally distributed, if the sample means $(\bar{X}_s)$ are normally distributed, then, of course, the statistic

$$z = \frac{x}{\sigma} = \frac{(X - \mu)}{\sigma_x}$$

has a normal distribution; the statistic

$$t = \frac{x}{\hat{\sigma}_{\bar{x}}} = \frac{X - \mu}{\hat{\sigma}_{\bar{x}}}$$

has a t distribution.

The difference between the t and the z cases may be illustrated as follows. When using the normal distribution, our interpretation of a confidence interval would reflect the fact that if many such intervals were to be computed, they would have differing midpoints simply because these midpoints are the sample means themselves, which would vary. But for any given large-size sample, these intervals would all be the same *width* because z is a constant depending upon the confidence coefficient selected. When using the t distribution in cases where the population standard deviation is unknown and the sample size is small, *both the midpoints and the widths* of the confidence intervals would vary if many such intervals were constructed. This is true because t is not a constant and varies for different size samples and for the different degrees of freedom. Also, $\hat{\sigma}_x$ varies from sample to sample.

An example. Assume that we are interested in knowing or estimating the average annual family expenditures for clothing in an area of relatively homogeneous single-family dwellings. A random sample of ten homes is selected; and on the basis of prolonged and costly interviews with the "head of the household," satisfactory total annual family clothing expenditure estimates are obtained. The $\bar{X}$ of these figures is \$838, with a standard deviation $(\hat{\sigma}_x)$ of \$110.

Question: What is the 95 percent confidence interval of the parameter, that is, μ, the true average annual expenditure of the area?

Data:

$n = 10$; $\bar{X} = \$838$; $\hat{\sigma}_x = \$110$
Confidence coefficient $= .95$
$t_{.95} = 2.262$; $\mu = ?$

Confidence Interval:

$$\bar{X} \pm t_{.95}\hat{\sigma}_{\bar{x}}$$

where

$$\hat{\sigma}_{\bar{x}} = \frac{\hat{\sigma}_x}{\sqrt{n}}$$
$$\hat{\sigma}_{\bar{x}} = \frac{\$110}{\sqrt{10}}$$
$$\hat{\sigma}_{\bar{x}} \simeq \$35$$

From Table D.1 in Appendix D we discover that $t_{.95}$ *for* $n - 1$ degrees of freedom is 2.262. This t value may be found by reading down the left column until 9 and reading across under the .05 column. If we had wanted $t_{.90}$, our table value would be 1.833, and for $t_{.99}$ the value would be 3.250. Students should check these last two values for practice in reading the table.

The .95 confidence interval is as follows:

$$\bar{X} \pm t_{.95}\hat{\sigma}_{\bar{x}}$$
$$\$838 \pm 2.262\ (\$35)$$
$$\$838 \pm \$79$$
$$\$759 - \$917$$

Interpretation: If many such samples were selected and if on the basis of this information a confidence interval were computed for each in the same manner, 95 percent of these intervals would include the true mean of the population. In reality, we act as though our interval, \$759–\$917, is one of the 95 percent. (See Figure 9.18.)

Comparison of z *and* t: In order to emphasize the contrast between these two theoretical distributions, Table 9.9 has been constructed. That is, in order to include the same percent of the area under the curve, we must go out further on the t scale than on the z scale. Also, although the z value will remain the same for a given confidence coefficient for *any large-size* sample, the t value will change with each size of sample for the same confidence coefficient. To illustrate, assume the following: $n = 15$; confidence coefficient = .95.

TABLE 9.9

Confidence Coefficient	z Value (n = Large)	t Value ($n = 10$)	t Value ($n = 5$)
.90	1.645	1.833	2.132
.95	1.96	2.262	2.776
.99	2.58	3.250	4.604

FIGURE 9.18

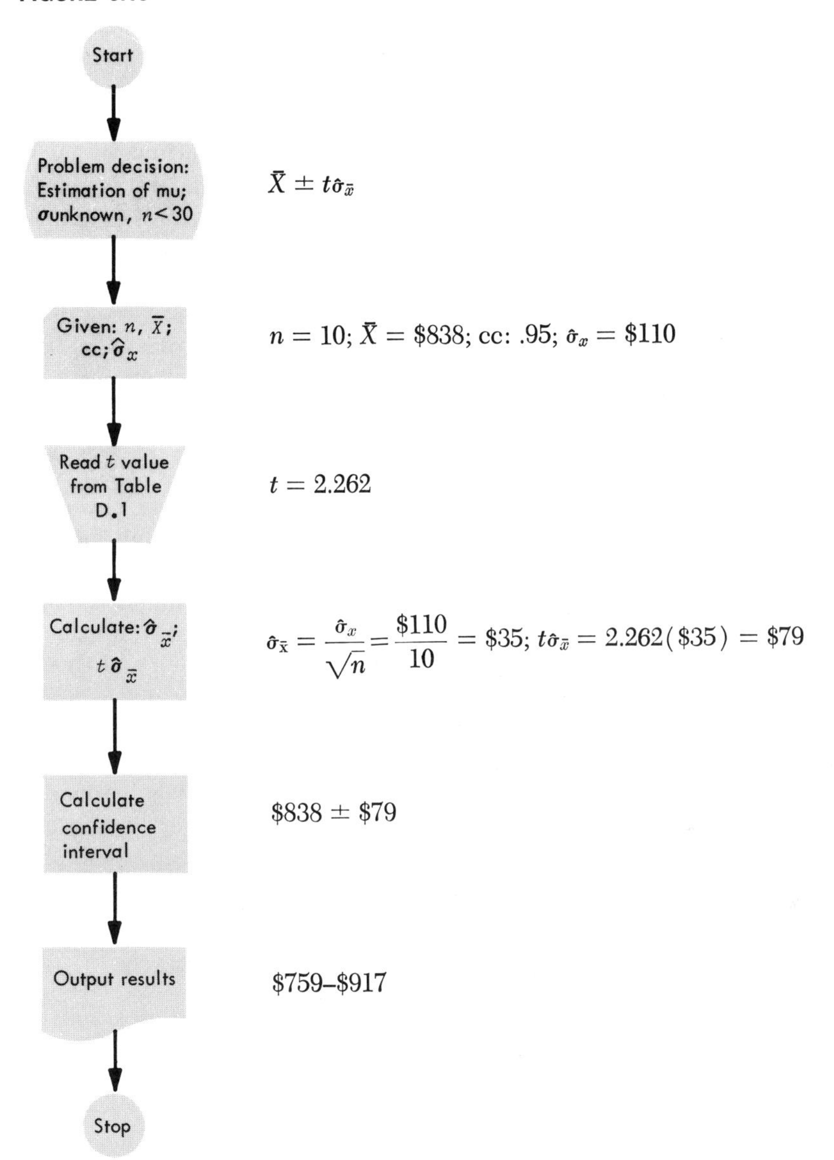

From Table D.1 in Appendix D, we can see that $t_{.95}$ for 14 degrees of freedom is 2.145. If n were 20 and the confidence interval remained at .95, then $t_{.95}$ would be 2.093. For large samples the z values would remain the same, that is, 1.96 for the .95 confidence interval. The student should check these values to be satisfied that this is true.

To test an hypothesis concerning a sample mean ($\bar{X}$), using either the normal distribution or the t distribution assumes that the sampling distribution of the $\bar{X}$ is normal. Therefore the only distinction to be made between the use of the normal curve and the t distribution is that the normal distribution is used with σ_x and the t distribution is used with the estimated standard deviation of the population ($\hat{\sigma}_x$). However, when $n \geqq 30$, the estimated standard deviation of the population approximately equals the population standard deviation, and the distribution approximates a normal curve. This can be seen in Figure 9.14. That is, as the sample size becomes larger and larger, the estimated standard deviation of the population approaches the true standard deviation of the population and the distribution of t approaches the normal probability distribution.

DEGREES OF FREEDOM

It should be noted here that the term "degrees of freedom" means the number of unrestricted chances for variation in the sampling measurements made. That is, the number of degrees of freedom in any given sampling situation is the number of measurements made minus the number of restrictions imposed in the random sampling process. For example, assume that a random sample of 30 observations is obtained and that the estimate of μ is desired. The mean is a constant for any given random sample. Consequently, $n - 1$ of the sample measurements could have been any values at all; but the one remaining value must then be that value which when added to the other $n - 1$ values equals n times the mean. Therefore, we say that one degree of freedom is lost in this case. That is, the restriction is the fixed value of the sample mean. (See Chapter 5.)

EFFECT OF SAMPLE SIZE ON PRECISION AND ACCURACY

In our examples of estimating a universe proportion or a universe mean on the basis of sample evidence, it became obvious that the precision of the estimate can be sharpened by giving up some probability of being right. On the other hand, the probability of being right can be increased by giving up some precision,[2] that is, by increasing the interval *range* in which it is inferred that the true universe lies. *The only way to increase both precision and reliability is to increase the size of the sample.* This is apparent from the fact that the sample size (n) appears in the denominator of both expressions for the standard error, which, of course,

[2] Precision is used here to relate to the width of the confidence interval, whereas reliability (or accuracy) relates to the confidence coefficient.

means that as the sample size increases, the standard error decreases. Consequently, for a given sample size the manner in which the conclusions or results are stated *always* involves striking a compromise between the desire for precision and the desire for reliability. In order to satisfy both demands more fully, the sample size must be increased. Clearly, this generally involves increasing costs. In any business problem the executive must weigh *alternative* costs. Perhaps, in the long run, it may be less expensive if more items are observed!

It is extremely important to note that doubling the sample size *does not* double the precision of the estimate. This follows from our formulas for the standard error, which indicate that this value is proportional to the *square root* of the sample size. Consequently, to double the precision of the estimate of the parameter (i.e., reduce the standard error by a factor of 2), the sample size must be increased *not* by a factor of 2 but by its square, 4. This is the concept of the *principle of decreasing variation,* that is, the size of the sampling error decreases as the square root of the number of items in the sample increases.

Effect of sample requirements on *n* for measurement data

Let us return to our previous example of the credit manager of the large department store. The data given in that problem were:

$\bar{X} = \$58.14$ (average delinquent charge account)
$n = 100;\ \hat{\sigma}_x = \$15.30;\ \hat{\sigma}_{\bar{x}} = \$1.53$
Confidence coefficient $= .95;\ z = 1.96$
$\bar{X} \pm z\hat{\sigma}_{\bar{x}} = \$58.14 \pm 1.96\ (\$1.53) = \$55.14 - \$61.14$
$z\hat{\sigma}_{\bar{x}} = 1.96\ (\$1.53) = \$3.00$

That is, the "sampling error" is $\pm\$3.00$ the way our results are stated. Assume that we desire to reduce this figure to $\pm\$1.50$, that is, cut the sampling error in half. Further, let us keep the same confidence coefficient or reliability. What effect would these demands have on the size of the sample required?

Where

$$d = z\hat{\sigma}_{\bar{x}} \qquad \text{and} \qquad \hat{\sigma}_{\bar{x}} = \frac{\hat{\sigma}_x}{\sqrt{n}}$$

then

$$d = z \cdot \frac{\hat{\sigma}_x}{\sqrt{n}} \tag{9.14}$$

Substituting given values in the latter formula,

$$\frac{1.50}{1.96} = \frac{15.30}{\sqrt{n}}$$

$$\sqrt{n} = \frac{15.30}{.7653}$$

Therefore, $n \simeq 400$. Thus, it is abundantly clear that if we wish to reduce the "sampling error" by one half, the original sample size must be increased by the square of 2, or $(2)^2(100) = 400$. (See Figure 9.19.) Alternatively, we might use the following:

$$n = \frac{z^2 \hat{\sigma}_x{}^2}{d^2} \tag{9.15}$$

where d is the desired "sampling error," that is, $\pm\$1.50$. Then,

$$n = \frac{(1.96)^2(15.30)^2}{(1.50)^2}$$

$$n = 400$$

In actual practice, 300 new observations probably would be taken and then combined with the original 100 items.

To illustrate further the impact on the needed sample size of heavy demands for high reliability, let us assume in our first example that we wish "to be right 99 percent of the time" instead of using the .95 confidence coefficient. Assume we are satisfied with a "sampling error" of $\pm\$3.00$. What do these requirements do to the sample size?

$$d = \$3.00;\ \text{confidence coefficient} = .9973$$

Then

$$z = 3.0$$

and

$$\frac{d}{z} = \frac{\hat{\sigma}_x}{\sqrt{n}}$$

Substituting,

$$\frac{3.00}{3.0} = \frac{15.30}{\sqrt{n}}$$

$$\sqrt{n} = 15.30$$

Therefore, $n \simeq 234$.

We readily see that by placing more stringent demands on the sample, that is, insisting on a higher probability, the number of observations needed is increased significantly (234 compared with 100). Conversely, if we were to give up some probability of being right and not insist on a very

FIGURE 9.19

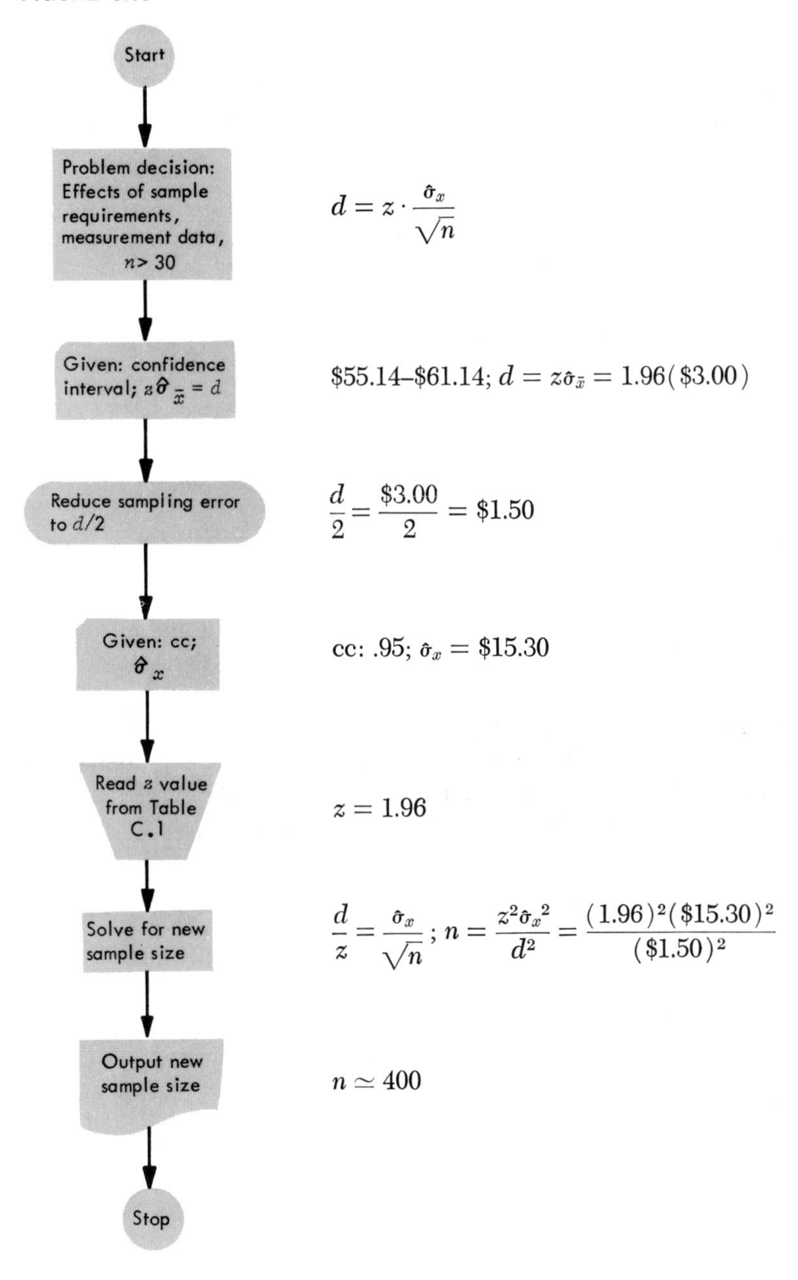
Start
Problem decision: Effects of sample requirements, measurement data, $n > 30$
$d = z \cdot \frac{\hat{\sigma}_x}{\sqrt{n}}$
Given: confidence interval; $z\hat{\sigma}_{\bar{x}} = d$
$55.14–$61.14; $d = z\hat{\sigma}_{\bar{x}} = 1.96(\$3.00)$
Reduce sampling error to $d/2$
$\frac{d}{2} = \frac{\$3.00}{2} = \$1.50$
Given: cc; $\hat{\sigma}_x$
cc: .95; $\hat{\sigma}_x = \$15.30$
Read z value from Table C.1
$z = 1.96$
Solve for new sample size
$\frac{d}{z} = \frac{\hat{\sigma}_x}{\sqrt{n}}; n = \frac{z^2\hat{\sigma}_x{}^2}{d^2} = \frac{(1.96)^2(\$15.30)^2}{(\$1.50)^2}$
Output new sample size
$n \simeq 400$
Stop

small "sampling error," the number of items needed in the sample would be reduced accordingly.

For example, if the credit manager indicated that a statement about the delinquent accounts would be satisfactory that was within ±$5.00 and was correct 90 percent of the time, our sample size would reflect these more lenient demands.

Where

$$d = \pm\$5.00;\ \text{confidence coefficient} = .90;\ z = 1.645$$

and

$$\frac{d}{z} = \frac{\hat{\sigma}_x}{\sqrt{n}}$$

then

$$\frac{5.00}{1.645} = \frac{15.30}{\sqrt{n}}$$

$$\sqrt{n} = \frac{15.30}{3.04}$$

$$n = \left(\frac{15.30}{3.04}\right)^2$$

Therefore, $n \simeq 25$.

In summary then, it is obvious that the size of the sample is directly related to the demands placed on it. The more insistent the researcher is for a relatively small sampling error and a high degree of reliability, the more observations will be needed. The reverse is also true.

Effect of sample requirements on *n* for attribute data

Let us recall the previous example of the magazine publisher where the basic information is

$$n = 200;\ p = 60\ \text{percent};\ \hat{\sigma}_p = 3.46\ \text{percent}$$

$$\text{Confidence coefficient} = .90;\ z = 1.645$$

$$\text{Confidence interval} = 60 \pm 1.645(3.46), \qquad \text{or 54–66 percent}$$

The sampling error is ±6 percent. Suppose the publisher wishes to reduce this error to ±3 percent, that is, reduce the error by one half. Further, let us assume the same confidence coefficient. What effect would these demands have on the sample size where

$$d = z\hat{\sigma}_p \qquad \text{and} \qquad \hat{\sigma}_p = \sqrt{\frac{p(100-p)}{n}}$$

See Figure 9.20.

Then,

$$d = z\sqrt{\frac{p(100 - p)}{n}}$$

and

$$d^2 = (z^2)\frac{[p(100 - p)]}{n}$$

$$n = \frac{z^2[p(100 - p)]}{d^2} \qquad (9.16)$$

Substituting, where $d = z\hat{\sigma}_p = 3$ percent,

$$n = \frac{(1.645)^2[60(100 - 60)]}{(3)^2}$$

$$n \simeq 722$$

To illustrate the impact of the demand for greater reliability, assume that we change the confidence coefficient to .95 but are satisfied with $d = 3$ percent.

Then, from Formula (9.16):

$$n = \frac{(1.96)^2[60(100 - 60)]}{(3)^2}$$

$$n \simeq 1025$$

Therefore the sample size again—just as with the mean—reflects the demands placed on it. Generally, the more we expect from the sample, the larger it must be.

Perhaps it is not realistic to use $p = 60$ percent in the above examples because, after all, p will vary from sample to sample. However, we must keep in mind that the example was used merely to illustrate the impact of the demands of the sample on n. It may be more realistic to use $\pi_0 =$ 50 percent in estimating the sample size because n will be at its largest when this is true. The reason is that $(\pi_0)(100 - \pi_0)$ is largest when $\pi_0 =$ 50 than when it is any other percentage. Consequently, to assure adequate sample size and meet the requirements placed on it, we should use $\pi_0 = 50$ percent. In our example, then, for confidence coefficient $= .90$ and $d = \pm 3$ percent:

$$n = \frac{(z)^2[\pi_0(100 - \pi_0)]}{(d)^2}$$

$$n = \frac{(1.645)^2[50(100 - 50)]}{(3)^2}$$

$$n \simeq 752 \qquad (9.17)$$

FIGURE 9.20

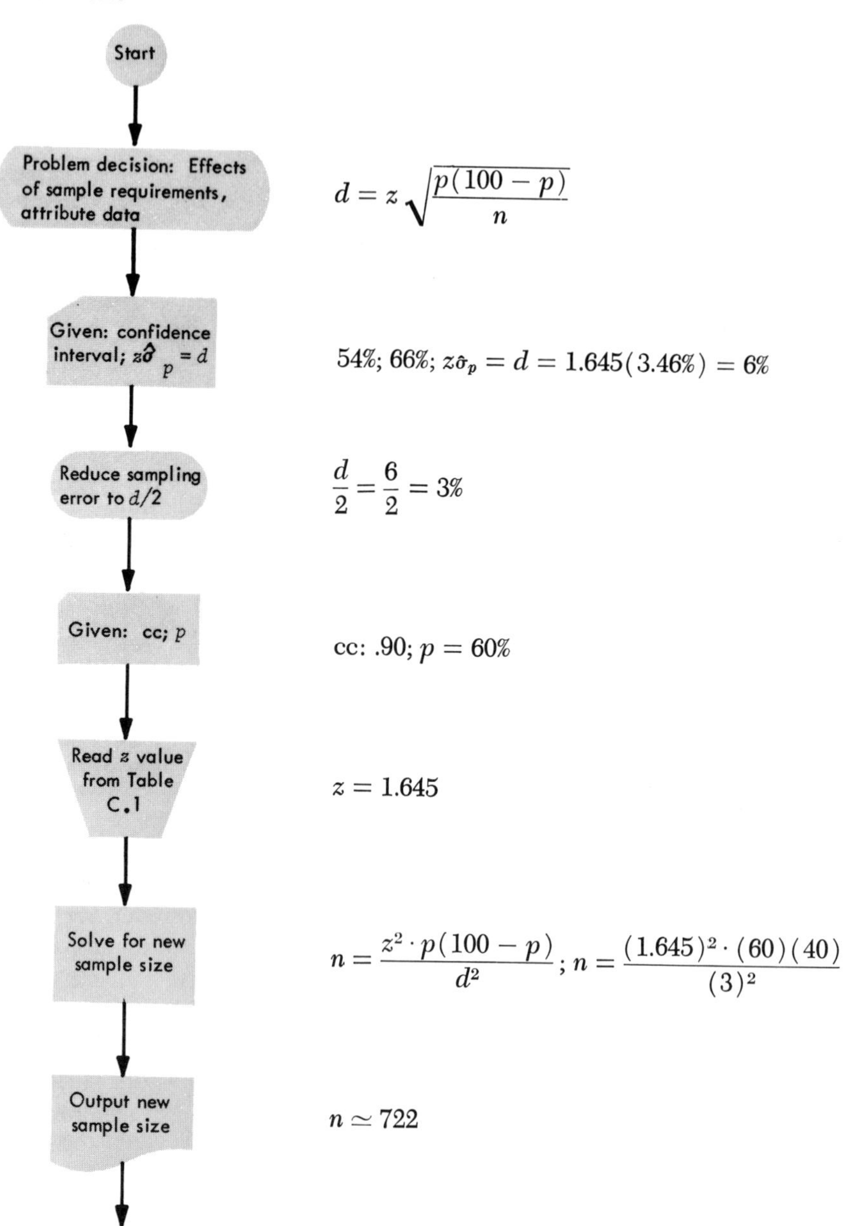
Start
Problem decision: Effects of sample requirements, attribute data
$d = z\sqrt{\frac{p(100-p)}{n}}$
Given: confidence interval; $z\hat{\sigma}_p = d$
54%; 66%; $z\hat{\sigma}_p = d = 1.645(3.46\%) = 6\%$
Reduce sampling error to $d/2$
$\frac{d}{2} = \frac{6}{2} = 3\%$
Given: cc; p
cc: .90; $p = 60\%$
Read z value from Table C.1
$z = 1.645$
Solve for new sample size
$n = \frac{z^2 \cdot p(100-p)}{d^2}$; $n = \frac{(1.645)^2 \cdot (60)(40)}{(3)^2}$
Output new sample size
$n \simeq 722$
Stop

Notice that n is larger here than the one determined above under similar demands; this is true because we used $\pi_0 = 50$ instead of the sample percentage 60 to insure n being adequate size.

Chapters 10 and 11 take up the other major problem of statistical inference. In this chapter, we have been considering theoretical sampling distributions and the problem of estimating a parameter on the basis of a statistic, that is, the sample evidence. We are now ready to be introduced to several other useful tools, namely, tests of hypotheses and operating characteristic curves (or power functions.)

SYMBOLS

π, population percentage
p, sample percentage
n, sample size
s_x, sample standard deviation
σ_x, or σ, population standard deviation
$\hat{\sigma}_x$, estimated population standard deviation for variable X

$\sigma_{\bar{x}} = \frac{\sigma_x}{\sqrt{n}}$, true standard error of mean

$\hat{\sigma}_{\bar{x}} = \frac{\hat{\sigma}_x}{\sqrt{n}} = \frac{s_x}{\sqrt{n-1}}$, estimated standard error of mean

σ_x^2, population variance for variable X
σ_x, population standard deviation for variable X
$\hat{\sigma}_x$, estimated population standard deviation for variable X
s_x, sample standard deviation for variable X

$$\hat{\sigma}_x = \sqrt{\frac{\Sigma f(X - \bar{X})^2}{n-1}} = s_x\sqrt{\frac{n}{n-1}}$$

$\bar{\bar{X}}$, grand mean of sampling distribution of sample means

$\sigma_{\bar{x}} = \sqrt{\frac{\sigma^2}{n} \cdot \frac{N}{N-1}\left(1 - \frac{n}{N}\right)}$, true standard error of mean

$\sigma_p = \sqrt{\frac{\pi(100-\pi)}{n}}$, true standard error of percentage

$\hat{\sigma}_p = \sqrt{\frac{p(100-p)}{n}}$, estimated standard error of percentage

$\sqrt{\frac{N-n}{N-1}}$, finite correction factor

$n = \frac{z^2\hat{\sigma}_x^{\,2}}{d^2}$, estimated sample size measurement data

$n = \frac{z^2 \cdot p(100-p)}{d^2}$, estimated sample size, attribute data

$$z = \frac{x}{\sigma_x} = \frac{X - \mu}{\sigma_x}$$

$$t = \frac{x}{\hat{\sigma}_x} = \frac{X - \mu}{\hat{\sigma}_x}$$

$$t = \frac{X - \mu}{\hat{\sigma}_{\bar{x}}}$$

PROBLEMS

9.1. Precisely and concisely define the following:

a. Point estimate.
b. Confidence interval.
c. Confidence coefficient.
d. Sampling distribution.
e. Parameter.
f. Statistic.
g. Standard error of the mean.

9.2. Compare and contrast attribute and measurement universes.

9.3. Given: $\mu = 120$; $\sigma = 20$; $n = 100$.

Required:

a. What is the probability that $\bar{X} \leq 118$?
b. What is the probability that $\bar{X} \geq 125$?

9.4. In an attempt to estimate the average amount of money spent each month by university students in a given university city, a market research firm selected a random sample of 200 students. The results were $\bar{X} = \$158$, the average amount spent, and $\hat{\sigma} = \$50$, the estimated population standard deviation of expenditures for all students. Answer the following:

a. What is your point estimate of average expenditures? Interpret this measure. What probability statement can you make as applied to this estimate?
b. What is the .95 confidence interval estimate of μ? Interpret this interval.
c. What is the .90 confidence interval estimate of μ?
d. What would be the .95 confidence interval if the sample size had been 100 instead of 200 but $\bar{X}$ and $\hat{\sigma}$ were the same?
e. How could we narrow our .95 confidence interval estimate?
f. What size sample would we need to select if we wanted to use a .99 confidence coefficient and reduce the sampling error of (*b*) by one half?

9.5. In an attempt to estimate the percentage of people in the Ann Arbor market area who prefer self-service hardware stores, a research firm conducted a probability sample survey of 300 families within the trading area. The results of the survey showed:

Attribute	Number of Respondents
Prefer self-service	210
Prefer use of clerks	90
	300

Required:

a. What is the estimated standard error of the percentage of families that prefer self-service hardware stores? Interpret this measure.
b. Using a .95 confidence coefficient, what is your interval estimate of the true proportion of respondents that prefer self-service? Interpret this interval.

9.6. M. Gauss, the advertising manager of the Alpha Manufacturing Corporation, claims that a particular mass-produced product of the company has an average life of five years. This statement is based on the evidence of a simulated test of a random sample of ten items which had a mean life of five years and a standard deviation of six months.

Required:

a. L. Bayes, the company statistician, questions Gauss' claim. On what grounds do you think Bayes has the audacity to be critical of the validity of this statement?

b. Help Bayes provide a better estimate of the average life of the product on the basis of the statistics of the sample.

9.7. Absenteeism remains a problem in some industries and has a bad effect on team productivity in some firms, resulting in higher costs. The Sigma Steel Corporation has a large number of production employees and generally considers a 10 percent daily absence rate as normal. One morning, R. Groan, the supervisor, noted that out of 625 production employees, 100 did not report for work. Is it probable that this absentee rate can be blamed on outside factors? (Assume an infinite universe.)

9.8. Given: $\bar{X} = 1{,}000$ hours; $s = 80$ hours; $n = 100$. What is your estimate of the universe mean, using a .95 confidence coefficient?

9.9. The average weekly earnings for a random sample of 225 skilled employees in a selected firm employing a large number of workers is $120. The sample standard deviation is $30.

Required:

a. What is your point estimate of the average weekly earnings of all skilled employees in this firm?

b. Using a .90 confidence coefficient, what is your interval estimate of μ? Interpret this interval.

c. The total weekly earnings for this week were stated to be $1,550,000 for 10,000 skilled employees. If you were an auditor inspecting these accounts, would you be suspicious of the discrepancy between the sample and the stated grand total?

9.10. The editor of a newspaper with a large circulation wishes to know what proportion of the paper's subscribers actually read the editorials. The newspaper has an independent research agency conduct a readership survey employing a probability sample from the list of subscribers. It is found that out of 600 subscribers, 90 read the editorials. Between what limits might the editor expect the true proportion to lie? Use a .99 confidence coefficient and interpret you interval.

9.11. The average monthly starting salary of a nationwide probability sample of 625 midyear MBA graduates was found to be $850, with a standard deviation of $30. Assuming that the job market outlook does not change significantly by June, what is the .95 confidence interval estimate of the true average starting MBA salary in the nation? Interpret this interval.

9.12. During the regular monthly audit in the payroll office of a large university a random sample of 300 records was spot-checked for accuracy. It was found that 30 vouchers contained errors. What inference can we make, using a .95 confidence coefficient?

9.13. Rule 2–3 of "The Rules of Golf" states that the velocity of a ball may not exceed 250 feet per second (fps) when tested by the United States Golf

Association (USGA) on a special machine used to measure this particular characteristic. The USGA allows a reasonable tolerance of ±2 percent. Assume that a random sample of 100 Long Drive golf balls was tested and that the average velocity was measured as 260 fps, with a sample standard deviation of 2 fps. Using a .99 confidence interval, what would be your estimate of the true mean fps of the total production run of these Long Drive balls? Would you say this particular production run meets Rule 2–3?

9.14. The True Shot Manufacturing Company is producing a new high-compression golf ball. A random sample of ten new balls is tested in order to estimate the true compression characteristics. The results of the sample are $X = 125$ and $s = 5$. What is your estimate of μ?

9.15. Manufacturers of golf balls are concerned with an important scientific concept called the coefficient of restitution, which is usually called "e." It is defined simply as the ratio of the relative velocity of the ball and the club after impact to the relative velocity before impact. Assume that a random sample of 64 golf balls is subjected to a test and it is found that the average "e" is .65, with a sample standard deviation of .16. What is your estimate of the universe value of "e," using a .90 confidence coefficient?

9.16. In order to meet the requirements of "The Rules of Golf," a golf ball must be no smaller than 1.680 inches in diameter. Assume that a random sample of ten balls is selected from a production run and measured. The results are $\bar{X} = 1.71$ inches and $s = .12$ inches. Using a .99 confidence coefficient, what is your estimate of the appropriate parameter? Would you say that this particular run meets the diameter standards established by the rules?

9.17. "The Rules of Golf" also states that a golf ball should be no heavier than 1.620 ounces. Assume that a random sample of 100 balls is selected from a production run and weighed. The sample results are $\bar{X} = 1.694$ ounces and $s = .09$ ounces. Using a .95 confidence coefficient, what is your estimate of the appropriate parameter? Would you say that this particular run meets the weight standards established by the rules?

9.18. A random sample of 225 families in a city indicates that they average 150 minutes of television viewing per day, with a standard deviation of 30 minutes. What are the 90 percent confidence limits for the true average viewing time of all families in this city?

9.19. A monthly sample of 625 families in a large metropolitan area indicates that on the average, 10 percent of the unemployed persons have been out of a job for at least 6 six weeks or more. Using a 95 percent confidence coefficient, estimate the parameter for the area. Interpret your interval estimate.

9.20. On the basis of a random sample 9 fines paid in the local traffic court, it is concluded at the 90 percent confidence level that the true average fine paid for traffic offenses lies between \$5 and \$8. What was the value of the estimated population standard deviation used in determining this interval?

9.21. A has averaged 80 strokes per round of golf for several years and notes that B has averaged 83 strokes for the last 10 rounds of 18 holes. The standard deviation of B's last 10 rounds is 4 strokes. How many strokes handicap per round can A give B and still feel 90 percent certain of winning in the long run?

9.22. On the basis of four brief tests, an applicant for a secretarial position makes the following record:

Dictation:	*Typing:*
$\bar{X} = 110$ words per minute (wpm)	$\bar{X} = 70$ wpm
$\hat{\sigma}_x = 10$ wpm	$\sigma_x = 5$ wpm

Required:

a. What would you estimate the applicant's true mean dictation and typing speeds to be, using a 90 percent confidence coefficient?

b. The applicant had indicated on the application blank prior to taking any tests that the applicant had a dictation speed of 120 wpm and a typing speed of 75 wpm. Do these estimates seem reasonable, based on your evidence?

9.23. "About 4 out of every 10 Michigan motorists—nearly 200,000 in 1975—are driving illegally," a high state official proclaimed.

A spot check of 500 operator license renewals showed that nearly 40 percent had expired, the official disclosed. "One applicant had been driving for almost two years on an invalid license," the official stated.

"On the basis of this spot check, I estimate that about 200,000 motorists driving in Michigan probably are not aware that their licenses have expired," the official concluded.

Required:

a. On the basis of your knowledge of sampling theory, comment on this actual news release.

b. What suggestions could you offer for improving the official's "estimate"?

9.24. Assume that a buyer wishes to submit a bid on some government surplus property which has been stored for some time under somewhat less than perfect conditions. The buyer wishes to come within ±5 percent of the true percentage of usable items. Also, the buyer wants to be sure about 90 percent of the time. The government will state only that the percentage of usable items is somewhere between 25 and 75 percent, but buyers may take samples to form their own opinion.

Required:

a. What size sample should the buyer select to *assure* an estimate with the desired precision?

b. The buyer later decides to assume more risk and tells the statistician that "if we come within ±10 percent, I can still make a reasonable profit." Assume the same confidence coefficient as in (*a*). What size sample would you suggest?

c. Can the precision of the samples in (*a*) and (*b*) be evaluated if the seller's estimate of the 25–75 percent is incorrect? Why, or why not?

9.25. Given: $d = \pm 5$ percent; $\pi_0 = 50$ percent; $z = 1.96$.

Required:

a. Determine n.

b. Determine n if $d = \pm 2.5$ percent and other given data remain the same.
c. Determine n if $d = \pm 5$ percent, $\pi_0 = 50$ percent, and $z = 2.58$.
d. Determine n if $d = \pm 2.5$ percent, $\pi_0 = 50$ percent, and $z = 2.58$.

9.26. Given: $\hat{\sigma}_x = 40$ units; $\bar{X} = 140$ units; $n = 100$.

Required:

a. Determine the confidence interval for the true parameter if the confidence coefficient = .9973.
b. Assume that you wish to reduce the sampling error in (*a*) to one fourth but keep the same confidence coefficient. Determine the new n.
c. Assume that you wish to reduce the sampling error in (*a*) to one fourth but also change the confidence coefficient to .90. Determine n.
d. Assume that the sampling error of (*a*) is permitted to be doubled and the confidence coefficient is reduced to .68268. Determine n.
e. Are the sample sizes equal in the above cases? Why, or why not?

9.27. As director of marketing research for the Dutch Chocolate Company, your task is to measure consumer reaction to a newly created home-baking chocolate by having potential users run a comparison test with a sample of its probable principle competitor, brand X.

Required:

a. Assume that you intend to run a market test for a random sample of consumers in the Ann Arbor area. You wish to determine potential users' preference within a ±6 percentage points with a confidence coefficient of 90 percent. Assuming that a priori one half of the consumers favor your product, how large a sample should you plan on working with?
b. Now let's assume that you wish to cut the sampling error to one half what was allowed in (*a*) above. That is, you wish to estimate the potential users' preference within a ±3 percentage points. What size sample would you need if you maintained a 90 percent reliability?
c. What size sample would you need if the confidence coefficient of part (*b*) were changed to 99.73 and the other data remained the same? What has happened to your sample size requirements? What about probable survey costs?
d. If your a priori guess about the consumer preference for your product is in error, what consequences does this forebode? Explain carefully.
e. If your sample in part (*c*) yields a sample proportion of percent favoring your product, are you safe in assuming that there probably is an actual preference for the Dutch Chocolate Company's new product? Explain.

9.28. A new enzyme is being developed and considered for production uses by the Now Chemical Corporation. The hope is that the new enzyme when used in a certain manufacturing process will act as a catalytic agent in increasing the output of acceptable chemicals. The important measure of the efficiency of the new enzyme was conceived as the ratio of "actual yield to the theoretical yield," the latter computed on the basis of past experience with an "old" enzyme. This measure is called the Yield Efficiency Production Index (YEPI). A YEPI of 105.7 would be interpreted to mean that 5.7 percent more acceptable chemicals were obtained from the batch than past experience would have predicted. The new enzyme was used in 50 batches of chemicals and an average yield of 132.8 with a standard deviation of 21.0 was computed.

Required:

a. Determine a 99 percent confidence interval for the true YEPI.

b. Would you conclude that the true YEPI obtained by the use of the new enzyme is above 100? Explain. Is it likely to be above 140?

c. One obviously nonstatistical member of the management team evaluating this test concluded: ". . . in my judgment, the standard deviation of 21.0 represents almost two thirds of the difference between 132.8 and 100.0. Therefore, there is no statistical evidence to support the contention that the new enzyme increases the output of acceptable chemicals." As a sophisticated statistician, how would you respond to this "argument"?

9.29. It was proposed that the board of directors of the Gamma Corporation include a "public" member. A random sample of 1,000 stockholders of the corporation were surveyed to determine their opinion on this proposal. Sixty-two percent of the stockholders favored the proposal while 38 percent were against it.

Required:

a. Construct a 90 percent confidence interval for the true proportion of stockholders who are in favor of having a public member on the board.

b. What is the meaning of this interval?

9.30. An efficiency expert of the Goodoldstat Corporation made 100 random observations on some production workers. In 42 observations the efficiency expert discovered that the worker was not busy, that is, the worker was waiting for a part or something.

Required:

a. Construct an 80 percent confidence interval for the proportion of the time the worker is idle.

b. Assume that Goodoldstat Corporation has 1,200 such workers; would your interval change given this additional information?

9.31. In a probability sample of 200 families in the city of Bogeyville, the mean disposable family income was estimated to be $8,250 with a standard deviation of $650. In a similar survey in Double Eagle City, the figures were $9,100 and $700 respectively.

Required:

a. Construct a 90 percent confidence interval for the family disposable incomes in Bogeyville.

b. Construct a 90 percent confidence interval for the family disposable incomes in Double Eagle City.

c. On the basis of the statistical techniques studied in this chapter, what would be your *intuitive* guess as to the differences in actual incomes in the two cities?

9.32. A probability sample of 400 citizens in a large community indicated that 240 favored that the planning commission grant approval for the construction of Briarwood Shopping Center. Use these data to establish the

90 percent confidence limits for the true proportion of the population who support the center.

9.33. Experience with skilled workers in the auto industry shows that the time required for the completion of a given job is approximately normally distributed with an estimated population standard deviation of ten minutes.

Required:

a. How large a sample would one need to select randomly if you wished to estimate the true population mean with no more sampling error than a plus or minus two minutes? Use a confidence coefficient of .95450.

b. What size sample would be required if the maximum error of estimate were to be within plus or minus four minutes with a probability of .95450 of being correct?

c. Given that the sample size you determined in (*b*) above yields a sample mean of 70 minutes, what are the .90 confidence limits for mu?

9.34. *a.* $\bar{X}$ is to mu as s is to what?

b. $\bar{X}$ is to mu as p is to what?

c. σ is to $\sigma_{\bar{x}}$ as $\hat{\sigma}$ is to what?

9.35. If $p = 50$ percent, and other things are equal, then the standard error of the percentage will be larger than if $p = 80$ percent. Is this statement correct? Why or why not?

9.36. The student council manages a student lounge in a heavy traffic area on campus. One member of the council believes strongly (an advertising major) that an attractive advertisement on a vending machine will increase sales. To test this claim, it was found that without the ad the average (mean) sales were $200 for 30 days, with an estimated population standard deviation of $10. The mean sales for the next 30 days when the ad was used turned out to be $205. Based on the last sample, how large would the estimated population standard deviation have to be in order to accept the hypothesis that the ad did not, in fact, increase the sales significantly? Use an alpha = .05. (Students should try to solve the problem intuitively but if not, then refer to Chapter 10.)

9.37. It is estimated that the average life of an American-made auto is eight years with a standard deviation of two years. If the average auto travels 12,000 miles per year and costs 10 cents per mile to run, what percent of the autos costs more than $7,200?

9.38. A probability sample of 26 college graduates earned an average (mean) income of $2,500 more per year than the average earnings of noncollege students. The sample standard deviation was $250. What are the 90 percent confidence limits of the mean amount by which college graduates' annual income exceeds the nongraduate group?

9.39. Use the information in Problem 9.38 and the following additional facts: (1) a college education costs approximately $12,000 for four years, and (2) the income forgone during the four years while attending college averages about $20,000 total. How long after graduation from college does it take the extra education to "begin paying off" in terms of dollars? Be .90 certain that it does not take longer than your answer.

9.40. Given the information at right concerning the annual taxable incomes of a group of five individuals, answer the questions below.

Individual	Income
A	$ 8,000
B	7,000
C	11,000
D	10,000
E	9,000

a. Construct a table showing all possible combinations of samples of two items drawn from the above population. Assume a simple random sample is the type used and that the first item selected is not replaced prior to drawing the second item.

b. Construct a table showing the frequency array and the probability of occurrence of the sample means from your table in (*a*) above.

c. What is the probability that the mean income of a sample of two items drawn from this population will be $8,000; $9,000; or some value other than $8,000 or $9,000?

9.41. Using the information generated by your response to Problem 9.40, demonstrate that the standard error of the mean is equal to the standard deviation of all possible sample means about the mean of the population.

9.42. Referring to Problem 9.40, how did you utilize the concept of a probability distribution of the sample mean (sampling distribution) in arriving at your responses to part (*c*)?

9.43. Again refer to Problem 9.40; answer the following:

a. Would the sampling distribution associated with a simple random sample of three individuals from the above population differ from the sampling distribution of the sample mean associated with the random sample of size 2 as you used? Explain why.

b. If the sample in Problem 9.40 were a nonprobability type rather than a simple random sample would your answers to that problem still be applicable? Why, or why not?

RELATED READINGS

Barnard, G. A.; Kiefer, J. C.; LeCam, L. M.; and Savage, L. J. "Statistical Inference." *The Future of Statistics,* pp. 139–62. New York: Academic Press, 1968.

Frier, J. Brown. "The Estimation of Knowledge by Multiple Choice Tests." *The American Statistician,* vol. 22 (October 1968), pp. 35–36.

Gardner, Martin. "A New Pencil-and-Paper Game Based on Inductive Reasoning." *Scientific American* (November 1969), pp. 140–46.

Guenther, William C. "Shortest Confidence Intervals." *The American Statistician,* vol. 23 (February 1969) pp. 22–25.

Hartley, H. O., and Rao, J. K. N. "A New Estimation Theory for Sample Surveys, II." *New Developments in Survey Sampling,* pp. 147–69. New York: John Wiley & Sons, Inc., 1969.

Springer, C. H.; Herlihy, R. E.; and Beggs, R. I. *Statistical Inference,* chaps. 6–8. Mathematics for Management Series, vol. III. Homewood, Ill.: Richard D. Irwin, Inc., 1966.

10

Tests of hypotheses

Once to every man and nation comes the moment to decide.

J. R. Lowell

In utilizing the theory of statistical inference, we select a random sample and compute a particular statistic. On the basis of the value of that statistic, we form some conclusion (estimation), or we make a decision in connection with the population from which the sample was drawn (tests of hypotheses). Normally, this decision involves the choice between two alternative courses of action. This is done with the aid of our mathematical model, which is a theoretical probability distribution. It is the sampling distribution of the statistic with which we are dealing. That is, it is the *theoretical* distribution which would result if we were to draw many samples of a given size and compute the corresponding statistics. The *one*—and it is only one—sample and one statistic with which we are dealing in the "real world" is then considered in the light of that theoretical distribution.

This is done in order to *estimate* the true value of the parameter (i.e., the population parameter) corresponding to the statistic (i.e., the sample value). Or we use the theory to *test the hypothesis* that the statistic is a member of that sampling distribution with parameters stipulated by that prior hypothesis. Obviously, the parameters of the theoretical sampling distribution are determined by the characteristics (parameters) of the hypothesized population. Therefore, this procedure we are about to encounter really is, in effect, testing an hypothesis about the parent universe.

Although situations requiring the procedures of estimation arise frequently, sampling problems also persist where the major task is to decide whether or not different samples came from the same universe, or if a given sample came from a known or hypothesized population. The latter, of course, is accomplished by testing to determine if there is a significant difference between the result obtained by sampling and an established, or assumed, standard. Clearly, this hypothesized standard might very well be the mean or proportion of a given universe.

If we discover a "relatively small difference" between the mean (or proportion) of a sample and the established universe mean (or proportion), it is clear from our knowledge of sampling variation that the observed difference easily may have occurred by mere chance. If this is true, we conclude that the "difference is not significant," that is, it could have arisen simply as a result of sampling variation. On the other hand, if there is a "relatively large difference" between a statistic and the hypothesized parameter, it is probable that the sample did not come from the population with the assumed characteristic. In this situation, we conclude that the "difference is significant," that is, caused by factors other than pure chance. It remains for us in the examples which follow to quantify the differences "relatively small" and "relatively large" as used above.

Application of statistical decision-making rules to select the appropriate alternative normally is referred to as testing an hypothesis, the reason being that each alternative course of action may be viewed as an hypothesis or solution to the problem. In our tests, we usually begin with the assumption that a sample did indeed come from an hypothesized universe. This assumption is called the *null hypothesis* or *zero hypothesis* (H_0), which except for chance variations of the sample would lead to zero differences. Then, observed differences between the sample and the hypothesized population are studied to evaluate the probability that they resulted by chance alone. Previously, the statistician must have decided on precisely how small the chance must be before the difference is arbitrarily called significant and the null hypothesis is rejected. That is, we conclude that the sample came from a universe that has a parameter different from our hypothesized value. For example, if we determine that the probability is as low as 6 out of 100 (.06), the statistician may not reject the null hypothesis. But if the probability of a chance occurrence is 5 out of 100 (.05) or less, the statistician may begin to believe that the chance factor is probably not strong enough to explain the observed difference. Some factor other than chance has been operating to cause the difference. In this latter case the statistician concludes that the difference between the sample result and the hypothesized universe is significant; therefore, the null hypothesis is rejected. In some cases the risk of rejecting the null hypothesis is great or perhaps too costly, and the researcher may not consider an observed difference to be significant unless the probability of chance variation is as low as 1 out of 100 (.01), or even 1 out of 1,000 (.001).

The choice as to the level of significance that will be chosen in reaching a decision that an observed difference is significant (from a sampling point of view) is largely a matter of judgment. Unless there is complete knowledge about the universe, no one can say a particular result could not have arisen by chance. Obviously, if complete knowledge about a population existed, there would be no reason to draw a sample! *Consequently, there is always the possibility that a sample result did arise by chance, even though the probability may be small.* The theoretical normal curve, which describes the sampling distribution (e.g., the probability distribution of sample means) approaches the base line ever more closely as the curve spreads out toward infinity. However, the curve never actually touches the base line. In other words, *in theory*, there is no final limit to the possibility of a chance occurrence. In the real world, for example, a gambler conceivably may throw a 7 on each of 100 successive throws of a pair of dice; however, the chance that this will happen is infinitesimally tiny, that is $(1/6)^{100}$. In real life an opponent probably would decide by the fifth or sixth consecutive throw of a 7 that *some factor other than chance* was affecting the outcomes! Statistically, one would say that the "difference is significant; it is too great to have occurred by chance."

Briefly, then, prior to using a statistical decision-making rule, we begin with a statement of the acceptable alternative courses of action. Simultaneously, criteria are established for choosing between these alternative courses of action. The so-called "statistical rule" simply is employed in the last step when the sample evidence is evaluated in accordance with the criteria for choosing a course of action.

KINDS OF TESTS OF HYPOTHESES FOR VARIOUS ALTERNATIVES

Basically, there are three kinds of problems which will concern us in our tests of hypotheses about a population mean or proportion. They include (1) two-sided tests, (2) one-sided alternatives to the right, and (3) one-sided alternatives to the left. (See Figures 10.1 and 10.2.)

Where one is concerned about the problem of detecting whether or not a parameter has either increased or decreased, the two-sided test is the appropriate one. If the researcher is interested in detecting only if the sample came from a population that has as its parameter one that is larger than the hypothesized parameter, then the one-sided test to the right is useful. The one-sided test to the left is used where we are concerned about discovering a decrease in the parameter. That is, does the sample come from a population whose parameter is less than the hypothesized value?

CONCEPT OF THE NULL HYPOTHESIS

A statistical null hypothesis is a statement about a population, and the test is used to decide whether or not to accept the hypothesis. The procedure of the test is to use sample data to invalidate or accept the null

FIGURE 10.1
Three kinds of tests of hypotheses for the various alternatives concerning a population mean

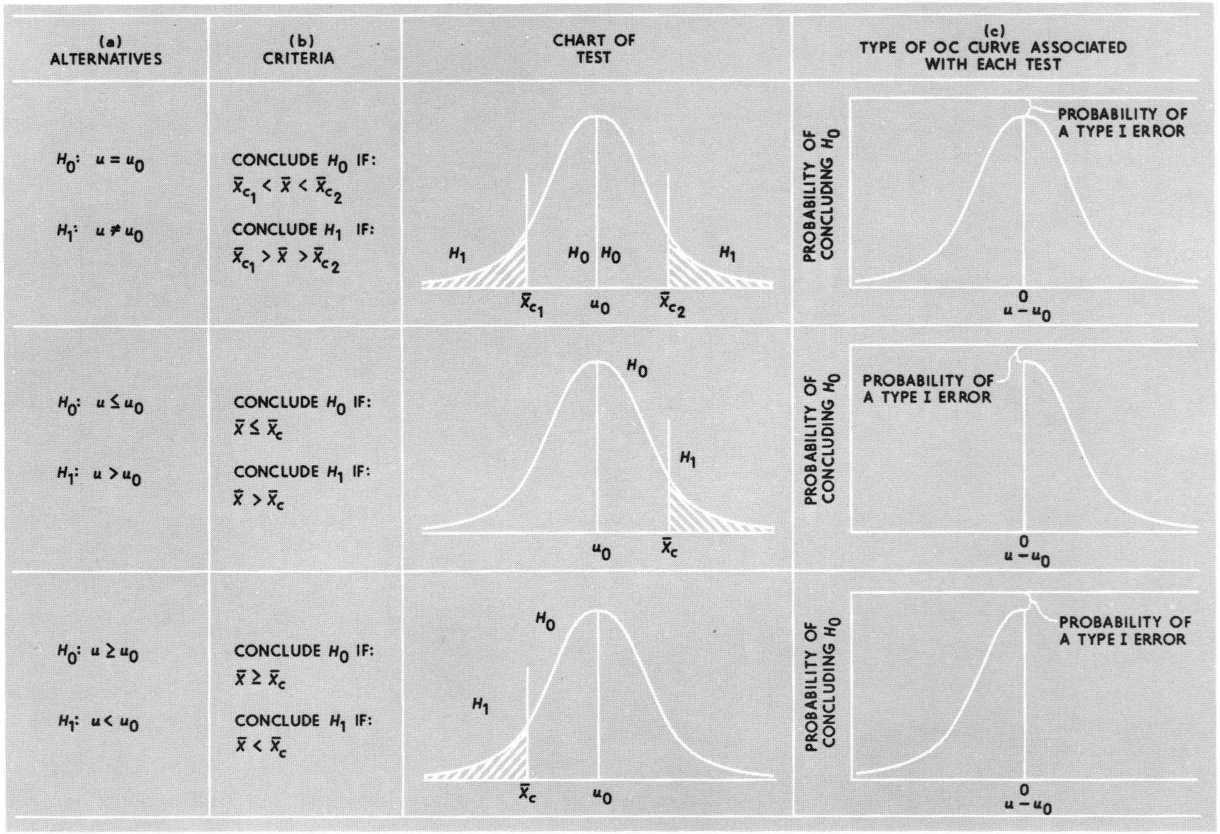

(a) ALTERNATIVES	(b) CRITERIA	CHART OF TEST	(c) TYPE OF OC CURVE ASSOCIATED WITH EACH TEST
H_0: $u = u_0$ H_1: $u \neq u_0$	CONCLUDE H_0 IF: $\bar{X}_{c_1} < \bar{X} < \bar{X}_{c_2}$ CONCLUDE H_1 IF: $\bar{X}_{c_1} > \bar{X} > \bar{X}_{c_2}$	H_1 H_0 H_0 H_1 $\bar{X}_{c_1}$ u_0 $\bar{X}_{c_2}$	PROBABILITY OF CONCLUDING H_0 PROBABILITY OF A TYPE I ERROR 0 $u - u_0$
H_0: $u \leq u_0$ H_1: $u > u_0$	CONCLUDE H_0 IF: $\bar{X} \leq \bar{X}_c$ CONCLUDE H_1 IF: $\bar{X} > \bar{X}_c$	H_0 H_1 u_0 $\bar{X}_c$	PROBABILITY OF CONCLUDING H_0 PROBABILITY OF A TYPE I ERROR 0 $u - u_0$
H_0: $u \geq u_0$ H_1: $u < u_0$	CONCLUDE H_0 IF: $\bar{X} \geq \bar{X}_c$ CONCLUDE H_1 IF: $\bar{X} < \bar{X}_c$	H_0 H_1 $\bar{X}_c$ u_0	PROBABILITY OF CONCLUDING H_0 PROBABILITY OF A TYPE I ERROR 0 $u - u_0$

FIGURE 10.2
Three kinds of tests of hypotheses for the various alternatives concerning a population proportion

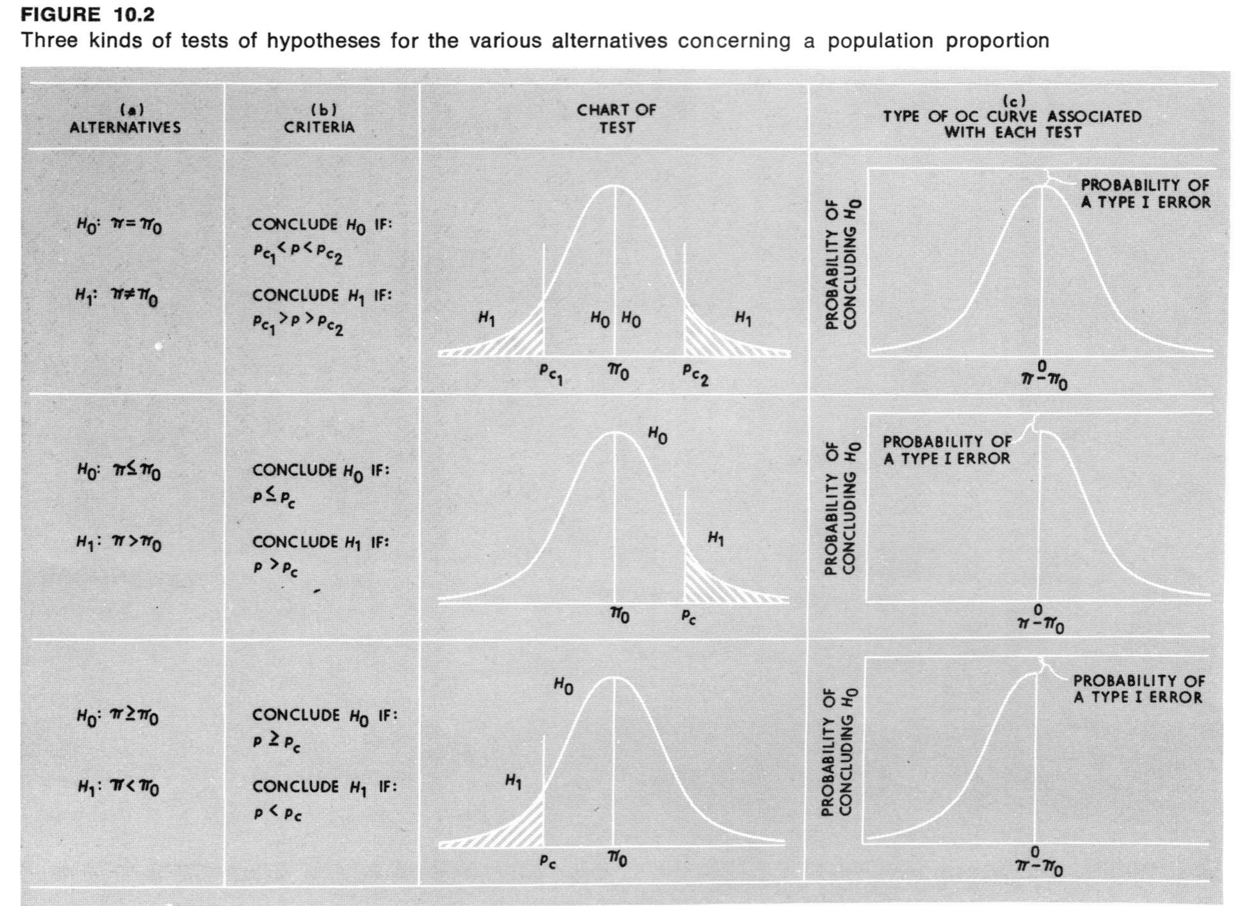

hypothesis in light of the established criteria. *Frequently, but not always, the null hypothesis represents the contrary of a prediction from a theory. That is, the investigator sets up as the null hypothesis to be tested the conclusion he or she hopes to disprove.* In general, the method is appropriate in a situation in which the analyst expects that on the basis of experimental results to be obtained, he or she may or may not be able to draw a certain inference. The inference is such that it can be put as the alternative to a statistical hypothesis; this latter then is the *null* hypothesis. In effect, the analyst proceeds in such a manner that in order to secure widespread acceptance of the conclusions, the analyst insists on a small a priori probability that he or she will falsely reject the null hypothesis, that is, accept the alternative hypothesis when H_0 is true. This a priori probability is, of course, the alpha risk.

It should be clear, and perhaps it is unfortunate, but the choice of a statistical test to use in testing an hypothesis must be made from the point of view of the investigator *before the experiment is completed.* Once we have the results of an experiment, we may all have 20/20 vision. That is to say, tests can always be designed which would result in the rejection of the hypothesis, whereas other tests would not. In this sense, hindsight always produces better vision, but such action does not reflect the facts of the real world. Analysts who allow the results of the experiment to determine their choice of a test place themselves in a position similar to that of the bridge expert who proclaims *after* the hand has been played precisely what the approach should have been.

The principle of the null or zero hypothesis relates to the concept of a *significant difference. In terms of sampling theory, a difference is considered statistically significant if it is too large to have occurred by random variations.* Using the null hypothesis concept, one assumes that the population difference between the observed statistic and the hypothesized parameter is zero. Any difference that did occur between the sample statistic (e.g., $\bar{X}$) and the hypothesized parameter (μ_0) is caused by random variations. That is, we would say the difference is not significant, and it probably occurred by chance error of sampling.

In this connection, it also should be noted that an *important difference* is the variation between the hypothesized parameter (e.g., μ_0) and the alternative value that the universe parameter (μ_1) might assume. If this difference exists and the conclusion is drawn on the basis of a random sample that the hypothesized universe was the one being sampled, a definitely objectionable error of inference would be made. It is desirable to select a test that will lead to the rejection of the null hypothesis, that is, conclude that the difference is significant, if such a difference is important! Managerial ability or judgment must dictate to the statistician what is to be considered an important difference.

TYPE I AND TYPE II ERRORS: AN ANALOGY WITH LIE DETECTOR TESTS

In any test of hypotheses, there are two kinds of errors we might make. First, on the basis of the sample evidence, it would be a mistake to reject a null hypothesis if it were in fact true. This is commonly called a Type I error and is sometimes referred to as the "producer's risk" in acceptance sampling. If the hypothesis were not true, accepting it would be a mistake. This is called a Type II error, that is, accepting the null hypothesis when, indeed, the alternative hypothesis is true. *Of course, it is possible to make only one of these errors in any one problem, but there is no way to determine which error, if either, is likely to be made.* The probability of a Type I error is equal to the alpha risk, whereas the probability of a Type II error, called the beta risk, depends upon how wrong the null hypothesis really is. In Chapter 11 we shall illustrate how the probabilities of these two kinds of errors are controlled within management-stated limits. Here, our purpose is only one of introduction; the analogy below may clarify the concepts.

The familiar "lie detector" test depends, in part, upon the extent to which the suspect's pulse rate rises above his or her normal rate (e.g., 72) when responding to pertinent questions. It is a medical fact that there are manifold reasons why a person's pulse rate might rise above his or her norm. Therefore, some rate above this standard (72) must be regarded as being significantly different—too large to have occurred by chance.

What happens if the rate is set too close to 72? In this case, some honest people may be unjustly accused of lying. Statistically speaking, too many Type I errors are made, that is, rejection of a true null hypothesis—the null hypothesis (H_0) being that the suspect is not lying, and the alternative hypothesis (H_1) is that the person is lying.

What happens if the rate is set too far above 72? Here, some liars would go undetected. That is, we commit a Type II error—accepting a false null hypothesis. Obviously, our society has been more willing to risk making this kind of error.

TESTS OF HYPOTHESES PROCEDURE

Fundamentally, any test of hypotheses involves about four basic steps: (1) the researcher frames the hypotheses, (2) the sampling plan is determined, (3) the sample evidence is evaluated in the light of established criteria, and (4) a decision is reached between the alternative courses of action. (See Figure 10.3.)

Clearly, in any problem analysts must establish the alternative courses of action; that is, they must state a realistic null hypothesis (H_0) and an

FIGURE 10.3
Test of hypotheses procedure

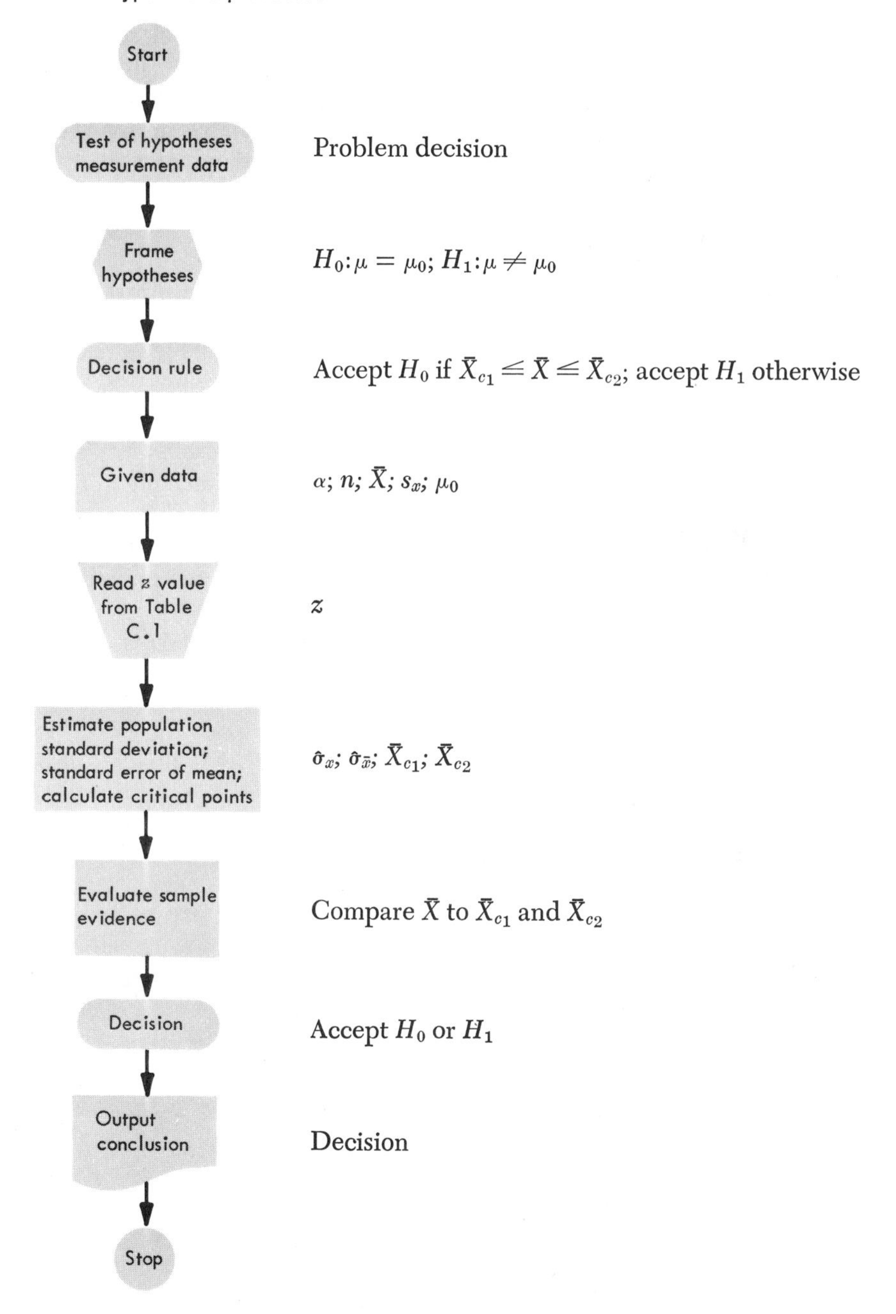

alternative hypothesis (H_1). Then, on the basis of some stated risks which management is willing to assume (α and β), the statistician determines the sampling plan. Involved here also is the establishment of some objective criteria for choosing between H_0 and H_1. Once the sample has been drawn and an appropriate statistic computed, the sample evidence is evaluated in the light of the established criteria. Finally, on the basis of this evidence a decision is reached and a course of action is chosen.

CHOICE OF THE LEVEL OF SIGNIFICANCE (α)

The choice of a level of significance, the alpha risk, always involves a compromise between two kinds of danger. First, there is the risk of concluding that a difference is significant when it really is not. This, of course, is called a Type I error when we reject a true null hypothesis. Obviously, the probability of committing a Type I error can be reduced by the selection of a strict level of significance, that is, a very low probability such as an α of .01 or .001. However, if this is done, the risk of making a Type II error increases commensurably. This follows because the acceptance interval is broad for a strict level of significance and may include differences caused by other than random factors. This increases the probability of accepting the null hypothesis when it is false.

On the other hand, if a less demanding level of significance is chosen

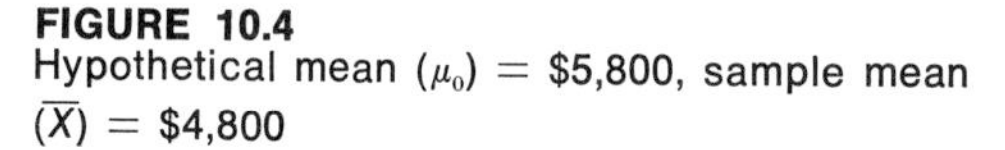

FIGURE 10.4
Hypothetical mean (μ_0) = \$5,800, sample mean ($\bar{X}$) = \$4,800

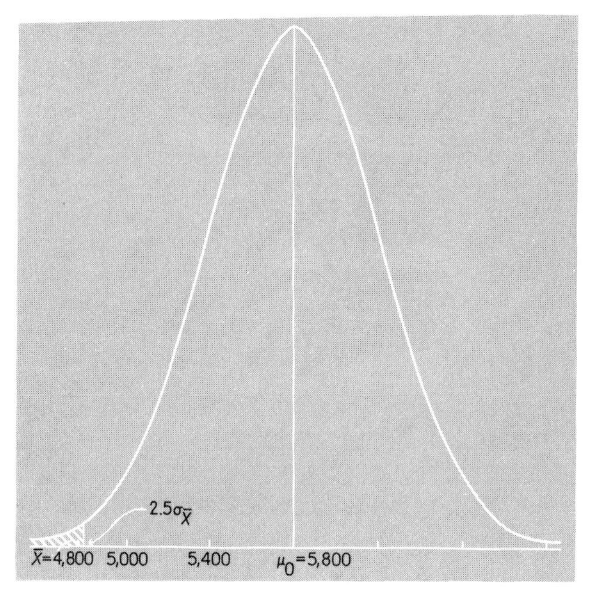

FIGURE 10.5
Distribution of sample means of size $n = 100$, showing acceptance and rejection regions

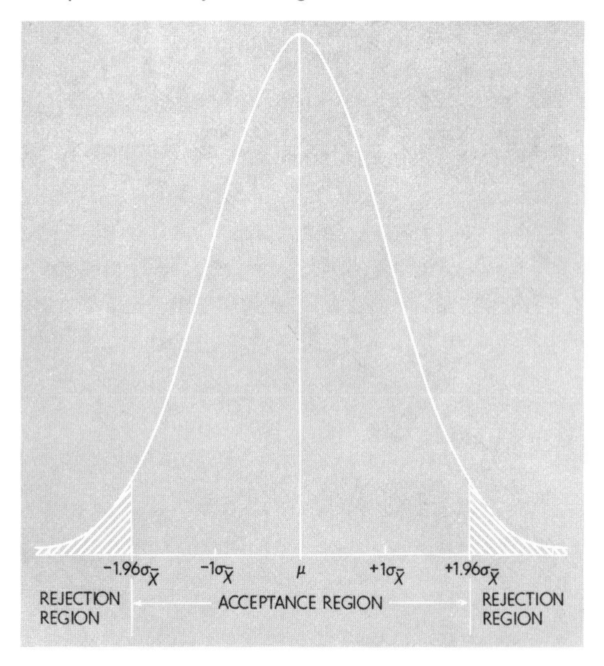

(e.g., use α of .05 instead of, say, .01 or .001), the probability of making a Type II error *decreases,* but the probability of making a Type I error *increases. The only way to reduce the probability of both kinds of errors is to increase the sample size.* Of course, neither error can ever be completely eliminated.

DECISION-MAKING RULE: POPULATION MEAN, LARGE SAMPLE, TWO-SIDED ALTERNATIVE

Assume that on the basis of retail sales in an area a market analyst states the hypothesis that average income per household after taxes is $5,800. For illustration purposes, let us further assume that we know the population standard deviation, σ_x, of incomes in the area is known to be $4,000. Suppose the market analyst selects 100 households at random and determines that their average income is $4,800. How might the original hypothesis that $\mu = \$5,800$ be tested, assuming a willingness to accept an alpha risk of .05? (See Figures 10.4, 10.5, and 10.6.)

We know that for a large-size sample the sampling distribution of sample means can be considered to be normal. In addition, we know that

FIGURE 10.6
Test of hypothesis re population mean; $n > 30$; 2-sided test; σ is known

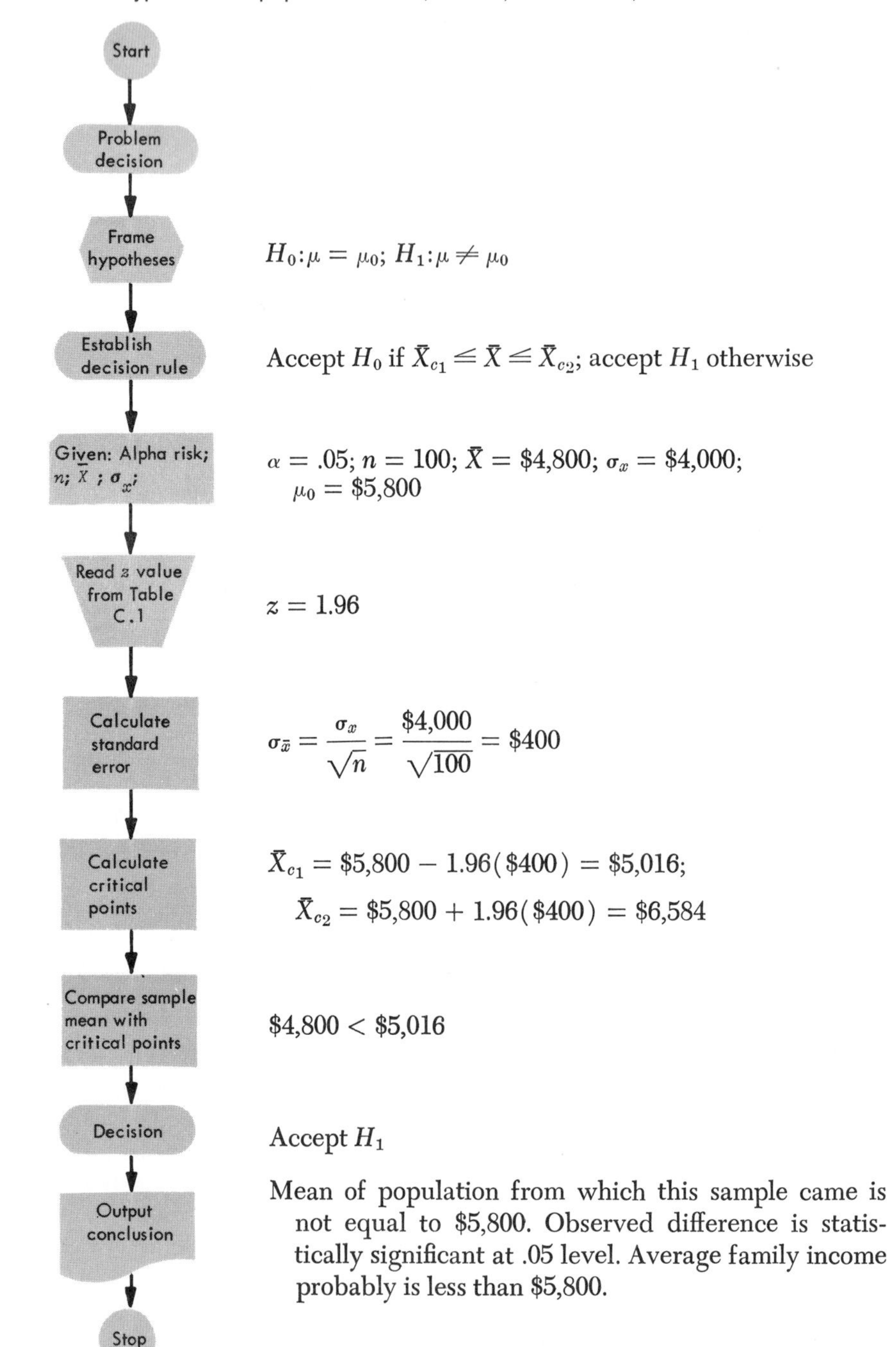

the standard error of the mean, the standard deviation of sample means, is (see Formula [9.1])

$$\sigma_{\bar{x}} = \frac{\sigma_x}{\sqrt{n}} \tag{10.1}$$

What are the alternative courses of action? First, we frame the hypotheses, as follows:

$H_0: \mu = \mu_0$. The mean of the universe from which the sample came equals the hypothesized parameter, \$5,800. The difference we did observe, that is, \$5,800 − \$4,800, is not significant; it probably occurred by chance.

$H_1: \mu \neq \mu_0$. The mean of the universe from which the sample came does not equal the hypothesized parameter. The observed difference is significant, that is, probably too great to have occurred by chance.

What are the criteria for choosing between these alternative courses of action? Conclude H_0 if

$$\mu_0 - z\sigma_{\bar{x}} \leq \bar{X} \leq \mu_0 + z\sigma_{\bar{x}}$$

Conclude H_1 otherwise.

Where z is the standard normal deviate associated with the specific α risk of concluding H_1 when H_0 is true (Figure 10.7).

Computations: Given

$$n = 100;\ \bar{X} = \$4{,}800;\ \sigma_x = \$4{,}000;\ \alpha = .05$$

$$\sigma_{\bar{x}} = \frac{\sigma_x}{\sqrt{n}}$$

FIGURE 10.7

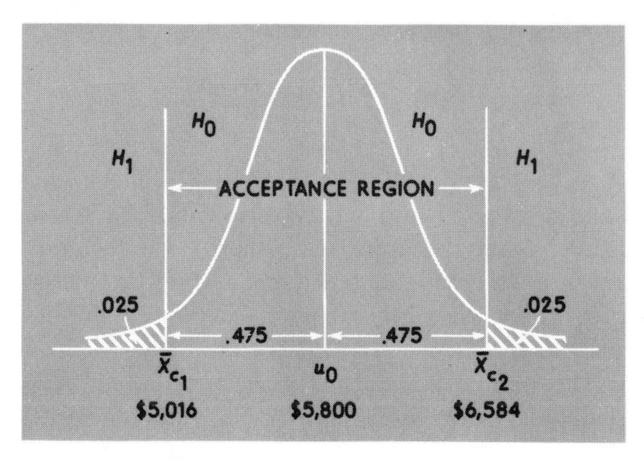

$$\sigma_{\bar{x}} = \frac{\$4{,}000}{\sqrt{100}} = \$400$$

$$\bar{X}_{c1} = \mu_0 - z\sigma_{\bar{x}} = \$5{,}800 - 1.96(\$400) = \$5{,}016$$
$$\bar{X}_{c2} = \mu_0 + z\sigma_{\bar{x}} = \$5{,}800 + 1.96(\$400) = \$6{,}584$$

What is our conclusion? Evaluating the sample evidence in the light of the established criteria, we reject the null hypothesis (H_0) because $\bar{X} =$ \$4,800 falls to the left of $\bar{X}_{c1}$. This means that the average income per household is something less than we have hypothesized, or that the mean of the universe from which this sample came probably does not equal \$5,800. The probability of obtaining a sample mean as small or smaller than \$4,800 from a universe that has as its true mean \$5,800 is less than .05. We could be wrong, but probability theory is on our side. We *might* be guilty of making a Type I error; however, that has been controlled at .05, the alpha risk. It should be noted that we have ignored attempting to control the probability of a Type II error at this point. Later, when discussing operating characteristic curves, this serious omission will be dealt with. At the moment Table 10.1 may be helpful in summarizing this concept.

TABLE 10.1

	Decision	
Alternative	Accept H_0	Accept H_1
If H_0 is true..........	No error	Type I error $= \alpha$
If H_1 is true..........	Type II error $= \beta$	No error

Alternative method

An alternative method, and one which is shorter but perhaps not as illuminating to the beginner, is to compute

$$z_c = \frac{\bar{X} - \mu_0}{\sigma_{\bar{x}}} = \frac{\$4{,}800 - \$5{,}800}{\$400} = -2.5$$

Because $z_c = -2.5 > z = -1.96$ we reject H_0. In effect, we substitute z_c values for $\bar{X}$ and $z_{\alpha/2}$ for $\bar{X}_{c1}$ and $\bar{X}_{c2}$. We also would have rejected H_0 had $+z_c > +z = 1.96$. While this approach is shorter, the technique illustrated first better prepares the student for understanding the material in Chapter 11. Instructors not planning to cover Chapter 11 might elect to use this technique.

Ordinarily, the standard deviation of the population, σ_x, is not known and we must depend upon the sample for an estimate of it. Thus, if the standard deviation of our sample s_x were computed to be \$3,861, our estimate of the population standard deviation is (see Formula [9.4])

$$\hat{\sigma}_x = s_x \sqrt{\frac{n}{n-1}} \tag{10.2}$$

or

$$\hat{\sigma}_x = \$3{,}861 \sqrt{\frac{100}{100-1}} \simeq \$3{,}880 \text{ or rounded to } \$3{,}900$$

With this estimate of the universe variation, we can then estimate the true standard error of the mean, $\hat{\sigma}_{\bar{x}}$, as follows (see Formula [9.2]):

$$\hat{\sigma}_{\bar{x}} = \frac{\hat{\sigma}_x}{\sqrt{n}}$$

$$\hat{\sigma}_{\bar{x}} = \frac{\$3{,}900}{\sqrt{100}} = \$390 \tag{10.3}$$

The remainder of the test is as before.

$$z_c = \frac{\$4{,}800 - \$5{,}800}{\$390} = -2.56$$

Therefore, at the .05 level of significance, we again reject the null hypothesis that the mean of the population from which this sample came is \$5,800. Indeed, we again conclude on the basis of our sample evidence and the established criteria that the universe parameter does not equal \$5,800.

DECISION-MAKING RULE: POPULATION MEAN, LARGE SAMPLE, ONE-SIDED ALTERNATIVE, RIGHT

Let us now consider essentially the same basic data but assume that we are interested in knowing *only* if the average income per household *is something greater than \$5800.* Here, we would place all of the alpha risk on the right side of our theoretical sampling distribution, and the procedure would be as given below. Assume our sample evidence were as follows (see Figures 10.8 and 10.9):

$$\bar{X} = \$6{,}200;\ s_x = \$3{,}564;\ \alpha = .05;\ n = 100$$

What are the alternative courses of action? They may be stated in this manner:

$H_0: \mu \leq \mu_0$. The mean of the universe from which this sample came is equal to or less than the hypothesized parameter, \$5,800. The difference we did observe, that is, \$6,200 − \$5,800, is not significant; it probably occurred by chance.

$H_1: \mu > \mu_0$. The mean of the universe from which this sample came is greater than the hypothesized parameter, \$5,800. The difference we did observe is significant, that is, probably too great to have occurred by chance.

What are the criteria for choosing between these alternative courses of action? Conclude H_0 if

$$\bar{X} \leq \mu_0 + z\hat{\sigma}_{\bar{x}}$$

Conclude H_1 if

$$\bar{X} > \mu_0 + z\hat{\sigma}_{\bar{x}}$$

Computations: Given

$$n = 100; \bar{X} = \$6{,}200; s_x = \$3{,}564; \alpha = .05; z = 1.645$$

from Formula (10.3)

$$\hat{\sigma}_{\bar{x}} = \frac{\sigma_x}{\sqrt{n}}$$

and from Formula (10.2)

$$\hat{\sigma}_x = s_x \sqrt{\frac{n}{n-1}}$$

Substituting,

$$\hat{\sigma}_x = \$3{,}564 \sqrt{\frac{100}{100-1}}$$

$$\hat{\sigma}_x \simeq \$3{,}600$$

Then,

$$\hat{\sigma}_{\bar{x}} = \frac{\$3{,}600}{\sqrt{100}}$$

$$\hat{\sigma}_{\bar{x}} = \$360$$

Because this is a one-sided test, $z = 1.645$ for $\alpha = .05$, as indicated in Figure 10.8.

What is our conclusion? Evaluating the sample evidence in the light of the established criteria, we accept the null hypothesis (H_0) because $\bar{X} = \$6{,}200$ falls to the left of our critical point, $\bar{X}_c$. This means that the average income per household is equal to or less than \$5,800. Inductively, we act as though the true universe parameter is something less than \$5,800. Theory tells us that the probability of randomly selecting a sample whose mean is as large or larger than \$6,200 if $\mu = \$5{,}800$ is less than .05. In

FIGURE 10.8

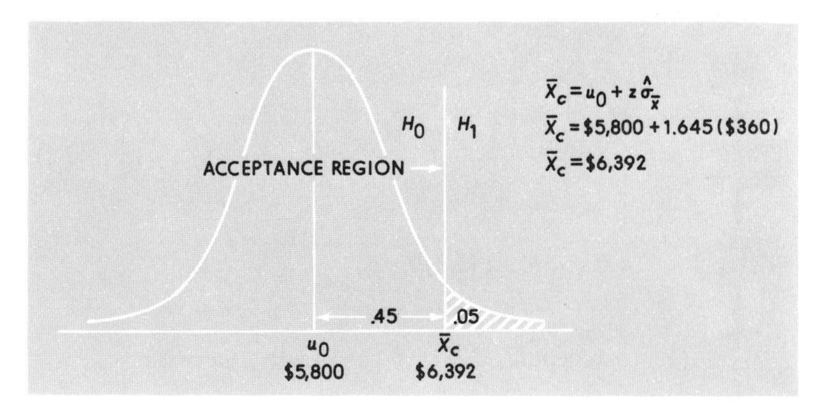

fact, given the alpha risk of .05, we would have had to obtain a mean greater than $6,392 to reject H_0.[1]

DECISION-MAKING RULE: POPULATION MEAN, LARGE SAMPLE, ONE-SIDED ALTERNATIVE, LEFT

Here, we are concerned about the hypothesis that the average income per household is $5,800 or less. This time the alpha risk is on the left side of our theoretical sampling distribution and the procedure is as follows. Assume our sample evidence is

$$\bar{X} = \$5{,}100;\ s_x = \$3{,}466;\ \alpha = .05;\ n = 100$$

What are the alternative courses of action? They may be stated in this manner:

$H_0: \mu \geqq \mu_0$. The mean of the universe from which this sample came is equal to or greater than the hypothesized parameter, $5,800. The difference we did observe, that is, $5,800 – $5,100, is not significant; it probably occurred by chance.

$H_1: \mu < \mu_0$. The mean of the universe from which this sample came is less than the hypothesized parameter, $5,800. The difference we did observe is significant, that is, probably too great to have occurred by chance.

---

[1] Using the shorter approach:

$$z_c = \frac{\$6{,}200 - \$5{,}800}{\$360} = 1.11$$

Because $1.11 < 1.645$ we accept H_0.

FIGURE 10.9
Test of hypothesis re population mean; $n > 30$; one-sided test, right

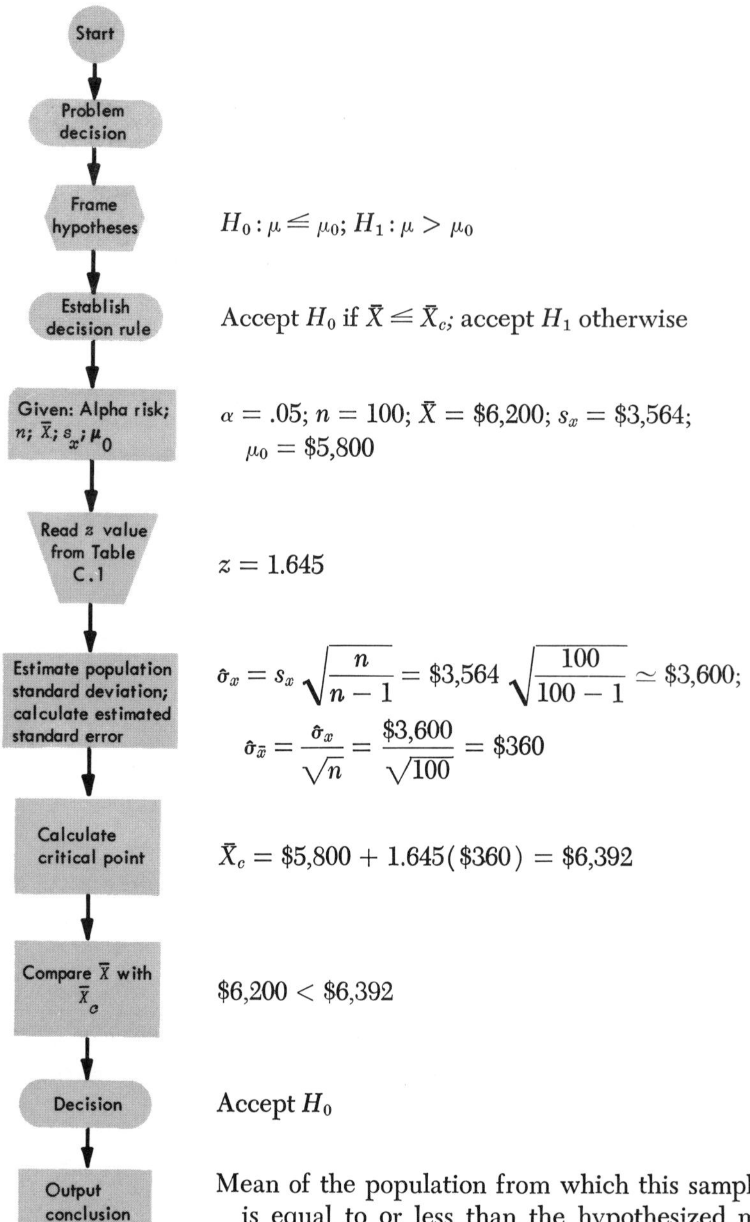

$H_0: \mu \leqq \mu_0$; $H_1: \mu > \mu_0$

Accept H_0 if $\bar{X} \leqq \bar{X}_c$; accept H_1 otherwise

$\alpha = .05$; $n = 100$; $\bar{X} = \$6{,}200$; $s_x = \$3{,}564$;
$\mu_0 = \$5{,}800$

$z = 1.645$

$$\hat{\sigma}_x = s_x \sqrt{\frac{n}{n-1}} = \$3{,}564 \sqrt{\frac{100}{100-1}} \simeq \$3{,}600;$$

$$\hat{\sigma}_{\bar{x}} = \frac{\hat{\sigma}_x}{\sqrt{n}} = \frac{\$3{,}600}{\sqrt{100}} = \$360$$

$\bar{X}_c = \$5{,}800 + 1.645(\$360) = \$6{,}392$

$\$6{,}200 < \$6{,}392$

Accept H_0

Mean of the population from which this sample came is equal to or less than the hypothesized mean of \$5,800. Observed difference is due to chance error of sampling.

What are the criteria for choosing between these alternative courses of action? Conclude H_0 if

$$\bar{X} \geqq \mu_0 - z\hat{\sigma}_{\bar{x}}$$

Conclude H_1 if

$$\bar{X} < \mu_0 - z\hat{\sigma}_{\bar{x}}$$

Computations: Given

$$n = 100;\ \bar{X} = \$5{,}100;\ s_x = \$3{,}466;\ \alpha = .05;\ z = 1.645$$

using Formula (10.2),

$$\hat{\sigma}_x = s_x\sqrt{\frac{n}{n-1}}$$

$$\hat{\sigma}_x = \$3{,}466\sqrt{\frac{100}{100-1}}$$

$$\hat{\sigma}_x \simeq \$3{,}500$$

and using Formula (10.3),

$$\hat{\sigma}_{\bar{x}} = \frac{\hat{\sigma}_x}{\sqrt{n}}$$

Then,

$$\hat{\sigma}_{\bar{x}} = \frac{\$3{,}500}{\sqrt{100}}$$

$$\hat{\sigma}_{\bar{x}} = \$350$$

Again, because this is a one-sided test, $z = 1.645$ for $\alpha = .05$, as indicated in Figure 10.10.

FIGURE 10.10

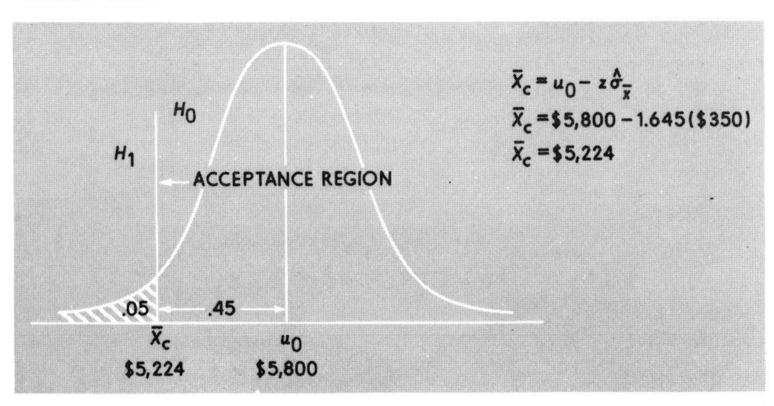

What is our conclusion? Examining our sample evidence in the light of the established criteria, we reject the null hypothesis (accept H_1) because $\bar{X} = \$5{,}100$ falls to the left of $\bar{X}_c$. This means that the average income per household is something less than \$5,800. The difference we observed is too great to have occurred through chance error of sampling. We may support this decision by probability theory, as follows:

$$z_c = \frac{X - \mu_0}{\hat{\sigma}_{\bar{x}}} = \frac{\$5{,}100 - \$5{,}800}{\$350} = -2.0$$

The area under a normal probability curve associated with a $z = -2.0$ is .47725 (see Appendix C, Table C.1); .50000 − .47725 = .02275. Therefore, only two times out of a hundred would we obtain a sample mean as small as or smaller than the one we observed simply by mere chance if the true population mean were \$5,800. Because this is less than our .05 alpha risk, we reject H_0.[2] (See Figure 10.11.)

DECISION-MAKING RULE: POPULATION MEAN, SMALL SAMPLE, TWO-SIDED ALTERNATIVE

As we noted in Chapter 9, sometimes it is convenient to work with small samples, for example, less than 30 observations. We discovered then that when $n < 30$ and *we do not know the population standard deviation*, the sampling distribution is approximated by Student's t distribution rather than the normal curve.

The Sigma Manufacturing Company wishes to promote some sort of a statement concerning the average life of its product. In fact, the sales manager wishes to advertise to the effect that the average life is five years. This enthusiasm is based on the following sample evidence and simulated tests of the product under varying conditions of use:

$$\bar{X} = 55 \text{ months (average life)}$$
$$s_x = 6 \text{ months}$$
$$n = 25$$

What are the hypotheses to be tested? They may be stated as follows:

$H_0: \mu = \mu_0$. The mean of the population from which this sample came is equal to the hypothesized parameter, that is, five years. The difference we observed (60 − 55) is not significant; it probably could have occurred by chance.

[2] Alternatively,

$$z_c = \frac{\$5{,}100 - \$5{,}800}{\$350} = -2.0$$

Because $-2.0 > -1.645$ we accept H_1.

FIGURE 10.11
Test of hypothesis re population mean; $n > 30$ one-sided test, left

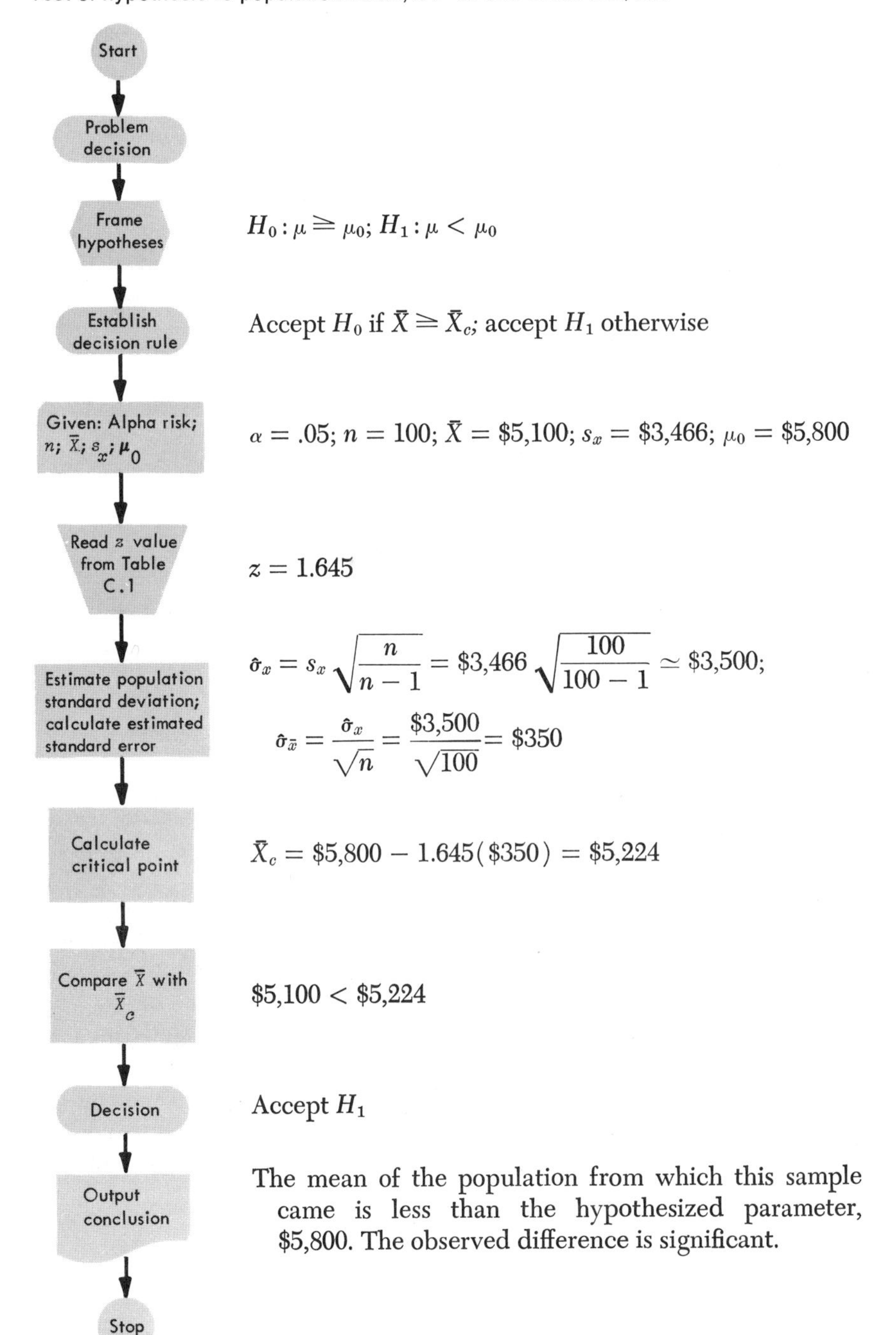

$H_1: \mu \neq \mu_0$. The mean of the population from which this sample came is not equal to the hypothesized parameter. The observed difference probably is too great to have occurred by chance.

What are the criteria for choosing between these hypotheses? Conclude H_0 if

$$\mu_0 - t\hat{\sigma}_{\bar{x}} \leq \bar{X} \leq \mu_0 + t\hat{\sigma}_{\bar{x}}$$

Conclude H_1 otherwise.
Computations: Given

$$n = 25;\ \bar{X} = 55 \text{ months};\ s_x = 6 \text{ months};\ \alpha = .01;\ t_{.01} = 2.797$$

using Formula (10.2),

$$\hat{\sigma}_x = s_x \sqrt{\frac{n}{n-1}}$$

$$\hat{\sigma}_x = 6 \sqrt{\frac{25}{25-1}}$$

$$\hat{\sigma}_x = 6.1 \text{ months}$$

and Formula (10.3),

$$\hat{\sigma}_{\bar{x}} = \frac{\hat{\sigma}_x}{\sqrt{n}}$$

$$\hat{\sigma}_{\bar{x}} = \frac{6.1}{\sqrt{25}}$$

$$\hat{\sigma}_{\bar{x}} = 1.22 \text{ months}$$

Referring to Table D.1 in Appendix D we note that for 24 degrees of freedom the .01 value of t is 2.797. The diagram in Figure 10.12 shows the critical points.

What is our conclusion? Because $\bar{X} = 55$ and is to the left of $\bar{X}_{c1}$, we reject the null hypothesis and accept H_1. That is, the mean of the population from which this sample came does not equal five years, as we have hypothesized. Only one time in 200 would we obtain a sample mean as small as or smaller than the one we observed if the true mean really were 60 months. We behave as though this rare event did not take place and conclude H_1. On the basis of this sample evidence, it would not be appropriate for the manager to pursue his intended promotional campaign. Further tests should be made and, perhaps, the new evidence combined with the old to form a larger sample[3] (see Figure 10.13).

[3] Again,

$$t_c = \frac{55 - 60}{1.22} = -4.09$$

Because $-4.09 > -2.797$ we accept H_1.

FIGURE 10.12

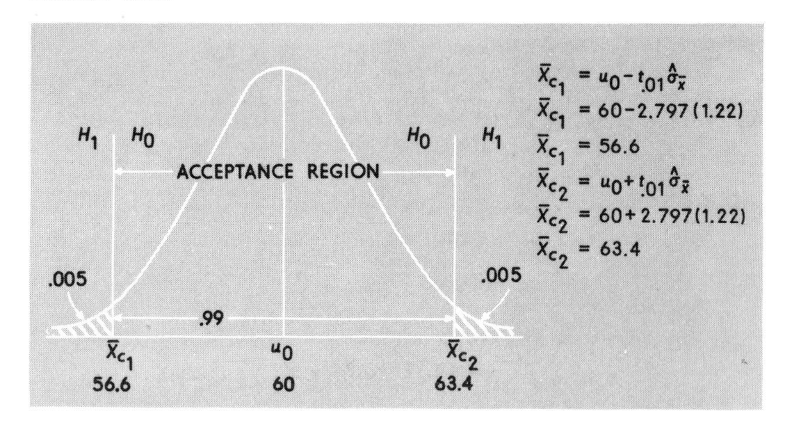

DECISION-MAKING RULE: POPULATION MEAN, SMALL SAMPLE, ONE-SIDED ALTERNATIVE, RIGHT

Let us assume that the sales manager now wishes to take another sample and attempt to determine if the average life of this product of Sigma Manufacturing Company under simulated conditions of use is something *greater than five years.*[4] Normally, we would analyze this problem placing all of the alpha risk on the right side of the sampling distribution. Assume that our sample produces the following:

$$n = 25; \bar{X} = 63 \text{ months}; s_x = 10 \text{ months}; \alpha = .01$$

What are the alternative hypotheses? They may be stated as

$H_0: \mu \leq \mu_0.$ The mean of the universe from which this sample came is equal to or less than the hypothesized parameter, five years. The difference we observed, that is, 63 − 60, is not significant; it probably could have occurred by chance.

$H_1: \mu > \mu_0.$ The mean of the universe from which this sample came is greater than the hypothesized parameter, five years. The difference we observed probably is too great to have occurred by chance.

What are the criteria for choosing between these two hypotheses? Conclude H_0 if

[4] From the point of view of the company, it *may* be more meaningful to consider this as a two-tailed test. However, we shall consider the problem useful as a right-tailed test and later as a left-tailed one. Actually, management's interest probably will be concentrated on whether or not the mean would be *at least* five years rather than exactly five years. In other words, conceivably this could be an ideal one-tailed test.

FIGURE 10.13
Test of hypothesis re population mean; $n < 30$; two-sided test; σ unknown

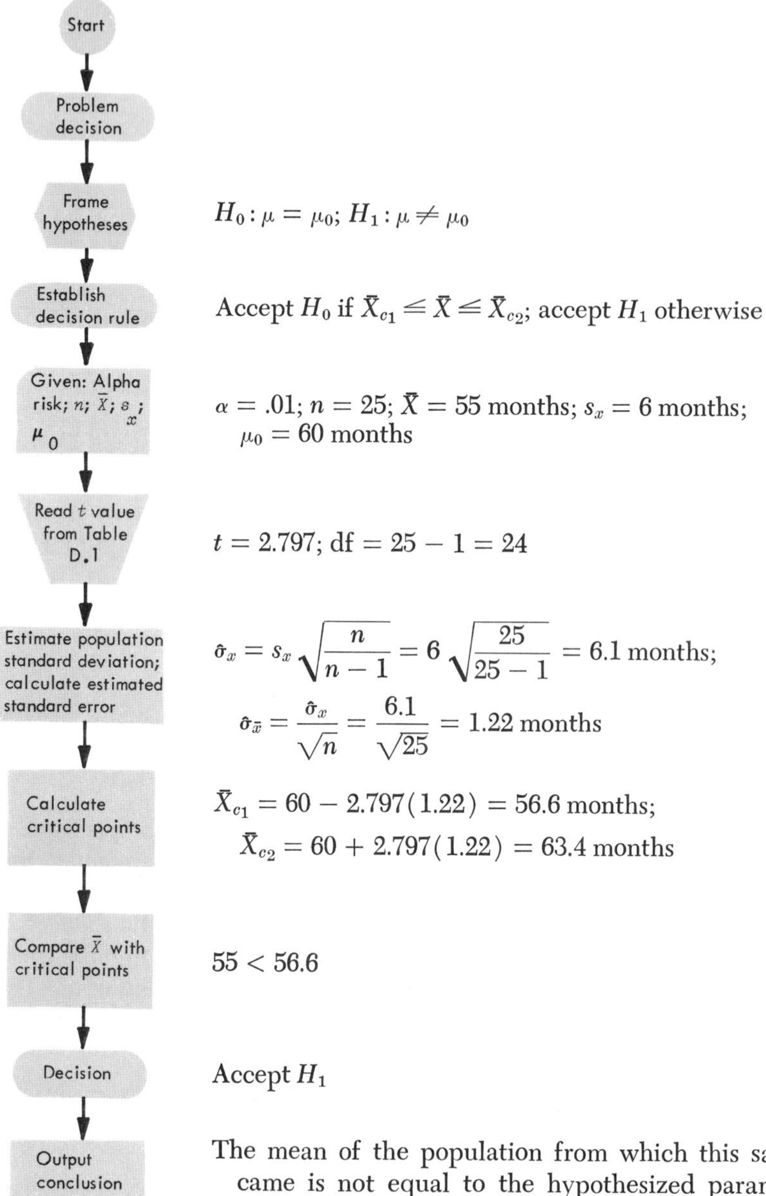

$$\bar{X} \leq \mu_0 + t_{.01}\hat{\sigma}_{\bar{x}}$$

Conclude H_1 if

$$\bar{X} > \mu_0 + t_{.01}\hat{\sigma}_{\bar{x}}$$

Computations: Given

$$n = 25; \bar{X} = 63 \text{ months}; s_x = 10 \text{ months}; \alpha = .01; t_{.01} = 2.492$$

$$\hat{\sigma}_x = 10\sqrt{\frac{25}{25-1}}$$

$$\hat{\sigma}_x = 10.2 \text{ months}$$

$$\hat{\sigma}_{\bar{x}} = \frac{\hat{\sigma}_x}{\sqrt{n}}$$

$$\hat{\sigma}_{\bar{x}} = \frac{10.2}{\sqrt{25}}$$

$$\hat{\sigma}_{\bar{x}} = 2.04 \text{ months}$$

What is our conclusion? Because $\bar{X} = 63$ months falls to the left of $\bar{X}_c$, we accept the null hypothesis, which states that the mean of the universe from which this sample came is equal to or less than five years. Therefore the sales manager should not advertise that the average life of this product is *greater* than five years. It may well be true, but the sample evidence does not support the proposition.[5] (See Figures 10.14 and 10.15.)

DECISION-MAKING RULE: POPULATION MEAN, SMALL SAMPLE, ONE-SIDED ALTERNATIVE, LEFT

Assume that the sales manager of the Sigma Manufacturing Company then decided to lower the sights and asked: "Can we legitimately state that our product will last at least four years under conditions of normal use? That is, can we offer a four-year guarantee?" Given the following new sample evidence:

$$n = 25; \bar{X} = 45 \text{ months}; s_x = 3 \text{ months}; \alpha = .01$$

What are the alternative hypotheses? They may be stated as

$H_0: \mu \geq \mu_0$. The mean of the universe from which this sample came is equal to or greater than the hypothesized parameter, four

[5] Again using the short method:

$$t_c = \frac{63 - 60}{2.04} = 1.47$$

Because $1.47 < 2.492$ we accept H_0.

FIGURE 10.14
Test of hypothesis re population mean; $n < 30$; one-sided test, right

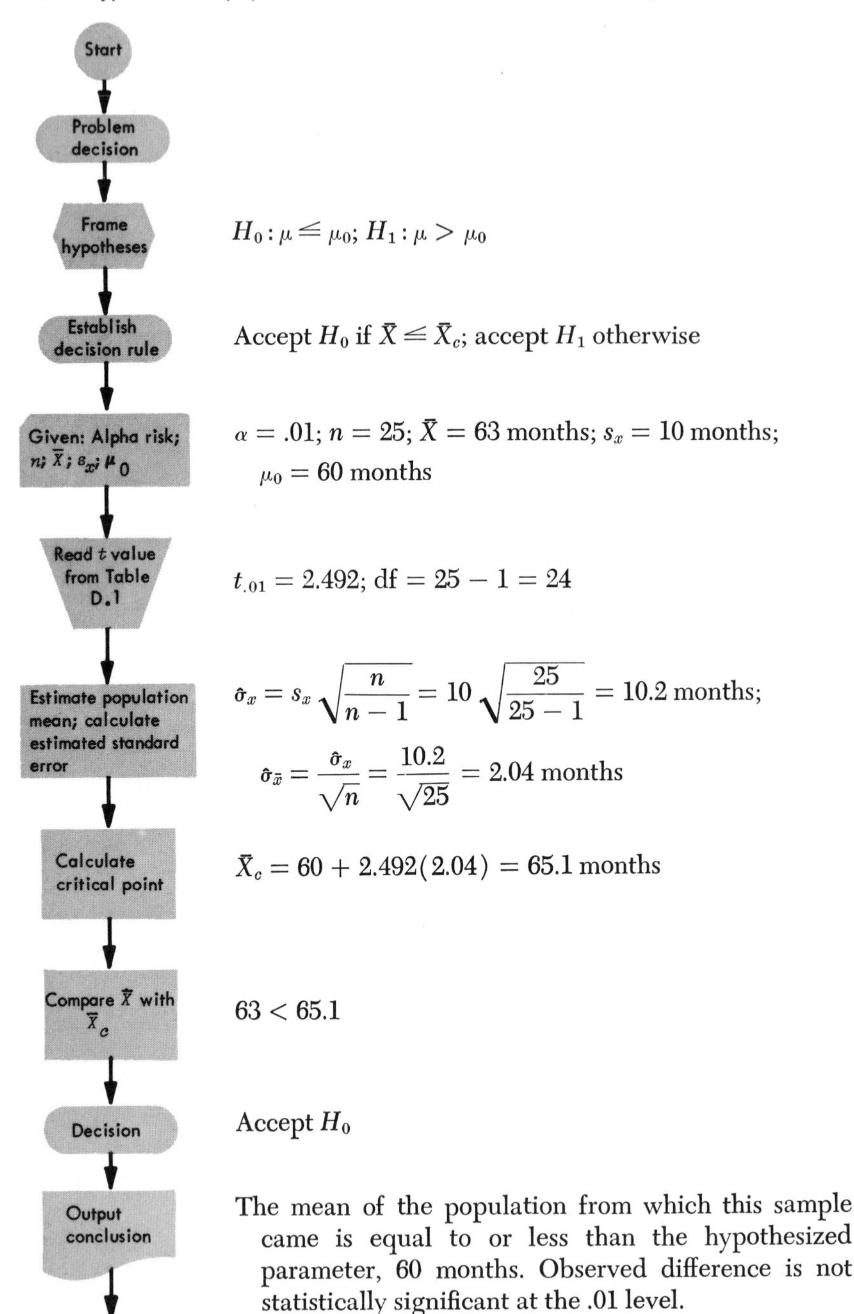

FIGURE 10.15

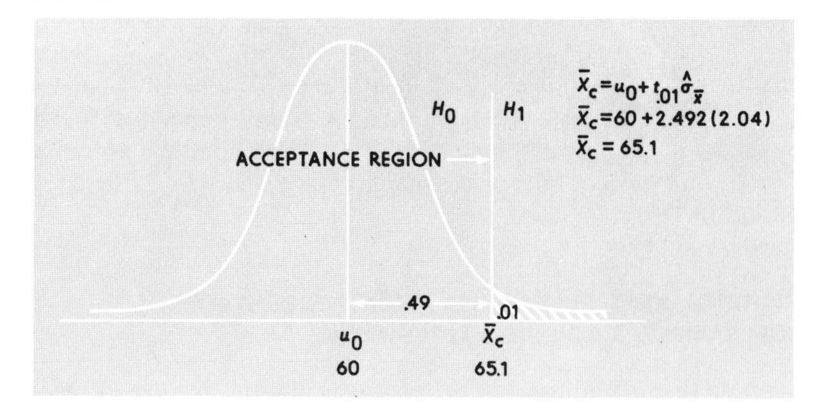

years. The difference we observed, that is, 48 − 45, probably could have occurred by chance.

$H_1: \mu < \mu_0$. The mean of the universe from which this sample came is less than the hypothesized parameter. The observed difference is significant, that is, probably too large to have occurred by chance.

What are the criteria for choosing between these alternative hypotheses? Conclude H_0 if

$$\bar{X} \geqq \mu_0 - t_{.01}\hat{\sigma}_{\bar{x}}$$

Conclude H_1 if

$$\bar{X} < \mu_0 - t_{.01}\hat{\sigma}_{\bar{x}}$$

Computations: Given

$$n = 25;\ \bar{X} = 45 \text{ months};\ s_x = 3 \text{ months};\ \alpha = .01;\ t_{.01} = -2.492$$

using Formula (10.2),

$$\hat{\sigma}_x = s_x \sqrt{\frac{n}{n-1}}$$

$$\hat{\sigma}_x = 3 \sqrt{\frac{25}{25-1}}$$

$$\hat{\sigma}_x = 3.06 \text{ months}$$

$$\hat{\sigma}_{\bar{x}} = \frac{\hat{\sigma}_x}{\sqrt{n}}$$

$$\hat{\sigma}_{\bar{x}} = \frac{3.06}{\sqrt{25}}$$

$$\hat{\sigma}_{\bar{x}} = .61 \text{ months}$$

What is our conclusion? Because $\bar{X} = 45$ and falls to the left of $\bar{X}_c$, we reject the null hypothesis (accept H_1). Again, on the basis of this evidence, it would be unsound judgment to use the recommended promotion policy. Probability theory indicates that there is only one chance out of one hundred that we would obtain a sample average of less than 46.5 months if the universe average really is 48 months. We assume that this rare event did not take place; therefore, we reject H_0.[6] (See Figures 10.16 and 10.17.)

DECISION-MAKING RULE: POPULATION PERCENTAGE, LARGE SAMPLE, TWO-SIDED ALTERNATIVE

Suppose that in addition to the hypothesis that the average income per family in the area was $5,800, the market analyst also stated the proposition that 15 percent of the families have incomes after taxes of $10,000 or more. As we learned above the first of these is an hypothesis that $\mu =$ $5,800, and the second is an hypothesis that $\pi = 15$ percent. We shall consider the latter hypothesis now.

If the sample size is large and π is not near zero or 100 percent, the normal curve may be used to test the above hypothesis.

Assume a sample of 100 families in the area yields the fact that only 7 percent have incomes of $10,000 or more. Using $\alpha = .05$, test the hypothesis that $\pi = 15$ percent.

What are the alternative hypotheses? They may be stated as

FIGURE 10.16

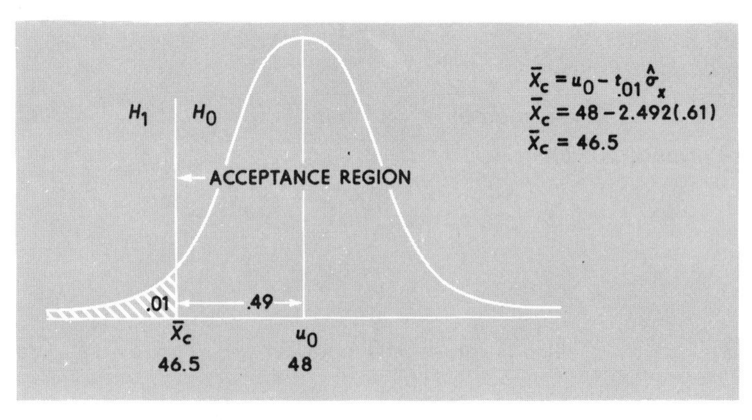

[6] Again,

$$t_c = \frac{45 - 48}{.61} = -4.91$$

Because $-4.91 > -2.492$ we accept H_1.

FIGURE 10.17
Test of hypothesis re population mean; $n < 30$; one-sided test, left

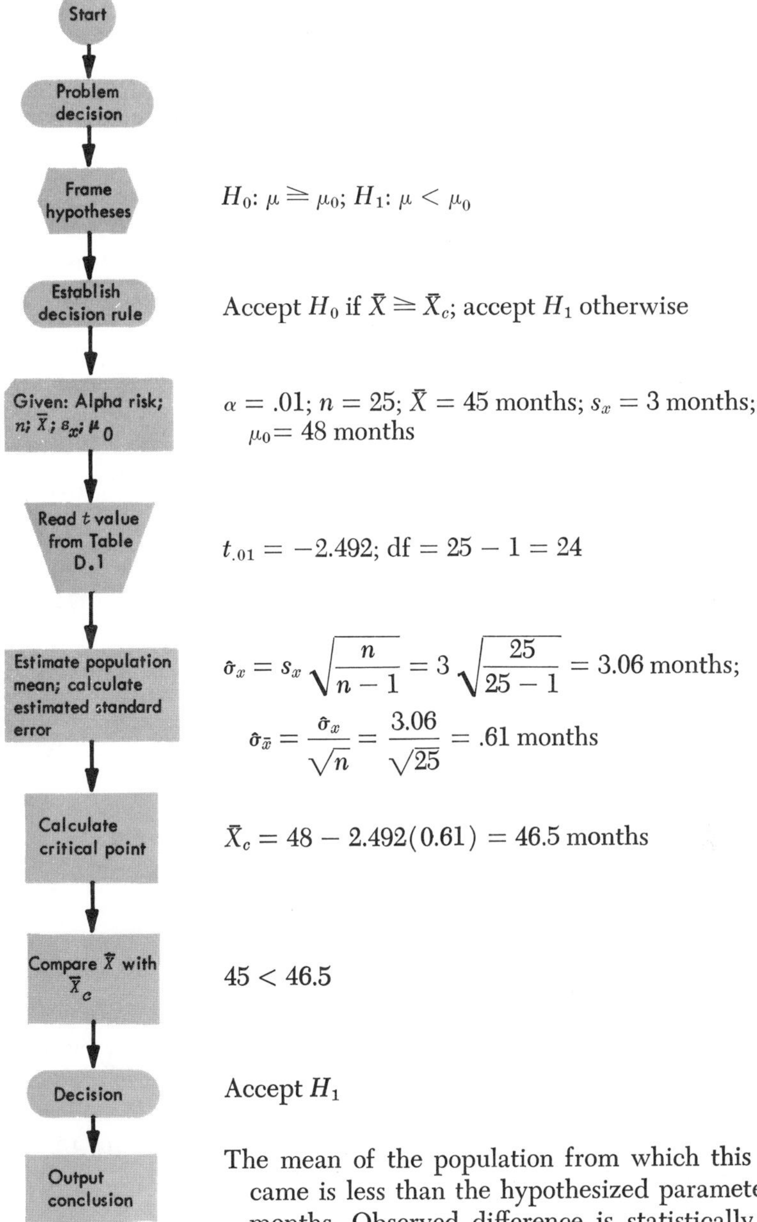

$H_0: \pi = \pi_0$. The percentage of families in the universe from which this sample came that have incomes of $10,000 or more is equal to the hypothesized parameter, that is, 15 percent. The difference we observed, that is, 15 − 7, is not significant; it probably could have occurred by chance.

$H_1: \pi \neq \pi_0$. The percentage of families in the universe from which this sample came that have incomes of $10,000 or more is not equal to the hypothesized parameter. The difference we observed is significant; it probably is too great to have occurred by chance.

What are the criteria for choosing between these alternative hypotheses? Conclude H_0 if

$$\pi_0 - z\sigma_p \leq p \leq \pi_0 + z\sigma_p$$

Conclude H_1 otherwise.

Computations: Given

$$n = 100;\ \pi_0 = 15 \text{ percent};\ \alpha = .05;\ z = 1.96$$

If $\pi = 15$ percent, the true standard error of the percentage, that is, the standard deviation of all possible sample percentages of size $n = 100$, is (see Formula [9.7])

$$\sigma_p = \sqrt{\frac{\pi_0(100 - \pi_0)}{n}} \tag{10.4}$$

$$\sigma_p = \sqrt{\frac{15(100 - 15)}{100}}$$

$$\sigma_p \simeq 3.6 \text{ percent}$$

$$p_{c1} = \pi_0 - z\sigma_p$$

$$p_{c1} = 15 - 1.96(3.6)$$

$$p_{c1} = 7.9, \quad \text{or} \quad 8 \text{ percent}$$

$$p_{c2} = \pi_0 + z\sigma_p$$

$$p_{c2} = 15 + 1.96(3.6)$$

$$p_{c2} = 22.1, \quad \text{or} \quad 22 \text{ percent}$$

(See Figure 10.18.)

What is our conclusion? Because $p = 7$ falls to the left of p_{c1}, we reject H_0 (i.e., accept H_1) and conclude that the true percentage of families with incomes of $10,000 or more is not equal to 15; indeed, it is probably less. In fact, only 2.5 times out of 100 would we receive a sample percentage less than eight if the true parameter really is 15. Therefore, we say the null hypothesis is untenable.[7] (See Figures 10.18 and 10.19.)

---

[7] Again,

$$z_c = \frac{7 - 15}{3.6} = -2.22$$

Because $-2.22 > -1.96$ we accept H_1.

FIGURE 10.18

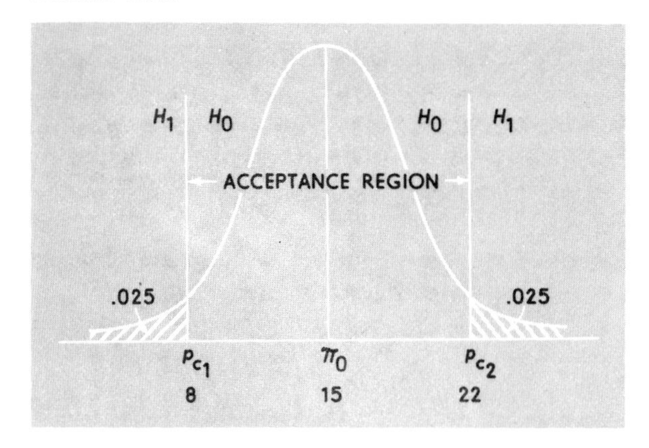

DECISION-MAKING RULE: POPULATION PERCENTAGE, LARGE SAMPLE, ONE-SIDED ALTERNATIVE, RIGHT

Assume that we are interested in knowing only that the percentage of families with incomes of $10,000 or more has increased during a given time period. Given

$$n = 100;\ \mathrm{p} = 12;\ \alpha = .01;\ \pi_0 = 10 \text{ percent}$$

Test the hypothesis that $\pi \leq \pi_0$.

What are the alternative hypotheses? They may be stated as

$H_0 : \pi \leq \pi_0$. The percentage of families with incomes of $10,000 or more is equal to or less than the hypothesized parameter, that is, 10 percent. The difference we observed (12 − 10) is not significant and probably could have occurred by chance.

$H_1 : \pi > \pi_0$. The percentage of families with incomes of $10,000 or more is greater than the hypothesized parameter. The difference we observed is significant and probably is too large to have occurred by chance.

What are the criteria for choosing between these alternative hypotheses? Conclude H_0 if

$$p \leq \pi_0 + z\sigma_p$$

Conclude H_1 if

$$p > \pi_0 + z\sigma_p$$

Computations: Using Formula (10.4),

FIGURE 10.19
Test of hypothesis re population percentage; large sample; two-sided test

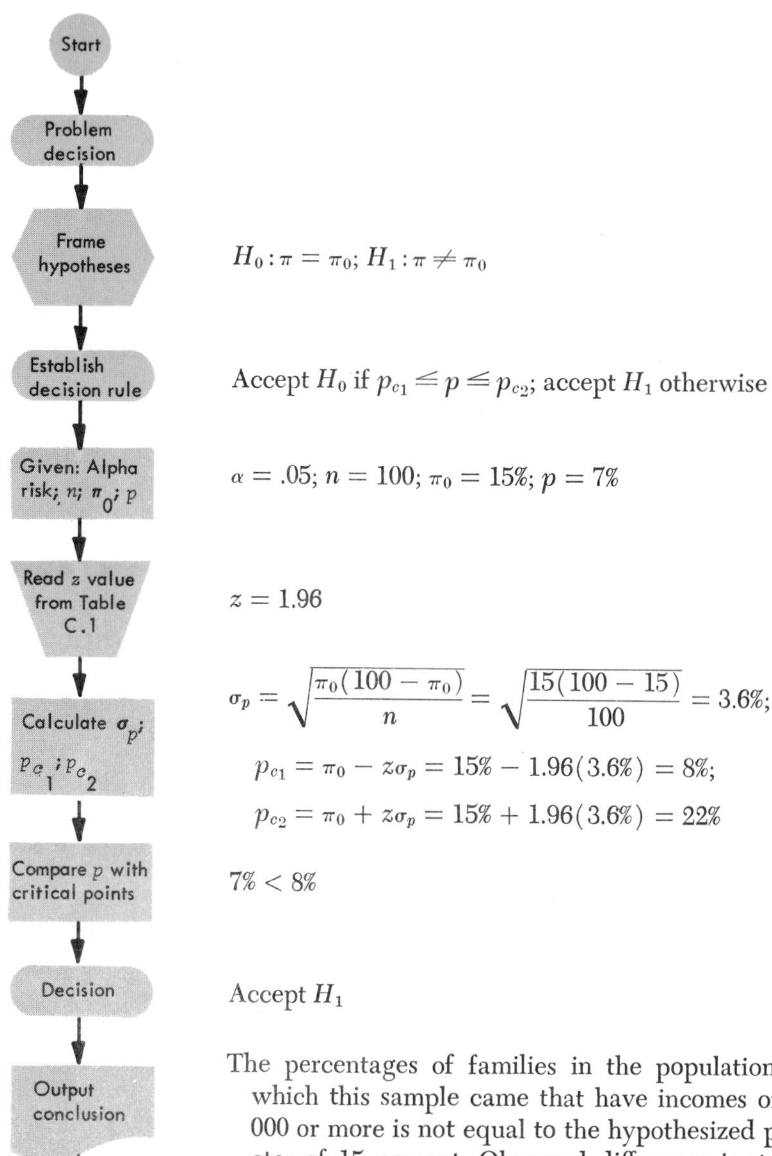

$H_0: \pi = \pi_0$; $H_1: \pi \neq \pi_0$

Accept H_0 if $p_{c1} \leqq p \leqq p_{c2}$; accept H_1 otherwise

$\alpha = .05$; $n = 100$; $\pi_0 = 15\%$; $p = 7\%$

$z = 1.96$

$$\sigma_p = \sqrt{\frac{\pi_0(100 - \pi_0)}{n}} = \sqrt{\frac{15(100 - 15)}{100}} = 3.6\%;$$

$$p_{c1} = \pi_0 - z\sigma_p = 15\% - 1.96(3.6\%) = 8\%;$$

$$p_{c2} = \pi_0 + z\sigma_p = 15\% + 1.96(3.6\%) = 22\%$$

$7\% < 8\%$

Accept H_1

The percentages of families in the population from which this sample came that have incomes of $10,000 or more is not equal to the hypothesized parameter of 15 percent. Observed difference is statistically significant at the .05 level.

$$\sigma_p = \sqrt{\frac{\pi_0(100 - \pi_0)}{n}}$$

$$\sigma_p = \sqrt{\frac{10(100 - 10)}{100}}$$

$$\sigma_p = 3 \text{ percent}$$

Clearly, this is a one-sided test for $\alpha = .01$, then $z = 2.33$, because the area between π_0 and p_c is equal to .49.

What is our conclusion? Because $p = 12$ falls within the acceptance region, we accept the null hypothesis (H_0). This means that we conclude the percentage of families with incomes of $10,000 or more is equal to or less than 10.[8] (See Figures 10.20 and 10.21.)

DECISION-MAKING RULE: POPULATION PERCENTAGE, LARGE SAMPLE, ONE-SIDED ALTERNATIVE, LEFT

Now, let us assume that we are interested in knowing only that the percentage of families with incomes of $10,000 or more has *decreased* during a given time period. Given:

$$n = 425;\ p = 12 \text{ percent};\ \alpha = .01;\ \pi_0 = 15 \text{ percent}$$

Test the hypothesis $\pi \geqq \pi_0$.

What are the alternative hypotheses? They may be stated as

H_0: $\pi \geqq \pi_0$. The percentage of families with incomes of $10,000 or more is equal to or greater than the hypothesized parameter, 15 percent. The difference we observed (15 − 12) is not significant; it probably could have occurred by chance.

H_1: $\pi < \pi_0$. The proportion of families with incomes of $10,000 or more is less than the hypothesized parameter, 15 percent. The observed difference is significant, that is, probably too great to have occurred by chance.

What are the criteria for choosing between these alternative hypotheses?

---

[8] Also,

$$z_c = \frac{12 - 10}{3} = .67$$

Because $.67 < 2.33$ we accept H_0.

FIGURE 10.20

Test of hypothesis re population percentage; large sample; one-sided test, right

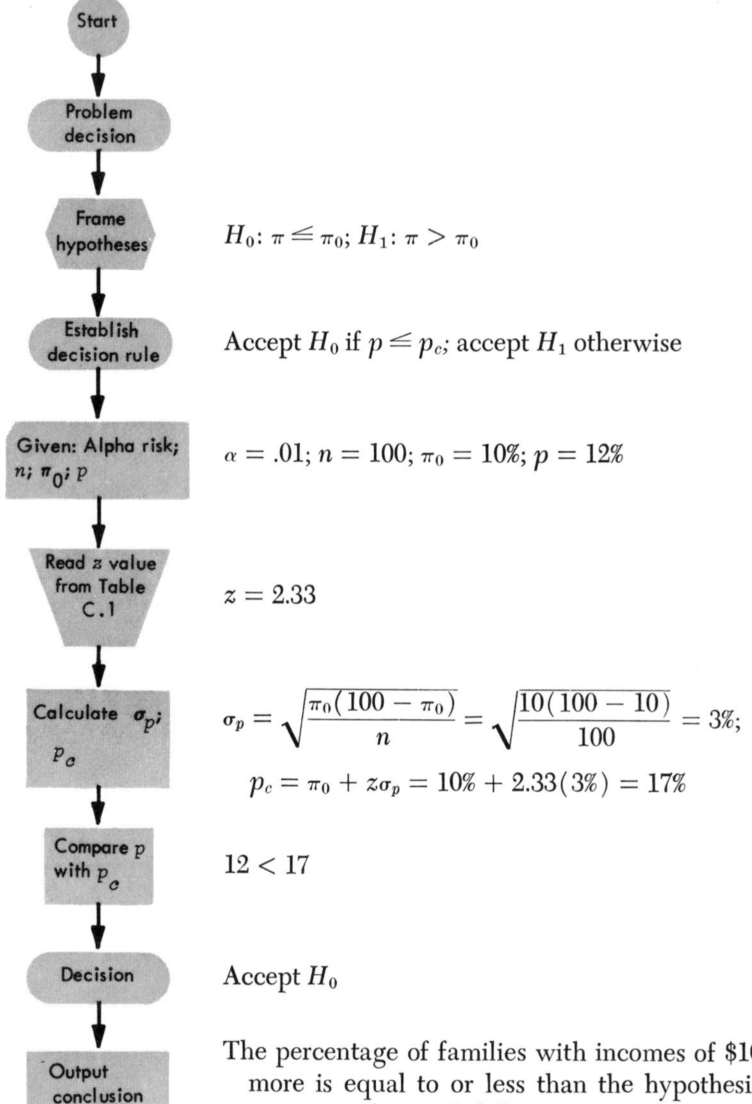

H_0: $\pi \leqq \pi_0$; H_1: $\pi > \pi_0$

Accept H_0 if $p \leqq p_c$; accept H_1 otherwise

$\alpha = .01$; $n = 100$; $\pi_0 = 10\%$; $p = 12\%$

$z = 2.33$

$$\sigma_p = \sqrt{\frac{\pi_0(100 - \pi_0)}{n}} = \sqrt{\frac{10(100 - 10)}{100}} = 3\%;$$

$$p_c = \pi_0 + z\sigma_p = 10\% + 2.33(3\%) = 17\%$$

$12 < 17$

Accept H_0

The percentage of families with incomes of $10,000 or more is equal to or less than the hypothesized parameter. Observed difference is not statistically significant at the .01 level.

FIGURE 10.21

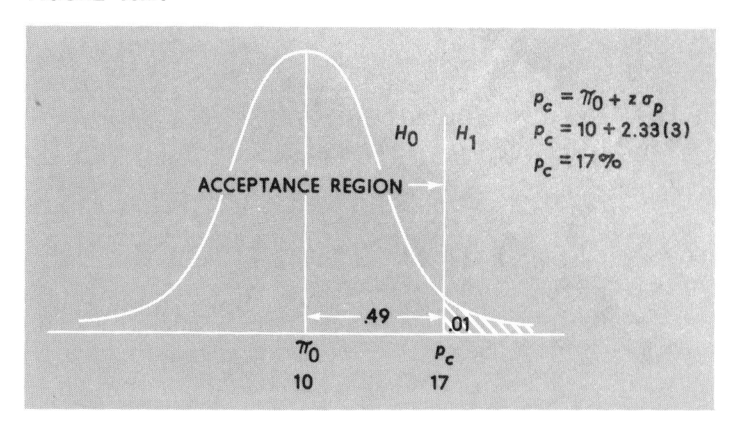

Conclude H_0 if

$$p \geqq \pi_0 - z\sigma_p$$

Conclude H_1 if

$$p < \pi_0 - z\sigma_p$$

Computations: Given

$$n = 425;\ p = 12 \text{ percent};\ \pi_0 = 15 \text{ percent};\ \alpha = .01;\ z = 2.33$$

using Formula (10.4),

$$\sigma_p = \sqrt{\frac{\pi_0(100 - \pi_0)}{n}}$$

$$\sigma_p = \sqrt{\frac{15(100 - 14)}{425}}$$

$$\sigma_p = 1.7 \text{ percent}$$

(See Figure 10.23.)

What is our conclusion? Because $p = 12$ falls to the right of p_c, we accept the null hypothesis, H_0. That is, the percentage of families with incomes of \$10,000 or more has not decreased in the given period. The difference we observed probably could have occurred by chance.[9] (See Figures 10.22 and 10.23.)

---

[9] Again,

$$z_c = \frac{12 - 15}{1.7} = -1.76$$

Because $-1.76 < -2.33$ we accept H_0.

FIGURE 10.22
Test of hypothesis re population percentage; large sample; one-sided test, left

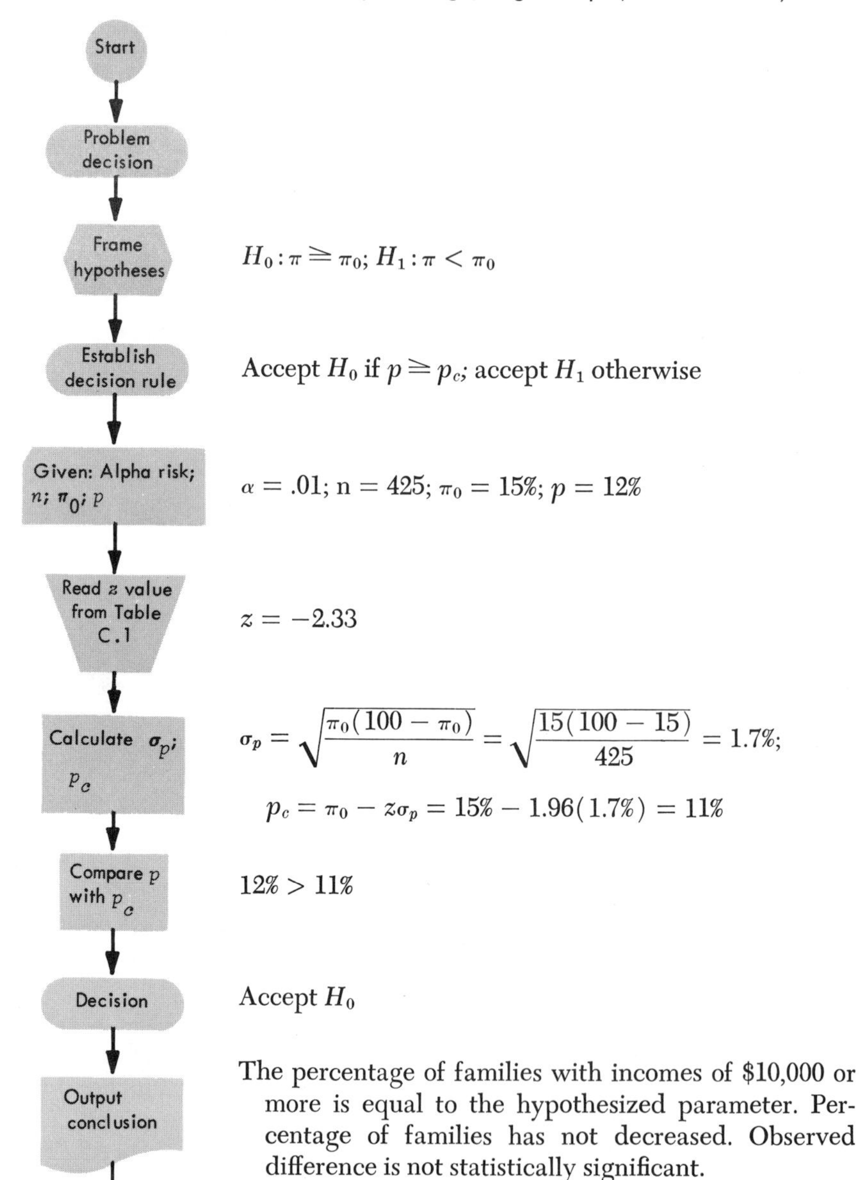

FIGURE 10.23

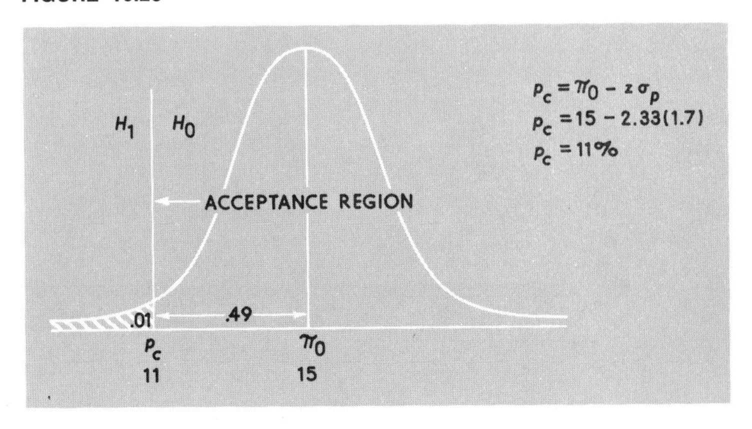

TESTING THE DIFFERENCE BETWEEN TWO MEANS[10]

Another kind of problem which finds many examples in the real world is testing whether two means are significantly different. Such problems merely involve the sampling distribution of the *differences* between means. That is, we need to know how $\bar{X}_1 - \bar{X}_2$ varies in repeated experiments. If a large number of experiments were carried out, a large number $\bar{X}_1 - \bar{X}_2$ differences would occur. Just as is true with one mean, it is not required that we carry out these experiments over and over because the form of the limiting distribution can be worked out mathematically. The sampling distribution of $\bar{X}_1 - \bar{X}_2$ will tend to be normally distributed if $\bar{X}_1$ and $\bar{X}_2$ possess independent normal distributions with means μ_1 and μ_2 and standard deviations σ_{x1} and σ_{x2}. (See Figure 10.24.) The mean of this sampling distribution will be $\mu_1 - \mu_2$ and a standard deviation expressed as

$$\sigma_{\bar{x}1 - \bar{x}2} = \sqrt{\sigma^2_{\bar{x}1} + \sigma^2_{\bar{x}2}} = \sqrt{\frac{\sigma^2_{x1}}{n_1} + \frac{\sigma^2_{x2}}{n_2}} \tag{10.5}$$

To illustrate, test the hypothesis that there is no difference between the mean weights of two samples of a production run of golf balls being manufactured to given specifications. Assume the following:

$$n_1 = 100 \qquad n_2 = 100$$
$$\bar{X}_1 = 1.620 \text{ ounces} \qquad \bar{X}_2 = 1.650 \text{ ounces}$$
$$\hat{\sigma}^2_{x1} = .4 \qquad \hat{\sigma}^2_{x2} = .5$$

---

[10] Theoretically, this method is not precise unless $n_1 = n_2$, although it does serve as a good approximation. The inexactness lies in the fact that in the real world, σ^2_{x1} and σ^2_{x2} frequently must be estimated. It can be shown that the use of correlation or analysis of variance techniques may provide better results where precision is of supreme importance.

FIGURE 10.24
Measurement data

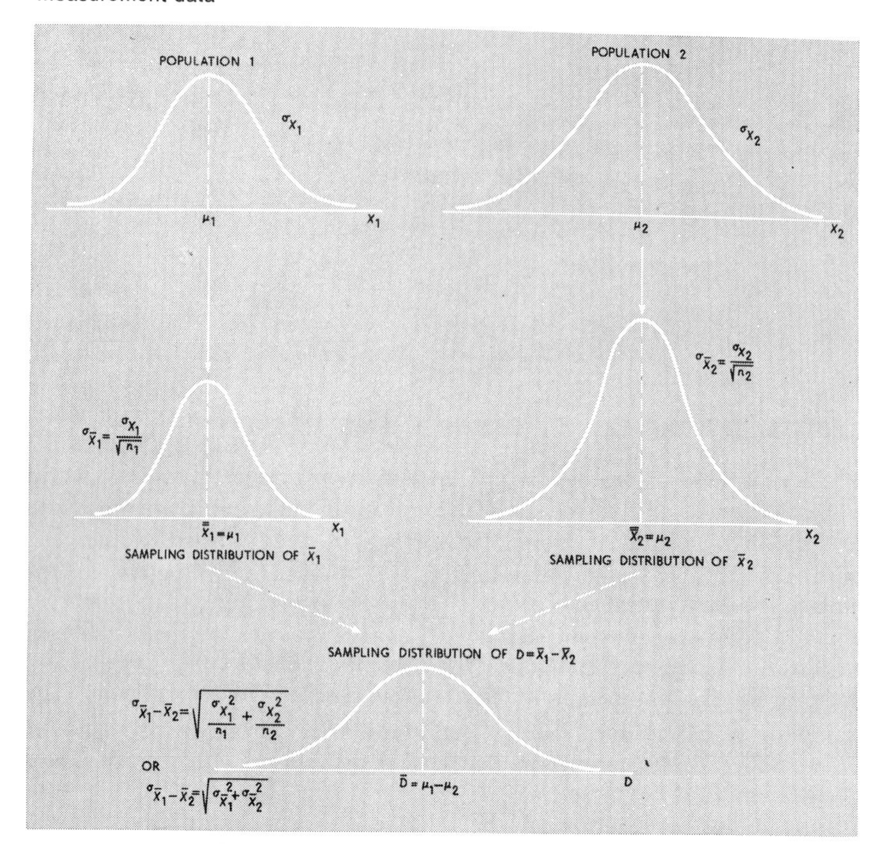

We set up the hypotheses:

$$H_0: \mu_1 - \mu_2 = 0, \quad \text{or} \quad \mu_1 = \mu_2$$
$$H_1: \mu_1 - \mu_2 \neq 0, \quad \text{or} \quad \mu_1 \neq \mu_2$$

Our criteria for choosing between H_0 and H_1, given $\alpha = .05$, would be: Conclude H_0 if

$$\bar{X}_{c1} \leq (\bar{X}_1 - \bar{X}_2) \leq \bar{X}_{c2}$$

Conclude H_1 otherwise.

What is needed is a measure of dispersion of the sampling distribution of the differences between $\bar{X}_1$ and $\bar{X}_2$. Unfortunately, σ_{x1}^2 and σ_{x2}^2 are unknown, so we must substitute $\hat{\sigma}_{x1}^2$ and $\hat{\sigma}_{x2}^2$ for these values. Using Formula (10.5)

$$\hat{\sigma}_{\bar{x}1-\bar{x}2} = \sqrt{\frac{\hat{\sigma}^2_{x1}}{n_1} + \frac{\hat{\sigma}^2_{x2}}{n_2}}$$

we find

$$\hat{\sigma}_{\bar{x}1-\bar{x}2} = \sqrt{\frac{.4}{100} + \frac{.5}{100}} = \sqrt{.009} = .094$$

Graphically, we have as shown in Figure 10.25, where

$$\bar{X}_{c_1} = 0 - z\hat{\sigma}_{\bar{x}1-\bar{x}2} = 0 - 1.96(.094) = -.18$$

and

$$\bar{X}_{c_2} = 0 + z\hat{\sigma}_{\bar{x}1-\bar{x}2} = 0 + 1.96(.094) = .18$$

Therefore the 2 samples of 100 each yielded the value $\bar{X}_1 - \bar{X}_2 = 1.62 - 1.650 = -.030$. This value falls within our acceptance region, and we accept our null hypothesis.[11]

Just as in all other tests of hypothesis problems, the acceptance of H_0 does not in any way imply that the null hypothesis is true. We do behave inductively; the sample evidence indicates that a significant difference does not exist and that the observed difference probably is due to chance. Unless further evidence is given which is contrary to our findings, for all practical purposes we assume there is no difference in the parameters μ_1 and μ_2. Students should not be tricked into believing that they have *proved* that the null hypothesis is true, because unless we examine all items of each universe, we cannot really be certain. (See Figures 10.25 and 10.26.)

FIGURE 10.25

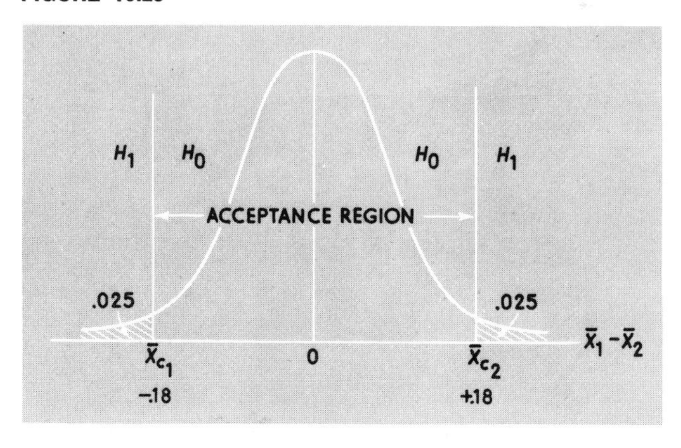

[11] Again,

$$z_c = \frac{(1.62 - 1.650) - 0}{.094} = -.32$$

Because $-.32 < -1.96$ we accept H_0.

FIGURE 10.26
Test of hypothesis whether two sample means came from the sample population; two-sided test; large sample

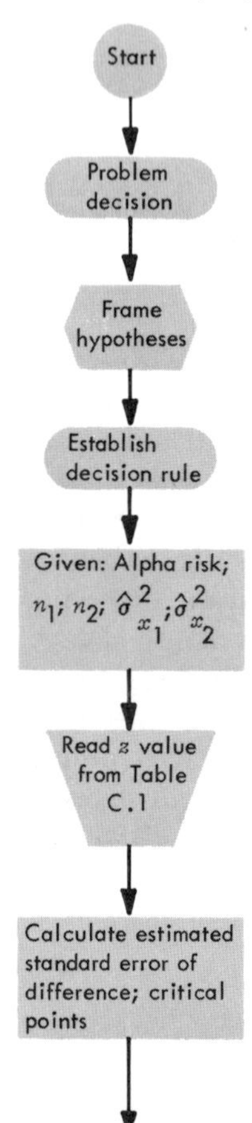

$H_0: \mu_1 - \mu_2 = 0$; or $\mu_1 = \mu_2$;
$H_1: \mu_1 - \mu_2 \neq 0$; or $\mu_1 \neq \mu_2$

Accept H_0 if $\bar{X}_{c1} \leqq (\bar{X}_1 - \bar{X}_2) \leqq \bar{X}_{c2}$;
accept H_1 otherwise

$\alpha = .05$; $n_1 = 100$; $n_2 = 100$; $\bar{X}_1 = 1.620$ ounces;
$\bar{X}_2 = 1.650$ ounces; $\hat{\sigma}^2_{x1} = .4$; $\hat{\sigma}^2_{x2} = .5$

$z = 1.96$

$$\hat{\sigma}_{\bar{x}_1 - \bar{x}_2} = \frac{\sigma^2_{x1}}{n_1} + \frac{\sigma^2_{x2}}{n_2} = \sqrt{\frac{.4}{100} + \frac{.5}{100}} = .094 \text{ ounces;}$$

$\bar{X}_{c2} = 0 - 1.96(.094) = -.18$; $\bar{X}_{c2} =$
$0 + 1.96(.094) = .18$

$\bar{X}_1 - \bar{X}_2 = 1.620 - 1.650 = -.030$;
$-.18 \leqq -.030 \leqq .18$

Accept H_0

Difference between two means is not statistically significant at the .05 level.

TESTING THE DIFFERENCE BETWEEN TWO PERCENTAGES[12]

The task of determining whether two universes differ with respect to a certain attribute is of some importance and occurs rather frequently in market research work. For example, is there any difference in the television viewing habits of men and women? Do smoking habits vary between males and females? Or is there a difference in the shopping characteristics of people of different races?

Problems of this type can be treated as testing the hypotheses that

$$H_0: \pi_1 - \pi_2 = 0, \quad \text{or} \quad \pi_1 = \pi_2$$
$$H_1: \pi_1 - \pi_2 \neq 0, \quad \text{or} \quad \pi_1 \neq \pi_2$$

where π_1 and π_2 are the population percentages of the given attribute. Where n_1 and n_2 denote the sample sizes, then p_1 and p_2 are the sample percentages. The variable we are interested in, then, is $p_1 - p_2$, which corresponds to using $\bar{X}_1 - \bar{X}_2$ in our example where we tested $H_0: \mu_1 = \mu_2$. The methodology for solving this problem is the same because p_1 and p_2 can be viewed as two independent normal random variables. The theory here is that $p_1 - p_2$ can be considered as being approximately normally distributed with mean $\pi_1 - \pi_2 = 0$ and with a standard deviation of the sampling distribution given by

$$\sigma_{p1-p2} = \sqrt{\frac{\pi_1(100 - \pi_1)}{n_1} + \frac{\pi_2(100 - \pi_2)}{n_2}} \tag{10.6}$$

or

$$\hat{\sigma}_{p1-p2} = \sqrt{(p)(100 - p)(1/n_1 + 1/n_2)} \tag{10.7}$$

(See Figure 10.27.) Because the values of π_1 and π_2 are unknown, they are estimated by using sample values. Although the parameters are unknown, they are assumed to be equal under the hypothesis $H_0: \pi_1 = \pi_2$. If this common value is denoted by p, then we can estimate p by the value obtained from the sample percentage of the combined data.

For example, in the southeast section of a large city, 120 out of 200 taxpayers are opposed to a proposed city income tax plan. In the northwest section of the same city, only 110 out of 200 are opposed to the same plan. Test to see if the rate of opposition is the same for these two sections of the city, using a .05 level of significance.

Calculations yield:

$$p_1 = 120/200 \times 100 = 60 \text{ percent}$$
$$p_2 = 110/200 \times 100 = 55 \text{ percent}$$

where

$$n_1 = n_2 = 200$$

[12] See note 10 on page 297.

FIGURE 10.27
Attribute data

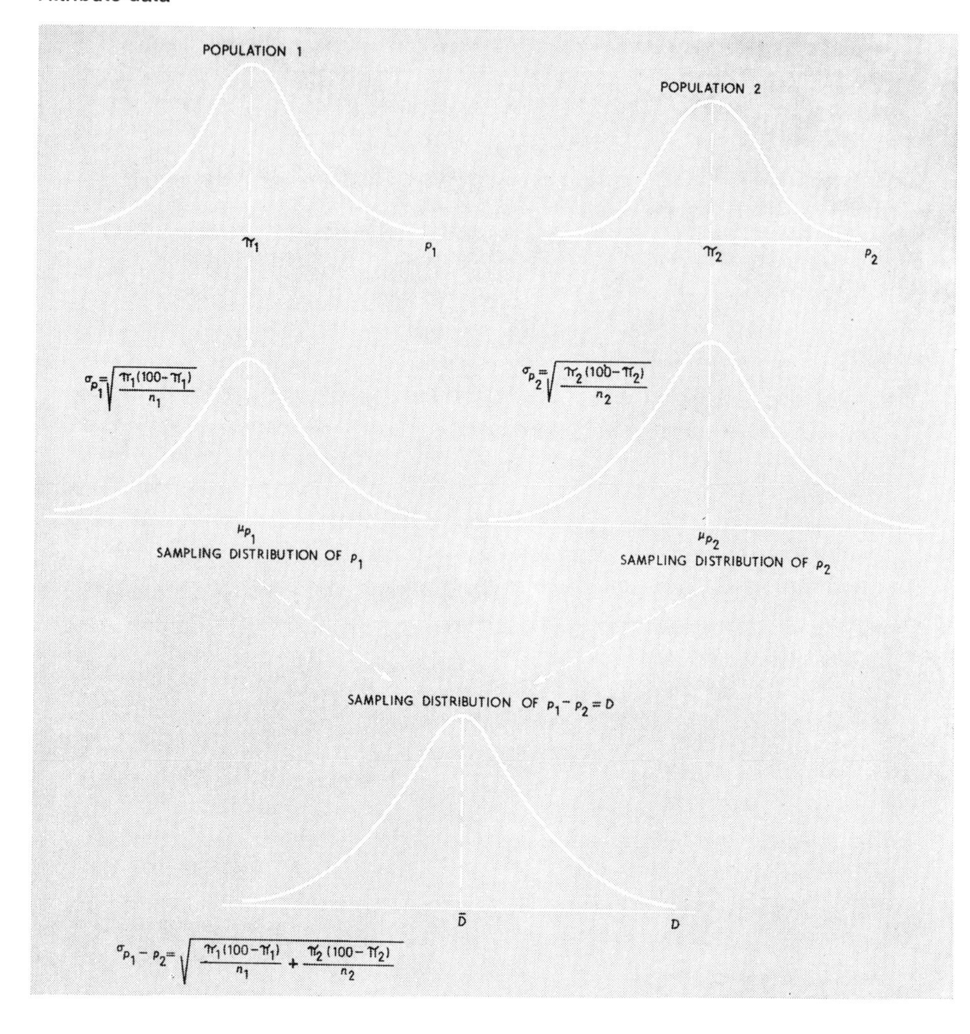

and

$$p = \frac{230}{400} \times 100 = 58 \text{ percent}$$

By replacing p_1 and p_2 in the formula for the standard error of the difference between two percentages ($\sigma_{p_1 - p_2}$), we obtain (see Formula [10.7])

$$\hat{\sigma}_{p_1 - p_2} = \sqrt{(58)(42)(1/200 + 1/200)} \simeq 4.9 \text{ percent}$$

Graphically, this problem appears as shown in Figure 10.29, where

$$p_{c_1} = 0 - z\hat{\sigma}_{p_1 - p_2} = 0 - 1.96(4.9) = -9.6$$

FIGURE 10.28
Test of hypothesis whether two sample percentages came from the same population; two-sided test; large sample

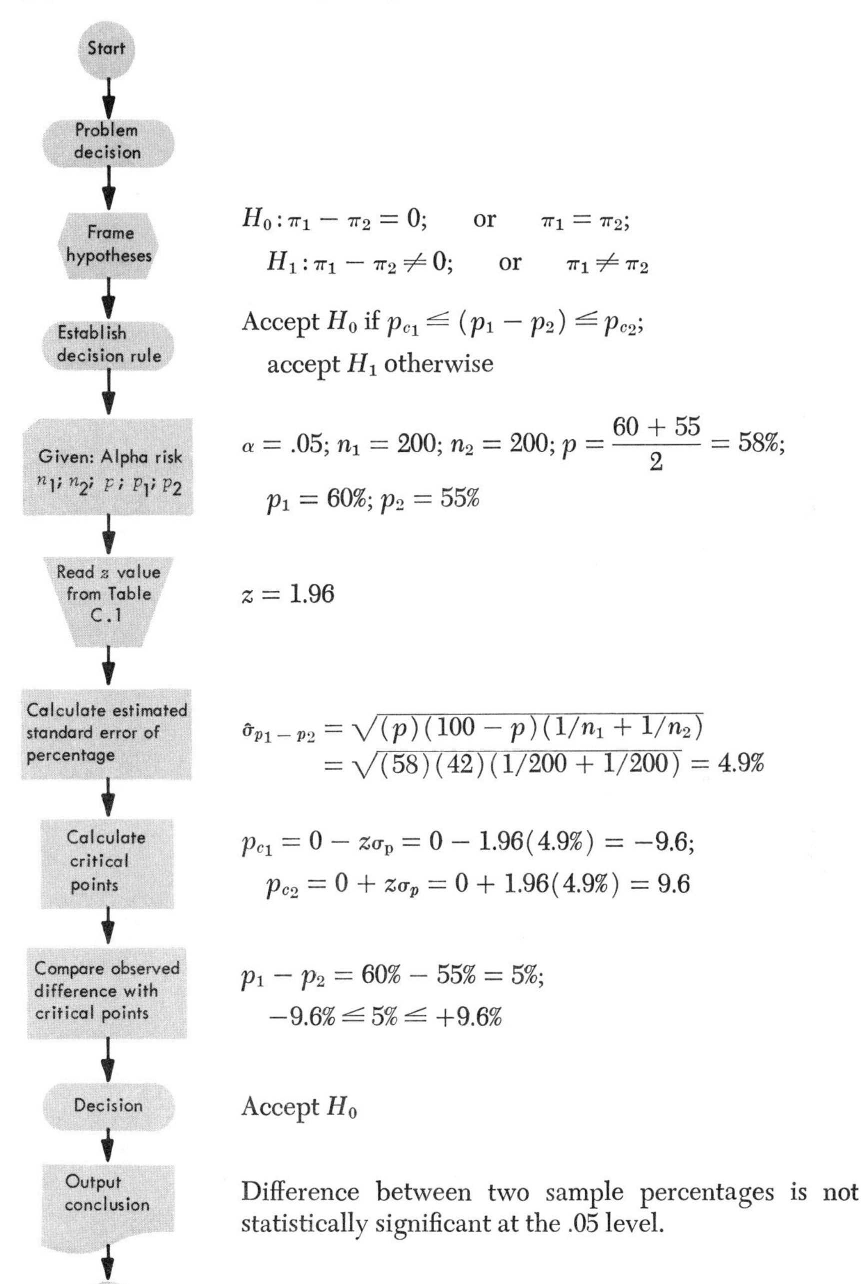

FIGURE 10.29

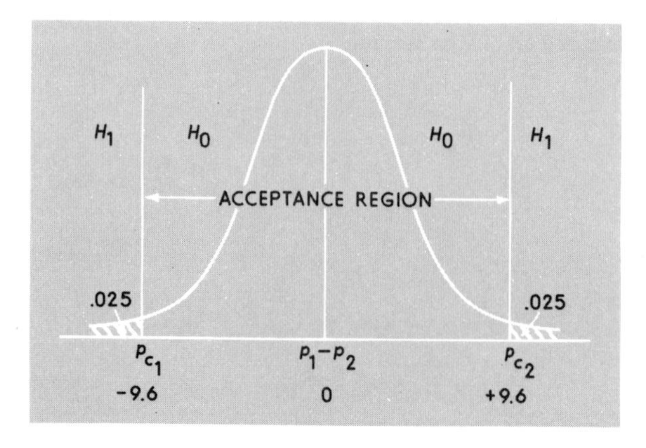

and

$$p_{c2} = 0 + z\hat{\sigma}_{p1-p2} = 0 + 1.96(4.9) = 9.6$$

Since $p_1 - p_2 = 60 - 55 = 5$ percent falls within acceptance region, we accept the null hypothesis that $\pi_1 = \pi_0$ and conclude that the difference we observed is not significant; it probably could have occurred by chance. It would appear that there are no real differences in reactions to the proposed income tax in the two sections of the city. Again, the student is cautioned that the null hypothesis has not been *proved* correct; however, unless further evidence is presented, we can assume that there exists no difference in reactions.[13] (See Figures 10.28 and 10.29.)

SYMBOLS

H_0, null hypothesis
H_1, alternative hypothesis
μ_0, hypothesized mean
π_0, hypothesized percentage
μ_1, alternative hypothesized value of mean
π_1, alternative hypothesized value of percentage
$\sigma_{\bar{x}}$, true standard error of mean
$\hat{\sigma}_{\bar{x}}$, estimated standard error of mean
σ_p, true standard error of percentage
$\hat{\sigma}_p$, estimated standard error of percentage
$\bar{X}_c$, $\bar{X}_{c_1}$, $\bar{X}_{c_2}$, critical points for measurement problem

---

[13] Again,

$$z_c = \frac{(60 - 55) - 0}{4.9} = 1.22$$

Because $1.22 < 1.96$ we accept H_0.

p_c, p_{c1}, p_{c2}, critical points for attribute problem
α, alpha risk, probability of a Type I error
β, beta risk, probability of a Type II error

$\hat{\sigma}_x = s_x \sqrt{\frac{n}{n-1}}$, estimated population standard deviation

s, s_x, sample standard deviation for variable X
$\sigma_{\bar{x}_1 - \bar{x}_2}$, true standard error of the difference between two means
$\hat{\sigma}_{\bar{x}_1 - \bar{x}_2}$, estimated standard error of the difference between two means
$\sigma_{p_1 - p_2}$, true standard error of the difference between two percentages
$\hat{\sigma}_{p_1 - p_2}$, estimated standard error of the difference between two percentages

PROBLEMS

10.1. A widely known trade journal claimed to have derived data which showed that the personal income per family in a given city was \$6,540. This estimate was questioned by L. B. Neyman of the Reliable Research Organization (RRO), who had just completed an economic status survey based on 100 randomly selected personal interviews with families in the city. Neyman discovered that the average personal income per family was \$6,080, with a sample standard deviation of \$900. On the basis of RRO data, test the hypothesis that $\mu = \$6{,}540$. Use $\alpha = .05$. What are your conclusions?

10.2. The daily newspaper of a city of 45,000 people, in a by-line article, states that 60 percent of the people in the town own their homes. J. La Place, a research assistant in the city manager's office, questions this hypothesis that $\pi = 60$ percent. La Place takes a sample of 300 assessment cards from the files of the city assessor. The 300 cards show:

Attribute	Number
Own their homes	170
Rent their homes	130
Total	300

On the basis of these results, should La Place accept or reject the hypothesis that $\pi = 60$ percent? Show how you arrived at your answer, using $\alpha = .05$.

10.3. The Pascal Public Opinion Poll (PPOP) claims that 66 percent of the voters in a forthcoming national election favor candidate K, and that the remaining 34 percent is distributed among three others. Test the hypothesis that the PPOP claim is reasonable if another independent survey conducted by Galton Associates, Inc., produced the following results, based on a nationwide probability sample. Use a .05 level of significance.

Attribute	Number of Respondents
Favor candidate K	408
Favor candidate G	92
Favor candidate T	56
Favor candidate R	44
Total	600

10.4. Dr. C. Pearson, the director of admissions at a large midwestern university, claims that the average score on the Graduate Management Admission Test (GMAT) by students accepted at the school for the next semester is 550. Given the population standard deviation of 96, a .05 level of significance, and a random sample of 64 students' GMAT test scores, determine the acceptance region for a two-sided test. Assume the sample of 64 yields an average of 548. Is Dr. Pearson's hypothesis tenable?

10.5. In which of the following should you reject the hypothesized value of the parameter, using a two-sided test?

a. $n = 10$; $\hat{\sigma} = 2.0$; $\alpha = .01$; $\mu_0 = 40$; $\bar{X} = 39$.
b. $n = 25$; $\sigma = 3.0$; $\alpha = .01$; $\mu_0 = 70$; $\bar{X} = 65$.
c. $n = 100$; $\hat{\sigma} = 2.5$ $\alpha = .05$; $\mu_0 = 100$; $\bar{X} = 103$.
d. $n = 225$; $\alpha = .05$; $\pi_0 = 60$ percent; $p = 53$ percent.
e. $n = 2{,}500$; $\alpha = .10$; $\pi = 50$ percent; $p = 48$ percent.

10.6. The Wald Food Wholesale Company, Inc. has found that for years the credit department has managed to keep outstanding accounts receivables at an average age of 27 days. This figure is used as a bench mark by which to judge the efficacy of its credit policy and collection efforts. Recently, a regular monthly check revealed that a random sample of 64 accounts yielded an average of 29 days, with a standard deviation of four days.

a. L. Yule, the credit manager, asks you whether this difference is significant. (Use $\alpha = .05$.) How would you respond from a statistical point of view?
b. What is the probability that a sample mean would vary by more than two days from the bench mark or standard?
c. Construct a .99 confidence interval for the mean of the universe, based on the sample evidence. What does this suggest relative to recent credit and collection experience?

10.7. "The Rules of Golf" simply states that a golf ball must "be no smaller than 1.680 inches in diameter." The Ace Manufacturing Company is producing a new type of golf ball, and one of the tests used is to insure that the golf balls meet the above rule. Assume that 100 randomly selected golf balls from a given production run are measured and the average diameter is 1.674 inches, with a sample standard deviation of .06 inches. Using $\alpha = .01$, test the hypothesis that $\mu \geqq \mu_0$. Do these balls meet the rule standards?

10.8. Another characteristic of a golf ball controlled by the rule makers pertains to weight. The rule says that the ball should be no heavier than 1.620 ounces. The Ace Manufacturing Company wishes to insure keeping within this standard. Assume that 12 golf balls are selected per hour from each hour's production run and weighed. In one particular hour, it was found that the mean weight was 1.650 ounces. The population standard deviation has remained rather constant at .06 ounces. Test the hypothesis that $\mu \leqq \mu_0$, using $\alpha = .01$. Does this hour's production meet the rule standards?

10.9. Rule 2–3 of "The Rules to Golf" states that the velocity of a ball may not exceed 250 feet per second (fps) when tested by the United States Golf Association (USGA) on a machine designed to measure accurately this characteristic. The USGA permits a reasonable tolerance of 2 percent. Assume that a sample of 225 golf balls was tested by the USGA and it was found that the average velocity was 260 fps, with a sample standard deviation of 3 fps. Using a .05 level of significance, test the hypothesis that $\mu \leq \mu_0$, using $\alpha = .01$. Does this hour's production meet the rule standards? USGA? Would your answer have been the same if you had used a .01 level of significance?

10.10. The USGA maintains a constant vigil on the possibility that current technology could soon make the present golf ball regulations obsolete. For example, new materials might be used that would make nonsense of the present velocity rule. Therefore the USGA is concerned with a scientific concept called the coefficient of restitution. This characteristic is usually labeled "e" and is merely the ratio of the relative velocity of the golf ball and the club after impact to their relative velocity before impact. To illustrate, if a club strikes the ball with a velocity of 200 fps, this figure is a relative velocity *prior* to impact, because the ball is standing still (zero velocity). Assume that the ball leaves the club after impact at a velocity of 250 fps and that the club is slowed down to 120 fps. The *after-impact* relative velocity is 250 minus 120, or 130 fps. Therefore, "e" has the value of 130/200, or .65, which is a fairly typical figure.

Assume that a velocity test is made on 100 randomly selected Birdie Balls from a large lot manufactured by the Remlap Golf Equipment Company. It is found that the average "e" is .70. The estimated population standard deviation is .10. Answer the following:

a. Using $\alpha = .10$, test the hypothesis that $\mu = \mu_0 = .65$.
b. Using $\alpha = .02$, test the hypothesis that $\mu \geq \mu_0$ where $\mu_0 = .65$.
c. Using $\alpha = .05$, test the hypothesis that $\mu \leq \mu_0$ where $\mu_0 = .65$.

10.11. Given the following information concerning the diameters of two different production runs of the same brand of golf balls:

$n_1 = n_2 = 100$ $\qquad \bar{X}_2 = 1.669$ inches
$\bar{X}_1 = 1.680$ inches $\qquad s_2^2 = .06$ inches
$s_1^2 = .04$ inches

Test the hypothesis that $\mu_1 = \mu_2$ with a .01 level of significance.

10.12. The same two samples of golf balls given in Problem 10.11 were tested for weight. The following results were obtained:

$n_1 = n_2 = 100$ $\qquad \bar{X}_2 = 1.625$ ounces
$\bar{X}_1 = 1.620$ ounces $\qquad s_2^2 = .03$ ounces
$s_1^2 = .06$ ounces

Test the hypothesis that $\mu_1 = \mu_2$ at the .05 level of significance.

10.13. In a nationwide investigation, it was found that 46 out of 460 long-time smokers eventually had a lung ailment. Only 9 out of 460 nonsmokers had a similar ailment. Using $\alpha = .10$, test the hypothesis that $\pi_1 = \pi_2$.

10.14. In one section of a large city, 200 out of 250 residents own the dwellings in which they live. In another section, 120 out of 200 residents own their homes. Using a .05 level of significance, can we conclude that there is no difference between these two sections of the city in terms of homeownership?

10.15. Explain the distinction between proving an hypothesis and testing an hypothesis.

10.16. The city assessor's office takes two samples of home evaluations (i.e., estimates of "true market value") from the results of a recent citywide reassessment project. Sample 1 is taken from area 1 and Sample 2 from area 2. The summarized results are:

Sample 1	Sample 2
$n_1 = 50$	$n_2 = 50$
$\bar{X}_1 = \$30{,}000$	$\bar{X}_2 = \$32{,}000$
$\hat{\sigma}^2_{x_1} = \$1{,}500$	$\hat{\sigma}^2_{x_2} = \$2{,}500$

Is the estimated "true market value" of the homes in area 2 significantly different from that in area 1? Use a .01 level of significance.

10.17. Two manufacturing processes are to be compared. Given the following:

$$n_1 = 100 \qquad n_2 = 100$$
$$\bar{X}_1 = 110 \qquad \bar{X}_2 = 115$$
$$\hat{\sigma}^2_{x_1} = 400 \qquad \hat{\sigma}^2_{x_2} = 225$$

Is there a significant difference between the two processes, using $\alpha = .05$?

10.18. It is hypothesized that more than 10 percent of the families in the United States plan to buy new automobiles this year. A sample of 2,000 families interviewed in a nationwide survey indicated that 160 planned to purchase new cars this year. Do you accept the hypothesis at the .01 level of significance?

10.19. It is hypothesized that more than 60 percent of the durable goods manufacturers in the United States plan to spend more on new plant and equipment this year than they did in the previous year. In a sample of 1,000 firms, what is the smallest number than can plan to increase such expenditures to permit acceptance of the hypothesis at the .10 level of significance?

10.20. Ggollek's, a large breakfast cereal producer, had decided to introduce a new product. The management requests the market research unit to determine which of two packages might have the greatest customer appeal. Using a nationwide panel of 2,000 families, each is given two entirely different packages with identical contents but identified as brand X and brand Y. The following results are reported to the statistician:

Favor brand X	825
Favor brand Y	750
No preference	425
Total	2,000

Is there a significant difference between customer preference for X or Y? Use a .01 level of significance.

10.21. A consumer research agency reported to a car manufacturer that a random sample of auto owners (different samples for each brand) revealed that for minor repairs, brand A had an average repair cost of $55.42 while brand B had an average of $52.78. (In each case the costs referred to the same model and year.) You have been asked to evaluate these data by responding to the following:

a. Would you assume that this is a measurement or attribute type problem? Why?

b. What might explain the observed differences between the two means?

c. As applied to this case, what might be meant by the term "statistical significance?"

d. If you could request additional information from the research agency to determine the significance of these data what kind of information would you ask for? (Generally, additional information costs money and therefore useless information is to be avoided.)

e. Assume that the manufacturer of brand A has just hired you as a market analyst. Given the information above (including what you just requested in item [*d*]) is there anything that management must provide that would help you?

10.22. Assume that on the basis of a probability sample we want to test the hypothesis that the average selling price for one-half acre residential lots in Ann Arbor, Michigan, is $15,000. No computations are necessary but answer the following:

a. Explain how we might commit a Type I error given the above null hypothesis.

b. Explain how we might commit a Type II error given the above null hypothesis.

10.23. The way in which we state the null hypothesis influences whether an error is Type I or Type II. For example, if we assume that M. Bogey is being interviewed for a research position with a large corporation. Bogey's probable supervisor is concerned about Bogey's competence as a researcher.

a. If the null hypothesis is stated as: "M. Bogey is a competent researcher," explain under what conditions might the supervisor be committing Type I and Type II errors.

b. If the null hypothesis is stated as: "M. Bogey is not a competent researcher," indicate how the Type I error of (*a*) above is now a Type II error and vice versa.

10.24. Refer to Problem 9.31. Given the background of Chapter 10 how would you respond to 9.31(*c*) in a statistical sense? Use a .05 level of significance.

10.25. Suppose that the null hypothesis is that we have an honest and true dime. We formulate a decision rule that indicates the null hypothesis is to be rejected if three tosses of the coin produce either three heads or three tails.

a. If the coin is in fact honest and true, what is the probability of a Type I error?

b. If the coin is in fact an honest and true dime, what is the probability of a Type II error?

10.26. In a probability sample poll of eligible voters in a large city, 50 out of 200 women favor some sort of revenue-sharing plan. On the other hand, 60 out of 200 men interviewed favored this plan. Using a .05 level of significance, is there a real difference of opinion in this city between women and men voters on the revenue-sharing plan?

10.27. A supplier is producing some product for the government according to some rigid specifications. In the recent past, the supplier has been able to produce these items with a standard deviation of 25 units. The government adopts the following decision-making rule: Select a probability sample of 100 items from a shipment and determine the mean (of the characteristic specified by the government). If this mean is 1,650 units or more, accept the total lot; if the mean is less than 1,650 units, reject.

a. It is calculated that the probability of rejecting the total lot is .98 if the true mean equals 1,645 units. Is this an alpha risk, a beta risk, or neither?.

b. Clearly, then, given the information in (*a*), the probability of accepting the shipment is .02. Is this an alpha risk, beta risk, or neither?

10.28. A large number of snowmobile batteries were subjected to a simulated test to estimate their average life. It turned out that the sample mean was 30 months of life. If 60 percent of them average between 28 and 32 months and 25 batteries lasted longer than 34 months, how many were in the sample tested?

10.29. In naming a new subcompact, an auto company official who suggested the name, claims that at least 55 percent of the potential customers will prefer "Prancer" to any other name suggested. In a national probability sample of consumers, 560 out of 1,000 interviewees stated a preference for "Prancer." Do you accept this company official's claim at the .10 level of significance?

10.30. Carefully explain the distinction between statistical estimation and statistical tests of hypothesis.

10.31. Using a two-sided test, would you accept the null hypothesis for an observed $p = 48$ percent in a random sample of 100 observations with alpha of .05 if $\pi_0 = 40$ percent?

10.32. In employing a decision-making rule involving a .01 risk of making a Type I error, management concluded that a parameter had changed.

a. If the risk of making a Type I error had been stated as .02, would management have reached the same conclusion? Why?

b. If the risk of making a Type I error had been stated as .001, would management have reached the same conclusion? Why?

10.33. A large midwestern state university has 3,247 faculty members. A random sample is taken of the faculty to determine their views on a student-faculty committee's proposed controversial judicial system. Out of the sample of 300, the proposal was rejected by 210 faculty members. In a similar survey of the student body, a sample of 300 students (representing 35,000) yielded a rejection rseponse from 225. Using a .10 level of significance, is there a real difference between the two rejection percentages?

10.34. The mean score of the Graduate Management Admission Test (GMAT) for a sample of 100 students admitted to an MBA program at a given university is 575, with a standard deviation of 25. The mean score of a sample of 100 students who actually enrolled in this university was 580 with a standard deviation of 30. Is there a significant difference between the two groups at the .05 level?

RELATED READINGS

ARNOLD, B. C. "Hypothesis Testing with a Preliminary Test of Significance." *Journal of American Statistical Association,* vol. 65, no. 332 (December 1970), pp. 1590–96.

BISCHOFF, CHARLES W. "Hypothesis Testing and the Demand for Capital Goods." *Review of Economics and Statistics,* vol. 50 (August 1969), pp. 354–68.

FREUND, JOHN E. *Mathematical Statistics,* chaps. 11–12. 2d ed. Englewood Cliffs, N.J.: Prentice-Hall, Inc., 1971.

LEHMAN, E. L. *Testing Statistical Hypotheses.* New York: John Wiley & Sons, Inc., 1959.

LINDGREN, B. W. *Statistical Theory,* chap. 7. New York: Macmillan Co., 1962.

MCCOLLOUGH, CELESTE, and VAN ATTA, LOCHE. *Statistical Concepts,* chaps. 5, 12, and 14–15. New York: McGraw-Hill Book Co., 1963.

MOOD, AEXANDER M., and GRAYBILL, FRANKLIN A. *Introduction to the Theory of Statistics,* chap. 9. 3d ed. New York: McGraw-Hill Book Co., 1974.

11

Operating characteristic curves: Power functions[1]

Uncertainty and expectations are the joys of life.

Congreve, Love for Love

THE PROBABILITY of rejecting a true hypothesis or accepting a false hypothesis was alluded to in Chapter 10. That is, it would be an error to reject a true statement about a population. This, we learned previously, is called an error of Type I. It would also be a mistake to accept a false hypothesis and thereby commit an error of Type II.

Curves used to depict the probability of rejecting the null hypothesis are called power curves (see Figure 11.1). That is, they show the probability of a Type I error, which can occur only when the hypothesis is true, and the probability of a Type II error, which can occur at every point *except* where the true parameter minus the hypothesized parameter equals zero. One minus the height of the curve gives the probability of a Type II error, except at the point where the hypothesis is true. At this latter point a Type II error cannot be made. Power curves are so called because they indicate a decision-making rule's ability (or power) to discriminate between the alternative courses of action or conclusions.

Operating characteristic curves (called OC curves) have a similar application, but they indicate the probability of accepting the zero hypothesis (see Figure 11.2). An *OC* curve tells us what our decision-making rule (or test) will do for all possible values that the true population parameter may assume. The curve shows the probability of a Type I error which, of course, can occur only when the hypothesis is true, for example, when $\mu - \mu_0 = 0$. (The probability is found by subtracting the height of the curve at this specific point from one.) An *OC* curve also

[1] This material may be considered optional. Although it is the feeling of the author that a first course should *introduce* students to this subject, the chapter may be omitted without loss of continuity.

FIGURE 11.1
Power curve

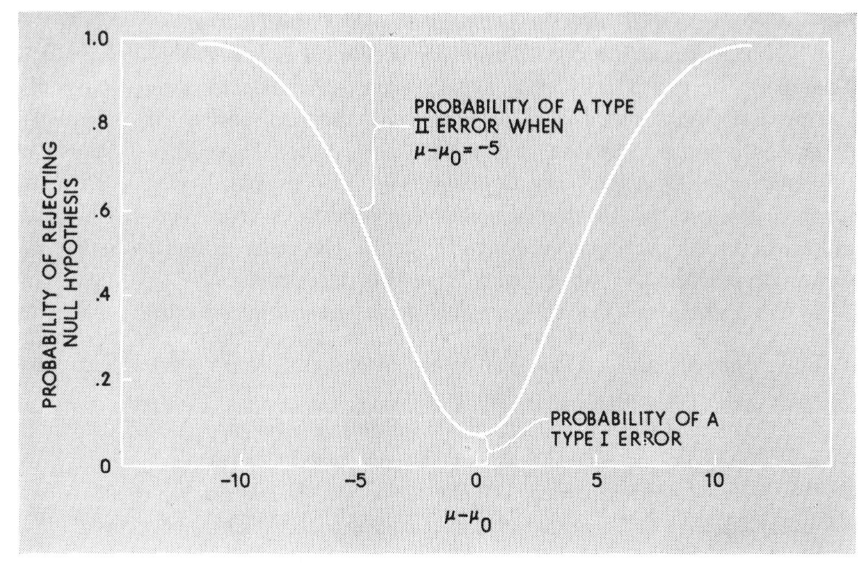

Note: Null hypothesis is $H_0 : \mu = \mu_0$.
Alternative hypothesis is $H_1 : \mu \neq \mu_0$.

FIGURE 11.2
Operating characteristic curve

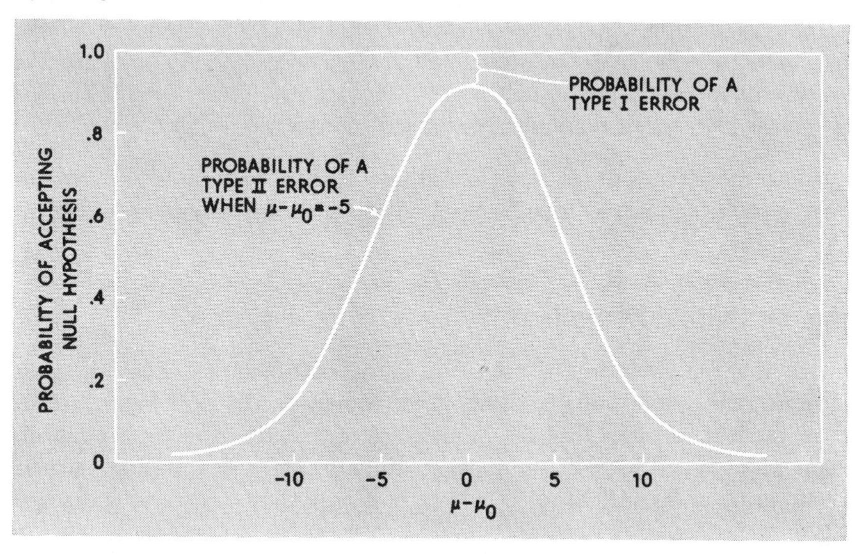

Note: $H_0 : \mu = \mu_0$
$H_1 : \mu \neq \mu_0$

shows the probability of a Type II error, which can occur at every point *except* where $\mu = \mu_0$. *At every point except where* $\mu - \mu_0 = 0$, *the height of the OC curve indicates the probability of a Type II error.*

The hazard associated with an error of Type I is called the alpha risk.[2] Therefore the probability of a Type I error is easily computed—it equals the preassigned alpha risk. Unfortunately, the probability of a Type II error (called the beta risk) cannot be determined so readily because it depends upon *how false the hypothesis is.* The probability of a Type II error decreases the further the null hypothesis is from the truth. The closer the null hypothesis comes to the truth, the more difficult it is to discriminate and the probability of a Type II error increases.

In these prefatory remarks, several points need emphasizing:

1. Power curves and *OC* curves relate to the problem of what happens in the *long run,* not to a specific event.
2. They are really unrelated to the confidence interval type of question which may be the more commonly encountered situation.
3. When the statistician speaks of the alpha and beta risks we must recognize that they are somewhat arbitrarily selected. They represent only two points of many which might be chosen, through which the power curve or *OC* curve must pass to satisfy the test or decision-making rule. Once the curve has been determined, we are no longer concerned with α and β as such but rather with the ability of the test to discriminate in the long run.
4. These two types of curves must be interpreted with caution. It may *appear* that the true mean of the population is variable because it does change on the x axis (see Figures 11.1 and 11.2). *However,* μ, *the population mean, is constant, and these curves merely indicate the discriminating power of our decision-making rule (test) for the various values which that unknown parameter (a constant) may assume.*

With these concepts in mind, let us now turn to two kinds of problems where such techniques may be applied for both measurement and attribute data.

CALCULATION OF PROBABILITY OF A TYPE II ERROR

As stated in Chapter 10, a test of a hypothesis includes choosing a test statistic and determining an acceptance region. If the null hypothesis is true, the test specifies that the probability of rejecting the H_0 is equal to some preassigned level of significance, the alpha risk. Naturally, it would be an error of Type I to reject a true null hypothesis. Also, as indicated

[2] The alpha risk may be looked upon as a sort of "unconfident coefficient" because it indicates the percent of time we expect to be wrong.

previously, it would be an error of Type II to accept a false null hypothesis. This is the beta risk. Until this point, our method of testing has controlled the chance of making a Type I error. Let us consider the chance of accepting the null hypothesis when the alternative course of action really should be taken.

If we declare the null hypothesis false, we imply some idea of what alternatives might exist. To illustrate, if for a normal population with a standard deviation of 10 units, we are testing the hypothesis that the true mean of the distribution equals 50, feasible alternative values of the mean might be 40, 60, 35, and so on. *In fact, the true mean might be any value but 50.* However, if it so happens that the true mean is 50, then there is a certain probability of rejecting the null hypothesis. If the true mean is 40, then there is some other probability of rejecting the hypothesis. If the true mean is 70, then there exists still another probability of rejection of the null hypothesis.

This probability of rejecting the null hypothesis is called the *power of a test.* Naturally, the *power* depends upon which alternative actually is true. In the real world of practical application we do not *know* which alternative is true. We usually are interested in the power of a test for several possible values of the mean. Of course, if the null hypothesis is not true, we would like the probability of rejecting to be as large as possible. In most real-life situations we must balance the probabilities of these two kinds of errors against one another. As mentioned above, the probability of a Type I error is equal to the size of the rejection region. This area equals the alpha risk. The probability of a Type II error is somewhat more complicated to evaluate because it depends upon how false the null hypothesis really is! *The further the null hypothesis is from the truth, the less likely we are to accept it. The closer the null hypothesis is to the fact, the more likely we are to accept a false statement.*

Illustration: Assume the facts below:

$$\mu_0 = 100 \text{ units; } \alpha = .05; \ \sigma_{\bar{x}} = 2.0 \text{ units}$$
$$\sigma = 20 \text{ units; } n = 100$$

Assume further that we are testing the following:

$$H_0 : \mu = \mu_0.$$
$$H_1 : \mu \neq \mu_0$$

Our criteria for choosing between these alternatives for this two-sided test are:

$$\text{Conclude } H_0 \text{ if } \mu_0 - z\sigma_{\bar{x}} \leq \bar{X} \leq \mu_0 + z\sigma_{\bar{x}}$$
$$\text{Conclude } H_1 \text{ otherwise}$$

Figure 11.3 can help us understand the concept of a Type II error. If the true mean of our population being sampled really is 100 units, prob-

FIGURE 11.3

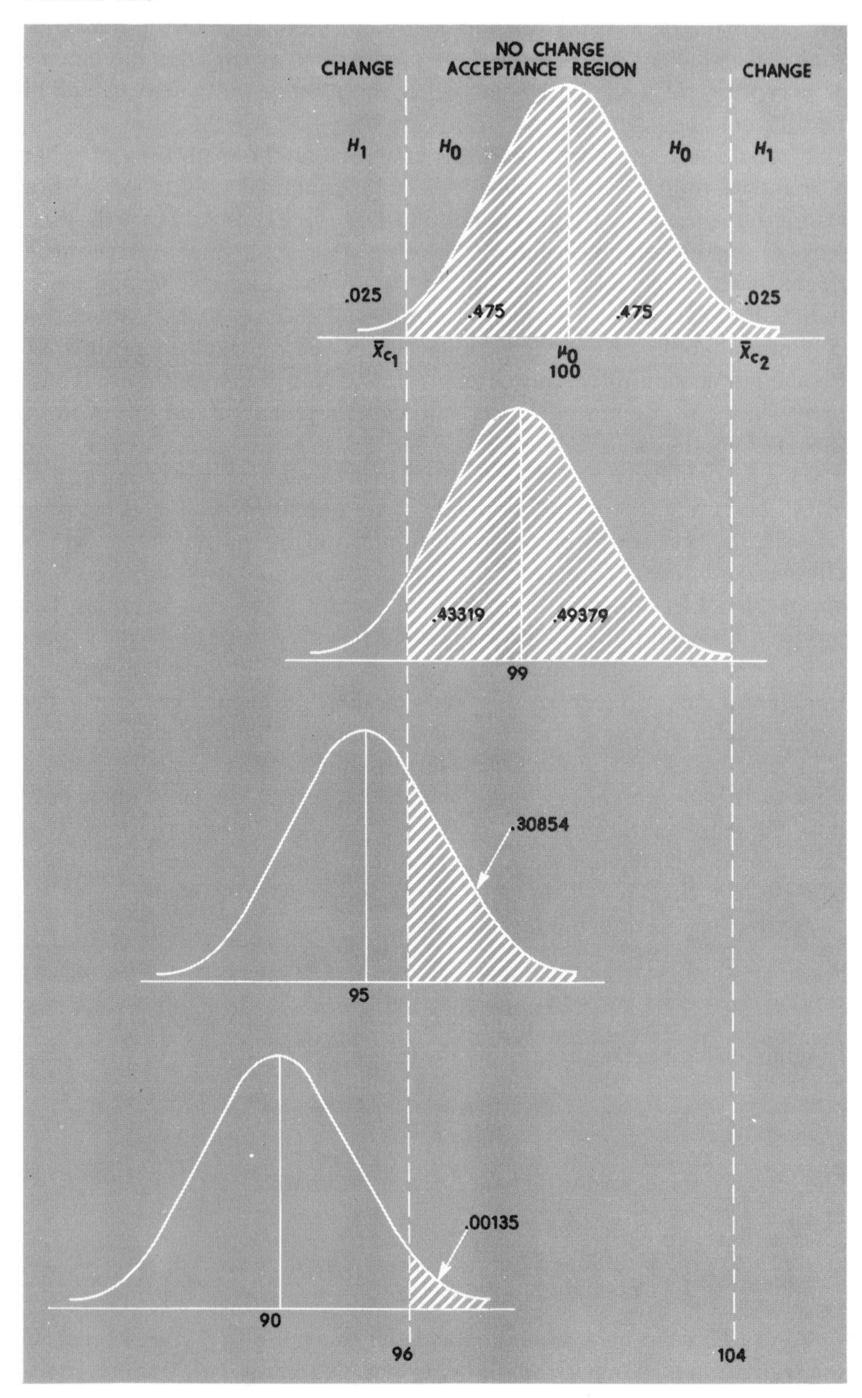
NO CHANGE
ACCEPTANCE REGION
CHANGE
CHANGE
H_1
H_0
H_0
H_1
.025
.475
.475
.025
$\bar{X}_{c_1}$
μ_0
100
$\bar{X}_{c_2}$
.43319
.49379
99
.30854
95
.00135
90
96
104

ability of accepting H_0 is .95 because of the given alpha. If this is the situation, we *cannot* make a Type II error. By definition a Type II error is to accept a false hypothesis, and naturally we cannot do so if H_0 is true.

However, if the true mean is 99 instead of as we have hypothesized, the probability of a Type II error is much greater than if the parameter is 95, 94, 93, 92, 91, or 90. As we might expect, the probability of a Type II error (accept a false null hypothesis) decreases rapidly the further H_0 is from the truth. This can be seen by observing that as we move further from the truth the less area of the curve remains in our acceptance region.

Let us compute the probabilities of a Type II error for the alternative values of the mean as given in Figure 11.3. If $\mu = 99$, then the probability of accepting the null hypothesis is:

$$z = \frac{96 - 99}{2} = -1.5 \qquad \text{Area } .43319$$

$$z = \frac{104 - 99}{2} = 2.5 \qquad \underline{.49379}$$

By addition, the answer is: .92698

If $\mu = 95$, then the probability of accepting the null hypothesis is:

$$z = \frac{104 - 95}{2} = 4.5 \qquad \text{Area } .50000$$

$$z = \frac{96 - 95}{2} = .5 \qquad \underline{.19146}$$

By subtraction, the answer is: .30854

If $\mu = 90$, then the probability of accepting the null hypothesis is:

$$z = \frac{104 - 90}{2} = 7.0 \qquad \text{Area } .50000$$

$$z = \frac{96 - 90}{2} = 3.0 \qquad \underline{.49865}$$

By subtraction, the answer is: .00135

It should be obvious to the student that we could determine probabilities for any alternative value that the true mean might assume. Table 11.1 shows the values for a few more alternatives. The student may wish to check some of these values for his or her own understanding. These values of Table 11.1 may be used to construct either a power curve or an operating characteristic curve for the above test, and they would appear much like Figures 11.1 and 11.2 respectively.

TABLE 11.1

If the True Mean Equals		Power Curve (probability of Rejecting H_0)		OC Curve (probability of Accepting H_0)
85		1.000		·000
90		.999		.001
91		.994		.006
92		.977		.023
93		.933		.067
94		.841		.159
95		.691		.309
96		.500		.500
97		.308		.692
98		.160		.840
99		.073		.927
100	Only place you can make Type I	.050	Note: You cannot make a Type II error here	.950
101		.073		.927
102		.160		.840
103		.308		.692
104		.500		.500
105		.691		.309
106		.841		.159
107		.933		.067
108		.977		.023
109		.994		.006
110		.999		.001
115		1.000		.000

CONTROLLING α AND β

Up to this juncture in the tests of hypothesis, our concern has been with α only. That is, the probability of a Type I error was controlled at some preassigned level of significance. Now the problem of controlling both α and β at arbitrary levels is considered. In applying this reasoning, several questions need answering. First, given the risk management is willing to assume in rejecting a true hypothesis (α) and the risk of accepting the false hypothesis (β), what size sample is needed to protect against these types of errors? Second, what criteria will be used to choose between the two alternative courses of action? We shall now consider situations for both mean and proportion differences.

Mean differences: Two-sided measurement data

Let us consider the following situation. A firm is turning out packages of dry cereal in family-sized boxes. The contents should weigh 16 ounces. Management wants to set up an inspection scheme to control the process.

That is, it does not want the machine process to place too much cereal in each box (thereby reducing profits), nor does it want to cheat the customers (i.e., boxes contain less than the stated 16 ounces).[3] What can the company statistician do to help?

First, the statistician can suggest a *type* of sampling design. Second, the statistician can determine the proper sample *size*. Third, the statistician can set up a *rule* to decide whether or not the process is functioning satisfactorily.

Let us further assume the following:

$$\mu_0 = 16.0 \text{ ounces}, \mu_1 = 15.75 \text{ ounces}, \mu_2 = 16.25 \text{ ounces}$$
$$\alpha = .02, \beta_1 = .05, \beta_2 = .05$$

We also know from past performances of these machines that the standard deviation of the contents of the package is .4 ounce. Obviously, management is more eager to avoid making a Type I error, that is, stopping the machine-filling process unnecessarily.

The statistician would probably suggest selecting a simple random sample of n boxes from each hour's production. However, actually determining sample size and the decision-making rule requires more detail. From the information above, we can see that management wants to take a 2 percent risk (α) of stopping the process when in fact it is performing according to desired standards. Management also wishes to protect against concluding that the machines are working satisfactorily when indeed they are not. In fact, management wants to take only a 5 percent risk(β_1) of concluding that the boxes contain 16 ounces on the average when the true mean really is 15.75 ounces. In addition, a 5 percent risk (β_2) is taken of concluding that the boxes contain 16 ounces on the average when in fact the true mean content is 16.25 ounces. Figure 11.4 graphically assists us in structuring the problem mathematically. The primary virtue in sketching such a detailed graph is to reduce the "story" part of the problem to a workable solution without getting mixed up in terminology. One should not be too technical about the precise distances on the x axis. In reality, if $\beta_1 \neq \beta_2$ or if $\mu_0 - \mu_1 \neq \mu_2 - \mu_0$, then one should draw two separate charts. In practice this is not always done even though combining the two separate problems does tend to distort the x scale. The important thing to remember is that this graphic approach is merely a method of reducing the problem to an orderly procedure in order that the statistician can develop a proper sampling plan. The graph never is presented as the solution to the problem even though it is extremely helpful in understanding the risks associated with the specifications required.

---

[3] This is termed a two-tailed test. Problems arise where a one-tailed test is useful. In this problem, for example, management might only have been interested in protecting against filling the boxes too full. Here the one-sided alternative may not be so realistic.

FIGURE 11.4

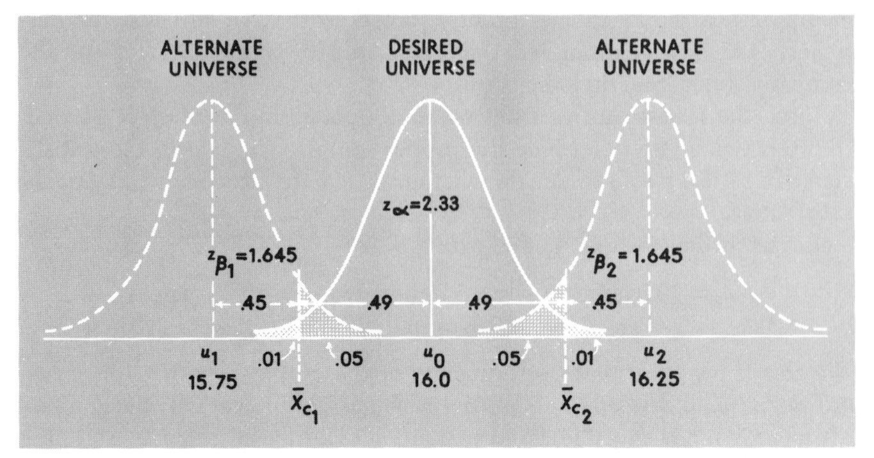

Note: z_α = standard normal deviate associated with alpha risk (see Table C.1, Appendix C)
z_{β_1} = standard normal deviate associated with beta risk with alternate universe μ_1
z_{β_2} = standard normal deviate associated with beta risk with alternate universe μ_2
$\bar{X}_{c_1}$ = critical point
$\bar{X}_{c_2}$ = critical point
Alternative courses of action:
$H_0 : \mu = \mu_0$ (null hypothesis, process satisfactory)
$H_1 : \mu \neq \mu_0$ (alternate hypothesis, stop process and investigate problem)

Figure 11.5 presents the solution procedure in a flowchart form. Perhaps, the junction point 30 needs some explanation. It merely says that once a decision has been reached by comparing the sample mean of the weights of 40 boxes from one hour's production with the critical points, we go back to the process of selecting the next hour's sample. This continues until we change the specifications or risks, which means a new sampling plan must be developed.

What size sample should be selected each hour? The following formula may be utilized to answer the question:

$$\mu_0 - z_\alpha \frac{\sigma_x}{\sqrt{n}} = \mu_1 + z_\beta \frac{\sigma_x}{\sqrt{n}}$$

then,

$$\mu_0 - \mu_1 = z_\alpha \frac{\sigma_x}{\sqrt{n}} + z_\beta \frac{\sigma_x}{\sqrt{n}}$$

and

$$\mu_0 - \mu_1 = \frac{\sigma_x}{\sqrt{n}} (z_\alpha + z_\beta)$$

$$\mu_0 - \mu_1 = \frac{\sigma_x(z_\alpha + z_\beta)}{\sqrt{n}}$$

FIGURE 11.5
Determining sampling plan by controlling alpha and beta, measurement data, two-sided test

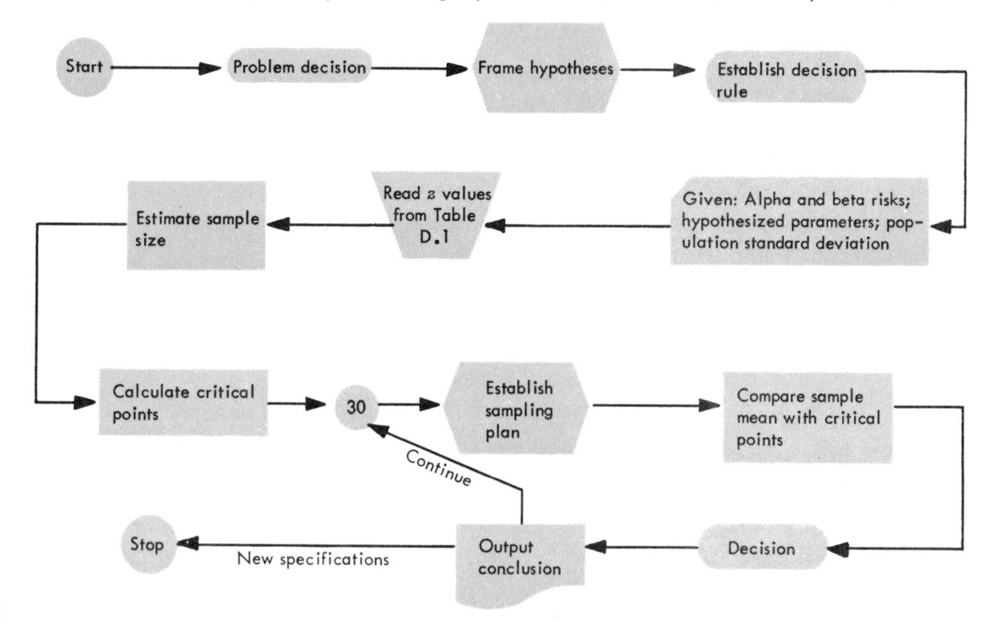

and

$$\sqrt{n} = \frac{\sigma_x(z_\alpha + z_\beta)}{\mu_0 - \mu_1}$$

where

$$n = \left[\frac{\sigma_x(z_\alpha + z_\beta)}{\mu_0 - \mu_1}\right]^2 \qquad (11.1)$$

If the standard deviation of the universe is unknown, the formula for determining n is

$$n = \left[\frac{\hat{\sigma}_x(z_a + z_\beta)}{\mu_1 - \mu_0}\right]^2 + \frac{z_\alpha^2}{2} \qquad (11.2)$$

substituting in Formula (11.1):

$$\begin{aligned} n &= \left[\frac{.4(2.33 + 1.645)}{16.0 - 15.75}\right]^2 \\ &= (6.36)^2 \\ &\simeq 40 \end{aligned}$$

That is, 40 boxes will be selected at random from each hour's production and weighed to determine their mean contents. From Figure 11.4, it may be noted the alternative courses of action are

$$H_0 : \mu = \mu_0 \qquad \text{(Process is OK)}$$
$$H_1 : \mu = \mu_0 \qquad \text{(Process needs checking)}$$

Rational criteria must now be determined in order to choose between these two alternate courses of action. Therefore, we need to compute $\bar{X}_{c1}$ and $\bar{X}_{c2}$, the critical points. They may be found thus:[4]

$$\bar{X}_{c1} = \mu_1 + \left(\frac{z_{\beta_1}}{z_\alpha + z_{\beta_1}}\right)(\mu_0 - \mu_1) \tag{11.3}$$

$$= 15.75 + \left(\frac{1.645}{2.33 + 1.645}\right)(16.0 - 15.75)$$

$$= 15.85$$

$$\bar{X}_{c2} = \mu_2 - \left(\frac{z_{\beta_2}}{z_\alpha + z_{\beta_2}}\right)(\mu_2 - \mu_0) \tag{11.4}$$

$$= 16.25 - \left(\frac{1.645}{2.33 + 1.645}\right)(16.25 - 16.0)$$

$$= 16.15$$

Consequently, the sampling plan and decision-making rule are:

1. Select a simple random sample of 40 boxes from each hour's production.
2. Determine the mean contents of these 40 boxes.
3. If $\bar{X}$ is between 15.85 ounces and 16.15 ounces, the process is allowed to continue.
4. If $\bar{X} \leqq 15.85$ ounces or if $\bar{X} \geqq 16.15$ ounces, H_1 is accepted and the process is stopped to determine what is wrong.

When $\mu_2 - \mu_0 = \mu_0 - \mu_1$ and when $\beta_1 = \beta_2$, then we need compute n for only one side because the sample size would be the same for both alternative universes. If $\mu_2 - \mu_0 \neq \mu_0 - \mu_1$ or $\beta_1 \neq \beta_2$, then we usually take the larger sample size or perhaps combine the two. An alternative formula for determining n is

$$\Delta = (z_\alpha + z_{\beta_1})\left(\frac{\sigma_x}{\sqrt{n}}\right) \tag{11.5}$$

[4] Alternative formulas are
$$\bar{X}_{c_1} = \mu_1 + (z_{\beta 1})(\sigma_{\bar{x}})$$
$$\bar{X}_{c_2} = \mu_2 - (z_{\beta 2})(\sigma_{\bar{x}})$$
$$\bar{X}_{c_1} = \mu_0 - (z_\alpha)(\sigma_{\bar{x}})$$
$$\bar{X}_{c_2} = \mu_0 + (z_\alpha)(\sigma_{\bar{x}})$$

where

$$\Delta = \mu_0 - \mu_1$$

which then may be written

$$\sqrt{n} = \frac{(z_\alpha + z_{\beta_1})\sigma_{\bar{x}}}{\mu_0 - \mu_1}$$

$$n = \left[\frac{(z_\alpha + z_{\beta_1})\sigma_{\bar{x}}}{\mu_0 - \mu_1}\right]^2 \qquad (11.6)$$

which is the same as Formula (11.1).

Assume that this plan is now operative and 40 boxes of cereal are chosen at random and their mean contents found to be 16.3 ounces. What is the decision? According to the alternative courses of action and the criteria established, the procedure is to stop the machines and attempt to correct the filling process. In addition, we would weigh that particular hour's production individually and correct any errors. If the $\bar{X}$ weight of the 40 boxes were 15.9 ounces, we would assume the process is in control and allow it to continue with no changes in the setting of the machines. (See Figures 11.4 and 11.5.)

Realistically, of course, we must set the alpha and beta risks and the allowable error (in this problem $\pm$ .25 ounces) in such a manner that it does not cost more to stop the process when the rule indicates than it would either (1) to place too much cereal in a few boxes and thereby reduce profits or (2) not to fill each box with something less than precisely 16 ounces and thereby suffer customer ill will (and perhaps legal action by the government). For example, it may be desirable to set μ_2 as high as 17 on the basis that we would prefer to have too much in a few boxes rather than interrupt production too frequently. *In practice, we would compare the power or OC curve for this decision-making rule with curves for other tests and select the rule(test) which is the most discriminating.* That is, we want the power of the test to be the highest possible (see Figures 11.6 and 11.7 and Table 11.2).

We may wish to check to see if our sampling plan actually provides the required protection indicated by management.

If the true mean of the population being sampled is 15.75 ounces, what is the probability of a Type II error? That is, what is the probability of accepting a false hypothesis?

We must determine the area that falls within our acceptance region. For example, we need to know the area beyond 15.85 ounces. (See Figure 11.8.)

Then

$$z = \frac{\bar{X}_{c1} - \mu_1}{\sigma_{\bar{x}}} \qquad (11.7)$$

TABLE 11.2

Alternative Courses of Action if We Conclude:	Decision		
	When $\mu = 16.0$	When $\mu = 15.75$	When $\mu = 16.25$
H_0 (process is OK)	Correct decision (.98)	Incorrect decision (.05)	Incorrect decision (.05)
H_1 (process mean is too small)	Incorrect decision (.01)	Correct decision (.95)	———
H_2 (process mean is too large)	Incorrect decision (.01)	———	Correct decision (.95)

FIGURE 11.6
Power curve based on data of dry cereal problem

If μ:	*Probability of Rejecting Null Hypothesis*
15.00	1.00
15.85	.50
16.00	.02
16.15	.50
17.00	1.00

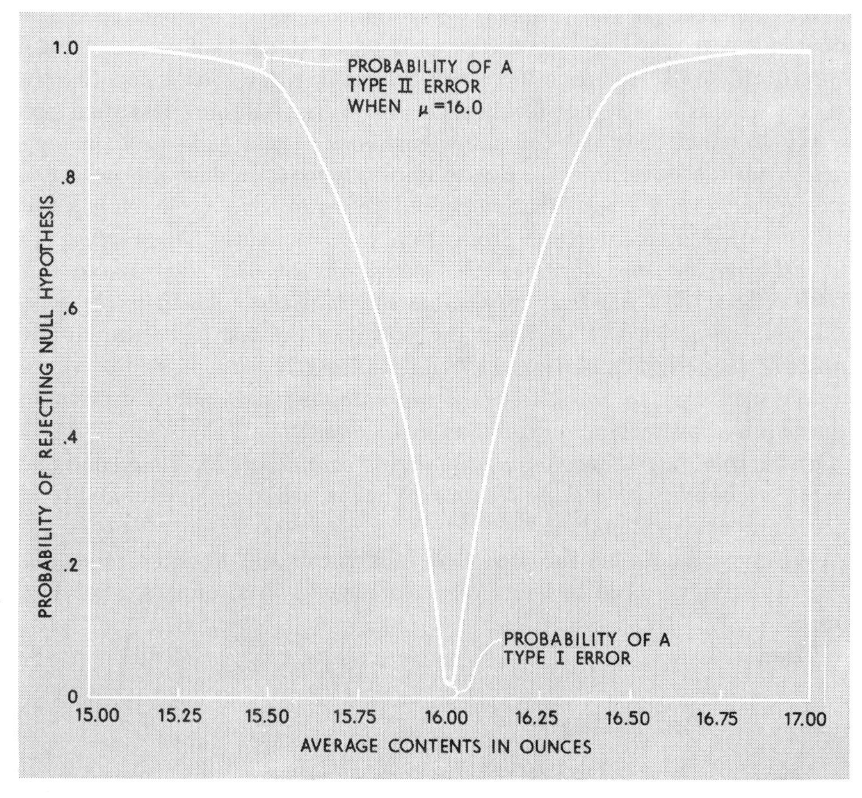

FIGURE 11.7
Operating characteristic curve based on data of dry cereal problem

If μ:	*Probability of Accepting Null Hypothesis*
15.00	.00
15.85	.50
16.00	.98
16.15	.50
17.00	.00

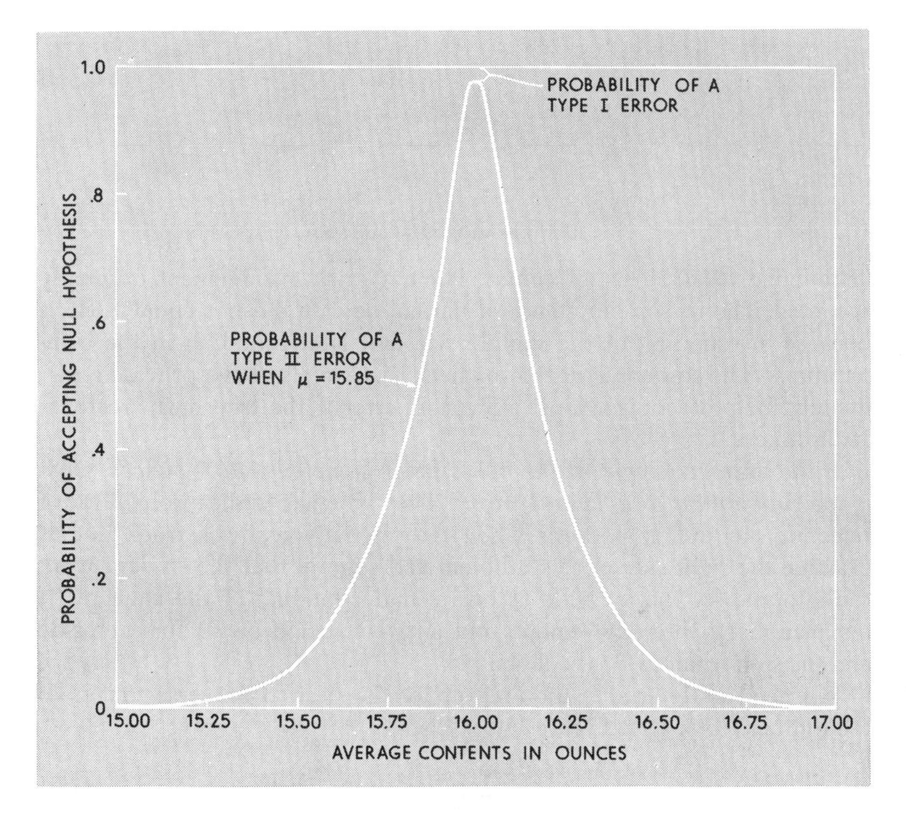

where

$$z = \frac{15.85 - 15.75}{\frac{.4}{\sqrt{40}}}$$

$$z = \frac{.10}{.06}$$

$$z = 1.66$$

The area between 15.75 and 15.85 then is found to be .45154. The area to the right of $z = 1.66$ or to the right of 15.85 is

FIGURE 11.8

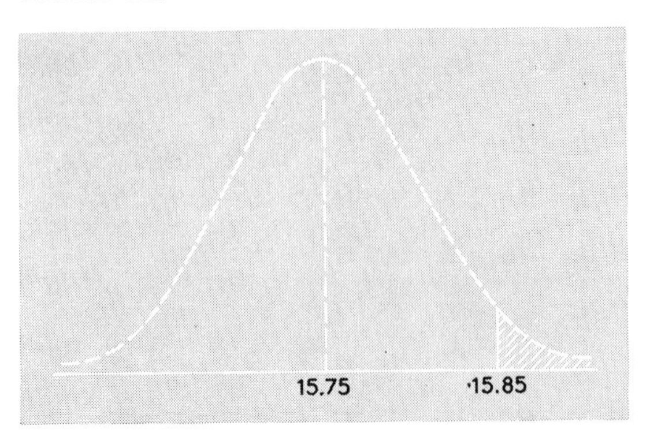

$$.50000 - .45154 = .04846$$

or rounded to .05. This, of course, is our β_1 risk management originally accepted. The reason our standard normal deviate did not equal 1.645 is because of rounding of the sample size and later rounding in the computations. The student is urged to check that we also have provided adequately for protection against a Type II error if the true mean really is 16.25 ounces.

If the average weight of the boxes being sampled is 16.0 ounces, what is the probability of a Type I error? That is, what is the probability of rejecting our null hypothesis when the alternate really is true? Let us examine the right side of this problem and suggest that the student work the example for the left side to insure understanding of the concept. If our plan meets the requirements, our answers should be .01 for each side and the sum equal to .02 the alpha risk.

We need to determine the area under the desired universe that is to the right of 16.15. (See Figure 11.9.) Then,

$$z = \frac{\bar{X}_{c2} - \mu_0}{\sigma_{\bar{x}}} \tag{11.8}$$

where

$$z = \frac{16.15 - 16.0}{\frac{.4}{\sqrt{40}}}$$

$$z = \frac{.15}{.06}$$

$$z = 2.5$$

The area between 16.0 and 16.15 then is found to be .49379. The area to

FIGURE 11.9
Desired universe

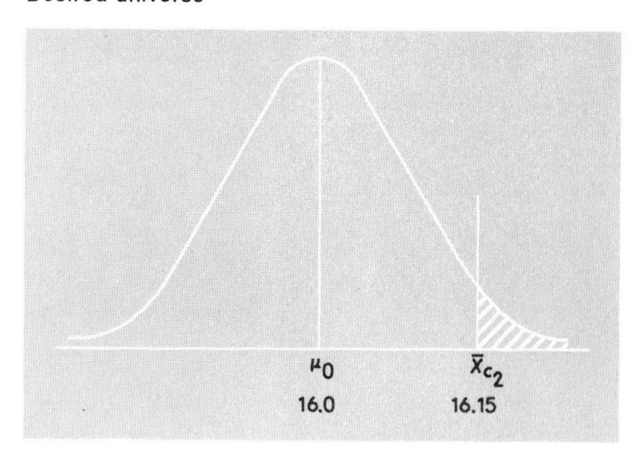

the right of 16.15 then is .50000 − .49379 = .00621, or rounded to .01 which equals alpha divided by 2.

It should be noted that if $\beta_1 \neq \beta_2$, several things must be kept in mind. First, in determining the sample size you should look upon this as two distinct problems unless $\mu_2 - \mu_0 = \mu_0 - \mu_1$. Otherwise complications arise in computing areas and in visualizing any movement along the x axis. Of course, if $\beta_1 = \beta_2$, then you merely use the larger sample of the two possibilities. If $\beta_1 \neq \beta_2$ and you use the larger sample, then you also have changed one of the beta risks. Second, if β_1 is more strict than β_2, then $\bar{X}_{c1}$ should be computed as more demanding than $\bar{X}_{c2}$. (That is, $\bar{X}_{c1}$ would reflect the different sample size.) Let us illustrate the points made above. (See Figures 11.10 and 11.11.)

FIGURE 11.10

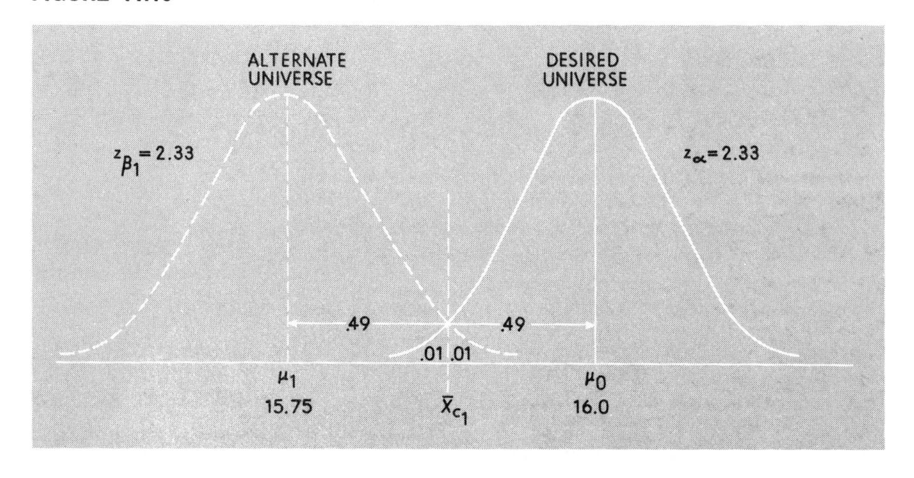

FIGURE 11.11

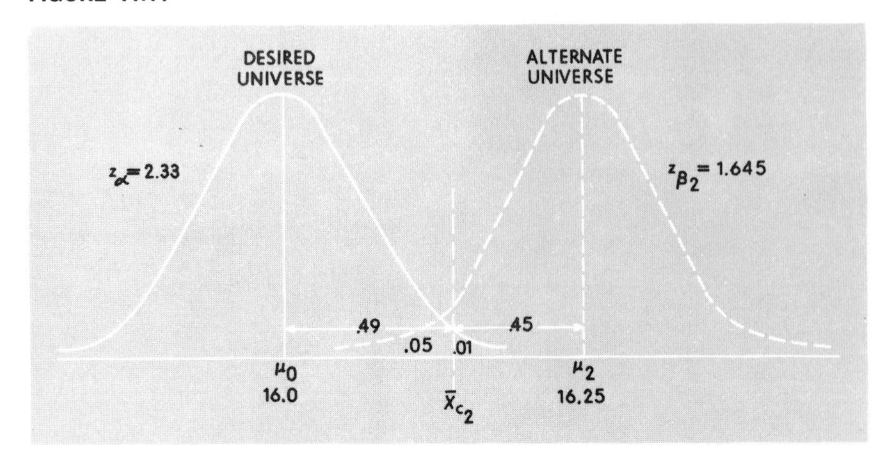

Assume the following:

$$\mu_0 = 16.0 \text{ ounces}; \mu_1 = 15.75 \text{ ounces}; \mu_2 = 16.25 \text{ ounces};$$
$$\alpha = .02; \beta_1 = .01; \beta_2 = .05; \sigma = .4 \text{ ounces}$$

Note the only change in our information is that now $\beta_1 \neq \beta_2$ as was the case previously. What affect does this have on our sampling plan?

Viewing these as separate problems our computations are as follows:

$$n_1 = \left[\frac{.4(2.33 + 2.33)}{16.0 - 15.75}\right]^2 \qquad n_2 = \left[\frac{.4(2.33 + 1.645)}{16.25 - 16.0}\right]^2$$

$$n_1 = \left(\frac{1.864}{.25}\right)^2 \qquad n_2 = \left(\frac{1.59}{.25}\right)^2$$

$$n_1 \simeq 56 \qquad n_2 \simeq 40$$

$$\bar{X}_{c1} = \mu_1 + z_\alpha \sigma_{\bar{x}}$$

$$\bar{X}_{c_1} = 15.75 + 2.33\left(\frac{.4}{\sqrt{56}}\right) \qquad \bar{X}_{c_2} = 16.25 - 1.645\left(\frac{.4}{\sqrt{40}}\right)$$

$$\bar{X}_{c_1} = 15.875 \qquad \bar{X}_{c_2} = 16.15$$

We now illustrate that the β_2 risk would be changed when $\beta_1 \neq \beta_2$ if one were to use the larger of the two samples and to view it as one problem. Then, the calculation of our second critical point would change to:

$$\bar{X}_{c_2} = 16.25 - 1.645\left(\frac{.4}{\sqrt{56}}\right)$$

$$\bar{X}_{c_2} = 16.16$$

Now, what is the probability of a Type II error if $\mu = 16.25$ ounces? Viewing this as we did as though we were solving *two* separate problems the answer would be .05. However, since we used the larger sample size here β_2 is changed.

$$z = \frac{\mu_2 - \bar{X}_{c2}}{\sigma_{\bar{x}}}$$

$$z = \frac{16.25 - 16.16}{\frac{.4}{\sqrt{56}}}$$

$$z = \frac{.09}{.05}$$

$$z = 1.80$$

The area associated with this standard normal deviate value is .46407. The area to the left of $\bar{X}_{c2}$ and in the acceptance region then is $.50000 - .46407 = .03593$. It is clear then that by viewing the problem as one, we have changed the β_2 risk considerably.

In summary, when $\beta_1 \neq \beta_2$ then you should handle the situation either as separate problems or recognize that by using the larger sample you have altered the probability of a Type II error.

Mean differences: One-sided measurement data

Assume that a wholesaler has found that for years the credit manager has been able to keep outstanding accounts receivables at an average age of 21 days with a standard deviation of 4 days. This figure is considered to be good, and management wishes to make certain that the average age does not increase significantly. Suppose the credit manager requests your help. Given the following additional information, what type of a monthly sampling plan would you recommend? The credit manager states that only "1 chance in 100" is wanted when indicating to top management that the average age is 21 days when it really is as large as 23. The credit manager is willing to take "5 chances out of a 100" of reporting that the average age of accounts receivables has increased when actually it has not.

To summarize our data:

$$\mu_0 = 21 \text{ days}; \mu_1 = 23 \text{ days}; \alpha = .05; \beta = .01; \sigma_x = 4$$

Graphically, the problem appears as shown in Figure 11.12.

What size monthly sample of accounts receivables should be observed?

Because $\mu_1 > \mu_0$ we use:

$$n = \left[\frac{\sigma_{\bar{x}}(z_\alpha + z_\beta)}{\mu_1 - \mu_0}\right]^2 \qquad (11.9)$$

FIGURE 11.12

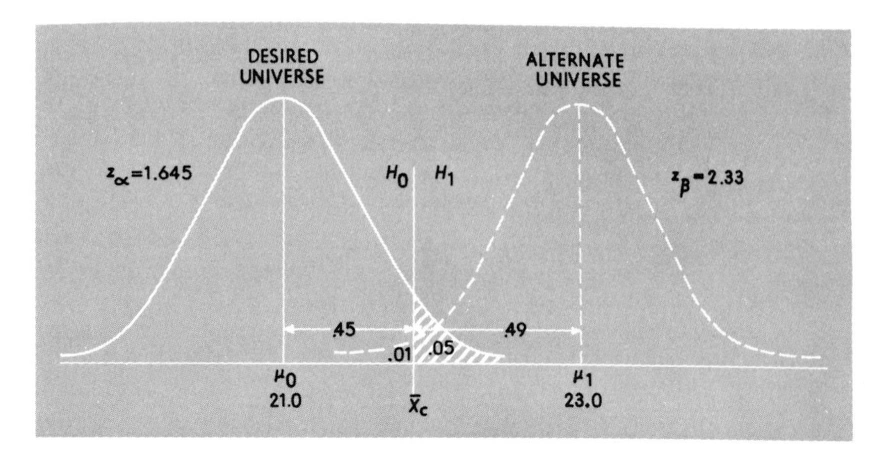

$$n = \left[\frac{4(1.645 + 2.33)}{23 - 21}\right]^2$$

$$n = (7.95)^2$$

$$n \simeq 63$$

What is the value of our critical point?

$$\bar{X}_c = \mu_0 + \left(\frac{z_\alpha}{z_\alpha + z_\beta}\right)(\mu_1 - \mu_0) \tag{11.10}$$

$$\bar{X}_c = 21.0 + \frac{1.645}{1.645 + 2.33}(23.0 - 21.0)$$

$$\bar{X}_c = 21.8$$

What are the alternative courses of action?

$H_0 : \mu \leq \mu_0$ (Average age has not increased)
$H_1 : \mu > \mu_0$ (Average age has increased)

What are the criteria for choosing between H_0 and H_1?

Conclude H_0 if $\bar{X} \leq \bar{X}_c$
Conclude H_1 if $\bar{X} > \bar{X}_c$

Consequently our monthly sampling plan would be as follows:

1. Select a random sample of 63 accounts receivables every month.
2. Determine $\bar{X}$ the average age of the accounts.
3. If $\bar{X}$ is $\leq$21.8 days, we conclude the credit and collection policies are effective and adequate.

4. If $\bar{X} > 21.8$, we conclude that the average age of the accounts receivables has increased and we will investigate to determine the cause.

Assume that this plan has been put into effect and that we obtain the following results for the past three months:

$$\bar{X}_1 = 20.9 \text{ days}$$
$$\bar{X}_2 = 21.9 \text{ days}$$
$$\bar{X}_3 = 22.5 \text{ days}$$

What might we reasonably conclude? While the average age of the accounts receivables was near our tolerable limit for two months, the sample for the third month indicates a probable increase has been taking place. Given management's stated objectives, a review of recent credit and collections policies seems in order.

Does the above sampling plan provide the credit manager with the protection indicated? If the true average age of the accounts receivables being sampled really is 23.0 days, what is the probability of a Type II error? That is, what is the probability of concluding that the credit and collections policies are OK when in fact the average age has increased? We must determine the area under the alternate universe curve that is to the left of 21.8. (See Figure 11.13.) Then,

$$z = \frac{\mu_1 - \bar{X}_c}{\sigma_{\bar{x}}} \tag{11.11}$$

where

$$z = \frac{23.0 - 21.8}{\frac{4}{\sqrt{63}}}$$

$$z = \frac{1.2}{.504}$$

$$z = 2.38$$

The area between 23.0 and 21.8 is found to be .49134. The area to the left of 21.8 is .50000 − .49134 = .00866, or rounded to .01 the beta risk.

If the true average age of the accounts receivables really is 21.0 days, what is the probability of a Type I error? That is, what is the probability of concluding that the credit and collections policies are in need of investigation when actually the average age is OK according to management's standards? Our answer should equal the alpha risk if our sampling plan meets the demands of the credit manager.

We need to determine the area under the desired universe curve that is to the right of 21.8. (See Figure 11.14.) Then,

FIGURE 11.13
Alternate universe

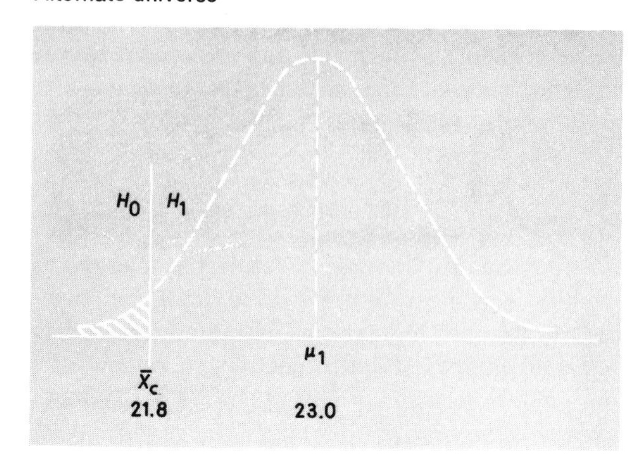

$$z = \frac{\bar{X}_c - \mu_0}{\sigma_{\bar{x}}} \tag{11.12}$$

where

$$z = \frac{21.8 - 21.0}{\dfrac{4}{\sqrt{63}}}$$

$$z = \frac{.8}{.504}$$

$$z = 1.59$$

FIGURE 11.14
Desired universe

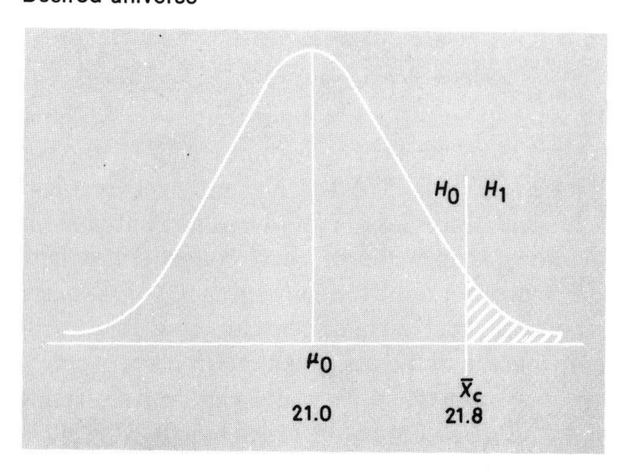

The area between 21.0 and 21.8 is found to be .44408. The area to the right of 21.8 then is .50000 − .44408 = .05592. Except for rounding errors the plan seems to be satisfactory.

Proportion differences

Similar reasoning may be employed for attribute data, that is, a situation where we are concerned with a proportion or percentage and the standard error of the percentage.

Assume a council member in a metropolitan city engages a competent research agency to give assistance in how to vote on the issue of urban renewal. The council member wants to vote according to the wishes of the majority of constituents—a commendable attitude indeed! The politician also indicates that sample opinions of constituents are desired to determine their reaction, favorable or unfavorable, to urban renewal. That is, π is the percentage of constituents favoring the program and is the decision parameter. The politician further indicates the desire to focus attention on two possible values of the population percentage: (1) $\pi_0 = 40$ percent, at which point it would be preferable to vote no; and (2) $\pi_1 = 55$ percent, at which point it would be preferable to vote yes.

Suppose the politician wants to take a 5 percent chance (the beta risk) of voting no if $\pi = 55$ percent. If $\pi = 40$ percent, the politician wants to take only a 1 percent risk of voting yes. That is, the probability of a Type I error, rejecting a true hypothesis, is fixed at .01. The probability of accepting a false hypothesis, a Type II error, is fixed at .05. The politician apparently feels it would be more embarrassing to vote for urban renewal if really the true proportion of constituents favoring the plan is less than a clear majority than it would be to vote no if a majority favor it (see Figures 11.15 and 11.16).

The alternative courses of action are

$$H_0: \pi \leq \pi_0 \qquad \text{(Vote no)}$$
$$H_1: \pi > \pi_0 \qquad \text{(Vote yes)}$$

What size sample should the research agency select, and what criteria should be employed to choose between these alternative courses of action?

The sample size n may be estimated thus:

$$n = \left[\frac{z_\alpha\sqrt{\pi_0(100 - \pi_0)} + z_\beta\sqrt{\pi_1(100 - \pi_1)}}{\pi_1 - \pi_0}\right]^2 \qquad (11.13)$$

$$n = \left[\frac{2.33\sqrt{(40)(60)} + 1.645\sqrt{(55)(45)}}{55 - 40}\right]^2$$

$$n = 171$$

FIGURE 11.15
Controlling α and β for π

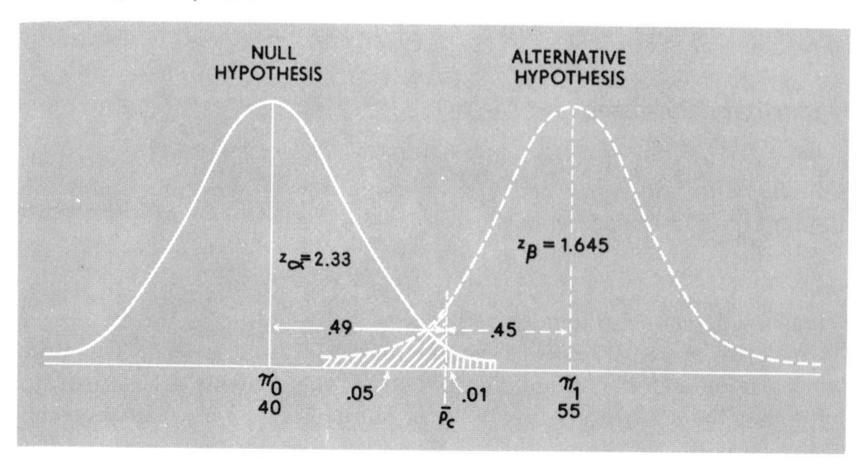

An alternative formula is

$$\pi_1 - z_\beta \sqrt{\frac{\pi_1(100 - \pi_1)}{n}} = \pi_0 + z_\alpha \sqrt{\frac{\pi_0(100 - \pi_0)}{n}} \qquad (11.14)$$

which may be transposed to:

$$\pi_1 - \pi_0 = z_\alpha \sqrt{\frac{\pi_0(100 - \pi_0)}{n}} + z_\beta \sqrt{\frac{\pi_1(100 - \pi_1)}{n}}$$

$$\sqrt{n} = \frac{z_\alpha \sqrt{\pi_0(100 - \pi_0)} + z_\beta \sqrt{\pi_1(100 - \pi_1)}}{\pi_1 - \pi_0}$$

$$n = \left[\frac{z_\alpha \sqrt{\pi_0(100 - \pi_0)} + z_\beta \sqrt{\pi_1(100 - \pi_1)}}{\pi_1 - \pi_0}\right]^2$$

which is the same as Formula (11.13).

One hundred seventy-one constituents selected in a random manner would be interviewed to determine their position on urban renewal. The proportion or percentage favoring would be calculated, and on the basis of this sample a voting decision would be made. What criteria must be used to make an intelligent choice between H_0 and H_1? The critical point, p_c, must be found:

$$p_c = \pi_1 - z_\beta \sqrt{\frac{\pi_1(100 - \pi_1)}{n}} \qquad (11.15)$$

$$p_c = 55 - 1.645 \sqrt{\frac{(55)(45)}{171}}$$

FIGURE 11.16
Determining sampling plan by controlling alpha and beta, attribute data, one-sided test

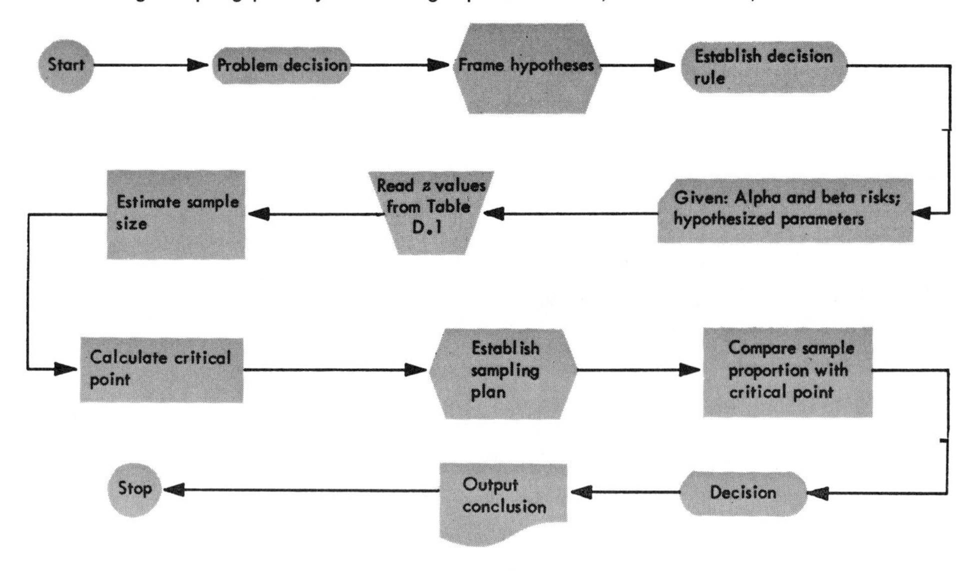

$$= 55 - 6.25$$

$$= 48.75, \qquad \text{or rounded to } 49\%$$

The criteria for choosing between H_0 and H_1 are these:

1. Select a simple random sample of 171 constituents and determine their position concerning urban renewal. (Actually, 175 or 200 items might be chosen, since this represents an estimate of n.)
2. If the percentage favoring urban renewal is $\leq$ 49 percent, accept H_0 and instruct the politician to vote no.
3. If the percentage favoring urban renewal is $>$ 49 percent, reject H_0 (i.e., accept H_1), and instruct the politician to vote yes.

Assume that a sample of 171 is taken and 87 indicate favoring urban renewal. How should the politician vote? The sample percentage is $87/171 \cdot 100 = 51$ percent. Therefore, $51 > 49$, and H_1 is accepted; and the politician is advised to vote yes. The research agency could be wrong, but the probability of both types of error has been controlled at .01 and .05. *That is, using these methods, only one time out of 100 would the politician be advised to vote yes when a majority of constituents were against such a program.* Also, only five times out of 100 would the politician be instructed to vote no when really a majority of the voters favored urban renewal (see Figures 11.17 and 11.18).

FIGURE 11.17
Power curve based on data of urban renewal problem

If π (%):	*Probability of Rejecting Null Hypothesis*
00	.00
40	.01
49	.50
55	.95
100	1.00

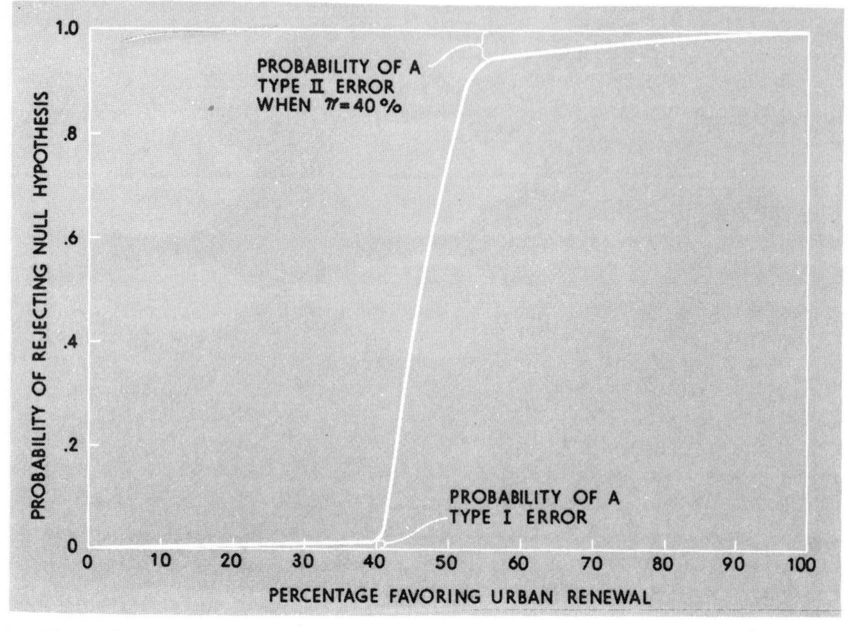

$H_0 = \pi \leq \pi_0$ $\quad\pi_0 = 40\%$
$H_1 = \pi > \pi_0$

Again, let us check on our sampling plan. If the proportion of voters favoring urban renewal really is 40 percent, what is the probability of a Type I error? That is, what is the probability of accepting our alternate hypothesis (instruct politician to vote yes) when in fact the null hypothesis is true?

We must determine the area under the null hypothesis curve that is to the right of 49 percent (p_c). If our plan is satisfactory, the answer should be equal to the stated alpha risk. (See Figure 11.19.)

Then,

$$z = \frac{p_c - \pi_0}{\sigma_p} \tag{11.16}$$

FIGURE 11.18
Operating characteristic curve based on data of urban renewal problem

If π (%):	*Probability of Accepting Null Hypothesis*
00	1.00
40	.99
49	.50
55	.05
100	.00

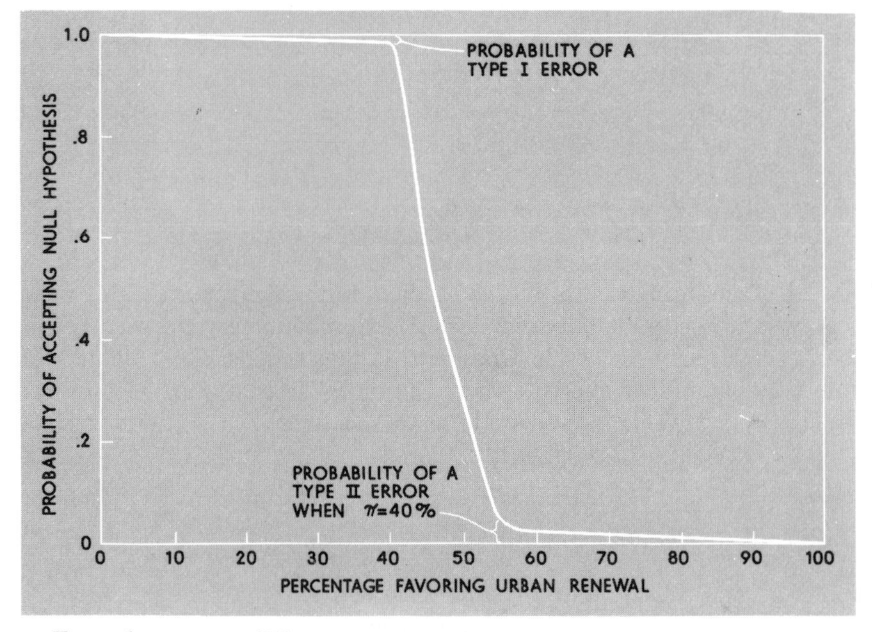

$H_0 = \pi \leq \pi_0$ $\quad \pi_0 = 40\%$
$H_1 = \pi > \pi_0$

where

$$z = \frac{49 - 40}{\sqrt{\frac{(40)(60)}{171}}}$$

$$z = \frac{9}{3.74}$$

$$z = 2.40$$

The area between 40 and 49 percent then is .49180. The area to the right of 49 percent then is .50000 − .49180 = .00820, or rounded to .01 which is our alpha risk.

FIGURE 11.19

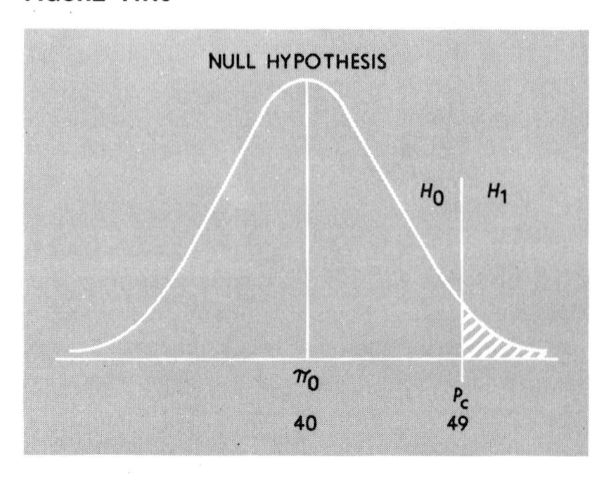

If the true proportion of the voters favoring urban renewal is 44 percent, what is the probability of a Type II error? Our answer should equal the .05 beta risk. What is the probability of instructing our politician to vote no when indeed the vote should be yes? Here we must determine the area under the alternate hypothesis curve that is to the left of our critical point. (See Figure 11.20.) Then,

$$z = \frac{\pi_1 - p_c}{\hat{\sigma}_p} \tag{11.17}$$

$$z = \frac{55 - 49}{\sqrt{\frac{(55)(45)}{171}}}$$

$$z = \frac{6}{3.8}$$

$$z = 1.59$$

The area between 55 and 49 is found to be .44408. The area to the left of 49 then is .50000 − .44408 = .05592 which except for rounding errors is equal to the beta risk.

Attribute data: Power of discrimination

An advertising agency hypothesized that 20 percent of the readers of a certain magazine could recall having seen a particular multicolor advertisement. In a random sample of 400 readers, only 60 could recall the advertisement at all. The difference from the expected value was stated

FIGURE 11.20

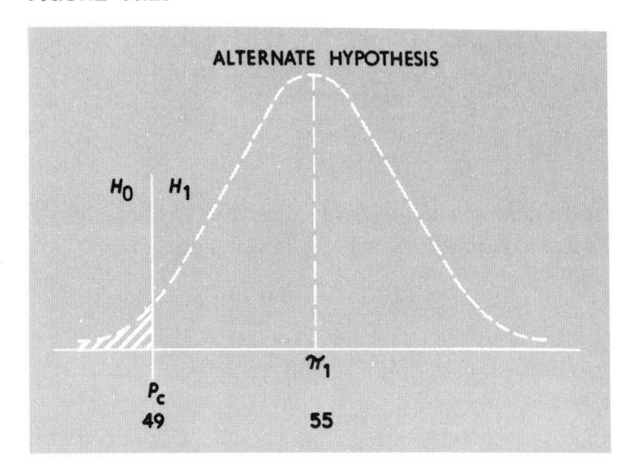

by the agency statistician as being significant at the .05 level. What did the statistician mean by this statement? Simply, that the difference between the sample value (15 percent) and the hypothesized value (20 percent) is too great to have occurred by chance error of sampling. Only 5 times out of 100 would we observe a sample value as small as 15 percent if the true proportion is as we hypothesized. We would reject the original hypothesis. (See Figure 11.21.)

If we were to describe the power of discrimination of the above test, we might utilize the concept of the power function. What would be the values we might show on the axes of an *OC* curve describing this test? The *y* axis of an *OC* curve indicates the probability of accepting the hypothesis that the true parameter is 20 percent. The *x* axis shows the

FIGURE 11.21

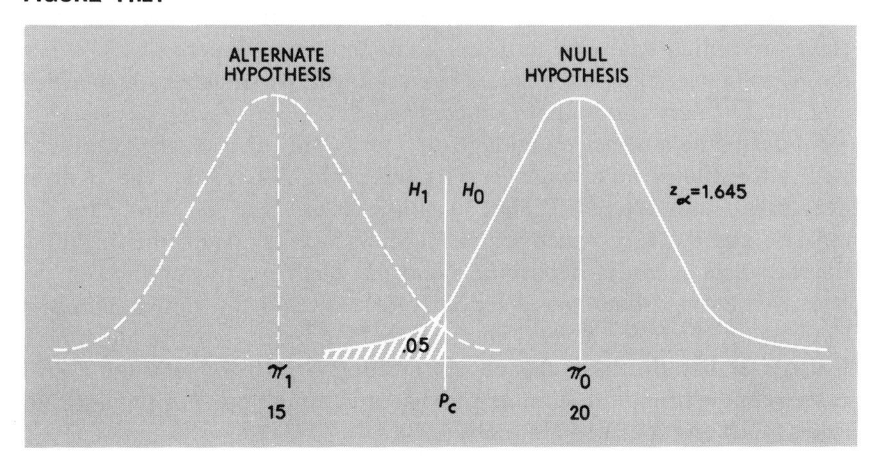

alternative values that the parameter might assume. Let us calculate one value on the *OC* curve and interpret its meaning. Given,

$$p_c = 20 - 1.645\sqrt{\frac{(20)(80)}{400}}$$

$$p_c = 16.71$$

If the true proportion is 15 percent, then the probability of accepting a false hypothesis or committing a Type II error is:

$$z = \frac{16.71 - 15.0}{\sqrt{\frac{(15)(85)}{400}}}$$

$$z = \frac{1.71}{1.79}$$

$$z = .96$$

The area associated with that standard normal deviate is .33147. The probability of a Type II error if the true proportion is really 15 percent then is

$$.50000 - .33147 = .16853$$

In other words, if this is the situation in the population, we would not likely accept the null hypothesis very often—only 17 times out of 100.

The power curve (or the *OC* curve) tells us that the probability of a Type II error depends upon the value of the unknown parameter. It is clear, then, that in stipulating the characteristics of the desired universe in order to construct a decision-making rule, the probability of a Type II error must be related to a specific value which the true (but unknown) parameter may assume. For example, we may indicate that the probability of a Type I error be 10 percent (α) and the probability of a Type II error (accepting a false hypothesis) be no more than 5 percent (β) when the *hypothetical* parameter differs from the true parameter by a specified amount. In this way *both* risks are controlled.

In designing a decision-making rule (test), the analyst recognizes that with a fixed level of significance (α) the probability of a Type II error decreases as n is increased. That is, as n increases, the standard error decreases, and the acceptance region becomes smaller. With this in mind, the statistician can (1) determine α and (2) control β by varying n.

Power curves show the probability of rejecting the null hypothesis. Comparing them for several decision-making rules, we can determine which test has the best chance of rejecting a false hypothesis. Power curves can be used to determine the discriminating power of the adopted rule against any specified alternative.

OC curves indicate the probability of accepting the null hypothesis. They are related to the power curve (for the same test) in that the height of the *OC* curve is one minus the height of the power curve for the same alternative value of the unknown parameter.

CAUTION

Inference, the process of estimating parameters from sample data, is the basis for much of our information concerning our community and our nation. Here the process, for the most part, has been limited to inference from large samples taken from large populations. There is a host of modifications that must be recognized when samples and/or populations are small.

There is no simple formula for determining sample size, but a crude approximation of sample size can be made by use of the formulas in this chapter. However, in order to determine the size of sample necessary to obtain results of specified precision, we need to be acquainted with more advanced work in sampling theory.

SYMBOLS

α, alpha risk, probability of a Type I error
β, beta risk, probability of a Type II error
H_0, null hypothesis
H_1, alternate hypothesis
$\bar{X}_{c1}$, $\bar{X}_{c2}$, critical points, measurement data
p_{c1}, p_{c2}, critical points, attribute data
z_α, z_{β_1}, z_{β_2}, standard normal deviates
μ_0, μ_1, μ_2, hypothesized parameters, measurement data
π_0, π_1, hypothesized parameters, attribute data

PROBLEMS

11.1 Precisely and concisely define the following:
a. Null hypothesis.
b. Alpha risk.
c. Beta risk.
d. Type I and Type II errors.
e. *OC* curve.

11.2. The Sorob Corporation manufactures "Eagle" brand golf balls, among many other golf products. The Sorob Corporation subjects these balls to many quality tests in order to insure meeting USGA standards. One such test deals with maintaining a diameter no smaller than 1.680 inches. Assume that L. Gauss, the company statistician, wishes to control the diameters of the "Eagle" brand golf balls by taking a random sample of n number of balls from each two-hour production run and measuring the balls carefully. According to the USGA, the balls should "not be smaller than 1.680 inches." Gauss wishes to avoid producing golf balls with diameters less than 1.675 inches; in fact, Gauss is willing to assume a beta risk of .01. The alpha risk assumed is .05. The estimated population standard deviation is .010. Answer the following:

a. What sampling *plan* do you suggest for Gauss? Draw a diagram to explain and justify your response.

b. What is the meaning of α and β for this problem?

c. Assume that a sample of size n (which you determined in [*a*] above) is chosen and the sample diameters average 1.679 inches. Does this lot of "Eagle" brand balls meet Gauss' and the USGA's standards? Why, or why not?

d. Calculate the probability of a Type II error if $\mu = 1.679$.

e. If β were .001, would your response be the same?

11.3. The USGA also is concerned about the weight of golf balls, with the rules stating that "the ball should be no heavier than 1.620 ounces." Therefore the Sorob Corporation is also interested in controlling this factor in the production of "Eagle" brand golf balls. (See Problem 11.2.) Here, L. Gauss wishes to avoid producing golf balls whose weight is greater than 1.620 ounces. Gauss is willing to take a .01 risk of concluding that the balls are 1.625 ounces or heavier, when indeed they are not. Gauss is also willing to assume a .05 risk of concluding that the balls are 1.620 ounces or less, when indeed they are not. Assume that a previous population standard deviation of 0.016 ounces is useful here.

a. What does the alpha risk mean here? What is P (Type I error)?

b. What does the beta risk mean here? What is P (Type II error)?

c. What sampling plan do you suggest for Gauss? Draw a diagram to illustrate your answers.

d. Assume that a sample of size n (which you determined in [*c*] above) is selected and that $\bar{X} = 1.621$ ounces. Does this lot of "Eagle" brand balls meet the desired specifications? Why, or why not?

e. If $\mu = 1.624$, calculate the probability of accepting a false hypothesis.

f. If β were .001, would your responses be the same?

11.4. It was hypothesized that 30 percent of the readers of the *New York Times* in a selected suburb could recall having seen a particular full-page advertisement. In a probability sample of 500 readers, only 110 could recall the advertisement at all. This difference from the "expected value" was said to be "significant at the .05 level."

a. Describe what is meant by this statement.

b. A statistician might utilize the concept of a power curve to describe the power of discrimination of such a test. Describe the values shown on the axes of the *OC* curve.

c. Calculate one value on the *OC* curve. What does it mean?

11.5. In the Office of Staff Benefits of the Fischer Manufacturing Corporation, 5 percent erroneous entries in employee records are considered typical. What sampling plan would you suggest that J. B. Petty adopt for the regular monthly audit if management wishes to protect against an 8 percent level of error? Use $\alpha = .05$ and $\beta = .10$.

11.6. Rule 2–3 of "The Rules of Golf" state that "the velocity of a ball may not exceed 250 fps when tested by the USGA (allowing a reasonable 2 percent tolerance)." The Sam Daens Golf Equipment Company, manufacturer of "Double Eagle" brand golf balls, maintains careful control of its production in order to meet this and other standards of the USGA. T. Shewart, Daens's statistician, wishes to design a sampling plan which will afford adequate protection in meeting the velocity standard. Shewart wishes to insure that the velocity of the "Double Eagle" is neither greater than nor less than 250 fps. Assume that the following control criteria are set up:

$$\mu_0 = 250 \text{ fps} \qquad \mu_1 = 245 \text{ fps} \qquad \mu_2 = 255 \text{ fps}$$
$$\alpha = .05 \qquad \beta_1 = .02 \qquad \beta_2 = .01$$
$$\hat{\sigma} = 5.75 \text{ fps}$$

Answer the following:

a. Explain the meaning of α, β_1, and β_2 for this problem. Which risk is Shewhart most interested in protecting against? Why? Why is Shewhart concerned about a velocity less than 250 fps, given the rule above?

b. What sampling plan do you suggest for Shewhart? Draw a diagram to illustrate your responses.

c. Assume that a random sample of size n (which you determined in [*b*] above) is selected and that $\bar{X} = 251$ fps. Does the "Double Eagle" brand meet the velocity standard? Why, or why not?

d. Draw an *OC* curve indicating this test.

e. Calculate the probability of accepting a false hypothesis when $\mu = 251$ and when $\mu = 248$.

f. If β_2 were .05, would this alter your responses? How?

11.7. Assume that L. Delta, a politician in a metropolitan city, engages a competent public opinion research agency to help reach a rational decision on the issue of a city income tax. The politician wants to vote according to the wishes of a majority of constituents. In this case the decision parameter is π, the proportion of eligible voters favoring a given city income tax proposal. Delta wants to focus attention on two possible values of the population proportion: (*a*) $\pi_0 = 45$ percent, at which point it would be preferable to vote no; and (*b*) $\pi_1 = 55$ percent, at which point Delta would vote yes. Assume that the politician wishes to take a .02 risk of voting no if $\pi = 55$ percent. If $\pi = 45$ percent, Delta is to take only a .01 risk of voting yes. Answer the following:

a. What is the meaning of α and β in this problem? What is the probability of a Type I error? Of a Type II error?

b. What sampling plan would you suggest the research agency follow, given the above constraints? Draw a diagram to illustrate your responses.

c. Assume that a probability sample of size n (which you determined in [*b*] above) is selected and $p = 53$ percent. How would you instruct Delta to vote, given the desire to vote with the majority of constituents and given the constraints?

d. Draw an *OC* curve indicating this test.

e. If $\pi_1 = 60$ percent, woud your responses be different? How?

f. Calculate the probability of accepting a false hypothesis if $\pi = 46$ percent.

11.8. Refer to Problem 10.31. If $\pi = 45$ percent, what is the probability of a Type II error? Draw a diagram to illustrate your answer.

11.9. Given the following diagram, respond to the following questions:

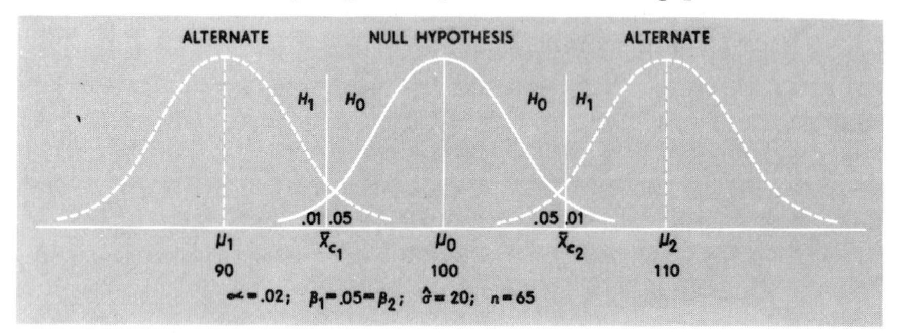

a. On the basis of the above diagram, the important difference is how many units?

b. Is management more anxious to protect against an error of Type I or Type II?

c. If α is changed to .05 and $\beta_1 = \beta_2 = .02$, will the sample size be larger or smaller than 65?

d. If α is changed to .01 and $\beta_1 = \beta_2 = .02$, will the sample size be larger or smaller than 65?

e. If μ_2 is changed to 105 but all other information remains the same, then will we need a larger or a smaller sample than 65?

f. What is an observed difference between $\bar{X}$ and μ_0 called?

g. If mu $= \mu_0$, what type of error may not be made?

h. Given the original diagram and information, if $\mu = \mu_2 = 110$ the probability of a Type II error is controlled at what level?

i. If $\mu = \mu_1 = 90$, what type of error may not be made?

j. When one rejects the null hypothesis, one assumes there is a significant difference between the sample result and what other value?

k. What is the probability of rejecting the null hypothesis called?

l. The height of the *OC* curve indicates the probability of a Type II error at every point except at one point. What is this point?

11.10. Given the following information:

$\pi_0 = 90$ percent	$\pi_1 = 80$ percent	$n = 200$
$\alpha = .01$	$\beta = .05$	$p_c = 85$ percent

a. Calculate the probability of a Type II error if $\pi = 86$ percent. Draw a diagram to illustrate your response.

b. Demonstrate that the sample size of 200 provides at least the protection required by the alpha risk of .01.

c. Demonstrate that the sample size of 200 provides at least the protection required by the beta risk of .05.

11.11. We wish to test the hypothesis that the mean test score on the Graduate Management Admission Test (GMAT) made by students applying to a large state university in a given year is 575. Using an estimated population standard deviation of 60, a .10 level of significance, and a properly drawn sample of 25 test scores at this university, answer the following:

a. Using a two-sided test what are the hypotheses to be tested?

b. What is the acceptance region?

c. Assume that a random sample of 25 GMAT scores provides an average of 550. Is the original hypothesis tenable? Why or why not?

d. If the true mean GMAT score of those students applying in a given year is 560, what is the probability of a Type II error?

e. If the true mean is 575 what is the probability of a Type I error? Of a Type II error?

11.12. Given the following:

$$\pi_0 = 60 \text{ percent} \qquad \pi_1 = 55 \text{ percent}$$
$$\alpha = .10 \qquad \beta = .05$$

a. Draw a diagram that mathematically structures this problem.

b. What are the hypotheses to be tested?

c. What are the criteria for choosing between these hypotheses?

d. What size sample would be suggested to protect against the specified risks?

e. If $p = 59$ percent, what is your decision?

f. Demonstrate that the sample size of (*d*) provides at least the protection required by the alpha risk of .10.

g. Demonstrate that the sample size of (*d*) provides at least the protection required by the beta risk of .05.

11.13. Refer to Problem 10.27.

a. If the true mean is 1,645, what is the probability of a Type II error?

b. If the true mean is 1,650, what is the probability of accepting shipment?

11.14. The Camelot Canning Corporation (CCC) employs a machine to fill cans of size No. 303 with peas. The label on the can states that the net weight of the peas is 16.5 ounces. Management of CCC does not wish to underfill the cans for fear of government action with respect to mislabeling its product, nor does it wish to overfill them for obvious reasons. The machine does not, of course, place precisely the same net weight in every can; in fact, this distribution of net weights tends to be normally distributed about 16.5 ounces. The mean of the probability distribution, that is, the average net weight of the peas placed in each can, can be controlled by altering the setting of the filling machine. Because of the machine's relative accuracy in filling cans to a specified weight the population standard deviation is estimated from past experience to be .40 ounces. The managemnet of CCC wants to develop a sampling inspection plan that will control the filling process at some specified weights with known risks. Management states that they want to take only a .02 alpha risk of stopping the machine-filling process when in fact it is performing according to desired standards. They also wish to protect against concluding that the machines are working satisfactorily when, indeed, they are not. In fact, CCC management wants to take only a .01 risk of concluding that the cans contain 16.5 ounces, on the average, when the true mean weight really is 16.0 ounces. In addition, a .05 percent risk is acceptable in concluding that the cans contain 16.5 ounces, on the average, when in fact the true mean weight is really 17.5 ounces. Answer carefully the following questions: (Draw a neat diagram to structure the problem.)

a. What are the hypotheses to be tested? State your response in symbols and in words.

b. What are the numerical values of μ_1, μ_2, μ_0?

c. What are the values of z_α, z_{β_1}, z_{β_2}?

d. What are the values of the critical points?

e. What size sample should be taken when periodic checks are made on this filling process?

f. What are the criteria you suggest be employed in choosing between these hypotheses?

g. Assume that the sampling plan that you suggested above is operative and that a random sample of size n cans is selected from this hour's filling process. Their mean net weight is found to be 17.30 ounces. What should be done?

11.15. In statistical inference problems, both the definition of a population and the specific characteristic to be studied depend primarily on the nature of the task being considered. For example, the following might be useful to help emphasize the importance of the concepts involved. The problem is broadly classified as a "low-cost housing supply problem."

1. *Problem definition:* To determine the effective supply of low-cost apartment units in Ann Arbor, Michigan.
2. *Population definition:* All apartment units in Ann Arbor.
3. *Population characteristic of interest:* Whether or not the apartment is available (binomial situation).

Complete the following broadly conceived problems in a similar manner using the three headings given above.

a. Acceptance of an order of TV tubes.

b. Production of male adult belts (estimation problem).

c. Establishment of a corporate retirement plan for the Pearson Company.

d. Purchases of consumer durable goods in United States (estimation problem).

e. Relationship of marginal propensity to consume and family incomes in the United States.

11.16. A large discount electronics supply company takes four inventories per year in each store. If there is a shortage or overage between the physical check and the records of a particular store's inventory in corporate headquarters, further investigation is made. Let us examine the data for store No. 72. Corporate management seeks a statistical decision rule to relate observations to the appropriate action. The parameter to be investigated is the population mean—the average value per physical unit of inventory. Assume the following information, answer the questions below:

1. Average value per physical unit of inventory for store No. 72 according to corporate headquarter's records: $60.
2. Estimated population standard deviation: $10.
3. Importance difference: ±$5.
4. Alpha risk, .10; beta_1 risk, .02; beta_2 risk, .02.

Required:

a. Explain the meaning of the alpha risk for this problem.

b. Explain the meaning of the beta_1 and beta_2 risks for this problem.

c. Draw a carefully conceived diagram to structure this problem graphically in order to solve it in an orderly fashion.

d. What is the null hypothesis to be tested? What is the alternate hypothesis to be tested? (State in symbols and words.)

e. What would be your sampling plan for store No. 72?

f. What are the values of the critical or decision points?

g. Assume that a sample of the size you recommend in (*e*) is taken of the inventory in store No. 72 and that the average value per physical unit of inventory is $59 ($\bar{X}_{72}$). What do you advise corporate management to do?

h. If the true population average value per physical unit in store No. 72 is $61, what is the probability of a Type I error? A Type II error?

i. Demonstrate that your sampling plan will provide for the protection that corporate management seeks to maintain with respect to the stated risks.

11.17. Given the following information:

$$\pi_0 = 45 \text{ percent} \qquad \pi_1 = 55 \text{ percent} \qquad n = 500$$
$$\alpha = .01 \qquad \beta = .02 \qquad p_c = 50.2 \text{ percent}$$

Required:

a. Calculate the probability of a Type II error if $\pi = 46$ percent. Draw a diagram and show all computations.

b. Demonstrate that the sample size of 500 provides the protection required by alpha $= .01$.

11.18. A marketing decision must be made on whether or not to market a new consumer product. A random sample of consumers is to be taken, and the decision between marketing the product or not to market the product is to be made on the basis of the sample results from this test market panel.

Management and the statistician decide to concentrate on the proportion of family units who say that they prefer this product after using it (a free sample) for a week; this proportion for all family units is the decision parameter. Management wishes to focus attention on two values of the population proportion. (1) if pi = 20 percent it would be preferable not to market the new product; and (2) if pi = 30 percent it would be desirable to market the product.

It is also determined that management is willing to take a greater risk (.05) of not marketing the product if pi = 30 percent than the risk they are willing to assume (.01) of marketing the product if pi = 20 percent.

Answer the following questions:

a. Sketch a diagram carefully and label it properly to illustrate this problem.

b. What is the null hypothesis to be tested? State your response in symbols and in words relating to this problem.

c. What is the alternative hypothesis to be tested? State your response in symbols and in words relating to this problem.

d. What is the meaning of the alpha risk in this problem?

e. What is the meaning of the beta risk in this problem?

f. What sampling plan would you suggest to protect against these risks?

g. Sketch the *OC* curve for the above test.

11.19. The Columbia Gas & Electric Company is concerned with the waiting time from the moment a customer calls the company by phone until the customer is connected with a service representative. The performance of service is described by the distribution of delay time. The problem at hand is whether current performance is the same as in the recent past or whether it has changed. If it has worsened, management wishes to know this in order that appropriate remedial action can be taken. An improvement in the service also would be investigated to see if it can be made permanent. If current performance is the same as in the past, however, management probably would not take any action at this time since past performance is considered satisfactory.

It was decided that the adequacy of performance in this particular situation could be reasonably measured by the average delay time. In the recent past, this has been 20 seconds.

Suppose that management states that it wishes to assume a maximum risk of .05 of concluding that the average delay time has changed when it has not. Management further states that it wants to take a .02 risk of concluding that the average delay time has not changed when in fact it actually has decreased by five seconds. Management also wishes to take only a .01 risk of concluding that the average delay time has not changed when it actually has increased by five seconds. Past experience indicates that the estimated population standard deviation is ten seconds.

a. Sketch a diagram to illustrate this problem.
b. What is the null hypothesis to be tested? State your answer in symbols and in words.
c. What are the alternate hypotheses to be tested? State your answer in symbols and in words.
d. What is the meaning of the alpha risk in this problem?
e. Explain the meaning of the beta risks in this problem.
f. What sampling plan would you suggest to protect against these risks?

11.20. A wholesaler offers a shipment of 500,000 assorted items, which normally sell for from $2.50 to $4.95, to the head of a chain of jewelry stores at a discount from the regular wholesale price. The jewelry store could feature these items at a special pre-Christmas sale at prices from $1 to $1.79 and still make a substantial profit because of the large number of items.

The jewelry executive estimates that 90 percent of all items bought must be sold in order to avoid a large leftover stock of merchandise. The potential buyer wishes to take only a .01 risk (alpha) of missing an opportunity to make a profit on the sale of the merchandise.

The jewelry executive estimates that if only 80 percent of the items can be sold, the venture should not be undertaken. However, the executive is willing to take a .05 risk (beta) of buying the merchandise when actually according to the executive's plans the items should not be purchased.

Therefore, before buying the large assortment of items, the jewelry executive wants to experiment with their salability, in effect, test-market the items this spring. Consequently, the executive persuades the wholesaler to sell a small number of items (some multiple of 100) so that consumer response can be tested before buying the whole shipment.

Question: Can you help the jewelry executive?

11.21. A manufacturer desires to include a three-year warranty with the sale of a product. Past experience indicates that the "average life" of the product can be described effectively by a normal distribution with an estimated population standard deviation of six months.

A cost study of past warranty claims shows that the most profitable economic level would be achieved if the average life of the product is at such a value that only 2 percent of the customers experience less than three years of product life. The cost analysis also indicates that protection must be provided against a 10 percent level of customer warranty claims.

Question: What sampling plan might you suggest for product control in order to implement the three-year warranty policy if the alpha risk is to be controlled at .10 and the beta risk at .05?

RELATED READINGS

BOEHM, GEORGE A. W. "Helping the Executive Make up His Mind." *Fortune,* vol. 65 (April 1962), pp. 128–31, 218–24.

BURCK, GILBERT. "Management Will Never Be the Same Again." *Fortune,* vol. 70 (August 1964), pp. 125–26, 199–204.

COHEN, JACOB. *Statistical Power Analysis for the Behavioral Sciences,* chaps. 1 and 2. New York: Academic Press, 1969.

DYCKMAN, T. R.; SMIDT, S.; and MCADAMS, A. K. *Management Decision Making under Uncertainty,* chaps. 16 and 17. New York: Macmillan Co., 1969.

FEINBERG, WILLIAM E. "Teaching the Type I and Type II Errors: The Judicial Process," *The American Statistician,* June 1971, pp. 30–32.

LEABO, DICK A. "Sample Requirements and Sample Size: A Pedagogical Aid," *FORTRAN Applications in Business Administration,* vol. II, pp. 233–40. Ann Arbor: Graduate School of Business Administration, University of Michigan, 1971.

LEHMAN, E. L. *Testing Statistical Hypotheses.* New York: John Wiley & Sons, Inc., 1959.

MCDANIEL, HERMAN, ed. *Applications of Decision Tables.* New York: Brandon/Systems Press, Inc., 1970.

MOOD, ALEXANDER M., and GRAYBILL, FRANKLIN A. *Introduction to the Theory of Statistics,* chap. 9. 3d ed. New York: McGraw-Hill Book Co., 1974.

SAVAGE, LEONARD J. *The Foundations of Statistics,* chaps. 16–17. New York: John Wiley & Sons, Inc., 1954.

Part V

Measuring economic changes

Thus this field of study [index numbers], almost alone in the domain of the social sciences, may truly be called an exact science—if it be permissible to designate as a science the theoretical foundations of a useful art.

Professor Irving Fisher, Yale University

12

Index numbers

All sciences are characterized by a close approach to exact measurement.

William Trufant Foster

BUSINESS MANAGEMENT has come to depend more and more on index numbers as indicators of changes in market conditions. In earlier days, industry moved from boom to depression and back again without fully realizing what was happening, except at the extremes. This was partly because cyclical changes never have been uniform throughout an economy; some industries have resisted change more than others, and some have even moved against the general trend. And within each industry, some firms have enjoyed prosperity, while others have been approaching bankruptcy. But today, variations in the indexes of sales, prices, production, new orders, profits per dollar of sales, and many others are closely watched; and frequently, they make headline news. This is especially true of such indexes as those of stock prices and the Consumer Price Index (CPI). The latter is widely used in so-called "escalator" clauses in collectively bargained labor contracts. These clauses provide for adjustments of wages according to some formula using the increase or decrease in the CPI to determine the increase or decrease in the wages of workers covered by the contract.[1] The CPI even has been applied to adjust alimony payments! Some financial institutions are offering escalator provisions for their certificates of deposit (CDs) which could yield more than the fixed-

[1] As of early 1975, it was estimated that over 5.1 million workers covered by collective bargaining contracts had their wages tied to changes in the CPI. In addition, another 44 million persons now find their incomes affected by the index, many as a result of statutory action, for example, social security beneficiaries, Civil Service employees, retired military personnel and postal workers. The number of these escalator clauses is increasing and also show signs of changing. For example, in the spring of 1974 settlements in the aluminum industry provided for annual automatic cost-of-living adjustments in pension benefit levels geared to a rise in the CPI.

rate longer term CDs. Interest rates on some deposits for four years have been geared to vary with year-to-year changes in the CPI.

Nevertheless, though index numbers are widely employed, they are far from perfect. They are generally based on scanty samples, and the mathematical theories on which they depend are not at all rigorous. This is *not* to say that they are *intentionally* misleading. It merely calls attention to the fact that many of them are sampling approximations and that further improvements are not likely to be made until the business public becomes aware of their limitations. As will appear later in the text, broader samples and more frequent revisions are needed. In the early 1970s the President's Commission on Federal Statistics was busy reviewing the operation of the governmental statistical system. Perhaps, some of their recommendations will strengthen an already vigorous and viable data collection system.

As one part of his economic program to combat inflation and the dollar problems abroad, President Nixon on August 15, 1971 announced his order freezing wages, prices, and rents for 90 days and proclaimed a "national emergency" in imposing a 10 percent surcharge on foreign imports. In this general discussion of index numbers it is worth noting that President Nixon's decision to invoke provisions of the Economic Stabilization Act of 1970 was said to be tied to a drop in two key economic indicators and a persistent rise in another. The indexes involved are the Industrial Production Index, the balance of payments, and the CPI. The Federal Reserve Board said that production dropped .8 of 1 percent at 106.0 (1967 = 100) in July 1971. This placed the Industrial Production Index 5.3 percent below the 1969 high and 3.3 percent above the November 1970 low point. At the same time the Department of Commerce stated that the balance of payments deficit during the second quarter of 1971 was the worst in U.S. economic history. Consumer prices during this time also had shown a trend upward at a pace beyond the "creeping inflation" rate. Just how important the levels of these indexes were in shaping public policy is difficult to determine. That the indicators did play some sort of a role seems unquestioned. Where indexes are used to form public policy it seems axiomatic that the average person should become informed about some of the key features of these economic indicators. For example, in addition to the applications of the CPI stated above, it is likely that automatic adjustments in social security benefits might be geared to the movements of the CPI. The price index surely played a major role in determining pay and price restraints imposed in Phase 2 and later periods of Nixon's program.

Undoubtedly, the relatively deep recession (the worst since 1948) and the double-digit inflation of 1974 were instrumental in bringing about a renewed interest in the general public of the measurement of price changes, industrial output, investment commitments, and so on, through the interpretation of index numbers. It is the view of this author that every well-informed citizen should be familiar with the major indexes discussed in this chapter as well as the economic indicators covered in Chapter 13. These statistical measures play an important role in the determination of governmental fiscal and monetary policies that affect each one of us. It certainly behooves us to be better informed about their uses and limitations. Therefore, let us consider some of the general methods of constructing these indicators of economic activity.

SIMPLE INDEX NUMBERS

Generally speaking, index numbers are percentages indicating changes in values, quantities, prices, or other market variables, each item being compared with the corresponding figure in a selected base. Typically, they measure changes over periods of time, though sometimes they present current comparisons; for example, they might compare costs in certain cities or firms. Simple index numbers deal with like denominations, such as pig iron, cotton, or population. Composite index numbers, which combine various categories such as bread, meat, and kilowatt-hours, are considered later.

Naturally, it is difficult to draw the line between informal percentages and formal index numbers. For example, the average price of cotton received by farmers in the United States in the years 1910–14 was 12.4 cents a pound; and in the years 1947–49, it was about 31.2 cents.[2] If the former price is taken as 100 percent, the latter is 252 percent (i.e., 31.2/12.4), which would be considered a mere percentage.[3] But if a continous series of monthly or annual percentages were given, they would be designated as index numbers, or collectively as an index. Obviously, custom alone determines the dividing line segregating ordinary percentages; the essential difference is merely in the degree of continuity or complexity.

An index of population

As an example of a series of percentages that barely qualifies as a simple index, the following figures may be cited. They consist of Bureau of the Census reports on population in continental United States, includ-

---

[2] See U.S. Department of Agriculture, *The Agricultural Situation*, December, 1963.

[3] The omission of a percent sign (%) in writing index numbers leads to seeming errors unless the sign or its decimal equivalent (.01) is mentally or actually supplied. Thus the ratio of index numbers 31.2/12.4 is literally 2.52. Times 100 percent, this becomes 252 percent, which means 252 (.01), and 252 is seemingly incorrect unless the percent sign or its equivalent is understood. The product of index numbers, 125 times 248, is the index number 310. To make the operation explicit or more precise, $(1.25)(2.48) = 3.10 = (3.10)(100\%) = 310\%$ or 310, as an index number with the percent sign understood.

ing armed forces overseas, for the decennial census years 1900–1970, in thousands: 75,995; 91,972; 105,711; 122,775; 131,669; 150,697; 178,464; 205,395. Reduced to percentages of the population in 1900, the percent sign being omitted, they become, 100, 121, 139, 162, 173, 198, 235, and 270.

The utility of such index numbers is evident. They show much more clearly than the original data the growth of population over the first half of the present century. They indicate, for example, that for every 100 persons in 1900, there were 270 persons in 1970; that is, 205,395/75,995 = 270, expressed as a percentage, the sign % being understood.

CHOICE OF BASE

The choice of a base period or item is somewhat arbitrary.[4] It is influenced by the comparisons that are to be emphasized. Even for a given set of comparisons, such as the population data just quoted, the base need not be the initial item. Expressed with 1970 as the base, they appear as the index, 37, 45, 51, 60, 64, 73, 87, and 100, which presents the same picture of growth as before, though perhaps not quite so clearly. They are rounded to the nearest whole number rather than stated as a decimal, inasmuch as the purpose—like that of a chart—is to give a general impression of the change in question, and meticulous accuracy detracts from the effect.

Other considerations may be involved in the choice of a time series base. Not only should it represent a natural basis of comparison with respect to time but it preferably should be representative with respect to trend. It would hardly do, for example, to take November 1975 as the revised base of an index of consumers' goods prices, since the reported item (165.1) was then eight points above the year's average (157.0). Nor would June have been any better, since the reported item then (160.6) was about two points above the average. Only in political statistics, where a biased yet technically accurate statement is sought, would such bases be tolerated.

Hence, for many index numbers, particularly of highly variable items, the base is often an extended period, such as the monthly or annual average of the three years 1957–59, a period following the sharp postwar inflation. Earlier bases covered the years 1935–39 and 1947–49; and an even earlier base, still utilized in the mid-70s by political supporters of the farm bloc, is 1910–14. Most current index numbers calculated and

---

[4] In 1976 the base period for most index numbers published by the federal government was 1967 = 100. The base period for almost all government statistical series is set by the Office of Management and Budget (OMB). When the revised CPI and a few others are introduced in 1977 it likely will have a new base period which at this time has not been determined.

published by the federal government use 1967 as the base year. A base period such as 1967 makes it easier for everyone to realize actual changes in the various indexes. For example, when the "cost of living" as measured by the price indexes goes up or down compared with 1967, most people will more readily remember past prices than they could remember what happened in the previous base period of 1957–59. Except for rounding differences, changing the base does not alter the percentage change between index figures over time.

It should be noted that when an extended period is taken as the base, the items within that period when expressed as separate percentages normally average 100. For example, if the average population of the seven census years given above (133,710, in thousands) is taken as the base, the index numbers, attained by dividing each census figure by the average become 57, 69, 79, 92, 98, 113, 133, and 154, which average 100 (when not rounded). Such a base, though not advisable in this case, is particularly suitable when the comparisons are intended to show which items are above and which are below the average, as in comparing population or productivity measures.

It should be obvious that because index numbers are essentially ratios, they may consistently be multiplied or divided by any designated factor, even though the base may then disappear. Such a procedure is necessary when the base is changed, or when two similar indexes covering overlapping periods of time are to be "spliced" into one continuous series.

THE "VALUE" INDEX

The most important data entering into the computation of market index numbers are classified as value, quantity, and price. These are generally composite indexes, and the latter two invoke certain problems of weighting when quantities and prices involving incommensurable units are to be combined. But a value index, in which incommensurable physical units are rendered commensurable in terms of the common value unit, the dollar, involves only simple percentages, like a simple index.

The value index may be illustrated by an index of retail sales based on national samples. Such an index is published in the *Federal Reserve Bulletin.* In 1976 the base period was described as 1967 = 100, and each monthly index number was essentially a percentage ratio of total reported sales in a given month compared with average monthly sales in the base period. Of course, the data on which such index numbers are based involve minor adjustments from time to time, due to variations in the composition of the sample, and other requirements. But essentially, they are simple percentage comparisons.

The annual index of retail sales for the period of 1971–75 was given in the April 1975 *Federal Reserve Bulletin* as follows (1967 = 100):

1971	122
1972	142
1973	157
1974	172
1975p	178

p = preliminary.

Expressed in percentages of the two years (1974–75 average is 175), these index numbers become (1974–75 = 100):

1971	70
1972	81
1973	90
1974	98
1975p	102

p = preliminary.

The computation of the original index is in principle as elementary as this readjustment of the base. In other words, the base is shifted by merely dividing by the 1974–75 average index, which is 175.

Value index "deflated"

A value index such as that of retail sales obviously is a composite of two distinctive variables: the quantity of goods sold and the average prices at which they are sold. If an index of prices is available approximating the type of goods marketed, an index of quantities sold may be obtained by dividing the sales index by the price index—a principle that may now be briefly illustrated.

It will be assumed that the Consumer Price Index approximates closely enough for illustrative purposes the prices of goods sold in retail stores. We then have the following comparable index for the years 1971–1975, each on the base 1967 taken as 100. The first, V, is the value index of retail sales noted above, and the second, P, is consumer prices (see Table 12.1) reduced to the same base:

V: 122, 142, 157, 172, 178
P: 121, 125, 133, 148, 157

Then the value index divided by the price index, year by year (V/P), expressed as a percentage, is the quantity index:[5]

[5] That a value index number divided by an appropriate price index number gives a quantity index number depends upon the theoretical assumption that what is true of individual items is also true of the aggregation of items combined as index numbers—for example, 100 bushels of wheat (q) at $2.20 per bushel ($p$) are worth $220 ($v$). Obviously, $pq = v$, $q = v/p$, and $p = v/q$. The same equations written as capital letters refer to appropriate index numbers, that is, $PQ = V$, $Q = V/P$, and $P = V/Q$. In practice, however, applied to indexes as commonly computed, these equations are found to be only approximately true.

$$Q = 101, 114, 118, 116, 113$$

Because both the value index and the price index have the same base, the quantity index will have this base so that the quantity index here reflects approximately the increasing *volume* of retail sales during the five years. This volume became nearly half again as large as in the base year, but it was not nearly so large as it would appear to be if judged by value of sales alone.

The computation by which the effect of rising prices is removed from the value index is sometimes described as deflating—a term that came into use when both prices and production were soaring during World War I. But when prices are declining, the quantity index obtained by "deflating" will rise in comparison with values.[6]

The deflated retail sales quoted above, or comparable figures, are often called sales in dollars of the base period, in this case of 1967. That is, they indicate theoretical changes in value of sales as they would have been if prices of the base period had remained at a constant average while other variables were unchanged.

THE VALUE OF THE DOLLAR

The reciprocal of the index of consumer prices is commonly used as a measure of the "value of the dollar," that is, of its purchasing power. Obviously, if prices rise, the dollar loses purchasing power, and its "value" declines; conversely, if prices fall, its "value" rises.[7] In strict theory a broader index than consumer prices should be used for measuring the value of the dollar, because this index represents only purchases by moderate-income receivers.

The computation of the "value of the dollar" may be illustrated as follows. In 1975 the index of consumer prices stood at 157.0 (1967 = 100) (see Table 12.1); hence the value of the dollar in terms of that base was (100/157.0) = 64. That is, from the base period to 1975—about eight years—the physical volume that could be purchased with a dollar was 36 percent less. Compared with that base, it was an 64-cent dollar. But compared with the pre-World War II years, 1935–39, when prices according to the present index averaged 41.9, prices in 1975 were (157.0/41.9), or 374.7. In terms of that base, the value of the dollar is therefore

---

[6] The value index, V, divided by an appropriate price index, P, will give a quantity index, Q, even though V and P have different bases. In that case, Q is a valid series of ratios and may be reduced to a desired base. However, the weightings of P, as noted later, should be appropriate for the period covered.

[7] When discussing the value of the dollar one is reminded of the banker who went to his M.D. for a medical checkup. The doctor said: "You are as sound as a dollar." Whereupon the banker exclaimed: "As bad as all that!" And then he fainted dead away!

TABLE 12.1
Consumer Price Index, United States, 1934–75 annual average, 1967 = 100

Year	Index	Year	Index	Year	Index
1934	40.1	1948	72.1	1962	90.6
1935	41.1	1949	71.4	1963	91.7
1936	41.5	1950	72.1	1964	92.9
1937	43.0	1951	77.8	1965	94.5
1938	42.2	1952	79.5	1966	97.2
1939	41.6	1953	80.1	1967	100.0
1940	42.0	1954	80.5	1968	104.2
1941	44.1	1955	80.2	1969	109.8
1942	48.8	1956	81.4	1970	116.3
1943	51.8	1957	84.3	1971	121.3
1944	52.7	1958	86.6	1972	125.3
1945	53.9	1959	87.3	1973	133.1
1946	58.5	1960	88.7	1974	147.7
1947	66.9	1961	89.6	1975p	157.0

p = preliminary.
Source: U.S. Department of Labor, Bureau of Labor Statistics, *Economic Report of the President,* February 1975, p. 300.

(100/374.7) 100 = 27, commonly interpreted as a 27-cent dollar. For some purposes the BLS index of wholesale prices is to be preferred to the Consumer Price Index as an indicator of the "value of the dollar." This is especially true for retailers' dollars in their trade. On the other hand, the gross national product deflators give the best detailed "value of the dollar." For example, some of the deflators (1958 = 100) for segments of GNP and their reciprocals in 1974 were as shown in Table 12.2.

It is important to understand that the value of the dollar is merely a relative matter, depending upon the period with which comparison is made. Financial writers sometimes venture opinions concerning what the real value of the dollar (i.e., its purchasing power) is or ought to be, but no authoritative agreement has been reached. Experience shows, how-

TABLE 12.2

Item	Deflator	"Value of the Dollar"
Gross national product	170.1	$0.59
Durable goods	123.6	0.81
Nondurable goods	169.9	0.59
Services	173.4	0.58
Gross private domestic investment	165.6	0.60

Source: U.S. Department of Labor, Bureau of Labor Statistics, *Economic Report of the President,* February 1975, p. 252.

ever, that rapid changes are disturbing to the economy and therefore should be avoided, if possible.

COMPOSITE INDEXES

As has been shown, indexes of value, such as retail sales, gross national product, and other series of data expressed in dollars are computed as percentages, as if they were simply indexes. Yet they are usually composite; that is, they combine a variety of physical units in such groups as food, clothing, and utilities. But when multiplied by price (pq), these physical units are additive because the dollar serves the same purpose for marketed goods that the pound serves for weights.

The basis of addition in value indexes may be elucidated by a simple illustration. If 20 bales of cotton and 10 tons of pig iron are to be added, obviously the nominal sum, 30, is worthless. If the items are reduced to a common physical unit, becoming 10,000 pounds of cotton and 20,000 pounds of iron, the aggregate 30,000 pounds, or 15 tons, may have meaning in a freight depot but not in a market report. But if their values in a specific market are added, as 20 bales of cotton at $150 a bale and 10 tons of pig iron at $60 a ton, the total, $3,600, expresses accurately their importance in the given market. The total is *dollars' worth* of goods, an obvious market measurement. Values, however, shift from time to time. It is this inconstancy of markets that complicates the making of index numbers.

Composite quantity and price indexes, as previously noted, invoke difficult theoretical problems of weighting. As we have seen, we cannot consistently add diverse physical quantities, such as 20 bales of cotton and 10 tons of pig iron, nor can we average prices when the quantities those prices represent are of varying size.

IMPLICITLY WEIGHTED INDEXES

In general, the inherent problem of index number construction is that of choice of weights to be used. Sometimes, such indexes are computed and the weights are implicit (unstated). In Table 12.3, part A illustrates the method by which an index is computed by merely considering prices. Part B illustrates the method by which an index is computed by merely adding relatives. The former assumes that each commodity is of equal importance, and the latter assumes that the amount which one dollar will buy in the base period is the proper weight. Also, of course, part B has all the weaknesses of arithmetically averaging ratios. (As noted in Chapter 5, the geometric mean should be used instead of the arithmetic mean.) In general, neither method is satisfactory. Thus, we must look to a somewhat more realistic method for computing index numbers.

TABLE 12.3
Simple aggregative and simple average of relative index numbers (data: three leading grain crops)

	1967 = 100 Mean Price per Bushel p_0	Relative p_0/p_0	1971 Price per Bushel p_n	Relative p_n/p_0	1975 Price per Bushel p_n	Relative p_n/p_0
A. *Simple Aggregative*						
Corn	$1.64		$1.16		$1.23	
Wheat	2.14		1.95		2.09	
Oats	.85		.62		.75	
	$4.63		$3.73		$4.07	
Index: $\Sigma p_n/\Sigma p_0$ =		100.0		80.6		87.9
B. *Simple Average of Relatives*						
Corn	$1.64	100.0	$1.16	70.7	$1.23	75.0
Wheat	2.14	100.0	1.95	91.1	2.09	97.7
Oats	.85	100.0	.62	72.9	.75	88.2
p_n/p_0		300.0		234.7		260.9
Index: $\Sigma(p_n/p_0)/\Sigma(p_0p_0)$ =		100		78.2		87.0

METHOD OF WEIGHTED AGGREGATES

A solution to the difficulty of combining various commodities has been found in the so-called "method of weighted aggregates" or, more simply, the *aggregative method.* It is not an entirely consistent method. Certain theoretical objections may be raised against it, but it gives useful approximation in practice. As far as a composite quantity index is concerned, it is related in principle to the process of "deflating," described above. In that process the value index, comprising the variables, quantity and price, is divided—item by correlative item—by an appropriate price index. As has been seen, this results in a quantity index ($V/P = Q$), expressing changes in only the volumes of goods sold. It is "sales at constant dollars."

Of course, it is not practical to construct both a value index and a price index to obtain a quantity index. But "sales at constant dollars" can be obtained simply by holding prices constant, that is, by ignoring actual price changes and multiplying physical quantities of sales by prices as of the first period, or of some other period considered typical. This gives a series of values which vary only with quantities sold. It can be reduced to percentages, that is, index numbers, by reference to a selected base. The prices held constant serve the purpose of weights. The period from which they are selected is called the weights base, which is not necessarily identical with the percentage base—the period taken as 100.

The construction of a composite price index usually proceeds on the same principle applied in reverse. Prices of the pertinent commodities are collected and are multiplied by typical quantity weights held constant throughout. This gives a value series of sales of constant quantities, reflecting only price changes, combined by use of so-called "quantity weights."

In practice, many variations in procedure may be found necessary. These will be considered later. But this so-called *aggregative method* of combining quantities and prices to form composite indexes is fundamental to a practical understanding of the nature of index numbers and is easily mastered.

INDEX FORMULAS—AGGREGATIVE METHOD

The aggregative method, utilized in computing market index numbers of value, quantity, and price, is readily expressed in abbreviated formulas. In such formulas, p and q stand for a price and correlative quantity, and pq is the value, v; for example, 10 pounds (q) of bread at \$0.20 (p) a pound equals \$2 (v). A subscript *zero* usually designates a p, q, or v in the base period, though subscript w is sometimes used for this purpose, particularly if the weights are taken from some other period than the percentage base. Successive prices, quantities, and values may be written $p_0, p_1, p_2, \ldots, p_n$ or in condensed form, 1 or n may be utilized. In that case, subscript 1 or n denotes any other "given" period than the base.

By use of these symbols, a series of actual values, added in each time period, is written:[8]

$$V = \frac{\Sigma p_n q_n}{\Sigma p_0 q_0} \tag{12.1}$$

$$Q = \frac{\Sigma p_0 q_n}{\Sigma p_0 q_0} \tag{12.2}$$

and

$$P = \frac{\Sigma p_n q_0}{\Sigma p_0 q_0} \tag{12.3}$$

[8] That averaging prices is more complex than averaging values or quantities is readily seen when successive sales of the same commodity are combined. Suppose two sales of identical goods are four yards @ \$1.50 and two yards @ \$2, a total value of \$10. The average quantity is $(4 + 2)/2 = 3$, and the average value is $(6 + 4)/2 = 5$. But the average price is not $(1.50 + 2.00)/2$, but $5/3$ (or the totals $10/6$)—not the simple average, \$1.75, but the weighted average, \$1.67. Considerations such as this have led to the conclusion that in strict theory a price index number should be based on value divided by an acceptable estimate of quantity. But such a procedure is not practical because it requires excessive data, and the usual method gives a satisfactory approximation.

It was formerly assumed that in strict theory, P should be V/Q; and where complete data are available, this ratio may be used as a check on weighted aggregative price index.

Aggregative method illustrated

The aggregative method just described is illustrated by the use of very simple data in Table 12.4. It is assumed that a Chicago speculator, dealing chiefly in grains, requires a price index of the three grain crops in which the speculator is interested, and for which estimates of visible supply are readily available. Currently, the speculator requires a monthly index; but for earlier years, annual figures are sufficient. These are computed here from estimated rounded data, on the base period 1967, extended to 1974 and 1975.

In part A—Quantity Index—the average price of each crop for the base period is first given. This is the price weight (P_w), which is here also the price in the percentage base period (p_0). The three columns following give production in millions of bushels. The production data in the base

TABLE 12.4
Aggregative method: Quantity and price index numbers, 1967 as weight and percentage base (data: three leading grain crops)

A. *Quantity Index from Formula (12.2)*

	1967			
	Mean Price* $p_w = p_0$	Mean Crop† q_0	1974 Crop† q_1	1975 Crop† q_2
Corn	1.64	3,066	3,455	3,403
Wheat	2.14	1,251	1,004	947
Oats	.85	1,294	1,163	1,308
$\Sigma p_0 q_n =$		8,805	8,803	8,719
Index: $\Sigma p_0 q_n / \Sigma p_0 q_0 =$		100.0	100.0	99.0

B. *Price Index from Formula (12.3)*

	1967			
	Mean Crop† $q_w = q_0$	Price* p_0	1974 Price* p_1	1975 Price* p_2
Corn	3,066	1.64	1.16	1.23
Wheat	1,251	2.14	1.95	2.09
Oats	1,294	.85	.62	.75
$\Sigma p_r q_0 =$		8,805.28	6,798.29	7,356.27
Index: $\Sigma p_n q_0 \Sigma p_0 q_0 =$		100.0	77.2	83.5

* Dollars per bushel.
† Millions of bushels.

period are reduced to annual averages so as to make them comparable with annual crops following. For a monthly average of crops marketed, the base data would be expressed in monthly averages.

Columns q_0, q_1, and q_2 (collectively designated as q_n) are each totaled as values ($\Sigma p_0 q_n$) at base prices. These values, rounded to millions of dollars, represent variable production at typical constant prices, thus giving due weight to the variable market measure of each crop but eliminating the variable year-to-year price by holding it constant. The values are then reduced to percentages—as for a single index—by dividing through by the annual average in the base period. In terms of market measurements in successive years, the combined three crops remained constant, then declined to 99.0 as compared with the base period. Over a long period, as would be expected, such indexes have shown an upward trend.

In part B—Price Index—the procedure is reversed. Average production in the base period (q_w or q_0) is listed. It could be entered in terms of annual totals without changing the results, or in terms of any numbers having the same ratios to each other. If such an index was to be continued over a long period of time, each q_0 could be divided by $\Sigma p_0 q_0$ and expressed as the percentage figures 34.824, 14.205, and 14.694, thus eliminating a series of divisions, as noted below.[9]

In succeeding columns, annual average prices are given. The prices here listed are supposed to be based on receipts by farmers, but published reports are apparently somewhat confused in this respect. The columns p_0, p_1, and p_2 (collectively p_n)are then totaled as values of crops in the base period. Thus the year-to-year price variable is retained, while production is held constant at typical ratios. Index numbers are then obtained, as before, by dividing through by the base value and expressing the quotients as percentages. But as noted above, if each q_0 had been divided by this base value, the percentage index numbers would have been obtained directly—a considerable saving of time in a long series of computations.

INDEX FORMULAS—WEIGHTED RELATIVES METHOD

In the actual construction of index numbers, the simplicity of the aggregative method may become complex by reason of various difficulties encountered in the collection and analysis of data. One of these sources of complexity has given rise to what has variously been described as the "method of weighted relatives," "weighted average of relatives," or the

[9] This is merely another form of the formula. That is,

$$\frac{\Sigma p_n q_0}{\Sigma p_0 q_0} = \Sigma \left(\frac{q_0}{\Sigma p_0 q_0} \cdot p_n \right)$$

"relative method." This method arose at first out of certain theoretical considerations now outmoded; but practically, its advantage lies in the fact that simple index numbers—the so-called "relatives"—of each series of data are computed first. This makes it possible to see at a glance how the various elements of the index are moving—which ones are changing most decisively and which are most erratic.

Before considering an illustration of the method it may be well to note that it is actually or approximately the aggregative method in disguise. This may be seen by its formulas, which are as follows for quantities and prices where specific values, p_0q_0, are weights:

$$Q = \frac{\Sigma\left(\frac{q_n}{q_0}\right)(p_0q_0)}{\Sigma p_0q_0} \tag{12.4}$$

$$P = \frac{\Sigma\left(\frac{p_n}{p_0}\right)(p_0q_0)}{\Sigma p_0q_0} \tag{12.5}$$

Inspection of these formulas will show that the numerator of Q reduces to p_0q_n, and of P to p_nq_0, making the fractions identical with the aggregative formulas. And even if p_0 or q_0 of the weights is selected from an appropriate weights base other than the percentage base (written as p_w and q_w), the weighted relatives may be expected to approximate the results obtained by the aggregative method. The significance of the formulas will, however, become more apparent when illustrated by Table 12.4, which utilizes the simple data of Table 12.3 .

Relative method illustrated

In table 12.5 the data of the preceding example are set down in a new order. In part A the quantities and in part B the prices are written as simple index numbers, the so-called "relatives." The first column in each part lists the value weights—the value of each crop in the base period, or numbers having the same ratios as these values. Several reasons may be adduced to justify the use of value weights, but the most obvious is that their use here reduces the computation to the aggregate method—or, as has been noted, to an approximation of it.

The items in each column of relatives are multiplied, respectively, by their weights and summed. As before, the value sums are reduced to percentages of the base period, that is, to index numbers. Except for variable results of rounding, these index numbers are the same as those obtained by the aggregate method.

As in the preceding example, the computations of a long series may be shortcut by reducing the value weights. In this case, it is sufficient to

TABLE 12.5
Weighted relatives method: Quantity and price index numbers (data and bases as in Table 12.4)

	1967 Value (millions of dollars) $p_w = p_0q_0$	1967 q_0/q_0	1967 p_0/p_0	1974 q_1/q_0	1974 p_1/p_0	1975 q_2/q_0	1975 p_2/p_0
				Simple Index Numbers			
A. *Quantity Index from Formula (12.4)*							
Corn	5,028	100.0		112.7		111.0	
Wheat	2,677	100.0		80.3		75.7	
Oats	1,100	100.0		89.9		101.1	
	$\Sigma(q_n/q_0)v_w =$	8,805.0		8,805.1		8,719.7	
	Index: $[\Sigma(q_n/q_0)v_w]/\Sigma v_w =$		100.0		100.0		99.0
B. *Price Index from Formula (12.5)*							
Corn	5,028		100.0		70.7		75.0
Wheat	2,677		100.0		91.1		97.7
Oats	1,100		100.0		72.9		88.2
	$\Sigma(q_n/q_0)\Sigma v_w =$		8,805.0		6,795.4		7,356.6
	Index: $[\Sigma(p_n/p_0)v_w]/\Sigma v_w =$		100.0		77.2		83.6

substitute for each its percentage of the total value (i.e., 57.104, 30.403, and 12.493). Each column of relatives will then give directly the required index number. This is illustrated in Table 12.6. If the average annual values of each crop in the base period are used as weights the results are unchanged. But the percentage weights are the most convenient if the index is to be extended through many time intervals or the base is a new base each period.

An example of the usefulness of the weighted average of relatives method is that of a university which is faced with the problem of preparing a budget for the next biennium. Suppose the university has a general

TABLE 12.6
Weighted relatives method: Price index, 1975, using percentage value weights (data and bases as in Table 12.5)

	1967				1975		
	Price per Bushel p_0	Relative p_0/p_0	Percentage of Value W	Product $W \cdot p_0/p_0$	Price per Bushel p_n	Relative p_n/p_0	Product $W \cdot p_n/p_0$
Corn	1.64	100	57.1	5,710.0	1.23	75.0	4,282.5
Wheat	2.14	100	30.4	3,040.0	2.09	97.7	2,970.0
Oats	.85	100	12.5	1,250.0	.75	88.2	1,102.5
			100.0	10,000.0			8,355.0
			Index: $\frac{\Sigma(W \cdot p_n/p_0)}{\Sigma W}$	= 100			83.6

expense account made up of expenditures for travel, fuel, other utilities, supplies, and maintenance, and over the years has experienced a percentage value of expenditure as follows:

Travel	5.4
Fuel	32.6
Other utilities	16.8
Supplies	35.4
Maintenance	9.8
	100.0

Further, assume that the university officials have evidence that indicates that category price changes in the next biennium compared with the present are likely to be:

Travel	up 10%
Fuel	up 5%
Other utilities	up 7%
Supplies	up 1%
Maintenance	up 3%

The university then can estimate its general expense money needs by use of the weighted average of relatives method shown in Table 12.7. In this illustration the general expense budget for the future biennium should be 104 percent of that for the present biennium.

PROBLEMS OF CONSTRUCTION

The making of index numbers as illustrated thus far has purposely been limited to the bare essentials of theory and practice. The usual method, depending upon actual values or weighted aggregates of value, with one of the two basic variables held constant, has been emphasized.

TABLE 12.7
Weighted relatives method: Price index, future biennium compared with present biennium general expense—University A (data: hypothetical)

Items	Percentage Value W	Relative p_n/p_0	Product $W \cdot p_n/p_0$
Travel	5.4	110	594
Fuel	32.6	105	3,420
Other utilities	16.8	107	1,800
Supplies	35.4	101	3,580
Maintenance	9.8	103	1,010
Σ	100.0		10,404

$$\text{Index: } \frac{\Sigma(W \cdot p_n/p_0)}{\Sigma W} = 104$$

In practice, data for many commodities may be required, often running into the hundreds. Great care must be exercised in selecting commodities that remain constant in quality over the period of time covered, and revisions must be made to provide for changes in buying patterns and for new products. Sometimes, it will be found more convenient to construct a *chain* index in which each year (or other period) is first compared with the preceding year as percentage base, and the resulting percentages are combined in one index by successive multiplication.

For example, if a given index for 1974/1973 (i.e., 1974 on a 1973 base) is 102.4, while 1975/1974 is 99.2, and 1976/1975 is 104.6, then the index is:

Year	Index
1973	100.0
1974	102.4
1975	102.4 (99.2) = 101.6
1976	101.6 (104.6) = 106.3

and so on through successive years. The reverse of this process breaks up a given index into *unit relatives,* corresponding to the original links in the chain. The chaining process is convenient when new items and new weights are frequently introduced.

Variations in method that arise in meeting specific problems of construction may be illustrated by reference to the Federal Reserve Board *Index of Industrial Production* (a quantity index) and the United States Department of Labor *Consumer Price Index,* to which reference has already been made. Both indexes are reported currently in the *Federal Reserve Bulletin* and the *Survey of Current Business.* Inspection of these and other government publications, as well as those of the Industrial Conference Board, the *New York Times,* and *Business Week,* will give an idea of the multiplicity of data and index numbers now available to the public.

Fisher's "ideal" index

The traditional methods of constructing index numbers have been subject to criticism by Professor Irving Fisher largely on the basis that the use of a *single* weight may cause an inconsistency in a series. His criticisms suggested a method of overcoming this problem when adequate data are available.

One serious limitation to the application of Fisher's technique is that it is necessary to have both quantity and price data. Fisher then suggests that indexes of quantity, price, and value be calculated. The value index (price times quantity) is constructed in a manner similar to that used in the common aggregative method, *except* that the aggregates are the sum of the actual prices times actual quantities, no weights being employed.

Consequently, the resulting aggregates reflect a combination of price and quantity changes, or, more specifically, gross income derived from the recorded sales. In constructing the price and quantity indexes, the weights are chosen from the base period; hence the indexes are referred to as base-weighted (P_b and Q_b).

Essentially, Fisher's method is a correction that is applied to these base-weighted indexes. Fisher maintains that some correction is desirable because for any given year the product of that particular year's price index and the same year's quantity index usually differs from the value index. Naturally, the product should equal the value index, just as any price times quantity should represent value. This comparison normally is referred to as the factor's test.

Fisher's procedure applies this test by dividing each year's value index by that year's price index:

$$(V \div P_b) = Q_r \tag{12.6}$$

to see if it checks with the given quantity index. If the indexes agree, the base-weighted index is taken as final; however, if there is disagreement, the result of the above division is taken as a second estimate of the quantity index and is called Q_r. A second estimate is found by dividing each year's value index by its quantity index:

$$(V \div Q_b) = P_r \tag{12.7}$$

And the revised price indexes (P_r) are called reverse-weighted indexes because they are identical with the figures that would result if the indexes were calculated with weights chosen from the given year instead of the base year.

Once the second estimates of P and Q are obtained for each year in the series, the data are averaged to obtain Fisher's final result as

$$P = \sqrt{P_b \times P_r} \tag{12.8}$$

$$Q = \sqrt{Q_b \times Q_r} \tag{12.9}$$

or, in effect, using the geometric mean. If the factor's test is satisfied, P times Q will precisely equal V, using the geometric mean, but will only be approximated if the arithmetic mean is used.

There are several good reasons why Fisher's "ideal" method of constructing index numbers is not in general use today. First, the cost involved in gathering price *and* quantity data becomes almost prohibitive. Second, the accuracy would not be improved sufficiently to justify the added cost. Third, although the method does remove an inconsistency, it is not entirely free of theoretical criticisms. Finally, as Croxton, Cowden, and Bloch have said, "it seems impossible to say specifically what it measures, other than to say that it is the geometric mean of the Laspeyres and the Paasche index numbers."[10]

SOME IMPORTANT INDEXES

Index of Industrial Production

This is the major index compiled by the Federal Reserve Board directly measuring the physical volume of mine and factory production in the United States. Although it is not as comprehensive as the gross national product, a value index, it is more specific. Almost any industrial producer can compare his or her progress with that of competitors, as well as with the growth of industry in the United States and the world.

Index numbers are reported in specific lines and combined into groups and a national total. The weighted relatives method is employed. The monthly items are given both with and without seasonal adjustment. In a few cases where direct production figures are not available, man-hours have been substituted. This is particularly useful in cases where several related products are manufactured by the same organization. It helps to simplify also the difficult problem of classification.

The base of the index, as published now, is 1967 = 100. An earlier revision revealed that the index had underestimated progress, due to a neglect of new products continually appearing on the market.

Consumer Price Index

As an illustration of the computation of a price index, the index of consumers' prices—formerly called the cost-of-living index—may be briefly described.

The monthly Consumer Price Index (CPI) is compiled by the Bureau of Labor Statistics (BLS) and currently is designed to measure the purchasing power of the urban consumer's dollar. According to the BLS the CPI serves the following two major functions: (1) it is a yardstick for revising wages, salaries, and other income payments to keep in step with rising prices; and (2) it is an indicator of the rate of inflation in the American economy.[11] Largely because of changes in consumer buying

---

[10] Frederick Croxton, Dudley J. Cowden, and Ben W. Bloch, *Practical Business Statistics*, 4th ed. (Englewood Cliffs, N.J.: Prentice-Hall, Inc., 1969), p. 305.

[11] See U.S. Department of Labor, *The Consumer Price Index*, October 7, 1963, *The Consumer Price Index: History and Techniques*, Bulletin No. 1517; *Consumer Prices in the United States, 1959–68*, Bulletin No. 1647, 1970; and Julius Shiskin, "Updating the Consumer Price Index—An Overview," *Monthly Labor Review*, July 1974, pp. 3–20. The present index measures price changes for approximately 400 items se-

habits it has become necessary to revise and update this index periodically. The index initially appeared in print in 1919 and the first revision was in 1940. Following World War II the CPI was overhauled in 1946 and again in 1953. By the late 1950s it became apparent that the weights needed to be reevaluated and the index was revised in 1964. Currently, the BLS is in the middle of another major updating of the CPI with the work scheduled for completion in 1977. Effective April 1977 the BLS will publish two CPIs: (1) an improved index for urban wage earners and clerical workers to meet the requirements of collective bargaining; and (2) an index for all urban households, which will provide a new comprehensive measure of the rate of inflation for the economy. Along with these revisions came changes in the nomenclature of this index. Until 1945 the BLS referred to The Cost-of-Living Index for the United States. In 1945 the name was changed to Consumers' Price Index for Moderate Income Families in Large Cities. Since 1964 the title used has been Consumer Price Index for Urban Wage Earners and Clerical Workers. Effective in 1977 the titles of *two* indexes will be Consumer Price Index for All Urban Households (the new index) and an updated Consumer Price Index for Urban Wage Earners and Clerical Workers. The change in the titles over the years shows the attempt of the BLS to accurately reflect the scope and coverage of the index.

These revisions in such a major index are costly. For example, the 1952 update took three years and cost $4 million; the 1960–64 revisions took five years at an expense of $6.5 million. The 1970s revisions are expected to take eight years and cost $40 million! However, given the economic significance of this indicator to our national policies, this expense is not so high when considered in light of the applications. If one considers the use of the index only as it is applied in collective bargaining escalator clauses, then the price for preparing the monthly index *including* the revisions amount to around 70–90 cents per $1,000 increase in payments to individuals. When the other applications of the CPI are considered then the costs are even less.

Current revisions. Each revision involves the development and use of a greater amount of relevant data plus a review of the statistical and economic concepts crucial to the construction of the CPI. The latest revisions are no exception with the major elements of the update of the 1970s including:[12]

---

lected to represent price movements for all goods and services purchased by urban wage earners and clerical workers, including families and single persons. Some critics claim that the CPI is too slow in recognizing changes in quality; therefore, these individuals maintain that the CPI overstates price rises or the rate of inflation. Still others complain that the index does not reflect accurately, or fast enough, the changes in consumer buying habits.

[12] See Shiskin, "Updating the Consumer Price Index," pp. 8–12.

1. A new survey of 20,000 families on a quarterly basis plus a panel survey of another 20,000 households (who kept two-week diaries of expenditure patterns) to determine:
   a. The proportion of spending for food, shelter, medical care, and so on, which are then used in assigning index weights; and
   b. The specific goods and services to be included in the market basket that is then priced each month.
2. A new sample of stores will be included to better reflect where people buy.
3. The BLS has modernized the conceptual framework to make the CPI more relevant and current in line with economic conditions and to improve sampling techniques.
4. Along with item 3 above the BLS will publish two separate indexes that will be much more in line with the applications. One to be used in collective bargaining contracts, the Consumer Price Index for Urban Wage Earners and Clerical Workers, and the new Consumer Price Index for All Urban Households as a more general measure of inflation.
5. Present goals are that the newly revised CPI will have sampling errors that are substantially lower than previously.
6. The BLS is attempting to improve the methods for the handling of quality changes. This is a major problem in the construction of a price index because both the quality of products and the consumption patterns are constantly changing. That is, how does one separate out the price change from changes in quality? Automobiles represent the classic example of this problem each new model year.

What does the CPI measure? One big problem in interpreting any index number is in deciding precisely what it does and does not measure. What the CPI does is to compare the cost of the market basket of goods and services in a given month relative to what the same items cost a month ago, a year ago, or many years ago. For example, given the current base year (1967) the market basket could have been purchased for $100. If in March 1976 the CPI was 150.0 it means that the same market basket of goods and services would then cost $150. This represents a price increase of 50 percent during the period 1967 to March 1976. Consumers usually shift their patterns of expenditures to avoid these price increases by substituting, for example, less expensive cuts of meat, and so on. Actually, the index does not reflect these substitutions or postponements of purchases. Rather the CPI assumes that the consumer purchases the same combination of goods and services each month as defined by the market basket. This is the major reason the CPI is called a *price index* and *not* a cost-of-living index. However, the press and the public in general usually refer to the CPI as the latter. A true cost-of-living index would include such items as income and social security taxes whereas a price index excludes such items because such costs are not directly associated with retail prices of specific goods and services. One reason for the constant revisions in the index is because the CPI cannot immediately reflect changes in consumer expenditure patterns. The CPI also cannot adjust quickly to the introduction of new products—they just will not be in the market basket immediately.

The methods used in compiling the CPI are much more detailed than this overall summary reveals. Information concerning the increase in prices of individual items and groups of items from one month to the next is desired, and this leads to many separate computations. For example, pork prices are based on three representative items: pork chops, hams, and bacon. The typical quantity weight—the index weight, q_w—times the price in a given month is called the expenditure weight, $q_w p_1$. This is compared with the expenditure weight of the preceding month, $q_w p_0$, for each item and for the group, as shown in Table 12.8. This indicates an increase of about 2 percent for pork $(.73/33.00 = 2.2 \text{ percent})$ from a given month to the next, or an index of $33.73/33.00 = 102.2$ (percent) if the first of the two months is tentatively considered the base.[13] The latter form may be described as, in effect, a weighted average of the price relatives 103, 101, and 102.

It would be an endless task to describe all the details of computation of the Consumer Price Index and other indexes in common use. Nor would it be profitable to consider such details unless we were employed to work on a given index. From the executive's point of view the important features are the principles on which they are constructed and the purposes they serve. All the major indexes are used to estimate the changing trends in the cyclic movements of markets. And the Consumer Price Index is now widely used in wage contracts of management and unions. In general, those companies which have collectively bargained contracts

TABLE 12.8

	$q_w p_0$	Price Change (percent)	Price Relative (percent)	$q_w p_1$
Pork chops	$15.00	+3	103	$15.45
Hams	8.00	+1	101	8.08
Bacon	10.00	+2	102	10.20
	$33.00			$33.73
Index	100.0			102.2

[13] A series of such month-to-month "relatives" may be "chained" by successive multiplications to obtain a continuous index and reduced to a required base (see page 368).

with unions that include a so-called "escalator" clause adjust the hourly rate of the workers quarterly. Some such formula as a one-cent-an-hour increase in pay for an increase of .5 in the index is common. Frequently, April, July, October, and January are the months in which the adjustment is made.

There are other important indexes which are constantly used by government and industry to assess the economic climate. The U.S. Department of Agriculture gathers data on prices received by farmers and prices paid by farmers, and computes the index for each. Both the Index of Prices Paid by Farmers and the Index of Prices Received by Farmers have the base period 1910–14. For some time an effort has been made completely to revise the indexes; but to date, little progress has been made. When the Index of Prices Received is divided by the Index of Prices Paid, the result is the political football known as the "parity ratio." An average price, 1910–14 = 100, multiplied by the Index of Prices Paid yields a product sometimes called the "parity price." The indexes are used by economists, government agencies, and farm groups in a host of ways, and it is to be hoped that they will be revised soon.

Wholesale Price Index

The Wholesale Price Index (WPI), also prepared by the Bureau of Labor Statistics (BLS), is designed to reflect the general rate and direction of the composite price movements in primary markets and the specific rates and directions of price movements for individual commodities or groups of commodities.[14] It is designed to measure "real" price changes not affected by such things as quality changes, quantities, or terms of sale. The term wholesale refers to sales in large lots, not to prices received by wholesalers, jobbers, or distributors. This is important and sometimes causes misleading interpretations of the WPI. The quoted prices used in constructing the index refer to the first important commercial transaction for each commodity. Later transactions for the same item are not included; but as raw materials are transformed into semifinished and finished goods, they are represented according to their importance in primary markets. Prices of approximately 2,200 items are included in this index. The WPI has been used in many ways that are similar to the applications of the CPI but it does not measure retail prices.

Most of the indexes mentioned in this chapter are published in such documents as the *Monthly Labor Review, Economic Indicators, Federal Reserve Bulletin,* or the *Economic Report of the President.*

---

[14] For details of the weighting and construction of the WPI see U.S. Department of Commerce, Office of Business Economics, *Business Statistics;* and U.S. Department of Labor, *Wholesale Prices and Price Indexes, 1962,* Bulletin No. 1411 (June 1965).

The Dow Jones averages

Perhaps, the best-known financial index is the Dow Jones 30 Industrials. When people ask, "What did the stock market do today?" they usually are referring to the D-J averages as an indicator of the movement of all common stocks. This indicator is discussed here because of its widespread use even though it might be misleading to view a single indicator as a measure of what is happening in the market. There are other stock market barometers but none that receive the publicity of the D-J 30 Industrials. Some other well known indicators include: the New York Times Industrials, Standard and Poor's 500 Industrials, D-J 20 Railroad Stocks, D-J 15 Utility Stocks, D-J 40 Bonds, and the Associated Press Industrials.

The D-J 30 Industrials index covers only a few of the large companies and some analysts criticize it for failing to mirror the movement of the hundreds of other stocks listed on the New York Stock Exchange. The supporters of the index argue that these 30 companies account for about one third of the market value of all stocks listed. They also point out that the stocks of these 30 industrials are so widely held that they do represent a broad spectrum of the investing public. Dow Jones reminds its critics that the major function of the averages is to provide a "general rather than a precise measure of the fluctuations in the securities markets and to reflect the historical continuity of security movements."[15] The initial Dow Jones industrial index was published in 1896 and was based on the movements of 12 common stocks. Later the index was expanded to include 20 issues and in 1928 the list included 30 stocks.[16]

In a manner similar to the methods used by the Bureau of Labor Statistics to modify the market basket of price indexes, the D-J index has had substitutions as some stocks become nonrepresentative of a substantial sector of our industrial society. When this happens a new stock is included in the index in a manner similar to the adjustment made for stock splits. Let us consider the problem of stock splits as they effect the calculations.

Sometimes the fluctuations of the D-J 30 Industrials (currently just over 900) are incorrectly interpreted as dollar-per-share prices. A 9-point rise in the index would represent a 1 percent change at the 900 level. If the average price of all common stocks listed on the New York Stock Exchange is around $50, a comparable percentage change in the latter would be a 50-cent rise. The major reason for this disparity stems from stock

---

[15] See Maurice L. Farrell, *The Dow Jones Investor's Handbook* (New York: Dow Jones & Co., Inc., 1968), p. 12.

[16] For a current list of stocks included in the index see Farrell, p. 13.

splits. If the board of directors believe that the price of the company's stock has risen too high for broad investor appeal, they might elect to split the higher priced shares into more numerous low-priced shares. Assume that a share of common stock is selling for $150 and is split three-for-one. The new price is now $50 but there are three times as many shares outstanding. Therefore, no investor's equity has been changed by this arithmetic, but unless an adjustment is made in an index like the Dow Jones the averages will be distorted.

Assume that we have a very simple stock market index based on the average of four separate issues selling at $100, $50, $30, and $20. Their average price is: ($100 + $50 + $30 + $20)/4 = $50. Now assume that a decision was made to split the $100 stock on a two-for-one basis. If we recomputed the average without any adjustment for the split the new figure would be: ($50 + $50 + $30 + $20)/4 = $37.50, down 25 percent from the average of $50. This is misleading and incorrect because the only thing that changed was the number of shares outstanding. An adjustment is made in order to keep the average at $50 in order that the historical record of the index is preserved. Dow Jones changes the divisor to make this adjustment when stock splits (or substitutions) occur in any of the 30 industrials. In our example, the new divisor would be 3 instead of 4 in order to maintain the average at $50.

The D-J 30 Industrials is not a complicated index of the types discussed previously; however, its *use* is similar. Strictly speaking, the measure is simply an average adjusted for stock splits and substitutions. Technically, the D-J 30 Industrials does not have a base period as ordinary index numbers do. The continuity and comparisons are maintained by changing the divisor. Because 30 common stock prices are involved, one might reasonably assume that the divisor used to determine the average is also 30. That was true initially but because of substitutions and splits the Dow Jones divisor has fallen sharply since the original indicator was published. By 1939 the divisor was 15.1, in 1950 it was 8.92, and since 1971 it has been around 2.

INDEXING

Professor Milton Friedman of the University of Chicago has recently caused some discussion among economists and the BLS suggesting that the *entire* economy of the United States should be subjected to what is called "indexing." Brazil is the most noteworthy example of the application of this procedure. Under indexing, when the CPI rises so do not only wages and salaries but also taxes, rents, interest rates, and most other items. The theory behind indexing and its major objective is to keep all or most of the economy in step with price changes. Presumably, the adjustments would be made in both directions, both up and down. A fundamental question that needs some research is what would be the effects on inflation, unemployment, real economic growth, distribution of income, and a whole host of other economic considerations. It is an interesting proposal, and the debate on the topic really is just getting started. Students interested in pursuing this topic further are urged to consult Yang's article listed in the Related Readings at the end of the chapter.

SUMMARY

As was noted earlier, the best way to become familiar with the growing number of available index numbers is to inspect the pages of publications which publish them. It is not possible to describe them here in detail. They are subject to constant revision—something that is very desirable because often sampling is inadequate and weights and bases are obsolescent. If one understands the simple method of weighted aggregations, the purposes which the index is designed to serve should be sufficiently clear. Geometric means, harmonic means, and a variety of theoretical tests formerly used have now largely passed out of use. In the future when reporting has been greatly expanded, such tests may be revived; but for the present, they may be ignored.

The number of applications and uses of index numbers probably runs into the thousands; however, a few of the most important purposes of these indicators might be listed here. They include:

1. *A measure of economic and social change.* The Index of Industrial Production of the Federal Reserve Board is a widely known index number of this type. An index of Gross National Product (GNP) can be used to show changes in the economic growth of a country, state, or area.
2. *A measure of inflation.* The problem of reporting *real* changes in output, sales, production, and so on, is handled by making an application of such index numbers as the Consumers Price Index (CPI), the Wholesale Price Index (WPI), the GNP deflators, and specific cost or price indexes such as the Construction Cost Index. Under this heading index numbers might be used both to measure the rate of price increases as well as "deflating" another index to remove the influence of price rises.
3. *A measure of seasonal variation.* All governmental and private index numbers that are affected by seasonal patterns usually are "adjusted" for this repetitive factor. By adjusting for seasonal variation these indexes remove the influences caused by changes in custom, habit, or

weather. Chapter 13 makes use of the seasonal index concept in short-term forecasting.

4. *Facilitating comparisons between two series.* Index numbers that have comparable base periods might be compared to examine the trends of each. Basically, the index number is a percentage and one can note the change in one or more series since the base period.
5. *A means of adjusting wages, prices, contracts, alimony payments by escalator clauses.* This is a method of protection against inflation and is an attempt to maintain a rising real income. Chapters 13 and 14 use index numbers as a way of demonstrating the fitting of several trend curves. Chapter 13 also discusses index numbers as a forecasting tool.

SYMBOLS

V, a value index
P, a price index
Q, a quantity index
p_0, price in a base period
p_n, price in a given period
q_0, quantity in a base period
q_n, quantity in a given period

PROBLEMS

12.1. *a.* Using the following simplified data, compute indexes of price, quantity, and value, 1967 = 100, using the weighted aggregative method.

1967		1971		1975	
p	q	p	q	p	q
4	2	4	2	5	3
8	5	9	6	9	6
5	4	6	4	7	4

b. Compute indexes of price, quantity, and value, 1967 = 100, using the data above and the weighted relatives method.

12.2. Suppose that an index of price for March 1976 is 102.5 as compared with the index of price for February 1976. The index of price for February 1976, 1967 = 100, is 125.4. What is the index of price for March 1976, 1967 = 100?

12.3. Suppose B. Jones began work at the Glassblowers shop in city A on October 20, 1965, at a weekly wage of $85. The price index in city A for October 1965 was 114.6 (1957–59 = 100). Jones's weekly pay in November 1975 was $110. The price index for city A, November 1975 was 125.3. What was the change in Jones's real income, 1957–59 = 100? What was the change in Jones's real income, October 1965 = 100?

12.4. Suppose that the average price of four commodities (A, B, C, and D) for the period 1970 was $A = \$0.60$, $B = \$0.20$, $C = \$0.35$, and $D = \$1.10$. Further assume that a recent survey shows that typical monthly consumption of the commodities is $A = 5$, $B = 10$, $C = 12$, and $D = 4$. March 1976 prices of the commodities were $A = \$0.80$, $B = \$0.30$, $C = \$0.20$, and $D = \$1.50$. What is the March 1976 price index, using the weighted aggregative method ($1970 = 100$)?

12.5.

Federal Reserve Bank credit and member bank reserves, 1929–74 (averages of daily figures, millions of dollars)

Year	Total Federal Reserve Bank Credit Outstanding	Total Member Bank Reserves	Year	Total Federal Reserve Bank Credit Outstanding	Total Member Bank Reserves
1929	1,643	2,395	1952	27,299	21,180
1930	1,273	2,415	1953	27,107	19,920
1931	1,950	2,069	1954	26,317	19,279
1932	2,192	2,435	1955	26,853	19,240
1933	2,699	2,588	1956	27,156	19,535
1934	2,472	4,037	1957	26,186	19,420
1935	2,494	5,716	1958	28,412	18,899
1936	2,498	6,665	1959	29,435	18,932
1937	2,628	6,879	1960	29,060	19,283
1938	2,618	8,745	1961	31,217	20,118
1939	2,612	11,473	1962	33,218	20,040
1940	2,305	14,049	1963	36,610	20,699
1941	2,404	12,812	1964	39,873	21,609
1942	6,035	13,152	1965	43,853	22,719
1943	11,914	12,749	1966	46,864	23,830
1944	19,612	14,168	1967	51,268	25,260
1945	24,744	16,027	1968	56,610	27,221
1946	24,746	16,517	1969	64,100	28,031
1947	22,858	17,261	1970	66,708	29,265
1948	23,978	19,990	1971	74,255	31,329
1949	19,012	16,291	1972	76,851	31,353
1950	21,606	17,391	1973	85,642	35,068
1951	25,446	20,310	1974	93,993	36,960

Source: U.S. Department of Labor, Bureau of Labor Statistics, *Economic Report of the President,* February 1975, p. 314.

a. Present these two series as simple index numbers, using that base period which will facilitate the sort of comparison you believe will be interesting and significant.

b. Present the data, or your index numbers, in acceptable graphic form.

c. Write a brief statement interpreting the two series.

d. Justify the base period you choose. Would 1933 be a good base year? Why, or why not? What about 1941–45 as the base period?

12.6.

Seasonally adjusted average daily figures (billions of dollars) of the money supply in the United States, 1959–74 (December)

Year	M_1	M_2	Year	M_1	M_2
1959	143.4	210.9	1967	186.9	349.6
1960	144.2	217.1	1968	201.7	382.3
1961	148.7	228.6	1969	208.7	392.2
1962	150.9	242.8	1970	221.4	425.3
1963	156.5	258.9	1971	235.3	473.1
1964	163.7	277.1	1972	255.8	525.7
1965	171.3	301.3	1973	271.5	572.2
1966	175.4	317.8	1974	283.8	613.9

Note: M_1 = currency plus demand deposits; $M_2 = M_1$ + time deposits at commercial banks other than large CD's.

Source: Board of Governors of the Federal Reserve System, *Economic Report of the President,* February 1975, p. 310.

a. Present these two series as simple index numbers, using that base period which will facilitate the sort of comparison you believe will be interesting or significant.

b. Write a brief paragraph summarizing the economic implications of the supply of money as depicted by your indexes.

c. Justify the selection of your base period.

12.7. Given the following data for May 1975, determine the "value of the dollar" as measured by each series.

Wholesale prices (1967 = 100) 159.3
Consumer prices (1967 = 100) 173.2

12.8. Refer to the following sources, and answer the questions below: *Economic Indicators,* the *Survey of Current Business, Business Week,* the *New York Times,* the *Monthly Labor Review, Barron's, First National City Bank Letter,* the *Morgan Guaranty Survey,* and the *Cleveland Trust Bulletin.*

a. List five index series measuring physical production or consumption.

b. List five index series measuring retail or wholesale prices.

c. List five index series measuring trade volume.

d. List five index series measuring transportation activity.

12.9. Assume that one index of economic activity shows a rise of 2 percent from November to December of a given year. Suppose that another index purporting to measure the same thing records a decline of 1 percent. Without benefit of any additional information, what possible explanations can you suggest for the lack of agreement in the direction and degree of change in business activity as measured by these indexes?

12.10.

Sources of personal income by selected types, 1954–74 (billions of dollars)

Year	Wages and Salaries	Proprietors' and Rental Income*	Dividends
1954	197	54	9
1955	211	56	11
1956	228	57	11
1957	239	59	12
1958	240	62	12
1959	258	62	13
1960	271	62	13
1961	278	64	14
1962	296	67	15
1963	311	68	17
1964	334	70	18
1965	359	76	20
1966	395	79	21
1967	423	83	21
1968	465	86	24
1969	510	91	24
1970	542	91	25
1971	574	95	25
1972	627	103	27
1973	692	124	30
1974	751	123	33

* Includes business and professional income, farm income, and rental income of unincorporated enterprises.

Source: U.S. Department of Labor, Bureau of Labor Statistics, *Economic Report of the President,* February 1975, pp. 266–67.

a. You are asked to "deflate" one of the above national income series, selecting the one to which the Consumer Price Index most appropriately applies. (Use Table 12.1)

b. What determines whether value series should be adjusted, as above, prior to their further analysis or use?

c. Would you have any reservations about using the Consumer Price Index to "deflate" any of the above series? Why, or why not?

12.11. A university spent $2 million for a new classroom building in 1972. A relevant index of construction costs is as follows:

Year	Index of Construction Costs
1972	88
1973	90
1974	95
1975	98
1976	100

On the basis of the above data, estimate the reproduction cost of the building in 1976.

12.12. Average weekly wages of semiskilled workers in the Golf-O-Matic Corporation were $185.25 in 1975; in 1976 they were $192. A cost-of-living index relevant to employees in this city was 103.1 in 1975 and 107.1 in 1976. What was the change in "real" average weekly wages from 1975 to 1976?

12.13. In March 1975 the Index of Prices Received by Farmers was 165 and the Index of Prices Paid by Farmers was 179. What is the parity ratio, calculated from these indexes?

12.14. "If we want to compare two index numbers, we must make certain that they have the same base period." Defend or refute this statement.

12.15. If the Consumer Price Index for two different cities happens to be the same, does this mean that the cities experienced the same dollar increase in the cost of living? Why, or why not? Does it mean that they had the same percentage change in prices? Why, or why not?

12.16. How much will the purchasing power of the dollar fall if prices rise 25 percent? Thirty percent?

12.17. Given the following Index of Industrial Production with a base period 1967 = 100, shift the base to 1974 and calculate the new values.

Year	Index (1967 = 100)
1965	91
1966	99
1967	100
1968	106
1969	111
1970	107
1971	107
1972	115
1973	126
1974	125

12.18. Given the following relevant data, calculate an index of physical volume of sales (1976 = 100).

Year	Sales (millions of dollars)	Price Index
1970	110	94
1971	125	97
1972	139	98
1973	140	100
1974	145	103
1975	148	103
1976	150	104

12.19. The Wholesale Price Index (WPI) for selected years (1967 = 100) is presented below. Calculate an index of the purchasing power of the dollar as measured by this index, and interpret your answer.

Year	WPI (1967 = 100)
1965	96.6
1966	99.8
1967	100.0
1968	102.5
1969	106.5
1970	110.4
1971	113.9
1972	119.1
1973	134.7
1974	160.1

12.20. Assume that an index is at 100 in 1972; it rises 3 percent in 1973, falls 1 percent in 1974, and rises 2 percent in 1975 and 3 percent in 1976. Calculate the index for the five years, using 1972 as the base year.

12.21. The U.S. Department of Labor reported that ". . . the Wholesale Price Index (WPI) rose .6 of 1 percent in January."

a. Would you classify this as an important change?

b. What is the annual rate of wholesale price increase based on the January WPI?

c. Could part of the January rise, as measured by the WPI, be caused by sampling error? Explain.

12.22. Given the following relevant information, calculate an index of physical volume of sales with a base of 1973 = 100.

Year	Sales (millions)	Price Index
1973	$105	100
1974	121	110
1975	150	125
1976	169	130

12.23. Assume that an index of price for 1974 is 160.0 calculated on the basis of weighted aggregative method. The base period for this price index is 1967 = 100. On the basis of the following information, calculate the price index for 1975 and 1976.

Commodity	1974		1975		1976	
	Price	Quantity	Price	Quantity	Price	Quantity
C_1	$3.00	20	$3.50	25	$5.00	25
C_2	4.00	10	5.00	10	7.00	12
C_3	5.00	20	6.50	20	6.50	22

12.24. Refer to Table 12.1 in the text for the annual values of the Consumer Price Index (CPI).

a. State the following salary data in terms of 1967 dollars.

Year	Salary	Year	Salary
1967	$15,500	1972	$19,300
1968	16,700	1973	20,300
1969	17,500	1974	21,300
1970	17,800	1975	22,200
1971	18,500	1976	23,600

b. Compute an index of real salary using 1967 = 100.

c. Change the base of the index of real salary from 1967 = 100 to 1976 = 100.

12.25. Given a Construction Cost Index of 150 in July 1976 with a base of 1967 = 100, what would it cost in July 1976 to replace a home built in 1967 at a cost of $50,000?

12.26. Given the following data respond to the questions below.

Total operations appropriations to higher education in Michigan, selected years (thousands of current dollars)

Academic Year	Appropriations	Academic Year	Appropriations
1951–52	$ 34,539	1968–69	$226,645
1960–61	98,016	1969–70	252,665
1961–62	98,584	1970–71	284,066
1962–63	104,082	1971–72	312,745
1963–64	109,831	1972–73	345,524
1964–65	131,158	1973–74	381,732
1965–66	159,929	1974–75*	423,600
1966–67	196,425	1975–76*	465,500
1967–68	204,587	1976–77*	500,000

* Estimated.

a. Using the Consumer Price Index (CPI) as a measure of inflation, what has happened to the level of real appropriations between 1951–52 and 1974–75? (Use the CPI for 1952 and 1975.)

b. How many dollars must be appropriated in 1976–77 to have the same purchasing power of 1951–52?

12.27. Refer to Table 12.1 and the CPI. Shift the base of this index to 1940 = 100. Write a brief paragraph on the inflation that is measured by this index with the new base.

12.28. Using one of the sources given in Chapter 2 compare the rate of inflation as measured by the GNP deflator and the CPI. Which do you think is the better measure?

12.29. Calculate an index for each of the following series using 1967 = 100.

Year	Corporation Profits Plus Depreciation	Dividend Payments	Undistributed Profits
1967	$ 89.6	$21.4	$25.3
1968	94.6	23.6	24.2
1969	96.8	24.3	20.5
1970	95.2	24.7	14.6
1971	106.5	25.0	21.1
1972	124.0	27.3	30.3
1973	144.1	29.6	43.3
1974	161.7	32.7	52.4
1975p	143.5	33.8	28.5

p = preliminary.
Billions of dollars, quarterly data at seasonally adjusted annual rates.

a. What is the meaning of the phrase "billions of dollars, quarterly data at seasonally adjusted annual rates"?
b. Write a brief paragraph outlining what has happened to each series since 1967.
c. Using the CPI as a deflator state each series in constant dollars.
d. Calculate an index for each of the series in part (*c*). Compare these series with the original one you computed. Write a brief summary paragraph describing the changes that took place since 1967.

RELATED READINGS

BECHTER, DAN M., and PICKETT, MARGARET S. "The Wholesale and Consumer Price Indexes: What's the Connection?" *Monthly Review*, Federal Reserve Bank of Kansas City (June 1973), pp. 3–9.

FISHER, IRVING. *The Making of Index Numbers.* New York: Houghton Mifflin Co., 1923.

———. *The Purchasing Power of Money.* Rev. ed. New York: Macmillan Co., 1920. These two books by Professor Fisher are considered classics.

HICKMAN, B. G. "Diffusion, Acceleration, and Business Cycles." *American Economic Review*, vol. 49, no. 4 (September 1959), pp. 535–65.

KENDRICK, JOHN W. "Productivity, Costs and Prices." *Wages, Prices, Profits and Productivity* (ed. CHARLES A. MYERS), chap. 11. New York: American Assembly, Columbia University, 1959. This chapter defines productivity concepts and indicates sources of measurement of productivity and the need for statistical improvements.

KARNOSKY, DENIS S. "A Primer on the Consumer Price Index." *Review*, Federal Reserve Bank of St. Louis (July 1974), pp. 2–7.

LOWENSTERN, HENRY. "Adjusting Wages to Living Costs: A Historical Note." *Monthly Labor Review* (July 1974), pp. 21–26.

MCGRAW-HILL. *Business Week Index: How It Is Made, What It Tells.* New York: McGraw-Hill Book Co.

MOORE, GEOFFREY H. *Measuring Recessions.* Occasional Paper 61. New York: National Bureau of Economic Research, Inc., 1958.

———. *Statistical Indicators of Cyclical Revivals and Recessions.* Occasional Paper 31. New York: National Bureau of Economic Research, Inc., 1950.

MUNDLAK, Y., and RAZIN, A. "Aggregation, Index Numbers and the Measurement of Technical Change." *Review of Economics and Statistics,* vol. 50 (May 1969), pp. 166–75.

SHISKIN, JULIUS. "Updating the Consumer Price Index." *Monthly Labor Review* (July 1974), pp. 3–20.

THIELGES, BERNARD A. "Computation of a Composition Quantity Index," *Fortran Applications in Business,* vol. II, pp. 439–48. Ann Arbor: Graduate School of Business Administration, University of Michigan, 1971.

U.S. DEPARTMENT OF LABOR, BUREAU OF LABOR STATISTICS. *The Consumer Price Index.* Washington, D.C.: U.S. Government Printing Office (September 1967).

———. *Consumer Prices in the United States, 1959–68.* Bulletin 1647. Washington, D.C.: U.S. Government Printing Office, 1970.

———. *Major BLS Programs: A Summary of Their Characteristics.* Washington, D.C.: U.S. Government Printing Office, 1967.

———. *Wholesale Price Indexes: Specifications for Individual Commodities.* Washington, D.C.: U.S. Government Printing Office (June 1963).

———. "Output per Man-Hour Measures: Industries," chap. 23. *Handbook of Methods for Surveys and Studies,* BLS Bulletin No. 1458. Washington, D.C.: U.S. Government Printing Office, 1967.

YANG, JAI-HOON. "The Case for and against Indexation: An Attempt at Perspective." *Review,* Federal Reserve Bank of St. Louis (October 1974), pp. 2–11.

13

Time series analysis—I: Trend

> The choice before man is not whether to engage in forecasting or to abstain from it, but whether to base expectations on "hunches" or on lessons carefully distilled from experience.
>
> *Arthur F. Burns*

THE STUDY of the movement of a series of data through time is known as time series analysis. Data such as sales are affected at any point in time by the composite force of trend, cycle, seasonal, and irregular factors. That is, at any given time, the sales of an organization are influenced by the long-term growth of the firm, by the business cycle or business conditions, by the seasonal nature of the demand, and possibly by some random shock. In order to appraise their effect on sales, estimates of these forces may be obtained by averaging methods. Of the four forces, trend is the longest in duration, the business cycle generally is of a shorter period; the seasonal pattern tends to repeat every 12 months; and the irregular fluctuations may be shorter than either cycle or seasonal. By averaging the original data over a relatively long period, the effects of the shorter forces of cycle, seasonal, and irregular are smoothed out and an estimate of trend remains.

This chapter concentrates on the trend factor while Chapter 14 discusses seasonal and cyclical variations. Both linear and nonlinear trends computations are illustrated here. Topics to be treated in this chapter include: computation of trend by freehand method, semiaverages method, and the least-squares technique. The latter is the most widely used and can be applied for both the original data and the logarithms. In addition, the problem of converting an annual trend equation into monthly terms for short-term forecasting is demonstrated. Later nonlinear curves such as the parabola, exponential, Pearl-Reed, Gompertz, and Phillips curves are discussed.

ANATOMY OF A TIME SERIES

Quarterly or monthly indicators of business activity, particularly in the field of industrial production and distribution, generally exhibit certain clearly marked variabilities. One of these is a computed secular or long-term trend (Y_c or T). Business as a whole tends to grow at the rate of 2 or 3 percent a year, though with many interruptions. Individual business firms and industries, of course, show many divergencies from this average. Index numbers of wholesale and consumer prices also exhibit secular trends, both upward and downward, though over a century or more they may return to an earlier level. For example, wholesale prices in the United States in the 1930s averaged nearly the same as the 1830s, though varying widely in the intervening years.[1]

At the other extreme, variabilities limited to each separate year may be found. Agricultural activity generally rises in the summer and declines in the winter. Retail trade in some lines shows marked activity just prior to Easter and Christmas. The prices of fruits and vegetables, as a rule, respond inversely to the marketing seasons. In fact, nearly all lines of business activity and most firms have their characteristic responses to the changing seasons (S).

The so-called "business cycle" (C), an irregular, wavelike variability affecting most lines of business, has been the subject of intensive statistical analysis and theoretical discussion. In recent years, it seems to be losing the slight degree of regularity that once characterized it. Nevertheless the swings of excessive activity alternating with slack times continue to be a feature of the business picture and must be recognized.

The course of business is subject to many seemingly chance factors varying from minor strikes to major wars, and including the mutability of nature and the vagaries of demand. Such factors cannot be reduced to definite rules, and they are generally lumped together as irregular (I) or accidental. (See Figure 13.1 for a typical representation of each of the components.)

Method of combining factors

By way of generalization, and introducing the method used here, data (Y) representing an economic time series may be described as the resultant force of four factors, symbolized thus:

$$Y = T \cdot S \cdot C \cdot I \qquad (13.1)$$

and the task of the statistician dealing with such a series is to segregate each insofar as this is possible. When annual data are analyzed, as in the

[1] Cf. U.S. Department of Commerce, *Historical Statistics of the United States, 1780–1945* (Washington, D.C.: U.S. Government Printing Office), pp. 223–34.

FIGURE 13.1
The anatomy of a time series

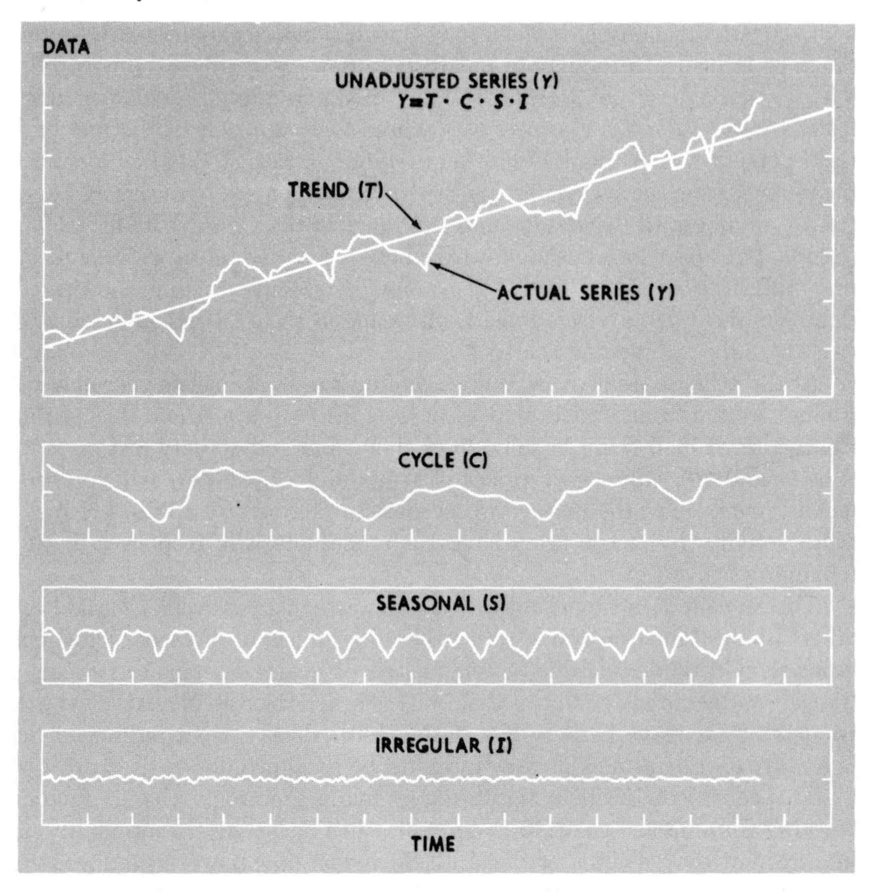

section that follows, the seasonal force is not reflected in the series, and the factors are

$$Y = T \cdot C \cdot I \tag{13.2}$$

In the following sections the four factors will be considered. The discussion of the long-term growth (or decline) will be restricted to the most elementary and probably the most useful type, the linear or straight-line trend. In a later section the fitting of curvilinear trends will be discussed.

From a mathematical point of view, there exist many ways by which the four major components may be combined to form the resultant force. For example, one can use an additive assumption to represent a multiplication assumption by the use of logarithms of the data as the input to the model. That is,

$$\log Y = \log T + \log S + \log I + \log C \qquad (13.3)$$

Each method, of course, produces a different series; therefore, it remains to justify our selection, that is, that the factors are multiplicative.

First, however, let us consider some alternative methods of combining the data. For example, an economic time series may be the resultant, among others, of one of the following:

$$Y = T + S + C + I$$
$$Y = T + S \cdot C \cdot I$$
$$Y = T + S + C \cdot I$$

In a mathematical sense, each of the above combinations is possible. Why, then, do we choose the multiplication method? It appears that in order to answer this question, we might ask another question: Does the method (any one of the four mentioned above) *make sense* in terms of the nature of the factors?

First of all, let us remember that the primary purpose for assuming *any* combination is to provide a *realistic and reasonable* basis for analyzing the actual data. The differences in the three methods listed above and the multiplication method can be illustrated best by considering the seasonal element. Assume that the sales of XYZ Company, Inc. are usually higher in June than they are in May of any given year. In 1976, June sales were $100,000, and May sales were $80,000. The difference between these two months was considered to be a result of the seasonal nature of the sales of XYZ Company, Inc. As statisticians of this firm, do we inform management that sales in June were 25 percent above May or that they were $20,000 higher? For 1976, of course, it makes no difference which way we report the change. However, if in 1977, May sales amount to $110,000, what is our forecast for June 1977, which will reflect the seasonal component? If we regarded June sales as being $20,000 higher than May sales, our estimate for June 1977, would be $130,000. If on the basis of 1976 we expect a 25 percent seasonal rise from May to June, our forecast for June 1977 would be $137,500.

In the first estimate ($130,000) the statisticians considered the seasonal difference as being a *plus* $20,000 and *added* it to the May figure. For the second estimate ($137,500), the statisticians considered the seasonal differential to be 25 percent and *multiplied* the May data by the seasonal index, 125. Which method is more appropriate? Although the answer depends upon the experience of the individual firm, in general the multiplicative assumption is the more reasonable one. That is, it seems more realistic to assume that equal percentage changes provide a better approximation than do equal changes in dollar amounts. Finally, although it is not a justification for our selection, the fact remains that most business analysts use the combination $Y = T \cdot S \cdot C \cdot I$ and find it appropriate (Formula [13.1]).

General comments

Perhaps the nature of these factors may be better appreciated if we symbolize their relationships. This understanding of the nature of the composite force is necessary for executives who are called upon to make decisions which reflect some sort of forecast. It seems reasonable, then, that these forecasts should be based on an understanding of the economic effects of such factors.

If we symbolize the actual monthly sales data of a firm by $Y = TSCI$, the annual figure may be stated as $Y = TCI$ because seasonal variations are not reflected in yearly totals. Further, if we divide annual data, TCI, by an estimate of trend (T and later symbolized by Y_c), a measure of the effect of CI is the quotient, later to be identified as the residual. An estimate of C is obtained by taking a moving average of the CIs, which will smooth out the irregular element because of the latter's shorter duration. Estimates of the seasonal force may be similarly computed, using monthly data or annual monthly averages. For example, if we divide $TCSI$ by a measure of TC, we eliminate trend and cycle, and the seasonal and irregular remain. Again, by using a moving average of several SIs, the irregular will be canceled, leaving an estimate of S.

Actual computation of these measures involves many more steps; however, this section is inserted to assist the student to follow the logic of the detail of the computations.

Time series analysis is important to anyone interested in studying the movements of economic data. By isolating or removing individual forces, the impact of each may be assessed. *In our discussion of the subject, we have restricted the application to forecasting. This is done in the interest of conciseness and clarity, and is not intended to minimize the significance of other uses of time series analysis.* It should be pointed out with respect to forecasting that trend, cyclical, or seasonal estimates utilized should be looked upon as a useful *basis for a first approximation* to a projection into the future, and not as a precise value.

METHODS OF ESTIMATING SECULAR TREND

Determining trend by inspection

Freehand method. Once the data have been charted, we can fit a trend by inspection. This is done by drawing a line through the data which *in* *the statistician's opinion* best describes the average long-term growth. Such a technique certainly has simplicity in its favor; however, it is subjective. A trend line is intended to depict the growth factor involved, and a freehand line is *what the statistician sees* as the secular pattern. This method should be employed only by those individuals with adequate economic background concerning the particular time series. Even then, the freehand method is limited in its application.

Selected points method. A similar technique of determining trend is the selected points method. This process, too, suffers by its subjective nature. If we consider the trend involved in the series to be that of a straight line, we need select only two points. Although this results in a slight refinement of the freehand method, the selected points method is still limited in its application.

The values of the two points selected, p_1 and p_2, are generally taken at the beginning and end of the time series. These values are two points on the appropriate trend line. To secure a measure of the slope (b) of the straight line, the difference between the Y values of these two points is divided by t, the number of years or other time interval separating them. The formula is

$$b = \frac{p_2 - p_1}{t} \tag{13.4}$$

The trend values are obtained by adding b for each successive time period to p_1 and subtracting b for each preceding time interval. The equation for a selected points straight-line trend is

$$\hat{Y}_c = p_1 + b(X_n - X_1) \tag{13.5}$$

where X_n indicates any given year or time period and X_1 represents the year associated with p_1.

To illustrate, a trend line using this method is fitted to the sales data given in Table 13.1. The trend values in the last column of Table 13.1 were estimated by selecting point p_1 as 1971, and the trend value taken as the actual sales. Point p_2 might be taken as 1975 with a trend value of 516; t then is $1975 - 1971 = 4$ and b, the slope of the trend line using Formula (13.4) is

$$b = \frac{p_2 - p_1}{t} = \frac{516 - 469}{4} = \frac{47}{4} = 11.75$$

Substituting in the trend equation, Formula (13.5):

$$\hat{Y}_c = p_1 + b(X_n - X_1)$$

we may calculate trend values for each year. The trend value for 1975 is

$$\hat{Y}_{75} = 469 + 11.75(1975 - 1971) = 516$$

TABLE 13.1
Computation of trend by selected points method sales of General Brands, Inc., 1970–76 (millions of dollars)

Year	Actual Sales Y	Trend Values Y_c
1970	436	457
1971	469	469
1972	483	481
1973	488	493
1974	514	504
1975	516	516
1976	528	528

Source: Hypothetical data.

For 1976 the trend value is

$$\hat{Y}_{76} = 469 + 11.75(1976 - 1971) = 528$$

Values for other years might be determined in a similar manner and are shown in Table 13.1. It must be kept in mind that this subjective method provides no more than a first approximation to trend. Somewhat more precise, at least less subjective, techniques of fitting trend are described below.

Determining trend by computation

Semiaverages method. This method is a more objective approach than the selected points technique. If we divide the series to be analyzed into two parts and take the arithmetic mean of each half, then a straight line passing through these points provides a trend line. Such a procedure is called the method of semiaverages.

Table 13.2 illustrates the procedure. Here, we have an odd number of years, namely, seven. We have three choices to divide such a series: (1) we can add one half of the middle year (1973) to each half, (2) we can add this value to each half, or (3) we can omit it. We have chosen the second method. Once we can locate the values of the semiaverages, we plot them in the middle of their respective time periods—in our illustration, 1971, and 1972, and 1974, 1975.

This method of semiaverages, although not as subjective as the inspection technique, is not completely free from criticism. Because we used the mean of each half of the series, any extreme will greatly affect the points. Such a trend will not be an accurate picture of the growth element in the series. In addition, values so obtained are not precise enough for predicative purposes or for trend elimination. Consequently,

TABLE 13.2
Computation of trend by semi-averages method, sales of General Brands, Inc., 1970–76 (millions of dollars)

Year	Sales	Semiaverage
1970	436	
1971	469	469*
1972	483	
1973	488	
1974	514	
1975	516	512†
1976	528	

* 469 = (436 + 469 + 483 + 488)/4 (plotted between 1971 and 1972).
† 512 = (488 + 514 + 516 + 528)/4 (plotted between 1974 and 1975).
Source: Hypothetical data.

we look to a mathematical method to furnish an equation and a general expression of the long-term trend.

***Least-squares method* (*short method*).** The formula for a straight line can be written $\hat{Y} = a + bx$, with a and b called the constants. The value of a is merely the height of the trend line at the Y intercept or the height of the line at the origin. That is, $\hat{Y}_c = a$ when $x = 0$. The other constant, b, represents the slope of the trend line; and when b is positive, the slope is upward; and when b is negative, the trend is downward, indicating a decline. The constant b is the increase in Y_c which is associated with each one-unit change in x. In Table 13.3, this one-unit change is each passing year. (See also Figure 13.2.)

The least-squares method of estimating trend will provide the same result even if different analysts compute the line independently. The estimates of trend computed in this manner are such that the sum of the *squared* deviations of the actual data from the estimated values is a *minimum*. Consequently, the term "least squares" is attached to this method. Each statistician may draw a slightly different trend line using the freehand method, but the slope and location of the trend line will be the same in each case when computed by the least-squares technique.

Table 13.3 illustrates the short least-squares method. Here, we have an odd number of years; so, to minimize calculations, 1973 is arbitrarily chosen as the origin. The basic data refer to annual sales of General Brands, Inc. from 1970 through 1976. Column 2 is simply Column 1 transferred into deviations from the origin in one-year units. That is, 1972 is −1 unit from 1973 (the origin), 1971 is −2, 1974 is +1, and so on. This is done only to reduce computations. The four-digit figures in Column 1 could be used directly, but that is unnecessary and cumbersome.

TABLE 13.3
Computation of a straight-line trend by method of least squares, odd number of years; sales of General Brands, Inc., 1970–76 (millions of dollars)

Year (1)	x Deviations in One-Year Units (2)	Y Sales (3)	xY (4)	x^2 (5)	$\hat{Y}_c$ Long-Term Trend (6)
1970	−3	436	−1,308	9	447.7
1971	−2	469	−938	4	462.0
1972	−1	483	−483	1	476.3
1973	0	488	0	0	490.6
1974	1	514	514	1	504.9
1975	2	516	1,032	4	519.2
1976	3	528	1,584	9	533.5
	$\Sigma x = 0$	$\Sigma Y = 3{,}434$	$\Sigma xY = 401$	$\Sigma x^2 = 28$	$\Sigma \hat{Y}_c = 3{,}434.2$

$a = \Sigma Y/n$
$= 3{,}434/7$
$= 490.6$

$b = \Sigma xY/\Sigma x^2$
$= 401/28$
$= 14.32142$

$\hat{Y}_c = a + bx$
$\hat{Y}_c = 490.6 + 14.3x$
Origin July 1, 1973
X in terms of years
Y in millions of dollars

Note: When the data are available, it is preferred to have more than seven years of information entering into the calculations in order to eliminate the cyclical effects.

In order to determine the values of a and b, we refer to the so-called "normal" equations. There are two normal equations for a first-degree curve, or straight line; three normal equations for a parabola, or a second-degree curve; four normal equations for a cubic. In all cases the number of normal equations is dependent upon the number of constants to be determined. We are concerned only with the first-degree curve.

The two normal equations are

$$\Sigma Y = na + b\Sigma x \tag{13.6}$$

$$\Sigma xY = a\Sigma x + b\Sigma x^2 \tag{13.7}$$

and when the origin is selected so that $\Sigma x = 0$, the equations are simplified because the terms containing Σx drop out. The normal equations then become

$$\Sigma Y = na \tag{13.8}$$

$$\Sigma xY = b\Sigma x^2 \tag{13.9}$$

In order to determine the constants, these may be arranged into

$$a = \Sigma Y/n \tag{13.10}$$

$$b = \Sigma xY/\Sigma x^2 \tag{13.11}$$

FIGURE 13.2
Schematic diagram of the straight-line trend formula

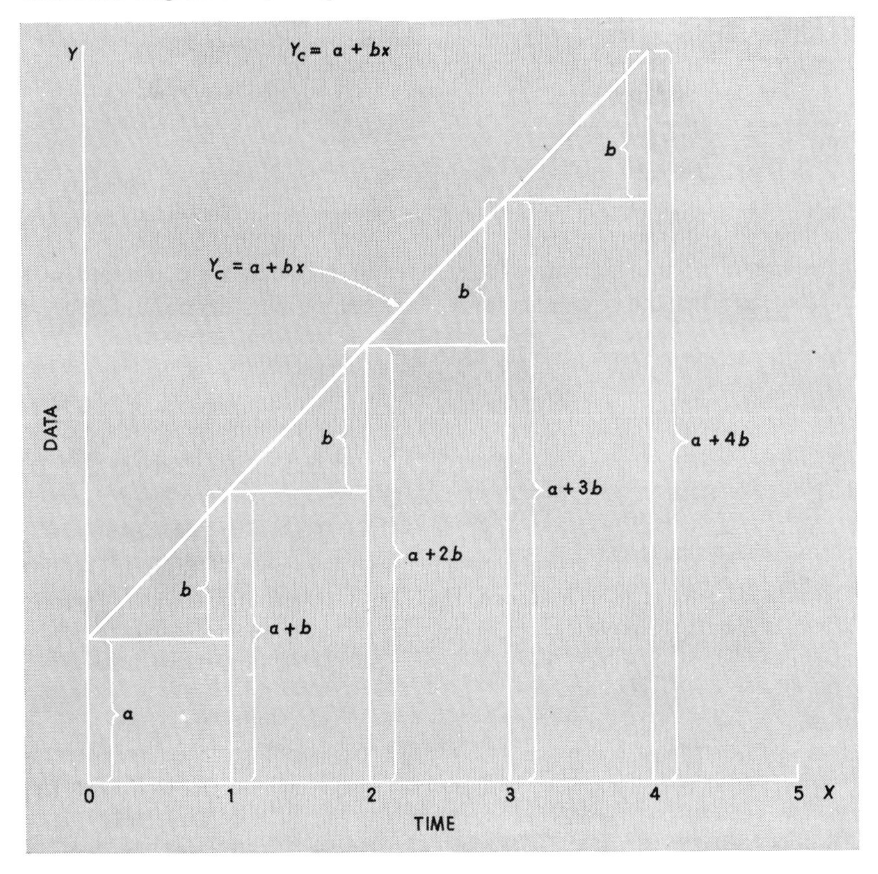

The equation for T or $\hat{Y}_c$ for each of the years in Table 13.3 may be written thus:

$$
\begin{aligned}
1970&: a + b(-3) = \hat{Y}_c \\
1971&: a + b(-2) = \hat{Y}_c \\
&\quad \vdots \qquad \vdots \qquad \vdots \\
1976&: a + b(3) \quad = \hat{Y}_c
\end{aligned}
$$

in which a and b are constants, the x's are $-3, -2, -1, 0, +1, +2, +3$; $\Sigma x = 0$; and the Y_c's stand for successive trend items yet to be determined. If the seven equations are added, we have (note that Σbx is written $b\Sigma x$, since b is a constant whereas x is a variable):

$$na + b\Sigma x = \Sigma \hat{Y}_c$$

or in general, since Σx is here necessarily zero, and because $\Sigma \hat{Y}_c$ must equal ΣY to give the trend the same average graphic height as the data,

$$na = \Sigma Y$$

and

$$a = \frac{\Sigma Y}{n} = \bar{Y} \tag{13.12}$$

The second normal equation for determining b may be computed in a similar way, but the equations for T or $\hat{Y}_c$ are multiplied by x and become

$$\begin{aligned} 1970&: ax + bx^2 = x\hat{Y}_c \\ 1971&: ax + bx^2 = x\hat{Y}_c \\ &\quad \cdot \qquad \cdot \qquad \cdot \\ &\quad \cdot \qquad \cdot \qquad \cdot \\ &\quad \cdot \qquad \cdot \qquad \cdot \\ 1976&: ax + bx^2 = x\hat{Y}_c \end{aligned}$$

in which the terms have the same meaning as before. If these equations are added, and if it is assumed that $\Sigma x\hat{Y}_c = \Sigma xY$ and that, as before, $\Sigma ax = a\Sigma x = 0$, we have

$$b\Sigma x^2 = \Sigma xY$$

and

$$b = \frac{\Sigma xY}{\Sigma x^2} \tag{13.13}$$

Also see Chapter 15.

Returning to Table 13.3, we can now follow the calculations for a and b. The constant a is equal to $\bar{Y}$, and b is equal to the sum of the cross products (Column 4) divided by Σx^2 (Column 5). Column 6 represents the growth factor in General Brands sales data from 1970 through 1976. The trend equation is

$\hat{Y}_c = 490.6 + 14.3x$
Origin: July 1, 1973
X in terms of years
Y in millions of dollars

The interpretation given to a is that it is the value of $\hat{Y}_c$ when $x = 0$; b is the slope of the trend line, that is, the amount of change in Y with each passing year; in this example, b indicates that on the average, sales of this firm increase by $14.3 million with each passing year. Utilizing this information, we may forecast sales for 1977. To do this, we determine how many deviating units 1977 is from the origin. It is +4 years which is the value of x for 1977. Substituting in the trend equation, we obtain

$$\hat{Y}_c = 490.6 + 14.3(4) = \$547.8 \text{ million}$$

That is, *on the basis of trend alone,* we would forecast sales of General Brands, Inc., in 1977 to be almost $548 million. However, as we shall discover later, other forces may operate to pull the actual figure above or below this trend value. (See Figure 13.3 also.)

Table 13.4 indicates a procedure for determining trend by the least-squares method when n is even. In this situation the origin may be chosen as January 1, 1974. Therefore, in transposing Column 1 into more workable form (Column 2), we can see that July 1, 1973, is −1 from the origin; July 1, 1972, is −3, and so on. The procedure for determining the trend values and the constants is exactly the same as in Table 13.3. If fiscal-year data were employed, the origin would have been July 1, 1974. In both examples, however, in order to use the short form of the normal equations, Σx must equal zero. It must be noted that (b) is now in terms of half-year units. In order to estimate the average annual growth *due to trend* from 1971–76, one must multiply $6.0 million by 2.

CONVERTING AN ANNUAL TREND EQUATION INTO MONTHLY TERMS

In making long-range forecasts of trend, annual equations such as discussed above are useful. However, it is frequently necessary for firms to

FIGURE 13.3
Sales of General Brands, Inc., 1970–76, trend line computed by least-squares method

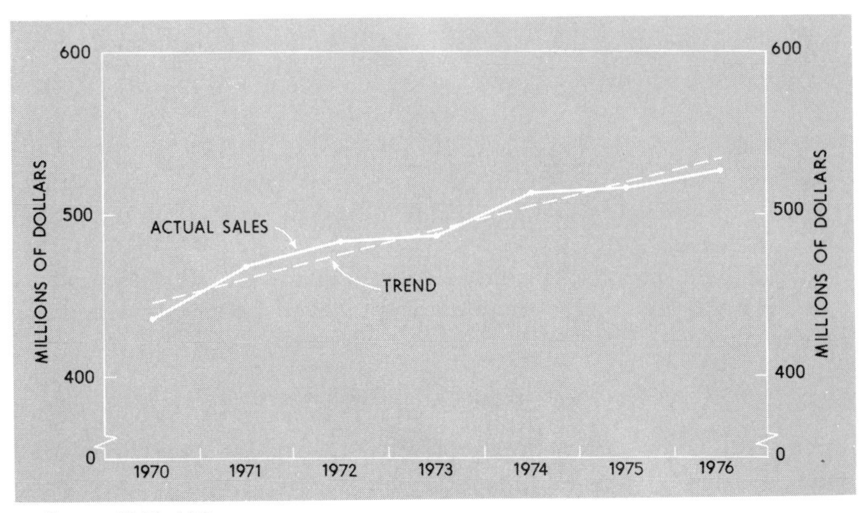

Source: Table 13.3.

TABLE 13.4
Computation of a straight-line trend by method of least squares, even number of years; sales of General Brands, Inc., 1971–76

Year (1)	x Deviations in Half-Year Units (2)	Y Sales in Millions of Dollars (3)	xY (4)	x^2 (5)	$\hat{Y}_c$ Long-Term Trend (6)
1971	−5	469	−2,345	25	469.7
1972	−3	483	−1,449	9	481.7
1973	−1	488	− 488	1	493.7
1974	1	514	514	1	505.7
1975	3	516	1,548	9	517.7
1976	5	528	2,640	25	529.7
	$\Sigma x = 0$	$\Sigma Y = 2{,}998$	$\Sigma xY = 420$	$\Sigma x^2 = 70$	$\Sigma \hat{Y}_c = 2{,}998.2$

$$a = \frac{\Sigma Y}{n} \qquad b = \frac{\Sigma xY}{\Sigma x^2} \qquad \hat{Y}_c = a + bx$$

$$a = \frac{2{,}998}{6} = 499.7 \qquad b = \frac{420}{70} = 6.0$$

$\hat{Y}_c = 499.7 + 6.0x$
Origin: January 1, 1974
X in terms of half-year units
Y in millions of dollars

Source: Hypothetical.

make short-term forecasts for a specific month or season. Short-term forecasts may be useful in scheduling purchasing for inventories, production, or budgeting control during these months. In order to utilize our trend equation for a short-term forecast, we must reduce it to monthly terms. The analysts could compute trend using monthly data; however, this is time consuming and sometimes confusing.

We shall illustrate the procedure by recalling the information from Table 13.3:

$\hat{Y}_c = 490.6 + 14.3x$
Origin: July 1, 1973
X in terms of years
Y in terms of millions of dollars

Table 13.5 indicates that our first step is to divide both of the constants a and b by 12 because they now represent *annual* increments. The equation then becomes

$$\hat{Y}_c = 40.9 + 1.2x$$

but b is still unsatisfactory because it refers to *annual* changes in *monthly* sales.

TABLE 13.5
Converting an annual trend equation into monthly terms (data based on Table 13.3)

$\hat{Y}_c = 490.6 + 14.32142x$
Origin: July 1, 1973
X in terms of years
Y in millions of dollars
} Annual equation

$b = 14.32142/12 = 1.19345$
$a = 490.6/12 = 40.9$
$b = 1.19345/12 = .09945$

Shifting origin to July 15, 1973:
$\hat{Y}_c = 40.9 + (.09945)(.5) = 40.95$

$\hat{Y}_c = 40.95 + .09945x$
Origin: July 15, 1973
X in terms of months
Y in millions of dollars
} Monthly equation

That is, b refers to the amount of change which occurs between, for example, January 1973 and January 1974. In order to utilize the formula in a short-term forecast, we need a b value which reflects the change in sales from January to February *of the same year,* or from one month to the next *of the same year.* This simple adjustment may be completed by again dividing b by 12.[2] The equation now is

$$\hat{Y}_c = 40.9 + .1x$$

which indicates that sales *due to trend* would normally be $100,000 higher from one month to the next within the same year. That is, b now is in terms of *monthly* increments in *monthly* sales.

Just as annual data are centered at the middle of the year (calendar year at July 1 and fiscal-year data at December 1), monthly data should be centered at the middle of their time period. This requires one more adjustment in the trend equation. As it now stands, the origin is July 1, 1973. The origin must be moved forward to July 15, 1973, the middle of the month. This step is achieved by substituting .5 for x in the equation; that is, the new origin is one half of a month from the old origin. This results in

$$\hat{Y}_c = 40.9 + .1(.5) = \$40.95 \text{ million}$$

Because $Y_c = a$ at the origin, the new value of a is 40.95, and the equation becomes

[2] The same result is obtained if we divide the original b value by 144 instead of by 12 twice.

$$\hat{Y}_c = 40.95 + .09945x$$

Origin: July 15, 1973
X in terms of months
Y in millions of dollars

Warnings

Students should not be misled by the mathematical preciseness of forecasts made from trend projections. First of all, conditions which prevailed during the period to which the calculations refer may change. Technological advances, changing consumer preferences, business conditions, population shifts, political developments, and other factors may alter the long-term growth of a firm's sales. These methods are useful as a *guide* for management *when* the individual making the forecast has adequate knowledge of the data and the effects of shifting economic conditions upon sales. Used as a *tool* in assisting management to reach intelligent decisions concerning plant expansion, budgetary control, purchasing of raw materials, inventories, and so forth, trend analysis can be an invaluable aid. If used by the economic neophyte, trend estimates can be dangerously misleading. In 1874, Mark Twain, in his classic *Life on the Mississippi,* summed up the nonsense side of *blindly* extrapolating trend:

> In the space of one hundred and seventy-six years the lower Mississippi has shortened itself two hundred and forty-two miles. That is an average of a trifle over one mile and a third per year. Therefore, any calm person, who is not blind or idiotic, can see that in the Old Oölitic Silurian Period, just a million years ago next November, the Lower Mississippi River was upward of one million three hundred thousand miles long, and stuck out over the Gulf of Mexico like a fishing-rod. And by the same token any person can see that seven hundred and forty-two years from now the Lower Mississippi will be only a mile and three quarters long, and Cairo and New Orleans will have joined their streets together, and be plodding comfortably along under a single mayor and a mutual board of aldermen. There is something fascinating about science. One gets such wholesale returns of conjecture out of such a trifling investment of fact.

NONLINEAR GROWTH CURVES

Until this point, it has been assumed that the trend pattern depicting the data was best described by the formula for a straight line. We also noted that trends are utilized to project what appears to be the composite direction of change into the future. In addition, secular trend lines are used as a base which averages out minor irregularities to make the general direction of change become obvious. The analyst also may compare given items as percentages of trend, that which is assumed to be *normal,* in order to evaluate seasonal and cyclical fluctuations.

Selecting the appropriate formula and curve that accurately and effectively represents the data is a statistical problem. Once we are able to decide what *shape* of a curve best describes the pattern of the data, our next task is to *fit* the selected curve to the data. A trend always depicts a relationship between two variables; in many instances, one of these is time. The basic data are shown on the Y axis of the chart while time is always measured on the X scale. Because there are many possible relationships among economic time series, the alternative shapes of trend lines are almost infinite. We can develop a mathematical formula to describe almost any relationship; fortunately, only a relatively few types of trends are needed to portray most economic data. Generally, the formulas are not too complex which also simplifies the task somewhat.

In some situations, the essential functional relationship between the variables clearly indicates the shape of the trend. Normally, this is not the case and the analyst must select the appropriate trend largely upon the basis of the analyst's knowledge of the data and inspection of the graphic pattern. Some of the types of curves usually appropriate for business and economic data are shown in Figures 13.4 and 13.5. The problem of deciding which curve to use generally dictates that our first step is the preparation of a chart which shows the relationship. In the choice of curves, the analyst must keep in mind the primary objective of trend

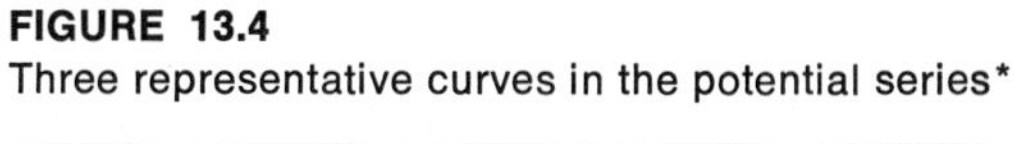

FIGURE 13.4
Three representative curves in the potential series*

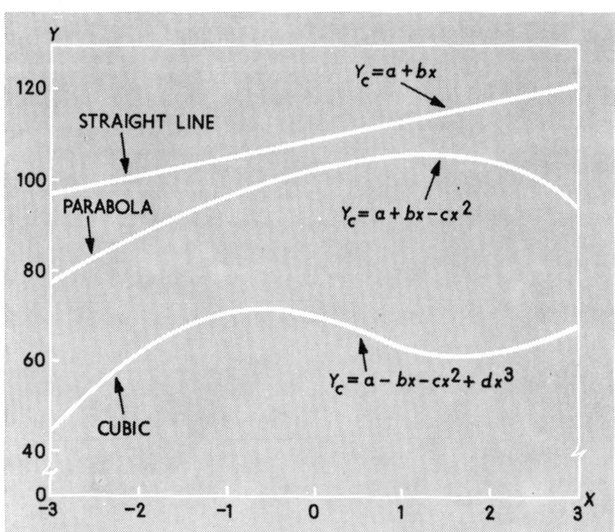

* The straight line and parabola belong to a group of trends called the *potential series*. These curves become increasingly complex by the addition of terms in higher powers of x, as dx^3. If the maximum power of x is 3, then the curve is called a cubic.

FIGURE 13.5
Three representative curves in the exponential series

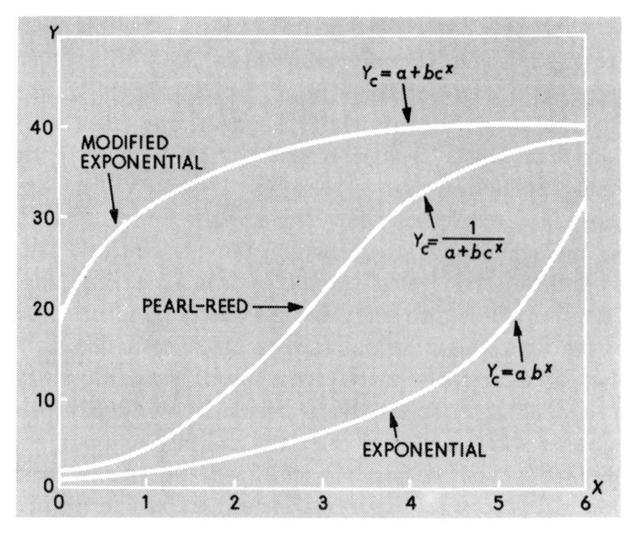

fitting, namely, the accurate description of the covariation. Let us consider fitting a few of the nonlinear trends useful in business and economics. While the straight-line trend is the most widely used, there are situations where its use is clearly wrong. A chart of the economic data might indicate a gradual change in the direction of the trend, or perhaps, the pattern tends to flatten out as it approaches a higher or lower level. Some form of a parabola might be appropriate in the first instance, where an exponential or modified exponential might be suggested for the latter case. Let us now consider a few of the nonlinear trends that are applicable to business and economic data.

The straight-line trends discussed above belong to a large family of simple polynomials. The general expression for these curves is

$$\hat{Y}_c = a + bx + cx^2 + dx^3 + ex^4 \cdots \text{and so on} \tag{13.14}$$

As stated previously, a is equal to the Y-intercept at the origin; b is the general slope of the line that indicates the rate of change in the data; c, d, and e represent the degree to which the curve is bent or changes directions. The straight-line trend moves in only one direction, either upward or downward, but it never alters its course. The distinguishing feature of the curves beyond the first degree is that the trend changes direction.

Within this family of polynomial trends, the straight line is the *special case* where all the constants beyond b are zero. It is called a curve of the first degree. The degree of power of the trend relates to the exponent of x in the final term. The formula for the straight-line trend has only the first two terms on the right side of the equality sign; the second-degree parabolic trend is

$$\hat{Y}_c = a + bx + cx^2 \tag{13.15}$$

and the cubic takes the following form:

$$\hat{Y}_c = a + bx + cx^2 + dx^3 \tag{13.16}$$

It is clear that the straight line obviously is not the appropriate representation of the growth pattern of every time series. Nonetheless, it actually is the most useful and undoubtedly the most widely employed of all trend lines. Whenever a curvilinear trend is used to depict the general drift of the data, the analyst must exercise care not to fit too high a degree parabola. The higher exponents, particularly at the extremes, produce a marked departure from linearity. This characteristic limits the usefulness of such curves in forecasting because one is almost certain to confuse trend and cycle.

As we will discover below, polynomial trends are easy to fit, and in theory they can be extended to any degree. As we reach a higher order the variation around the trend line gets smaller. That is, the $\Sigma(Y - \hat{Y}_c)^2$ becomes smaller. Theoretically, one might keep using a higher degree curve until the number of constants in the equation equals the number of observations. At this point the trend will coincide with the data, but because so many degrees of freedom are lost the result is statistically meaningless.

Within the range of the data, polynomial trends frequently provide an excellent fit. From a logical point of view though it is difficult at times to justify these curves for use in economic forecasting. For example, even the simplest polynomial, the straight-line trend, will if extended far enough become negative in one direction and technically increase without limit in the other. This is true of the third-degree curve, too. The second-degree parabola will increase or decrease without limit at both ends. Naturally, the fourth-degree polynomial will do likewise. The facts are that most economic time series do not (indeed cannot) have negative values. They also tend to reach an upper or lower bound in one or both directions.

Parabola

The parabola belongs to a group of trends (see Figure 13.4) which become increasingly complex by the addition of terms in higher powers of X. If the maximum power of X is 2, the curve is a second-degree parabola; if 3, the curve is a cubic; if 4, it is a quartic, and so on. Mathematically, this group of curves is called the potential series; some curves in the exponential series are shown in Figure 13.5.

The second-degree parabola then may be described by the equation,

$$\hat{Y}_c = a + bX + cX^2 \tag{13.17}$$

but if time centering and a regular sequence of X intervals[3] are assumed, the formula becomes as in Formula (13.15):

$$\hat{Y}_c = a + bx + cx^2$$

Again, x is the symbol for the centered deviations, that is, taken from the mean of the X series. Similar to the equation for a straight line, this formula includes an element and a constant for each term. The elements are the successive powers of x, which are combined with the constants, a, b, and c to describe the curve to represent the data. As described under the least-squares method (page 395), a represents the height of the line at the origin where $x = 0$; b is a measure of the slope of the curve at the origin; and c is an additional constant that determines the degree of curvilinearity. Fitting a second-degree parabola involves the computation of the values of the three constants.

In order to fit the parabola to our data we refer again to the normal equations.[4] Because we have three constants to solve for, we need three normal equations:

$$\begin{aligned} na + b\Sigma X + c\Sigma X^2 &= \Sigma Y \\ a\Sigma X + b\Sigma X^2 + c\Sigma X^3 &= \Sigma XY \\ a\Sigma X^2 + b\Sigma X^3 + c\Sigma X^4 &= \Sigma X^2 Y \end{aligned}$$

However, if we look upon the X observations as deviations from their mean—time centering process—the variable may be written as x where,

$$x = X - \bar{X} \tag{13.18}$$

and Σx, Σx^3 will each equal zero. The centered normal equations then become:

$$\begin{aligned} \Sigma Y &= na + b\Sigma x + c\Sigma x^2 \\ \Sigma xY &= a\Sigma x + b\Sigma x^2 + c\Sigma x^3 \\ \Sigma x^2 Y &= a\Sigma x^2 + b\Sigma x^3 + c\Sigma x^4 \end{aligned}$$

By algebraic manipulation the second equation provides a formula for b ($a\Sigma x$ and $c\Sigma x^3$ drop out because Σx and $\Sigma x^3 = 0$):

$$\Sigma xY = b\Sigma x^2$$

$$b = \frac{\Sigma xY}{\Sigma x^2} \tag{13.19}$$

[3] That is, we have a Y value for each successive time interval, for example, month or year.

[4] See Chapter 15 also.

Also, the first time centered normal equation becomes:

$$na = \Sigma Y - b\Sigma x - c\Sigma x^2$$

$$a = \frac{\Sigma Y - c\Sigma x^2}{n} \qquad (13.20)$$

and the third normal equation might be used to solve for *c:*

$$c\Sigma x^4 = \Sigma x^2 Y - a\Sigma x^2 - b\Sigma x^3$$

$$c\Sigma x^4 = \Sigma x^2 Y - \frac{(\Sigma Y - c\Sigma x^2)}{n}\Sigma x^2$$

$$nc\Sigma x^4 = n\Sigma x^2 Y - \Sigma x^2 \Sigma Y + c\Sigma x^2 \Sigma x^2$$

$$nc\Sigma x^4 - c\Sigma x^2 \Sigma x^2 = n\Sigma x^2 Y - \Sigma x^2 \Sigma Y$$

$$c = \frac{n\Sigma x^2 Y - \Sigma x^2 \Sigma Y}{n\Sigma x^4 - \Sigma x^2 \Sigma x^2} \qquad (13.21)$$

Of course, by substituting the required summations, which may be easily obtained from the data (see Table 13.6), in the appropriate formulas, the values of the constants can be determined.

The procedure for fitting a second-degree parabola is illustrated in Table 13.6; the data represent the U.S. Department of Labor Index of Wholesale Prices, on an annual basis, 1940 through 1974. When the data are plotted (see Figure 13.6), a degree of curvilinearity is obvious. Therefore, the parabola seems appropriate to indicate the general movement of these prices during this time. Although recent years (1960–74) suggest a cubic might be appropriate.

Computing the parabolic trend

In computing the trend values, the time scale has been centered and the values of ΣxY, Σx^2, $\Sigma x^2 Y$, and Σx^4 are obtained. (See Table 13.6.) The second-degree parabolic trend equation fitted to these data is:

$\hat{Y}_c = 91.82945 + 2.06862x - 0.02830x^2$
Origin: July 1, 1957
x in terms of years
Y: 1967 = 100

This equation may then be solved for all of the values of x. The trend fitted may be checked in that the sum of the trend items ($\Sigma \hat{Y}_c$) should approximately equal the sum of the data (ΣY). A second, and generally useful, check involves plotting the data and the trend. (See Figure 13.6.) Finally, one may check the calculations by substituting the constants in the third normal equation:

$$\Sigma x^2 Y = a\Sigma x^2 + b\Sigma x^3 + c\Sigma x^4$$

$$309{,}311 = (91.82945)(3{,}570) + (2.06862)(0.0) + (-0.02830)(654{,}378)$$

TABLE 13.6
Fitting a parabolic trend (data: Index of Wholesale Prices, 1940–1974 [1967 = 100]; see Figure 13.6)

Year X	Index Y	Time Centered x	xY	x^2	x^2Y
1940	40	−17	−680	289	9,000
1941	45	−16	−720	256	11,520
1942	51	−15	−765	225	11,475
1943	53	−14	−742	196	10,388
1944	54	−13	−702	169	9,126
1945	55	−12	−769	144	7,920
1946	62	−11	−682	121	7,502
1947	76	−10	−760	100	7,600
1948	83	− 9	−747	81	6,723
1949	79	− 8	−632	64	5,056
1950	82	− 7	−574	49	4,018
1951	91	− 6	−546	36	3,276
1952	89	− 5	−445	25	2,225
1953	87	− 4	−348	16	1,392
1954	88	− 3	−264	9	792
1955	88	− 2	−176	4	352
1956	91	− 1	− 91	1	91
1957	93	0	0	0	0
1958	95	1	95	1	95
1959	95	2	190	4	380
1960	95	3	285	9	855
1961	94	4	376	16	1,504
1962	95	5	475	25	2,375
1963	94	6	564	36	3,384
1964	95	7	665	49	4,655
1965	97	8	776	64	6,208
1966	100	9	900	81	8,100
1967	100	10	1,000	100	10,000
1968	102	11	1,122	121	12,342
1969	106	12	1,272	144	12,448
1970	110	13	1,430	169	18,590
1971	114	14	1,596	196	22,344
1972	119	15	1,785	225	26,775
1973	135	16	2,160	256	34,560
1974	160	17	2,337	289	46,240
Totals	3,113*	0	7,385	3,570	309,311

* Difference due to rounding.
$b = \Sigma xY/\Sigma x^2 = 7{,}385/3{,}570 = 2.0682.$

$$c = (n\Sigma x^2Y - \Sigma x^2\Sigma Y)/(n\Sigma x^4 - \Sigma x^2\Sigma x^2) = \frac{(35)(309{,}311) - (3{,}570)(3{,}113)}{(35)(654{,}378) - (3{,}570)(3{,}570)} = -0.02830.$$

The small discrepancy in this last check (309,312 instead of 309,311) is the consequence of rounding the constants.

Where the constants a and b are positive, the second-degree parabola will have an upward drift that is convex to the base. If both constants are negative, the trend will drift downward and approximately concave

x^4	a	+	bx	−	cx^2	=	$\hat{Y}_c$
83,521	91.82945		−35.16654		−8.17870		48.484
65,536	↓		−33.09792		−7.24480		51.487
50,625			−31.03293		−6.36750		54.429
38,416			−28.96068		−5.54680		57.322
28,561			−26.89206		−4.78270		60.155
20,736			−24.82344		−4.07520		62.931
14,641			−22.75482		−3.42430		65.650
10,000			−20.68620		−2.83000		68.313
6,561			−18.61758		−2.29230		70.920
4,096			−16.54896		−1.81120		73.469
2,401			−14.48038		−1.38670		75.962
1,296			−12.41172		−1.01880		78.399
625			−10.34310		−0.70750		80.779
256			− 8.27448		−0.45280		83.102
81			− 6.20586		−0.25470		85.369
16			− 4.13724		−0.11320		87.579
1			− 2.06862		−0.02830		89.733
0			0.0		0.0		91.829
1			2.06862		−0.02830		93.870
16			4.13724		−0.11320		95.853
81			6.20586		−0.25470		97.781
256			8.27448		−0.45280		99.651
625			10.34310		−0.70750		101.465
1,296			12.41172		−1.01880		103.222
2,401			14.48038		−1.38670		104.923
4,096			16.54896		−1.81120		106.567
6,561			18.61758		−2.29230		108.155
10,000			20.68620		−2.83000		109.686
14,641			22.75482		−3.42430		111.160
20,736			24.82344		−4.07520		112.578
28,561			26.89206		−4.78270		113.939
38,416			28.96068		−5.54680		115.243
50,625			31.03293		−6.36750		116.495
65,536			33.09792		−7.24480		117.683
83,521	↓		35.16654		−8.17870		118.817
654,378	3,214.03075						3,115.068*

$$a = (\Sigma Y - c\Sigma x^2)/n = \frac{(3{,}113) - (-0.02830)(3{,}570)}{35} = 91.82945.$$

$$\hat{Y}_c = 91.82945 + 2.06862x - 0.02830x^2.$$

to the base. An interesting point concerning the value of b in a second-degree formula is that if the data are time centered, this measure of the slope of the line is identical with the b value of a straight-line trend.

The fourth edition of this book fitted a parabola to the same data for the years 1940–70. Actually, the parabolic trend fitted that time period

FIGURE 13.6
Fitting a parabolic trend (data: Index of Wholesale Prices, All Commodities, 1940–74 [1967 = 100]; see Table 13.6)

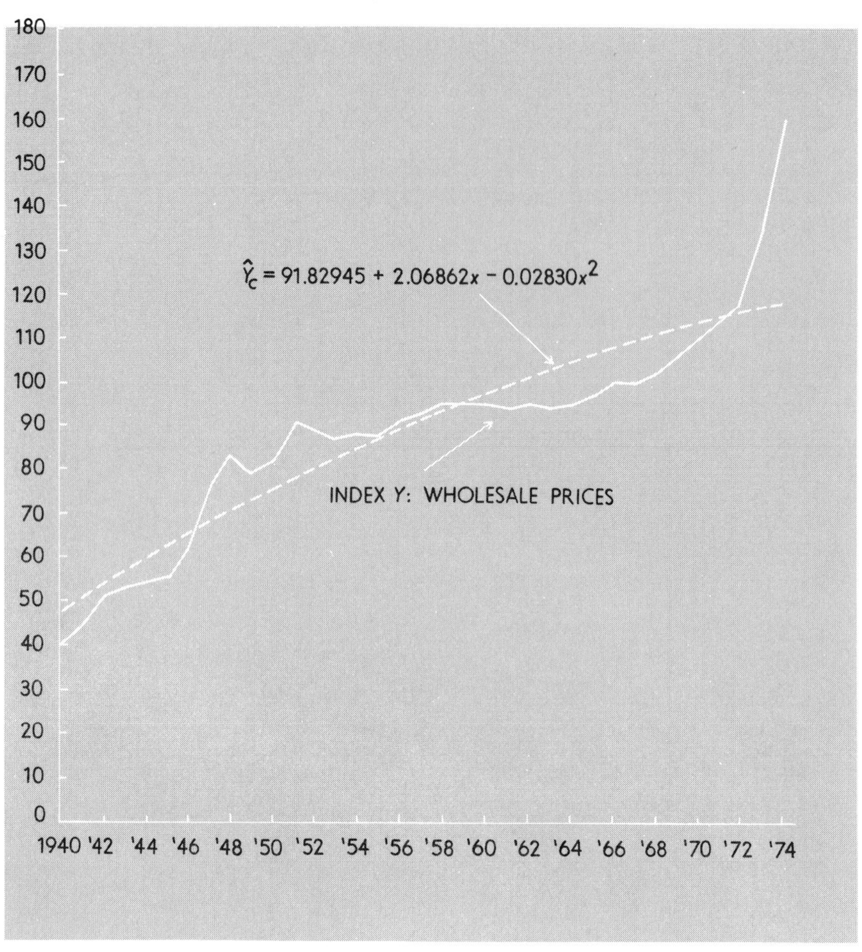

much better because of the significant inflation that occurred during the years 1971–74. In Figure 13.6 we may note that the Index of Wholesale Prices are considerable above the computed trend line. Until 1973 the parabola was a good fit to the data. Several lessons might be learned here. First, that extrapolation of a trend far into the future is treacherous business. Second, the double-digit inflation of 1974 becomes clear and the calculations of the parabolic trend demonstrates the impact of a few extreme items.

Test for parabolic trend

If a series of data actually form a parabolic pattern, the second differences will all be the same. Where the second differences are "more or less the same," the analyst might reasonably conclude that the parabola will be a good fit. To illustrate the test, let us consider the parabola $\hat{Y}_c = 1 + 2x + x^2$, and let us calculate $\hat{Y}_c$ values for $x = 0, 1, 2 \ldots 10$. (See Table 13.7.)

TABLE 13.7
Test of parabolic trend

x	$\hat{Y}_c$	First Differences	Second Differences
0	1	—	
		3	
1	4		2
		5	
2	9		2
		7	
3	16		2
		9	
4	25		2
		11	
5	36		2
		13	
6	49		2
		15	
7	64		2
		17	
8	81		2
		19	
9	100		2
		21	
10	121		

EXPONENTIAL TRENDS

Trends of the exponential series (Figure 13.5) tend to be characteristic of a time series in its early growth stages. This is particularly true of a series that is representing a rapidly expanding industry, for example, the electric power production in the 1920s, population, or any other similar growth pattern. Another example is the Index of Production in the Chemicals, Petroleum, and Rubber Products Industry, 1948–74. (See Figure 13.7 and Table 13.8.) A good way of detecting the appropriateness of an exponential curve is to plot the original data on ratio or semilogarithmic paper. (See Figure 13.8.) If the composite change (up or down) appears as an approximate straight line, then the exponential trend probably is the most descriptive of the series. Perhaps, the exponential trend is

FIGURE 13.7
Fitting the exponential trend (data: Index of Production in the Chemicals, Petroleum, & Rubber Products Industry, 1948–1974; [1967 = 100; see Table 13.8])

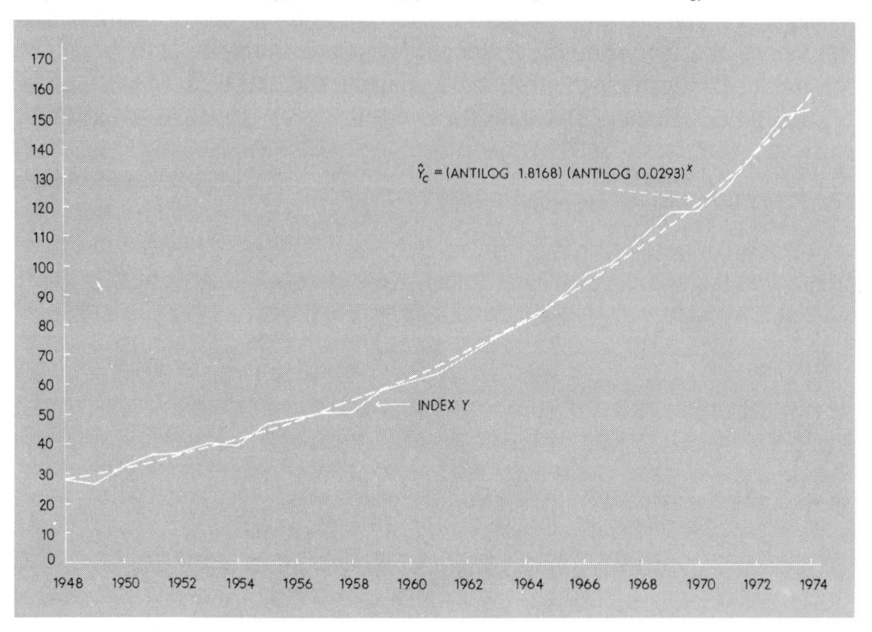

not frequently used; however, it is significant as an introduction to a set of trends of which the exponential is an element.

Fitting the exponential trend

The procedure for fitting an exponential trend is not complicated because we merely fit a straight-line trend to the logarithms of the data. The computations are carried on exactly as if we were using the original data. Once the curve is so defined we simply take the antilogarithms as the trend values.

The procedure[5] is illustrated in Table 13.8. (Also see Figures 13.7 and 13.8.) The formula for the exponential curve is $Y_c = ab^x$. The exponential trend fitted to the Index of Production in the Chemicals, Petroleum, and Rubber Products Industry, 1948–74, indicates that we might expect about a 6.9 percent *average annual rate* of growth in this index. As stated above the procedure is the same as fitting an arithmetic straight line to the data except that the logarithms (base 10) are used. Let us trace the computations for the first row of Table 13.8. The first two col-

[5] A shortcut procedure for fitting an exponential trend may be accomplished by a method of selected points.

TABLE 13.8
Fitting the exponential trend (data: Index of Production in the Chemicals, Petroleum, and Rubber Products Industry, 1948–1974 [1967 = 100; see Figure 13.7 and Table 13.8 below])

Year X (1)	Index Y (2)	Log Y (3)	Time Centered x (4)	x^2 (5)	xY (6)	Log $\hat{Y}_c$ (7)	Trend of Data $\hat{Y}_c$ (8)
1948	27	1.4314	−13	169	−18.6082	1.4359	27.3
1949	26	1.4150	−12	144	−16.9800	1.4652	29.2
1950	32	1.5051	−11	121	−16.5561	1.4945	31.2
1951	36	1.5563	−10	100	−15.5630	1.5238	33.4
1952	37	1.5682	−9	81	−14.1138	1.5531	35.7
1953	40	1.6021	−8	64	−12.8168	1.5824	38.2
1954	39	1.5911	−7	49	−11.1377	1.6117	40.9
1955	46	1.6628	−6	36	−9.9768	1.6410	44.8
1956	48	1.6812	−5	25	−8.4060	1.6703	46.8
1957	50	1.6990	−4	16	−6.7960	1.6996	50.1
1958	50	1.6990	−3	9	−5.0970	1.7582	53.6
1959	57	1.7559	−2	4	−3.5118	1.7582	57.3
1960	60	1.7782	−1	1	−1.7782	1.7875	61.3
1961	63	1.7993	0	0	0.0000	1.8168	65.6
1962	69	1.8388	1	1	1.8388	1.8461	70.2
1963	75	1.8751	2	4	3.7502	1.8754	75.1
1964	80	1.9031	3	9	5.7093	1.9047	80.3
1965	87	1.9395	4	16	7.7580	1.9340	85.9
1966	96	1.9823	5	25	9.9115	1.9633	91.9
1967	100	2.0000	6	36	12.0000	1.9926	98.3
1968	109	2.0374	7	49	14.2618	2.0219	105.2
1969	117	2.0682	8	64	16.5456	2.0512	112.5
1970	118	2.0719	9	81	18.6471	2.0805	120.4
1971	125	2.0969	10	100	20.9690	2.1098	128.8
1972	138	2.1399	11	121	23.5389	2.1391	137.7
1973	149	2.1732	12	144	26.0784	2.1684	147.4
1974	152	2.1818	13	169	28.3634	2.1977	157.6
Totals	2,026*	49.0527	0	1,638	48.0306		2,027.0*

* Difference due to rounding error.

$a = \Sigma Y/n = 49.0527/27 = 1.8168.$
$b = \Sigma xY/\Sigma x^2 = 48.0306/1{,}638 = 0.0293.$

Trend of logs: $\hat{Y}_c = a + bx = 1.8168 + 0.0293$
or $\log \hat{Y}_c = \log a + (\log b)(x)$

Trend of data:
$\hat{Y}_c = ab^x = (\text{antilog } 1.8168)(\text{antilog } 0.0293)^x$
$\hat{Y}_c = (65.59)(1.069)^x$
Which means that this index tends to rise 6.9 percent per year or the average annual *rate* of growth is 6.9 percent.

umns give the original data by year. Column (3) gives the logs of the original data, for example, log of 27 = 1.4314. (See Appendix B for the logs.) Column (4) is the usual centering process to reduce the computations. Note that the year 1948 is 13 units from the origin (1961) and hence the minus sign. Column (5) is the product of (−13) and the log

FIGURE 13.8
Fitting the exponential trend (data: Index of Production in the Chemicals, Petroleum, and Rubber Products Industry, 1948–1974 [1967 = 100; see Table 13.8 and Figure 13.7])

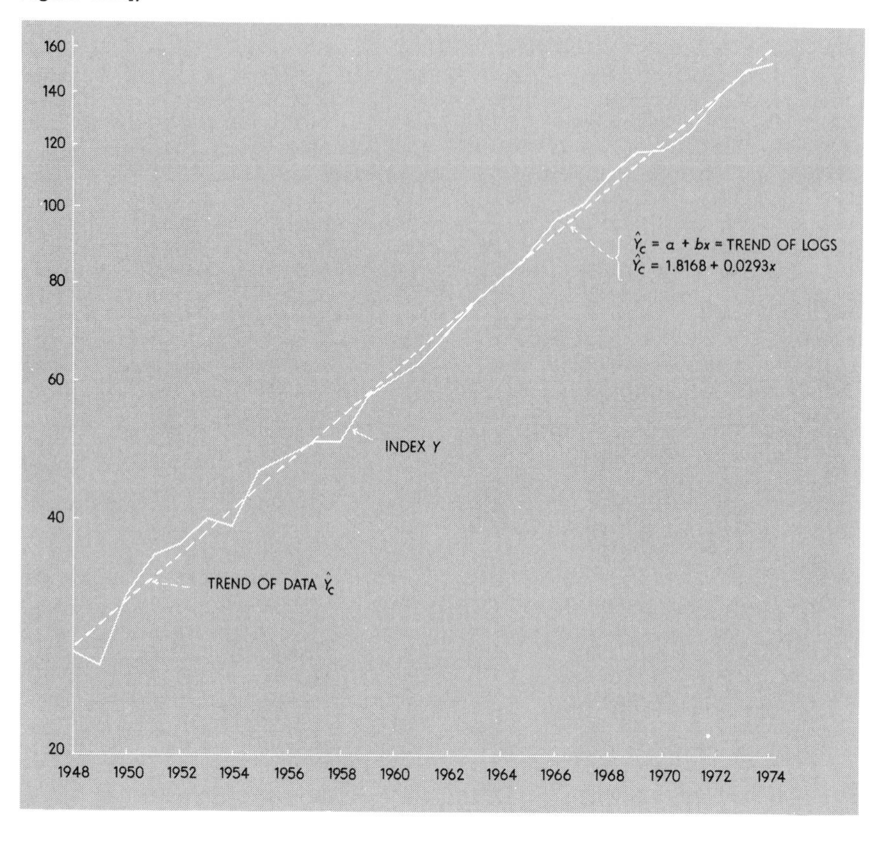

of 27 or (1.4314). The figure −18.6082 is recorded in Column (6). Column (7) is the log of the calculated trend determined by the following equation:

$$\log \hat{Y}_c = 1.8168 + 0.0293\,(-13) = 1.4359$$

The final column in Table 13.8 is the antilog of 1.4359 or 27.3. This estimate compares well with the actual index value of 27 for 1948. We can see from these data, and from Figures 13.7 and 13.8, that the exponential trend is a good fit to the original data. The difference between the two figures is that one is plotted on arithmetic paper and the other on semilog to show rates of change.

The modified exponential curve appears as an inverted exponential trend. (See Figure 13.5.) The trend pattern of an economic time series that tends to level off and approach the horizontal axis may be described by a modified exponential curve. The formula for this trend line is $\hat{Y}_c = a + bc^x$ and is fitted in a manner similar to the previous example.

Pearl-Reed and Gompertz curves

Both the Pearl-Reed and Gompertz growth curves appear as an elongated S. In terms of absolute growth these curves are useful to depict the pattern of a mature industry. For example, the steel industry (using almost any useful measure) shows: (*a*) an early period when growth increments were small, (*b*) an intermediate period of rapid expansion, and (*c*) an approach to maturity. If these curves are plotted on a log scale, the *relative* growth appears to be increasing at a decreasing rate. Long-term changes in business and economic activity generally fit this description, too.

Normally, if both the earlier and latest stages of growth are to be represented by a trend pattern, the modified exponential type of curve usually is not appropriate. In such cases a further variation is necessary, and in most cases either the Pearl-Reed growth curve or a simple form of a Gompertz trend can be useful. Both of these curves are rather complex; however, their general nature can be considered here. The Pearl-Reed curve may be defined as:

$$Y = \frac{1}{a + bc^x} \qquad (13.22)$$

and the equation for the simple Gompertz curve is

$$\log Y = a + bc^x \qquad (13.23)$$

where a and b are in terms of logs of the original data.

These two particular "growth curves" are useful for depicting past trends and probable future tendencies because they represent the so-called law of growth or S curve. That is, an industry, population, or a product might tend to grow at a nearly constant *percentage* rate during its youth; however, as the normal maturity develops the rate of change tends to decline. Therefore, on an arithmetic scale such curves tend to take the shape of the elongated S, while using a ratio or logarithmic scale the curve tends to flatten out and looks like this  for a growing series and a declining series like this  . It should be noted that unless the series is reflecting a declining *rate* of growth or decline a growth function cannot be fitted to the data. The auto industry is an excellent example of this type of growth. During Henry Ford's early years of production the series tended to grow slowly at a relatively constant percentage increase. Later as more paved roads became available and the product more reliable the growth increased significantly. Currently, the auto industry seems to be tapering off in terms of its growth. The steel industry is another such example. Generally, the older, more mature industries fit into this category. The elongated S curve might be a good description for the life cycle of many products.

THE PHILLIPS CURVE

Perhaps, the Phillips curve could just as well be discussed in Chapter 15 when the topic under consideration is regression analysis. It is mentioned here simply because of the nonlinear shape of the relationship between the unemployment rate and the percentage changes in consumer prices on a year-to-year basis. During periods of relatively high unemployment the Phillips curve comes into prominence. When prices were rising at a rapid rate in 1969 and 1970 the press gave considerable mention to this statistical and economic phenomenon. The curve portrays the serious public policy problem of premature inflation. It shows dramatically how unemployment is high when the nation's rate of inflation is low but drops sharply as prices rise. Its creator is Professor A. W. H. Phillips, formerly of the University of London, who currently is on the faculty of the Australian National University.

Figure 13.9 indicates the relationship for a few recent years that reflect both inflationary and deflationary times. The chart reflects the public policy dilemma of the trade-off between price stability and employment. The social costs of relative price stability apparently have been rising unemployment and retarded economic growth. Unemployment has always been more of a problem for blacks and other races than for whites. (See Table 13.9 and Figure 13.9.) In 1958 and 1961, for example, unemployment of blacks exceeded 12 percent while the rate for whites was about 6 percent; less than one half of that for blacks. As prices rose relatively slowly in the early 1960s, unemployment rates for both blacks and whites fell sharply. By the end of 1968 the unemployment rates for both groups were at about one half of the rate in 1961. Nonetheless, the unemployment rate for blacks was well over 6 percent nationally and started to climb again in 1969. By the end of 1970 the rate for blacks was over 8 percent. In 1974 while the inflation rate was at the double-digit level (11.0 percent) for the first time, the unemployment rate for blacks was almost twice (9.0 percent) as compared to 5.0 percent for whites. 1975 seemed to continue this relationship. Paradoxically, from the point of the Phillips curve, this was at the same time when consumer prices were rising at the fastest rate in recent economic history. It might also be noted that the data of Table 13.9 refer to annual averages for unemployment and year-to-year changes for the Consumer Price Index. The annual averages tend to conceal short-term variations and suffer from the usual problem of using one number to represent several. Possibly, it might be

FIGURE 13.9
The Phillips curve (fitted by freehand method)

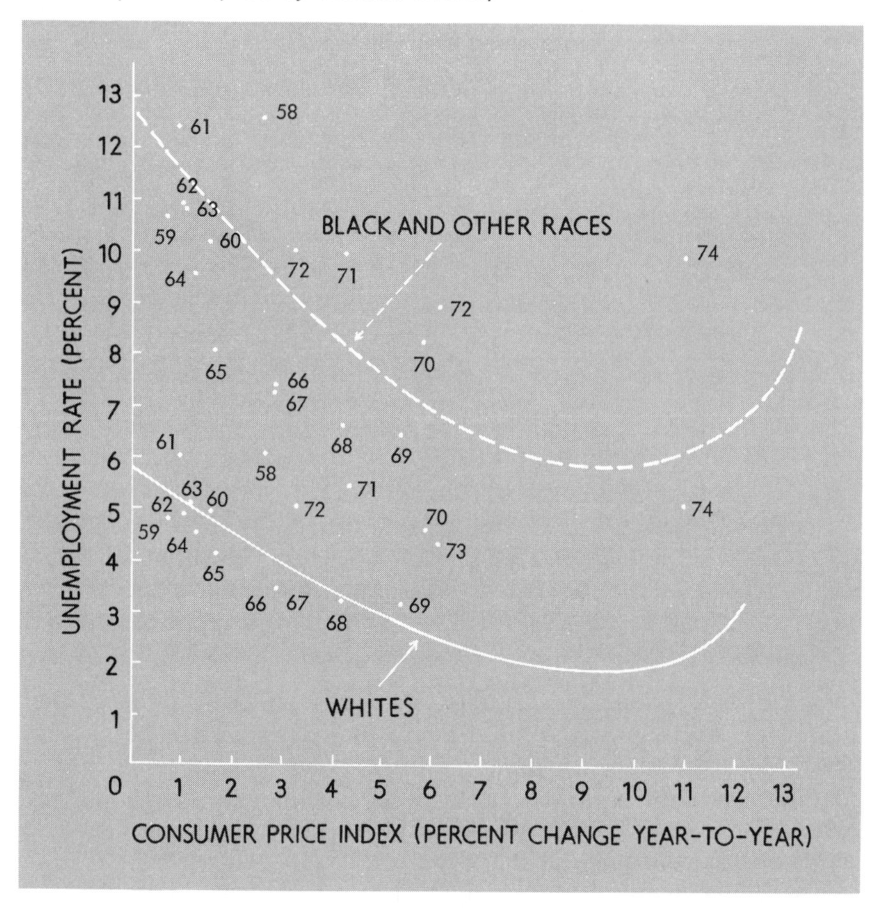

more meaningful to view the relationship portrayed by the Phillips curve by using monthly data in order not to smooth out the fluctuations.

Some economists agree that the Phillips curve does not tell the whole story with respect to the relationship of prices to unemployment. They argue that it gives only a short-run view and that the problem of unemployment versus inflation must be looked at in historical perspective. Unfortunately, the unemployed person in the short-run is the one who suffers. These economists suggest that the Phillips curve implies that there is no alternative to inflation but unemployment and that it pays little attention to the public dislike of either alternative. However, most economists probably would agree with one lesson of the Phillips curve. That is, if inflationary pressures subside to below the 1.1 percent rate,

TABLE 13.9
Relationship of unemployment and consumer prices

Year	Unemployment Rate Percent		Consumer Price Index (percent change year to year)
	White	Black	
1958	6.1	12.6	2.7
1959	4.8	10.7	0.8
1960	4.9	10.2	1.6
1961	6.0	12.4	1.0
1962	4.9	10.9	1.1
1963	5.0	10.8	1.2
1964	4.6	9.6	1.3
1965	4.1	8.1	1.7
1966	3.4	7.3	2.9
1967	3.4	7.4	2.9
1968	3.2	6.7	4.2
1969	3.1	6.4	5.4
1970	4.5	8.2	5.9
1971	5.4	9.9	4.3
1972	5.0	10.0	3.3
1973	4.3	8.9	6.2
1974	5.0	9.9	11.0

Source: U.S. Government Printing Office, *Economic Report of the President,* February, 1975, pp. 279 and 302.

then the black unemployment rate may well soar beyond the 12 percent level. Well-known economist Pierre Rinfret stated publicly in 1968 that "Any national policy designed to moderate inflation—and heaven knows we need it—must face up to the problem of the black trade-off. What this means is that an attack on inflation must be coupled with an attack on black unemployment."

The Phillips curve has been applied in other comparisons, too. For example, probably the most reflective measure of the supply of labor relative to the demand is the unemployment rate. (This is the figure that receives widespread publicity in the press and TV each month.) The higher this rate, the larger is the excess of the quantity of labor on the market than is demanded by all employers in the aggregate. Therefore, theoretically an efficient labor market would suggest, *ceteris paribus,* that higher unemployment rates would correlate with slower wage growth. The converse, lower rates of unemployment being associated with faster wage increases, would also follow from this reasoning. Of course, labor markets are not efficient in the strict sense. That is, market information is not perfect and the markets are not homogeneous. Therefore, one cannot expect a perfect relationship between wage changes and the unemployment rate. However, there probably is a negative relationship between wage changes and changes in the unemployment rate. Phillips found a close relationship between the changes of current wages and the unemployment rate. That is, the larger the increases in wages the lower the unemployment rate, and conversely. Studies in the United States during the 1960s confirmed this relationship in a mature economy. Recently, because of the double-digit inflation, these relationships in the United States have not held. Wages in current dollars have continued to rise throughout a high unemployment rate in the late 1960s and mid-1970s. Either the Phillips curve has shifted or the relationship of these two variables has changed. Two of the most logical speculations as to why this relationship no longer seems to hold are: (*a*) changes in the composition of the labor market, and (*b*) expectations about the rate of inflation in the future. Some analysts even have gone so far as to question whether or not a Phillips curve really exists in the long run. These people believe in a so-called "natural rate of unemployment" from which any deviation will tend to lead to accelerating wage increases or decreases. Actually, another logical explanation for the current deviations from the Phillips curve is that there probably is more than one labor market, each slightly different in composition, which produces a dispersion of unemployment across markets which might shift the relationship.

Some studies have been reported in the mid-1970s that suggest a statistically significant relationship between wage rates and unemployment rate changes for about one third of the states. These relationships fitted the conventional theory. Only manufacturing wage changes were used in the computations; however, a more inclusive wage measure did suggest similar results. Until such time as the Phillips theory is rejected economists and other business analysts probably will be discussing this tool of statistical analysis. In the meantime it is worthwhile to recognize the applications and limitations of the Phillips curve.

PRECAUTIONS

There is no substitute for a sound knowledge of the data under observation. This is especially true in trend fitting and correlation and regression analysis. Subjective judgment enters into the selection of the type of trend to be fitted and used. A useful first step in both cases is to plot the data and to study the graph. This permits the analyst to select a curve that adequately describes the pattern of the economic series.

Once the trend has been properly fitted, the analyst usually extrapolates to make a forecast. Care must be exercised here because events may change to alter the pattern in the long run. Perhaps, a new trend really is being established at another level. Davies has said that ". . . trend is like an established habit; it probably will continue, but other forces may interrupt or change it." Consequently, any extrapolation of a trend line should be done for only a relatively short period of time. A useful example of extrapolation can be found in Bureau of Census estimates of population from one census year to the next (ten years later). Such estimates generally prove to be accurate enough for most applications.[6]

The point to remember is that the fitting of a trend, much like the application of many other statistical procedures, is not simply a mathematical process. In many respects the mathematical mechanics are secondary to the judgment of the analyst which is based upon an appreciation of the data and the problem at hand. Many factors usually enter into the establishing of a trend pattern of most economic time series. Complex interrelationships in our economic system normally involve such factors as the state of technology, changes in governmental policies, the level of economic activity, and a whole host of other forces.

Nonetheless, recognizing the limitations and dangers, trend extrapolation is useful in business. Very often it can be looked upon as what may be normal or as a bench mark. The analysis of business cycles is aided by this means. From the point of view of the individual firm, trend analysis of external series, for example, bank debits, population, retail sales, prices, and so on, can be equally helpful as the study of the firm's internal data relating to sales, production, and the like.

Chapter 14 discusses other aspects of time series analysis including the measurement of cyclical and seasonal variations. Also, the leading economic indicators of the National Bureau of Economic Research (NBER) are discussed in detail.

SYMBOLS

$\hat{Y}_c$ or T, computed or estimated secular or long-term trend
S, computed seasonal variations
C, computed cyclical variations
I, irregular or random fluctuations
p_1, p_2, values of two selected points
b, slope of a straight line
t, difference between the Y values of $p_2 - p_1$
X_n, given year or time period
X_1, year or time period associated with p_1
a, a constant indicating height of line at the Y intercept, or height of line at the origin
b, a constant indicating the slope of the line; increase in $\hat{Y}_c$ associated with each unit change in x
x, centered time period

[6] An historic example of an error resulting from assuming that a trend would continue without change was made by Abraham Lincoln in his second message to Congress. Lincoln's advisors assumed continuation of the population increase rate of 1790 to 1860 and predicted the U.S. population would reach 251,689,914 in 1930! Aside from the extrapolation error—which was bad enough—the estimation to the last digit was ridiculous.

PROBLEMS

13.1. The table on the right is a list of employees in nonagricultural establishments, in tens of thousands.

Year	Y
1969	529
1970	514
1971	534
1972	544
1973	542
1974	558
1975	570

Required: What is the value of–

a. a?

b. b?

c. ΣxY?

d. Σx^2?

e. ΣY?

f. $\Sigma \hat{Y}_c$?

g. What is the trend equation with origin 1972, X units = one year?

13.2. Suppose

$$\hat{Y}_c = 100 + 16x$$

X units = one year

Origin: July 1, 1972

Required:

a. What is the trend equation with origin July 1, 1976, X units = one year?

b. What is the trend equation with origin April 1, 1976, X units = one-half year?

13.3. Using a source such as the *Statistical Abstract of the United States, The National Income and Product Accounts of the U.S., 1929–65, Long Term Economic Growth, 1860–1965,* select a series whose growth pattern may best be described by the following:

a. Modified exponential.

b. Exponential.

c. Pearl-Reed.

d. Second-degree parabola.

Fit the second-degree parabola to the series it best represents. Do the same for the exponential trend.

13.4. Plot U.S. Gross National Product (GNP) in constant dollars from 1910 to date on (*a*) a ratio-scale chart and (*b*) an arithmetic-scale chart. How is the trend pattern best described? Fit an appropriate trend to these data.

13.5. The LeBeaux Instrument Company of Ann Arbor was organized in 1969. Two years later it became a public company. The company produces air sampling instruments, membrane air filters, electrophoresis equipment (used to separate the component parts of a mixture by applying an electric current), chromatography equipment, and several other newer related products. Shown below are LeBeaux sales data since 1969. Fit an appropriate trend line to the data after plotting the information on graph paper. Interpret your results.

Year	Sales (thousands of dollars)	Year	Sales (thousands of dollars)
1969	50	1973	760
1970	80	1974	1,050
1971	225	1975	1,720
1972	420	1976	2,250

13.6. Assume that the trend component of an economic time series is nonlinear. Several basic approaches to the analysis of this trend are possible; what might these be?

13.7. Calculate a trend line for the sales of General Mills Corp.:

Year	Sales (millions)	Year	Sales (millions)
1961	$576	1969	$ 885
1962	546	1970	1,092
1963	524	1971	1,185
1964	541	1972	1,901
1965	559	1973	1,662
1966	525	1974	2,000
1967	603	1975	2,309
1968	669		

13.8. Fit an appropriate trend line to the following population data:

Year	U.S. Population (millions)	Year	U.S. Population (millions)
1961	184	1969	203
1962	187	1970	205
1963	189	1971	207
1964	192	1972	209
1965	195	1973	210
1966	197	1974	212
1967	199	1975	214
1968	201		

13.9. Given the data on the right, determine the trend by using the selected points method.

Corporate profits plus capital consumption allowances, 1961–75 (billions of dollars)

Year		Year	
1961	53.5	1969	96.8
1962	61.3	1970	95.2
1963	64.8	1971	106.5
1964	72.3	1972	124.0
1965	82.9	1973	144.1
1966	89.5	1974	161.7
1967	89.6	1975	167.8
1968	94.6		

Source: *Economic Indicators,* April, 1975, p. 7.

13.10. Given:

$$\hat{Y}_c = 72 + 288x$$

Origin: July 1, 1968
X in terms of years
Y in millions of dollars

Convert the above into a monthly trend equation.

13.11. There are two major types of consumer credit outstanding in this country: (1) instalment credit, which includes loans extended for the purchase of automobiles, other durable goods, repairs and modernization of homes, and personal loans–all of which are repayable in "easy monthly payments"; and (2) non-instalment credit, which embraces charge accounts, service credit (electric, telephone, doctor bills, etc.), and single payment loans, which you will surely be reminded of if payment is not received when due.

The volume of both types of credit has increased materially in recent years:

Consumer credit in the United States (billions of dollars)

End of Year	Instalment Credit	Noninstalment Credit
1953	23.0	8.4
1954	23.6	9.0
1955	29.0	10.0
1956	31.7	10.6
1957	33.9	11.1
1958	33.6	11.5
1959	39.2	12.3
1960	43.0	13.2
1961	43.5	14.2
1962	48.0	15.1
1963	54.2	16.3
1964	60.5	17.9
1965	71.3	19.0
1966	76.2	20.0
1967	79.4	21.4
1968	87.7	23.0
1969	97.1	24.0
1970	102.1	25.1
1971	111.3	27.1
1972	127.3	30.2
1973	147.4	33.0
1974	156.1	34.0
1975	157.0	33.6

Source: *Economic Indicators,* Council of Economic Advisors and *Federal Reserve Bulletin.*

Part I.

a. Fit straight-line trends to the two series by the least-squares method.
b. Are the two b values comparable? What does each indicate?

c. Present the two series in a chart and show the trend of each.

d. State your conclusions relative to the long-term growth of the two types of consumer credit.

Part II. (See Problem 14.3.)

13.12. Given the following trend information:

$\hat{Y}_c = 30 + 2.16x$	Origin: July 1, 1965
Y in millions of dollars	x in terms of years

Convert this equation into monthly terms. Be sure your response is in the most practical form.

13.13. Fit the most appropriate trend to the following data:

Percent of external to internal sources of funds, corporate nonfinancial sector, 1965–1974

Year	Percent	Year	Percent
1965	.35	1970	.72
1966	.41	1971	.66
1967	.52	1972	.72
1968	.48	1973	.77
1969	.68	1974	.85

13.14. Given the following data, respond to the questions below.

Gross internal funds as a percent of capital expenditures, nonfinancial corporate sector, 1960–1974

Year	Gross Internal Funds (billions of dollars)	Capital Expenditures (billions of dollars)	Percent
1960	$34.4	$ 38.7	88.9
1961	35.6	36.3	98.0
1962	41.8	43.6	95.9
1963	43.9	45.2	97.1
1964	50.5	51.6	97.9
1965	56.6	62.3	90.9
1966	61.2	76.5	80.0
1967	61.4	71.4	86.0
1968	61.7	75.0	82.3
1969	60.7	83.7	72.5
1970	59.4	84.0	70.7
1971	68.0	87.2	78.0
1972	78.7	102.5	76.8
1973	84.6	121.5	69.6
1974	81.4	125.8	64.7

Source: *Flow of Funds*, Board of Governors of the Federal Reserve System.

a. Plot each of these series on a ratio or logarithmic chart.

b. Fit a trend line to each series.
c. Write a brief paragraph reflecting the financial implications of these trends.

13.15.

Debt/equity ratios, U.S. manufacturing corporations, 1955–1974

Year	D/E Ratio	Year	D/E Ratio
1955	.209	1965	.273
1956	.232	1966	.308
1957	.246	1967	.345
1958	.244	1968	.370
1959	.239	1969	.402
1960	.245	1970	.437
1961	.250	1971	.444
1962	.251	1972	.435
1963	.253	1973	.439
1964	.254	1974	.469

Source: SEC/FTC *Quarterly Financial Report.*

a. Plot this series on an appropriate chart.
b. Fit the trend line to the data and draw in this trend on your chart.
c. Write a brief paragraph of the financial implications of this trend.

RELATED READINGS

BELLMAN, R., and ROTH, R. "Curve Fittings by Segmented Straight Lines." *Journal of the American Statistical Association,* vol. 64, no. 327 (1969), pp. 1079–84.

BRY, GERHARD, and BOSCHAN, CHARLOTTE. "Interpretation and Analysis of Time-Series Scatters." *The American Statistician,* vol. 25 no. 2 (April 1971), pp. 29–33.

CUNNYNHAM, J. *The Spectral Analysis of Economic Time Series.* Bureau of Census Working Paper No. 14. Washington, D.C., 1963.

DUESENBERRY, JAMES; ECKSTEIN, O.; and GROMM, G. "A Simulation of the U.S. Economy in Recession." *Econometrica,* vol. 28 (1960), pp. 749–809.

FIRST NATIONAL CITY BANK, NEW YORK CITY. "Has Mr. Phillips Thrown Policymakers a Curve?" *Monthly Economic Letter* (February 1971), pp. 10–12.

GRANGER, C. W. J., and HATANAKA, M. *Spectral Analysis of Economic Time Series.* Princeton, N.J.: Princeton University Press, 1957.

HUSBY, RALPH D. "A Nonlinear Consumption Function Estimated from Time-Series and Cross-Section Data." *Review of Economics and Statistics,* vol. 53, no. 1 (February 1971), pp. 76–79.

MORTENSEN, D. T. "Job Search, the Duration of Unemployment, and the Phillips Curve." *American Economic Review,* vol. 60, no. 5 (December 1970), pp. 847–62.

PARZEN, E. "An Approach to Empirical Time Series Analysis." *Radio Science,* vol. 68D, no. 9 (1964), pp. 937–51.

———. *Time Series Analysis Papers.* San Francisco: Holden-Day, 1967.

PHILLIPS, A. W. H. "The Relationship between Unemployment and the Rate of Change of Money Wage Rates in the United Kingdom, 1862–1957." *Economica,* November 1958, pp. 283–99.

14

Time series analysis—II: Cyclical and seasonal variations

We cannot properly know things as they are unless we know how they came to be what they are.

C. L. Becker

WE NOW turn to the other major forces that affect economic time series, namely, cyclical variations and seasonals. While trend has a smooth pattern the cycles tend to be more unpredictable and irregular. On the other hand, seasonal patterns tend to be reoccurring and rather stable over reasonable periods of time. Consequently, when making short-term forecasts the impact of seasonal variations upon trend estimates should be recognized. The potential impact of a cycle on an economic time series generally can be best handled through a complete analysis of general business conditions at the moment rather than in using the historical record as a pattern to be followed. The problems of analyzing economic conditions are not the main subject for this text; however, in helping the analyst evaluate future cyclical variations this chapter does discuss in some detail the economic indicators of the National Bureau of Economic Research (NBER). These indexes are followed very closely by economists and business forecasters and therefore represent a topic worthy of the student's time. Let us first turn to the problems of estimating cyclical variations.

METHODS OF ESTIMATING CYCLICAL VARIATIONS

Although an application of the data in Table 14.1 may have limited value in a forecast, a better appreciation of the *impact* of cycle and irregular may be obtained by adjusting mathematically for trend. Column 4 reflects the fluctuations *from* the long-term trend of General Brands, Inc. sales data which apparently result from either cycle or irregular forces.

TABLE 14.1
Adjustment of sales of General Brands, Inc. for trend, fiscal years, 1970–1976

Year (1)	Sales (millions of dollars) TCI (2)	Trend (millions of dollars) T or $\hat{Y}_c$ (3)	Adjusted Sales (percent) $TCI/T = CI$ (4)
1970	436	447.6	97.4
1971	469	461.9	101.5
1972	483	476.2	101.4
1973	488	490.5	99.5
1974	514	504.8	101.8
1975	516	519.1	99.4
1976	528	533.4	99.0

(See Chapter 13 for the trend computations.) Column 2 contains the actual sales figures; and from our earlier assumption, it is the product of trend, cycle, and irregular, TCI. By dividing the annual data, TCI/T, we obtain CI. (Note: T is used interchangeably with $\hat{Y}_c$.) That is, by dividing the actual data by an estimate of trend, what remains is an estimate of the *impact* of CI. Column 4 is the result of this process and is expressed as an index or percent. That is, T or $\hat{Y}_c$ (trend) is looked upon as the norm and equal to 100 percent. Any deviations from this norm are considered to arise from cyclical and irregular fluctuations. To illustrate, sales of General Brands, Inc. were 2.6 percent *below* normal (trend) in 1970 and 1.8 percent *above* normal in 1974 (see Figure 14.1, too). Presumably,

FIGURE 14.1
Index of cyclical and irregular forces, General Brands, Inc. sales, 1970–1976, adjusted for trend

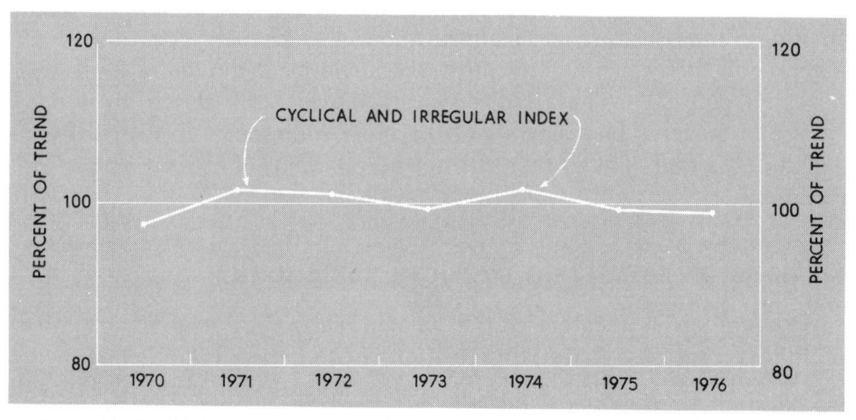

Source: Table 14.1.

these differences were caused by the cyclical pattern or some irregular force. Only a person familiar with the economic situation and the characteristics of the company could single out these forces. A moving average of Column 4 may tend to smooth out the irregular influence, but the procedure holds very little interest to the analyst because it is more academic than useful, and it is not considered here.

National Bureau of Economic Research indicators

The National Bureau of Economic Research (NBER) has been engaged in the study of business cycles for years, and it has had a continuing interest in identifying economic time series that provide an intelligent basis for appraising current business conditions. (See Table 14.2.) Again, the advent of high-speed electronic computers has facilitated the effectiveness of careful study of so many economic indicators. In the NBER's search for answers to questions relating to an evaluation of the current position of the economy, Wesley C. Mitchell, Arthur F. Burns, and later Geoffrey H. Moore analyzed more than 800 statistical series. Burns and Mitchell confirmed the feeling that no "ideal" single indicator existed. In fact, they listed 21 indicators with one or more superior characteristics in terms of analyzing current business conditions. In 1950, Moore revised this list; and in 1963 a new list of 26 indicators was released by the NBER.

In 1967, the NBER published a new volume that extended the list of indicators. (See Geoffrey H. Moore and Julius Shiskin, *Indicators of Business Expansions and Contractions* [National Bureau of Economic Research, 1967].) This latest list includes 37 leading series, 25 roughly coincident, 11 lagging, and 16 unclassified by timing, or 89 in all; 73 are monthly and 16 are quarterly. This list includes 13 series not on the "list of 26" of 1963 and omits 5 series. Also 14 series previously unclassified by timing are now included. A short list of 25 series, drawn from the full list, is presented in Table 14.2. This latest more selective list includes 12 leading, 7 roughly coincident, and 6 lagging series; 21 are monthly and 4 are quarterly. The NBER employs a dual classification pattern which groups the business indicators by cyclical timing and economic process. (See Table 14.3.) The principal basis of classification is the economic process with the cyclical timing being the secondary criterion. The economic processes are divided into minor ones that tend to have distinct differences in cyclical timing. The historical timing record of each series at business cycle peaks and troughs is the major basis in the timing classification.

Leading series. Therefore, some of the leading series—those that precede a business downturn or upturn—relate to future output rather than current production, whereas others measure profit *expectations.* Some em-

TABLE 14.2
National Bureau of Economic Research 25 indicators of economic activity

(1) Series	(2) Business Cycle Peak or Trough	(3) Business Cycle Turns Covered	(4) Leads	(5) Lags	(6) Rough Coinci- dences*	(7) Business Cycle Turns Skipped	(8) Median Lead (−) or Lag (+) Mos.
Leading Series:							
A. Employment and unemployment:							
1. Average workweek, production workers,	P	9	7	1	1(0)	1	− 6
manufacturing	T	10	6	1	3(2)	1	− 4
2. Nonagricultural placements, BES	P	5	4	0	0(0)	1	−11
	T	5	4	1	4(0)	0	− 1
B. Formation of business enterprises:							
3. Index of net business formation	P	5	4	0	0(0)	1	−20
	T	5	4	0	3(1)	0	− 3
C. New investment commitments:							
4. New orders, durable goods	P	10	8	0	1(0)	2	− 8
	T	10	8	0	6(1)	1	− 2
5. Contracts and orders, plant and equip-	P	4	4	0	0(0)	0	− 8
ment	T	4	3	1	2(0)	0	− 3
6. New building permits, private housing	P	11	8	1	1(0)	2	−13
units	T	11	9	0	4(1)	1	− 5
D. Inventory investment and purchasing:							
7. Change in book value, manufacturing,	P	5	5	0	0(0)	0	−14
and trade inventories	T	5	4	0	2(1)	0	− 6
E. Prices, costs, and profits:							
8. Industrial materials prices	P	10	8	1	3(0)	2	− 6
	T	11	5	1	6(4)	2	0
9. Stock prices, 500 common stocks	P	22	17	2	9(1)	2	− 4
	T	22	16	3	5(1)	2	− 4
10. Corporation profits after taxes	P	10	7	0	4(2)	1	− 6
	T	10	6	2	7(2)	0	− 2
11. Ratio, price to unit labor cost, manufac-	P	10	9	1	4(0)	0	−11
turing	T	11	8	2	6(1)	0	− 3
12. Change in consumer installment debt	P	7	6	0	1(0)	1	−12
	T	7	5	1	3(0)	1	− 4

Series							
Roughly Coincident:							
A. Employment and unemployment:							
1. Employees in nonagricultural establishments	P	7	5	1	5(1)	0	− 2
	T	7	1	1	7(5)	0	0
2. Unemployment rate, total	P	7	4	1	2(1)	1	−4
	T	7	0	5	6(2)	0	+2
B. Production income, consumption, trade:							
3. GNP in constant dollars	P	8	2	2	6(2)	2	0
	T	9	5	1	3(1)	2	−3
4. Industrial production	P	10	5	2	5(3)	0	0
	T	11	4	1	8(6)	0	0
5. Personal income	P	9	2	5	5(1)	1	+1
	T	10	8	0	7(1)	1	−2
6. Manufacturing and trade sales	P	4	3	0	2(1)	0	−4
	T	4	1	0	4(3)	0	0
7. Sales of retail stores	P	10	2	4	3(0)	4	+1
	T	11	3	2	4(1)	5	0
Lagging Series:							
A. Employment and unemployment:							
1. Unemployment (15 and over weeks)	P	4	1	2	3(1)	0	+1
	T	4	0	4	2(0)	0	+2
B. Fixed capital investments:							
2. Business expenditures, new plant and equipment	P	10	1	5	9(4)	0	0
	T	10	1	8	7(1)	0	+2
C. Inventories:							
3. Book value, manufacturing and trade	P	5	1	4	3(0)	0	+2
	T	5	1	4	4(0)	0	+2
D. Prices, credit, and interest rates:							
4. Labor cost per unit of output, manufacturing	P	10	0	7	0(0)	3	+8
	T	11	0	7	1(0)	4	+9
5. Commercial and industrial loans outstanding	P	6	1	3	3(0)	2	+2
	T	6	0	4	3(0)	2	+2
6. Bank rates on short-term business loans	P	10	2	6	2(1)	1	+5
	T	11	0	9	3(0)	2	+5

* Includes exact coincidences shown in parentheses and leads and lags of three months or less.

Source: Geoffrey H. Moore and Julius Shiskin, *Indicators of Business Expansions and Contractions* (New York: National Bureau of Economic Research, Inc., 1967), pp. 49–101.

TABLE 14.3

Gross classification of NBER cyclical indicators by economic process and cyclical timing (number of series in parentheses)

Economic Process / Cyclical Timing	I. EMPLOYMENT AND UNEMPLOYMENT (15 series)	II. PRODUCTION, INCOME, CONSUMPTION AND TRADE (8 series)	III. FIXED CAPITAL INVESTMENT (14 series)	IV. INVENTORIES AND INVENTORY INVESTMENT (9 series)	V. PRICES, COSTS, AND PROFITS (10 series)	VI. MONEY AND CREDIT (17 series)
LEADING INDICATORS (37 series)	Marginal employment adjustments (6 series)		Formation of business enterprises (2 series) New investment commitments (8 series)	Inventory investment and purchasing (7 series)	Sensitive commodity prices (1 series) Stock prices (1 series) Profits and profit margins (4 series)	Flows of money and credit (6 series) Credit difficulties (2 series)
ROUGHLY COINCIDENT INDICATORS (25 series)	Job vacancies (2 series) Comprehensive employment (3 series) Comprehensive unemployment (3 series)	Comprehensive production (3 series) Comprehensive income (2 series) Comprehensive consumption and trade (3 series)	Backlog of investment commitments (2 series)		Comprehensive wholesale prices (2 series)	Bank reserves (1 series) Money market interest rates (4 series)
LAGGING INDICATORS (11 series)	Long-duration unemployment (1 series)		Investment expenditures (2 series)	Inventories (2 series)	Unit labor costs (2 series)	Outstanding debt (2 series) Interest rates on business loans and mortgages (2 series)

Source: U.S. Department of Commerce, Bureau of Census, *Business Conditions Digest,* January 1971, p. 2.

ployment and inventory series are also in this group because they reflect early responses of the economy to changing business conditions or prospects. The reason these 12 series tend to lead the business cycle is that many reflect investment commitments and expectations for the future; therefore, they represent economic forecasts and, as such, provide one of the earliest positive signs of the direction of economic activity.

Coincidence series. Those economic indicators termed by the NBER as "roughly coincident" are fairly comprehensive measures of the current volume of *aggregate* economic activity. For example, included in this group are gross national product, industrial production, nonagricultural employment, unemployment, bank debits, retail sales, and personal income. The primary reasons the coinciders tend to change with the general movements of the economy are that all of the series (1) directly measure broad phases of business activity or (2) have something to do with financing production and distribution.

Because of the relatively unusually smooth curve of each series on this coincident list, the *turning points* in the level of economic activity for the economy as a whole normally can be identified much easier than by studying the leading series curves.

Lagging series. Generally, the six lagging series move after the coincident group. Several of the series–labor cost per dollar of real GNP and bank rates on short-term business loans–measure costs which are slow to respond to changing economic conditions. However, these factors do influence subsequent changes in expectations and investment decisions.

Summary. Table 14.2 indicates the lead or lag time in months of the 25 indicators. All of these series are graphically presented in a monthly publication of the Bureau of the Census of the U.S. Department of Commerce called *Business Conditions Digest.* The NBER series are selected to summarize and interpret the complex interactions that make up the business cycle, and their record as indicators has been impressive. Nonetheless the business analyst must recognize that judgment still plays an important role in interpreting any economic data while also recognizing the importance of these indicators as valuable tools of short-term forecasts of business activity.

1. Several findings of the NBER in this area seem worth summarizing.[1] Since 1854, expansions of aggregative economic activity have lasted longer than contractions. Expansions have averaged about 30 months and contractions about 20 months. Although there is considerable variation about these averages and economic contractions at times are longer than expansions, there exist reasons for concluding that as a

[1] See the list of NBER studies provided in the Related Readings at the end of this chapter.

consequence of shift in the structure of the economy, expansions are becoming longer than contractions.
2. Rates of growth in aggregate economic activity during expansions have been more uniform than rates of contraction. This conclusion suggests that a more reliable forecast can be made of the rate of growth at the initial stages of an expansion than of the rate of decline in a similar stage of contraction.
3. Expansion rates have been faster in the initial stages—especially during the first six months—than in the late stages of economic expansion.
4. The severity of the preceding recession or contraction affects the rate of growth during the initial stages of an expansion. Usually, after severe contractions the rates of growth during the initial stages of expansion are faster than those following a mild recession.
5. Recoveries generally exceed previous peaks much more quickly after a mild recession than after a sharp contraction.

METHODS OF ESTIMATING SEASONAL VARIATIONS

In long-term forecasting, for example, in long-range planning such as plant expansion, capital requirements, and so on, estimates of trend and cycle are useful. However, more precise estimates are required for short-term forecasts. Here, estimates of the seasonal pattern of a firm's sales prove helpful in placing orders for raw materials or scheduling production for a specific month or season. As indicated earlier, most firms' sales are subject to changes in the seasons; consequently, an analysis of the seasonal *pattern* is required in short-range forecasting.

Ratio-to-moving-average method

Various methods of computing a seasonal index are available, but to date the most commonly used is the ratio-to-moving-average method. An index of seasonal variation for one year is a *specific* seasonal; an average of several specific seasonals provides a measure of *typical* seasonal variation. If data for enough years are used, this technique has the advantage of smoothing out cyclical and irregular fluctuations. However, we must also remove trend from the original data. The ratio-to-moving-average method is the best available for averaging out C and I and eliminating T, thereby permitting the analyst to calculate a *typical* seasonal index.

Basically, five steps are involved in the computation of a seasonal index. Referring to Table 14.4, they are:

1. After arranging the data by months and by years, a 12-month moving total, Column 2, is computed. For example, 436 = sum of data for the 12 months of 1970; 441 = 436 + 35 (January 1971) − 30 (January

TABLE 14.4
Hypothetical example of computation of specific seasonal index for monthly sales of General Brands, Inc., 1970–1976, by ratio-to-moving-average method

Year and Month	(1) Original Data: Monthly Sales (millions of dollars), *TCSI*	(2) 12-Month Moving Total	(3) Centered 24-Month Moving Total, 24(*TC*)	(4) Specific Seasonal Relative as a Percentage 24(*TCSI*) ÷ 24(*TC*)
1970:				
January	30			
February	35			
March	38			
April	39			
May	42			
June	45			
July	50	436	877	137.0
August	35	441	885	94.9
September	32	444	889	86.4
October	30	445	893	80.6
November	31	448	901	82.6
December	29	453	910	76.5
1971:				
January	35	457	916	91.7
February	38	459	918	99.3
March	39	459	919	101.8
April	42	460	922	109.3
May	47	462	926	121.8
June	49	464	933	126.0
July	52	469	939	132.9
August	35	470	940	89.4
September	33	470	941	84.2
October	32	471	942	81.5
November	33	471	943	84.0
December	34	472	945	86.3
.	.	.	.	.
.	.	.	.	.
.	.	.	.	.
1976:				
January	39	514	1,030	90.9
February	40	515	1,029	93.3
March	43	518	1,033	99.9
April	46	521	1,039	106.3
May	53	525	1,046	121.6
June	54	529	1,054	123.0
July	56			
August	40			
September	39			
October	38			
November	39			
December	41			

1970); 444 = 441 + 38 (February 1971) − 35 (February 1970); and so on. This step practically eliminates S.

2. When the items averaged are an even number, as in the case at hand, a minor difficulty arises. The moving average is to be compared with the data, but the first item computed centers at neither June nor July, but at the midpoint of the year, just between the months, the irregular lengths of the months being disregarded. The second moving average centers between July and August. Actually, no serious error will result if we arbitrarily center each moving average at its seventh month. But custom dictates that another method be used.
3. Column 3 indicates this method, namely, centering Column 2. This is done by taking a 24-month moving total and centering this sum between the respective two 12-month moving totals of Column 2. The result is the same as if the estimate of *TC* were multiplied by 24. In effect, this provides an estimate of the impact of *TC* because step 1 smooths out *SI*. This may seem confusing; that is, step 1 removes S when our goal is to discover a seasonal pattern! This is necessary, of course, because it is impossible to compute a seasonal index directly by the averaging process. Seasonal variations are shorter in duration than *C* or *T* and would be ironed out. Therefore, we must first compute *TC* and then divide the original data by this estimate.
4. This brings us to our fourth step and Column 4. For example, July 1970 specific seasonal; 137.0 = (24)(50)/877 × 100; August 1970: (24)(35)/885 × 100 = 94.9. Here a ratio is computed of the original data (*TCSI*) to the estimate of *TC*, which results in *specific* seasonal relatives and an estimate of the impact of *SI*.
5. Table 14.5 indicates the final step, that is, smoothing out the irregular factor and determining our *typical* seasonal index. Here the 72 specific seasonal relatives are arrayed from high to low, and a modified mean is calculated. For example, January: 365.7 = 91.9 + 91.7 + 91.2 + 90.9; 91.4 = 365.7/4. This irons out *I* and leaves a typical seasonal pattern for sales of General Brands, Inc. Figure 14.2 indicates the seasonal index graphically. It may be noted that the peak months are May, June, and July, with September and October being the lowest.

Typical seasonal index. In order to have a seasonal pattern emerge from the calculations, the specific seasonal relatives for each month must be considered. Instead of merely selecting a specific year and saying this pattern is typical, the *SI*s are averaged to reduce the influence of *I*. In the example of Table 14.5, the irregular years are omitted from the calculations in determining the typical pattern. That is, the extreme items are omitted, both high and low, and the average thus obtained is a modified mean. One final adjustment is required. Because the average *normal* monthly index is looked upon as being equal to 100 percent, the total of

FIGURE 14.2
Seasonal index of General Brands, Inc. sales

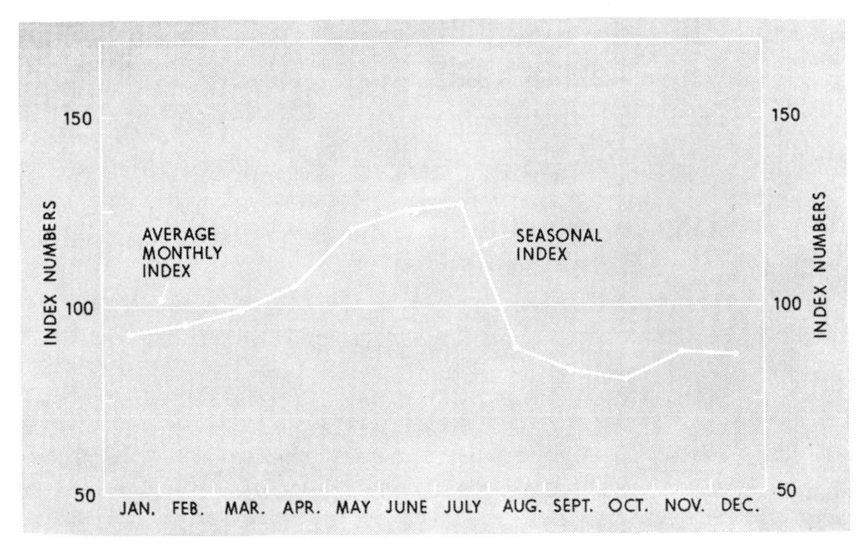

Source: Table 14.5.

the 12 modified means should be 1,200. Because of rounding, the actual total is too small; therefore a correction factor must be applied. This multiplier is obtained by dividing 1,200 by 1,198.3. In this instance, because the correction factor is greater than one, it results in slightly increasing each typical index. The final row in Table 14.5 is the typical seasonal index and may be considered to be the seasonal pattern of General Brands, Inc., that is, *typical* for *any* year.

Exponential smoothing

Naturally, the various electronic computers available today are being used to make sales forecasts in many firms. One method that makes effective use of the high-speed computers is called *exponential smoothing*. This is simply a technique of weighting past sales data and the seasonal component in a method that minimizes the forecasting error. Mathematically, exponential smoothing is nothing more or less than a weighted moving average. The aspect of this technique that gives it this special name is the fact that the weighting factors are chosen on the basis of a geometric progression. As a result, each term in the sequence is discounted according to its position in the order giving recent past events heavier weight to these items than is given to previous items. In effect, this approaches reality in that things that happened today are given more attention than events of a few days, weeks, or months ago. Currently, work is in process

TABLE 14.5
Computation of typical seasonal index for General Brands, Inc.

Rank	Jan.	Feb.	Mar.	Apr.	May	June	July	Aug.	Sept.	Oct.	Nov.	Dec.	Total
1	93.4	99.3	101.8	109.3	121.8	127.3	137.0	94.9	86.4	83.9	92.5	92.2	
2	91.9	96.1	100.8	106.3	121.7	126.2	134.8	90.8	84.2	81.5	89.5	89.1	
3	91.7	95.7	99.9	105.8	121.6	126.0	132.9	90.6	84.0	81.3	88.6	88.4	
4	91.2	95.3	99.6	105.4	120.6	125.3	131.3	89.4	83.8	81.2	84.0	86.3	
5	90.9	93.9	98.3	103.9	119.9	124.2	129.8	89.0	81.2	80.6	82.6	86.0	
6	90.8	93.3	96.1	102.9	119.8	123.0	128.0	87.9	80.3	80.1	81.2	76.5	
Modified total	365.7	381.0	398.6	421.4	483.8	501.7	528.8	359.8	333.2	324.6	344.7	349.8	
Modified mean	91.4	95.2	99.6	105.4	121.0	125.4	132.2	90.0	83.3	81.2	86.2	87.5	1,198.3
Seasonal index	91.5	95.3	99.7	105.5	121.1	125.5	132.3	90.1	83.4	81.3	86.3	87.5	1,199.5

Correction factor: $\frac{1,200.0}{1,198.3} = 1.00141867.$

to develop and improve computer-based sales forecasting methods by weighting the cycle component as well as evaluating the effect of promotional campaigns.

This forecasting procedure has been described quite clearly in such references as: Robert Brown, *Statistical Forecasting for Inventory Control* (McGraw-Hill, 1959); Peter R. Winters, "Forecasting Sales by Exponentially Weighted Moving Averages," *Management Science*, vol. VI, no. 3 (April 1960), pp. 324–42; and Robert Brown, *Smoothing, Forecasting and Prediction of Discrete Time Series* (Prentice-Hall, 1963). Some of the major advantages of exponential smoothing over more traditional methods are: (*a*) it operates efficiently on historical data; (*b*) it requires a minimum of carry-forward of data from one time period to the next; (*c*) it can be modified to accommodate data that contain distinct trend or seasonal patterns; and (*d*) it can be performed at modest cost either manually or by an electronic computer.

In terms of forecasting sales data based on historical evidence, the basic formula for first-order exponential smoothing might be expressed as:

Expected Sales for the Next Period = w (Actual Sales during the Previous Period) + $(1 - w)$ (Expected Sales for the Previous Period)

Clearly, four pieces of missing information are required to use the above equation:

a) Expected sales for period t made in period $t - 1$; $(\hat{X}_{t-1})$;
b) Actual sales for the previous period (X_t);
c) A weighting factor (w) whose value lies between zero and one; and
d) Expected sales forecast in period t for the next period $(\hat{X}_{t+1})$.

The relationship of the weighting factor and actual sales data and the forecasted sales may be stated in terms of:

$$\hat{X}_{t+1} = \hat{X}_{t-1} + w(X_t - \hat{X}_{t-1})$$

This equation merely adjusts the previous forecast by the use of the weighting factor. For example, if the previous forecast were too low, the new estimate will be equal to the last forecast adjusted upward to include some fraction of the forecast error. Obviously, the reverse would take place if the forecast proved too high.

Given the above, any new forecast results in a simple weighted average of the actual sales and the previous forecast. The equation becomes

$$\hat{X}_{t+1} = wX_t + (1 - w)\hat{X}_{t-1}$$

Perhaps, a few words about the role of the weighting factor might be appropriate at this point.[2] Clearly, if the value assigned to w is near zero,

[2] Some authors prefer the term smoothing constant to weighting factor. These same authors then use the symbol α instead of w.

the influence of the most recent *actual* sales will be relatively small. Forecasts under this situation would tend to vary little from period to period. On the other hand, if w is near one, the influence of previous sales data will be great and the forecasts will show greater fluctuations from one period to the next. The constant w, therefore, determines the relative influence that most recent actual sales will have on the new forecast. Where little historical data are available, a relatively high value (.4 or greater) of w makes the forecasts responsive to current conditions. Brown and Winters have indicated that generally the value of the weighting factor is relatively low and seldom exceeds .3. In fact, Brown has indicated that ". . . a value of .1 for w will function effectively in many situations." There is a commonsense justification for using a relatively small weighting factor because it minimizes any erratic fluctuations caused by irregular factors but at the same time allowing for a modest refinement based on current conditions. Some experts have been critical of exponential smoothing procedures on the grounds that they are empirically inadequate when major policy decisions undergo change. They also object to the assumption that differing weights should be assigned observations depending solely upon their relative recency. There are methods for determining the "optimum value" of w, and the reader is referred to the previously cited works for illustrations.

First-order exponential smoothing. To illustrate the technique of exponential smoothing let us take a close look at a first-order progression and then extend our discussion to the concept of a second-order case. The reader will recall from the early study of mathematics (also Chapter 13) that a geometric progression approaches the exponential trend as the number of terms in the progression approaches infinity. Let us assume that we have the simple geometric progression of

$$1, g, g^2, g^3, g^4, \ldots, g^{n-1}$$

If g is positive and less than 1, then each term becomes progressively smaller. For example, if $g = 1/4$ the sequence becomes

$$1, 1/4, 1/16, 1/64, 1/256, \ldots, (1/4)^{n-1}$$

Also, remember that earlier we said that exponential smoothing really is nothing more than a weighted average. If we used our geometric progression as the basis for the weighting of the items in a weighted average,[3] we would have the equation

$$\bar{X}_i = \frac{1X_i + gX_{i-1} + g^2X_{i-2} + g^3X_{i-3} + g^4X_{i-4} + \ldots + g^{n-1}X_{i-n-1}}{1 + g + g^2 + g^3 + g^4 + \ldots + g^{n-1}} \qquad (14.1)$$

[3] Recall that the formula for a weighted mean is $\bar{X}w_i = \frac{\Sigma(w_ix_i)}{\Sigma w_i}$.

The numerator is, of course, the items weighted by the geometric progression while the denominator is the *sum* of the terms in the geometric progression. (The latter replaces the sum of the weights in the usual formula for a weighted mean.) The sum of a geometric progression may be expressed mathematically as:

$$\Sigma GP = \frac{X_1(1-g)^n}{(1-g)} \tag{14.2}$$

and in our simple geometric progression $X_1 = 1$ the formula becomes

$$\Sigma GP = \frac{1-g^n}{1-g} \tag{14.3}$$

When $n \rightarrow \infty$ then g^n becomes quite small when g is less than 1. As a matter of fact, under these conditions $g^n \rightarrow 0$ in which case the numerator $1 - g^n$ approaches 1. Therefore, for a geometric progression that has an infinite number of terms the sum is

$$\Sigma GP\infty = \lim_{n\rightarrow\infty} \frac{1-g^n}{1-g} = \frac{1}{1-g} \text{ when } |g| < 1 \tag{14.4}$$

Now going back to our previous formula (14.1) for a weighted mean and substituting $1/(1-g)$ in place of the sum of the geometric progression in our numerator we obtain

$$\bar{X}_i = \frac{X_i + gX_{i-1} + g^2X_{i-2} + g^3X_{i-3} + g^4X_{i-4} + \ldots + g^{n-1}X_{i-n-1}}{1/(1-g)} \tag{14.5}$$

which reduces to

$$\bar{X}_i = (1-g)(X_i + gX_{i-1} + g^2X_{i-2} + g^3X_{i-3} + g^4X_{i-4} + \ldots + g^{n-1}X_{i-n} \tag{14.6}$$

which in a more general form can be expressed as below once $\bar{X}_i$ has been determined by multiplying both sides of the above equation by g we obtain

$$\bar{X}_i = (1-g)X_{i+1} + g\bar{X}_i \tag{14.7}$$

From this formula one can compute the value of the last exponentially weighted average plus the latest term in the sequence. As indicated previously this is nothing more than a weighted moving average, the distinction that merits special attention and a separate name stems from the special weighting technique. The exponentially moving average also *moves* in a different way. To compute an ordinary moving average we subtract the "oldest" term and add the latest one. In the exponentially smoothed average old terms are never dropped; they merely receive less weight than the new terms. Theoretically, no term is ever deleted; however, the weighting factors tend to reduce the impact rapidly. Perhaps, this is a good justification for the use of the exponential smoothing method because it actually becomes closer to reality. That is, items that are more recent (and therefore probably more vivid in our memory) are given more weight than items that are much "older."

The ultimate step in using the formula for exponential smoothing is to use the value of $\bar{X}_i$ as the estimate of the *next* term in the sequence, that is,

$$\hat{X}_{i+1} = \bar{X}_i \tag{14.8}$$

The objective of exponential smoothing is to produce a moving average that reflects significant fluctuations but ignores small random variations. The technique is useful to a broad range of business problems and is easily computed. One can change the number of terms (n) and the weighting factor (w) quickly. If you wish to fit a curve to a stable series that smooths out random or irregular fluctuations then a small value (.01) of the weighting factor (w) is chosen. However, if you wish to reflect a rapid response to an actual change in the series then a larger value of w is called for, for example, .04 or .05. This ease of changing the weighting system makes it convenient to estimate two series, namely, one stable and another responsive and then use the computer to determine which mode is desired at any point in time. The value of the weighting or smoothing constant is arrived at largely by experiment and experience. Under so-called "normal" conditions a relatively small w might be in order; however, if important changes are expected or predicted than w probably should be increased. Once a new pattern emerges then w might be decreased again.

While no attempt will be made to justify it here, mathematically it turns out that an exponentially smoothed average with a weighting factor w is statistically similar (approximately equal variability) to a moving average containing $(2/w) - 1$ terms. Table 14.6 indicates this relationship for a few selected values of w.

TABLE 14.6

Number of Terms of a Moving Average*	Weighting Factor (w) of Exponentially Weighted Average
199	.01
39	.05
19	.10
9	.20
6	.30
4	.40
3	.50

* Equally weighted with the number of terms rounded.

Let us examine how closely a nine-term moving average approximates an exponentially smoothed average with a weighting factor of .20. Consider the following sequence with a rising trend

$$X_i = 4, 5, 6, 7, 8, 9, 10, 11, 12, 13, 14, \ldots, 20$$

The computations in Table 14.7 indicate the similarity of the two methods, that is, moving averages and exponential smoothing. The forecasting rules used in the computations of Table 14.7 are

$$\hat{X}_i = \bar{X}_{i+1}$$
$$\hat{X}_i = .20X_{i-1} + (1 - .20)X_{i-2}$$

The table shows the major differences, for example, the loss of data in the moving average technique and the complete estimates of the exponential smoothing method. Because the series is relatively stable a weighting factor of $w = .20$ is not unreasonable. However, given the rising trend of the series we see that a first-order exponential smoothing is inadequate as the estimates seem to be worse as we move to the larger values of X_i.

The forecaster must keep in mind that no matter what the value of the weighting factor (w) is used the exponentially smoothed averages always lag behind a constantly rising or falling linear trend. The forecast will then contain a cumulative error unless the trend force can be estimated. If trend can be determined then one can correct the forecast

TABLE 14.7
Estimating X_i based on a nine-term moving average and exponential smoothing with $w = .20$

i	X_i	$\hat{X}_i$ (nine-term moving average)	$\hat{X}_i$ Exponential Smoothing $\hat{X}_i = .20X_{i-1} + (1 - .20)X_{i-2}$
1	4		[4]
2	5		(.2) (5) + (.8) (4) = 4.2
3	6		(.2) (6) + (.8) (4.2) = 4.56
4	7		5.05
5	8		. 5.64
6	9		6.31
7	10		. 7.05
8	11		7.84
9	12		8.67
10	13	8	. 9.54
11	14	9	10.43
12	15	10	11.34
13	16	11	12.27
14	17	12	13.22
15	18	13	14.16
16	19	14	15.13
17	20	15	(.2) (20) + (.8) (15.13) = 16.10

to eliminate the lag. The technique that can be used is called second-order exponential smoothing. This technique is merely an extension of what we have been discussing, and the new formula might be

$$\hat{X}_{i+1} = \bar{X}_i + (\text{trend correction}) \tag{14.9}$$

In effect, the originally smoothed averages are smoothed again using the same weighting factor. The difference between the first and second smoothed average is the trend factor. The forecasting equation then becomes

$$\bar{X}_{i+1} = \bar{X}_i + (\bar{X}_i - \bar{\bar{X}}_i) \tag{14.10}$$

where $\bar{\bar{X}}_1$ is the second-order exponentially smoothed average. $\bar{\bar{X}}_i$ is found by smoothing the first-order averages by the formula

$$\bar{\bar{X}}_i = w\bar{X}_i + (1 - w)\bar{\bar{X}}_{i-1} \tag{14.11}$$

A second-order exponentially smoothed average is merely a first-order smoothed again! The difference between the two is the trend correction.

Table 14.8 uses the exponential smoothing rule

$$\hat{X}_{i+1} = \bar{X}_i + (\bar{X}_i - \bar{\bar{X}}_i)$$

where $\bar{\bar{X}}_i$ is the second-order exponentially smoothed average. $\bar{\bar{X}}_i$ is found by smoothing the first-order averages by the Formula (14.11):

$$\bar{\bar{X}}_i = \bar{X}_i + (1 - w)\bar{\bar{X}}_i$$

Columns (1) and (2) of Table 14.8 are taken from Table 14.7 except that $\bar{X}_i$ is used instead of $\hat{X}_i$. Also, because $\hat{X}_i = \bar{X}_{i-1}$ the indexing also changes. We can see that our new estimates are doing a much better job of tracking the trend as our error becomes smaller.

Why stop with the second-order exponential smoothing technique if it brings us closer to the actual data than the first order? If the trend is *nonlinear* then a third-order smoothing might be useful. The approach is much the same as just described.

Higher order exponential smoothing. There are several major problems that arise when higher order exponential smoothing is used instead of simple order (first) to forecast the movement of an economic time series. First, sometimes the forecast will reflect excessive linear growth from the use of the higher order exponential. However, some analysts feel that this problem, if it exists, can be overcome adequately by using the simple exponential smoothing techniques that assume a constant level in the series. Second, the selection of the value of the weights for smoothing the economic time series is even more crucial in higher order procedures. Some research has found that *both* simple and higher order exponential smoothing cause forecasts to become more unstable as the

TABLE 14.8
Estimating X_t by second-order exponential smoothing (columns [1] and [2] are from Table 14.7)

X_t	$\bar{X}_t$	$\bar{\bar{X}}_t$	$X_{t+1} = \bar{X}_t + (\bar{X}_t - \bar{\bar{X}}_t)$
4	[4.2]	[4.2]	4.2 + (4.2 − 4.2) = 4.2
5	4.56	4.27*	4.56 + (4.56 − 4.27) = 4.85†
6	5.05	4.43	5.05 + (5.05 − 4.43) = 5.67
7	5.64	4.67	6.61
8	6.31	5.00	. 7.62
9	7.05	5.41	8.69
10	7.84	5.90	. 9.78
11	8.67	6.45	10.83
12	9.54	7.07	. 12.01
13	10.43	7.74	13.12
14	11.34	8.46	14.22
15	12.27	9.22	15.32
16	13.22	10.02	16.42
17	14.16	10.85	17.47
18	15.13	11.71	18.55
19	16.10	12.59	19.61
20	17.08	13.49	17.08 + (17.08 − 13.49) = 20.67

* $\bar{\bar{X}}_t = w\bar{X}_t + (1 - w)\bar{\bar{X}}_{t-1} = (.20)(4.56) + (.80)(4.2) = 4.27$. Other values of $\bar{\bar{X}}_t$ are similarly computed.

† Other values of $\hat{X}_{t+1}$ are similarly computed. The readers should do several of these computations to confirm their knowledge of the procedures.

value of the weights increases. Apparently, this is primarily the result of the influence of an irregular factor. In more recent years, a technique called *adaptive exponential smoothing* has been developed to minimize some of the problems alluded to above. To date, the results of this research have not found their way into the elementary textbook level and they are beyond the scope of this book. It should be noted, however, that adaptive smoothing procedures, frequently used with spectral analysis to detect a strong seasonal, depict a set of models (rather than a single model forecast) where the constant *adapts* to conform to the changes in the data. The one major advantage of such a method is that it will provide a more accurate forecast of an economic series that tends to have some abrupt irregular (random) shifts in direction.

Conclusions. In some cases, the use of the technique of exponential smoothing results in a smaller forecasting error; however, the selection of an optimal weighting factor is crucial. Where forecasts are required for relatively staple, high-volume items, the market analyst would certainly want to consider using this technique. On the other hand, exponential smoothing as a technique for prediction has very limited application for sales items that have many wide variations in expected demand. *As is* *the case with any method employed to forecast the future, the prediction is no better than the data used no matter how elaborate or complicated the mathematical procedure.*

There is a great deal more to this topic than implied in this brief discussion. Our purpose here is to introduce the student to the subject. Answers to questions such as why does the second-order exponentially smoothed average approximate a linear trend, when to use higher order smoothing, and many other matters are not covered because they generally require more time and space than can be allotted in an elementary statistics text. Students that are seriously interested in the topic and who feel they might use the technique in their research are strongly urged to study one or more of the readings cited or to elect a course that deals specifically with the subject of exponential smoothing.

As one expert has stated, "It is far better to be approximately correct than precisely wrong." Too often the mathematically oriented people forget this point in their zeal to apply their newly discovered tools.

APPLICATION OF TIME SERIES ANALYSIS

Why do business analysts break down an economic time series into its major components? Frequently, this is necessary to analyze carefully the effects of the major forces comprising the time series model. For example, the analyst studies the various components:

1. To analyze the effects of trend, cycle, or seasonal characteristics by themselves, or
2. To remove the influence of one force while analyzing the others, and/or
3. To provide a *basis* for sound forecasting and thereby facilitate intelligent executive planning and control.

Although the final section concentrates on item 3 above, Table 14.9 indicates the procedure for *removing* the influence of the seasonal upon a firm's sales. Utilizing the typical seasonal from Table 14.5, S is removed by dividing the original monthly data ($TSCI$) by this index. This leaves a sales figure which reflects the composite force of trend, cycle, and irregular. That is, Column 3 of Table 14.9 indicates that sales of General Brands, Inc. would *normally* have been $42.6 million in January 1976; however, the seasonal characteristics of this firm were in operation to *depress* actual sales $3.7 million below their "expected level." An examination of Column 3 reveals that there is much less fluctuation in the deseasonalized data than in Column 1, the original data. This is true because the effects of the seasonal have been removed. The variations which do remain are considered to be caused by trend, cycle, and irregular.

TABLE 14.9
Adjustment of General Brands, Inc. sales for seasonal variation, 1976

Month	(1) Sales (millions of dollars) $TCSI$	(2) Seasonal Index S	(3) Adjusted Sales (millions of dollars) $TCSI/S = TCI$
January	39	91.5	42.6
February	40	95.3	42.0
March	43	99.7	43.2
April	46	105.5	43.6
May	53	121.1	43.8
June	54	125.5	43.0
July	56	132.3	42.3
August	40	90.1	44.4
September	39	83.4	46.8
October	38	81.3	46.7
November	39	86.3	45.2
December	41	87.5	46.8

Long-term forecasting

When an organization is considering making a costly investment in plant and equipment or expansion of facilities through new construction, it is necessary to make *some* forecast of the future. In making long-term forecasts of a year or more in the future, the analysis of trend becomes essential.

Although we should not be misled by a mathematical formula, it does seem reasonable that where long-range forecasts are necessary, they should be based on facts, not intuition. Even a decision *not* to expand reflects some sort of forecast. So it seems logical that executives responsible for decisions such as these make them on as intelligent a basis as possible. In addition, planning for necessary financing, inventories, and raw-material purchases requires some form of a forecast. Effective budgetary control of an organization is geared to the sales forecast. Consequently, let us consider the data of General Brands, Inc. and prepare a forecast for 1977.

Referring to data from Table 13.3, it is recalled that the trend equation is

$$\hat{Y}_c = 490.6 + 14.3x$$
Origin: July 1, 1973
X in terms of years
Y in millions of dollars

Because 1977 is four years (four-step deviations) from the origin, the number 4 may be substituted for x in the above equation, as follows:

$$\hat{Y}_c = 490.6 + 14.3(4) = \$547.8 \text{ million}$$

The forecast for sales in 1977 of General Brands, Inc. would be $547.8 million *if* cycle and irregular forces were inoperative. Seasonal is of no concern in long-term forecasts using annual data. However, no organization is completely insulated from the effects of the business cycle; therefore the analysts must consider business conditions as well as trend in preparing a long-term forecast. It is impossible to predict the effect of irregular because of its random character, and it is therefore ignored here. For example, not many people can forecast the *effect and timing* of a tax cut, a soldiers' bonus, a major war, a flood, a fire, or an earthquake!

If an analysis of the general business situation reveals a composite forecast for an increase in business activity for 1977, for example, 5 percent above 1976, we may make another adjustment of our forecast:

$$\text{Trend} \times \text{Cycle} = \text{Forecast}$$

Therefore,

$$\$547.8 \text{ million} \times 1.05 = \$575.2 \text{ million}$$

The adjusted forecast, which now reflects TC, is approximately $575 million and is 10.9 percent above the 1976 level.

Short-term forecasting

This type of forecast proves necessary in planning purchases of raw materials, labor requirements, sales campaigns, and so on, for a specific month or season. In such a forecast the effects of the seasonal become significant, as well as trend and cycle. Because we are dealing with monthly data, we must refer to Table 13.5 and the trend equation in monthly terms. The required equation is

$$\hat{Y}_c = 40.95 + .1x$$

Origin: July 15, 1973
X in terms of months
Y in millions of dollars

Let us now forecast sales of General Brands, Inc. to assist in budgetary planning, first of all, for July, 1977; and second, for the summer season, April through July.

Because July 1977 is 48 months from the origin, this value is substituted in the equation for x:

$$\hat{Y}_{c(\text{July '77})} = 40.95 + .1(48) = \$45.75 \text{ million}$$

That is, *if* only trend forces were working on the time series (sales), one would reasonably expect sales of General Brands, Inc. in July 1977 to be approximately $46 million. However, because seasonal and cyclical variations do occur in monthly data, these forces must be recognized and their likely impact estimated.

Referring to Table 14.5, it is noted that the seasonal index for July is 132.3 (1.323 as a decimal fraction). Therefore the trend estimate, which also reflects the seasonal pattern, becomes

$$\text{Trend} \times \text{Seasonal} = \hat{Y}_c$$
$$\$45.75 \text{ million} \times 1.323 = \$60.53 \text{ million}$$

Assume July 1977 business activity, consequently, General Brands, Inc. sales are expected to be approximately 5 percent above the normal level for July. Therefore the above estimate is modified to include the impact of the cycle:

$$\text{Trend} \times \text{Seasonal} \times \text{Cycle} = \hat{Y}_c$$
$$\$45.75 \text{ million} \times 1.323 \times 1.05 = \$63.55 \text{ million}$$

An obvious conclusion from such a procedure is that the business analyst must constantly reappraise the situation and thereby be in a position to adjust the short-term forecast quickly and intelligently.

Utilizing the same procedure, let us prepare an estimate for the April–July season. First, estimates of trend are obtained for each month; adjust these estimates for seasonal variations, sum the monthly figures, and adjust for the changing business cycle.

April 1977 is 45 months from the origin; therefore the estimate reflecting TS is (S for April stated as a decimal is 1.055; see Table 14.5):

$$\hat{Y}_{c(\text{April '77})} = 40.95 + .1(45) = \$45.45 \text{ million}$$
$$\text{Trend} \times \text{Seasonal} = TS$$
$$\$45.45 \text{ million} \times 1.055 = \$47.95 \text{ million}$$

May 1977 is 46 months from the origin; therefore the estimate reflecting TS is (S for May stated as a decimal fraction is 1.211; see Table 14.5):

$$\hat{Y}_{c(\text{May '77})} = 40.95 + .1(46) = \$45.55 \text{ million}$$
$$\text{Trend} \times \text{Seasonal} = TS$$
$$\$45.55 \text{ million} \times 1.211 = \$55.16 \text{ million}$$

June 1977 is 47 months from the origin; therefore the estimate reflecting TS is (S for June stated as a decimal fraction is 1.256; see Table 14.5):

$$\hat{Y}_{c(\text{June '77})} = 40.95 + .1(47) = \$45.65 \text{ million}$$
$$\text{Trend} \times \text{Seasonal} = TS$$
$$\$45.65 \text{ million} \times 1.255 = \$57.29 \text{ million}$$

The July 1977 estimate computed above is $60.53 million; therefore the forecast for the April–July season is:

Month	Estimate of TS (millions of dollars)
April	47.95
May	55.16
June	57.29
July	60.53
	Σ = 220.93, or 221

Because it is expected that business conditions would be 5 percent above normal for this period, the above estimate is adjusted to reflect the impact of C, too:

$$\text{Trend} \times \text{Seasonal} \times \text{Cycle} = \hat{Y}_c$$
$$\$220.93 \text{ million} \times 1.05 = \$232.0 \text{ million}$$

Therefore the purchasing department, personnel department, and all other areas to be affected may be planning on making orders for inventories, etc., for April–July 1977 sales of approximately $232.0 million. This represents an increase of 11.1 percent over the same season of 1976.

SUMMARY

The analysis of time series has many applications in economic research. One important use indicated here is in planning for the business future, that is, forecasting, although we should not be misled by the mathematical preciseness of trend projection, for example, we should recognize the technique for what it is worth. Estimates obtained from an annual trend equation provide a *basis* for sound executive control. Certainly, some limitations must be considered. One restriction, the assumption that the trend forces will continue to operate in the future as they did in the past, is hazardous, particularly beyond the range of the data. There are other obvious drawbacks, but they still do not invalidate the method as a *foundation* for forecasting.

If a forecast is needed for a specific month or season, the annual trend equation converted into monthly terms is applicable. In addition, we must recognize the seasonal pattern in short-range forecasts; therefore the trend estimate must be multiplied by the seasonal index.

Such estimates as indicated above, where the annual trend equation is used, or where trend and seasonal are synthesized, should be looked upon as useful *first approximations.* Although much statistical research effort has gone into cycle theory, the most useful estimates of the impact of cyclical variations are obtained through careful analysis of business conditions. Consequently, these first approximations should be adjusted to reflect the effects of the changing business situation. Above all, the forecast must be stated in usable terms which are appropriate to the decisions to be made.

One final caution: Even though the analyst may be successful in predicting the effects of trend, cycle, and seasonal, some random shock may occur to invalidate the whole forecast. By their very nature, these irregular factors defy systematic prediction. Nonetheless, an understanding of the impact of trend, seasonal, and cycle can help the executive adjust his or her method of operations more effectively to such random shocks. One should also note that trend, cycle, and seasonal patterns may shift. Changing economic or political conditions, for example, a major war, may significantly alter the long-term trend or cyclical variations. Shifting the date of introducing new models of a product may change the seasonal pattern of a firm's sales.

Chapter 15 deals with simple linear correlation and regression analyses while Chapter 16 discusses the problems and techniques of multiple and partial correlation and regression. It should be noted that these methods also may be useful in economic and business forecasting. That is, correlation and regression analyses might be considered some additional tools available to assist the business analyst in an attempt to improve upon the predictive process.

SYMBOLS

Y, original data
$\hat{Y}_c$, estimated value
I, irregular component of a time series
C, cyclical component of a time series
S, seasonal component of a time series
T, trend component of a time series
$\hat{X}_{t-1}$, expected sales forecast for period t made in period $t-1$
$\hat{X}_{t+1}$, expected sales forecast for next period made in period t
X_t, actual sales for period t
w, weighting factor between 0 and 1
$\Sigma GP\infty$, sum of a geometric progression of an infinite number of terms.

PROBLEMS

14.1. On the basis of the data below, prepare (*a*) a long-term forecast of sales of Sigma Products, Inc., that is, for 1976; and (*b*) a forecast for the June, July, and August season of 1976. Interpret your forecasts.

$\hat{Y}_c = 543.2 + 20.1x$
Origin: July 1, 1969
X in terms of years
Y in terms of billions of dollars

Sigma Products, Inc. (seasonal index)

Month	Index	Month	Index
January	75.0	July	135.6
February	77.3	August	122.5
March	82.5	September	101.0
April	84.0	October	101.3
May	86.2	November	102.4
June	129.7	December	102.5

14.2 The table on the right is a list of employment in nonagricultural establishments, in tens of thousands.

Month and Year	Y	Typical Seasonal Index
1975:		
May	554	99.7
June	555	100.4
July	556	99.5
August	555	100.0
September	561	101.0
October	561	101.2
November	562	101.2
December	562	102.2
1976:		
January	563	98.5
February	565	98.3
March	567	98.9
April	569	99.1

Required:

Deseasonalize the data.

14.3. Refer to Problem 13.11, page 423.

a. Compute percentage deviations from trend for each of the two series, Instalment Credit and Noninstalment Credit, and explain the meaning of these deviations.

b. Present and compare the deviations in a separate chart.

14.4. The annual trend equation for the sales of Innisbrook Corporation is represented by the following:

$\hat{Y}_c = 576 + .00x$ $\quad$ $x =$ years

$Y =$ thousands of dollars $\quad$ Origin: July 1, 1965

a. Based on the past several years, monthly sales during January have been around $60,000. What is the typical seasonal relative for January?

b. If your seasonal relative for January was greater than 100, does this necessarily indicate that at least one of the remaining 11 months has a seasonal relative that is less than 100? Explain.

14.5. The limited partnership of Nassau & Nassau has had approximately the same annual increase in sales of $36,000 per year for the past 10 years. In 1972 the annual sales were a total of $312,000. Ignoring seasonal, irregular, and/or cyclical variations, what would be the expected monthly sales for December 1973?

14.6. The Dimeaskins Corporation had actual sales of $20,000 in October 1972. The predicted secular trend for this month was $30,000 with cyclical variations at 95 percent (a decline of 5 percent) and the seasonal index for October is 105. Comment on the corporation's performance for this month and its future effects. (Assume that the cyclical and seasonal indexes are correct.)

14.7. In the 18 months following the outbreak of war in Korea, average retail prices of all commodities moved upward and buying was marked by flurries of panic purchasing.

1950	Retail Sales (billions)	Index of Sea-sonal	Retail Prices 1947–49 = 100	1951	Retail Sales (billions)	Index of Sea-sonal	Retail Prices 1947–49 = 100
Jan. . .	$ 9.5	89	101	Jan. . .	$12.2	89	109
Feb. . . .	9.3	84	100	Feb. . . .	11.2	84	110
Mar. . . .	11.1	102	101	Mar. . . .	12.9	102	110
April . . .	11.1	97	101	Apr. . . .	11.9	97	110
May . . .	11.7	102	101	May . . .	12.7	102	111
June . . .	12.0	104	102	June . . .	12.7	104	111
July . . .	12.3	95	103	July . . .	11.5	95	111
Aug. . . .	12.7	101	104	Aug. . . .	12.5	101	111
Sept. . .	12.5	101	104	Sept. . . .	12.4	101	112
Oct. . . .	12.1	105	105	Oct. . . .	13.2	105	112
Nov. . . .	11.9	101	106	Nov. . . .	12.7	101	113
Dec. . . .	14.8	119	107	Dec. . . .	14.6	119	113

Source: Adapted from retail sales and retail price series appearing in *Survey of Current Business,* U.S. Department of Commerce, Washington, D.C.

a. You are asked to adjust the dollar volume of retail sales during this period for the influence of seasonal variation and to present your adjusted series in chart form. Can you identify any buying flurries?

b. Next, adjust the retail sales series derived in part (*a*) for changes in the price level and show this series on the same chart. How did the physical volume of retail sales change during the year and one half following June 1950?

14.8. Given that the seasonal index for the sales of a firm is 110 for a particular month, what does this mean?

14.9. "A key assumption in the 'classical' method of time series analysis is that each of the component movements in a time series can be isolated individually from a series." Do you agree with this statement? Does this assumption create any serious limitations to such analysis?

14.10. "A 12-month moving average of time series data removes trend and cycle." Do you agree? Why, or why not?

14.11. Clothing sales of the men's department in a large department store in a metropolitan area amounted to $5,000 in March; the seasonal index for March is 105, and the trend level of sales is $4,800. What is the measure of cyclical and irregular influences for March?

14.12. Frequently, gross national product data are stated in terms of "billions of dollars; quarterly data at seasonally adjusted annual rates." What does this statement mean?

14.13. Answer the following by a brief statement on each:

a. Why must short-term forecasts be more precise than long-term ones?

b. What is the major *objective* of seasonal analysis?

c. What is the statistical norm or bench mark which a given firm must use as a check against its position at any given time?

14.14. If the seasonal index of sales for a firm is 75 in the month of August, what would be your guess as to the percentage of the annual sales to be expected in August? Why?

14.15. What are the distinguishing differences between exponential smoothing and a moving average?

14.16. Distinguish between a first-order exponential smoothing and higher-order exponential smoothing.

14.17. How is the weighting factor (smoothing constant) determined in exponential smoothing forecasts?

14.18. Given the following monthly sales of the Nassau Press, Ltd. determine a *forecast* of the fourth-year sales by:

a. A month-to-month forecast using the data for the first three years. Calculate the absolute deviation for each month, that is, difference between the actual sales and your forecast.

b. Using the data for the first three years forecast sales using a two-month moving average. Again calculate the absolute deviation for each month.

c. Using a two-month weighted average where the current month is weighted 2 and the previous month 1 using the data for the first three years. Again calculate an absolute deviation for each month.

d. Using a smoothing constant or weighting factor of $w = .20$ do a first-order exponential smoothing forecast. Calculate the absolute deviation for each month.

e. Using the data from (*d*) above do a second-order exponential smoothing forecast. Again calculate the absolute deviation for each month.

f. Construct a table summarizing the results of these five forecasts by comparing maximum and minimum plus and minus deviations and the various absolute deviations. Which forecast is the best? Why?

	Sales in Thousands of Dollars			
Month	First Year	Second Year	Third Year	Fourth Year
January	1,825	1,820	2,015	2,060
February	1,510	1,730	1,750	1,770
March	1,795	2,040	2,060	2,130
April	1,780	2,100	2,020	2,190
May	1,765	1,960	2,055	2,050
June	1,700	1,945	1,890	2,185
July	1,760	2,005	1,950	2,195
August	1,820	1,790	2,010	2,130
September	1,630	1,925	1,770	2,040
October	1,790	1,860	1,940	2,150
November	1,800	2,020	1,990	2,260
December	1,560	1,930	2,000	1,970

RELATED READINGS

ANSCOMBE, R. J. "Rejection of Outliers." *Technometrics,* vol. 2, no. 2, pp. 123–47.

ASKOVITZ, S. I. "A Short-Cut Graphic Method for Fitting the Best Straight Line to a Series of Points according to the Criterion of Least Squares." *Journal of the American Statistical Association,* vol. 52, no. 277 (March 1957), pp. 13–17.

BOX, G. E. P., and JENKINS, G. M. *Time Series Analysis, Forecasting and Control.* San Francisco: Holden-Day, Inc., 1970.

BROWN, ROBERT G. *Smoothing, Forecasting, and Prediction of Discrete Time Series.* Englewood Cliffs, N.J.: Prentice-Hall, Inc., 1963.

BUTLER, WILLIAM F., and KAVESH, ROBERT A., eds. *How Business Eonomists Forecast.* Englewood Cliffs, N.J.:Prentice-Hall, Inc., 1966.

DUNN, DOUGLAS M. *Local Area Forecasting–An Adaptive Approach.* Unpublished Ph.D. dissertation, University of Michigan, Graduate School of Business Administration, 1970.

GIBSON, WELDON B. "Long Range Forecasting as a Management Tool." *Business Economics,* vol. 4 (September 1969), pp. 36–39.

JASZI, GEORGE. "Econometric Forecasting Does Work: At Least Reasonably Well." *Business Economics,* vol. 2 (Spring 1967), pp. 37–43.

MCLAUGHLIN, ROBERT L. *Time Series Forecasting.* Market Research Techniques Series, No. 6. Chicago: American Marketing Association, 1962.

MOORE, GEOFFREY H. *Business Cycle Indicators,* vol. I. New York: National Bureau of Economic Research, Inc., 1961. Contains contributions to the analysis of current business conditions.

———. *Business Cycle Indicators,* vol. II. New York: National Bureau of Economic Research, Inc., 1961. Contains basic data on cyclical indicators.

———. *Measuring Recessions.* Occasional Paper 61. New York: National Bureau of Economic Research, Inc., 1958.

———, and SHISKIN, JULIUS. *Indicators of Business Expansions and Contractions.* New York: National Bureau of Economic Research, Inc., 1967.

SHISKIN, JULIUS. *Signals of Recession and Recovery.* Occasional Paper 77. New York: National Bureau of Economic Research, Inc., 1961.

———. "The Census Bureau Seasonal Adjustment Program." *Business Economics,* vol. 4 (September 1969), pp. 71–73.

———, and EISENPRESS, HARRY. *Seasonal Adjustments by Electronic Computer Methods,* Technical Paper No. 12. New York: National Bureau of Economic Research, Inc., 1958.

TINBERGEN, JAN. *Statistical Testing of Business Cycle Theories, Vol. II: Business Cycles in the United States, 1919–1932.* New York: Agathon Press, 1968.

TRIGG, D. W., and LEACH, A. G. "Exponential Smoothing with an Adaptive Response Rate." *Operational Research Quarterly,* vol. 18, no. 1, pp. 53–59.

U.S. Department of Commerce. *Long Term Economic Growth, 1860–1970.* Washington, D.C.: U.S. Government Printing Office, 1971.

———. Business and Defense Services Administration. *Facts for Marketers and Measuring Markets.* Washington, D.C.: U.S. Government Printing Office, 1971.

———. Bureau of Domestic Commerce, *U.S. Industrial Outlook, 1975 with Projections to 1980.* Washington, D.C.: U.S. Government Printing Office, 1975.

———. Social and Economic Statistics Administration. *Social Indicators, 1973.* Washington, D.C.: U.S. Government Printing Office, 1973.

U.S. Department of Labor, Bureau of Labor Statistics. *Projections 1970, Inter-industry Relationships, Potential Demand, Employment.* Washington, D.C., U.S. Government Printing Office, 1969.

Winter, P.R. "Forecasting Sales by Exponentially Weighted Moving Averages." *Management Science,* vol. 6, no. 3, pp. 324–42.

Zarnowitz, Victor. *An Appraisal of Short-Term Economic Forecasts.* Occasional Paper 104. New York: National Bureau of Economic Research, Inc., 1967.

Part VI

Basic concepts of statistical inference—II

Science is simply common sense at its best—that is rigidly accurate in observation, and merciless to fallacy in logic.

Thomas Huxley

15

Simple linear correlation and regression analysis

THOUGH THE TECHNIQUES of correlation and regression analyses have made use of complex mathematics, the concepts are embodied in the most elementary observations of what goes on about us. For example, annual sales of a firm may vary with per capita income in the area or the level of employment, or the efficiency of employees may vary with scores on placement tests. Measures that thus tend to change in unison are said to exhibit a positive relationship or correlation. (See Figure 15.1.) On the other hand, the relation of accidents to age over 40 may be a negative association. Often, such correlations make prediction possible—as, for example, the probable efficiency of a prospective employee based on a test score.

It should be clear at this point that the analysis and interpretation of correlation and regression between two variables require a relationship which involves the pairing of observations or measurements. Thus, one might pair test scores (X) with efficiency scores (Y) of workers in a shop and attempt to detect and analyze the degree of correlation between the two variables.

It should be observed here that although the rate of technical progress has advanced rapidly, thereby making the use of desk calculators virtually obsolete for correlation and regression analysis, the basic *concepts* of this powerful tool remain unchanged. All of the computations in this chapter, as well as those in Chapter 16, are designed to accomplish an appreciation of these elementary principles rather than emphasize the techniques.

In the recent past, correlation analysis was used to imply all of the techniques currently associated with correlation and regression. Recently, more emphasis has been placed on the techniques of regression analysis, thereby requiring some distinction between the two terms.

Correlation relates to the general problem of measuring the degree of association, using such measures as the coefficient of correlation and the

FIGURE 15.1
Coincidental correlation

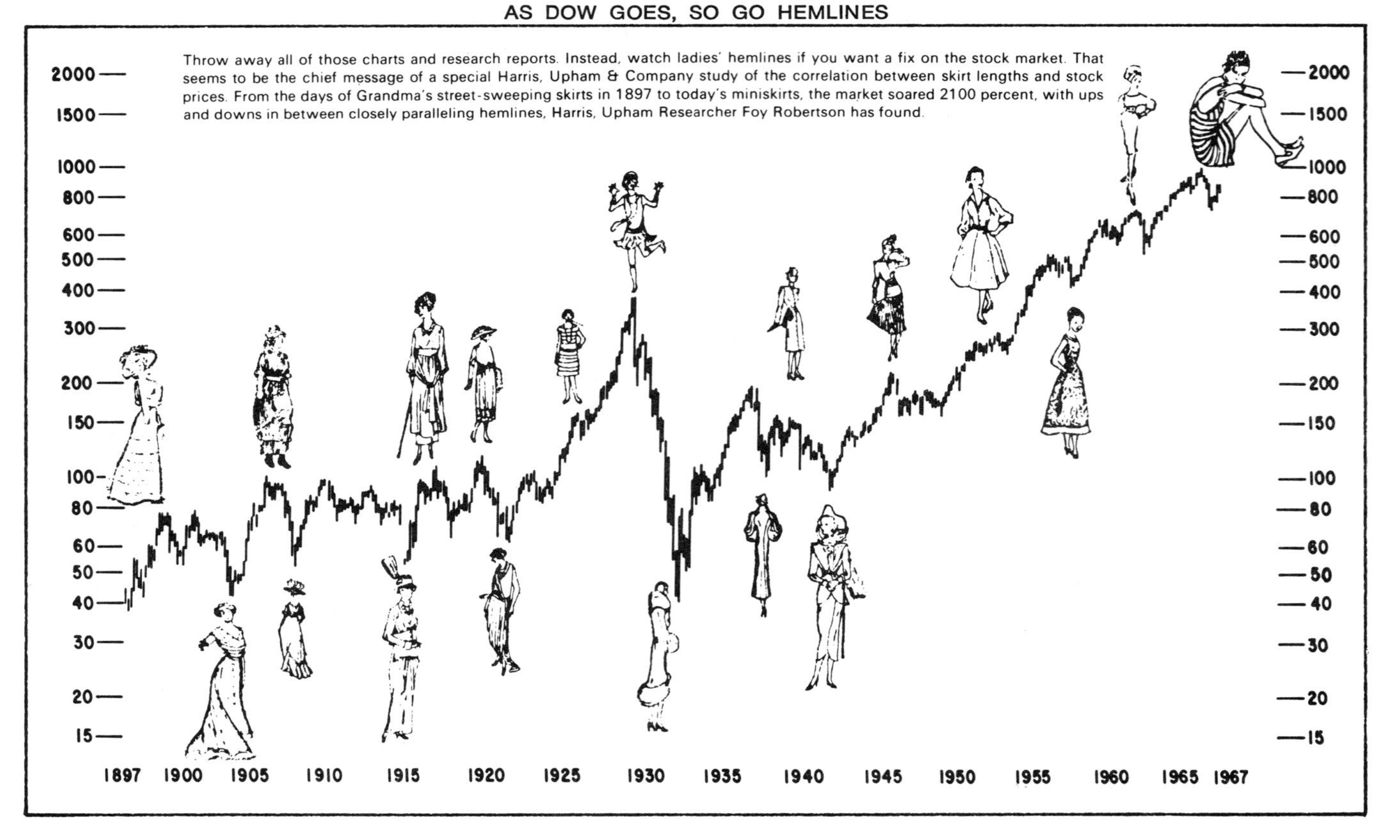

Source: *Detroit Free Press.*

coefficient of determination. In addition, the significance of the correlation is tested by using Fisher's F test or the t test. In some cases the standard error of the coefficient of correlation is determined. In any event, correlation analysis has as its *basic* objective the measurement of the degree of covariation existing between the variables.

Regression analysis utilizes the regression equation in attempting to predict the level of the dependent variable based on the level of some independent factor. In regression analysis, such measures as the net regression coefficient, the standard error of b, the standard error of estimate, and the regression equation are important. One might say that regression analysis deals with a broader and more interesting set of problems; however, correlation analysis is a useful auxiliary tool that acts as an aid in understanding regression.

The simplest economic relationship is a two-variable function, and for this reason the study of correlation and regression analysis normally begins with this model. Also, it is usually intelligent to limit the initial discussion to the linear model so the data and the computations do not get into the way of the student achieving an understanding of the concepts. For example, we may think of the quantity demanded of a given product as being a function of the price; or we may view absenteeism as a function of overtime; or production costs might be dependent upon units of output. The basic formula for a two-variable linear model is $\hat{Y}_c = a + bX$ where a and b are sample values of unknown parameters. $\hat{Y}_c$ is the estimate of the dependent variable given some level of X, the independent variable. For example, quantity demanded would be the dependent variable and price the independent. The constant a represents the height of the regression line at the Y-intercept. The slope of the line is measured by b. Both of these constants are known as regression coefficients. The above formula actually represents a sample line which might be a reflection of the true population relationship as described by $Y'_c = A + BX$.

Generally, the relationships among variables are not such that every observation falls *precisely* on the regression line; therefore, we introduce a random disturbance or error term into the equation which for a sample then becomes:

$$\hat{Y}_c = a + bX + e \tag{15.1}$$

The error term (e) might stem from three identifiable sources: (a) there might be a sampling error in the data; (b) the data may contain errors of specification; or (c) the data might be subject to measurement error. Numerically, the error term might be viewed as the residual or the difference between the predicted value and the true value of the dependent variable.

TABLE 15.1
The entrance test score (X) and efficiency test score (Y) of six employees of a company, January 1, 1976

Worker	Test X	Test Y
A	7	27
B	15	45
C	13	51
D	3	9
E	10	33
F	12	51

THE SCATTER DIAGRAM

So that the method of analysis will not be blurred by working with extensive data, a sample of six workers is used. Table 15.1 contains the entrance test score (X) and the corresponding efficiency score (Y) for each of the six workers.

When the two variables are plotted against each other, the result is called a *scatter diagram.* Figure 15.2 pictures this relationship. The chart is made by plotting as follows: $X = 7, Y = 27$; $X = 15, Y = 45$; $X = 13, Y = 51$; $X = 3, Y = 9$; $X = 10, Y = 33$; $X = 12, Y = 51$.

In correlation and regression analysis, it is customary to plot what is regarded as the independent variable on the X scale with values increasing from left to right, and the dependent variable on the Y scale with values increasing from bottom to top. It can be seen that there is a considerable degree of covariation in the two variables, a fact clearly indicated by the way in which the Y values generally increase as X values increase.

Figure 15.3 shows four possible patterns which should help in the understanding of the nature of linear relationships.

The fact that the data in Figure 15.2 do not fall into a precise pattern of a straight line as in Figure 15.3 A and B, in effect, states the first problem of correlation, which is the measurement of the extent to which covariation does exist. That is, to what extent do the test scores and efficiency scores of Table 15.1 vary together?

DECOMPOSITION OF VARIANCE

There are three kinds of variance of considerable importance in correlation and regression analysis. They include:

1. *The Total Variance.* This is the variation of the actual data, the dependent variable Y, from their mean, $\bar{Y}$. This variation is represented by

FIGURE 15.2
Scatter diagram of entrance test scores (X) and later efficiency scores (Y) of six workers

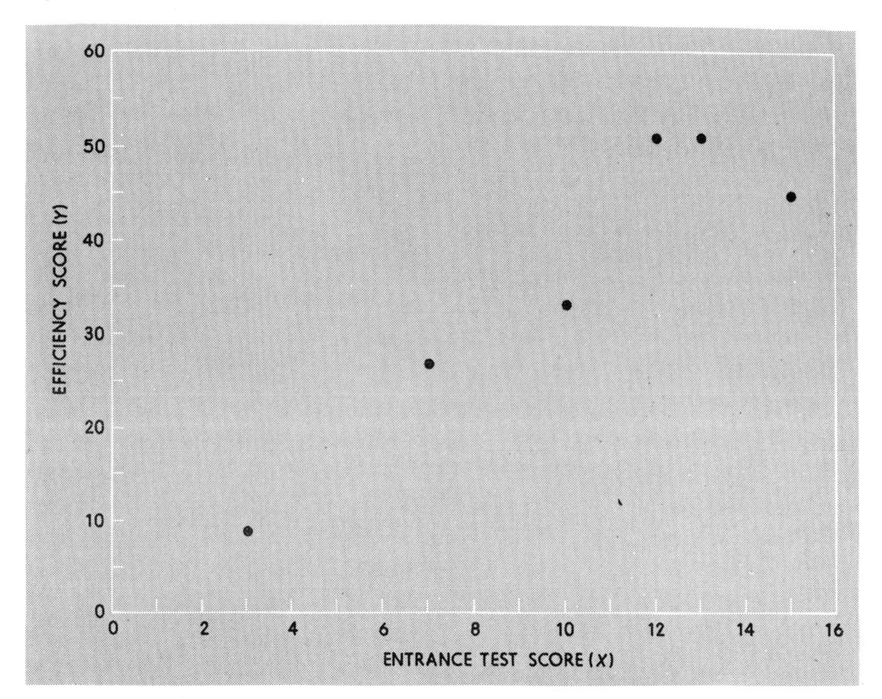

the standard deviation of the Y values, σ_y. The square of this measure, σ_y^2, is called the total variance. The deviations involved here are $\Sigma(Y - \bar{Y})^2$.

2. *The Unexplained Variance.* This is the variation of the dots in the scatter diagram from the regression line. This dispersion is measured by the square of the standard error of the estimate, σ_{yx}^2. That is, the deviations involved here are $\Sigma(Y - \hat{Y}_c)^2$.
3. *The Explained Variance.* This is the variation of the predicted values of the dependent variable ($\hat{Y}_c$) from their mean ($\bar{Y}_c$). Since $\Sigma Y = \Sigma \hat{Y}_c$, then $\bar{Y} = \bar{Y}_c$. This is called the explained deviation, $\sqrt{\Sigma(\hat{Y}_c - \bar{Y}_c)^2}$, and is described by the standard deviation of the predicted values, σ_{y_c}. The square of this measure, $\sigma_{y_c}^2$, is termed the explained variance.

Figure 15.4 presents these concepts by using only *one* dot representing an actual point, in order not to cloud the picture. The distance between points A and C represents the variation involved in determining the total variance to be explained. The distance between A and B graphically represents the variation in the dependent variable (Y) that is un-

FIGURE 15.3
Correlation patterns

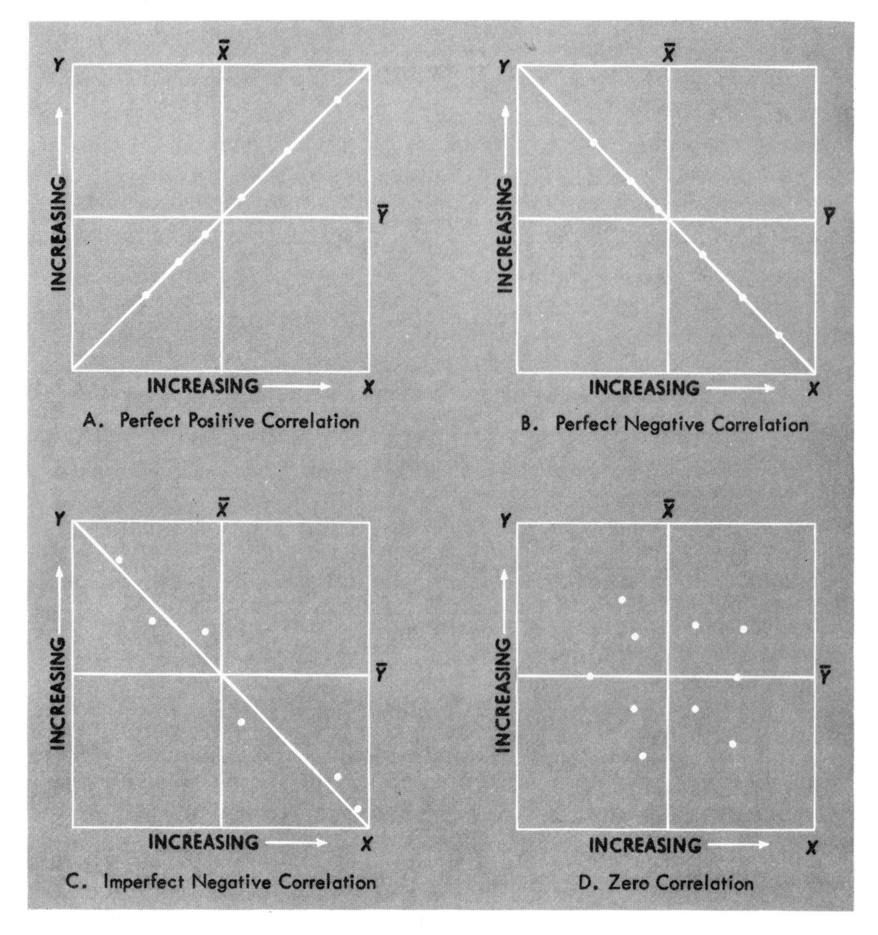

explained by variation in the independent variable (X). The unexplained variance, then, is the deviation of the dots from the regression line. To some extent, this measure indicates the degree of reliability of the regression equation. The distance between B and C represents the variation in the dependent variable that is accounted for by variation in the independent variable. Later, we discover that a ratio of the explained variance to the total variance provides a measure of the degree of correlation.

Some important relationships exist among these variances. For example,

$$\sigma_y^2 = \sigma_{yx}^2 + \sigma_{y_c}^2 \tag{15.2}$$

FIGURE 15.4
Decomposition of variance

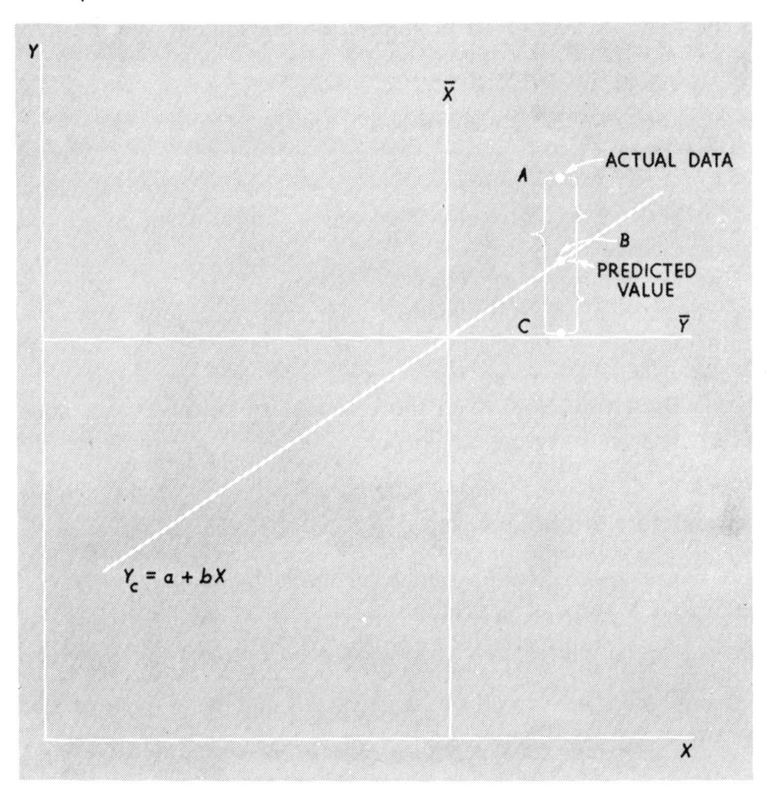

That is, the total variance equals the unexplained plus the explained. And

$$\rho^2 = \frac{\sigma^2_{y_c}}{\sigma^2_y} \qquad (15.3)$$

which is the population coefficient of determination and indicates the percentage of the variation in Y that is accounted for by variation in X. Also, if

$$\sigma^2_{yx} = \sigma^2_y \quad \text{then} \quad \sigma^2_{y_c} = 0$$

and, $\rho^2 = 0$, or no correlation exists. And if

$$\sigma^2_{yx} = 0 \quad \text{then} \quad \sigma^2_y = \sigma^2_{y_c}$$

then $\rho^2 = 1.0$, and the relationship is perfect. That is to say, X is a perfect predictor of Y.

CORRELATION, REGRESSION, AND PREDICTION

The methods employed in computing correlation may be most easily understood if they are related to the purpose which regression analysis usually serves. This purpose is prediction. Simple correlation essentially involves fitting a line or curve called the *regression* to data. Correlation may involve complex regression fitting; but here, we shall consider only the linear or straight-line regression. As here computed, it assumes that one or both of the correlative series is a sample from an approximately normal distribution.

The limited data used consist of scores (X) in preliminary entrance tests given to six machine operators who apply for jobs. These scores are compared with scores made in a later efficiency test (Y) with a view to determining whether the former were adequate for approximately predicting the latter. If so, then the preliminary entrance tests constitute a useful means of selecting employees.

Approximate prediction

A crude approximation might be made by dealing merely with the mean and a standard deviation (Table 15.2). If all the entrance test scores (X) were practically alike, that is, close to $\bar{X} = 10$, then it might be assumed that the later efficiency score of a given worker who made a reasonable X score would be likely to fall, about 68 times in 100, within the range

$$\bar{Y} \pm \hat{\sigma}_y = 36 \pm 16.4$$

TABLE 15.2
Approximate prediction based on means and a standard deviation of entrance test scores (X) and later efficiency scores (Y)

Worker	X	Y	Y^2
D	3	9	81
A	7	27	729
E	10	33	1,089
F	12	51	2,601
C	13	51	2,601
B	15	45	2,025
Sums	60	216	9,126
	$\bar{X} = 10$	$\bar{Y} = 36$	$\bar{Y}\Sigma Y = 7,776$
			$\Sigma y^2 = 1,350$

$$\hat{\sigma}_y = \sqrt{\frac{1,350}{6-1}} = 16.4$$

where

$$\hat{\sigma}_y = \sqrt{\frac{\Sigma y^2}{n-1}} = \sqrt{\frac{1{,}350}{5}} = 16.4 \qquad (15.4)$$

or between the limits 19.6 and 52.4. This is too wide a range for useful prediction, so a more accurate method is sought. (See Figure 15.5.)

A second approximation that promises some improvement consists of grouping the workers into two classes according to the merits of their entrance scores (X) and testing the prediction range for each group (Table 15.3). For a worker who made X score, at or below the average, the prediction range is

$$\bar{Y} \pm \hat{\sigma}_y = 23 \pm 12.5$$

where

$$\hat{\sigma}_y = \sqrt{\frac{\Sigma y^2}{n-1}} = \sqrt{\frac{312}{2}} = 12.5$$

FIGURE 15.5

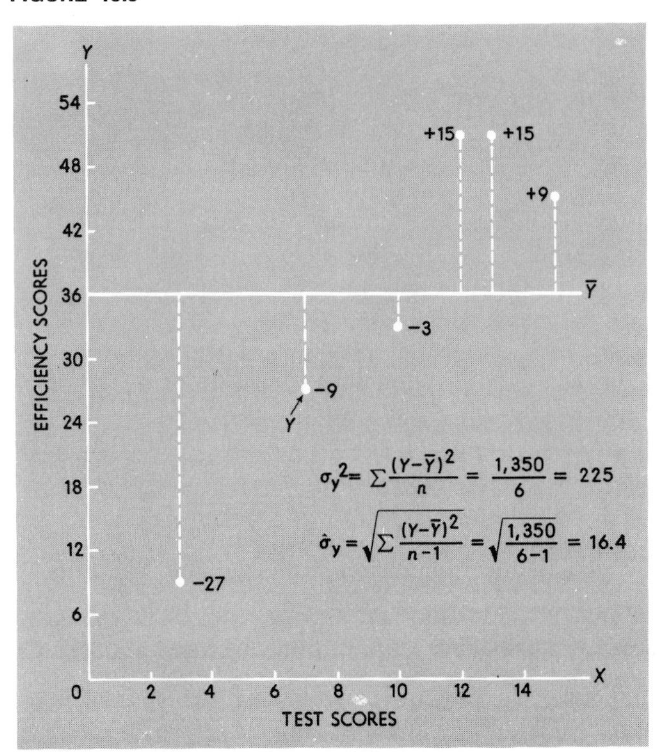

TABLE 15.3
Approximate prediction based on means and standard deviation of entrance test scores (X) and later efficiency scores (Y) grouping workers into two classes according to the merits of their entrance test scores

	Class Having Test Score Equal to or Less than the Average Test Score			Class Having Test Score above the Average Test Score		
	X	Y	Y^2	X	Y	Y^2
	3	9	81	12	51	2,601
	7	27	729	13	51	2,601
	10	33	1,089	15	45	2,025
Sums	20	69	1,899	40	147	7,227
		$\bar{Y} = 23$	$\bar{Y}\Sigma Y = 1,587$		$\bar{Y} = 49$	$\bar{Y}\Sigma Y = 7,203$
			$\Sigma y^2 = 312$			$\Sigma y^2 = 24$

$$\hat{\sigma}_y = \sqrt{\frac{312}{3-1}} = 12.5 \qquad \hat{\sigma}_y = \sqrt{\frac{24}{3-1}} = 3.5$$

$$\hat{\sigma}_y = \frac{(12.5 + 3.5)}{2} = 8.00$$

And for those above the average X score the range is

$$\bar{Y} \pm \hat{\sigma}_y = 49 \pm 3.5$$

where

$$\hat{\sigma}_y = \sqrt{\frac{\Sigma y^2}{n-1}} = \sqrt{\frac{24}{2}} = 3.5$$

The average prediction range is now ± 8 [i.e., $(12.5 + 3.5)/2$] as compared with ± 16.4 previously obtained—a considerable improvement.

This second approximation, however, suggests that further improvement might be made. More adequate data could be broken down into several groups for predictive purposes. The result would be several "moving averages," with a standard deviation range attached to each. But a smoothed moving average would be practically a regression line, which would include all given values of X and others that might be interpolated. And a unified standard deviation of the regression line as a whole—the so-called "standard error" of estimate σ_{yx}—could then be computed. This, in fact, is the method actually used in regression procedure; and in many cases, linear regressions fit the data satisfactorily.

Prediction by regression: Computation of least-squares line

By simultaneously solving the following two normal equations, the least-squares straight line may be fitted to data.

$$\Sigma Y = na + b\Sigma X \tag{15.5}$$
$$\Sigma XY = a\Sigma X + b\Sigma X^2 \tag{15.6}$$

That is, for a first-degree curve involving two constants in the regression equation, we need only two normal equations. For a parabola or a multiple regression, we would need three or more equations. We may determine ΣX, ΣY, ΣXY, and ΣX^2, and then compute the constants a and b by simultaneously solving the formulas above.

However, a simplified procedure may be used by the solution of the equations determined by the centering process. That is, if we look upon the observations as deviations from their means, the variables may be written as x and y, where

$$x = X - \bar{X} \qquad \text{and} \qquad y = Y - \bar{Y} \tag{15.7}$$

The normal equations then become

$$\Sigma y = na + b\Sigma x \tag{15.8}$$
$$\Sigma xy = a\Sigma x + b\Sigma x^2 \tag{15.9}$$

But in Chapter 5, we found that $\Sigma x = 0$ and $\Sigma y = 0$; therefore the *centered* equations for the constants are

$$a = 0 \tag{15.10}$$

$$b = \frac{\Sigma xy}{\Sigma x^2} \tag{15.11}$$

The first equation merely says that the regression line passes through the intersection of the axes at $\bar{X}$ and $\bar{Y}$. In actual practice the constant a is found by dividing the first original normal equation by n, as follows:

$$\Sigma Y = na + b\Sigma X$$

dividing by n:

$$\frac{\Sigma Y}{n} = \frac{na}{n} + \frac{b\Sigma X}{n}$$

Therefore,

$$\bar{Y} = a + b\bar{X}$$

and

$$\text{a} = \bar{Y} - b\bar{X} \tag{15.12}$$

The procedure in Table 15.4 follows that of trend fitting discussed in Chapters 13 and 14 and calls for only a brief explanation here. As in the

TABLE 15.4
Linear regression of entrance test scores (X) of six machine operators and later efficiency test (Y)

Worker	X	Y	X^2	XY	Y^2	$\hat{Y}_c$
A	7	27	49	189	729	25.5
B	15	45	225	675	2,025	53.5
C	13	51	169	663	2,601	46.5
D	3	9	9	27	81	11.5
E	10	33	100	330	1,089	36.0
F	12	51	144	612	2,601	43.0
Sums	60	216	696	2,496	9,126	216.0
	$\bar{X} = 10$	$\bar{Y} = 36$	$\bar{X}\Sigma X = 600$	$\bar{X}\Sigma Y = 2{,}160$	$\bar{Y}\Sigma Y = 7{,}776$	
			$\Sigma x^2 = 96$	$\Sigma xy = 336$	$\Sigma y^2 = 1{,}350$	

$$b = \frac{\Sigma xy}{\Sigma x^2} = \frac{336}{96} = 3.5$$

$$a = \bar{Y} - b\bar{X} = 36 - 3.5\,(10) = 1$$

$$\hat{Y}_c = 1 + 3.5X$$

$$r^2 = \frac{b\Sigma xy}{\Sigma y^2}$$

$$r^2 = \frac{(3.5)(336)}{1{,}350}$$

$$r^2 = \frac{1176}{1350} = .87$$

case of any linear trend, the centered sums[1] Σxy and Σx^2 are first required, since their ratio measures the slope of the regression line. Of course, in the case at hand, where both the means are round numbers, the deviations x and y could conveniently have been written thus:[2]

$$\begin{aligned}
x&: -3,\ 5,\ 3,\ -7,\ 0,\ 2;\ \Sigma x^2 = 96\\
y&: -9,\ 9,\ 15,\ -27,\ -3,\ 15;\ \Sigma y^2 = 1{,}350\\
xy&: 27,\ 45,\ 45,\ 189,\ 0,\ 30;\ \Sigma xy = 336
\end{aligned}$$

[1] Computation is facilitated by recognizing that

$$\begin{aligned}
\Sigma x^2 &= \Sigma(X-\bar{X})^2 = \Sigma X^2 - 2\bar{X}\Sigma X + n\bar{X}^2 = \Sigma X^2 - \bar{X}\Sigma X\\
\Sigma y^2 &= \Sigma(Y-\bar{Y})^2 = \Sigma Y^2 - 2\bar{Y}\Sigma Y + n\bar{Y}^2 = \Sigma Y^2 - \bar{Y}\Sigma Y\\
\Sigma xy &= \Sigma[(Y-\bar{Y})(X-\bar{X})] = \Sigma XY - \bar{Y}\Sigma X - \bar{X}\Sigma Y + n\bar{X}\bar{Y}\\
&= \Sigma XY - 2\bar{X}\Sigma Y + \bar{X}\Sigma Y = \Sigma XY - \bar{X}\Sigma Y
\end{aligned}$$

[2] In strict theory the units employed in correlations are x/s_x and y/s_y, as given below. These are complete abstractions rather than dollars, number of persons, etc. In fact, a term such as ΣX^2 cannot be logically interpreted except as it is taken as a function of abstract numbers, for what is the meaning of, say, \$10 or 500 persons, squared! It is seldom necessary, however, to work in terms of x/s_x and y/s_y. The actual procedures utilized in correlation give the same results as would be obtained by the more tedious methods of strict logic. Data of Table 15.2 in s units are

$$\begin{aligned}
x/s_x&: -1.25,\ -.75,\ -.50,\ .00,\ .75,\ 1.75;\ \Sigma(x/s_x)^2 = 6 = n\\
y/s_y&: -.60,\ -1.00,\ -1.00,\ .20,\ .60,\ 1.80;\ \Sigma(y/s_y)^2 = 6 = n
\end{aligned}$$

The cross products total 5.6, an average of .933, which, as shown later, is the product-moment coefficient of correlation, r. Note that $\Sigma(x^2/s_x^2) = \Sigma x^2/(\Sigma x^2/n) = n$.

But such a direct method is seldom economical. Generally, it is more convenient to cumulate the squares and cross products on a machine without writing the itemized columns. They appear in the example merely to demonstrate the method. Then the correction terms are employed, that is,

$$\Sigma x^2 = \Sigma X^2 - \bar{X}\Sigma X$$
$$\Sigma y^2 = \Sigma Y^2 - \bar{Y}\Sigma Y$$
$$\Sigma xy = \Sigma XY - \bar{X}\Sigma Y$$

or some variation of this method of centering.

The regression equation, to be used in prediction, namely,

$$\hat{Y}_c = a + bX = (\bar{Y} - b\bar{X}) + (\Sigma xy/\Sigma x^2)X$$

is computed as

$$\hat{Y}_c = 1 + 3.5X$$

which may be applied to other values of X within the limits of the data, or perhaps tentatively a little beyond. In the case here considered, for example, a new preliminary score of $X = 10$ would indicate a probable efficiency score of $\hat{Y}_c = 1 + 3.5(10) = 36$. But as a general rule it should be recognized that the validity of a prediction equation diminishes rapidly close to or beyond the limits of the data to which it is fitted. (See Figure 15.6.)

Figure 15.7 presents the relationship in three dimensions as a means of better understanding correlation and regression. It is called a bivariate normal distribution because of the assumption that the conditional distribution of X or of Y is always normal. We might view these as six separate distributions with the same standard deviation. Each distribution has different expected values (see Table 15.5 for the $\hat{Y}_c$ values). Along the X axis we measure the entrance test scores and along the Y axis is the efficiency score. Along the third axis are given the densities of the six distributions. (Actually, we do not know what these values are because the populations were not reported.) The expected values lie on a straight line through the constant a. This is the regression line and is sort of a dynamic mean. That is, on the condition that X has a certain value Y assumes a value on the regression line according to the relationship. The conditional probability distributions represent the fact that not always when X is 7, Y is 26 or that when X is 3, Y is 12. As a matter of fact these values are distributed normally about the expected value ($\hat{Y}_c$) with a standard deviation called the standard error of estimate (σ_{yx}).

Assumptions of correlation and regression analysis

Several theoretical assumptions are necessary regarding the simple linear model and the conditional probability distributions of the de-

FIGURE 15.6

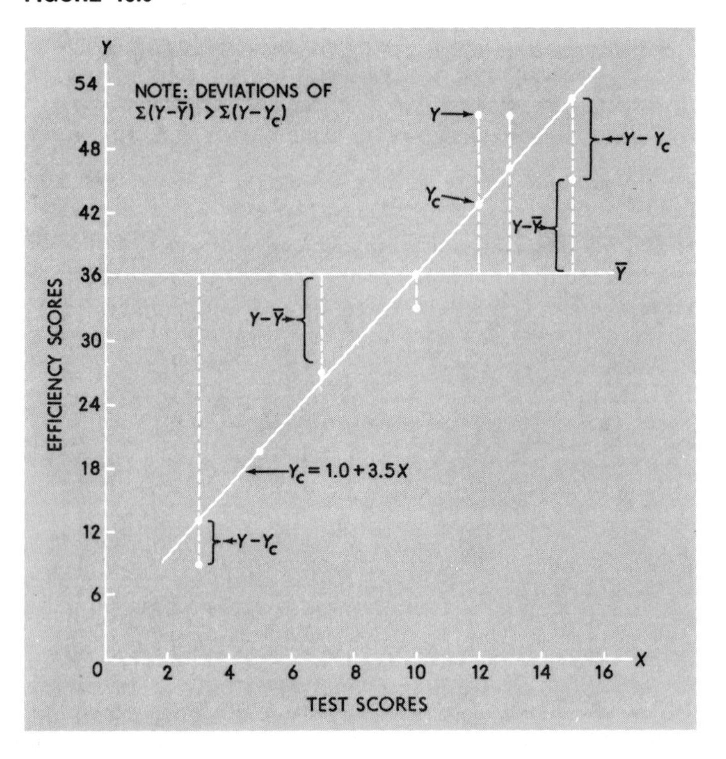

pendent variable. These assumptions are: (1) that successive observations are independent; (2) that the conditional probability distributions of Y given X have the same standard deviations no matter what is the value of X (see Figure 15.7); (3) that the means of these conditional probability distributions lie on the same straight line described by the equation $Y' = A + BX$ which is the expression for the true (population) regression line; (4) that the value of X is known in advance; and (5) that the conditional probability distributions for Y are normal.

The independence assumption is the usual one made for a random sample and is one of the reasons that correlation of time series presents some unusual problems. Observations over time usually are not completely independent; that is, the level of GNP in one quarter is not independent of what GNP was in the previous quarter. Some transformations must be made to overcome this assumption. The second assumption is required because if all standard deviations of all of the conditional probability distributions were not assumed to be equal then one could not make use of the standard error of the estimate in predicting values of Y based upon values of X. One would need to have a different standard error for each distribution. Estimation of these standard errors would be

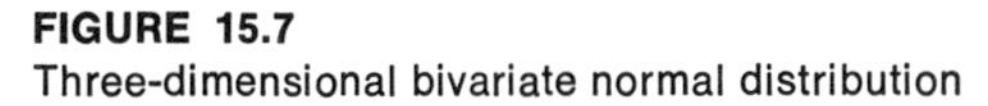

FIGURE 15.7
Three-dimensional bivariate normal distribution

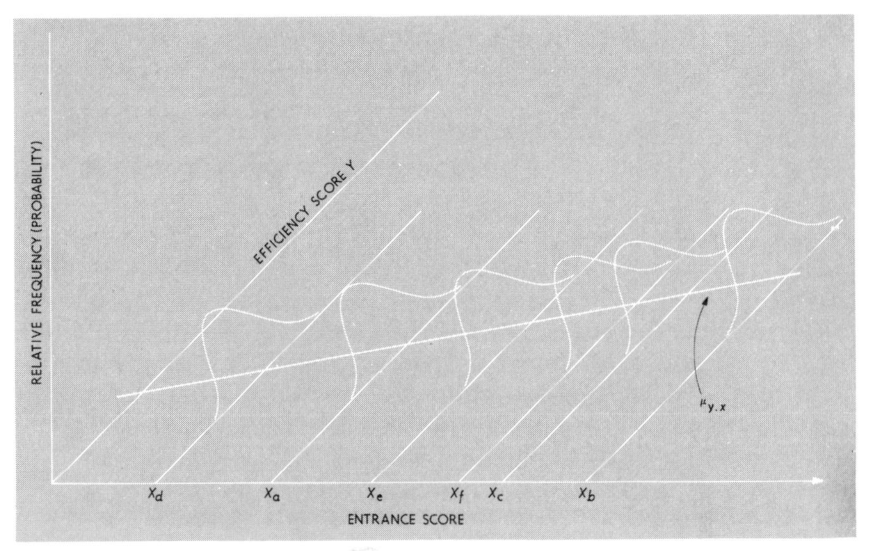

impracticable and probably impossible. The third assumption relates to the fact that the linear model requires that the expected values (the means) of all of the distributions lie on a straight line. The fourth assumption merely states that if you do not know the value of X no estimate of Y can be made. The final restriction on the simple linear regression model would provide inaccurate information if the conditional probability distributions are not normal. Another theoretical distribution might be more appropriate and might indicate that nonparametric correlation might be useful instead of the Pearsonian methods of this chapter. (For a discussion of nonparametric statistical methods please see Chapter 17.)

Standard error of estimate

The 68 percent probability prediction range, represented by the standard deviations previously computed, may now be estimated as a standard deviation of the differences $d = Y - \hat{Y}_c$, as computed in Table 15.5 and indicated graphically in Figure 15.8. If computed merely for the sample, this standard deviation (s_{yx}) is $\sqrt{\Sigma d^2/n}$, but for prediction purposes, allowance must be made for the universe of paired X's and Y's, of which the data are merely a small sample. This allowance is made by dividing Σd^2, where d is $Y - \hat{Y}_c$, by the degrees of freedom, $n - 2$. The two degrees lost are determined by the fact that such deviations appear only

TABLE 15.5
Direct computation of the standard error of the estimate $\hat{\sigma}_{yx}$

	Y	$\hat{Y}_c$	$Y - \hat{Y}_c$	$(Y - \hat{Y}_c)^2$
	27	25.5	1.5	2.25
	45	53.5	−8.5	72.25
	51	46.5	4.5	20.25
	9	11.5	−2.5	6.25
	33	36.0	−3.0	9.00
	51	43.0	8.0	64.00
Sums	216	216.0	0.0	174.00

$$\hat{\sigma}_{yx} = \sqrt{\frac{\Sigma(Y - \hat{Y}_c)^2}{n - 2}} = \sqrt{\frac{174.00}{4}} = \sqrt{43.5} = 6.6$$

after two X, Y points have tentatively indicated a regression, so that the count begins ordinarily when n is 3. This prediction standard deviation, as noted earlier, is called the estimated standard error of estimate $\hat{\sigma}_{yx}$. It is here found to be approximately 6.6 units vertically above and below the regression line; consequently, the prediction at $X = 10$, referred to above, may be written as the 68 percent probability range, $\hat{Y}_c = 36 \pm 6.6$, or between the limits 29.4 and 42.6.

It should be noted that $\Sigma(Y - \hat{Y}_c)^2$ is a minimum, since this is a least-squares line. That is, as measured by the squared deviations, the least-squares line is graphically closer to the data than any other line that might be drawn. If we let

$$Y'_c = A + BX$$

be the *population* regression line,[3] then

$$\Sigma(Y - \hat{Y}_c)^2 < \Sigma(Y - Y'_c)^2$$

If we knew Y'_c we would divide by n, but we must estimate Y'_c with $\hat{Y}_c$, and this leads to a smaller variation which is based on the estimates of a of A and b of B. Thus, we decrease the denominator to $n - 2$, losing one degree of freedom for each of our estimates a and b upon which $\hat{Y}_c$ is based.

Prediction based upon a regression line and the standard error of estimate proves to be a considerable improvement over the crude estimates based simply on Y standard deviations of one or two groups, as follows:

---

[3] Generally, in simple linear regression the population regression line is

$$Y'_c = \alpha + \beta X$$

where α and β are used to denote the Y intercept and slope, respectively. However, because α and β were used in Chapter 11 to indicate the probabilities of Type I and Type II errors, to avoid confusion we use A and B for the regression parameters.

FIGURE 15.8
Correlation of scores in preliminary tests (X) with later efficiency scores (Y) of six workers

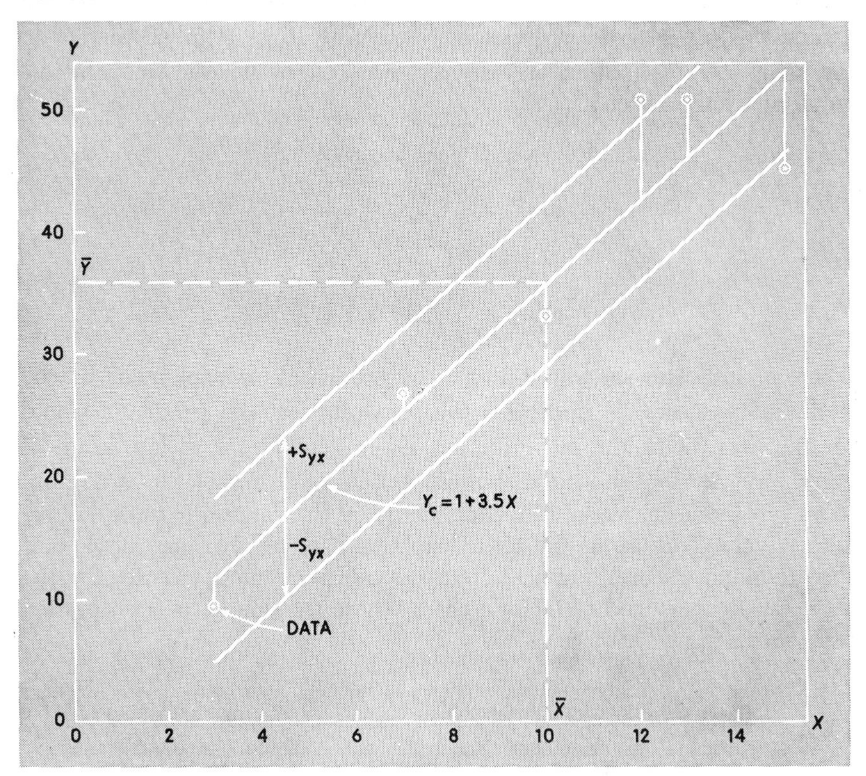

1. By Y standard deviation only: $\pm\ \hat{\sigma}_y \ = \pm\ 16.4$
2. Same, but by two groups: $\pm\ \hat{\sigma}_y \ = \pm\ \ 8.0$
3. By standard error of estimate: $\pm\ \hat{\sigma}_{yx} = \pm\ \ 6.6$

That the range of $\pm\hat{\sigma}_{yx}$ is expected to be less than the range of $\pm\hat{\sigma}_y$ when the regression has an appreciable slope is indicated by an equation that is true of a least-squares regression equation, namely,[4]

[4] It may be noted in Table 15.4 that Σy_c^2 as computed equaled $b\Sigma xy$, and that $\Sigma y^2 = \Sigma y_c + \Sigma d^2$. These equations, based on centered x and y, are readily proved thus: By definition, $y_c = bx$; hence, $\Sigma y_c^2 = b^2\Sigma x^2$. But $b = \Sigma xy/\Sigma x^2$; therefore, $\Sigma y^2 = b\Sigma xy$. Then, write $d = y - y_c$. Squaring, summing, and adding Σy_c^2 to each side of the equation gives

$$\Sigma y_c^2 + \Sigma d^2 = \Sigma y^2 - 2\Sigma yy_c + 2\Sigma y_c^2$$

But the last two terms cancel, since $y_c = bx$, $yy_c = bxy$, and $\Sigma yy_c = b\Sigma xy = \Sigma y_c^2$.

$$\Sigma d^2 = \Sigma y^2 - \Sigma y_c^2 \tag{15.14}$$

where in computing $\hat{\sigma}_{yx}$, the numerator of the variance, $\hat{\sigma}_y^2$, is reduced by a function (Σy_c^2) of the regression (see Table 15.4). Applied merely to the sample, the equation of centered squares, transposed, may be divided by n and written as

$$\frac{\Sigma y^2}{n} = \frac{\Sigma y_c^2}{n} + \frac{\Sigma d^2}{n}$$

or

$$\hat{\sigma}_y^2 = \hat{\sigma}_{yc}^2 + \hat{\sigma}_{yx}^2; \ (225 = 196 + 29)$$

where $\hat{\sigma}_y^2$ indicates the variability to be predicted or "explained" by the preliminary scores, $\hat{\sigma}_{yc}^2$ indicates the variability of the predictions, and $\hat{\sigma}_{yx}^2$ indicates the failure of prediction. Or in general, for various kinds of correlation, it may be said that these variances represent the variability to be explained, the variability explained, and the variability unexplained, respectively. The use of the term "explained" would be more obvious if the preliminary scores had been, let us say, experience in months. But the term can be stretched to include cases where the emphasis is on prediction. (See Figure 15.9.)

FIGURE 15.9

DESCRIPTIVE MEASURES

VARIABLE	n	MEAN	STD DEV	SE OF MEAN	MINIMUM	MAXIMUM
CPBT	16	79.281	25.613	6.4032	49.700	141.00
GPP	16	597.54	113.65	28.414	433.40	775.90

ANALYSIS OF VARIANCE OF CPBT n = 16

SOURCE	DF	SUM OF SQRS	MEAN SQUARE	F-STATISTIC	SIGNIF
REGRESSION	1	8200.4	8200.4	70.011	.0000
ERROR	14	1639.8	117.13		
TOTAL	15	9840.3			

MULTIPLE R= .91288 R-SQR= .83335 SE= 10.823

VARIABLE	PARTIAL	COEFFICIENT	STD ERROR	T-STATISTIC	SIGNIF
CONSTANT		-43.645	14.936	-2.9216	.0112
GPP	.91288	.20572	.24586 -1	8.3672	.0000

In summary then, our prediction of $\hat{Y}_c$ is as follows:

Using Y Only	Using X and Y Regression
$\hat{\sigma}_y = 16.4$	$\hat{\sigma}_{yx} = 6.6$
$\bar{Y} \pm \hat{\sigma}_y = 36 \pm 16.4$	$\bar{Y}_c \pm \hat{\sigma}_{yx} = 36 \pm 6.6$
20–52	29–43, where $\bar{X} = 10$

Given the regression of X and Y the coefficient of determination, r^2, is .87 and the coefficient of nondetermination, k^2, is .13. That is, only 13 percent of the variation in X is left unexplained by the variation in Y. Table 15.6 summarizes the analysis of variance for a simple linear relationship in general and then uses the data generated by the regression of X and Y discussed above.

TABLE 15.6
Analysis of variance for simple linear regression model

	General		Data from Regression of X and Y	
Source of Variation	Variation	Variance	Variation	Variance
Explained by regression	$\Sigma(\hat{Y}_c - \bar{Y}_c)^2$ or $b\Sigma xy$	$\sigma^2_{y_c}$	1,176.00	196.00
Unexplained by regression	$\Sigma(Y - \hat{Y}_c)^2$	σ^2_{yx}	174.00	29.00
Total	$\Sigma(Y - \bar{Y})^2$	σ^2_y	1,350.00	225.00

COEFFICIENT OF CORRELATION

Though the chief purpose of regression as applied in the field of business is prediction, there is a place for a coefficient or indicator that states in a single figure the degree of relationship—or predictability—that exists between two sets of data. For example, when production increases in the ascending phase of the business cycle, a percentage index, suggesting the likelihood that wholesale prices would also rise, would be convenient. In earlier times of freer markets the coefficient of correlation of production and prices was commonly

$$r = \frac{\Sigma xy}{n\sigma_x\sigma_y} = .80 \text{ or more} \tag{15.15}$$

representing a high probability that the two cycle movements would agree.

As thus utilized, the coefficient of correlation, r, is, *in effect*, a percentage figure; however, this kind of interpretation generally is reserved for the coefficient of determination, r^2. If there is perfect agreement between two series expressed in standard deviation units, then r, computed as described below, will be 1.00 (i.e., 100 percent). If there is exact disagreement, one moving up when the other moves down, the computed coefficient will be -1.00 (i.e., -100 percent). Various degrees of agreement or disagreement will register on the scale between these two extremes—a coefficient of zero meaning that no relationship is registered.

Pearsonian product moment *r*

The coefficient r was first popularized by the famous English statistician, Karl Pearson (1857–1936). The term "product moment" refers to what are commonly called the cross products which when centered are expressed as Σxy. When reduced to standard deviation units, the average cross products would constitute r as Pearson conceived it, but this is not the most convenient or the most logical approach.

Ratio of variances

The most obvious basis for a coefficient of correlation that would have the effect of a percentage is a comparison of the so-called "explained" (or predictable) variability with the total variability of the y series (see Table 15.7). In terms of variances, r could then be taken as

$$r^2 = \frac{\Sigma y_c^2}{\Sigma y^2}, \quad r = \sqrt{\frac{\Sigma y_c^2}{\Sigma y^2}} \qquad (15.16)$$

This squared coefficient, r^2, is sometimes called the coefficient of *determination.* By contrast, the corresponding ratio of unexplained variability,

TABLE 15.7
Direct computation of $\Sigma y_c^2 = \Sigma(\hat{Y}_c - \bar{Y}_c)$

	$\hat{Y}_c$	$\hat{Y}_c - \bar{Y}_c$	$(\hat{Y}_c - \bar{Y}_c)^2$
	25.5	−10.5	110.25
	53.5	17.5	306.25
	46.5	10.5	110.25
	11.5	−24.5	600.25
	36.0	.0	.00
	43.0	7.0	49.00
Sums	216.0	00.0	1,176.00

$$\sigma_{y_c}^2 = \frac{1,176}{6} = 196$$

$$r^2 = \frac{1,176}{1,350} = .871$$

$k^2 = \Sigma d^2/\Sigma y^2$, is called the coefficient of nondetermination,[5] and the square roots are the coefficient of correlation and of noncorrelation, respectively. To summarize:

Squares:

Coefficient of determination: $r^2 = \Sigma y_c^2/\Sigma y^2$

Coefficient of nondetermination: $k^2 = \Sigma d^2/\Sigma y^2$

$$r^2 + k^2 = \Sigma y^2/\Sigma y^2 = 1$$

Square roots:[6]

Coefficient of correlation: $r = \sigma_{yc}/\sigma_y$

Coefficient of noncorrelation: $k = \sigma_{yx}/\sigma_y$

$$r = \sqrt{1 - k^2}$$

where $d = Y - \hat{Y}_c$. The symbol k is sometimes called the coefficient of *alienation.* However, in practice the only terms and symbols commonly used are r, the coefficient of correlation, and the square coefficient, r^2.

It should be recalled at this point that in computing correlation, the basic statistics are Σx^2, Σxy, and Σy^2, as indicated in Table 15.4. The slope b is $\Sigma xy/\Sigma x^2$; the constant a is $\bar{Y} - b\bar{X}$; Σy_c^2 is $b\Sigma xy$; and $\Sigma d^2 = \Sigma y^2 - \Sigma y_c^2$. For prediction, the standard error of estimate, $\hat{\sigma}_{yx}$, is $\sqrt{\Sigma d^2/(n-2)}$. And the degree of covariation is often computed most conveniently by finding $\Sigma y_c^2 = b\Sigma xy$, as just stated, and dividing by Σy^2. That is,

$$r^2 = \frac{\Sigma y_c^2}{\Sigma y^2} = \frac{b\Sigma xy}{\Sigma y^2} \qquad (15.17)$$

and

$$r = \sqrt{\frac{b\Sigma xy}{\Sigma y^2}}$$

Various formulas for *r*

It is worth noting, however, that r^2 and r may be computed by a number of interchangeable formulas. To begin with, the product-moment formula, involving the use of centered x and y in standard deviation units, may be written

[5] This symbol k should not be confused with its common use as a constant. As here used, it was originally a name initial, k for Kelley, like F for Fisher.

[6] The statistics r and k are related as two sides of a right triangle with an hypotenuse of unity. The same relationship holds for σ_y and σ_d with hypotenuse σ_y. When r^2 is at its maximum of unity, obviously k is zero. c

$$r = \frac{\Sigma xy}{n\sigma_x\sigma_y} = \frac{\Sigma xy}{\sqrt{\Sigma x^2 \Sigma y^2}} \tag{15.18}$$

and r^2 may be written

$$r^2 = \frac{(\Sigma xy)^2}{\Sigma x^2 \Sigma y^2} = \frac{\Sigma xy}{\Sigma x^2} \cdot \frac{\Sigma xy}{\Sigma y^2} = b_{yx}b_{xy} \tag{15.19}$$

where b_{yx} (y on x) is the slope previously computed (i.e., $\Sigma xy/\Sigma x^2$), and b_{xy}–as discussed later–is the slope that would be obtained if Y were taken as the base (i.e., $\Sigma xy/\Sigma y^2$). From the various formulas thus obtained, the one that is most convenient for the problem at hand will, of course, be chosen.

The algebra of the transformation of these formulas may be indicated thus:

$$r = \Sigma\left(\frac{x}{\sigma_x}\right)\left(\frac{y}{\sigma_y}\right)/n = \frac{\Sigma xy}{n\sigma_x\sigma_y}$$

But

$$\sigma_x = \sqrt{\frac{\Sigma x^2}{n}}, \quad \sigma_y = \sqrt{\frac{\Sigma y^2}{n}}, \quad \sigma_x\sigma_y = \frac{1}{n}\sqrt{\Sigma x^2 \Sigma y^2}$$

Hence, substituting in the preceding transformation,

$$r = \frac{\Sigma xy}{\sqrt{\Sigma x^2 \Sigma y^2}} \tag{15.20}$$

which, written as a square, is readily transformed into the equation for r^2 first given above, thus:

$$r^2 = \frac{(\Sigma xy)^2}{\Sigma x^2 \Sigma y^2} = \frac{\Sigma xy}{\Sigma x^2} \cdot \frac{\Sigma xy}{\Sigma y^2} = \frac{b\Sigma xy}{\Sigma y^2} = \frac{\Sigma y_c^2}{\Sigma y^2}$$

Also, since $\Sigma y_c^2 = \Sigma_y^2 - \Sigma_y{}^2 - \Sigma_d{}^2$ the last form may be written

$$r^2 = \frac{\Sigma y^2 - \Sigma d^2}{\Sigma y^2} = 1 - k^2 \tag{15.21}$$

SIGNIFICANCE OF CORRELATION

As in many problems, the computation of a correlation coefficient raises the question of statistical significance. Is a given correlation, with its prediction formula, based upon a large enough sample so that the populations may be reasonably assumed to be correlated? The answer to this question can involve the use of tables of t and F.

The F table in question is general in nature, involving the probability of the ratio of two related mean squares occurring by chance. In the case of correlation the two related mean squares are Σy_c^2 and Σd^2, each divided by its own degrees of freedom. For Σy_c^2—the summed centered squares of the regression items—the degrees of freedom reduce to one. For it is rationalized that in this case we draw samples from a population of paired X, Y items lying anywhere on the extended trend line. The data—the frame of reference—furnish directly one of these points, namely, at the means, $\bar{X}$ and $\bar{Y}$. Hence, one other sample regression point is effective in determining the line, and no others are needed. Significance, therefore, may be indicated by the formula[7]

$$F = \frac{\dfrac{\Sigma y_c^2}{1}}{\dfrac{\Sigma d^2}{(n-2)}} = \frac{\Sigma y_c^2}{\Sigma d^2}(n-2) \tag{15.22}$$

which will increase with the validity of the correlation and the number of paired items.

The formula for F as given above may be transformed by writing the ratio $\dfrac{\Sigma y_c^2}{\Sigma d^2}$ as $\dfrac{r^2}{1-r^2}$, making the equation

$$F = \frac{r^2}{1-r^2}(n-2) \tag{15.23}$$

The transformation is readily verified if we write the ratio in question as

$$\frac{\dfrac{\Sigma y_c^2}{\Sigma y^2}}{\dfrac{\Sigma d^2}{\Sigma y^2}} = \frac{r^2}{k^2} = \frac{r^2}{1-r^2}$$

Quite often, the significance of linear correlation is measured by means of the statistic t, which is merely the square root of F.[8] But whether F or t

[7] If the fitted regression line is a parabola of the second degree, $\hat{Y}_c = a + bX + cX^2$ (or higher), where three or more points are required to determine the trend (three or more terms in the regression equation), these required points may be expressed by m, and the mean squares are $m - 1$ and $n - m$, respectively. Hence the general equation for F as here used is

$$F = \frac{\dfrac{\Sigma y_c^2}{(m-1)}}{\dfrac{\Sigma d^2}{(n-m)}} = \frac{\Sigma y_c^2 \cdot n - m}{\Sigma d^2 \cdot m - 1}$$

which is applicable to simple linear correlation (where $m = 2$).

[8]
$$F = \frac{r^2}{1-r^2} \cdot n - 2$$
$$t = \sqrt{F} = \frac{r\sqrt{n-2}}{\sqrt{1-r^2}}$$

is employed, significance at the 5 or 1 percent level is read from the appropriate table. What level is accepted as satisfactory depends somewhat upon the nature of the problem. It can be said, however, that the standard commonly accepted designates the 5 percent level as significant and the 1 percent level as highly significant.[9]

Reference to Table 15.7 will show that r^2 there computed is .871. Substituting this value in the F equation,

$$F = \frac{r^2}{1 - r^2}(n - 2) = \frac{.871}{1 - .871}(4) = 27.00$$

The tabulated 5 percent level for $n - 2$ degrees of freedom is 7.71 (see Appendix F), and the 1 percent level is 21.20. These are the F's that would be expected to arise by mere chance from similar samples of uncorrelated populations once in 20 times or once in 100 times, respectively. Hence the odds that an F of 27.00 has arise merely by chance are very small, and the correlation may be regarded as highly significant from the standpoint of statistical sampling.

The initial hypothesis is that correlation in the population is zero. A significant F is the basis for rejecting this hypothesis.

Another way of testing the significance of r is through the use of the table in Appendix G (page 683). This table provides the values of the correlation coefficient for different levels of significance and is utilized as follows. The left column is merely the sample size adjusted for the degrees of freedom, or $n - m$, where m is the number of constants in the regression equation. For our example $n = 6$, $m = 2$, $r = .93$; and reading from the table on page 683, we find that our correlation is significant at all three levels.

Still another technique for appraising the relationship makes use of the standard error of the coefficient of correlation (σ_r). The formula for this measure of dispersion of the sampling distribution of r is

$$\sigma_r = \frac{1 - r^2}{\sqrt{n - m}} \qquad (15.24)$$

Once the standard error has been computed, it is used in setting up a confidence interval for the population coefficient in a manner similar to the techniques of Chapter 9. Generally, it is not considered valid to use the σ_r unless the sample size is greater than 50 and $r < .90$. Otherwise, it is better to use the z transformation. Why? Because z tends to normalize the skewness of the distribution of the correlation measures. That is, the distribution of r in small samples from correlated populations is skewed, and its standard error is difficult to estimate. However, we can transform

[9] These levels are arbitrarily selected. Sometimes the 2 percent level is used, and at other times the 10 percent level.

sample r's into a quantity z whose sampling distribution is nearly normal. A good approximation of the standard error of z is

$$\sigma_z = \frac{1}{\sqrt{n-3}} \tag{15.25}$$

and for small-size samples, it is considered desirable to add the quantity

$$\frac{r}{2(n-1)}$$

to the value of z. Where z may be found,

$$z = 1.1513 \log \frac{1+r}{1-r} \tag{15.26}$$

or it may be determined from a nomograph.[10]

The sampling distribution of z is approximately normal even from correlated universes; consequently, the probability distribution of the differences of sample z's is almost normal. Therefore, we can use the z transformation (and Appendix I) to test whether two sample r's are significantly different or whether the difference occurred by chance.

INTERVAL ESTIMATE FOR *B*

Frequently, the analyst is interested in estimating the slope of the *population* regression line for its own sake, rather than merely as a means of prediction of a level of the dependent variable. The regression equation for the population is

$$Y'_c = A + BX$$

where A equals the universe equivalent of the sample a, and B is the parameter analogous to b, the slope of the sample line. We recall that

$$\hat{Y}_c = a + bx$$

really is a *sample* regression line and is only one of many possible; therefore, we may be interested in estimating the universe slope.

Our point estimate of B is the value of b, or 3.5 (see Table 15.4). Clearly, with a different sample, b will vary, and we should know something about the sampling distribution of the regression coefficient if we are to construct an interval estimate of B.

The mean of the sampling distribution of b is B; therefore, b is an unbiased estimate of B. (See Figure 15.10.)

---

[10] A form of line chart upon which appear scales for the variables involved in a particular formula in such a way that corresponding values for each variable lie on a straight line which intersects all of the scales. In nonelementary statistical work the nomograph is not as extensively used as are tables.

FIGURE 15.10
Sampling distribution of b where $\mu = B$

DESCRIPTIVE MEASURES

VARIABLE	n	MEAN	STD DEV	SE OF MEAN	MINIMUM	MAXIMUM
FDIFCPBT	15	5.9267	9.1297	2.3573	-10.900	23.500
FDIFGPP	15	21.573	17.864	4.6124	-19.900	45.100

ANALYSIS OF VARIANCE OF FDIFCPBT   n = 15

SOURCE	DF	SUM OF SQRS	MEAN SQUARE	F-STATISTIC	SIGNIF
REGRESSION	1	194.56	194.56	2.6012	.1308
ERROR	13	972.37	74.797		
TOTAL	14	1166.9			

MULTIPLE R= .40833   R-SQR= .16673   SE= 8.6485

VARIABLE	PARTIAL	COEFFICIENT	STD ERROR	T-STATISTIC	SIGNIF
CONSTANT		1.4246	3.5747	.39853	.6967
FDIFGPP	.40833	.20869	.12939	1.6128	.1308

The standard deviation of this sampling distribution of b, that is, the true standard error of the net regression coefficient, is

$$\sigma_b = \frac{\sigma_{yx}}{\sqrt{\Sigma X^2 - \dfrac{(\Sigma X)^2}{N}}} \tag{15.27}$$

where ΣX^2, $(\Sigma X)^2$, and N are *universe* values.

From the centering process we find that

$$\frac{(\Sigma X)^2}{N} = (\Sigma X)\frac{(\Sigma X)}{N} = \bar{X}\Sigma X$$

and

$$\Sigma X^2 - \bar{X}\Sigma X = \Sigma x^2$$

so the formula for the standard error of b might also be expressed as

$$\sigma_b = \frac{\sigma_{yx}}{\sqrt{\Sigma X^2 - \bar{X}\Sigma X}} \quad \text{or} \quad \sigma_b = \frac{\sigma_{yx}}{\sqrt{\Sigma x^2}} \tag{15.28}$$

Because the sampling distribution is nearly normal, the confidence interval for B is the usual form for large samples:

$$b \pm z\sigma_b$$

Because we do not know the true standard error of the estimate (σ_{yx}), we substitute our best estimate ($\hat{\sigma}_{yx}$) in our formula. This gives

$$\hat{\sigma}_b = \frac{\hat{\sigma}_{yx}}{\sqrt{\Sigma X^2 - \dfrac{(\Sigma X)^2}{n}}} \qquad (15.29)$$

where $\hat{\sigma}_b$ is a point estimate of σ_b and the data relate to the sample.

Using the data of Table 15.4, we have

$$\hat{\sigma}_b = \frac{6.6}{\sqrt{696 - \dfrac{(60)^2}{6}}}$$

$$\hat{\sigma}_b = .67$$

Because $n = 6$ and we substituted $\hat{\sigma}_{yx}$ for σ_{yx}, the t distribution is applicable, and our interval estimate of B is

$$b \pm t\hat{\sigma}_b$$

And using a .95 confidence coefficient, our interval is

$$3.5 \pm 2.776(.67)$$
$$1.64\text{–}5.36$$

That is, the range of the average productivity associated with each unit of X is between 1.64 and 5.36. On the average, the productivity increase associated with each unit change in X is 3.5.

DUAL REGRESSION AND PREDICTION

If Figure 15.8 (data of Table 15.4) were turned 90 degrees counterclockwise so that the Y scale as base is made to appear the independent variable (though read backwards), a new regression line would be required. That is, whereas on the original base (Y on X),

$$b_{yx} = \frac{\Sigma xy}{\Sigma x^2} = \frac{336}{96} = 3.5$$

$$r^2 = \frac{(b)\,(\Sigma xy)}{\Sigma y^2} = \frac{(3.5)\,(336)}{1{,}350} = .87 \qquad (15.30)$$

on the alternate Y base (X on Y),

$$b_{xy} = \frac{\Sigma xy}{\Sigma y^2} = \frac{336}{1{,}350} = .248$$

$$r^2 = \frac{(.248)(336)}{96} = .87$$

Note: The slope of the regression line (b) changes as well as the values of Σy^2 and Σx^2; that is, Σy^2 becomes Σx^2. Therefore, we note that while the slope of the regression line changes the value of the coefficient of determination remains the same (.87).

At first glance, this might be thought to be the same slope as before, viewed from a different angle. But if so, the two slopes b_{yx} and b_{xy} should be reciprocals, which is contrary to fact. In order to compare the two, the new slope b_{xy} may be expressed as read on the X axis by taking its reciprocal thus:

$$\frac{1}{b_{xy}} = \frac{1}{.248} = 4.03226$$

which could have been computed as $\Sigma xy/\Sigma y^2$. Comparison with b_{yx}, computed as $\Sigma xy/\Sigma x^2 = 3.5$, shows that it is steeper than the latter, which would be expected, since zero correlation would make each regression line horizontal to the axis from which it is computed.

In the case here discussed (Table 15.4), prediction is naturally from the preliminary tests (X) to measured efficiency (Y), but cases will sometimes arise where a reverse prediction is desired. So, for the sake of illustration, we shall assume that it is desirable to estimate the preliminary score of a worker whose efficiency score is 27.

It is hardly necessary to discuss in detail the method of reverse prediction, since it can most simply be done by mentally or actually interchanging the labels of the data (read X as Y and Y as X) and using the same formulas as before. This gives the reversed regression, or prediction equation base Y, as

$$\hat{X}_c = 3.022 + .248Y$$

which, for $Y = 27$, yields the predicted efficiency score of $\hat{X}_c = 9.74$. The estimated standard error is found to be $\hat{\sigma}_{xy} = 1.76$, making the 68 percent probability range 9.74 ± 1.76, or between the X limits, 7.98 and 11.50.

There is a certain paradoxical inconsistency involved in prediction as reversed by the dual regressions. It was noted earlier that a preliminary test of $X = 10$ resulted in an efficiency prediction score of $Y = 27$. But it had just been shown that the efficiency score of $Y = 27$ implies a predicted preliminary test of 9.74. The two results are apparently contradictory.

An examination of correlation procedure will show that the paradox of the dual prediction lines arises quite naturally out of the orthogonal, or perpendicular, relationship of a given X to a predicted Y, or the reverse. The requirement of this relationship is readily seen by reference to a chart such as Figure 15.8. When a regression is fitted by the least-squares method, it is natural to think of any given data point and associated regression point as located on the same ordinate, directly above the given X, and parallel to the Y axis. But when the Y axis is the base, the deviations are taken parallel to the X axis. Consequently, two different regressions are obtained.

The divergence of the two regression lines is obviously a function of the degree of correlation. If correlation is perfect ($r^2 = 1.00$), all data points are on the regression line, which is exactly the same whether computed on the X axis or the Y axis. But as correlation diminishes, the divergence of the two regressions increases. As was noted earlier, the maximum divergence comes when $r = 0$. This relationship between correlation and the divergence of the regressions is registered most directly in the equation $r^2 = b_{yx}b_{xy}$, which becomes zero when the two lines are perpendicular to each other, since each slope on its own axis is zero.

CORRELATION OF TIME SERIES: SOME WARNINGS

When data ordered in time are correlated, certain unique problems are encountered, especially if one wishes to apply any error formulas or tests of significance. The two major problems are: (1) the nonindependence of successive observations and (2) the effect of the relationship of long-term trend upon the correlation coefficient. No meaningful generalizations may be made unless these shortcomings are corrected. There are methods available which will in large part correct these conditions, but a detailed discussion is beyond the scope of this text. A few adjustments will be mentioned here and an example presented which illustrates the problems.

The formulas for reliability, that is, the standard error of estimate and the general appraisal of the correlation coefficient by F, are based upon the theory of random sampling and a normally distributed population. Chapter 3 stated that random sampling assumes that each observation in a sample is selected purely at random from all the items in the universe. That is, each possible item has an equal opportunity of being chosen. The theory further assumes that successive samples are chosen in such a manner that values found in one sample have no relation or connection with the values found in the next sample. However, it is apparent that the level of the Gross National Product (GNP), or any other economic time series, in any one given year is not completely independent of that of the previous year. Rather it is probably near the level established in the preceding year adjusted by new or changing factors. This internal correlation between members of a series of observations ordered in time is called *autocorrelation*.[11] This autocorrelation provides sort of a "built-in" association that may make any literal interpretation of the

[11] Some writers refer to this internal correlation between members of a series as *serial* correlation. To further confuse the issue, Durbin-Watson uses the term serial correlation to refer to the relationship between two independent variables.

coefficient of determination meaningless unless some adjustments are made of the raw data. The second problem with respect to the impact of trend on the relationship can be compensated for by one of the techniques mentioned below.

Correlation of actual time series data. If one correlates the actual raw data the relationship might still be useful for long-term forecasting. The regression equation will reflect the combined effects of trend and cycle. No reliability statement about either the regression equation or the coefficient of determination would be in order. The only recourse the analyst might have is to select only those time series that have a *logical* close relationship or to make one of the adjustments below.

Correlation of first differences. As a short-term forecasting tool the analyst may wish to correlate the first differences of the data. While absolute differences may be used, generally it is preferred to correlate the percent differences from year to year. This will eliminate all trend influences except for one year and avoids any possible error in the selection of trend. The percent differences are desired because the percentages usually have a uniform distribution. In addition, if absolute differences are used in the calculations the figures generally become larger in later years simply because of the growth factor. These extremes can effect the correlation measures importantly. However, sometimes the absolute first differences will produce *less* distortion as a measure of the rate of change than will relative first differences. This is particularly true if both series increase or decrease as a function of time.

Correlation of logarithms of data. If one is not certain of the type of trend in the data, or if the analyst is more interested in the relative relationship, then the use of the logs of the data might be considered. Generally, the common logs to the base 10 are used; however, sometimes using the natural logs is more effective in removing the multicollinearity effects. This is particularly true of some financial data, for example, price changes of warrants or common stocks. Using logs tends to remove most of the influence of any trend effects without having to identify the specific trend. In many respects, this technique is much like using relative first differences of the data. In fact, some analysts prefer to use the first differences of the logs in order to study the relative relationships. The analyst must remember that the b value will be stated in terms of logs so the antilog must be found in order to state the slope of the regression line in terms of a percentage.

Sometimes the use of logs will make it possible to use the linear model when the original data might be curvilinear in its relationship. For example, if $C = A \cdot B$ then $\log C = \log A + \log B$ which is linear. Also where the data values show great variation then the use of logs will tend to normalize the distribution. That is, if you have four observations with values of 10, 100, 100, and 1,000 the arithmetic mean is 302.5 which is skewed to the right. If we convert the original data to logs we obtain a normal distribution which helps us meet an important assumption of correlation analysis. It might be noted here that whether or not logs are used the analyst may plot the residuals (the difference between the actual data values and the estimated) as a histogram to determine if the distribution is normal to satisfy the linear model assumption.

Correlation of cycles. If the economist is interested in measuring the effects of short-run business cycle changes, it frequently is desirable to correlate the percentages of trend (i.e., CIs). When time series data unadjusted for trend are correlated, the coefficient of correlation reflects the relationship between the fluctuations and also that between the trends of the two series. Thus, if one is interested in measuring the cyclical association between two series, trend must be eliminated. Then the selection of the appropriate trend becomes crucial. The reliability of two or more cycles may be measured whenever deviations from trend approximate normal distributions. Then the cycles are considered direct measures of certain variable forces and the correlation indicates the degree to which one series covaries with another, *above what might have been expected by chance.* Many times the removal of the trend factor also eliminates any curvilinear effect that might be present. This makes the use of a linear relationship more appropriate in such cases.

Correlation of per capita data. If the study of *economic* relationships is most important, the correlation of per capita data is recommended, the reason being that population growth is the dominant component in many dependent variables of interest to the economist. Using per capital data tends to remove the strong influence of the growth of the economy due simply to the trend of population changes.

Time as separate independent variable. Where the influence of trend (time) is of interest the analyst might consider a multiple regression with time as one of the independent variables. The regression coefficients then reflect the separate influence of each independent variable, as well as trend, on the dependent variable.

The data below were correlated several times with and without some sort of an adjustment.[12] First the raw data of Table 15.8 were correlated, and no statement of the significance of r^2 is appropriate. All that can be said with respect to this relationship is that the coefficient of determination between Gross Private Product (GPP) and Corporate Profits before Taxes (CPBT) is about .83. The coefficient of correlation, r, is .91. Table 15.9 and Figure 15.11 summarize this relationship. A few words about MIDAS notation might be appropriate here. The descriptive portion of Table 15.9 seems clear enough except for the term "SE OF MEAN."

---

[12] The computer program used is called MIDAS and was written by Daniel Fox and Kenneth E. Guire of the staff of the Statistical Research Laboratory of the University of Michigan. Computations were run on the IBM 370/168.

TABLE 15.8
Correlation of time series data, Gross Private Product (X), and Corporate Profits, before Taxes (Y) (billions of dollars)

Year	X GPP*	Y CPBT
1959	433.4	52.1
1960	444.0	49.7
1961	452.3	50.3
1962	482.9	55.4
1963	503.2	59.4
1964	532.0	66.8
1965	567.0	77.8
1966	603.5	84.2
1967	617.5	79.8
1968	647.0	87.6
1969	664.9	84.9
1970	661.7	74.0
1971	685.6	83.6
1972	731.7	99.2
1973	776.9	122.7
1974p	757.0	141.0

* Gross Private Product (GPP) equals GNP less compensation of general government employees. Data presented here are for nonfarm enterprises.

p = preliminary.

Source: *Economic Report of the President*, February 1975, pp. 261 and 266.

This is the estimated standard error of the mean and in the symbols used in this chapter it is $\hat{\sigma}_{\bar{x}}$. The analysis of variance portion of Table 15.9 reveals a lot. First, we learn that the $r^2 = .833$ and the standard error of the estimate is $10.823 billion.[13] The F-statistic indicates that this relationship is statistically significant; however, one might question the appropriateness of this measure given the fact that the data are in the original form and no transformation was made. Also, the regression sum of squares (8,200.4) represents the explained variance and if this figure is divided by the degrees of freedom, that is, $m - 1$, which is $2 - 1 = 1$, we obtain the mean square. Similarly, the unexplained variance is 1,639.8 and it divided by its degrees of freedom $(n - m)$ gives us $1{,}639.8/(16 - 2) = 117.3$, its related mean square. Then if these two related mean squares are used we determine F:

$$8{,}200.4 \div 117.3 = 70.011$$

[13] MIDAS does not make a distinction between a simple coefficient of correlation (r) and a multiple (R). Consequently, in the output we note the term "multiple R = .91288." This really is r in our terminology.

TABLE 15.9
Correlation and regression output of the raw data of GPP (X) and CPBT (Y) using MIDAS

DESCRIPTIVE MEASURES

VARIABLE	n	MEAN	STD DEV	SE OF MEAN	MINIMUM	MAXIMUM
CPBT	16	79.281	25.613	6.4032	49.700	141.00
GPP	16	597.54	113.66	28.414	433.40	775.90

ANALYSIS OF VARIANCE OF CPBT n = 16

SOURCE	DF	SUM OF SQRS	MEAN SQUARE	F-STATISTIC	SIGNIF
REGRESSION	1	8200.4	8200.4	70.011	.0000
ERROR	14	1639.8	117.13		
TOTAL	15	9840.3			

MULTIPLE R= .91288 R-SQR= .83335 SE= 10.823

VARIABLE	PARTIAL	COEFFICIENT	STD ERROR	T-STATISTIC	SIGNIF
CONSTANT		-43.645	14.936	-2.9216	.0112
GPP	.91288	.20572	.24586 -1	8.3672	.0000

The other figure in the sum of squares column (9,840.3) is the total variance. If we take the ratio of the explained to the total variance we calculate the coefficient of determination (r^2):

$$8{,}200.4 \div 9{,}840.3 = .83335 \text{ (rounded)}$$

By taking the square root of .8335 we obtain $r = .91288$.

The final part of Table 15.9 shows the regression equation:

$$\hat{Y}_c = a + bX = -43.645 + .20572X$$

Again the standard errors and t-statistic are given even though the valid use of these terms might be in doubt here given the nature of the data. Technically, all we can say about this relationship is that the coefficient of determination is .83; without some data adjustment it is not realistic to be talking about sampling errors and statistical significance. Note that the partial coefficient of correlation (to be discussed in Chapter 16) is .91288 the same as the r. The reason being we have only one independent variable.

MIDAS uses the notation of a $-$ 1 to indicate that a zero has been omitted. For example, the standard error of the regression coefficient (b) is given as .24586 -1. This value really is .024586. Incidentally, the standard error of the regression (10.823) is computed by dividing the

FIGURE 15.11
Scatter diagram of the relationship of Gross Private Product (GPP) and Corporate Profits before Taxes (CPBT) (data in original form; billions of dollars)

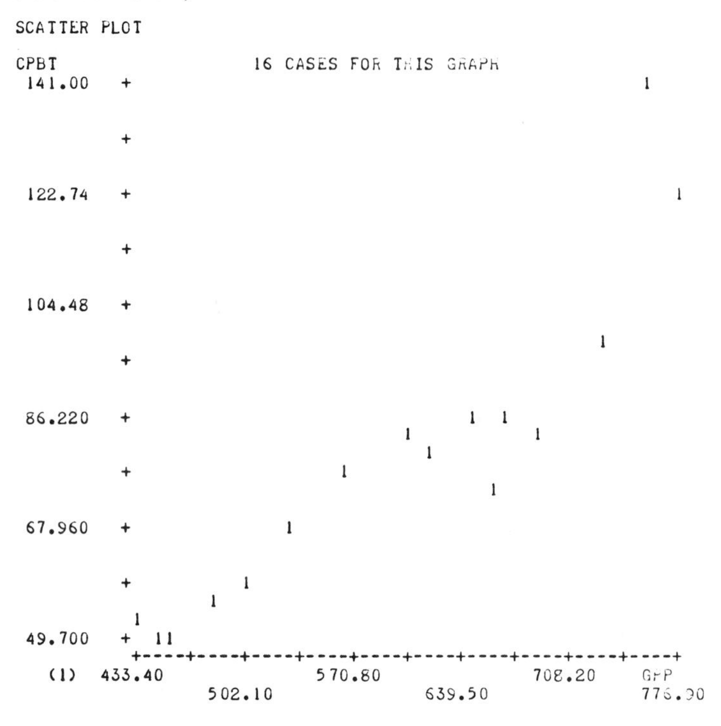

unexplained sum of squares (1,639.8) by $n - m$ degrees of freedom (16 − 2) and then taking the square root. That is,

$$\sqrt{\frac{1{,}639.8}{14}} = 10.823$$

A second relationship calculated made use of the absolute first differences of the data. Some very interesting things take place here. Table 15.10 and Figure 15.12 summarize the results. The most significant item to note is the dramatic drop in the coefficients. The coefficient of determination, r^2, now is .17 compared to .83 when raw data were used. The coefficient of correlation, r, is .41 when contrasted with .91 previously. From the standpoint of a general appraisal of the relationship using the F test, one might conclude that the correlation is not significant at the .10 level when first differences are used in the computations. That is, the correlation could have occurred by chance. When Figure 15.12 is compared with Figure 15.11, the decline in the relationship is obvious.

TABLE 15.10
Correlation and regression output of the first differences of GPP (*X*) and CPBT (*Y*) using MIDAS

DESCRIPTIVE MEASURES

VARIABLE	n	MEAN	STD DEV	SE OF MEAN	MINIMUM	MAXIMUM
FDIFCPBT	15	5.9267	9.1297	2.3573	-10.900	23.500
FDIFGPP	15	21.573	17.864	4.6124	-19.900	45.100

ANALYSIS OF VARIANCE OF FDIFCPBT   n = 15

SOURCE	DF	SUM OF SQRS	MEAN SQUARE	F-STATISTIC	SIGNIF
REGRESSION	1	194.56	194.56	2.6012	.1308
ERROR	13	972.37	74.797		
TOTAL	14	1166.9			

MULTIPLE R= .40833   R-SQR= .16673   SE= 8.6485

VARIABLE	PARTIAL	COEFFICIENT	STD ERROR	T-STATISTIC	SIGNIF
CONSTANT		1.4246	3.5747	.39853	.6967
FDIFGPP	.40833	.20869	.12939	1.6128	.1308

The number 2 in Figure 15.12 indicates that two points have the same value.

The third relationship summarized in Table 15.11 serves to introduce the subject of Chapter 16 and reflects the influence of the trend factor on the relationship between GPP and CPBT. That is, time is introduced into the model as a second independent variable. Its influence may be seen by referring to the coefficient of regression. (Note: the original data are used here.) According to these data, with each passing year CPBT tend to decrease by $2.0669 billion.

SOME ADDED CAUTIONS

Although the methods of regression analysis often make prediction possible, there are many pitfalls to be avoided in their application. Correlation of data ordered in time is a most useful technique, but we must recognize that the data are not randomly distributed and the laws of probability do not apply. Also, the beginner in applying correlation and regression techniques must be aware of some limitations brought about by: (*a*) violations of the theoretical assumptions; (*b*) the improper use of the regression equation and the various coefficients; and (*c*) the misinterpretation of these measures of correlation and regression analysis.

FIGURE 15.12
Scatter diagram of the relationship of Gross Private Product (GPP) and Corporate Profits before Taxes (CPBT), first differences (billions of dollars)

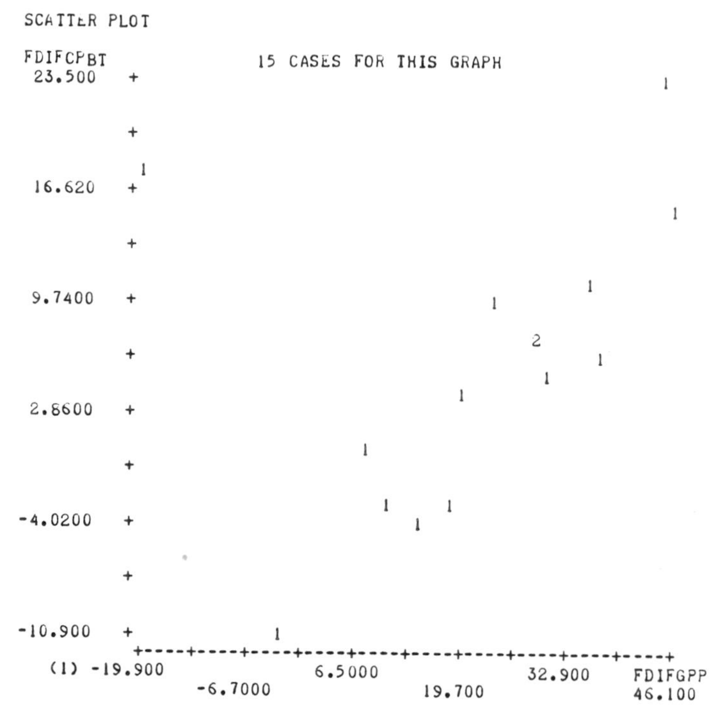

TABLE 15.11
Correlation and regression of Gross Private Product (GPP), time, and Corporate Profits before Taxes (CPBT) (billions of dollars)

ANALYSIS OF VARIANCE OF CPBT n = 16

SOURCE	DF	SUM OF SQRS	MEAN SQUARE	F-STATISTIC	SIGNIF
REGRESSION	2	8222.9	4111.4	33.047	.0000
ERROR	13	1617.4	124.41		
TOTAL	15	9840.3			

MULTIPLE R= .91413 R-SQR= .83564 SE= 11.154

VARIABLE	PARTIAL	COEFFICIENT	STD ERROR	T-STATISTIC	SIGNIF
CONSTANT		-77.410	80.950	-.95627	.3564
TIME	-.11703	-2.0669	4.8649	-.42486	.6779
GPP	.36891	.29163	.20376	1.4311	.1750

The reader also is cautioned that the variables correlated must be related in the real world on some basis other than pure mathematical. That is, you must have a logical basis for bringing these variables together. Further, one extreme item can completely change the results of a given relationship.

Effect of extreme items

The inclusion of one or two extreme items can change the aggregate relationship of a group of paired values from a large *r* to a small *r*, and vice versa. Table 15.12 demonstrates this. The inclusion of the A item results in a negative coefficient of .1217. When the A item is excluded,

TABLE 15.12
An illustration of the effect of one extreme item

	X	Y	X^2	XY	Y^2
A	10	70			
B	12	20			
C	13	24			
D	16	22			
E	18	16			
F	19	26			
G	20	30			
H	24	34			
I	25	34			
J	33	34			
Sums	190	310	4,044	5,774	11,660
	$\bar{X} = 19$	$\bar{Y} = 31$	$\bar{X}\Sigma X = 3{,}610$	$\bar{Y}\Sigma X = 5{,}890$	$\bar{Y}\Sigma Y = 9{,}610$
			$\Sigma x^2 = 434$	$\Sigma xy = -116$	$\Sigma y^2 = 2{,}050$

$$r = \frac{-116}{\sqrt{(434)(2{,}050)}} = \frac{-116}{943} = -.123;\ r^2 = -.015$$

	X	Y	X^2	XY	Y^2
A	Omitted				
B	12	20			
C	13	24			
D	16	22			
E	18	16			
F	19	26			
G	20	30			
H	24	34			
I	25	34			
J	33	34			
Sums	180	240	3,944	5,074	6,760
	$\bar{X} = 20$	$\bar{Y} = 26.6$	$\bar{X}\Sigma X = 3{,}600$	$\bar{Y}\Sigma X = 4{,}800$	$\bar{Y}\Sigma Y = 6{,}400$
			$\Sigma x^2 = 344$	$\Sigma xy = 274$	$\Sigma y^2 = 360$

$$r = \frac{274}{\sqrt{(344)(360)}} = \frac{274}{352} = .778;\ r^2 = .606$$

the coefficient is a positive .7784. The correlation of data for states sometimes takes on this aspect—one state being "out of line" with the others can destroy the true picture of the relationship for all states.

Two or more groups: Heterogeneity

Sometimes, data cannot be considered as one group because of a condition of heterogeneity. That is, the data may be made up of two groups which when considered separately show a very high correlation of the variables under consideration; but when erroneously considered to be of like character and correlated as one group, they show a very low correlation or, as illustrated in Table 15.13, show not only a low correlation coefficient but one with the opposite sign from those for the separate correlations.

Sample regression line

The method discussed here using the standard error of the mean ignores the fact that in many cases the regression line computed is a sample one. That is, the constant a and the regression coefficient b are for the sample. The population regression line might be described by $Y'_c = A + BX$. This error is measured by

$$S_{y_c} = \sqrt{1/n + (X - \bar{X})^2/\Sigma x^2}$$

Therefore, in our example we might have estimated $\hat{Y}_c$

$$\hat{Y}_c \pm tS_{y_c}$$

and then we add and subtract $t\hat{\sigma}_{yx}$. This refinement provides an interval within which the true (population) regression line probably lies. In most instances in dealing with social and economic statistics, the adjustment is debatable given the fact that the original data may not be that precise.

Serial correlation: The Durbin-Watson statistic

The term autocorrelation was introduced earlier to imply the internal correlation between members of a series of observations ordered in time. Some writers imply that autocorrelation and serial correlation are the same thing. Still others insist that autocorrelation be used to denote the correlation of members of a time series with themselves and serial correlation to signify the relationships of members of two different series. In the early 1950s Durbin and Watson described how to measure the possible existence of serial correlation (which they define as the correlation between two independent variables) in the unexplained variation from an

TABLE 15.13
An illustration of the effect of considering unlike groups separately

	X	Y	X^2	XY	Y^2
		One Group			
A	10	30			
B	12	32			
C	13	31			
D	14	36			
E	21	41			
F	28	14			
G	30	12			
H	31	16			
I	35	19			
J	36	19			
Sums	230	250	6,216	5,049	7,180
	$\bar{X} = 23$	$\bar{Y} = 25$	5,290	5,750	6,250
			$\Sigma x^2 = 926$	$\Sigma xy = -701$	$\Sigma y^2 = 930$

$$r = \frac{-701}{\sqrt{(926)(930)}} = -.7595;\ r^2 = -.5768$$

	X	Y	X^2	XY	Y^2
		Two Groups			
A	10	30			
B	12	32			
C	13	31			
D	14	36			
E	21	41			
Sums	70	170	1,050	2,452	5,862
	$\bar{X} = 14$	$\bar{Y} = 24$	$\bar{X}\Sigma X = 980$	$\bar{Y}\Sigma X = 2,380$	$\bar{Y}\Sigma Y = 5,780$
			$\Sigma x^2 = 70$	$\Sigma xy = 72$	$\Sigma y^2 = 82$

$$r = \frac{72}{\sqrt{(70)(82)}} = +.9473;\ r^2 = .8974$$

	X	Y	X^2	XY	Y^2
F	28	14			
G	30	12			
H	31	16			
I	35	19			
J	36	19			
Sums	160	80	5,166	2,597	1,318
	$\bar{X} = 32$	$\bar{Y} = 16$	$\bar{X}\Sigma Y = 5,120$	$\bar{Y}\Sigma X = 2,560$	$\bar{Y}\Sigma Y = 1,280$
			$\Sigma x^2 = 46$	$\Sigma xy = 37$	$\Sigma y^2 = 38$

$$r = \frac{37}{\sqrt{(46)(38)}} = +.8852;\ r^2 = .7836$$

equation fitted by the least-squares method.[14] They contend that if serial correlation does exist in the unexplained variation (residuals) this indicates that some force present in the dependent variable has not been explained by the regression. Sometimes the transformation of the variables, or the inclusion of additional independent variables (or one or more variables should be omitted), or a combination of these might help account for the serial correlation. Students interested in a more detailed description of how to compute the Durbin-Watson statistic and tables for interpretation are referred to the two articles in *Biometrika.*

When two independent variables are correlated the condition is known as *collinearity*. However, if correlation exists among more than two independent variables then the term used is *multicollinearity*. The general term applicable to this condition is serial correlation; however, as noted previously, some writers refer to serial and autocorrelation as being one and the same. This text makes a distinction between the two terms as stated above.

There seems to be some modest debate as to precisely what the Durbin-Watson statistic actually does measure. Some writers would claim that it measures autocorrelation, that is, the problem of nonindependence of successive observations in the same series. Others maintain that because the D-W statistic (d) is based upon the ratio of the sum of the squared differences in successive *residuals* that it probably measures both autocorrelation and serial correlation to some degree.

Before a correction can be made for serial correlation, or a transformation of the data used, we must be able to detect its presence. The D-W statistic is the best-known test that is applicable. Appendix L contains a table for testing for serial correlation. However, many computer programs will output the results of this test, and no table is required for interpretation. Figure 15.13 summarizes the D-W test but it needs some explanation. In the D-W test the critical values of the lower bound d_L and the upper bound d_u are given in the table for various values of the number of independent variables (k') and the sample size (n). If the computed D-W statistic is *less* than the lower bound the null hypothesis is rejected and one accepts the alternative that positive serial correlation does exist. If the D-W statistic computed is *greater than* the upper bound we do *not* reject the null hypothesis that there is no serial correlation among the variables. If the test statistic falls within the bounds of d_L and d_u the D-W test is inconclusive. However, if the computed D-W statistic is greater than $4 - d_L$ the null hypothesis is rejected. This means

[14] See J. Durbin and G. S. Watson, "Testing for Serial Correlation in Least Squares Regression," *Biometrika*, vol. 37, 1950, pp. 409–28; and *Biometrika*, vol. 38, 1951, pp. 159–78. Also see Donald L. McLagan, "A Non-econometrician's Guide to Econometrics," *Business Economics*, vol. 8, no. 3 (May 1973), pp. 38–45; and Robert A. Berman, "An Econometrician's Reactions to 'A Non-econometrician's Guide to Econometrics.' " *Business Economics*, vol. 9, no. 1 (January 1974), pp. 81–82.

FIGURE 15.13
Acceptance and rejection regions of the null hypothesis in the Durbin-Watson test for serial correlation

Values of the D-W Statistic d				
0 d_L	d_u	2	$(4 - d_u)$	$(4 - d_L)$
Decision				
Reject the null hypothesis; accept the conclusion of positive serial correlation	Neither accept nor reject the null hypothesis	Accept the null hypothesis	Neither accept nor reject the null hypothesis	Reject the null hypothesis; accept the conclusion of negative serial correlation
Note: Successive residuals are algebraically close together	*Note:* The test is inconclusive	*Note:* No serial correlation among residuals	*Note:* The test is inconclusive	*Note:* Successive residuals are algebraically far apart

H_0: No serial correlation among residuals.
H_1: Serial correlation among residuals exists.
An example using the table from Appendix L: Assume that $k' = 1$, that is, one independent variable; $n = 15$ and $d = 2.10$. Then from the table we can determine that $d_L = .95$ and that $d_u = 1.23$; and that $4 - d_u = 4 - 1.23 = 2.77$; that $4 - d_L = 4 - .95 = 3.05$. Therefore, because $d = 2.10 > 1.23$ and $2.10 < 2.77$ we accept the null hypothesis.

that we in effect accept the alternative hypothesis and conclude that there is negative serial correlation. That is, that the residuals are algebraically far apart. If the D-W computed statistic is less than $4 - d_u$ we do *not* reject the null hypothesis. Again, if the test statistic falls within the boundary of $4 - d_L$ and $4 - d_u$ the D-W test is not conclusive.

Generally, if the D-W statistic is "very low," then successive values of the residuals are algebraically very close to each other and may be taken as an indication of positive serial correlation. When the test statistic is "very large" then successive values of the residuals are algebraically far apart, and this may indicate negative serial correlation. A rule of thumb that some researchers follow is this: If the D-W statistic is between 1.5 and 2.5 serial correlation is not significant. If the D-W computed value is below 1.5 there might be positive serial correlation between residuals of the fitted data and the actual data. If the D-W statistic is greater than 2.5 negative serial correlation might be present. Both interpretations are dependent upon the sample size and the number of independent variables.

These are but a few of the many cautions that should be cited to students of correlation and regression analysis. However, an exhaustive treatment of the subject is not properly included in a book of this kind. The material in this chapter is intended to develop a critical attitude by the consumer of statistics as well as a general understanding of what correlation and regression mean.

Employed with caution, insight, and a great knowledge of the data to be correlated, the techniques of this chapter and Chapter 16 are some of the most useful tools available to those who work in the area of statistical and economic analysis. Probably the most important function of simple linear correlation and regression analysis is to indicate whether or not it might be appropriate and worthwhile to seek other independent variables. In the examples of this chapter some variation in the dependent variable was left unexplained by variations in the independent variable. Chapter 16 discusses the problem of building a model that includes two or more independent variables. In addition, the highly useful devices of partial correlation and regression and the parabolic regression are considered.

SYMBOLS

X, an independent variable
Y, a dependent variable
$\hat{\sigma}_y$, estimated standard deviation of the universe of variable Y
a, vertical distance of straight line from an origin, the Y intercept
b, slope of a straight line
$\hat{Y}_c = a + bX$, straight-line regression of Y on X
s_x, sample standard deviation of variable X
s_y, sample standard deviation of variable Y
σ_{yx}, true standard error of estimate
$\hat{\sigma}_{yx}$, estimated standard error of estimate
s_{yx}, sample standard error of estimate
$\sigma^2_{y_c}$, explained variance
σ^2_{yx} unexplained variance
σ^2_y, total variance
r, sample coefficient of correlation
r^2, sample coefficient of determination
ρ^2, population coefficient of determination
ρ, population coefficient of correlation
k^2, coefficient of alienation, or coefficient of nondetermination
F, the ratio of variances statistic
σ_r, true standard error of the coefficient of correlation
σ_z, true standard error of the variable z
σ_b, true standard error of the net regression coefficient
$\hat{\sigma}_b$, estimated standard error of the net regression coefficient
B, population regression coefficient, slope of line
A, population value of a
d, Durbin-Watson test statistic

PROBLEMS

15.1. The XYZ Company correlates product performance (Y) with test scores (X) of a sample of employees, using the data on the right. Answer the questions below:

Employee	X	Y
C	3	10
F	5	15
A	6	5
D	8	40
B	11	30
E	15	50

Required:

What is the value of–

a. a?
b. b?
c. ΣX^2?
d. ΣXY?
e. ΣY?
f. $\Sigma \hat{Y}_c$?
g. What is the regression equation?
h. What is the predicted performance of an individual with a test score of 20?

15.2. Use the data from Problem 15.1 above.

Required:

a. Compute the total variation, Σy^2.
b. Compute the unexplained variation, Σd^2.
c. Compute the explained variation, Σy_c^2.
d. Compute r.

15.3. S. Jones feels that there is a positive relationship between sales of Jones's company's product in the southeast states and per capita income in the prior year. Jones decides to correlate 1975 sales with 1974 per capita income. The data are as follows:

	1974 per Capita Income (thousands of dollars) X	1975 per Capita Sales (dollars) Y
Virginia	3.3	20
West Virginia	2.6	15
Kentucky	2.8	15
Tennessee	2.8	14
North Carolina	2.9	10
South Carolina	2.6	12
Georgia	3.0	16
Florida	3.4	20
Alabama	2.6	20
Mississippi	2.2	8
Louisiana	2.8	18
Arkansas	2.6	11

Source: Hypothetical.

Required:

What is the value of—

a. Σx^2?
b. Σy^2?
c. Σxy?
d. b?
e. $b\Sigma xy$?
f. r^2?
g. r?
h. a?
i. What is the regression equation?
j. Is r significant?
k. What is $\hat{\sigma}_{yx}$, the estimated standard error of the estimate?

15.4. Plot per capita personal savings against per capita disposable income in the United States annually from 1929 to date. Identify each point by year. Suggested sources for the data include *U.S. Income and Output* and *Economic Indicators,* published by the U.S. Department of Commerce. In addition:

a. Calculate the least-squares regression line for the data from 1929 through 1940. Draw the corresponding line on the chart and extend it to the limits of your chart in both directions.
b. What was the average per capita marginal propensity to save during the 1929–40 period? (Recall that the percentage of disposable income saved out of additional increments of such income defines the marginal propensity to save.)
c. Is the sign of the constant a meaningful? Why, or why not?
d. What happens to the ratio of the annual per capita savings to annual per capita income as the latter increases? What accounts for this phenomenon?
e. Do the same thing as (*a*) and (*b*) for the years 1946 to date. What conclusions can you validly draw when these two time periods are contrasted with respect to savings in the economy?
f. Why do you think it was suggested that the years 1941–45 be omitted from your calculations? Do you agree with the decision to omit these years?

15.5. Plot personal consumption expenditures against personal disposable income in the United States annually from 1929 to date. Calculate the least-squares regression line for the two periods 1929–40 and 1946 to date. Write a brief statement analyzing the data based upon your chart and calculations.

15.6.

Profits after taxes and cash dividends of all United States corporations (billions of dollars)

Year	Profits	Dividends	Year	Profits	Dividends
1951	20	9.0	1964	38	17.8
1952	17	9.0	1965	46	19.8
1953	18	9.2	1966	50	20.8
1954	17	9.8	1967	47	21.4
1955	23	11.2	1968	48	23.6
1956	24	12.1	1969	45	24.3
1957	22	12.6	1970	39	24.7
1958	22	11.6	1971	46	25.0
1959	29	12.6	1972	58	27.3
1960	27	13.4	1973	73	29.6
1961	27	13.8	1974	85	32.7
1962	31	15.2	1975p	83	33.7
1963	33	16.5			

p = preliminary.
Source: *Economic Indicators*, April 1975, p. 7.

Is there any relationship between these two series? Calculate whatever correlation and regression measures you deem necessary to support a brief statement concerning the implications of the data for the economy as a whole. What are the limitations of your analysis?

15.7.

Corporation profits after taxes and expenditures for new plant and equipment (billions of dollars)

Year	Profits	Expenditures for New Plant and Equipment	Year	Profits	Expenditures for New Plant and Equipment
1954	17	27	1965	46	54
1955	23	29	1966	50	64
1956	24	35	1967	47	65
1957	22	37	1968	48	68
1958	22	31	1969	45	76
1959	29	33	1970	39	80
1960	27	37	1971	46	81
1961	27	36	1972	58	88
1962	31	38	1973	73	100
1963	33	41	1974	85	112
1964	38	47	1975p	83	116

p = preliminary.
Source: *Economic Indicators*, April 1975, pp. 7 and 9.

Analyze and evaluate the relationship between these two series, using whatever correlation and regression techniques you deem appropriate. Are there any limitations concerning your analysis?

15.8.

Weekly hours worked by production workers in manufacturing industries in the United States and unemployment

Year	Average Weekly Hours	Unemployment (percent of civilian labor force)	Year	Average Weekly Hours	Unemployment (percent of civilian labor force)
1952	40.7	3.1	1964	40.7	5.2
1953	40.5	2.9	1965	41.2	4.5
1954	39.6	5.6	1966	41.3	3.8
1955	40.7	4.4	1967	40.6	3.8
1956	40.4	4.2	1968	40.7	3.6
1957	39.8	4.3	1969	40.6	3.5
1958	39.2	6.8	1970	39.8	4.9
1959	40.3	5.5	1971	39.9	5.9
1960	39.7	5.6	1972	40.6	5.6
1961	39.8	6.7	1973	40.7	4.9
1962	40.4	5.5	1974	40.0	5.6
1963	40.5	5.7	1975p	38.6	9.0

p = preliminary.
Source: *Economic Indicators,* April 1975, pp. 11 and 14; and *Economic Report of the President,* February 1975, p. 276.

Analyze and evaluate the relationship between these two series, using whatever correlation and regression analysis you deem appropriate. Are there any limitations to your analysis?

15.9

Bond yields and interest rates of prime commercial paper

Year	Yield High-Grade Municipal Bonds	Prime Commercial Paper: 4–6 Months
1960	3.73	3.85
1961	3,46	2.97
1962	3.18	3.26
1963	3.23	3.55
1964	3.22	3.97
1965	3.27	4.38
1966	3.82	5.55
1967	3.98	4.80
1968	4.51	5.90
1969	5.81	7.83
1970	6.51	7.72
1971	5.70	5.11
1972	5.27	4.69
1973	5.18	8.15
1974	6.09	9.87
1975p	6.80	6.10

p = preliminary.
Source: *Economic Indicators,* April 1975, p. 33.

Analyze and evaluate the relationship between these two series, using whatever correlation and regression techniques you deem appropriate. Are there any limitations to your analysis?

15.10. Given $\hat{Y}_c = 50 + .9X$. What is the meaning of .9 in the equation? of 50?

15.11. A market analyst for a major electronics manufacturer wishes to test the hypothesis that there is a relatively high correlation between the number of color television sets sold annually in a city and the population of that city. The analyst attempts to check on the validity of the hypothesis by analyzing the data for 50 cities chosen at random.

a. If the 50 cities included those of widely differing sizes, would this tend to distort the relationship as measured by the coefficient of correlation? Would it have an impact on the standard error of estimate?

b. If the per capita purchases of color television sets were constant for cities of all sizes, what would be the nature of the regression line?

c. How would the market analyst further proceed to explain quantitatively the fact that several cities of the same size may have experienced pronounced differences in color television set sales?

15.12. Precisely and concisely define the following:

a. The constants a and b

b. r^2

c. σ_{yx} and $\hat{\sigma}_{yx}$

d. σ_b and $\hat{\sigma}_b$

e. z transformation

f. σ_r

g. $\sigma^2_{y_c}$, σ^2_{yx}, and σ^2_y

15.13. There are three kinds of variance of importance to correlation and regression analysis. What are they, and what is their relationship?

15.14. Explain the paradox of dual regression and prediction in terms understandable to the nonstatistician.

15.15. What is the nature of the problem associated with heterogeneous data in terms of correlation and regression analysis?

15.16. Given $\hat{Y}_c = 100 + .60X$; $\hat{\sigma}_{yx} = 5$; $n = 40$.

a. What is the confidence interval for $\hat{Y}_c$ when $X = 200$ with a confidence coefficient of .95?

b. What is the confidence interval for $\hat{Y}_c$ when $X = 300$ with a confidence coefficient of .90?

15.17. Test the significance of $r^2 = .60$ when $n = 12$. Use .05 level as significant and the .01 level as highly significant.

15.18. Given the following:

Year	U.S. Population (millions)	General Mills Sales (millions of dollars)
1961	184	576
1962	187	546
1963	189	524
1964	192	541
1965	195	559
1966	197	525
1967	199	603
1968	201	669
1969	203	885
1970	205	1,022
1971	207	1,120
1972	209	1,316
1973	210	1,593
1974	212	2,000
1975	214	2,309

Source: *Economic Report of the President*, February 1975; and *Annual Report*, 1975, General Mills, Inc.

a. Plot both series on arithmetic graph paper.
b. Calculate the relationship using the raw data as given.
c. Calculate the relationship using relative first differences. Compare your measures with those in (*b*). What has happened to the relationship?
d. Calculate appropriate trend values for each series and then compute the CIs for each. Correlate the CIs and compare your results with those of (*b*) and (*c*) above.

15.19. Distinguish clearly between trend analysis and regression analysis.

15.20. Explain the theory of the criterion of the least-squares method as applied to trend analysis and regression analysis.

15.21. On the basis of a high positive relationship ($r^2 = .95$) which seemed to indicate that a high dividend/earnings ratio was accompanied by a high price/earnings ratio, a financial analyst made the following observation ". . . that companies which pay out larger shares of their earnings in dividends have stocks which are higher priced relative to their earnings" Is this conclusion justified on the basis of this relationship? Defend your position.

15.22. On the basis of a high negative relationship ($r^2 = .99$), it appeared that the correlation between grades in statistics and the time spent in studying is negative! Does this justify the conclusion ". . . that the smaller the number of hours causes the higher grade." If not, why not? Carefully defend your position.

15.23. Carefully defend or refute the following statement: "The standard error of the estimate usually is equal to or greater than the standard deviation of the dependent variable."

15.24. The correlation and regression analysis involving time series data presents some unique problems.

a. What are these problems?

b. How might the analyst adjust the raw data in order to compensate for these unique problems?

15.25. Given the following values of a variable Y: 5, 4, 4, 1, and 2, what is your best estimate of $\hat{Y}_c$? Why?

15.26. Given the following variable X values: 1, 2, 3, 4, and 5, which are related to the Y values of Problem 15.24 demonstrate that correlation and regression analysis may be useful in improving the prediction of Y_c. Be sure to show the relationships of the variances involved in the correlation.

15.27. Given the following random sample of 20 observations of the records of first-year graduate students, respond to the questions below.

	X Quantitative Test Score	Y First-Year Graduate GPA*
	18	1.1
	26	3.3
	30	3.9
	36	4.5
	40	6.5
	34	4.0
	19	1.2
	27	3.1
	37	5.4
	42	6.9
	48	7.1
	22	1.3
	29	3.5
	39	5.7
	52	8.1
	38	6.7
	46	7.9
	39	6.2
	54	8.0
	57	8.7
Sums	733	103.1

* GPA based on a 9.0 scale: ranging from A + = 9.0 to E = 0.0.

a. Calculate the relationship between these two variables.

b. Test the significance of the correlation using .05 level.

c. Given an X value of 60 what is the $\hat{Y}_c$ value? Construct a 95 percent confidence interval.

15.28. Given the following information from a firm's annual report (in millions of dollars)—

Year	X^* Capital Expenditures and Investments	Y Earnings before Income Taxes
1968	21.8	54.1
1969	18.0	60.5
1970	21.6	65.3
1971	26.4	73.6
1972	24.7	61.6
1973	36.4	77.9
1974	35.1	81.9
1975	26.7	55.6

* Lagged two years.

a. Determine the relationship of the above two series using the data as given.
b. Calculate the relative first differences of the above data and determine the relationship.
c. Compare the two relationships. What can you say about the correlation and regression?
d. If $X = 38.0$ for 1976, what is your estimate of $\hat{Y}_c$ for both regressions?

RELATED READINGS

BERKSON, JOSEPH. "Smoking and Lung Cancer." *The American Statistician,* vol. 17 (October 1963), pp. 15–22.

BOX, G. E. P., and PIERCE, D. A. "Distribution of Residual Autocorrelations in Autoregressive-Integrated Moving Average Time Series Models." *Journal of the American Statistical Association,* vol. 65, no. 332 (December 1970), pp. 1507–25.

COCKRANE, D., and ORCUTT, G. H. "Application of Least Square Regressions to Relationships Containing Auto-Correlated Error Terms." *Journal of the American Statistical Association,* vol. 44, no. 245 (March 1949), pp. 32–61.

DIXON, W. J. "Analysis of Extreme Values." *Annuals of Mathematical Statistics,* vol. 21, no. 115, pp. 480–506.

DRAPER, NORMAN, and SMITH, HARRY. *Applied Regression Analysis.* New York: John Wiley & Sons, Inc., 1966.

DUCHAN, ALAN I. "A Relationship between the *F* and *t* Statistics and the Simple Correlation Coefficients in Classical Least Squares Regression." *The American Statistician,* vol. 23 (June 1969), pp. 27–28.

EZEKIEL, MORDECAI, and FOX, KARL A. *Methods of Correlation and Regression Analysis.* 3d ed. New York: John Wiley & Sons, Inc., 1959.

FISHER, R. A. *Statistical Methods for Research Workers,* chaps. 5–7. 14th ed., revised and enlarged. New York: Hafner Publishing Co., Inc., 1973.

FREUND, F. J. "A Warning of Roundoff Errors in Regression." *The American Statistician,* vol. 17 (December 1963), pp. 13–16.

HALPERIN, MAX; RASTOGI, S. C.; HO, IRVIN; and YANG, Y. Y. "Shorter Confidence Bonds in Linear Regression." *Journal of the American Statistical Association,* vol. 62, no. 319 (September 1967), pp. 1050–67.

LEABO, DICK A. "The Declining Marginal Propensity to Save." *Business Economics,* vol. 6, no. 3 (May 1971), pp. 25–29.

OSTLE, BERNARD, and STECK, GEORGE S. "Correlation between Sample Means and Sample Ranges." *Journal of the American Statistical Association,* vol. 54, no. 286 (June 1959). This article is of particular interest to those persons interested in quality control.

PEARSON, KARL. "Mathematical Contributions to the Theory of Evolution—VII. On the Correlation of Characters Not Quantitatively Oriented." *Philosophical Transactions of the Royal Society of London,* Series A, vol. 195 (1901), pp. 1–47.

ROBINSON, WILLIAM. "Ecological Correlations and the Behavior of Individuals." *American Sociological Review,* vol. 15 (1950), pp. 351–57. Some dangers involved in generalizing from ecological relationships among individuals are vivdly described.

SIMON, HERBERT A. *Models of Men,* New York: John Wiley & Sons, Inc., 1957. This book attempts to establish the direction of causal relationships by means of statistical methods.

WAUGH, FREDERICK V., and FOX, KARL A. "Graphic Computation of $R_{1.23}$." *Journal of the American Statistical Association,* vol. 52, no. 280 (December 1957), pp. 479–81.

WECHSLER, HENRY; GROSSER, G. H.; and GREENBLATT, MILTON. "Research Evaluating Anti-depressant Medications on Hospitalized Mental Patients: A Survey of Published Reports during a Five-Year Period." *Journal of Nervous and Mental Disease,* vol. 141 (August 1965), p. 223 ff.

WINN, DARYL, and LEABO, DICK A. "Rates of Return, Concentration and Growth—Question of Disequilibrium." *The Journal of Law and Economics,* vol. 17, no. 1 (April 1974), The University of Chicago, pp. 97–115.

16

Introduction to multiple and partial correlation and regression analysis

IN THE ANALYSIS of economic data, and particularly in the application of correlation and regression analysis, it is rare to discover an association so obvious and so comprehensive that the coefficient of correlation represents a perfect relationship, that is, $r = \pm 1.0$. In such a situation, we recall, all of the variation in the dependent series (X_0) is accounted for or accompanied by variation in the independent variable (X_1). Simple coefficients of correlation such as $r = 1.0$, which would indicate that no further analysis appears necessary, are seldom observed. Instead, correlation and regression analysis, as discussed in Chapter 15, generally has its main virtue in delineating the limits of such a relationship and in emphasizing the need for further research.[1]

To illustrate the point, assume that we have measured the relationship between weekly hours worked (X_1) and quit rates (X_0) for a large group of production workers in a given corporation. A simple linear regression line would indicate the general nature of the degree of association between these two variables. It might also provide a basis for estimating quit rates by using weekly hours as the predictor. Generally, a single analysis of this type probably would not result in sufficiently accurate estimations. Therefore the logical question can be asked: How may we improve the prediction process? Immediately, one possibility suggests itself, that is, the identification of other conditions (independent variables) similarly related to quit rates. Then the problem is to combine these variables into a more efficient regression equation. Perhaps overtime, piece rates, work conditions affecting health, distance workers must

[1] In this chapter the dependent variable will be X_0 and the estimated value labeled $\hat{X}_0$. The independent variables will be $X_1, X_2, \ldots, X_n$. Recall that in the previous chapter the dependent variable was Y and the estimated value $\hat{Y}_c$.

travel to plant, and many others might also cause variations in quit rates. Theoretically, if we can identify *all* of the factors affecting quit rates and combine them into a new regression equation, our prediction should be perfect. In the "real world," it is difficult to isolate and measure all the forces; and frequently, an upper limit on the number of useful independent variables is rather quickly reached.

Further to illustrate the possibilities of such analysis, consider the following management problem. Assume that one problem faced by a firm is to discover conditions that are associated with voluntary quits and to predict possible continuance rates and necessary replacements on the basis of such analysis. To these ends, quit rates first are compared with actual average weekly hours worked. The dependent variable (X_0) is the actual observed quit rates, whereas the independent variable (X_1) is the actual average weekly hours worked. Assume that we obtain the following results from a probability sample of sets of paired values:

$$r = .60 \text{ (coefficient of correlation)}$$
$$r^2 = .36 \text{ (coefficient of determination)}$$
$$\hat{X}_0 = -.45 + .041X_1$$

This would indicate that 36 percent of the variation in quit rates is accompanied by variations in average weekly hours worked. Or we might look at the coefficient of nondetermination ($k^2 = 1 - r^2$) and state that 64 percent of the variations in quit rates is left unexplained by variations in average weekly hours worked. The regression equation shows that quit rates change by .04 units for each one-hour change in average weekly hours worked.

Clearly, the next phase of the problem is to provide a method by which the data with respect to the covariation of quit rates and hours worked may be combined with those relating to the coassociation of quit rates and some other factor. This, of course, is essentially the method of multiple and partial correlation and regression analysis. That is, multiple correlation and regression analysis involves the measurement of a relationship of a single dependent variable (X_0) and a number of independent variables in combination ($X_1 \ldots X_n$). Partial correlation and regression analysis measures the degree of association between two variables holding a third variable constant or at its average level. Whereas simple correlation ignores the impact of a third or more independent variables, partial correlation actually helps quantify the statement "other things being equal."

In simple correlation the prediction formula (regression equation) actually defines a trend line fitted to two series of data. In multiple correlation the principal problem is the computation of a *composite* trend line in which two or more independent variables are elements. With some minor modifications, this problem has its counterpart in the fitting of a parabolic trend. In the parabolic trend equation, $T = a + bX + cX^2$, it is necessary to estimate the constants a, b, and c. Although a merely positions the trend line relative to the Y axis, b and c are basically weights assigned to each series in order that the parabolic trend may come as close to the data as measured by the criterion of least squares. In multiple correlation the same objective is sought, but we use slightly different notation. The constant a is the height of the regression plane above the origin or the value of $\hat{X}_0$ when $X_1 \ldots X_n$ are zero, and the regression coefficients associated with the independent series X_1 and X_2 are designated as b_1 and b_2, respectively. The latter two constants are customarily described as *coefficients of net regression.* They play a prominent role in the process of partial analysis, as we shall discover later. The multiple linear regression equation is

$$\hat{X}_0 = a + b_1X_1 + b_2X_2 + \cdots + b_nX_n \tag{16.1}$$

BUILDING THE MULTIPLE CORRELATION AND REGRESSION MODEL

After the simple model dealing with only two variables has been formulated on a nonstatistical basis, we may wish to include other independent variables to help explain more of the variation in the dependent variable. How does the analyst identify appropriate independent variables? How many should be added? For the most part, answers to these questions flow from the skill and imagination of the person together with a technical knowledge of the dependent series. Another invaluable input is an awareness of how these variables are related in the real world. If our goal were to predict the sales of a major construction materials manufacturer, we might first think of the number of new housing starts as one potentially important independent variable. Other possible independent variables might be the rate of family formation, some index of general economic conditions, disposable income, level of consumer credit outstanding, interest rates for new home construction and for remodeling purposes, and many others. Figure 16.1 attempts to trace schematically the procedures used in the development of a multiple regression model. Once the business economist has concluded that correlation and regression techniques are appropriate to solve his or her problem, the next most important step is to build a model justifying the independent variables on a nonmathematical basis. The identification of the dependent variables is not as difficult. (In our example, it would be the sales data.) Then, of course, the regression is run on a computer and the results studied for possible model modifications. If some of the independent variables chosen are highly correlated, we might encounter the problems of multicollinearity or autocorrelation. (See Chapter 15 for a discussion of the

FIGURE 16.1
Schematic procedure for multiple correlation and regression model building

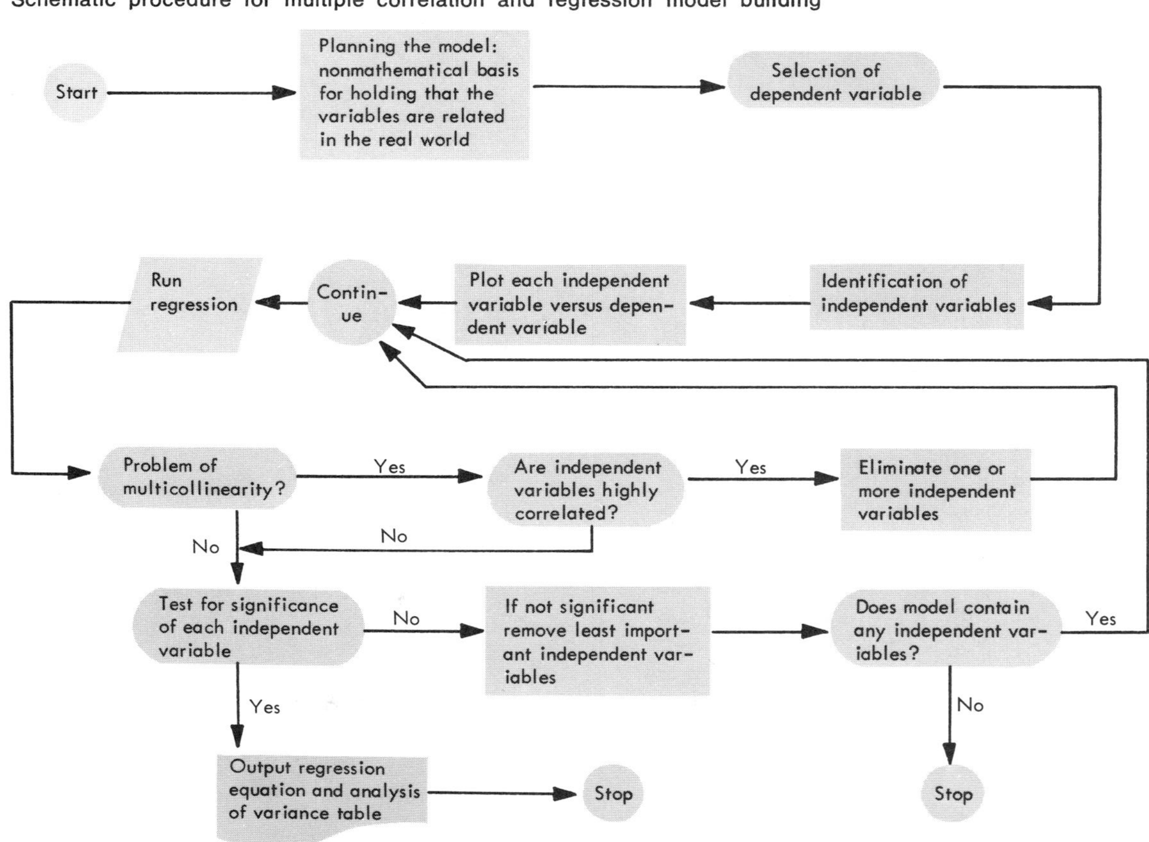

problem of correlating time series.) If this is the case, we then should eliminate one or more independent variables and rerun the regression. Independent variables also might be removed from the model on the basis of some F or t test.

Parenthetically, the reader is reminded that when we speak of "running the correlation" we are thinking of using some high-speed digital computing system. Most of the techniques mentioned so far in this text can be programmed for computer usage, and generally that is the way the computations are generated. However, illustrations and problems in this text can be worked "by hand" on a desk calculator if necessary. Clearly, the procedure of stepwise regression to be mentioned later is most adaptable for computer solution. In other instances, the use of the desk calculator has the advantage of demonstrating to the neophyte statistician precisely what goes on in some of the more involved computations. He or she also learns very quickly that the computer is a friend, indeed!

Returning to the problem of multicollinearity, it usually is defined as a condition in which there exists a linear relationship connecting the independent variables. The coefficients of the regression on such variables are then indeterminate, and the standard errors, in effect, become infinite.[2] Generally, the same term is used when the independent variables are subject to error. Collinearity may occur in sample data due to chance. Or, there may actually be a linear relationship between the variables in the population but due to sampling errors the observed values do not exhibit a linear relation. The regression line in such cases can be determined but is statistically unreliable in the sense that different samples might give entirely opposite results. Theoretically, the existence of one linear relationship is called collinearity and the condition of several is described as multicollinearity.

Selecting the best regression equation

Draper and Smith[3] refer to two contradictory criteria for rationally selecting a multiple regression equation for prediction purposes: (1) to make the equation most useful for prediction we want our model to include as many as possible related (to the dependent variable) independent variables to help explain the variation in the dependent variable; and (2) because of costs involved in the data collection stage we want to minimize the number of independent variables required. This concept

[2] See Maurice G. Kendall and William R. Buckland, *A Dictionary of Statistical Terms* (London: Oliver and Boyd, 1957), p. 191.

[3] See Norman Draper and Harry Smith, *Applied Regression Analysis* (New York: John Wiley & Sons, Inc., 1966), Chap. 6. Also refer to M. A. Elfroymson, "Multiple Regression Analysis" in A. Ralston and H. S. Wilf, *Mathematical Models for Digital Computers* (New York: John Wiley & Sons, Inc., 1962).

is related to the problem of statistical and economic efficiency discussed in Chapter 3 dealing with sampling theory. Sometimes we must compromise and give up some statistical efficiency in the interest of costs. The cost of collecting data for more independent variables may far outrun the benefit from their inclusion in the model. Statistical theory is not much help here because these problems must be solved on an economic basis. Clearly, the personal knowledge and judgment of the business economist about the variables are most important at this stage. There are at least six different procedures currently available for selecting the best regression equation.[4] However, because this is an elementary text we shall confine ourselves to the one most useful (and therefore, probably the "best") to business economists, namely, the procedure called stepwise regression. Figure 16.2 illustrates the basic logic of the stepwise procedure.

Stepwise regression. This helpful technique begins with a simple correlation matrix and enters into regression the independent variable most highly correlated with the dependent variable. Using the partial coefficients generated with respect to the other variables, the computer program then selects the next variable to enter the model. (Hence the reference to which step the variable entered.) It should be noted that in the stepwise regression an independent variable which may have entered the model at an earlier step by being the best single one, might later be rejected (on the basis of an F test). This can happen because of the relationship between the variable in question and other independent variables now in the model. Any independent variable judged not significant in terms of improving the regression equation is rejected. This stepwise procedure is continued until the program runs out of independent variables or when there are no more to be included or excluded.[5]

In effect, the stepwise regression selects one independent variable at a time (i.e., step by step) by computing the optimum coefficients for a linear mathematical equation. Optimum relates to the objective of minimizing the squared deviations of the predicted value of the dependent variable from the actual value of the dependent variable. This means that the best straight line (in terms of minimum squared error) is fitted to the data in n space with n being incremented by one each step. If the inclusion of the last variable causes the statistical significance of a previ-

---

[4] See Draper and Smith, *Applied Regression Analysis,* pp. 171–72.

[5] A readily available stepwise program was developed by UCLA and is called BMD2R as adapted by the Statistical Research Laboratory at the University of Michigan, Ann Arbor. For an application of this program see Dick A. Leabo, "Stepwise Regression Analysis Applied to Regional Economic Research," *Proceedings* of the Business and Economics Section, 130th Annual Meeting of the American Statistical Association, Detroit, December, 1970, pp. 454–58. The University of Michigan program MIDAS, referred to in Chapter 15, also has this capability and is easier to use than the BMD2R.

FIGURE 16.2
Schematic procedure of stepwise regression

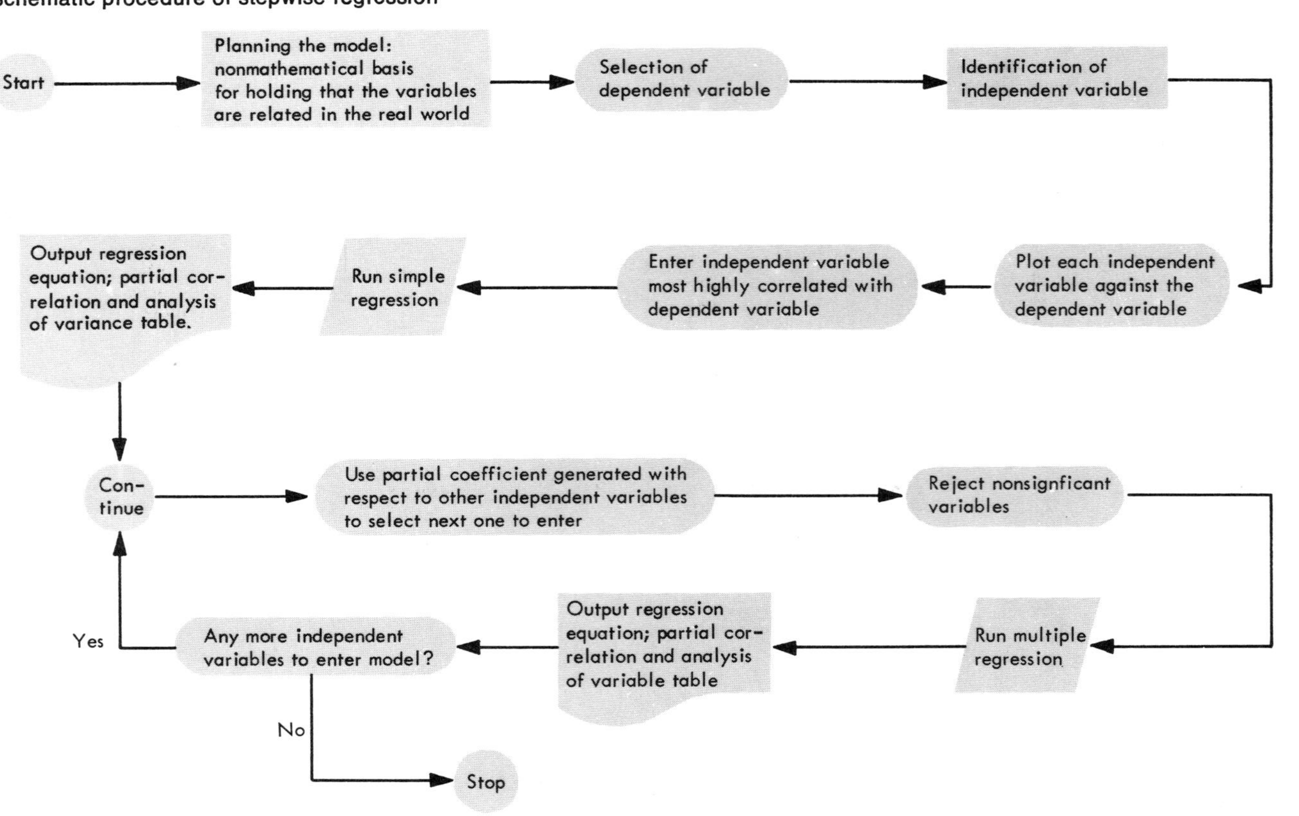

ously included independent variable to drop below a predescribed acceptable level, the next step deletes the now insignificant variable and recalculates the equation. Usually, most programs are written so the significance level for inclusion of a variable is .01 and .005 for deletion; however, these may be set at the discretion of the statistician.

Warning. If the selection of the independent variables can be done so easily and quickly by a high-speed computer, why would the analyst try any other technique? The reason is that the stepwise procedure should not be employed by an amateur unfamiliar with the many pitfalls of "machine selection" of variables. There is absolutely no substitute for a knowledge of the relation of the variables as they exist in the real world. The computer will only sort those variables given to it and can do nothing about independent variables not considered by the analyst. Once the analyst has made a critical evaluation of his or her model and relied on personal knowledge of the data to select the candidates for the independent variables, then the stepwise procedure is a valid and most helpful tool. In the hands of the amateur, it is much too easy to rely too heavily on the mechanical selection by the computer.

THE NORMAL EQUATIONS AND THE REGRESSION CONSTANTS

In order to combine the covariation among three or more variables, we must determine proper weights so that the simple regressions (i.e., X_1 on X_0 and X_2 on X_0, etc.) may be expressed in a composite equation similar to the one above. Basically, the problem is one of computing appropriate values for the coefficients of net regression ($b_1, b_2, \ldots, b_n$). These weights may be discovered through the use of normal equations in a manner comparable to the least-squares straight regression line discussed in Chapter 15. Chapter 15 indicates that the least-squares criterion is so designed that the method fits the regression line in such a way that the standard deviation of the differences between the estimated values and the actual data is a minimum. In other words, in terms of the squared deviations (about the regression line), the least-squares line is graphically closer to the actual data than any other line that may be drawn.

We recall from Chapter 15 that the value of the constant b in the simple linear regression equation can be discovered by using two normal equations so prepared that their solution assures the closeness of fit required. For the simple linear relationship, these two normal equations are

$$Na + b\Sigma X = \Sigma Y \tag{16.2}$$

$$a\Sigma X + b\Sigma X^2 = \Sigma XY \tag{16.3}$$

With centering, they become (see Chapters 13 and 15):

$$Na = \Sigma Y \tag{16.4}$$

$$b\Sigma x^2 = \Sigma xY \tag{16.5}$$

because the terms containing ΣX drop out when the origin is selected so that $\Sigma X = 0$.

Or for convenience, we write

$$a = \bar{Y} \tag{16.6}$$

$$b = \frac{\Sigma xY}{\Sigma x^2} \tag{16.7}$$

For use with multiple correlation and regression analysis, where we have three variables and therefore two coefficients of net regression, the required general normal equations are

$$Na + b_1\Sigma X_1 + b_2\Sigma X_2 = \Sigma Y \tag{16.8}$$

$$a\Sigma X_1 + b_1\Sigma X_1^2 + b_2\Sigma X_1X_2 = \Sigma X_1Y \tag{16.9}$$

$$a\Sigma X_2 + b_1\Sigma X_1X_2 + b_2\Sigma X_2^2 = \Sigma X_2Y \tag{16.10}$$

And since it is customary to use X_0 for the dependent variable in place of Y, the equations become

$$Na + b_1\Sigma X_1 + b_2\Sigma X_2 = \Sigma X_0$$

$$a\Sigma X_1 + b_1\Sigma X_1^2 + b_2\Sigma X_1X_2 = \Sigma X_1X_0$$

$$a\Sigma X_2 + b_1\Sigma X_1X_2 + b_2\Sigma X_2^2 = \Sigma X_2X_0$$

Although we obviously could solve the normal equations without centering, it reduces the calculations if we center them. Clearly, we keep adding normal equations as we add independent variables. Equally true, the task of simultaneously solving these equations becomes increasingly more tedious as the number of variables and the number of paired items (n) increases. Consequently, except for understanding the concepts of correlation and regression, it would be desirable to program the computations for use on the various electronic computers available. However, before we can intelligently program a multiple correlation problem, it would seem advantageous first to "learn what correlation is all about." In other words, *if* the only reason we are going through the mathematical methods is to learn a *technique,* then we are wasting our time. Computers usually can do large problems faster and cheaper; however, we must *understand the relationships* before we have the confidence to request computer assistance.

Returning to the three normal equations, we can abbreviate them somewhat by centering without destroying our understanding of the concepts. That is, it is possible to reduce the work of calculating the values of $b_1, b_2, \ldots, b_n$ by centering each variable or setting its origin at the mean of the series. In that event, for a problem of two independent variables and one dependent variable,

$$\Sigma(X_1 - \bar{X}_1) = \Sigma x_1 = 0$$
$$\Sigma(X_2 - \bar{X}_2) = \Sigma x_2 = 0$$
$$\Sigma(X_0 - \bar{X}_0) = \Sigma x_0 = 0$$

and our first normal equation, as well as the first terms in the other two equations, disappears. Then the coefficients of net regression are defined:

$$b_1\Sigma x_1^2 + b_2\Sigma x_1x_2 = \Sigma x_1x_0$$

$$b_1\Sigma x_1x_2 + b_2\Sigma x_2^2 = \Sigma x_2x_0$$

And if these centered equations are solved algebraically for b_1 and b_2, they become

$$b_1 = \frac{\Sigma x_2^2\Sigma x_1x_0 - \Sigma x_1x_2\Sigma x_2x_0}{\Sigma x_1^2\Sigma x_2^2 - (\Sigma x_1x_2)^2} \qquad (16.11)$$

$$b_2 = \frac{\Sigma x_1^2\Sigma x_2x_0 - \Sigma x_1x_2\Sigma x_1x_0}{\Sigma x_1^2\Sigma x_2^2 - (\Sigma x_1x_2)^2} \qquad (16.12)$$

And a is found from the first normal equation as

$$Na = \Sigma X_0 - b_1\Sigma X_1 - b_2\Sigma X_2$$
$$a = \bar{X}_0 - b_1\bar{X}_1 - b_2\bar{X}_2 \qquad (16.13)$$

MEASURE OF THE DEGREE OF ASSOCIATION

Generally, in measuring the degree of the covariation in a multiple relationship, we use the coefficient of multiple correlation, R. It has an interpretation similar to r in simple correlation in that it represents the square root of the ratio of explained variance to the total variance.

Or

$$R = \sqrt{\frac{\Sigma t^2}{\Sigma x_0{}^2}} \qquad (16.14)$$

where

$$\Sigma t^2 = b_1\Sigma x_1x_0 + b_2\Sigma x_2x_0 \qquad (16.15)$$

and is the aggregate *variance accounted for* by the two independent variables, and

$$\Sigma x_0^2 = \Sigma X_0^2 - \bar{X}_0 \Sigma X_0 \qquad (16.16)$$

and is the *total aggregative variance* to be accounted for. It follows that the unexplained aggregative variance is

$$\Sigma d^2 = \Sigma x_0^2 - \Sigma t^2 \qquad (16.17)$$

because the total variance is composed of the explained and unexplained; so

$$\Sigma x_0^2 = \Sigma t^2 + \Sigma d^2 \qquad (16.18)$$

or

$$\sigma_{x0}^2 = \sigma_t^2 + \sigma_d^2 \qquad (16.19)$$

The square of the multiple correlation coefficient is called the *multiple coefficient of determination,* and its symbol is R^2.

Therefore,

$$R^2 = \frac{\Sigma t^2}{\Sigma x_0^2} = \frac{\sigma_t^2}{\sigma_{x0}^2} \qquad (16.20)$$

and

$$R^2 = \frac{b_1 \Sigma x_1 x_0 + b_2 \Sigma x_2 x_0}{\Sigma x_0^2} \qquad (16.21)$$

STANDARD ERROR OF ESTIMATE

The explained variance is that included in the multiple regression equation, and the differences between X_0 and the regression line indicate the degree to which X_1 and X_2 *fail* to explain the variability in the dependent variable (X_0). The sample standard deviation of these residuals,

$$s_{0.12} = \sqrt{\frac{\Sigma d^2}{n}} \qquad (16.22)$$

is called the *standard error of estimate,* which provides some measure of the degree of reliability of the multiple regression equation, in addition to R^2.

Recall from Formula (16.17) that

$$\Sigma d^2 = \Sigma x_0^2 - \Sigma t^2$$

and the sample standard error of estimate squared is

$$s_{0.12}^2 = \frac{\Sigma d^2}{n} = s_{x0} - \sigma_t^2 \qquad (16.23)$$

and the estimated population standard error of estimate squared is

$$\hat{\sigma}^2_{0.12} = \frac{\Sigma d^2}{n} \cdot \frac{n}{n-m} \tag{16.24}$$

In actual practice, though, it is seldom desirable to calculate individual variances so that neither Σd^2 nor Σt^2 may be readily determined. In such cases the standard error of the estimate (for a three-variable multiple relationship) is more easily determined:

$$s_{0.12} = \sqrt{\frac{\Sigma x_0^2 - \Sigma t^2}{n}} = \sqrt{\frac{\Sigma x_0^2 - (b_1 \Sigma x_0 x_1 + b_2 \Sigma x_0 x_2)}{n}} \tag{16.25}$$

which can be expanded, if necessary by the addition of terms $(b_3 \Sigma x_0 x_3 + \ldots + b_n \Sigma x_0 x_n)$. This represents the variance of the residuals, which really is a measure of the degree to which the regression fails to explain all of the variation in X_0.

BETA REGRESSION COEFFICIENTS[6]

As indicated in the introductory section, the *b*'s are simply weights expressing the influence of the independent variables as they are combined into an estimate of the dependent variable. In using the *b* values, we do encounter some difficulties. The units of X_1 and X_2 may be different, and their respective variations also may differ greatly. As a result, the *b*'s do not indicate the actual comparative significance of each independent variable.

Consequently, we turn to a measure called the beta coefficient, which recognizes the dissimilarity of the units and their variances. We can either use the simple coefficient of correlations to determine the betas, where

$$r_{01} = \beta_1 + \beta_2 r_{12} \tag{16.26}$$

$$r_{02} = \beta_1 r_{12} + \beta_2 \tag{16.27}$$

or use the standard deviations of the various series,

$$\beta_1 = b_1 \left(\frac{\sigma_{x1}}{\sigma_{x0}} \right) \tag{16.28}$$

$$\beta_2 = b_2 \left(\frac{\sigma_{x2}}{\sigma_{x0}} \right), \text{ and so on.} \tag{16.29}$$

[6] The beta regression coefficients, symbolized by $\beta_1 \beta_2 \ldots \beta_n$, are *not* to be confused with the probability of a Type II error discussed in Chapter 11. They represent entirely different concepts; but traditionally, the same symbol has been used for both. The only defense of such action is that there should be no confusion because it is clear by the context in which the measures are used that the beta regression coefficients relate to regression analysis, whereas the beta risk relates to the probability of a Type II error.

Because these constants (betas) provide a more accurate idea of the relative importance of the independent variables, they are more useful than the *b*'s in regression analysis. In addition, we can measure the sampling error of the betas through the standard error of beta.

COMPUTATION OF THE MEASURES

Let us compute some of the above measures to obtain a better "feel" for the concepts involved. Assume that we are attempting to quantify the problem of identifying conditions that are associated with voluntary quits in the labor force of a large corporation, because on the basis of this analysis we want to estimate possible continuance rates and needed replacements. To these ends, quit rates are compared with actual weekly working hours and with compensated overtime. In order to simplify the computations and not require an IBM 360 system to do the calculations we shall work with a sample of 10 random observations. Even here, we shall discover that the calculator time is not small; this will serve to emphasize the need for computer assistance when, say, $n = 100$ and we have a dozen or more independent variables. In such situations the statistician very quickly gratefully seeks the aid of the electronic computer. The data in Table 16.1 are used in our illustration. It should be noted that if the

TABLE 16.1
Relationship of voluntary quit rates (X_0), weekly hours worked (X_1), and compensated overtime (X_2) in Corporation A ($n = 10$; $\bar{X}_0 = 1.0$; $\bar{X}_1 = 35$; $\bar{X}_2 = 6$)

Time Period		X_1 Weekly Hours	X_2 Overtime	X_0 Quit Rates	Z Check
1		26	3	.5	29.5
2		29	5	.6	34.6
3		27	2	.7	29.7
4		35	6	.8	41.8
5		31	8	.9	39.9
6		37	7	1.0	45.0
7		45	6	1.1	52.1
8		40	9	1.2	50.2
9		41	8	1.5	50.5
10		39	6	1.7	46.7
Σ		350	60	10.0	420.0√
Cross products (*P*)	1	12,628	2,186	367.5	15,181.5√
	2		404	64.5	2,654.5√
	0			11.34	443.34√
	Z				18,279.34√
Cross products, centered, (*np*)	1	3,780	860	175.0	4,815.0√
	2		440	45.0	1,345.0√
	0			13.40	233.40√
	Z				6,393.40√

regression is run leading X_1 and X_2 by one unit the coefficient of determination is higher (.76 compared to .61) and the standard error is lower.

The *P* block or the cross products are found by taking the squares and cross products. For example,

$$\Sigma X_1^2 = 12{,}628$$

which is the sum of $(26)^2 + (29)^2 + \cdots + (39)^2$. Also,

$$\Sigma X_1X_2 = 2{,}186$$

which is the sum of $(26)(3) + (29)(5) + \cdots + (39)(6)$. Also,

$$\Sigma X_1X_0 = 367.5$$

which is the sum of $(26)(.5) + (29)(.6) + \cdots + (39)(1.7)$. Also,

$$\Sigma X_1Z = 15{,}181.5$$

which is the sum of $(26)(29.5) + (29)(34.6) + \cdots + (39)(46.7)$, where

$$\Sigma X_1Z = \Sigma X_1^2 + \Sigma X_1X_2 + \Sigma X_1X_0$$

And

$$\Sigma X_2^2 = 404$$

which is the sum of $(3)^2 + (5)^2 + \cdots + (6)^2$. And

$$\Sigma X_1X_0 = 64.5$$

which is the sum of $(3)(.5) + (5)(.6) + \cdots + (6)(1.7)$. And

$$\Sigma X_2Z = 2{,}654.5$$

which is the sum of $(3)(29.5) + (5)(34.6) + \cdots + (6)(46.7)$, where

$$\Sigma X_2Z = \Sigma X_1X_2 + \Sigma X_2^2 + \Sigma X_2X_0$$

Also,

$$\Sigma X_0^2 = 11.34$$

which is the sum of $(.5)^2 + (.6)^2 + \cdots + (1.7)^2$. Also,

$$\Sigma X_0Z = 443.34$$

which is the sum of $(.5)(29.5) + (.6)(34.6) + \cdots + (1.7)(46.7)$, where

$$\Sigma X_0Z = \Sigma X_1X_0 + \Sigma X_2X_0 + \Sigma X_0^2$$

Finally,

$$\Sigma Z^2 = 18{,}279.34$$

which is the sum of $(29.5)^2 + (34.6)^2 + \cdots + (46.7)^2$, where

$$\Sigma Z^2 = \Sigma X_1Z + \Sigma X_2Z + \Sigma X_0Z$$

The centered cross products, *np* block, are found as follows:

$$\Sigma x_1^2 = (\Sigma X_1^2 - \bar{X}_1 \Sigma X_1)(n)$$
$$\Sigma x_1^2 = [12{,}628 - (35)(350)]10 = 3{,}780$$

$$\Sigma x_1 x_2 = (\Sigma X_1 X_2 - \bar{X}_1 \Sigma X_2)(n)$$
$$\Sigma x_1 x_2 = [2{,}186 - (35)(60)]10 = 860$$

$$\Sigma x_1 x_0 = (\Sigma X_1 X_0 - \bar{X}_1 \Sigma X_0)(n)$$
$$\Sigma x_1 x_0 = [367.5 - (35)(10.0)]10 = 175.0$$

$$\Sigma x_1 z = (\Sigma X_1 Z - \bar{X}_1 \Sigma Z)(n)$$
$$\Sigma x_1 z = [15{,}181.5 - (35)(420.0)]10 = 4{,}815.0$$

$$\Sigma x_2^2 = (\Sigma X_2^2 - \bar{X}_2 \Sigma X_2)(n)$$
$$\Sigma x_2^2 = [404 - (6)(60)]10 = 440$$

$$\Sigma x_2 x_0 = (\Sigma X_2 X_0 - \bar{X}_2 \Sigma X_0)(n)$$
$$\Sigma x_2 x_0 = [64.5 - (6)(10.0)]10 = 45.0$$

$$\Sigma x_2 z = (\Sigma X_2 Z_2 - \bar{X}_2 \Sigma Z)(n)$$
$$\Sigma x_2 z = [2{,}654.5 - (6)(420.0)]10 = 1{,}345.0$$

$$\Sigma x_0^2 = (\Sigma X_0^2 - \bar{X}_0 \Sigma X_0)(n)$$
$$\Sigma x_0^2 = [11.34 - (1.0)(10.0)]10 = 13.40$$

$$\Sigma x_0 z = (\Sigma X_0 Z - \bar{X}_0 \Sigma Z)(n)$$
$$\Sigma x_0 z = [443.34 - (1.0)(420.0)]10 = 233.40$$

Technically, the terms on the left side of the above equations still have (n) in them. That is, they should be $n\Sigma x_1^2$, $n\Sigma x_1 x_2$, and so on, because we multiplied by n but never removed it. The reason we do this is that in many problems, decimal numbers are thus omitted. Here, we could have ignored the multiplication by n, but because n is in *all* of the terms, we usually do not write it in the left side, since it will not affect our computations of the constants or the coefficient of correlation.

Let us continue with the calculations of the various measures. First, the net regression coefficients:

from Formula (16.11):

$$b_1 = \frac{\Sigma x_2^2 \Sigma x_1 x_0 - \Sigma x_1 x_2 \Sigma x_2 x_0}{\Sigma x_1^2 \Sigma x_2^2 - (\Sigma x_1 x_2)^2}$$

$$b_1 = \frac{(440)(175.0) - (860)(45.0)}{(3{,}780)(440) - (860)^2}$$

$$b_1 = .041468$$

from Formula (16.12):

$$b_2 = \frac{\Sigma x_1^2 \Sigma x_2 x_0 - \Sigma x_1 x_2 \Sigma x_1 x_0}{\Sigma x_1^2 \Sigma x_2^2 - (\Sigma x_1 x_2)^2}$$

$$b_2 = \frac{(3{,}780)(45.0) - (860)(175.0)}{(3{,}780)(440) - (860)^2}$$

$$b_2 = .021221$$

The constant a may be found by using Formula (16.13):

$$a = \bar{X}_0 - b_1\bar{X}_1 - b_2\bar{X}_2$$
$$a = 1.0 - (.041468)(35) - (.021221)(6)$$
$$a = -.579$$

Therefore the multiple regression equation is

$$\hat{X}_0 = -.579 + .041468X_1 + .021221X_2$$

which may be interpreted to mean that estimated voluntary quit rates, beginning with a negative value of a, increase by .041468 for each unit change in weekly hours and increase by .021221 for each unit change in compensated overtime.

Analyzed in this way, the multiple regression equation may be used to estimate voluntary quit rates (therefore, continuance rates and needed replacements) on the basis of weekly hours worked and compensated overtime, just as we used the simple regression equation in Chapter 15. Here, we have a composite relationship expressed in terms of a formula. Table 16.2 indicates the estimated values for the ten observations.

From Table 16.2, last two columns, we can see that the estimates are relatively close to the actual quit rates in most cases. However, since

TABLE 16.2
Calculation of estimated quit rates

Time	X_1	X_2	X_0	b_1X_1	b_2X_2	$\hat{X}_0$	$X_0 - \hat{X}_0$	$X_0/\hat{X}_0$ as Percent of Normal
1......	26	3	.5	1.078	.064	.56	−.06	89
2......	29	5	.6	1.203	.106	.73	−.13	82
3......	27	2	.7	1.120	.042	.58	+.12	125
4......	35	6	.8	1.451	.127	1.00	−.20	80
5......	31	8	.9	1.286	.170	.88	+.02	102
6......	37	7	1.0	1.534	.149	1.10	−.10	91
7......	45	6	1.1	1.866	.127	1.41	−.31	78
8......	40	9	1.2	1.659	.191	1.27	−.07	94
9......	41	8	1.5	1.700	.170	1.29	+.21	116
10......	39	6	1.7	1.617	.127	1.17	+.53	145
Σ......	350	60	10.0			9.99	+.01	

$$\hat{X}_0 = -.579 + .041468X_1 + .021221X_2$$

other factors besides weekly hours and overtime apparently account for some voluntary quits, a relatively large portion of the variance is left unexplained. For time period 1, the estimated value of the dependent variable is computed:

$$\hat{X}_0 = -.579 + .041468(26) + .021221(3)$$
$$\hat{X}_0 = .56$$

The degree of association may be measured by the coefficient of determination by using Formula (16.21):

$$R^2_{0.12} = \frac{b_1 \Sigma x_1 x_0 + b_2 \Sigma x_2 x_0}{\Sigma x_0^2}$$

$$R^2_{0.12} = \frac{(.041468)(175) + (.021221)(45.0)}{13.40}$$

$$R^2_{0.12} = .61$$

This means that 61 percent of the variation in voluntary quit rates is explained by variations in average weekly hours worked and compensated overtime. Thirty-nine percent of the variation in the dependent variable is left unexplained by these two independent variables. It should be noted that the subscript of the coefficient indicates that the numbers to the right of the decimal are the independent variables.

We also may have discovered the value of $R^2_{0.12}$ by using Σt^2, where (see Formula [16.15])

$$\Sigma t^2 = b_1 \Sigma x_1 x_0 + b_2 \Sigma x_2 x_0$$

and is the explained variance, whereas Σx_0^2 is the total variance to be explained. Then, from Formula (16.20),

$$R^2_{0.12} = \frac{\Sigma t^2}{\Sigma x_0^2}$$

$$R^2_{0.12} = \frac{8.21}{13.40}$$

$$R^2_{0.12} = .61$$

It was stated previously that the b's do not provide the relative importance of the independent variables. Therefore, we turn to the beta coefficients using Formula (16.28):

$$\beta_1 = b_1 \cdot \frac{\sigma_{x1}}{\sigma_{x0}},$$

where

$$\sigma_{x1} = 6.16; \sigma_{x0} = .36; \quad \text{and} \quad b_1 = .041468$$

So

$$\beta_1 = .041468 \left(\frac{6.16}{.36}\right)$$

$$\beta_1 = .709564$$

and using Formula (16.29):

$$\beta_2 = b_2 \cdot \frac{\sigma_{x2}}{\sigma_{x0}}$$

where

$$b_2 = .021221 \quad \text{and} \quad \sigma_{x2} = 2.0$$

Then,

$$\beta_2 = .021221 \left(\frac{2.0}{.36}\right)$$

$$\beta_2 = .117894$$

which, interpreted, means that average weekly hours are considerably more important in the determination of the level of the voluntary quit rates than is compensated overtime. Although, in this problem, we would have reached the same conclusion using the *b*'s, this is not always the case. In addition, the relative importance of the independent variables is not so clear using the *b*'s.

The statistical reliability of both the *b*'s and the β's may be found as follows:

$$\hat{\sigma}_\beta = \left[\frac{1 - R^2_{0.12}}{(1 - r^2_{21})(n - m)}\right]^{1/2} \tag{16.30}$$

where r^2_{21} is the simple coefficient of determination for weekly hours and overtime, n is the number of paired items, and m equals the number of constants in the multiple regression equation.

Then, we compute t values to determine the significance of each, where

$$t_1 = \frac{\beta_1}{\hat{\sigma}_{\beta 1}} \text{ and } \quad t_2 = \frac{\beta_2}{\hat{\sigma}_{\beta 2}} \tag{16.31}$$

Then, we refer to Appendix D and compare the computed t values with table values for $n - m$ degrees of freedom.

Another convenient way to measure the importance of the independent variables is by a "rule of thumb" which says that "if any of the regression coefficients are of a magnitude less than twice their standard error, they are not significant."[7]

A measure of the estimated standard error of the net regression coefficients (b's) also is available, where

$$\hat{\sigma}_{b0.12} = \frac{\hat{\sigma}_{x0}}{\hat{\sigma}_{x1}} \left[\frac{1 - R^2_{0.12}}{(n - m)(1 - R^2_{0.12})} \right]^{1/2} \qquad (16.32)$$

and is interpreted in a manner similar to the other standard errors encountered earlier.

Reliability of multiple coefficient of correlation

We may use Appendix Table F.1 to evaluate the significance of the correlation. Table F.1 is based on the ratio of the two variances involved. That is, the F value represents the ratio of the explained variance to the unexplained variance adjusted for the degrees of freedom lost. Such tables describe the coefficients that may be expected to occur by chance alone among samples of uncorrelated data once in 20 times (.05 level), once in 100 times (.01 level), and once in 1,000 times (.001 level).

For multiple correlation the calculated F_c is found:

$$F_c = \frac{R^2_{0.12}}{1 - R^2_{0.12}} \cdot \frac{n - m}{m - 1} \qquad (16.33)$$

And in the problem above,

$$F_c = \frac{.61}{1 - .61} \cdot \frac{10 - 3}{3 - 1}$$

$$F_c = 5.474$$

And from Table F.1,

$$F_{.05} = 4.737$$
$$F_{.01} = 9.547$$

Consequently, we conclude that the coefficient here computed ($R^2_{0.12}$ = .61) is significant, from a statistical sampling point of view. That is to say, the correlation did not occur by chance, and we assume that there is a relationship among the series.

In effect, we could test the significance of our coefficient as follows:

[7] See Frederick V. Waugh, "A Simplified Method of Determining Regression Constants," *Journal of the American Statistical Association*, vol. 30, no. 2, pp. 694–700.

H_0: Population $R^2_{0.12} = 0$, that is, there is no correlation in the universe; the observed $R^2_{0.12} = .61$ occurred by chance.
H_1: Population $R^2_{0.12} \neq 0$, that is, there is correlation in the universe; the observed $R^2_{0.12} = .61$ is significant.

Then, we use our F table to apply a critical ratio which we may subject to probability interpretation in order to decide between H_0 and H_1. The criterion for our decision at the .05 level is:

Conclude H_0 if

$$F_c \leq F_{.05}$$

Conclude H_1 if

$$F_c > F_{.05}$$

In our problem, since $F_c = 5.474 > F_{.05} = 4.737$, we conclude H_1. That is, we may say the correlation is significant but not *highly significant* because $F_c = 5.474 < F_{.01} = 9.547$.

Standard error of estimate

It is frequently useful to have another measure of reliability of the multiple regression equation in addition to the coefficient of multiple determination. A good measure is the *standard error of the estimate.* It may easily be calculated, using Formula (16.17):

$$\Sigma d^2 = \Sigma x_0^2 - \Sigma t^2$$

where

$$s^2_{0.12} = \frac{\Sigma d^2}{n}$$

and corrected for the error of sampling, we obtain Formula (16.24):

$$\hat{\sigma}^2_{0.12} = \frac{\Sigma d^2}{n} \cdot \frac{n}{n-m}$$

or when using the centering equations,

$$\hat{\sigma}^2_{0.12} = \frac{\Sigma x_0^2 - (b_1 \Sigma x_1 x_0 + b_2 \Sigma x_2 x_0)}{n-m} \qquad (16.34)$$

Note: Here, we must divide Σx_0^2, $\Sigma x_1 x_0$, and $\Sigma x_2 x_0$ by n first, because the terms really are $n\Sigma x_0^2$, $n\Sigma x_1 x_0$ and $n\Sigma x_2 x_0$. (See page 524.) Therefore,

$$\hat{\sigma}^2_{0.12} = \frac{1.340 - [(.041468)(17.50) + (.021221)(4.5)]}{10 - 3}$$

$$\hat{\sigma}^2_{0.12} = .074$$
$$\hat{\sigma}_{0.12} = .272$$

Interpreted, the standard error of the estimate may be applied as follows to our "quit rate" problem. When

$$X_1 = 31; X_2 = 8; \hat{X}_0 = .88$$

from our regression equation, that is, *on the average,* when $X_1 = 31$ and $X_2 = 8$, then the voluntary quit rates are estimated to be .88. The .95 confidence interval of the quit rate would be

$$\hat{X}_0 \pm t_{.05}\hat{\sigma}_{0.12}$$

or .88 ± 2.365 (.272), so the interval is .24–1.52. This seems to indicate a need for using additional independent variables in our correlation or taking a larger sample, or maybe both. That is, the interval is rather wide and indicates a low-degree predictable power of the regression equation.

Simple coefficient of correlation

When "running" a multiple correlation, it frequently is convenient and desirable to measure the simple relationships existing among the series. We can do this by referring to the *np* block in the multiple correlation. Notice that a diagonal line (3,780, 440, and 13.40) represents the centered squares (Σx_1^2, Σx_2^2, and Σx_0^2) and that the other items, except the check column, are centered cross products ($\Sigma x_1 x_2$, $\Sigma x_1 x_0$, and $\Sigma x_2 x_0$).

A convenient form for computation of the simple *r*'s is illustrated below, using the *np* block of the "quit rate" problem:

	X_1	X_2	X_0	Z
np block 1	3,780	860	175.0	4,815.0
2	*(860)*	440	45.0	1,345.0
0	*(175.0)*	*(45.0)*	13.40	233.40

Divide each row by the centered squared to obtain the *b*'s:

	X_1	X_2	X_0	Z
1	1.000	.228	.046	
2	*(1.954)*	1.000	.102	
0	*(13.060)*	*(3.358)*	1.000	

Then, multiplying the complementary b's (i.e., b_{21} by b_{12}, etc.):

$$r_{01}^2 = (.046)(13.06) = .60$$
$$r_{02}^2 = (.102)(3.358) = .34$$
$$r_{21}^2 = (.228)(1.954) = .44$$

This indicates that 60 percent (r_{01}^2) of the variation in voluntary quit rates is explained by variations in average weekly hours worked, and so on.

PARTIAL CORRELATION

Previously, the b's were referred to as *net* regression coefficients. This is an accurate description because these constants reflect the net regression of the dependent variable (X_0) upon the respective independent series after the impact of the remaining independent variables theoretically has been held constant. That is to say, b_1 is the measure of net regression of X_0 on X_1; b_2 is the measure of net regression X_0 on X_2; b_3 is the measure of net regression of X_0 on X_3; and so on, until b_n is the measure of net regression of X_0 on X_n. Naturally, the concept of these measures being a *net* value stems from the assumption that the effects of the other independent variables have been removed.

Partial correlation analysis is a direct result of this concept of net regression. As alluded to previously, this technique, which is a logical extension of multiple correlation and regression analysis, helps the analyst evaluate the association between a dependent variable and a *single* independent series, other factors remaining constant. It is by using this method of regression that the economist can quantify the phrase "other things being equal." Clearly, this approach is decidedly different from the simple correlation discussed in Chapter 15. Recall that there, we were concerned with the measurement of the covariation between two series while ignoring the influence of any other independent variables.

The objectives of partial correlation and regression analysis are similar to laboratory techniques of the natural sciences. In the so-called "exact" sciences, it is relatively easy to hold constant the impact of a number of factors while measuring the relationship between other variables. In business, for example, it may be useful to measure the degree of association between certain costs and profits of given products holding the demand constant. Fortunately, the b's or the β's provide us with a basis for doing just that.

Illustration

From our earlier example the net regression coefficient was $b_1 = .041468$, which is the net regression coefficient describing the regression of voluntary quit rates (X_0) on average weekly hours worked (X_1). An estimating equation may be readily set up by discovering the value of the constant *a:*

$$a = \bar{X}_0 - b_1\bar{X}_1$$
$$a = 1.0 - (.041468)(35)$$
$$a = -.451$$

Therefore the estimating equation becomes

$$\hat{X}_0 = -.451 + .041468X_1$$

where X_0 is the estimated quit rate, assuming *average* compensated overtime. Estimates may be made as usual merely by substituting appropriate weekly hour values for X_1.

Also, $b_2 = .021221$ is the net regression coefficient describing the regression of voluntary quit rates (X_0) on compensated overtime (X_2). The estimating equation may be found as above:

$$a = \bar{X}_0 - b_2\bar{X}_2$$
$$a = 1.0 - (.021221)(6)$$
$$a = .873$$

Therefore the estimating equation becomes

$$\hat{X}_0 = .873 + .021221X_2$$

where $\hat{X}_0$ is the estimated quit rates, assuming weekly hours worked remains at its average. Again, estimates may be made merely by substituting appropriate values of compensated overtime for X_2

Coefficient of partial correlation

The coefficient of partial correlation is symbolized by the characteristic (r) of simple correlation, but there always is a subscript indicating the variables being correlated and the series held constant. To illustrate: $r_{01.2}$ is the partial correlation coefficient where the series X_0 is correlated with X_1, holding X_2 constant. In general, the portion of the subscript to the left of the period designates the series being correlated, whereas the variables to the right of the period are those whose influences theoretically are held at their average level.

Calculation of the coefficient of partial correlation might proceed from that of the net regression coefficient. However, generally, it is more convenient to utilize a method that combines the simple correlation coefficients to determine the partial coefficient. Consequently, the coefficient of partial correlation between the dependent variable X_0 and the independent variable X_1, holding constant X_2, another independent series is

$$r_{01.2} = \frac{r_{01} - (r_{02})(r_{21})}{\sqrt{(1 - r_{02}^2)(1 - r_{21}^2)}} \tag{16.35}$$

Applying this measure to our example where the simple correlation coefficients are:

$$r_{01}^2 = .60;\ r_{01} = .77$$
$$r_{02}^2 = .34;\ r_{02} = .58$$
$$r_{21}^2 = .44;\ r_{21} = .66$$

The covariation of quit rates and weekly hours, holding overtime constant, is

$$r_{01.2} = \frac{.77 - (.58)(.66)}{\sqrt{(1 - .34)(1 - .44)}} = .64$$

This coefficient of partial correlation reflects the correlation of voluntary quit rates and weekly hours, assuming compensated overtime remaining constant at its average value.

Likewise, when weekly hours are held constant, the relationship of quit rates and overtime can be found:

$$r_{02.1} = \frac{r_{02} - (r_{01})(r_{21})}{\sqrt{(1 - r_{01}^2)(1 - r_{21}^2)}}$$

Therefore the measure of partial correlation is

$$r_{02.1} = \frac{.58 - (.77)(.66)}{\sqrt{(1 - .60)(1 - .44)}}$$

$$r_{02.1} = .155$$

which is the measure of the relationship between quit rates and overtime, holding weekly hours at their average value.

Meaning of partial correlation

These coefficients of partial correlation have a definite measurable importance in terms of the proportion of the total variance in the dependent variable that they account for, a meaning similar to that noted with respect to the coefficients of simple and multiple correlation. In the latter cases, we noted that the coefficient of determination (r^2 and R^2) indicated the proportion of the total variance in the dependent variable that is accounted for or explained by the variance of the regression equation. It can be shown that where the addition of an independent variable in the multiple correlation results in a reduction of the unexplained variance, this reduction, expressed as a percentage of total unexplained variance (before the inclusion of any given additional series), is equal to the coefficient of partial determination for that given series. Consequently, the

TABLE 16.3
Meaning of partial correlation: "Quit rate" problem (X_1 = weekly hours; X_2 = compensated overtime; X_0 = voluntary quit rates)

The Meaning of $r_{01.2} = .64$; $r^2_{01.2} = .41$	
1. The total variance to be accounted for (Σx_0^2), expressed as a percentage	100.0%
2. Variance accounted for by the simple coefficient of determination, r^2_{02} (see page 533)	34.0
3. Remainder of variance in simple correlation unaccounted for, 100.0 − 34.0	66.0
4. Variance accounted for by multiple coefficient of determination, $R^2_{0.12}$ (see page 526)	61.0
5. Remainder of variance in multiple correlation unaccounted for, 100.0 − 61.0	39.0
6. Reduction in the accounted-for variance by including r_{01} in the multiple correlation, 66.0 − 39.0	27.0
7. Percent reduction in unaccounted-for variance, 27.0/66.0 × 100	41.0
8. $r^2_{01.2} = .41$, or 41%	

TABLE 16.4
Meaning of partial correlation: "Quit rate" problem (X_1 = weekly hours; X_2 = compensated overtime; X_0 = voluntary quit rates)

The Meaning of $r_{02.1} = .16$; $r^2_{02.1} = .025$	
1. The total variance to be accounted for $(\Sigma x_0)^2$, expressed as a percentage	100.0%
2. Variance accounted for by the simple coefficient of determination, $r^2_{01} = .60$ (see page 531)	60.0
3. Remainder of variance in simple correlation unaccounted for, 100.0 − 60.0	40.0
4. Variance accounted for by multiple coefficient of determination, $R^2_{0.12}$ (see page 526)	61.0
5. Remainder of variance in multiple correlation unaccounted for, 100.0 − 61.0	39.0
6. Reduction in the accounted-for variance by including r_{02} in the multiple correlation, 40.0 − 39.0	1.0
7. Percent reduction in unaccounted-for variance, 1.0/40.0 × 100	2.5
8. $r^2_{02.1} = .025$, or 2.5%	

squared coefficient of partial correlation measures the percentage reduction in the unexplained variance that is attributable to covariation, with the given independent series represented by the coefficient. Tables 16.3 and 16.4 illustrate the above concept.

AN APPLICATION OF STEPWISE REGRESSION

In order to demonstrate the theory and application of stepwise regression, the computer results of the quit rate problem are now shown (Tables

16.5, 16.6, and 16.7). The data were run on the IBM 370/168 at the University of Michigan Computing Center using the MIDAS stepwise regression subprogram as developed by the Statistical Research Laboratory at Ann Arbor. While there is no intent to imply great precision by the number of decimal places, the computer output is presented just as the computer printout shows them. MIDAS prints out five significant digits rounding the last one. This should make it easier to follow the program. The dependent variable is listed as X0 (quit rates); X1 (weekly hours) and X2 (overtime) are the independent variables as they were in our original model. Table 16.5 of the output is self-explanatory identifying the problem and reproducing the data.

TABLE 16.5

```
M I D A S
STATISTICAL RESEARCH LABORATORY
UNIVERSITY OF MICHIGAN
13:00:48
SEP 23, 1975


< READ V=1-3 C=1-10 F=* FC=* L=X0QRATES,X1WKHRS,X2OTIME >

READ OBSERVATIONS
VARIABLES BY CASE

10 CASES READ FOR 3 VARIABLES


< WRIT V=* C=* FO=* >


WRITE OBSERVATIONS
VARIABLES BY CASE

 1.            2.            3.
 X0QRATES      X1WKHRS       X2OTIME


 .50000        26.000        3.0000

 .60000        29.000        5.0000

 .70000        27.000        2.0000

 .80000        35.000        6.0000

 .90000        31.000        8.0000

 1.0000        37.000        7.0000

 1.1000        45.000        6.0000

 1.2000        40.000        9.0000

 1.5000        41.000        8.0000

 1.7000        39.000        6.0000

10 CASES WRITTEN FOR 3 VARIABLES
```

Table 16.6 provides the descriptive measures of the mean, standard deviation, and the range. Also indicated are the standard error of the mean which might be used to make an interval estimate of each of the variables. For example, it is estimated that 95 percent of the workweeks are included within the interval $\bar{X}_1 \pm z\hat{\sigma}_{\bar{x}}$:

$$35.0 \pm 1.96(2.0494) = 30.9 \text{ to } 39.1$$

which is interpreted in the same manner as the confidence intervals calculated in Chapter 9. (Note: MIDAS prints everything in capital letters, like most computer output, therefore, $N = 10$ really refers to the sample size and normally would be given as n.)

The second portion of Table 16.6 provides the simple correlation matrix. That is, the simple coefficients of correlation (r) are presented along with the degrees of freedom (DF $= 8$), and the significance of these measures of the relationships among the variables. For example, the relationship between X0 quit rates and X1 weekly hours is described by $r = .7776$. The R @ .9900 $=$ .7646 indicates that the relationship is highly significant at the .01 level. The R @ .9500 $=$ .6319 indicates the size

TABLE 16.6

< DESC V=1-3 C=ALL >

DESCRIPTIVE MEASURES

VARIABLE	N	MINIMUM	MAXIMUM	MEAN	STD DEV
1.X0QRATES	10	.50000	1.7000	1.0000	.38586
2.X1WKHRS	10	26.000	45.000	35.000	6.4807
3.X2OTIME	10	2.0000	9.0000	6.0000	2.2111

< CORR V=1-3 C=ALL >

CORRELATION COEFFICIENTS

N= 10 DF= 8 R@.0500= .6319 R@.0100= .7646

VARIABLE			
1.X0QRATES	1.0000		
2.X1WKHRS	.7776	1.0000	
3.X2OTIME	.5860	.6668	1.0000
	1. X0QRATES	2. X1WKHRS	3. X2OTIME

of the coefficient of correlation needed to be judged significant at the .05 level. It is noted that $r_{02} = .5860$, the relationship between X0 quit rates and X2 overtime is not significant at the .05 level. The relationship of X2 overtime and X1 weekly hours (.6668) is significant but not highly significant at the .01 level. Of course, these computations agree with those done "by hand" on the desk calculator and shown on pages 531 and 533. That is, $r^2_{01} = (.7776)^2 = .60$; $r^2_{02} = (.5860)^2 = .34$; and $r^2_{21} = (.6668)^2 = .44$.

Table 16.7 provides the analysis of variance of the problem along with the regression equation and the measures of standard error. The explained variance and its related measures are listed in the row REGRESSION. The F test indicates that the relationship is statistically significant

TABLE 16.7

```
< REGR V=1,2-3 C=ALL >

LEAST SQUARES REGRESSION


ANALYSIS OF VARIANCE OF 1.XOQRATES   N= 10 OUT OF 10

     SOURCE              DF   SUM SQRS      MEAN SQR        F-STAT      SIGNIF

     REGRESSION           2   .82119        .41059          5.5399      .0361
     ERROR                7   .51881        .74116  -1
     TOTAL                9   1.3400

     MULT R= .78283   R-SQR= .61283  SE= .27224


     VARIABLE     PARTIAL     COEFF         STD ERROR      T-STAT      SIGNIF

     CONSTANT                 -.57871       .50921        -1.1365      .2932
   2.X1WKHRS      .64054      .41468  -1    .18791  -1    2.2069      .0631
   3.X2OTIME      .14411      .21221  -1    .55076  -1    .38531      .7114




< REGR V=1,2-3 C=ALL O=STANDARD >

LEAST SQUARES REGRESSION


ANALYSIS OF VARIANCE OF 1.XOQRATES   N= 10 OUT OF 10

     SOURCE              DF   SUM SQRS      MEAN SQR        F-STAT      SIGNIF

     REGRESSION           2   .61283        .30641          5.5399      .0361
     ERROR                7   .38717        .55310  -1
     TOTAL                9   1.0000

     MULT R= .78283   R-SQR= .61283  SE= .23518


     VARIABLE     PARTIAL     BETA WT       STD ERROR      T-STAT      SIGNIF

   2.X1WKHRS      .64054      .69648        .31560         2.2069      .0631
   3.X2OTIME      .14411      .12160        .31560         .38531      .7114
```

from a sampling point of view at the .04 level. The unexplained variance and its related measures are listed in the row ERROR. The total variance is the sum of these two. The coefficient of multiple determination ($R^2_{0.12}$) of .61283 agrees with the value previously calculated on page 526. The coefficient was calculated by MIDAS by dividing .82119 by the total sum of squares 1.3400. The standard error of the regression (.27224) was found by taking the square root of the unexplained sum of squares (.51881) divided by $n - m$ degrees of freedom. That is $(.51881)/10 - 1 = .07411$. The square root of .07411 is .27224.

The multiple regression equation is given directly below the coefficient of correlation as:

$$\hat{X}_0 = -.57871 + .041468X_1 + .021221X_2$$

which agrees with the equation given on page 525, Table 16.2. The coefficients of partial correlation are also given as $r_{01.2} = .64$ and $r_{02.1} = .14$ which agree with the values given on page 533, except for rounding errors.

The final section of Table 16.7, where the standardizing option is used, calculates the beta values in a manner similar to that shown on page 526. Again, the differences are largely due to rounding; however, the concept of betas used in MIDAS is slightly modified compared to the betas on page 521.

Table 16.8 makes use of the stepwise regression subprogram of MIDAS which is virtually the same as the UCLA BMD2R program widely used by many universities. The independent variable X1 weekly hours, which is the most highly correlated with the dependent variable X0 quit rates, enters the stepwise equation first. (See EQN = 1.) The significant levels used for inclusion and exclusion of variables were .10 and .20 respectively. That is, the significance level for inclusion of any variable must be less than .10 and greater than .20 for exclusion of any independent variables. The unique feature of the stepwise program is that once a variable is included it is still subjected to the inclusion significance test if another variable enters the model. That is, in some cases once a second or third variable enters the regression the initial variable included might then be judged not significant and therefore is excluded. Of course, this did not happen in our example. In fact, only one variable entered and then the stepwise program stopped because X2 overtime had a significance level of .7114 which is greater than .20 and therefore is excluded from the model. Because there are no more independent variables the program stops. The partial coefficients of correlation are the measures used to determine whether or not a second or third or fourth, and so on, variable enters the model. The initial independent variable to enter is selected on the basis of the simple coefficient of correlation (r) that is the highest in the correlation matrix. It is noted here that the simple relationship between X0 quit rates and X1 weekly hours is expressed by the coefficient

TABLE 16.8

```
< SEL V=1,2-3 C=ALL O=STEPWISE,FORWARD M=* L=.10,.20 >

SELECTION OF REGRESSION


ANALYSIS OF VARIANCE OF X0QRATES              N= 10          EQN= 1

SOURCE                 DF    SUM OF SQRS   MEAN SQUARE   F-STATISTIC SIGNIF

REGRESSION              1    .81019        .81019        12.233        .0081
ERROR                   8    .52981        .66227 -1
TOTAL                   9    1.3400

MULTIPLE R= .77757   R-SQR= .60462    SE= .25735


VARIABLE          PARTIAL    COEFFICIENT    STD ERROR     T-STATISTIC SIGNIF

CONSTANT                     -.62037        .47037        -1.3189       .2237
X1WKHRS           .77757     .46296 -1      .13236 -1     3.4976        .0081

REMAINING         PARTIAL    SIGNIF

X2OTIME           .14411     .7114


REGRESSION OF X0QRATES             FORWARD  SELECTION

EQN   R-SQR   STD ERR   # VAR     VARIABLE         PARTIAL   T-STAT      SIGNIF

1     .60462  .25735    1 IN      X1WKHRS          .77757    3.4976      .0081
```

of determination (.60462) which is not much smaller than the $R^2_{0.12}$ of .61283.

Table 16.9 shows the predicted values of the dependent variable (X_0) as well as the residuals. The latter being the difference between the actual values of X_0 and the predicted values. The residuals are then plotted against each variable in Figure 16.3. Normally it would be preferred to plot these on separate charts; however, they are combined here to indicate an option available on most computers and to save space here. The residuals represent the variation in the dependent variable *unexplained* by the regression. Plotting the residuals against the variables is a useful graphic tool to determine if there is a systematic relationship between the unexplained variation and the other variables. If there is a high relationship between the residuals and a particular variable the analyst might consider eliminating this independent variable from the regression model. If the residuals are plotted against time one can readily see the influence of the trend factor if present. On the basis of Figure 16.3 it does not appear that any systematic correlation is present in the data. An examination of the residuals in this manner can help us to examine our original assumptions about the model and to permit us to make any necessary modifications. While a detailed study of the residuals is not appropriate

FIGURE 16.3

```
COMMAND
?SCATTER V=4,1,2,3 OPTIONS=ONEGRAPH

SCATTER PLOT

RESIDUAL                 10 CASES FOR THIS GRAPH
 1.3842   +                                   3   2                1

          +

 .94429   +

          +

 .50434   +                                          2  13

          +3  2   1

 .64380 -1+           2   1                            3

          +X     3                       1     2               3

              1  2           X           2    3
-.37558   +

                     1           2   3
          +

-.81553   +                         1  3                            2
          +----+----+----+----+----+----+----+----+----+----+
  (1)  .50000          .98000            1.4600      QUITRATE
              .74000            1.2200               1.7000
  (2)  26.000          33.600            41.200      WEEKHOUR
              29.800            37.400               45.000
  (3)  2.0000          4.6000            7.6000      OVERTIME
              3.4000            6.2000               9.0000
```

here, a comment with respect to *outliers* might be in order. If one or more of the residual points are far greater than the others (at least three standard deviations from the mean of the residuals), one might be suspect of this nontypical data plot. One should not eliminate *any* outlier too rapidly because it might be providing information that the other points cannot. Perhaps, we made an error in our original data. Or possibly the outlier indicates an unusual combination of circumstances that might be of interest to the analyst.[8] At this point our computer run was terminated at a total cost of less than $2! Certainly, this is one of today's better bargains. How long would this same set of calculations and plots have taken an efficient statistician to do? The answer is too frightening even to contemplate. Clearly, if we had many observations and/or many independent

[8] See F. J. Anscombe and J. W. Tukey, "The Examination and Analysis of Residuals," *Technometrics*, vol. 5 (1963), pp. 141–60.

TABLE 16.9
List of the actual (observed) values of X_0 quit rates, the predicted values based on multiple regression model of X_0, X_1 and X_2, and the residuals $(X_0 - \hat{X}_0)$

Observation	Actual Values (observed) X_0	Predicted Values $\hat{X}_0$	Residuals $X_0 - \hat{X}_0$
1	.50000	.56312	−.06312
2	.60000	.72997	−.12997
3	.70000	.58337	.11663
4	.80000	1.00000	−.20000
5	.90000	.87657	.02343
6	1.00000	1.10416	−.10416
7	1.10000	1.41468	−.31468
8	1.20000	1.27100	−.07100
9	1.50000	1.29125	.20875
10	1.70000	1.16587	.53413

variables to work with this procedure would become unmanageable by the desk calculator.

Actually, the analysis was taken one step farther. Tables 16.10 and 16.11 reflect the attempt to determine the possible influence of any curvilinearity in the regression model. To do this two new variables were created, namely variables 5 and 6. These variables represent the square of the values of variables X_2 and X_3 respectively. Then the regression model was run again using X_2, X_3, X_5, and X_6 as the independent variables and X_0 the dependent variable (quit rates). Because the coefficient of multiple determination ($R^2_{0.12356} = .69391$) is higher than the previous relationship we can conclude that allowing for some curvilinearity has improved the model as a representation of the data. This is true even though the standard error of the estimate is slightly larger than previously. However, the level of significance has gone up to .1418 as opposed to .0361 with only three independent variables.

GRAPHIC MULTIPLE CORRELATION

Although the above introduction to the mathematical methods of multiple correlation serve as a useful springboard to further work in the area, a nonmathematical approach has been developed but is not described in many basic statistics texts. This is unfortunate because the graphic method of multiple correlation makes it much easier to comprehend the underlying concepts than does the mathematical approach. In addition, for brief problems such as the one described above, the graphic technique is much quicker, and if done properly, it can give adequate approximations of the regression in many situations. And although this method is

TABLE 16.10

```
< TRANS V5=V2*V2 >

VARIABLE TRANSFORMATION

        VARIABLE    TOTAL    VALID    MISS

     5.VAR 5           10       10       0


< TRANS V6=V3*V3 >

VARIABLE TRANSFORMATION

        VARIABLE    TOTAL    VALID    MISS

     6.VAR 6           10       10       0


< WRIT V=5-6 C=* FO=* >


WRITE OBSERVATIONS
VARIABLES BY CASE

 5.          6.
 VAR 5       VAR 6


 676.00      9.0000

 841.00      25.000

 729.00      4.0000

 1225.0      36.000

 961.00      64.000

 1369.0      49.000

 2025.0      36.000

 1600.0      81.000

 1681.0      64.000

 1521.0      36.000

10 CASES WRITTEN FOR 2 VARIABLES
```

a useful tool of analysis, its primary virtue is that it offers the statistical neophyte a thorough comprehension of correlation and regression analysis. There is no better way to obtain a feel for the data than by plotting the series as charts; it highlights the relationship and warns of any intercorrelations present. Only the statistician with a great deal of experience can achieve a similar appreciation of the association when computers are used. Naturally, once one has the feel of the data and a complete grasp of the concepts, the electronic computers remove the sheer drudgery of multiple correlation and regression analysis. By now, the reader should

TABLE 16.11

```
< REGR V=1,2,3,5,6 >

LEAST SQUARES REGRESSION


ANALYSIS OF VARIANCE OF 1.XOQRATES    N= 10 OUT OF 10
```

SOURCE	DF	SUM SQRS	MEAN SQR	F-STAT	SIGNIF
REGRESSION	4	.92983	.23246	2.8337	.1418
ERROR	5	.41017	.82033 -1		
TOTAL	9	1.3400			

```
MULT R= .83301   R-SQR= .69391  SE= .28641
```

VARIABLE	PARTIAL	COEFF	STD ERROR	T-STAT	SIGNIF
CONSTANT		-5.6499	4.7848	-1.1808	.2908
2.X1WKHRS	.48534	.37025	.29829	1.2412	.2696
3.X2OTIME	-.32614	-.22280	.28881	-.77144	.4753
5.VAR 5	-.44050	-.44492 -2	.40551 -2	-1.0972	.3226
6.VAR 6	.30319	.16000 -1	.22489 -1	.71145	.5086

detect that it is the function of both Chapter 15 and Chapter 16 to provide this *basic* understanding. Once the student has this needed background, he or she will be ready for more advanced work; and in time will be prepared to utilize effectively the assistance of computers.

For purposes of illustration, let us continue with the basic data of the "quit rate" problem. The variables are

$$X_1 = \text{average weekly hours worked}$$
$$X_2 = \text{compensated overtime}$$
$$X_0 = \text{voluntary quit rates}$$

Again, essentially, our objective is to discover a basis for estimating voluntary quit rates. We assume, of course, that a functional relationship exists among the factors. Remember that in the mathematical methods described above, our major goal was to develop a *formula* for estimation purposes. In the graphic approach, we really are merely attempting to identify a *curve* which depicts the relationship of the variables. Although the two techniques have slightly varying primary objectives, it should be noted that the graphic method may be applied to any situation where the mathematical approach is appropriate.

Once the data have been arranged in the usual correlation form, the next step is to plot a scatter diagram of the dependent variable (X_0) against one of the independent variables. We shall notice that the order of choice of the independent variables is not important; however, generally, it is traditional to number them in decreasing order of significance. Fulfillment of this customary requirement can be facilitated by plotting preliminary scatter diagrams between the dependent and each of the independent variables. Obviously, the dependent variable that would appear to have the highest relationship to X_0 would be termed X_1, and so on. Because we need to identify the dots, they should be labeled in all charts.

Figure 16.4 represents the simple relationship of voluntary quit rates (X_0) on average weekly hours worked (X_1). (See Table 16.12 for basic data.) Here, we want to explain variations in quit rates in terms of

FIGURE 16.4
Relationship of quit rates and weekly hours worked

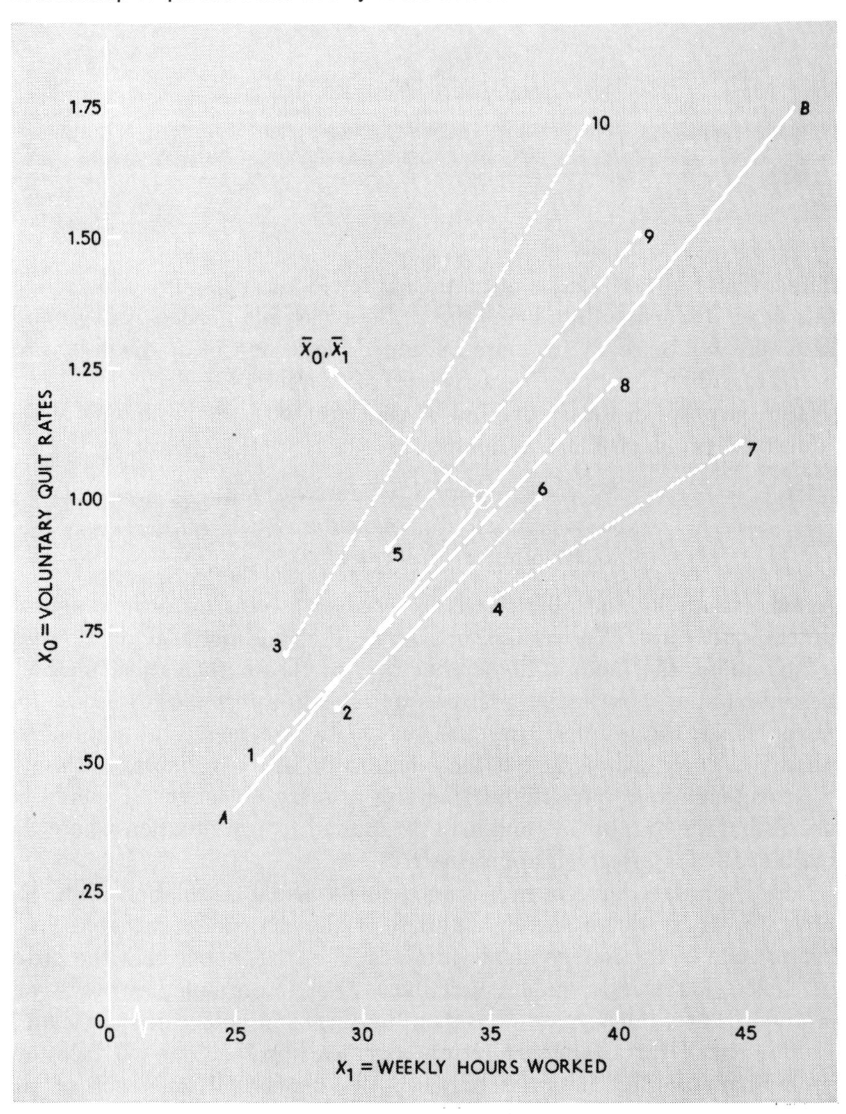

TABLE 16.12
Form for graphic multiple correlation: "Quit rate" problem

(1) Time Period	(2) X_0 Quit Rates	(3) X_1 Weekly Hours	(4) X_2 Compensated Overtime	(5) Expected Quit Rate Estimated on Basis of Regression of X_0 on X_1 (see Figure 16.4)	(6) $X_0 - \hat{X}_0$ (Column 2 – Column 5)	(7) Expected Additional Change in X_0 Due to X_2 (see Figure 16.5)	(8) Quit Rates as Estimated from X_1 and X_2 (Column 5 + Column 7)	(9) Actual Quit Rates Minus Expected Values (Column 2 – Column 8)
1......	.5	26	3	.46	.04	.12	.58	–.08
2......	.6	29	5	.65	–.05	.04	.69	–.09
3......	.7	27	2	.53	.17	.16	.69	.01
4......	.8	35	6	1.00	–.20	.00	1.00	–.20
5......	.9	31	8	.78	.12	–.08	.70	.20
6......	1.0	37	7	1.12	–.12	–.05	1.07	–.07
7......	1.1	45	6	1.61	–.51	.00	1.61	–.51
8......	1.2	40	9	1.31	–.11	–.13	1.18	.02
9......	1.5	41	8	1.38	.12	–.08	1.30	.20
10......	1.7	39	6	1.25	.45	.00	1.25	.45
Σ......	10.0	350	60		–.09			
$\bar{X}$'s.....	1.0	35	6					
Estimation:				10.09		–.02	10.07	

Column 9 = d; $\Sigma d^2 = .6025$

$$R^2_{0.12} = 1 - \frac{\Sigma d^2}{\Sigma X_0^2 - n\bar{X}_0^2}$$

$$R^2_{0.12} = 1 - \frac{.6025}{11.34 - (10)(1)^2} = .55$$

changes in weekly hours. Column 5 of Table 16.12 represents the expected quit rates estimated on the basis of the regression of X_0 on X_1. These values are read from Figure 16.4, the vertical axis. Care should be taken to read the values as precisely as possible.

For simplicity, a straight line was used in Figure 16.4. In drawing the freehand line, several guides are used. First, generally, a straight line should pass through the means of both series. The theoretical foundation for this lies in the concept of the least-squares principle. A second guide is that by definition, line *AB* is a line *of average* relationship and, as such, should be drawn so that approximately one half of the dots are on either side. It should be noted that it is accurate enough to approximate the slope of line *AB* because any error will be corrected in the following chart or charts. A third guide in determining the slope of *AB* is the "drift lines" —the dotted lines in Figure 16.4. Groups of two or more X_1 items that have nearly the same values are chosen. The following dots are grouped for our purposes: 3, 10; 5, 9; 1, 8; 2, 6; and 4, 7. Then each of these sets are connected by dotted drift lines, and the average slope of these drift lines is taken as the slope for line *AB*.

The use of these drift lines is the graphic method counterpart of estimating the net regression coefficients (slopes) in the mathematical method described above.

The regression line *AB* in Figure 16.4 represents the approximated average relationship between X_0, voluntary quit rates, and X_1, weekly hours worked. Consequently, we can estimate quit rates from this line. From Figure 16.4, we note that for the first time period, when $X_1 = 26$, $\hat{X}_0 = .46$. This figure .46 is then recorded in Column 5 of Table 16.12. Similarly, the values for the other nine periods are determined and recorded.

Because the dots on Figure 16.4 do not fall on the line *AB*, it is obvious that an important portion of the variation in quit rates is not explained by variations in weekly hours. Perhaps many other factors (additional independent variables) may be affecting labor turnover. Therefore, we need to know the amount of the variation in X_0 that is unaccounted for by fluctuations in X_1. Our next step is to subtract values in Column 5 from Column 2, that is, $X_0 - \hat{X}_0$, and record these values in Column 6 of Table 16.12. Although it is much easier to determine these values from the table, if we read from Figure 16.4 they represent the *vertical* deviations of the dots from line *AB*.

In order to explain more of the fluctuations in quit rates unexplained by weekly hours, we now turn to X_2. Figure 16.5 is constructed by using the data of Column 6 on the vertical axis and the horizontal axis representing X_2, compensated overtime. Notice that a heavy horizontal line is drawn depicting the zero value for the unexplained deviations from line *AB*. Actually, the zero line merely is the regression line drawn horizon-

FIGURE 16.5
Relationship of compensated overtime, X_2, and unexplained deviations from line *AB*

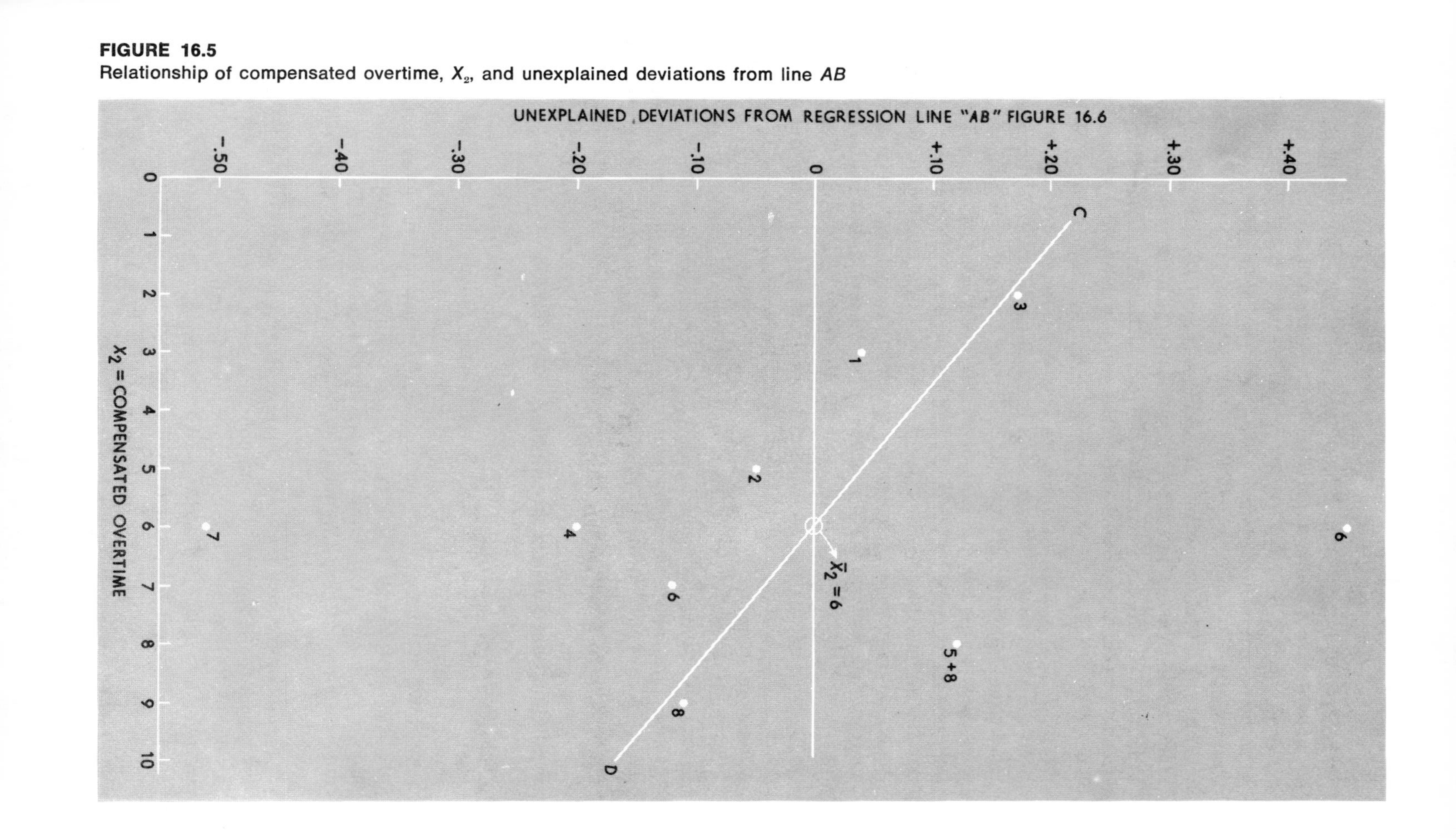

tally. Once again, a scatter diagram is plotted, this time using the data of Column 6 of Table 16.12 and Column 4, the remaining independent variable. Another freehand regression line, *CD*, is fitted to the data, as done previously in Figure 16.4. The line *CD* should pass through $\bar{X}_2$ at the zero line.

The total of Column 6 would equal zero if line *AB* had been a perfect fit; the mean then would be zero. This follows from the concept discussed in Chapters 5 and 6, namely, that the sum of the deviations of the items from their mean is zero. If the line *AB* is fitted carefully, the scatter about line *CD* is reduced. In any event, if the relationship is linear, line *AB* passes through $\bar{X}_0$ and $\bar{X}_1$, and line *CD* passes through $\bar{X}_2$ at the zero line.

In statistical terms the regression line *CD* in Figure 16.5 indicates the relationship between quit rates and compensated overtime (X_0 and X_2), after allowing for the influence of weekly hours (X_1). That is, line *CD* indicates the amount by which $\hat{X}_0$ will be higher or lower than the relationship of line *AB* represents. To illustrate the point, in Figure 16.4, with weekly hours (X_1) at 29, voluntary quit rates may be estimated at .65 if compensated overtime (X_2) remains at its average, 6. But in the second time period, when X_1 actually was 29, the level of overtime was sufficiently below its average (5 compared to $\bar{X}_2 = 6$) to pull actual quit rates down to .60 as compared to what would have been expected from the X_0 on X_1 relationship.

In the time period 9 the actual quit rates were higher than expected, based on the relationship of X_0 on X_1 (1.5 actual compared to 1.38 estimated). This is because compensated overtime was sufficiently above its mean (8 compared to 6) to raise actual quit rates above what would have been expected on the basis of weekly hours alone.

Prediction based on graphic method

Although the most important use of the graphic method is to assist the analyst in obtaining a feel for the relationship among the variables, Figures 16.4 through 16.6 may be used in prediction. In this application, it is worthwhile to note that generally the independent variable(s) must be predicted first; then, on the basis of these predictions the dependent variable is estimated. Consequently, we should remember that regression analysis is not a forecasting panacea; it merely facilitates the prediction process.

Prediction for time period 2. Assume that $X_1 = 40$ and $X_2 = 80$. The first step is to use Figure 16.4 and note that when weekly hours (X_1) have a value of 40, estimated voluntary quit rates are expected to be 1.30. Then we refer to Figure 16.5 and discover that when compensated overtime (X_2) is equal to 8, the estimated voluntary quit rates are $-.08$.

Therefore, our final estimate is $\hat{X}_0 = 1.30 - .08 = 1.22$.

FIGURE 16.6
Actual and estimated quit rates

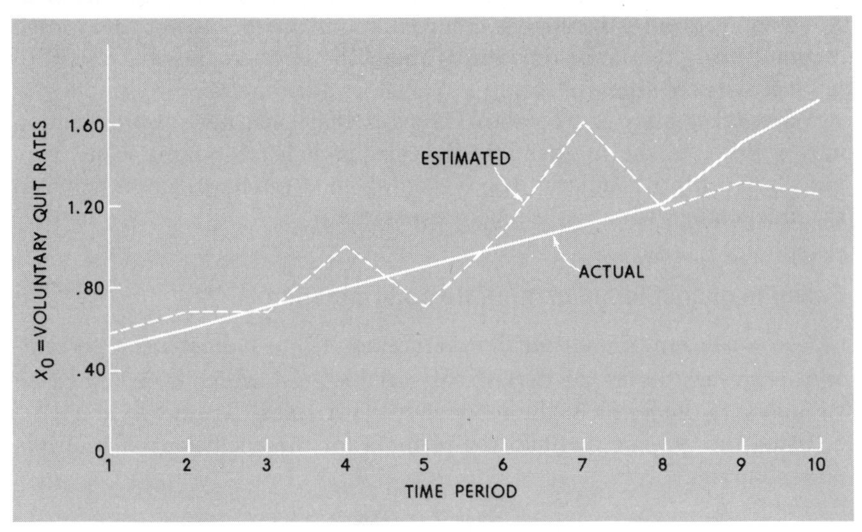

Procedure with three or more independent variables

If we had another independent variable (X_3) added to our relationship, the residuals in Column 9 of Table 16.12 would be obtained by reading off the deviations from the regression line *CD* in Figure 16.5, or by subtracting the values of Column 8 from those of Column 2. These values are then plotted against the values of X_3, and a new set of predictions is obtained. This new regression line would depict the partial relationship between X_0 and X_3, with X_1 and X_2 held constant.

Further refinements

Although the above procedure is only a first approximation to the regression curve, we could have improved our estimate by the method of successive approximations. To illustrate, in our original example using X_1, X_2, and X_0, the deviations from the regression line *CD* in Figure 16.5 could be recorded in Column 9, and then these values would be plotted about the regression line *AB* of Figure 16.4 against X_1. Generally, it is desirable for the analyst to use different-colored pencils here to identify easily the new sets of dots. Then a new regression line would be fitted on a freehand basis to the colored dots; this line should pass through the means of the two series. In effect, this new regression line is an improved estimate of the relationship of X_0 on X_1. The vertical deviations of the new scatter around the new regression line in Figure 16.4 would then be plotted about the regression line in Figure 16.5 against the same X_2 values.

Again, a new regression line is drawn (use another color) on Figure 16.5 and would represent the improved approximation to the relationship of X_0 on X_2. Naturally, the analyst can continue to repeat the above approximations, using the latest deviations about the latest regression line until no improvement seems necessary.

What takes place in the above steps is that each new approximation moves closer to the mathematically calculated least-squares value very quickly. This is particularly true if the degree of intercorrelation between the independent variables is low or nonexistent.

Estimate of coefficient of multiple determination

The coefficient of multiple determination ($R^2_{0.12}$) indicates the percentage of the variation in the dependent variable (X_0) which is explained by variations in the levels of the independent variables (X_1 and X_2).

Using the graphic method, the formula for the coefficient of multiple determination is

$$R^2_{0.12} = 1 - \frac{\Sigma d^2}{\Sigma X_0^2 - n\bar{X}_0^2}$$

where d equals the deviation of any point from the final regression in the last chart, that is, Figure 16.4 in our example, or the difference between the actual and the estimated value for the dependent variable (Column 9).

The value for our example is

$$R^2_{0.12} = 1 - \frac{.60}{11.34 - (10)(1)^2}$$

$$R^2_{0.12} = .55$$

Notice that this value is less than the one calculated by the mathematical method ($R^2_{0.12} = .61$). Had we used some further successive approximations, as described above, the graphic method would have produced virtually the same results.

The above formula indicates that $R^2_{0.12}$ is the ratio of the explained variance to the total variance. That is, the ratio of

$$\frac{\Sigma d^2}{\Sigma X_0^2 - n\bar{X}_0^2}$$

represents the variance in X_0 unexplained by the variance in X_1 and X_2.

NONLINEAR REGRESSION: A WARNING

In Chapter 14 it was discovered that not all trends can be described accurately with a straight line; neither can all regressions be fitted best unless the nonlinear models are considered. For example, the relationship between units of output and wages of production workers based on a straight piece rate might be depicted by a straight line. However, if the wage rate is an incentive type where the piece rate varies as output rises, then the relationship would be nonlinear. Output per worker and the length of workday also may be curvilinear because the production probably increases as the workday increases; but after a certain number of hours the output will "tail off" and not continue to increase at the same rate. Other examples might be that of interest rates and the volume of loans or of capital input and production. The number of cases where the curvilinear model is appropriate might be extended but the point should be obvious; namely, that not all relationships are best described by a straight line. Nonetheless, it should be kept in mind the more complex the relationship the more complicated is the mathematical formula for fitting the curve.[9] In addition, much like the higher order trend lines, the more complicated the regression equation the greater the risk one must assume in using the relationship for predictive purposes. A curve that turns down sharply or rises at an exponential rate will produce embarrassing results if events suddenly change that alters the relationship. Probably the best way to test for nonlinearity is to fit the regression by straight line and then by some curvilinear relationship. If the latter case has a higher coefficient of determination, then it probably is appropriate. The application will dictate whether or not the nonlinear model is useful. In many cases more of the variation in the dependent variable might be accounted for by the nonlinear model but it may not be practical to use in any prediction. There is a trade-off here between a more simplified model and a better statistical fit. Usually, the latter implies collecting more data at a cost that must be weighed against benefits. In the social sciences it is not always clear that more information is always possible or useful.

How does the analyst decide whether or not the linear or nonlinear case is more appropriate? Usually, in applying regression techniques the first step, whether the problem is a simple linear relationship or a more complicated model, is to plot the variables as scatter diagrams. Of course, a good knowledge of the series you are working with is taken for granted and should aid the statistician to reach a decision. Visual inspection and judgment are vital at this point. Careful analysis of the scatter plots might prove fruitful in explaining some of the interrelationships among the variables and also lead to the application of the proper statistical tools (which may or may not be correlation and regression). *It should be abundantly*

---

[9] Recently, a director of the operations research unit of a well-known consumer soft goods manufacturer stated in a professional speech ". . . that the most useful *single* statistical tool that I have is simple linear regression." Considering that the speaker is a Ph.D. in mathematics, the statement came as a modest surprise to his audience of applied statisticians.

clear that if we used sufficiently complex mathematical formulas we could make the curve pass through every data point. Unfortunately, the resulting loss of the degrees of freedom would be so great that the measures of correlation and regression are useless. There is no substitute for a sound knowledge of the data and good common sense in the application of statistical methods to any problem. This is especially true of correlation and regression. The selection of the appropriate regression curve has some similarity to that of the problem of identifying the proper trend line. In both instances familiarity with the data and a knowledge of how these variables are related in the real world are the best things discriminating analysts have going for them. No amount of mathematical manipulation can substitute for good judgment at this level of analysis. Students interested in learning more about this subject are referred to the Related Readings at the end of this chapter.

USE OF DUMMY VARIABLES

In relatively recent times the use of dummy variables in correlation and regression models has proven helpful when the factor is qualitative. Applications of this type are frequent in economic analysis and in marketing research. In the models discussed previously the independent variables could assume any value over a continuous range. It has been found useful to introduce a factor that has only two or a few more distinct levels. If we are comparing the outputs under two widely differing processes, the dummy variable takes the value of either 0 or 1, depending upon which process was used. In more general terms, we assign the factor the value of 1 if it "has the attribute" or 0 if it "does not have the attribute." In market research one might be interested in correlating brand preferences. The choices the analyst might be interested in are "your product" or "not your product." Mathematicians refer to a dummy variable as being a quantity written in a mathematical expression in the form of a variable although it represents a constant. They prefer to avoid the use of the term dummy variable as we have done; however, applied statisticians make use of the concept as indicated above. Generally, when dummy variables are used in the model the coefficient of determination is normally low. This is especially true if n is large. Again, students are urged to follow up on this topic on their own initiative by referring to one or more of the Related Readings.[10]

---

[10] See also the following excellent discussions of the topic: Jan Kmenta, *The Elements of Econometrics* (New York: Macmillan Publishing Co., Inc., 1971), chap. 11; J. Johnston, *Econometric Methods,* 2d ed. (New York: McGraw-Hill Book Co., 1972), pp. 176–207; and R. J. Wonnacott and T. J. Wonnacott, *Econometrics* (New York: John Wiley & Sons, Inc., 1970), pp. 68–76.

There are two primary types of research studies where dummy or binary variables might be useful. They include many consumer survey studies made with sample data of a cross-sectional nature, that is, collected at a single point in time. Some time series studies also can use dummy variables effectively. In fact, time itself frequently is coded and used as an independent variable. Acts of history such as labor strikes, war, recession periods, and so on, may be treated as qualitative variables where their effects or impact must somehow be measured or identified. Another important use of this technique in time series analysis relates to the seasonal force and its relationship to the dependent variable. Sometimes geographic areas are used in sales analysis as dummy variables in forecasting. It is common in some research projects in the social sciences for many economic or demographic variables to be categorized in a qualitative sense. For example, race, sex, religion, geographic region, occupation, education, and many others are used as categorical dummy variables because they are not so easily quantifiable.

An example: Earnings per share

Data for this example refer to the quarterly earnings per share of a hypothetical firm by quarters. Table 16.13 indicates the original data and provides a summary of the descriptive measures.[11] The usual procedure in this type of a situation is to define a *set* of dummy variables, with the number of variables used being one less than the number of categories. The reason is simple logic in the sense that $k - 1$ dichotomies contain all the data. That is, if the data are not in the first three categories we know the observations belong in the fourth class. In our example, the first quarter is omitted because if the data are not in the other three then we know they belong to the first quarter. When this is done the effects of the first quarter would show up in the regression constant a. Table 16.14 presents the correlation matrix for this relationship. Note that the correlation coefficient between time and earnings per share (EPS) is .7068 with a coefficient of determination (r^2) of .4996. While the relationship between EPS and the third (D3) and fourth (D4) quarters appears to be negative, the r for the second quarter (D2) is. 4675. (The notation R @ .9500 = .4044 is the .05 significance level and the R @ .9900 = .5151 is the .01 level. Values of the coefficient of correlation contained in this matrix must exceed these values to be judged significant.) Table 16.15 summarizes the multiple relationship of this problem where the dependent variable is EPS, time is the first independent variable, and D4, D3, and D2 relate to their respective quarters. $R^2 = .86448$ with a standard error of .099649.

[11] The data were run on the computer program MIDAS.

TABLE 16.13
Example of the use of four dummy variables

```
WRITE V=1-5 C=1-24 F=* FO=(5F7.2)


WRITE OBSERVATIONS
VARIABLES BY CASE
    .70    1.00    0.     0.     0.
    .81    2.00    0.     0.     1.00
    .60    3.00    0.     1.00   0.
    .63    4.00    1.00   0.     0.
    .54    5.00    0.     0.     0.
    .84    6.00    0.     0.     1.00
    .68    7.00    0.     1.00   0.
    .75    8.00    1.00   0.     0.
    .95    9.00    0.     0.     0.
   1.17   10.00    0.     0.     0.00
    .76   11.00    0.     1.00   0.
    .90   12.00    1.00   0.     0.
   1.11   13.00    0.     0.     0.
   1.25   14.00    0.     0.     1.00
    .78   15.00    0.     1.00   0.
    .86   16.00    1.00   0.     0.
   1.15   17.00    0.     0.     0.
   1.24   18.00    0.     0.     1.00
    .81   19.00    0.     1.00   0.
    .88   20.00    1.00   0.     0.
   1.28   21.00    0.     0.     0.
   1.45   22.00    0.     0.     1.00
   1.08   23.00    0.     0.00   0.
   1.14   24.00    1.00   0.     0.

    24 CASES WRITTEN FOR    5 VARIABLES


COMMAND
?DESC V=1-5 C=1-24


DESCRIPTIVE MEASURES

VARIABLE        N    MEAN       STD DEV   SE OF MEAN    MINIMUM     MAXIMUM

EARN           24   .93167     .24602    .50219  -1   .54000      1.4500
TIME           24   12.500     7.0711    1.4434       1.0000      24.000
D4             24   .25000     .44233    .90289  -1   0.          1.0000
D3             24   .25000     .44233    .90289  -1   0.          1.0000
D2             24   .25000     .44233    .90289  -1   0.          1.0000


COMMAND
?
```

With an F value of 30.299 the relationship is judged highly significant. (It should be questioned whether a significance test is appropriate here given the fact that the data are in time series form and no transformations were made.) The remainder of the table indicates the partial correlation coefficients and the coefficients of regression along with their standard errors.

Similarly we might use the various seasons as dummy variables, for example, winter, spring, summer, and fall. Again we would use only three

TABLE 16.14
Correlation matrix of the dummy variable example

CORR V=1-5 C=1-24

CORRELATION COEFFICIENTS

N= 24 DF= 22 R@.9500= .4044 R@.9900= .5151

VARIABLE					
EARN	1.0000				
TIME	.7068	1.0000			
D4	-.1718	.1251	1.0000		
D3	-.3516	.0417	-.3333	1.0000	
D2	.4675	-.0417	-.3333	-.3333	1.0000
	EARN	TIME	D4	D3	D2

COMMAND
?

of the four as variables with the effects of the omitted season being reflected in the constant a. The coefficients of regression, b_2, b_3, and b_4 would then reflect the other three seasons with the impact of b_1 (the omitted season) contained in the a value. Or, we might use various geographical regions in a similar manner.

TABLE 16.15

REGR V=1;2,3,4,5 C=1-24

LEAST SQUARES REGRESSION

ANALYSIS OF VARIANCE OF EARN N= 24

SOURCE	DF	SUM OF SQRS	MEAN SQUARE	F-STATISTIC	SIGNIF
REGRESSION	4	1.2035	.30087	30.299	.0000
ERROR	19	.08867	.99299 -2		
TOTAL	23	1.3921			

MULTIPLE R= .92977 R-SQR= .86448 SE= .99649 -1

VARIABLE	PARTIAL	COEFFICIENT	STD ERROR	T-STATISTIC	SIGNIF
CONSTANT		.65879	.52228 -1	12.614	.0000
TIME	.90083	.26929 -1	.29776 -2	9.0438	.0000
D4	-.56940	-.17579	.58222 -1	-3.0193	.0071
D3	-.66395	-.22386	.57840 -1	-3.8703	.0010
D2	.49937	.14474	.57609 -1	2.5124	.0212

RELATIONSHIP OF CORRELATION AND REGRESSION ANALYSIS TO THE ANALYSIS OF VARIANCE

Although the analysis of variance usually is not considered an appropriate topic for an elementary statistics text, several general comments and illustrations are in order. Basically, the analysis of variance procedure may be described as an *extension* of correlation and regression analysis. That extension involves a group of statistical procedures whose objective is the analysis of the squared standard deviations. In its simplest form the analysis of variance includes such problems as the significance of the difference between means. (See Chapter 10.) And until relatively recently (last 20–30 years), problems such as these were so handled or were solved through the use of correlation and regression analysis. Very quickly, correlation becomes rather cumbersome in these situations; and theoretically, the method of testing the significance of the difference between means does not give the same result as the analysis of variance unless $n_1 = n_2$. However, it should be noted that the use of the standard error of the difference between means does serve as a convenient approximation in many situations. Finally, both the correlation techniques and the methods of the significance of the difference between means emphasize the elementary principle of the analysis of variance. That principle is a comparison of the variability between groups (similar to correlation and regression) and the variance within groups.

In Chapter 10 the procedures for testing the significance of the *difference* between *two* statistics were illustrated. First, we considered tests involving two means, and later we studied procedures for analyzing the difference between two percentages. In Chapters 15 and 16 our attention has been directed toward the measurement and evaluation of the covariation between two or more series. Now we are ready to deal with the extension of these correlation methods. *The analysis of variability, particularly that variance represented by the squared standard deviation, is the purpose of the statistical methods under consideration in this section. In its simplest form then, analysis of variance deals with tests of the difference between two statistics, whereas correlation analysis primarily is a form of the study of variance between groups.* An important advantage of analysis of variance contrasted with more elementary techniques of comparing two statistics or two groups is that it can be extended to apply to many means and to many groups of data to test whether or not the random samples came from the same population. Besides these uses, the analysis of variance can be helpful in planning future sample designs. However, there is no intention to provide a complete description of the techniques in this basic text. Rather it is hoped that students might be motivated to pursue further their readings on the design of experiments and the analysis of variance and stimulate their imagination with respect to applications of these topics. Many excellent texts and articles are available on these subjects.[12]

An application[13]

The example chosen to demonstrate the application of the techniques of analysis of variance to sample data comes from the area of marketing management. The illustration relates to a packaging and pricing situation. While the data employed are hypothetical, actual "real world" market information may easily be substituted and the procedure remains the same. Therefore, while the conclusions based on these hypothetical data may not be true of a given market situation, the technique is still a valid and useful one. The fundamental proposition of importance is that the population variance may be estimated from our sample in several ways, and the analysis of these estimates can lead us to some interesting conclusions with respect to the population.

Basic data. Our illustration relates to market research. Let us assume that a manufacturer is planning to introduce a new nondurable consumer product to a national market. The questions needed to be answered by the vice president for marketing and the director of product planning include:

1. Will different package design and size appeal to different segmented markets in the United States?
2. Will the typical consumer fix different prices as reasonable to the same product in different packages?
3. Would a more expensive package create the impression of a higher value of the product?

To determine some empirical guide as to whether the different package designs suggest different values to the consumer, the following procedure was utilized. Four different package designs were tested in ten separate market areas. Each of the four packages were shown to a random sample of consumers in these ten test markets. *In order not to lead*

[12] See D. A. S. Fraser, *Statistics: An Introduction* (New York: John Wiley & Sons, Inc., 1958), chaps. 13–14; Henry Scheffe, "Alternative Models for the Analysis of Variance," *Annuals of Mathematical Statistics*, vol. 27 (1965), pp. 251–71; W. C. Guenther, *Analysis of Variance* (Englewood Cliffs, N.J.: Prentice-Hall, Inc., 1964); C. R. Hicks, *Fundamental Concepts in the Design of Experiments* (New York: Holt, Rinehart & Winston, Inc., 1964); and Alexander M. Mood and Franklin A. Graybill, *Introduction to the Theory of Statistics*, 2d ed. (New York: McGraw-Hill Book Co., 1963), chap. 14.

[13] This section is based upon a paper presented by the author at the 117th Annual Meeting of the American Statistical Association in Atlantic City. See Dick A. Leabo, "Use of Sample Survey Data and Analysis of Variance as a Guide to Packaging and Pricing Policies," American Statistical Association *Proceedings*, 117th Annual Meeting, pp. 468–69.

TABLE 16.16
Average prices assigned to the same product in different packages in different markets

Test Market	Package Design Y_1	Y_2	Y_3	Y_4	ΣY_r	$(\Sigma Y_r)^2$
A	66	42	54	77	239	57,121
B	83	65	89	53	290	84,100
C	61	29	61	61	212	44,944
D	52	59	80	41	232	53,824
E	60	37	59	72	228	51,984
F	78	48	52	49	227	51,529
G	70	53	48	56	227	51,529
H	87	61	72	82	302	91,204
I	92	80	67	50	289	83,521
J	56	74	40	43	213	45,369
ΣY_c	705	548	622	584	2,459	615,125
ΣY_c^2	51,423	32,390	40,720	35,934	160,467	
$(\Sigma Y_c)^2$	497,025	300,304	386,884	341,056	1,525,269	

the consumer into grading the packages, separate samples were chosen for each package in each market.

The figures in Table 16.16 are the average prices (arithmetic means) suggested for each package by each sample in each test area. For example, the mean price suggested in Test Market A for package Y_1 is 66 units. Another sample in the same city yielded a suggested average price of 42 units for the same product in package design Y_2, and so on.

Methodology. To perform this test, the hypothesis to be tested is that there is no significant difference (from a statistical sampling point of view) between packages or markets in terms of product value as suggested by consumers. *Basically this is our null hypothesis: differences in packaged design or differences in markets have no effect on the consumers' estimate of value (price) of the same product.*

Table 16.17 indicates the procedure. The data from Table 16.16 provide for the analysis. The ten test markets are coded from A, B, C, . . . , J, and the four package designs are labeled Y_1, Y_2, Y_3, and Y_4. In Table 16.17 the total variation is symbolized by Σy^2. The variation between test markets (rows) is Σt_r^2; the variation between package designs (columns) is Σt_c^2. The residual variation, that is, the variance unexplained by differences in test markets or package design, is Σd^2.

Conclusions. In order to test our hypothesis that the sample variances differ more than might be expected on the basis of random variation, we refer to the F distribution. We recall from Chapter 15 that the F distribu-

TABLE 16.17
Computation of mean squares for data of Table 16.16

$\Sigma t_r^2 = (615{,}125 \div 4) - 151{,}167 = 2{,}614$
$\Sigma y^2 = 160{,}467 - 151{,}167 = 9{,}300$
$\Sigma t_c^2 = (1{,}525{,}269 \div 10) - 151{,}167 = 1{,}360$
$\Sigma d^2 = 9{,}300 - 2{,}614 - 1{,}360 = 5{,}326$

Mean Squares Corrected for Degrees of Freedom

Whole table:	$\Sigma y^2 \div (n-1) = 9{,}300 \div (40-1) = 238.46$
Between columns:	$\Sigma t_c^2 \div (m_c - 1) = 1{,}360 \div (4-1) = 453.33$
Between rows:	$\Sigma t_r^2 \div (m_r - 1) = 2{,}614 \div (10-1) = 290.44$
Residuals:	$\Sigma y^2 - \Sigma t_c^2 - \Sigma t_r^2 = 5{,}326 \div (39-12) = 197.26$
	$(n-1) - (m_c - 1) - (m_r - 1)$
By columns:	$F_{c.r} = 453.33 \div 197.26 = 2.30 (F_{.05} = 2.96;\ F_{.01} = 4.60)$
By rows:	$F_{r.c} = 290.44 \div 197.26 = 1.47 (F_{.05} = 2.10;\ F_{.01} = 3.16)$

tion provides a probability interpretation of the ratio of two related mean squares occurring by chance alone.

With respect to differences in package designs, the hypotheses might be stated as follows:

H_0: Differences in package designs are not significant with respect to the consumers' estimates of the value of the new product.
H_1: Differences in package designs are significant with respect to the consumers' estimates of the value of the new product.

The criteria for rationally choosing between these alternatives are:

$$\text{Accept } H_0 \text{ if } F_{c.r} \leq F_{.05}$$
$$\text{Accept } H_1 \text{ if } F_{c.r} > F_{.05}$$

The Table of F (page 680) indicates that with respect to the variance by columns—differences in package design—the .05 level of significance is 2.96 and the .01 level is 4.60. *These are the F values that one might expect by mere chance, once in 20 times and once in a 100 times, respectively, if differences in package design have no effect on the consumers' estimates of value of the same product.* Our appropriate variance ratio, corrected for the loss of degrees of freedom, is $F_{c.r} = 2.30$. This is smaller than both table values of F; consequently, the null hypothesis (H_0) is accepted. That is, on the basis of our sample data, differences in package designs have no effect upon the consumers' estimates of the value (price) of the same product. There *may* be a difference nationally, but additional evidence would be required in order to reject the null hypothesis.

With respect to variances by rows—differences in markets—the hypotheses are:

H_0: Differences in market areas are not significant with respect to the consumers' estimates of the value of the new product.

H_1: Differences in market areas are significant with respect to the consumers' estimates of the value of the new product.

The criteria for reaching a decision are:

$$\text{Accept } H_0 \text{ if } F_{r.c} \leq F_{.05}$$
$$\text{Accept } H_1 \text{ if } F_{r.c} > F_{.05}$$

The Table of F indicates that with respect to the variance by rows—differences in market areas—the .05 level of significance is 2.10 and the .01 level is 3.16. Again, these are the F values that one might expect by mere chance, once in 20 times and once in 100 times, respectively, if differences in market areas are not important with respect to the consumers' estimates of the value of the new product. Our appropriate variance ratio, corrected for the loss of degrees of freedom, is $F_{r.c} = 1.47$. This figure is smaller than both table values of F; consequently, the null hypothesis (H_0) is accepted. That is, on the basis of our sample evidence we conclude that differences in market areas have no effect upon the consumers' estimates of the value (price) of the same product.

Computations: Dual analysis of variance

The procedure described above normally is referred to as dual analysis of variance because of the application of the technique to both columns and rows. Let us examine the nature of the computations and the meaning of the computed values.

First, Σy^2 is simply the *total variation* of the data as in any problem of correlation. That is, Σy^2 is the centered squares of the whole table minus the correction term, C. That is,

$$C = (\Sigma Y_c)^2 \div n$$
$$= (2{,}459)^2 \div 40 = 151{,}167$$

where n is the number of observations, 40; and

$$\Sigma y^2 = (\Sigma Y_c^2) - C$$
$$= 160{,}467 - 151{,}167 = 9{,}300$$

where the first term on the right side of the equation is the grand total of the sum of the individual squares, that is,

$$\Sigma(\Sigma Y_c^2) = 51{,}423 + 32{,}390 + 40{,}720 + 35{,}934$$
$$= 160{,}467$$

next, Σt_r^2 measures the variability of markets, each measured by their average price. That is,

$$\Sigma t_r^2 = [\Sigma(\Sigma Y_r)^2 \div d] - C$$
$$= (615{,}125 \div 4) - 151{,}167 = 2{,}614$$

where d is the number of package designs.

Third, Σt_c^2 measures the variability of package designs, each measured by the average price. That is,

$$\Sigma t_c^2 = [\Sigma(\Sigma Y_c)^2 \div m_r] - C$$
$$= (1{,}525{,}269 \div 10) - 151{,}167 = 1{,}360$$

where m_r is the number of test markets.

Finally, Σd^2 is the part of the total variability of the table unaccounted by market and package design variability.[14] That is,

$$\Sigma d^2 = \Sigma y^2 - \Sigma t_r^2 - \Sigma t_c^2$$
$$= 9{,}300 - 2{,}614 - 1{,}360$$
$$= 5{,}326$$

Degrees of freedom

Sometimes the determination of the degree of freedom is a confusing problem. It can easily be seen, however, that with respect to the whole table (Σy^2) the degrees of freedom are $n - 1$ because this is an ordinary standard deviation problem. With respect to the column mean squares (Σt_c^2) the degrees of freedom are $m_c - 1$, that is, the number of columns minus 1. For the row mean squares it is $m_r - 1$ or the number of rows minus 1. The degrees of freedom associated with the residual mean squares (Σd^2) can be found simply as the balance not required for columns and rows, that is, $(n - 1) - (m_c - 1) - (m_r - 1) = n - m_c - m_r + 1$. The number of residual degrees of freedom might be checked as $(m_c - 1)(m_r - 1)$.

The statistic $F_{c.r}$, relating to the variability by columns—that is, the variability of average price reflected by package designs in different markets—is found to be the ratio of the mean square by columns (Σt_c^2) to the residual mean squares (Σd^2). The critical values are read from the table by reference to the degrees of freedom involved, namely, 27 and 3. Therefore we look in column $n_1 = 3$ and row 27 to find the .05 and .01 levels of 2.96 and 4.60 respectively.

The symbol $F_{c.r}$ suggests a similarity to partial correlation discussed

[14] Dual residual variability (Σd^2) will not be the sum of the residual variabilities that can be obtained by columns and rows. The reason lies in the fact that column and row variabilities interact to help explain the actual total variability (Σy^2). Theoretically, one could conceive of a dual analysis of variance problem that yields no residual variability; however, the variation by columns and rows taken individually might be large.

above. In effect, dual analysis of variance between columns—average price suggested by different package designs—is computed after variability between rows—differences in markets—has been eliminated. For example, if the latter variability is eliminated by reducing each row price to average row price, Σt_r is reduced to zero and Σy^2, the total variance, is diminished by a like amount.

The significance of the variability by rows—differences in prices suggested by different markets—may be tested, too ($F_{r.c}$), in a similar manner.

SYMBOLS

X_0, dependent variable
$X_1, X_2, \ldots, X_n$, independent variables
$\hat{X}_0$, estimated value of dependent variable
a, the $\hat{X}_0$ intercept; value of $\hat{X}_0$ when $X_1, X_2, \ldots, X_n = 0$
$b_1, b_2, \ldots, b_n$, net regression coefficients; b_1 equals unit change in $\hat{X}_0$ associated with each unit change in X_1, if $X_2 \ldots X_n$ are held constant
r_{01}, r_{02}, r_{21}, simple correlation coefficients
$R^2_{0.12}$ coefficient of multiple determination; percent of variation in X_0 associated with variations in X_1 and X_2
$\beta_1, \beta_2, \ldots, \beta_n$, beta regression coefficients; expresses each series of data in terms of its own standard deviation
F, ratio of explained variance to unexplained variance, adjusted for degrees of freedom
m, number of constants in multiple regression equation
Σx^2_0, total variance
Σt^2, explained variance
Σd^2, unexplained variance (correlation)
$\sigma_{0.12}$, true standard error of estimate
$s_{0.12}$, sample standard error of estimate
$\hat{\sigma}_{0.12}$, estimated standard error of estimate
$\hat{\sigma}_{\beta_1}, \hat{\sigma}_{\beta_2}$, estimated standard error of betas
$\hat{\sigma}_b$, estimated standard error of b
$\hat{\sigma}_{x0}$, estimated standard deviation of population of X_0 series
$\hat{\sigma}_{x1}$, estimated standard deviation of population of X_1 series
$r_{01.2}$, partial correlation coefficient of X_0 series and X_1 series, holding X_2 constant
Σy^2, total variation
Σt^2_r variation between rows
Σt^2_c variation between columns
Σd^2, residual variation unexplained by differences in columns or rows (analysis of variance)

PROBLEMS

16.1. Recruitment and training of new employees is an expensive operation. Therefore a midwestern corporation, which has an excellent national reputation and is growing at a fast rate, is especially interested in hiring new employees who will show some loyalty to the company and remain with it for many years. What characteristics, besides general mental abilities and skill, tend to indicate employment stability in new personnel? The company makes a study of the records of current and former employees. The following data based on this study seem pertinent:

Employee	X_1 Emotional Instability Test Score	X_2 Community Activities Test Score	X_0 Tenure, in Years
A	18	12	5
B	15	11	3
C	11	10	7
D	16	14	11
E	10	15	4
F	8	12	2
G	21	20	15

a. What are the values of net regression coefficients b_1 and b_2? Interpret these measures.
b. What is the multiple regression equation? Indicate the use of the equation in prediction.
c. Compute and interpret the multiple coefficient of determination. Test for significance.
d. Compute and interpret the beta coefficients.
e. What is the value of the standard error of estimate for this problem, and how might it be used?
f. Calculate the simple correlation coefficients.
g. Calculate whatever measures you think might be useful in the study of partial correlation as applied to this problem. Interpret these values.
h. Illustrate the meaning of partial correlation for this problem, using a table similar to Table 16.3.

16.2. Apply the techniques of graphic multiple correlation to the above problem. Do whatever you think is necessary, and then prepare a brief paragraph summarizing the relationship.

16.3. Given the following:

$$\hat{X}_0 = -50 + 1.05X_1 + .78X_2$$
$$R^2_{0.12} = .58;\ \hat{\sigma}_{0.12} = .9 \text{ units};\ n = 25$$

a. Provide an estimate of the dependent variable when $X_1 = 50$ and $X_2 = 200$.
b. Test the significance of the correlation.

16.4. Given the following:

$$r^2_{01} = .42;\ r^2_{02} = .36;\ r^2_{21} = .28$$

Calculate and interpret the coefficients of partial correlation.

16.5. Precisely and concisely define the following:

a. Coefficient of multiple determination.
b. Standard error of estimate.
c. Standard error of beta.
d. F.
e. Σt^2.
f. Σx^2_0.
g. Net regression coefficients.
h. Beta coefficients.

16.6. A large department store serving the metropolitan area of a city of over four million people hires many part-time sales personnel. These clerks are standbys and supplement the regular crew. Various departments call the night before to hire additional sales help as needs are anticipated. There are many factors affecting the daily sales of the various departments. The personnel manager tried to isolate a few of these variables in an attempt to minimize the cost of hiring salespeople, yet at the same time maximizing sales opportunities. The following data were selected by the personnel manager as being useful as guidelines for hiring policies.

Time Period	X_1 Daily Advertising Expenditures (000 dollars)	X_2 Average Daily Measurement of Rain, Snow, or Sleet (in inches)	X_0 Average Daily Sales (000 dollars)
1	2	.05	28
2	1	1.02	20
3	4	.01	32
4	3	.02	29
5	5	.01	36
6	7	.04	33
7	9	1.11	26
8	6	.03	34
9	8	.51	32
10	10	.02	40

a. Evaluate the effectiveness of the data for this purpose, using whatever correlation and regression techniques you feel desirable.
b. Would you recommend that these variables continue to be used in the future? Why, or why not?
c. What other independent variables would you suggest as being useful here?

16.7. If you were the market analyst for a large furniture manufacturer, what economic variables might you suggest as being helpful in predicting industry and firm sales? Why?

16.8. Given $R^2_{0.123} = .46$, $n = 8$; test the significance of this measure.

16.9. Given the following:

$$\Sigma X_0 = 78;\ \Sigma X_1 = 114;\ \Sigma X_2 = 90;\ n = 10$$
$$\Sigma X_0^2 = 662;\ \Sigma X_1^2 = 1{,}422;\ \Sigma X_2^2 = 1{,}092$$
$$\Sigma X_1 X_0 = 957;\ \Sigma X_1 X_2 = 1{,}620;\ \Sigma X_2 X_0 = 1{,}120$$

a. Determine the regression equation.
b. Determine $R^2_{0.12}$.
c. Test the significance of $R^2_{0.12}$.

16.10. Defend or refute the following statement: "The standard error of the estimate usually is greater than the standard deviation of the dependent variable."

16.11. "The decision concerning the significance of the population coefficient of determination at a stated level of significance depends upon the value of $R^2_{0.12}$ and the size of the sample." Defend or refute this statement.

16.12. Given the following data with respect to a sample of the secretaries of the Tint-a-Lot Corporation:

Secretary	X_1 Years of Service	X_2 Index of Secretarial Competence	X_0 Weekly Wage
A	2	100	$ 80
B	3	110	90
C	2	90	75
D	1	85	70
E	4	120	125
F	6	125	135
G	8	130	150
H	5	115	110

a. Calculate the regression equation.
b. Calculate $R^2_{0.12}$.
c. Calculate $\hat{\sigma}_{0.12}$.
d. Calculate β_1 and β_2.
e. Calculate r^2_{01}; r^2_{02}; r^2_{21}.

16.13. Test the significance of $R^2_{0.12}$ in Problem 16.12 above. Test the significance of the simple correlation coefficients in Problem 16.12.

16.14. Assume the following data represent two series of aptitude test scores of comparable clerical personnel; determine whether the means of the series are significantly different at the .05 level.

(a)		(b)		(c)	
A_1	A_2	B_1	B_2	C_1	C_2
3	12	2	10	5	8
5	16	4	12	7	10
	18		12		12
	25		18		14

16.15. Random samples of the production efficiency scores of certain types of machine operators were obtained from different plants of a firm. Do the scores for the individual plants show a significant variation at the .01 level?

Employee Class	Number of Employees by Plant				
	P_1	P_2	P_3	P_4	P_5
E_1	2	3	1	1	1
E_2	3	3	1	4	5
E_3	3	2	6	3	3
E_4	1	2	1	1	1
E_5	1		1	1	

16.16. Analyze the following sets of uncorrelated data to determine if the means differ significantly at the .05 level:

(*a*)		(*b*)		(*c*)	
A_1	A_2	B_1	B_2	C_1	C_2
10	9	4	2	3	4
12	11	3	5	14	18
8	10	1	9	3	7
14	15	5	11	11	16
16	20	7	13	9	15

16.17. In estimating industrial production for a given state, preliminary and final figures were reported at two different times. Did the final figures differ significantly from the preliminary data? Use .05 level.

	Preliminary	Final
January	155	156
February	152	153
March	154	154
April	154	155
May	155	156
June	157	159
July	158	160
August	158	161
September	159	162
October	160	159
November	161	158
December	159	157

16.18. The Huracan Motor Company retails three leading sports-type automobiles: XH–1, XH–2, and XH–3. Four salespeople handle the entire selling function for the firm. The following data represent the number of autos sold in each of three months during 1975. Is there a preference for one auto over the other two?

Sales-person	April XH-1	April XH-2	April XH-3	June XH-1	June XH-2	June XH-3	December XH-1	December XH-2	December XH-3
S_1	8	4	3	6	2	4	1	2	1
S_2	4	10	5	8	5	7	2	1	1
S_3	2	6	5	6	6	6	1	1	2
S_4	2	4	3	9	7	5	1	2	1

16.19. Term paper: Using the available internal financial data, and any appropriate external information, develop a regression model(s) for the purpose of predicting the earnings or sales of a selected firm. Perhaps you may wish to consider using ratio analysis and correlation and regression analysis to predict earnings. It is suggested that you build your model and do the analyses directed toward management responsible for making decisions relating to investments in new plant, equipment, financing and credit policies. A suggested format for your report might be:

Contents

Findings in brief (1–2 typewritten pages)
Background of firm and environment
Discussion of variables: Reasons for inclusion or exclusion
Analysis and conclusions
Appendix: Computer printout
Bibliography

It is recommended that your report be limited to ten typewritten pages. Remember your report is to be read and acted upon by top management. Note Pascal's warning: It is much more difficult to write a short report than a long one. Be precise but concise!

RELATED READINGS

Andrews, Frank; Morgan, James; and Songuist, John. *Multiple Classification Analysis.* Ann Arbor: Institute for Social Research, Survey Research Center, 1966.

Box, G. E. P. "A Useful Method for Model Building." *Technometrics,* vol. 4 (1962), pp. 301–18.

———. "Some Notes on Nonlinear Estimation." Technical Report No. 25, Department of Statistics, University of Wisconsin, Madison, 1964.

Burnett, T. D., and Guthrie, D. "Estimation of Stationary Stochastic Regression Parameters." *Journal of the American Statistical Association,* vol. 65, no. 332 (December 1970), pp. 1547–53.

Cochran, W. G. "Some Effects of Errors of Measurement on Multiple Correlation." *Journal of the American Statistical Association,* vol. 65, no. 329 (March 1970), pp. 22–34.

Cramer, Elliot M. "Significance Tests of Models in Multiple Regression," *The American Statistician,* vol. 26, no. 4 (October 1972), pp. 23–25.

DRAPER, NORMAN, and SMITH, HARRY. *Applied Regression Analysis.* New York: John Wiley & Sons, Inc., 1966.

EFROYMSON, M. A. "Multiple Regression Analysis," in A. Ralston and H. S. Wilf, *Mathematical Methods for Digital Computers,* Article 17. New York: John Wiley & Sons, Inc., 1962.

GEARY, R. C., and LESSER, C. E. V. "Significance Tests in Multiple Regression." *The American Statistician,* vol. 22 (February 1968), pp. 20–21.

GUENTHER, W. C. *Analysis of Variance.* Englewood Cliffs, N.J.: Prentice-Hall, Inc., 1964.

HARDER, T., and PAPPI, F. U. "Multiple-Level Regression Analysis of Survey and Ecological Data." *Social Science Information,* vol. 7, no. 5, pp. 43–67.

HARTLEY, H. O. "The Estimation of Nonlinear Parameters by 'Internal Least Squares.'" *Biometrika,* vol. 35 (1964), pp. 31–45.

———, and BOOKER, A. "Nonlinear Least Squares Estimation." *Annuals of Mathematical Statistics,* vol. 36 (1965), pp. 638–50.

IVES, KENNETH H., and GIBBONS, JEAN D. "A Correlation Measure for Nominal Data." *The American Statistician,* vol. 21 (December 1967), pp. 16–17.

KARON, BETRAN P. "A Note on the Treatment of Age as a Variable in Regression Equations." *The American Statistician,* vol. 18 (June 1964), pp. 27–28.

KIEFF, JAMES R. JR. "Testing the Assumptions of a General Linear Model—Multiple Regression Analysis." *Fortran Applications in Business Administration,* vol. II, pp. 173–202. Ann Arbor: Graduate School of Business Administration, University of Michigan, 1971.

LEABO, DICK A. "Use of Sample Survey Data and Analysis of Variance as a Guide to Packaging and Pricing Policies." American Statistical Association, *Proceedings,* Business and Economics Section, 117th Annual Meeting (1957).

———. "The Declining Marginal Propensity to Save." *Business Economics,* vol. 6, no. 3, May 1971, National Association of Business Economists, pp. 25–29.

———. "Stepwise Regression Analysis Applied to Regional Economic Research." Working Paper No. 27. Division of Research, University of Michigan, Ann Arbor. A later version of this paper was published in *Proceedings,* Business and Economics Section, 130th Annual Meeting of the American Statistical Association (1970), pp. 544–48.

———. "Stochastic Analysis as Applied to the Multiple Linear Regression Equation." American Statistical Association, *Proceedings,* Business and Economics Section, 120th Annual Meeting (1960).

———. "Relationship of Michigan Personal Income and GNP." American Statistical Association, *Proceedings,* Business and Economics Section, 116th Annual Meeting (1956).

LONGLEY, JAMES W. "An Appraisal of Least Squares Programs for the Electronic Computer from the Point of View of the User." *Journal of the American Statistical Association,* vol. 62, no. 319 (September 1967), pp. 819–41.

MARQUARDT, D. W. "An Algorithm for Least Squares Estimation of Nonlinear Parameters." *Journal of the Society of Industrial and Applied Mathematics,* vol. 2 (1963), pp. 431–41.

SCHEFFE, HENRY. "Alternative Models for the Analysis of Variance." *Annuals of Mathematical Statistics,* vol. 27 (1965), pp. 251–71.

SHUMWAY, R. H. "Applied Regression and Analysis of Variance for Stationary Time Series." *Journal of the American Statistical Association,* vol. 65, no. 332 (December 1970), pp. 1527–46.

SONQUIST, JOHN A. *Multivariate Model Building.* Ann Arbor: Institute for Social Research, University of Michigan, 1970 (paperback).

———, and MORGAN, JAMES. *The Detection of Interaction Effects.* Ann Arbor: Institute for Social Research, University of Michigan, 1970 (paperback).

SRIKANTAN, K. S. "Canonical Association between Nominal Measurements." *Journal of the American Statistical Association,* vol. 65 (March 1970), pp. 284–92.

SUITS, DANIEL B. "The Use of Dummy Variables in Regression Equations." *Journal of the American Statistical Association,* vol. 52 (December 1957), pp. 548–51.

WALLIS, K. F. "Lagged Dependent Variables and Serially Correlated Errors: A Reappraisal of Three-Pass Least Squares." *Review of Economics and Statistics,* vol. 49 (1967), pp. 555–67.

Part VII

Nonparametric statistics: Data analysis with less restrictive assumptions

To be absolutely certain about something, one must know everything or nothing about it.

Olin Miller

17

Chi-square analysis and other selected nonparametric tests

If we begin with certainties, we shall end in doubts; but
If we begin with doubts, and are patient in them, we shall end in certainties.

Francis Bacon

THE STATISTICAL inference procedures discussed so far have been concerned with the value of a population parameter. That is, in some cases the methods assumed that the sample statistic came from a normally distributed population and could be used to estimate the parameter. These procedures might be properly called *parametric*. In some research designs it is not feasible or possible for the investigator to make rigid mathematical assumptions about the shape of the population being sampled. This limitation has brought about the development of a group of statistical methods called *nonparametric* or *distribution-free* tests.[1]

In recent years there has been an increasing emphasis in business research in the application of some of the methods of the behavioral sciences. Interest in the analytical tools and theories of psychology and sociology in particular has been strong in the areas of industrial relations, marketing, and organization theory. Applications in the area of accounting have been slower in coming and more limited but apparently are increasing. The trend seems to be for business researchers to make more, not less, use of some of the statistical methods and research findings from the behavioral sciences. As a result there has been a renewed interest in some

[1] Two excellent texts dealing with the subject are: Myles Hollander and Douglas A. Wolfe, *Nonparametric Statistical Methods* (New York: John Wiley & Sons, Inc., 1973), and E. L. Lehmann, *Nonparametric Statistics: Statistical Methods Based on Ranks* (San Francisco: Holden–Day, Inc., 1975.)

of the so-called nonparametric methods. The basic reason for this is that these tests can be made on data from the lower end of the scale of measurement. Nonparametric tests also do not require that the model make any assumptions about the population from which the sample observations came. Most nonparametric tests, however, do require that the sample observations are independently drawn and that the sample comes from the target population. Parametric methods require these and other stronger mathematical assumptions, and as a consequence are generally considered more robust from a statistical point of view.[2] Parametric tests require that the data be of a higher order of measurement. Some mathematical writers prefer to avoid the use of the term nonparametric and recommend instead the use of distribution-free statistics. (Most people view the terms as being synonyms.) The basis for this preference is that these tests of a hypothesis or confidence interval estimation procedures do not depend upon the form of the underlying distribution, hence, the term distribution-free statistics. For example, confidence intervals may be constructed from the median, based on binomial variation, which are valid for any continuous distribution. In terms of tests of hypotheses, the term distribution-free is implied to mean that the test of the null hypothesis is independent of the distribution. These writers prefer to avoid any confusion by restricting the term nonparametric to those cases where the hypotheses do not make an explicit statement about a parameter. However, most researchers in the social sciences only use the term nonparametric and find it descriptive and useful without causing any unusual confusion. Before proceeding any further let us pause briefly to refresh our memories about the four basic scales of measurement.

Scales of measurement

While this is not the place to discuss measurement theory and its problems, some fundamental distinctions among the levels might be in order. There are four levels of measurement that are important: nominal, ordinal, interval, and ratio. *Parametric statistical methods cannot be validly applied at the first two levels.* Unfortunately, sometimes the nonmathematically oriented researcher overlooks this fundamental fact. Whereas nonparametric tests may be applied at all four levels of measurement, most researchers would use the more powerful parametric tests where appropriate.

---

[2] For example, parametric tests require some degree of homoscedasticity of the data. This concept relates to dispersion as measured by the variance. If the variance of one variable is the same for all fixed values of another variable, the distribution is said to be homoscedastic in the first variate. The opposite case is said to be hetroscedastic. Parametric tests also assume that the measurements have been made in at least an interval scale level.

Nominal scale. This is the lower end of the measurement spectrum and sometimes is referred to as a classification scale. Data measurement is at its least powerful level when classifications, numbers, or symbols are used to categorize observations. For example, numbers or symbols can be classified as nominal data if they clearly indicate mutually exclusive subclasses. Auto license plates are one illustration of numbers and letters being used to identify groups, yet a nominal scale is involved. In Michigan, for example, the first of three alphabetic letters identifies the county; three digits follow the letters. All auto owners in Michigan residing in the same county could represent a subclass on a nominal scale. The same car would not be registered in more than one county (subclass). Nonparametric statistical tests may be made on data in nominal scale, but it is inappropriate to attempt to use the classical parametric tests discussed in previous chapters on such data.

Ordinal or ranking scale. If a classification is not only different but if the categories stand in some sort of relation to each other, we call this an ordinal scale. This is a common scale in consumer research and in many areas of research on human behavior. Perhaps, one of the easiest ordinal scales to comprehend is that of military rank. A captain is "higher than" a lieutenant who in turn is "higher than" a sergeant who outranks a corporal who outranks a private first-class who outranks a private. Clearly, the various grades are distinct mutually exclusive classifications, but they also stand in relation to one another. If the relation holds between some but not all of the sets of classification, it is referred to as a partially ordered scale. If the relation holds for all sets of classes so that a full rank ordering of classes is possible, it is an ordinal scale. A transformation of ordinal data that does not change the order of the information is permissible. This is one reason that rank correlation techniques are useful in evaluating some buyer behavior data. Products may be ranked by their order of preference by the consumer. Or if punishments for a crime are classified as "not severe," "severe," or "highly severe," this would be an example of an ordinal scale. Or, sometimes a sample respondent is given a statement to read and then asked whether he or she "agrees," "strongly agrees," "disagrees," "strongly disagrees," or has "no opinion." These classifications are near the weakest level of measurement, but at times the qualitative response is the only possible scale appropriate. Again, parametric tests are not applicable at this level.

Interval scale. If the distances between any two numbers on a scale are known, and if all the characteristics of an ordinal scale are met, then the researcher can say that the measurement is of a higher order, called an interval scale. The unit of measurement in an interval scale and the zero point are subjectively assigned. For example, temperature is measured by an interval scale (in fact two different scales, Centigrade and Fahrenheit), and the zero point differs in both. However, the two scales are related and contain the same type of information. Transformations may be made from one scale to the other—if you remember the relationship![3] When the conditions of interval scaling are met, both nonparametric and parametric tests can be applied to the data. Unfortunately, it is not always possible to transform ordinal data to a higher level thereby making the more powerful parametric tests useful.

Ratio scale. The final level of measurement is that of the ratio scale. All an interval scale needs to become a ratio scale is to have a true zero point. A ratio scale derives its value from the fact that the ratio of any two values is independent of the *units* in which they are measured. The unit of measurement does not distort the relationship. For example, two series stated in differing units may be shown on the same ratio chart. *Rates* of change are shown not absolute values.

Siegel and Bradley detail some advantages and disadvantages of nonparametric tests when compared with parametric methods. It is not particularly worthwhile here to argue the case for either approach; both procedures have their applications and limitations. Generally, some of the major strong points presented for the use of nonparametric tests are: (*a*) they are easier to learn, (*b*) the mathematical derivations are more readily understood by the nonmathematician, (*c*) they can be applied to data of a classificatory nature that are frequently found in social science research, (*d*) they are able to better handle ranked data, and (*e*) the computations are relatively simple. The third item (*c*) seems to be the most valid argument for their use. The more statistically robust parametric tests require that rigorous assumptions be met for a valid application. However, when all of those conditions are obtained and the measurement is at the interval level or higher, these tests are almost always preferred. Usually, nonparametric tests are, at this time at least, considered inappropriate for detecting interactions in analysis of variance problems, unless certain limiting assumptions about the additivity of the variances is made. If this requirement is met, then the parametric tests are applicable and preferred.

CHI-SQUARE ANALYSIS

Another theoretical distribution that is useful as a model to test an hypothesis about a population is called chi-square (χ^2). In our previous tests involving qualitative or attribute data, we were, in effect, comparing an *observed frequency* with an *expected frequency*. The Student t or normal probability distributions helped us to reach a rational decision. In the tests involving the significance of the difference between two statistics

---

[3] For those of us who are forgetful the relationship is:

$$\text{Fahrenheit} = 9/5 \ \text{Centrigrade} + 32°$$

(two sample means or two sample percentages), we were actually comparing two sets of observed frequencies. Many of the previous tests of hypothesis dealing with percentages or proportions are really special cases that might have been handled equally well by using the chi-square test —a more general approach to solving these and many more similar problems.

The chi-square test involves both a probability interpretation and a measure of reliability by comparing observed frequency distributions with theoretical (expected) distributions. This is done, of course, to determine whether or not the theoretical distribution is a good approximation of the actual situation. Introduced by Karl Pearson in 1900, the chi-square test probably is the best known and the most important of all nonparametric methods. Because the test makes no assumption about the population being sampled, it might be labeled a distribution-free method. However, there is a rigidly defined chi-square distribution with a mathematically expressed frequency function. (See Appendix J.)

Basic procedure

As in all cases involving a statistical statement about a population, we must first frame the null hypothesis. Usually, the null hypothesis is stated in such a manner that if it is true, the observed data should on the average tend to exhibit certain frequency characteristics. These frequency characteristics are the *expected frequencies* (f_e) that are computed. Then the sample data, the *observed frequencies* (f_o), are obtained. These two sets of characteristics are then compared, and the statistic χ^2 is computed by the following formula:

$$\chi_c^2 = \sum \frac{(f_o - f_e)^2}{f_e} \tag{17.1}$$

A shortcut formula for computing chi-square is available and might be useful in some situations. In any event, both formulas produce the same result. The shortcut formula is not used in the text examples because generally the arithmetic involved is small anyway, and there is no particular advantage in a more efficient formula. The basic chi-square equation may be modified to:

$$\chi_c^2 = \sum \frac{(f_o - 2f_of_e + f_e)^2}{f_e}$$

$$= \sum \frac{f_0^2}{f_e} - 2\sum \frac{f_of_e}{f_e} + \sum \frac{f_e^2}{f_e}$$

$$= \sum \frac{f_0^2}{f_e} - 2 \quad \Sigma f_o + \Sigma f_e \tag{17.2}$$

however,

$$\Sigma f_o = \Sigma f_e = n$$

therefore,

$$\chi_c^2 = \sum \frac{f_o^2}{f_e} - n \tag{17.3}$$

The question to be answered is how closely do the observed data approximate the expected frequencies of the theoretical distribution? The measurement of the closeness of the fit involves the differences between the actual and the expected frequencies as reflected by χ_c^2 the computed value of chi-square. This computed value of chi-square is then examined in light of the probability of the variation occurring by chance. Reference is usually made to a table or chart similar to those contained in Appendix J. *The test is to determine whether or not the computed value of chi-square differs significantly from zero.*

Value of chi-square

The computed value of chi-square (χ_c^2) depends upon the actual differences between the two sets of frequencies under consideration. If these two sets agree perfectly, that is, they are identical, the value of $\chi_c^2 = 0$. Clearly, this is the smallest value that chi-square might assume since it is the sum of squared numbers it can never be negative. *Even if our null hypothesis is true, we should not anticipate that the sample data will agree perfectly with our theoretical expectations.* (Recall the notion of the pattern of sample variation discussed in previous chapters.) However, by using the chi-square distribution in a manner similar to the way we previously applied the normal, F, and Student t distributions, we can determine the probability of chance occurrences of the values of chi-square as large as, or larger than, the observed value. In so doing, the analyst can make a rational decision whether to accept or reject the null hypothesis on the basis of probability theory.

Review of degrees of freedom concept

Generally, a convenient rule for determining the number of the degrees of freedom in any problem is equal to the number of variables. For a sample of n observations there are n degrees of freedom (df). The df are reduced by one for each restriction, that is, for each statistic computed and used as an estimator of a parameter. (See Chapters 5 and 8.) In the case of chi-square analysis there are two different distributions to be compared—the observed and the expected. *Consequently, the degrees of free-*

dom are reduced by the number of ways in which the computed frequencies are made to agree with the observed frequencies.

We recall that in estimating the population standard deviation where μ is unknown, the number of degrees of freedom is $n - 1$. This is true because in computing the deviations we use $(X - \bar{X})^2$ instead of $(X - \mu)^2$, the desired value. We reduce the *df* by one because $\bar{X}$ is substituted for μ.

Types of chi-square tests

Two main types of tests will be examined later: (1) tests of "goodness of fit" and (2) tests of the independence of principles of classification (contingency tables). Actually, there is a great deal of similarity between the two, and some writers merely treat the latter test as a special case of the former. Both tests involve the comparison of the two sets of frequencies. Each type also states the null hypothesis in such a manner that if it is true, the differences between the expected and observed frequencies are considered not significant.

For the "goodness of fit" test, the null hypothesis specifies some theoretical distribution as being descriptive of the population from which the sample came. Usually, a binomial, normal, uniform, or some other distribution is appropriate. From the selected theoretical distribution, the expected frequencies are computed. While the uniform distribution is the simplest to use in determining the expected frequencies, it probably is the most useful. This test enables the analyst to decide whether the sample observations may be regarded as being randomly drawn from a universe that has the characteristics specified by the theoretical distribution (the expected frequencies).

The test of the independence of principles of classification involves rational decision making as to whether a set of classification criteria is meaningful and effective. *The null hypothesis normally is stated that the principles of classification are independent, that is, there is no relationship between the criteria.*

Assumptions

There are five basic conditions which must be met in order for chi-square analysis to be validly applied. They are:

1. The experimental data (sample observations) must be independent of each other;
2. The sample data must be drawn at random from the target population;
3. The data should be expressed in original units for convenience of comparison, and not in percentage or ratio form;
4. The sample should contain at least 50 observations; and
5. There should be no less than five observations in any one cell.

The first two conditions are the same for the application of all standard error formulas and tests of significance. The third condition is no serious restriction because the analyst usually can convert the ratios back to their original units. *The fourth restriction concerning sample size is important because the chi-square distribution, like other statistics, might be erratic when the number in the sample is small.* The final point relates to the previous one but usually we are concerned about cell size for only the expected frequencies.

TESTS OF GOODNESS OF FIT

Uniform distribution

In order to illustrate the fundamentals of chi-square analysis, let us first examine an abstract population and then move to more "real life" applications. Assume a 6-sided die, numbered from 1 to 6, is tossed at random 60 times. The results of the experiment are as shown in Table 17.1.

TABLE 17.1
Sixty tosses of one six-sided die

Face Number	Frequency of Occurrence
1	8
2	11
3	9
4	13
5	14
6	5
	60

If the die were honest and the tosses independent, then our theoretical "expected" frequencies for each face would be ten. Obviously, these relate to the long run and would more nearly be approximated as n becomes large. Our test is to see whether or not these 60 tosses represent a random selection from our theoretical hypothesized population. Therefore, our null hypothesis (H_0) is that there is no difference in the frequencies except for chance error of sampling. Let us test this hypothesis using the .05 level of significance. The diagram of our test appears as Figure 17.1.

The criteria for reaching a decision are:

$$\text{Accept } H_0 \text{ if } \chi_c^2 \leq 11.07$$
$$\text{Reject } H_0 \text{ if } \chi_c^2 > 11.07$$

One degree of freedom is lost in this type of a problem; since $df = n - 1$ we have five degrees of freedom in this situation because there are

FIGURE 17.1
Chi-square probability distribution ($df = 6 - 1 = 5$; .05 level of significance)

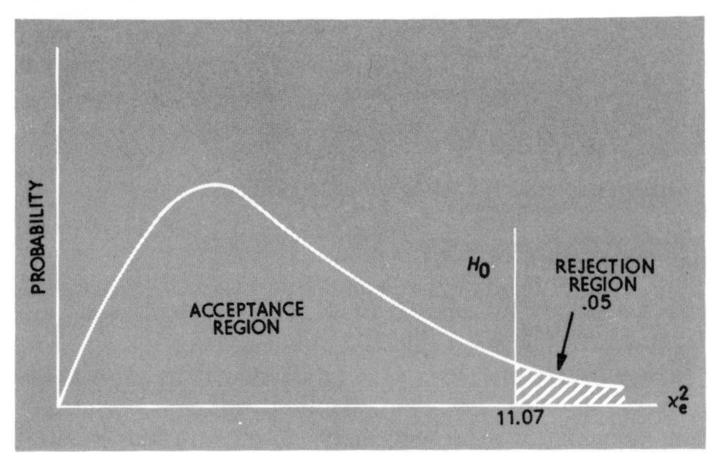

six classes. From the table in Appendix J, for 5 *df* at the .05 level we find the value of χ_e^2 as 11.070. This is the value of chi-square we would expect by mere chance once in 20 times if there is no difference in the two sets of frequencies (see Table 17.2).

Our $\chi_c^2 = 5.60$; therefore, the null hypothesis is accepted, that is, the difference is not significant from a statistical sampling point of view.

Because the probability associated with the appearance of any one face in any random toss is 1/6, *this example illustrates the test for a theoretical uniform distribution.* Let us now consider a binomial situation.

TABLE 17.2
Computation of χ_c^2 for 60 tosses of a die

(1) Face Number	(2) Observed Frequency f_o	(3) Expected Frequency f_e	(4) $f_o - f_e$	(5) $(f_o - f_e)^2$	(6) $\frac{(f_o - f_e)^2}{f_e}$
1	8	10	−2	4	.4
2	11	10	1	1	.1
3	9	10	−1	1	.1
4	13	10	3	9	.9
5	14	10	4	16	1.60
6	5	10	−5	25	2.50
	60	60			$\chi_c^2 = 5.60$

TABLE 17.3
One hundred random tosses of three coins

Number of Heads	Frequency of Tosses
0	15
1	35
2	40
3	10
	100

Binomial distribution

Assume 3 coins are tossed at random 100 times and we are interested in the number of heads (see Table 17.3).

The theoretical frequencies might be computed by expanding the binomial expression $(H + T)^n$. Since the probability of a head equals the probability of a tail, the expression may be written for this experiment as:

$$100(H + T)^3$$

or

$$100(H^3 + 3H^2T + 3HT^2 + T^3)$$
$$100[(.5)^3 + 3(.5)^2(.5) + 3(.5)(.5)^2 + (.5)^3]$$
$$100(.125 + .375 + .375 + .125)$$
$$12.5 + 37.5 + 37.5 + 12.5$$

Table 17.4 presents the observed and expected frequencies and the computation of χ_c^2. The null hypothesis is that these 100 tosses represent a random sample from the population. The diagram of the test appears as Figure 17.2.

TABLE 17.4
Computation of χ_c^2 for 100 tosses of three coins

(1) Number of Heads	(2) Observed Frequency f_o	(3) Expected Frequency f_e	(4) $f_o - f_e$	(5) $(f_o - f_e)^2$	(6) $\frac{(f_o - f_e)^2}{f_e}$
0	15	12.5	2.5	6.25	.50
1	35	37.5	−2.5	6.25	.17
2	40	37.5	2.5	6.25	.17
3	10	12.5	−2.5	6.25	.50
	100	100.0			$\chi_c^2 = 1.34$

FIGURE 17.2
Chi-square probability distribution ($df = 4 - 1 = 3$; .10 level of significance)

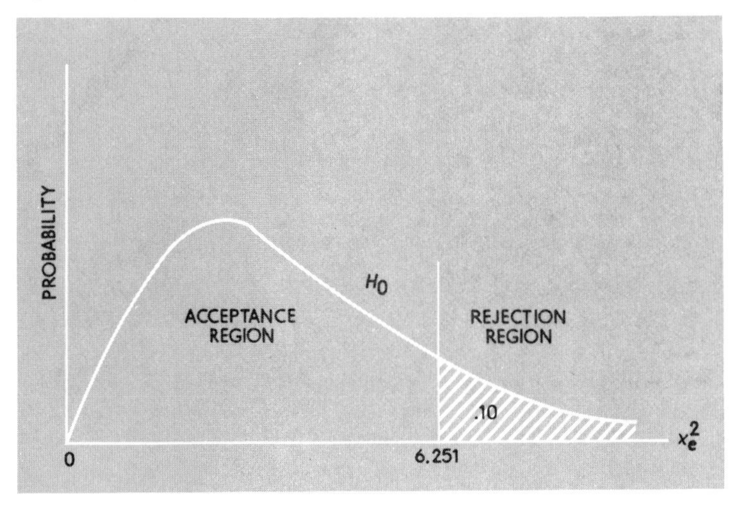

The criteria for reaching a decision are:

$$\text{Accept } H_0 \text{ if } \chi_c^2 \leq 6.251$$
$$\text{Reject } H_0 \text{ if } \chi_c^2 > 6.251$$

Our $\chi_c^2 = 1.34$; therefore, the null hypothesis is accepted; that is, the results merely represent sampling variations. The coins probably were honest and the tosses random (see Table 17.4).

Application of the test of goodness of fit: Normal distribution

Let us now examine a more practical application of the chi-square test of significance using the normal distribution as an approximation of the frequencies. The procedure is basically the same as for the uniform or binomial distributions except that the computations of the theoretical frequencies might be a little more involved. Let us make use of the data contained in Table 4.4 on page 73 that refer to the weekly wages of some skilled workers. Table 17.5 reflects the computations necessary to determine the theoretical frequencies.

Recall that the mean and the standard deviation are all that are needed to define a normal curve. Referring to Appendix C, especially Table C.2 for the values of the ordinates, we may complete the calculations. Recall

TABLE 17.5

Estimating the theoretical frequencies on the basis of the ordinates of the normal distribution (data on weekly wages of skilled workers taken from Table 4.4, p. 73; $\bar{X} = \$75$; $s = \$15$; $n = 200$)

Weekly Wage Class Mid-point	Observed Frequencies f_o	Deviations from Mean $(X - \bar{X})$	z Value	Ordinate Values Table C.2	Ratio to Sum of Ordinates	Theoretical Frequencies f_e
\$ 50	14	−25	−1.67	.0989	.067	13
60	40	−15	−1.00	.2420	.165	33
70	54	− 5	− .33	.3778	.257	51
80	46	+ 5	.33	.3778	.257	51
90	26	+15	1.00	.2420	.165	33
100	12	+25	1.67	.0989	.067	13
110	6	+35	2.33	.0264	.018	4
120	2	+45	3.00	.0044	.003	2†
Total	200			1.4682	.999*	200

* Will not total 1.000 because of rounding.

† Rounded up to make totals agree.

that in order to standardize any distribution we use the following formula:

$$z = \frac{(X - \bar{X})}{\hat{\sigma}}$$

where we divide the difference between the item and the arithmetic mean by the standard deviation ($\hat{\sigma}$). (Note: because the data are sample results we use $\hat{\sigma}$ as an estimate of σ. Because $n = 200$, $\hat{\sigma} \simeq \$15$.) Working through the first value of z we find

$$z = \frac{(50 - 75)}{15} = -1.67$$

Reading from Table C.2 we discover that the ordinate associated with a z value of 1.67 (we ignore the signs) is .0989. This same procedure is followed for each of the midpoints and the sum of the ordinates is determined. Then we find the ratio of each individual ordinate to the sum and multiply the resulting percentage times the sample size n. In this case $n = 200$ so $(.067)(200) = 13$. The other values of the theoretical frequencies are determined in a like manner.

Table 17.6 presents the computations of chi-square using the expected frequencies as determined by fitting the normal distribution to the data on weekly wages of skilled workers. The *null hypothesis is that the normal distribution is a good approximation of this wage distribution.* Obviously, the alternative hypothesis is that the normal distribution is not a good fit. The diagram of the chi-square test appears in Figure 17.3. The criteria for reaching a decision are:

$$\text{Accept } H_0 \text{ if } \chi_c^2 \leq 14.067$$
$$\text{Reject } H_0 \text{ if } \chi_c^2 \geq 14.067$$

TABLE 17.6
Computation of chi-square for weekly wage problem

(1) Weekly Wage Class Midpoint	(2) Observed Frequency f_o	(3) Expected Frequency f_e	(4) $f_o - f_e$	(5) $(f_o - f_e)^2$	(6) $(f_o - f_e)^2/f_e$
$ 50	14	13	1	1	.077
60	40	33	7	49	1.485
70	54	51	3	9	.176
80	46	51	−5	25	.490
90	26	33	−7	49	1.485
100	12	13	−1	1	.77
110	6	4	2	4	1.000
120	2	2	0	0	.000
					$\chi_c^2 = 4.790$

FIGURE 17.3
Chi-square probability distribution ($df = 8 - 1 = 7$; .05 level of significance)

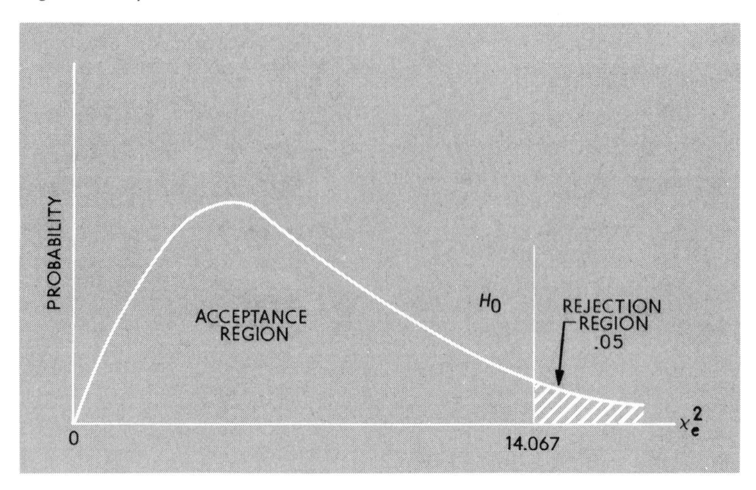

Our value of the computed chi-square is 4.790; therefore, the null hypothesis is accepted and we conclude that the normal curve is a good fit to this distribution of weekly wages of skilled workers. The critical value of chi-square of 14.067 was determined from Appendix J with 7 degrees of freedom. That is, we have 8 wage classes so $df = 8 - 1 = 7$ using the .05 level of significance we read the value 14.067. Because our computed value of chi-square 4.790 is less than 14.067 we accept H_0.

TESTS OF INDEPENDENCE OF PRINCIPLES OF CLASSIFICATION

Contingency tables: 2 X 2

In this type of chi-square test the decision relates to whether or not a set of criteria of classification is effective and/or meaningful. Are the principles of classification independent? A market research agency has felt that including a merchandise or money premium with a mailed questionnaire "substantially increases the proportion of returned questionnaires." It was decided to test this hypothesis on the next mail survey. One set of 500 questionnaires were mailed to a random sample of respondents; included as an inducement for cooperating was a newly minted 25-cent coin. Using appropriate follow-up procedures with the nonrespondents, 185 in this group eventually completed and returned the questionnaires. Identical questionnaires were sent to another random sample of 400 individuals; no premium was included. One hundred thirty-two respondents submitted questionnaires after similar follow-up procedures were used.

Question: Is the inclusion of the newly minted coin effective in raising the rate of return? Naturally, we could solve this problem utilizing the techniques described in Chapter 10 relating to the standard error of the difference between two percentages. Generally speaking, however, problems of this type can be dealt with more easily using the chi-square test for independence of principles of classification. The results should be identical, but in practice the chi-square test involves less arithmetic.

The two classifications here are (1) coin inclusion and (2) response. *Are they independent or does the inclusion of a coin effect response?* This yields two categories for each classification, therefore, we call this a 2×2 contingency table—two rows and two columns. Such a table has four cells but only one degree of freedom. That is, $df = (r - 1)(c - 1) = (2 - 1)(2 - 1) = 1$. *There is only one degree of freedom remaining because we place three restrictions on our computations.* They are: (1) the total, (2) the number of questionnaires with the coin, and (3) the number of responses. These restrictions fix our totals, and when one expected frequency is known, the other three are determined.

The null hypothesis is that the two principles of classification—inclusion of a newly minted 25-cent coin and response—are independent. We must compute the expected frequencies so all values in the four cells are proportional to the totals.

If the inclusion of the coin had no effect on response, we would expect the total questionnaires returned $(185 + 132 = 317)$ to be proportionately distributed between those with the coin and those without (see Table 17.7).

The expected frequencies for the *responding* individuals are computed as follows:

$$\frac{317}{900} \cdot 500 = 176 \text{ for those with coin included}$$

$$\frac{317}{900} \cdot 400 = 141 \text{ for those without the coin}$$

The expected frequencies for the *nonrespondents* are:

$$\frac{583}{900} \cdot 500 = 324 \text{ for those with the coin included}$$

$$\frac{583}{900} \cdot 400 = 259 \text{ for those without the coin}$$

See Table 17.8.

The null hypothesis is that the principles of classification are independent; that is, the inclusion of the coin does not effect response (from a statistical point of view). The diagram of the test is shown in Figure 17.4.

TABLE 17.7
Observed and expected frequencies—coin and questionnaire problem

Coin Included	Observed Frequencies			Expected Frequencies		
		Responded			Responded	
	Total	Yes	No	Total	Yes	No
Yes.......	500	185	315	500	176	324
No.......	400	132	268	400	141	259
Total......	900	317	583	900	317	583

TABLE 17.8
Computation of χ_c^2

(1) Cell (row-column)	(2) Observed Frequency f_o	(3) Expected Frequency f_e	(4) $f_o - f_e$	(5) $(f_o - f_e)^2$	(6) $\frac{(f_o - f_e)^2}{f_e}$
1 – 1..........	185	176	9	81	.46
1 – 2..........	315	324	−9	81	.25
2 – 1..........	132	141	−9	81	.57
2 – 2..........	268	259	9	81	.31
	900				$\chi_c^2 = 1.59$

FIGURE 17.4
Chi-square probability distribution ($df = (2 - 1)(2 - 1) = 1$; .10 level of significance)

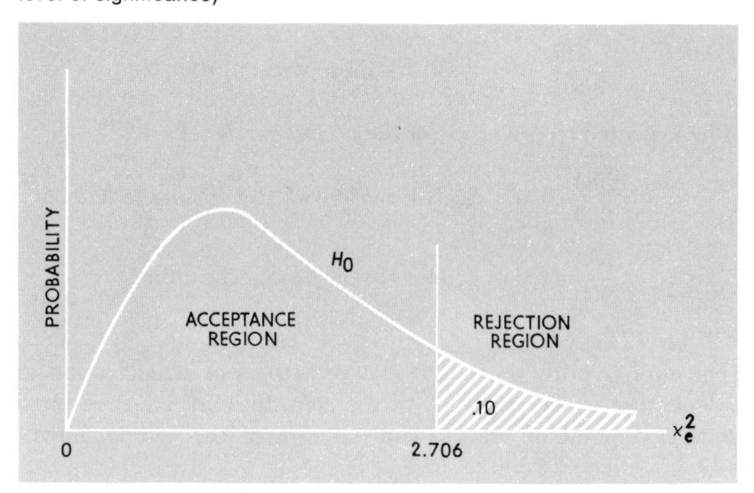

The criteria for reaching a decision are:

$$\text{Accept } H_0 \text{ if } \chi_c^2 \leq 2.706$$
$$\text{Reject } H_0 \text{ if } \chi_c^2 > 2.706$$

Therefore, we accept the null hypothesis that the coin inclusion has no effect on response.

Contingency tables: 4 X 3

Let us extend our analysis to a problem where we have four rows and three columns. Assume a national survey of consumers with respect to their plans to purchase a new automobile during 1976 reveals the figures shown in Table 17.9.

TABLE 17.9
Potential market—new automobiles, 1976 (by income level)

Disposable Family Income	Plan to Purchase	Undecided	Not Planning to Purchase	Total
Under $4,000	60	45	308	413
$4,000–$5,999	130	60	425	615
$6,000–$9,999	140	52	361	553
$10,000 and over	115	51	303	469
Total	445	208	1,397	2,050

It should be noted that if we were interested in testing the significance of a relationship between two or more classifications, we could use the analysis of variance techniques of Chapter 16. For example, if we also classified the families by the number of drivers, as well as income level, the relationship between these two factors as they effect new car purchase plans could be analyzed. The present problem may be conveniently handled by chi-square procedures (see Table 17.10).

TABLE 17.10
Computation of expected frequencies—new car purchase plans problem

(1) Disposable Income	(2) Plan to Purchase	(3) Undecided	(4) Not Planning to Purchase	(5) Total
Under $4,000	$413/2{,}050 \times 445 = 90$	$413/2{,}050 \times 208 = 42$	$413/2{,}050 \times 1{,}397 = 281$	413
$4,000–$5,999	$615/2{,}050 \times 445 = 134$	$615/2{,}050 \times 208 = 62$	$615/2{,}050 \times 1{,}397 = 419$	615
$6,000–$9,999	$553/2{,}050 \times 445 = 120$	$553/2{,}050 \times 208 = 56$	$553/2{,}050 \times 1{,}397 = 377$	553
$10,000 and over	$469/2{,}050 \times 445 = 102$	$469/2{,}050 \times 208 = 48$	$469/2{,}050 \times 1{,}397 = 319$*	469
Total	445	208	1,397	2,050

* Rounded to make totals agree.

The null hypothesis is that there is no difference in purchase plans by income level. Using the .01 level of significance and $df = (4-1)(3-1) = 6$, the criteria are:

$$\text{Accept } H_0 \text{ if } \chi_c^2 \leq 16.812$$
$$\text{Reject } H_0 \text{ if } \chi_c^2 > 16.812$$

See Table 17.11.

Because $\chi_c^2 = 20.41 > 16.812$, we reject the null hypothesis; that is, *we conclude that income level is important in considering purchase plans of new autos.* Given that almost 62 percent (12.80/20.41) of χ_c^2 is the result of the first income class, we may wish to check our decision by eliminating the "Under $4,000" class from our computations. Then there are only 1,637 responses to be analyzed. Tables 17.12 and 17.13 show the revised computations.

Therefore, because $\chi_c^2 = 4.10 < 13.277$, we accept the null hypothesis and conclude that differences in income levels are *not* important in effecting purchase plans for new cars. Apparently, once we eliminate the very lowest income level there are other factors that are more important. Perhaps such things as level of consumer durable goods debt outstanding, age of present car, and styling might be some factors the consumer also enters into his or her decision-making process. At least, this latter analysis casts some doubt on our original conclusion and we should continue our analysis of other variables. *Possibly, in this particular case the analysis of*

TABLE 17.11
Computation of χ_c^2 for new car purchase plans problem

(1) Cell (row-column)	(2)* Observed Frequency f_o	(3)† Expected Frequency f_e	(4) $f_o - f_e$	(5) $(f_o - f_e)^2$	(6) $\frac{(f_o - f_e)^2}{f_e}$
1 – 1	60	90	–30	900	10.00
1 – 2	45	42	3	9	.21
1 – 3	308	281	27	729	2.59
2 – 1	130	134	– 4	16	.12
2 – 2	60	62	– 2	4	.65
2 – 3	425	419	6	36	.09
3 – 1	140	120	20	400	3.33
3 – 2	52	56	– 4	16	.29
3 – 3	361	377	–16	256	.68
4 – 1	115	102	13	169	1.66
4 – 2	51	48	3	9	.19
4 – 3	303	319	–16	256	.80
	2,050	2,050			$\chi_c^2 = 20.41$

* From Table 17.9.
† From Table 17.10.

TABLE 17.12
Computation of expected frequencies—new car purchase plans problem, omitting first income level

(1) Disposable Income	(2) Plan to Purchase f_e	(3) Undecided f_e	(4) Do Not Plan to Purchase f_e	(5) Total
$4,000–$5,999......	615/1,637 × 385 = 145	615/1,637 × 163 = 61	615/1,637 × 1,089 = 409	615
$6,000–$9,999......	553/1,637 × 385 = 130	553/1,637 × 163 = 55	553/1,637 × 1,089 = 368	553
$10,000 and over....	469/1,637 × 385 = 110	469/1,637 × 163 = 47	469/1,637 × 1,089 = 312	469
Total	385	163	1,089	1,637

variance technique might be a more powerful tool since it can measure the interaction of income level and some other variables as they effect purchase plans.

TESTS OF SAMPLE DESIGN

Sometimes chi-square analysis is used to test the "goodness of fit" of a sample frequency distribution to some actual or theoretical population in order to evaluate the sample design. Because the population distribution is usually known, either by empirical data or assumption, the null hypothesis is that there exists no significant difference between the two distributions. That is, any difference we observed was due to chance error of sampling. Again we test the hypothesis by comparing the computed chi-square with the table value. We then decide on the basis of a probability

TABLE 17.13
Computation of χ_c^2 for new car purchase plans problem omitting lower income level

(1) Cell (row-column)	(2) Observed Frequency f_o	(3) Expected Frequency f_e	(4) $f_o - f_e$	(5) $(f_o - f_e)^2$	(6) $\frac{(f_o - f_e)^2}{f_e}$
1 – 1...........	130	145	–15	225	1.55
1 – 2...........	60	61	– 1	1	.02
1 – 3...........	425	409	16	256	.63
2 – 1...........	140	130	10	100	.78
2 – 2...........	52	55	– 3	9	.16
2 – 3...........	361	368	– 7	49	.13
3 – 1...........	115	110	5	25	.23
3 – 2...........	51	47	4	16	.34
3 – 3...........	303	312	– 9	81	.26
	1,637	1,637			$\chi_c^2 = 4.10$

$df = (3 - 1)(3 - 1) = 4$; χ_e^2 at .01 level = 13.277

interpretation, the chance that a χ^2_c as large or larger than the one we observed is likely to occur as a consequence of sample variation.

Assume a study is made where the number of persons in the household is an important variable. Suppose 2,000 families are interviewed and the distribution of the sample and the corresponding Census estimates for all United States households is as stated in Table 17.14.

There are $n - 1$ or 7 degrees of freedom in this problem. At the .01 level of significance $\chi^2_e = 18.475$ (see Table 17.15).

The null hypothesis is that there is no significant difference between the two distributions. That is, the sample comes from the distribution of all United States families in terms of number of persons in the household. On the basis of our calculations the χ^2_c is too large to have occurred

TABLE 17.14
Hypothetical example of relative distribution of 2,000 families in sample and all United States families by number of persons in household, 1976

Number of Persons in Household	Sample Families %	Sample Families No.	U.S. Families %
1	7.5	150	8.0
2	31.8	636	30.3
3	27.0	540	25.2
4	18.2	364	17.0
5	9.5	190	10.5
6	4.7	94	5.0
7	1.0	20	2.1
8 or more	.3	6	1.9
	100.0	2,000	100.0

TABLE 17.15
Computation of χ^2_c for comparative size of families in the sample and United States, 1976

(1) Number of Persons in Household	(2) Sample Frequency f_o	(3) U.S. Distribution f_e	(4) $f_o - f_e$	(5) $(f_o - f_e)^2$	(6) $\frac{(f_o - f_e)^2}{f_e}$
1	150	160	−10	100	.62
2	636	606	30	900	1.49
3	540	504	36	1,296	2.57
4	364	340	24	576	1.69
5	190	210	−20	400	1.90
6	94	100	− 6	36	.36
7	20	42	−22	484	11.52
8 or more	6	38	−32	1,024	26.95
Total	2,000	2,000			$\chi^2_c = 47.10$

through chance error of sampling alone. The sample is not representative of the larger household size. Perhaps, some additional interviews might be called for in at least the last two classes. The main differences between the sample and the population are: (1) oversampling of families in the 2, 3, and 4 size; and (2) undersampling of families of 7 and 8 or more persons. *Obviously, where size of family is related to the problem being studied, such discrepancies could introduce a bias and adjustments seem in order.*

SUMMARY AND WARNINGS: CHI-SQUARE ANALYSIS

The chi-square distribution as shown in most tables may be considered a *continuous* distribution. However, the number of possible combinations of frequencies is *finite*, and these can be represented by a *discrete* probability distribution. As is the case in many sampling problems, when the sample size is large, this approximation of a discrete distribution by a continuous one is generally termed adequate. The procedures described above are then held appropriate. As noted earlier, when the *expected* frequencies are small, some modification of the technique is in order.

As a general rule of thumb, it is desirable that one have five or more expected frequencies in each cell. If less than five expected frequencies appear in any cell, it is possible to combine classes in such a way to maintain the usefulness of the procedure. Also when the number of degrees of freedom remaining is greater than two, *one* cell with fewer than five expected frequencies may be permitted. In some cases it can be shown that even though the expected frequencies are below the requirement, the usefulness of the test may not be destroyed. The results do become inexact though and may cast doubt on the decision.

The suggestion that the expected frequencies should be at least five or more is a "statistically safe" rule. The only permissible exceptions might be when the other expected frequencies are large, the number of degrees of freedom remaining is at least 2, and the .05 level of significance is used. *Generally, if the .01 level of significance is appropriate then no exceptions are granted.*

Where the expected frequencies are too small but one does not wish to combine classes, Yates's correction might be used. This involves changing the *observed* frequencies in each cell by 0.5 in the manner illustrated by the formula:

$$\chi_c^2 = \sum \frac{(|f_o - f_e| - \frac{1}{2})^2}{f_e}$$

This reduces the computed value of chi-square and has a negligible effect on the computed value of chi-square when the frequencies are large. We read $|f_o - f_e|$ as the "absolute value of the difference between the observed and expected frequencies." That is, signs are ignored, only the *size* of the difference enters into the computations.

When the cells contain too few frequencies as outlined above, ignoring Yates's correction might lead to excessive rejection of the null hypothesis because the computed value of χ^2_c *is overstated.* On the other hand, Yates's correction tends to overcompensate for this and might result in excessive *acceptance* of the null hypothesis! What should the analyst do? As is the case with the application of any statistical technique, the "rule of reasonableness" is appropriate. That is, it might be reasonable to test the null hypothesis in our usual manner. If the hypothesis is accepted, we should be satisfied; if rejection is indicated, then recalculate chi-square using Yates's correction. If you still reject the null hypothesis, then be satisfied with the results. Only if the null hypothesis is rejected *without* Yates's correction but accepted when the adjustment is used, should the analyst consider a more exact test.

McNemar test for the significance of changes: Two related samples

A nonparametric test for studying the impact of those so-called "before and after" sample designs is the McNemar test for the significance of changes. Basically, this test is another application of the chi-square analysis with the frequencies determined by the before and after situations (using the identical respondents for both the "before and after" sampling of opinions). If the data are measured in nominal or ordinal scale then this test must be used to analyze the impact of changing conditions because no parametric test is appropriate. The McNemar test might be used to measure the effectiveness of a training session on employees, or the impact of a specific editorial, or the impact of a specific advertisement, or the effects of some reading materials mailed to a panel of respondents, or even to test the possible effects of changing residency on political views. The reader may recognize that all of the examples given are of the type where the respondent might serve as his or her control, and in which some nominal measure would be used to determine the impact of the changing conditions or the before and after differences. Because the same respondents are used in both situations the two samples are related.

To test the significance of any before or after changes we might first set up a table of frequencies to represent the two sets of responses from the same individuals. For example, we would group the responses (pluses and minuses) into one of four different cells. Where C_1, C_2, C_3, and C_4 are cells that contain all the responses representing the changing conditions or the before and after replies. Responses placed in cells C_1 and C_4 identify those respondents (or responses) that show changes; cells C_2 and C_3 reflect no change. Therefore, cells C_1 and C_4 represent the total num-

	After	
Before	−	+
+	C_1	C_2
−	C_3	C_4

ber of respondents (or responses) that changed. The expected values of (*a*) the cases that changed in one direction would be $\frac{1}{2}(C_1 + C_4)$ and (*b*) the cases that changed in the other direction also would be $\frac{1}{2}(C_1 + C_4)$. The null hypothesis is:

H_0: The expected frequencies in both cells C_1 and C_4 are $\frac{1}{2}(C_1 + C_4)$
H_1: The expected frequencies in both cells C_1 and C_4 are $\neq \frac{1}{2}(C_1 + C_4)$

Recall the basic formula for calculating the value of chi-square from (17.1) is:

$$\chi_c^2 = \sum \frac{(f_o - f_e)^2}{f_e}$$

In the McNemar test we are interested in the frequencies in cells C_1 and C_4 only, then using the notation above we obtain

$$\chi_c^2 = \frac{(C_1 - C_4)^2}{(C_1 + C_4)} \quad \text{with } df = 1 \tag{17.4}$$

Given the null hypothesis, the sampling distribution of the test statistic as indicated by Formula (17.4) tends to follow the chi-square distribution with one degree of freedom. However, we need to modify this formula because we are using a continuous distribution (chi-square) to approximate a discrete one. If the expected frequencies are small then using the chi-square distribution to approximate a discrete distribution, then the fit may not be good. Yates's correction for continuity helps to remove or reduce this possible source of error. With this correction, Formula (17.4) then becomes

$$\chi_c^2 = \frac{(|C_1 - C_4| - 1)^2}{(C_1 + C_4)} \quad \text{again with } df = 1 \tag{17.5}$$

Formula (17.5) tells us to subtract 1 from the *absolute* value of the difference between C_1 and C_4 before squaring. That is, by taking the absolute value we ignore signs of differences. We then refer to Appendix J to determine whether or not the computed value of chi-square could have occurred by chance error of sampling. The criteria for accepting or rejecting the null hypothesis at the .05 level with $df = 1$ are

$$\text{Accept } H_0 \text{ if } \chi_c^2 \leq 3.841$$
$$\text{Accept } H_1 \text{ if } \chi_c^2 > 3.841$$

Which indicates that if we accept the null hypothesis then our conclusion is that the before and after effect differences are not significant. On the other hand, if our test leads us to accept H_1 then we conclude that the after effects are statistically significant. The illustration below is an application of the McNemar test for the significance of changes.

An example. Assume that the League of Women Voters (LWV) is interested in voter reaction and the impact of some materials they send out before each election in a city of 100,000 population. The particular issue the LWV is concerned about is whether or not a specific school millage will be supported by the electorate. A study is made to determine voter reaction to this issue prior to the LWV sending out some explanatory materials to all eligible voters. A sample of 50 eligible voters is surveyed to measure their reactions to the proposed millage both before and after the materials were mailed. The study data were grouped according to the form below:

Before the Mailing	After the Mailing	
	(−) Against Proposal	(+) Favor Proposal
(+)	C_1	C_2
(−)	C_3	C_4

Where an individual was classified in category C_1 if the voter changed from a + (favor) to a − (against); the voter was counted in cell C_4 if he or she changed from a − (against) to a + (favor). If no change is detected in voter reaction to the proposal then he or she is tallied in either C_2 or C_3.

After the materials had been mailed to the eligible voters and the individuals had some time to digest the information about the millage proposal the data were summarized as follows:

Before the Mailing	After the Mailing	
	(−) Against Proposal	(+) Favor Proposal
(+)	6	27
(−)	16	15

(Note: some voters were undecided both before and after the mailings; the McNemar test is interested in the changes only so these individuals are not included in any of the cells.)

The *hypotheses* to be tested in this example are

H_0: The expected frequencies in both cells C_1 and C_4 are equal to $\frac{1}{2}(C_1 + C_4)$. That is, there is no difference in voter preference either before or after the mailing of materials by the LWV.

H_1: The expected frequencies in both cells C_1 and C_4 are *not* equal to $\frac{1}{2}(C_1 + C_4)$. That is, there is a statistically significant difference observed in terms of voter preference after the mailing by the LWV.

The *criteria* for accepting or for rejecting the null hypothesis at the .01 level of significance with $df = 1$ are:

$$\text{Accept } H_0 \text{ if: } \chi_c^2 \leq 6.635$$
$$\text{Accept } H_1 \text{ if: } \chi_c^2 > 6.635$$

The computations of the chi-square value for these data are (from Formula [17.5])

$$\chi_c^2 = \frac{(|C_1 - C_4| - 1)^2}{(C_1 + C_4)} \quad \text{with } df = 1$$

$$= \frac{(|6 - 15| - 1)^2}{(6 + 15)}$$

$$= \frac{8^2}{21}$$

$$= 3.048$$

According to the criteria set up above we would accept the null hypothesis and conclude that the mailings by the LWV did not make a statistically significant difference in changing voter preferences with respect to the proposed millage. The LWV may then wish to evaluate the type of information sent to eligible voters. If no change in voter reactions are obtained by sending the information then the LWV may wish to allocate this expenditure of funds to some other form of voter information transmission. Or, the LWV may wish to reexamine the approach of the materials sent by mail to see if advantages and disadvantages of such proposals can be made clearer to the typical voter.

Some writers may question the use of the Yates's correction for continuity here given the fact that none of the frequencies was less than the usually accepted rule of thumb of less than 5. In this case then the binomial test could be used. However, in either case, using the binomial or without the Yates correction, the conclusion would be the same. Given the fact that the definition of what is a "relatively small frequency" is arbitrarily set at 5 the McNemar test was used as a conservative approach.

Summary of procedure. The basic steps in applying the McNemar test are: (*a*) group the observed frequencies in a fourfold table of the type suggested above; (*b*) determine the *expected* frequencies in each cell C_1 and C_4 by $E = \frac{1}{2}(C_1 + C_4)$; (*c*) if the expected frequencies are less than 5 then use the binomial test rather than the McNemar test; (*d*) if the expected frequencies are 5 or larger, compute the value of chi-square using Formula (17.5); (*e*) establish the hypotheses to be tested and the criteria for choosing between them; (*f*) calculate the value of chi-square; and (*g*) compare this value with the expected value of chi-square given $df = 1$ at chosen level of significance and make decision.

Kolmogorov-Smirnov one-sample test

In the sense that the Kolmogorov-Smirnov (KS) one-sample test really is a test of goodness of fit, it actually is related to the chi-square analysis discussed above. Simply, the KS test is concerned with the degree of agreement between the set of sample observations and some *specified* theoretical distribution. This powerful test answers the question: did the sample values come from a population having the specified distribution? Unlike the previous chi-square tests we covered the KS test deals with specifying the theoretical cumulative frequency distribution (CFD_e) and comparing these with the observed cumulative frequency distribution (CFD_o). Just as we did in the chi-square test of goodness of fit, the KS one-sample test states under the null hypothesis (H_0) what the theoretical cumulative frequency distribution represents. The point of greatest divergence of the CFD_e and the CFD_o is calculated and evaluated in probability terms. That is, the sampling distribution of the differences between these two distributions provides us with the necessary probability measure. In this limited sense then the evaluation is no different than in any situation, parametric or nonparametric, in which we use probability theory to help us make a decision.

If the KS one-sample test is similar to the test of goodness of fit as measured by the chi-square distribution, why would we be interested in still another statistical tool? Why not just use the chi-square test and not worry about the KS test? The reason is that in actuality the KS test is more powerful from a statistical point of view, and for very small samples the chi-square test is not appropriate. The statistical efficiency of the KS test stems from the fact that it treats individual observations separately and as a consequence, unlike the chi-square test, does not lose information by the combining of categories. We recall that when using small samples and the chi-square test, it is necessary to combine categories before computing the statistic χ^2_c. Under such circumstances, the KS one-sample test is more powerful (more efficient) than the chi-square test for the goodness of fit. Perhaps, a reexamination of the weekly wage data of Table 4.4 might illustrate the power and application of the KS test. (Table 17.5 contains the same data but not as a cumulative frequency.)

Probably, at this point a summary of the basic procedure might be helpful in understanding the test. Initially, the analyst must specify the CFD_e that is hypothesized under H_0. Second, the observed data should be arranged in a cumulative distribution giving us CFD_o, pairing each interval of the observed values with a comparable interval of the theoretical frequencies. Third, for each step on the CFDs subtract the observed from the expected frequencies. Four, determine the statistic D, the *maximum deviation*, by using Formula (17.6). Fifth, refer to Appendix M and determine the probability (two-tailed test) that is associated with the occurrence of the null hypothesis of values as large (or larger) as the observed D. That is, what is the probability of observing a value as large as D merely by chance error of sampling. Six, if that probability is equal to or less than the preselected level of significance (alpha) then we reject the H_0. If that probability is greater than α then we accept the H_0.

An example. Referring to Tables 4.4 and 17.5 we can construct the necessary table to generate the CFD_e and the CFD_o as well as to calculate the value of D. The reader should detect that the first row of frequencies (f) are the actual sample observations grouped by class limits and are taken from Table 4.4. These data have been used in prior tests to illustrate the relationship of some of the statistical techniques discussed in this text. Row two contains the expected cumulative frequencies (CFD_e) while row three shows the observed cumulative frequencies (CFD_o). The final row are the *absolute* differences (signs ignored) of each sample observation from its paired theoretical (expected) value.

Our next step is to determine the value of D using

$$D = \text{maximum } |f_o - f_e| \tag{17.6}$$

From Table 17.16 we can see that the *maximum* value of these deviations (row three) is 11. We should note that the KS test focuses its attention on the maximum deviation, hence, the calculation of the statistic D.

The hypotheses to be tested are

H_0: That the CFD_o is identical with the normal probability distribution CFD_e

H_1: That the CFD_o is not identical with the normal probability distribution CFD_e

and the criteria for selecting the most appropriate hypothesis are

$$\text{Accept } H_0 \text{ if the probability of } D \geqq 11 \text{ is } > \alpha$$
$$\text{Accept } H_1 \text{ if the probability of } D \geqq 11 \text{ is } < \alpha$$

TABLE 17.16
Distribution of weekly wages, skilled workers (see Tables 4.4 and 17.5)

	Class Midpoint							
	$50	$60	$ 70	$ 80	$ 90	$100	$110	$120
f = number of workers in that class (see Table 4.4)	14	40	54	46	26	12	6	2
f_e = theoretical cumulative distribution of workers under H_0 (normal distribution) (see Table 17.5) (CFD_e)	13	46	97	148	181	194	198	200
f_o = observed cumulative distribution of workers (CFD_o)	14	54	108	154	180	192	198	200
$f_o - f_e$	1	8	11	6	1	2	0	0

If we assume a level of significance (α) of .01 and refer to Appendix M we find that for $n = 200$ that the probability of a $D \geqq 11$ is less than .01. Therefore, the alternate hypothesis, H_1, is accepted and we conclude that the distribution of weekly wages for unskilled workers in this plant is not normal. Notice that using the chi-square test (see Table 17.6) we found the difference to be *not* significant at the .05 level. Using the KS test above we found that the difference is significant at the .01 level which suggests that the power of this test is greater than the chi-square test for the same data. This is especially true for small samples where the chi-square test definitely is less powerful and for some very small samples the chi-square test may not even be appropriate because of the small number of expected frequencies in some categories.

THE SIGN TEST: TWO RELATED SAMPLES

In some research in marketing or industrial relations, quantitative measurement is not always possible or economically feasible. Many times in these situations it is possible to rank pairs of data where two related samples are involved. If the researcher is attempting to control two conditions and then establish that they have differing effects, the sign test is useful. This simple test gets its name from the fact that plus and minus signs are its raw data. *The test is based on the differences between two pairs of observations and it ignores magnitudes.* The fact that the two samples are related and not independent makes the application of a parametric test inappropriate. Probably the only assumption necessary for this test is that the variable being studied has a continuous distribution. Recall that a distribution is said to be continuous when the variate may take values in a continuous range. (A discrete distribution is one where the variate can take only a discontinuous set of values.) Unlike most parametric tests, the sign test does not make any assumption about the shape of the distribution (of differences), nor does it assume that the observations came from the same population.

Procedure. We may designate the first item in each pair of observations as X_1 and the second item as X_2. The value of the difference $(X_1 - X_2)$ will be either a plus, a minus, or zero. The sign test ignores all differences where the value is zero. The remaining set of pairs of observations will be the sample, n. The null hypothesis can be developed from intuition. That is, focusing our attention on the signs of the differences we would expect that the number of pairs of observations that have $X_1 > X_2$ to equal the number of pairs of observations where $X_1 < X_2$. That is, our null hypothesis assumes that we might expect an equal number of pluses and minuses. Given the background of probability theory we cannot reasonably expect the number of signs to be precisely equal. However, the hypothesis that the conditions are not different is rejected if too few signs of one kind occur. What the analyst must decide in advance is just what is "too few." It should be obvious that the binomial distribution can be of some assistance here. Because zero differences are ignored, there are only two possible outcomes, either a plus or a minus. Therefore, the probability of a plus equals the probability of a minus which equals .5. We know from earlier work that the mean and standard deviation can be determined for the binomial as follows: The expected value of pluses is equal to n times the proportion of pluses in the population; and the standard deviation is the square root of n times the proportion times one minus the population proportion. In symbols,

$$E(p) = n\pi \qquad \text{and} \qquad \sigma = \sqrt{n\pi(1 - \pi)}$$

where p equals the number of pluses and π is the population parameter. The standard normal deviate for samples of $n > 30$ we have

$$z_c = \frac{p - .5 - E(p)}{\sigma}$$

Note that if there are more minuses than pluses we should substitute m in the above equations. π equals the population proportion $= .5$.

Illustration. Married students in a large midwestern state university were interviewed, husbands and wives separately, to determine their reaction to a proposed faculty-student designed universitywide judiciary council. The proposed judiciary council would have authority to discipline students (or faculty) and would take precedence over any such body in

TABLE 17.17
Married students' reaction to proposed universitywide judiciary council

Student Couple No.	Rating H	Rating W	Sign $X_h - X_w$	Student Couple No.	Rating H	Rating W	Sign $X_h - X_w$
1	8	6	+	21	3	5	–
2	3	4	–	22	5	7	–
3	2	1	+	23	4	2	+
4	1	2	–	24	1	1	0
5	7	6	+	25	6	8	–
6	5	4	+	26	3	6	–
7	6	5	+	27	5	3	+
8	8	9	–	28	6	7	–
9	3	5	–	29	2	2	0
10	3	4	–	30	3	4	–
11	3	3	0	31	6	4	+
12	6	8	–	32	2	3	–
13	4	3	+	33	8	9	–
14	3	5	–	34	4	9	–
15	2	5	–	35	4	6	–
16	3	2	+	36	5	4	+
17	1	3	–	37	6	8	–
18	5	4	+	38	6	6	0
19	2	3	–	39	6	7	–
20	1	3	–	40	7	8	–

Note: Rating of 10 means that the proposal is perfect: rating of 1 means that it is considered useless.

any individual school or college or department. The evaluation of the proposal was on a ten-point scale with a value of ten being the best. That is, assigning a value of ten indicates that the person felt the proposal was perfect. Table 17.17 summarizes the results of the survey.

Computations.

1. Establish the hypotheses:
   $H_0: P(X_h > X_w) = P(X_h < X_w)$. There is no difference in the ratings by husbands or the wives.
   $H_1: P(X_h > X_w) < P(X_h < X_w)$. There is a significant difference in the ratings of husbands and wives. The median of the differences is negative; more wives rate the proposal higher than husbands.
2. Criteria for choosing between hypotheses:

$$\alpha = .05 \qquad n = 36 \qquad z = 1.645$$

$$\text{Accept } H_0 \text{ if } z_c \leq 1.645$$

$$\text{Accept } H_1 \text{ if } z_c > 1.645$$

3. Calculations:

$$m = \text{number of minuses} = 25 \qquad \pi = .5$$

$$E(m) = n\pi = (36)(.5) = 18$$

$$\sigma = \sqrt{n\pi(1-\pi)} = \sqrt{(36)(.5)(1-.5)} = 3$$

$$z_c = \frac{m - \frac{1}{2} - E(m)}{\sigma} = \frac{(25 - \frac{1}{2} - 18)}{3} = 2.16$$

4. Decision:
Accept H_1 because $2.16 > 1.645$. The sign test shows a significant difference between the ratings of husbands and wives.

WILCOXON MATCHED-PAIRS SIGNED-RANKS TEST: TWO RELATED SAMPLES

Another test that is useful when two samples are not independent is the Wilcoxon matched-pairs signed-ranks test. It is similar to the previous test except that instead of merely recording the direction of the differences between pairs of observations this nonparametric test recognizes magnitude as well. Consequently, it is considered to be a more powerful test. It is not unusual for qualitative data to be observed in such a manner that the researcher is able to determine which item of a pair is greater (i.e., affix a plus or minus sign as was done in the sign test) and then to rank the differences from the lowest to the highest. When the data are in this form, the Wilcoxon test is appropriate.

Procedure. The absolute differences of the set of matched pairs of observations are determined first. Then the items are ranked from the lowest absolute difference to the highest, ignoring the sign. Finally, the less frequent sign (plus or minus) is ranked. Again if the absolute difference is zero, the pair of items is dropped from the sample. If there are ties in ranks, all the tied values of the absolute differences are given the same rank. A statistic T is computed as the sum of the positive or negative ranks on the basis of the smaller total. A standard normal deviate is computed to determine the probability of observing such a T value and a decision is reached.

Illustration. Let us use the same sample survey data of the previous example. Let d_i stand for the absolute differences between X_h and X_w in order that each matched pair has a value. Again, if $d_i = 0$, we will drop the pair from the sample and reduce n accordingly. The null hypothesis assumes that the sum of ranks of the pluses is equal to the sum of the minuses. For samples of $n > 30$ the sampling distribution of T is approximately normal with

$$\mu_T = \frac{n(n+1)}{4} \tag{17.7}$$

$$\sigma_T = \sqrt{\frac{n(n+1)(2n+1)}{24}} \tag{17.8}$$

$$z_c = \frac{T - \mu_T}{\sigma_T} \tag{17.9}$$

Computations.

1. Establish the hypotheses:
   H_0: The sum of the positive-sign ranks (husbands) is the same as the sum of the negative-sign ranks (wives) for the population.
   H_1: The sum of the positive-sign ranks (husbands) is less than the sum of the negative-sign ranks (wives). The wives rate the proposal higher than the husbands.
2. Criteria for choosing between hypotheses:

$$\alpha = .05 \qquad n = 36 \qquad z = 1.645$$
$$\text{Accept } H_0 \text{ if } z_c < \pm 1.645$$
$$\text{Accept } H_1 \text{ if } z_c > \pm 1.645$$

3. Calculations (see Table 17.18):

$$\mu_T = \frac{n(n+1)}{4} = \frac{36(36+1)}{4} = 333$$

$$\sigma_T = \sqrt{\frac{n(n+1)(2n+1)}{24}} = \sqrt{\frac{(36)(37)(73)}{24}} = 64$$

$$z_c = \frac{T - \mu_T}{\sigma_T} = \frac{186 - 333}{64} = -2.29$$

4. Decision:
   Accept H_1 because $-2.29 > -1.645$. The Wilcoxon test shows a significant difference in the ratings of husbands and wives with the latter rating the proposal higher.

MANN-WHITNEY *U* TEST: TWO INDEPENDENT SAMPLES

If only an ordinal level of measurement has been attained then the Mann-Whitney U test makes a good substitute for the parametric test of the significance of the difference between two means (see Chapter 10). It is also useful to the researcher who wants to relax some of the assumptions of the parametric test. This nonparametric test can be used to determine whether or not two independent samples are drawn from the

TABLE 17.18
Computation of statistic T for Wilcoxon matched-pairs signed-ranks test of married student data (see Table 17.17)

Student Couple No. d_i	Rank* of d_i	Rank with Less Frequent Sign +	Student Couple No. d_i	Rank of d_i	Rank with Less Frequent Sign +
1.....2	+26.5	26.5	21.....2	−26.5	
2.....1	−10.0		22.....2	−26.5	
3.....1	+10.0	10.0	23.....2	+26.5	26.5
4.....1	−10.0		24.....0	Eliminated	
5.....1	+10.0	10.0	25.....2	−26.5	
6.....1	+10.0	10.0	26.....3	−34.5	
7.....1	+10.0	10.0	27.....2	+26.5	26.5
8.....1	−10.0		28.....1	−10.0	
9.....2	−26.5		29.....0	Eliminated	
10.....1	−10.0		30.....1	−10.0	
11.....0	Eliminated		31.....2	+26.5	26.5
12.....2	−26.5		32.....1	−10.0	
13.....1	+10.0	10.0	33.....1	−10.0	
14.....2	−26.5		34.....5	−36.0	
15.....3	−34.5		35.....2	−26.5	
16.....1	+10.0	10.0	36.....1	+10.0	10.0
17.....2	−26.5		37.....2	−26.5	
18.....1	+10.0	10.0	38.....0	Eliminated	
19.....1	−10.0		39.....1	−10.0	
20.....2	−26.5		40.....1	−10.0	
		$d_i = X_h - X_w$	$T = 186.0$		

* Ranked without regard to sign from low to high; sign added after ranking.

$$\frac{1+2+3+\cdots 19}{19} = 10.0; \quad \frac{20+21+22+\cdots 33}{14} = 26.5$$

same population or from two populations having the same mean. The Mann-Whitney U test normally is a two-sided test.

Procedure. As we did in Chapter 10, let us have n_1 = items in sample No. 1 and n_2 = items in sample No. 2. We will also let $N^* = n_1 + n_2$. Our initial step is to form an array of times from lowest to highest in an algebraic sense, that is, where the lowest rank is assigned to the largest negative numbers (see Tables 17.19 and 17.20). The researcher should identify each item in terms of the sample it represents in order to calculate R_1 or R_2. R_1 is the sum of the ranks assigned to items in sample 1 and R_2 is the sum of the ranks assigned to items in sample 2. Either one may be used in the test but one must be careful in interpreting the hypotheses.

The statistic U is found by

$$U = n_1 n_2 + \frac{n_1(n_1 + 1)}{2} - R_1 \tag{17.10}$$

or if the items of sample 2 are used,

$$U = n_1 n_2 + \frac{n_2(n_2 + 1)}{2} - R_2 \tag{17.11}$$

Mann and Whitney (see Related Readings at the end of this chapter) have shown that as n_1 and n_2 increase in size, the sampling distribution of U rapidly approaches the normal distribution with the following parameters:

$$\mu_u = \frac{n_1 n_2}{2} \tag{17.12}$$

and,

$$\sigma_u = \sqrt{\frac{(n_1)(n_2)(n_1 + n_2 + 1)}{12}} \tag{17.13}$$

However if ties occur, the variability of the set of ranks is changed and a correction must be made in the standard deviation of the sampling distribution of U. It is

$$\sigma_u = \sqrt{\left(\frac{n_1 n_2}{N^*(N^* - 1)}\right)\left(\frac{N^{*3} - N^*}{12} - \sum T\right)} \tag{17.14}$$

TABLE 17.19
Weights of samples of two types of golf balls produced by differing processes (ounces)

$n_1 = 20$ (Balata Cover)	$n_2 = 20$ (Surlyn Cover)
1.620	1.599
1.623	1.598
1.624	1.600
1.619	1.610
1.618	1.615
1.622	1.614
1.625	1.628
1.621	1.629
1.617	1.630
1.626	1.631
1.620	1.650
1.621	1.648
1.619	1.630
1.620	1.632
1.626	1.636
1.622	1.620
1.623	1.619
1.621	1.620
1.619	1.624
1.618	1.622

$N^* = 40$

TABLE 17.20
Computation of R_1 and R_2

Array of Sample Weights	Sample	Rank	R_1	R_2
1.598	2	1		1
1.599	2	2		2
1.600	2	3		3
1.610	2	4		4
1.614	2	5		5
1.615	2	6		6
1.617	1	7	7	
1.618	1	8.5	8.5	
1.618	1	8.5	8.5	
1.619	2	11.5		11.5
1.619	1	11.5	11.5	
1.619	1	11.5	11.5	
1.619	1	11.5	11.5	
1.620	1	16	16	
1.620	1	16	16	
1.620	1	16	16	
1.620	2	16		16
1.620	2	16		16
1.621	1	20	20	
1.621	1	20	20	
1.621	1	20	20	
1.622	1	23	23	
1.622	1	23	23	
1.622	1	23	23	
1.623	1	25.5	25.5	
1.623	1	25.5	25.5	
1.624	2	27.5		27.5
1.624	1	27.5	27.5	
1.625	1	29	29	
1.626	1	30.5	30.5	
1.626	1	30.5	30.5	
1.628	2	32		32
1.629	2	33		33
1.630	2	34.5		34.5
1.630	2	34.5		34.5
1.631	2	36		36
1.632	2	37		37
1.636	2	38		38
1.648	2	39		39
1.650	2	40		40
			404.0	426.0

where $N^* = n_1 + n_2$ and $T = \dfrac{t^3 - t}{12}$ and t is the number of sample observations tied for a given rank. We may test the hypothesis by computing

$$z_c = \frac{U - \mu_u}{\sigma_u}$$

The adjustment of the standard deviation of the sampling distribution of U is justified because the Mann-Whitney U test assumes that the weights of the sampled golf balls represent a distribution that has underlying continuity. Under this assumption, and with the most precise measurement possible, in theory the probability of a tie is zero. Using the normally crude measurements of the behavioral sciences or even those of this illustration, ties frequently occur. In reality these two observations or more that are tied really are different but our measurement devices cannot detect these differences.

Tied values are given the average of the tied ranks, for example, if two items are tied for first and second ranks their assigned rank is $(1 + 2)/2 = 1.5$. Ties that occur within the same sample cause no theoretical problem in applying the test but ties from different samples suggest the use of the correction. In most cases the correction factor is negligible, but the researcher is urged to make adjustment to be on the safe side.

Illustration. A firm manufactures golf balls under two sets of conditions. On one brand the cover is the balata material coming from the bully-gum trees found in the valleys of the Amazon River in Brazil. This is the material used on most of the better golf balls that are not solid or made of synthetics. A new process has been developed, called Surlyn, that covers the ball with a plastic covering that is harder to cut. The U.S. Golf Association rule is that balls should be no heavier than 1.620 ounces. Consequently, any balls that do not meet this standard must be sold at a lower price. Twenty golf balls are selected at random from the two different production runs and each is weighed with the results recorded in Table 17.19. The question is: Do the two types of balls differ significantly in terms of weight at the .05 level?

Perhaps, this problem could be solved by the use of a parametric test of the difference between two means. However, the assumption of normality required by that test might be debated and the test results doubted. In addition, the measurements might be in question.

Computations.

1. Establish the hypotheses:

$$\begin{array}{ll} H_0: \mu_1 = \mu_2 & \text{No difference in weight} \\ H_1: \mu_1 \neq \mu_2 & \text{Difference is significant} \end{array}$$

2. Criteria for choosing between the hypotheses:

$$\begin{array}{l} \text{Accept } H_0 \text{ if } -1.96 \leq z_c \leq +1.96 \\ \text{Accept } H_1 \text{ otherwise} \end{array}$$

3. Calculations:

$$n_1 = 20 \quad n_2 = 20 \quad \alpha = .05$$

$$\mu_u = \frac{n_1 n_2}{2} = \frac{(20)(20)}{2} = 200$$

$$U = n_1 n_2 + \frac{n_1(n_2 + 1)}{2} - R_1$$

$$U = (20)(20) + \frac{20(20 + 1)}{2} - 404 = 206$$

$$T = \frac{t^3 - t}{12}$$

$$T = \frac{2^3 - 2}{12} + \frac{4^3 - 4}{12} + \frac{5^3 - 5}{12} + \frac{3^3 - 3}{12} + \frac{3^3 - 3}{12}$$

$$+ \frac{2^3 - 2}{12} + \frac{2^3 - 2}{12} + \frac{2^3 - 2}{12} + \frac{2^3 - 2}{12} = 21.5$$

$$\sigma_u = \sqrt{\left(\frac{n_1 n_2}{N^*(N^* - 1)}\right)\left(\frac{N^{*3} - N^*}{12} - \sum T\right)}$$

$$\sigma_u = \sqrt{\left(\frac{(20)(20)}{40(40 - 1)}\right)\left(\frac{40^3 - 40}{12} - 21.5\right)} = 11.83$$

$$z_c = \frac{U - \mu_u}{\sigma_u} = \frac{206 - 200}{11.83} = .51$$

4. Decision:
Because $.51 < 1.96$ we accept H_0. Difference in weights between the two types of golf balls is not significant.

RANDOMIZATION TEST FOR TWO INDEPENDENT SAMPLES

A good substitute for the parametric t-test of the significance of the difference between two means of two independent samples is the randomization test. This is especially true when n_1 and n_2 are small, that is, less than 30. When n_1 and n_2 are large the Mann-Whitney U test may be regarded as a nonparametric randomization test applied to the ranks of the observations and a good approximation to the randomization test. Also, when the samples are large, the computations necessary for the randomization test become tedious; however, if certain conditions are met these computations may be avoided. The parametric t-test we recall also requires an assumption of a normal distribution and equal variances of the populations. The randomization test does not require these rigid assumptions. Because the test does use the numerical values of the observations the measurements must be at interval scale level or higher.

Under the null hypothesis (H_0) of the randomization test the observations of samples n_1 and n_2 are assumed to come from the same population (universe). That is, only chance is involved, and there really is no statistical significant difference between the two universes. The basic procedure of the test for a small sample is: (1) Determine the number of possible outcomes in the region of rejection which may be found by $\alpha(p_{n_1}^{n_2+n_1})$. This is merely the level of significance times the number of combinations of n things taken x at a time, where $n_2 + n_1 = n$ and $n_1 = x$; (2). Identify as belonging to the region of rejection that number of the most extreme possible outcomes (i.e., the largest difference between the sum of the values in n_1 and the sum of the values in n_2). For a two-tailed test only one half of the number of extremes are in one direction and the other half are in the opposite direction. For a one-tail test all of the expected number of extremes are in the predicted direction. (3) If the observed values are one of the expected or possible outcomes listed in the region of rejection we accept the H_1 at the given level of significance. That is, we reject the H_0. Again, where the samples are too large to conveniently enumerate all of the possible outcomes in the region of rejection we might use the formula for the t-test:

$$t = \frac{\bar{A} - \bar{B}}{\sqrt{\dfrac{\Sigma(B - \bar{B})^2 + \Sigma(A - \bar{A})^2}{n_1 + n_2 - 2}\left(\dfrac{1}{n_1} + \dfrac{1}{n_2}\right)}} \tag{17.15}$$

which tends to be distributed as an approximation to the Student t-distribution with degrees of freedom equal to $n_1 + n_2 - 2$. Therefore, the probability interpretation of this statistic may be made by referring to Appendix D and the table of t. Where ΣA and ΣB are the sum of the values of the observations in samples n_1 and n_2 respectively. And, where $\bar{A}$ and $\bar{B}$ represent the arithmetic means of these two samples. Incidentally, n_1 does not need to equal n_2 in order to apply the randomization test. It should be noted that even though Formula (17.15) is used we are using it here as a *nonparametric* test because the assumption of normality required by the parametric t-test is not necessary here. The use of this formula does require that the following three conditions be met: (1) measurement is in *at least* interval scale, (2) the kurtosis of the combined samples is small, and (3) that the ratio of $n_1 n_2$ lies between ⅕ and 5. That is, that the larger sample is not more than five times the smaller one. If this latter condition is met then the randomization distribution of all possible outcomes is closely approximated by the Student t-distribution.

In the interest of saving time and space the example to follow will assume that all three of these conditions are met and therefore, the computations of all possible outcomes (as done with a small sample) will be omitted and we can use Formula (17.15). Again, it should be noted that an alternative to the randomization test that does not require these three conditions be met is the application of the Mann-Whitney U test discussed above.

An example. Let us make use of the same data that were studied under the illustration of the Mann-Whitney test. (See Table 17.19 for the data.) What is needed are: ΣA, ΣB, $\bar{A}$, and $\bar{B}$. We may then calculate the statistic t and test the following hypotheses:

H_0: Differences between the two populations from which n_1 and n_2 were drawn are not statistically significant.

H_1: Differences between the two populations from which n_1 and n_2 were drawn is too great to have occurred by error of sampling, that is, the differences are statistically significant.

The criteria for choosing between these two conclusions are

$$\text{Accept } H_0 \text{ if: } t \leq t_{.05}$$
$$\text{Accept } H_1 \text{ if: } t > t_{.05}$$

For this problem then, $\Sigma A = 32.424$, that is, the sum of the values of the observations in n_1; $\Sigma B = 32.455$, that is, the sum of the values of the observations in n_2; $\bar{A} = 1.621$ and $\bar{B} = 1.623$. We may then calculate t using Formula (17.15)

$$t = \frac{1.621 - 1.623}{\sqrt{\dfrac{(32.455 - 1.623)^2 + (32.424 - 1.621)^2}{20 + 20 - 2}\left(\dfrac{1}{20} + \dfrac{1}{20}\right)}}$$

$$t = -.003$$

From Table D.1 we can see that this t value is not statistically significant and we accept H_0. This is the same conclusion reached by the Mann-Whitney U test earlier. Again we conclude that the differences in weights of these two samples are not statistically significant from a sampling point of view.

ONE-SAMPLE RUNS TEST FOR RANDOMNESS

The concepts of a random sample and randomness were discussed in Chapters 7 and 8. Recall that a random sample refers to the method of selection of the sample items in such a manner that each possible sample has a fixed and determinate probability of selection. Randomness is obtained when the ordering process of a set of objects is carried out in such a way that every possible order is equally probable. In order for the researcher to make valid inferences about a population based upon sample observations, the items must be randomly chosen. Sometimes in the social sciences the execution of a random sample is difficult to maintain and the randomness of selection might be suspect. The one-sample runs test is designed to test the hypothesis that the sample is random. The test is based on the order or sequence in which the observations were made. A *run* is defined as a sequence of identical symbols (occurrences) that are preceded and followed by different symbols or by none at all. For example, assume that we toss a coin ten times and record the sequence of heads (H) or tails (T) as follows:

H H H H H T T T T T

If R represents the number of runs we have $R = 2$. Intuitively, this seems like too few runs to be a random process for either a fair coin or a fair tosser. Something seems to have interfered with the randomness. On the other hand if we had the following order of heads and tails

H T H T H T H T H T

we have $R = 10$ which seems too perfect to have occurred by chance.

Similar concepts can be applied to research in business and economics. Assume that a survey of student opinions is being taken with respect to a specific issue. Sex is an important variable, therefore, male and female students are to be interviewed in approximately the same proportion as they exist in the population (the total enrolled number of students in the university). Assume that one interviewer's records show the sequence of the sex of the students the interviewer contacted as

M M M M M M M M M M

Clearly, the interviewer interviewed only male students and the randomness of the sequence might be questioned. Or if the interviewer's records were

F M F M F M F M F M

they might also be suspect. Let us turn to an example to demonstrate the application of the one-sample runs test.

Procedure. Let R equal the number of runs, n_1 equal the number of observations of one kind, and n_2 equal the number of observations of another kind. Generally, the one-sample runs test is a two-tailed test. A one-sided test could be used but it implies that the direction of the deviation from randomness is predicted in advance. If either of n_1 or n_2 is greater than 20, we may use the normal distribution to approximate the sampling distribution of R with

$$\mu_R = \frac{2n_1n_2}{n_1 + n_2} + 1 \tag{17.16}$$

$$\sigma_R = \sqrt{\frac{2n_1n_2(2n_1n_2 - n_1 - n_2)}{(n_1 + n_2)^2 + (n_1 + n_2 - 1)}} \tag{17.17}$$

and

$$z_c = \frac{R - \mu_R}{\sigma_R}$$

Illustration. Assume that we test for randomness the 100 trials given in Chapter 7, page 147. (Use a significance level of .05.) In this example a brief table of random numbers was generated by drawing from a container tags numbered from 0 through 9. The procedure for the selection of the tags to preserve randomness is described on page 146. The following runs are contained in the 100 draws:

3 1 99 44 3 6 22 9 1 3 8 6 0 8 1 0 5 7 1 7 1 22 8 6 1 9 1 3 4 8 4
77 11 4 3 7 9 7 8 44 9 6 8 4 2 1 2 5 8 7 4 3 9 5 9 7 4 8 0 6 3 6
4 33 5 9 1 5 8 4 1 8 9 1 3 6 9 2 7 1 5 1 8 9 0 2 4 9 3 5 7 3

Using the 10 digits 0, 1, 2, 3, . . . , 9, one would expect about as many even-numbered digits to be selected as odd because the $P(\text{even}) = P(\text{odd}) = .5$. Therefore, let n_1 equal even-numbered digits and n_2 equal the odd.

Computations.

1. Establish the hypotheses:
   H_0: The order of selection of the digits is random
   H_1: The order of selection of the digits is not random
2. Criteria for choosing between the hypotheses:

   Accept H_0 if $-1.96 \leq z_c \leq +1.96$
   Accept H_1 otherwise

3. Calculations:

$$n_1 = 43 \quad n_2 = 57 \quad R = 92$$
$$\alpha = .05 \quad z = 1.96$$

$$\mu_R = \frac{2n_1n_2}{n_1 + n_2} + 1 = \frac{2(43)(57)}{43 + 57} + 1 = 50$$

$$\sigma_R = \sqrt{\frac{2n_1n_2[(2n_1n_2 - n_1 - n_2)]}{(n_1 + n_2)^2 + (n_1 + n_2 - 1)}}$$

$$= \sqrt{\frac{2(43)(57)[2(43)(57) - 43 - 57]}{(43 + 57)^2 + (43 + 57 - 1)}} = 152.7$$

$$z_c = \frac{R - \mu_R}{\sigma_R} = \frac{92 - 50}{152.7} = .28$$

4. Decision:
   Because $.28 < 1.96$ we accept H_0 and conclude that the order of the selection of the digits is random.

RANK CORRELATION

In some situations, an adaptation of the Pearson parametric method of correlation discussed in Chapter 15 can be applied to the rankings of the data instead of to the original units. The coefficient of rank correlation (r_r) requires that both variables be measured in ordinal scale or higher. The observations may be ranked from low to high or reverse; it makes no difference. Obviously, if the dependent variable is ranked from the largest to the smallest values and the independent variable is ranked from the smallest to the largest, then r_r will be negative. Using rankings in correlation instead of the original data is analogous to the calculation of the median or quartiles. The basis for this statement is that each sample value is given emphasis with respect to its *position* in the array rather than its *magnitude*. This makes rank correlation useful for brand preference analysis and wherever rankings are important.

Procedure. The formula for determining the degree of rank association between two variables is:[4]

$$r_r = 1 - \frac{6\Sigma d^2}{n(n^2 - 1)} \tag{17.18}$$

where d is the difference between the correlative ranks of each set of paired values $(X - Y)$. The regression line passes through the average rank, $(n + 1)/2$, on each scale and has a slope of r_r. Unless many ties exist, the coefficient of rank correlation should be only slightly lower than the Pearson coefficient. If the data in the original units are irregular (several extreme values), rank correlation might be preferred over the parametric approach. The reason being that under the conditions of Chapter 15 an extreme item in each variable will be given an extraordinary weight. Each extreme observation will appear as a square and there will be one cross product. Rank correlation removes this emphasis on absolute values much the same as in the calculation of the median or quartiles. In the past this technique has had limited application because of the problems associated with large sets of data. In today's computer age this no longer is a factor.

---

[4] For a derivation of this formula and the rationale for the constants, see Sidney Siegel, *Nonparametric Statistics for the Behavioral Sciences* (New York: McGraw-Hill Book Co., 1956), pp. 203–4.

Illustration. Let us assume we wish to correlate first-year graduate grade point averages (GPA) with aptitude test scores, with the latter being the independent variable (X). We wish to determine the degree of association between aptitude and academic performance. Table 17.21 shows the calculations.

If n is 10 or greater, the significance of the coefficient of rank correlation may be tested by

$$t = r_r \sqrt{\frac{n-2}{1-r_r^2}} \qquad (17.19)$$

For a large n (10 or greater for this test), the value is distributed as Student t with $df = n - 2$. Then the significance may be determined by referring to Appendix D. For this illustration,

$$t = .89 \sqrt{\frac{10-2}{1-.79}} = 5.48$$

Table D.1 shows that for $df = 10 - 2 = 8$, at t value as large as 5.48 is significant at the .001 level for a one-tail test.

TABLE 17.21
Rank correlation (data: original data ranked from high to low)

Student	X Aptitude Test Score Rank	Y GPA Rank	$d = X - Y$ Rank Differences	$d^2 = (X - Y)^2$
A	1	2	−1	1
B	4	3	1	1
C	10	9	1	1
D	8	10	−2	4
E	5	5	0	0
F	7	6	1	1
G	3	1	2	4
H	2	4	−2	4
I	6	7	−1	1
J	9	8	1	1
Sums	55	55	0	18

$$r_r = 1 - \frac{6\Sigma d^2}{n(n^2-1)} = 1 - \frac{6(18)}{10(10^2-1)} \simeq .89$$

$$a = (n+1)(1-r_r) \div 2 = (10+1)(1-.89)/2 = .605$$

$$b = r_r = .89$$

$$Y_c = .605 + .89X$$

$$\sigma_r = \frac{1-r_r^2}{\sqrt{n}} = \frac{1-.79}{\sqrt{10}} \simeq .06$$

OTHER NONPARAMETRIC TESTS

Several other nonparametric tests that might be useful to the social scientist or business analyst will be briefly discussed below; however, the computational details of application will be omitted. The interested person should consult one of the texts mentioned in footnote 1 of this chapter.

Friedman two-way analysis of variance by ranks: Related samples

In order to test the null hypothesis that k samples have been drawn from the same population when the samples are related, the Friedman two-way analysis of variance by ranks test is most helpful. For this test it makes no difference whether or not the items are ranked from high to low or from low to high. The test does assume that the measurements which are used to indicate ranks must be at least an ordinal scale. Because the test is designed for matched samples, the number of observations is the same in each of the k samples. The matching may be done in various ways, for example, age, education, aptitude test scores, motivation, social-economic status, and many other variables that might be of interest to the researcher. For example, the behavioral scientist might be interested in studying the same group of subjects under each of many conditions or controls. Or the analyst might be analyzing several samples, each consisting of k matched subjects or items, and then the analyst might randomly assign one subject to each condition or control of interest. If one were measuring the effectiveness of four different management training programs, one might assign one person at random from each of n sets of $k = 4$ trainees to each program. Then the task would be to test the data to determine if the different programs produce significantly different results as measured by some condition of interest to management.

For this test the ranks of the subjects or items are the data used not the actual measured condition. If the null hypothesis is true then we would expect the distribution of ranks by each condition would be a matter of chance. That is, we would expect the ranks to appear in all training programs with about equal frequency. That is, in terms of the variable management is studying, it makes no difference which training program the trainee enters. If the employees' scores are independent of the training program then the set of ranks in each training program (each sample) probably comes from the discrete (discontinuous) rectangular distribution of ranks and the rank totals for each program would be about equal. If the employees' scores were dependent on which training program they entered, then the ranks and rank totals would vary significantly and the null hypothesis would be rejected.

The Friedman test uses a statistic χ_r^2 to apply this tool of analysis when a parametric test is inappropriate because of the level of measurement. When the number of conditions or groups (matched samples) are not too small then χ_r^2 is distributed approximately as chi-square with $df = k - 1$, when[5]

$$\chi_r^2 = \frac{12}{Nk(k+1)} \sum_{j=1}^{k} (R_j)^2 - 3N(k+1) \qquad (17.20)$$

where N = the number of rows (samples or groups); k = the number of columns (conditions or programs); R_j = the sum of ranks in the jth column. If the number of rows and columns are not too small (at least ten) then Appendix J and the table of the percentage points of the chi-square distribution is applicable; however, if they are small then some special tables must be used.[6]

Tukey's quick test: Independent samples[7]

In response to an apparent need, Tukey introduced in 1959 his quick test of the hypothesis that two random variables are identically distributed against the alternative hypothesis that the two means are not equal to each other. The need for such a test stemmed from the fact that an easier to apply tool was in demand. That is, the Mann-Whitney U test is likely to be *statistically* more powerful in a similar situation. However, because the Mann-Whitney test is more complicated it tends to go unused at times when it would be appropriate to use. Hence, Tukey refers to the *practical* power of a test, that is, the product of the statistical power by the subjective probability that the procedure will be used. On this basis the Tukey quick test will be used more frequently. Tukey's description of his test is limited to three sentences:[8]

> Given two groups of measurements, taken under conditions (treatments, etc.) A and B, we feel the more confident of our identification of the direction of difference the less the two groups overlap one another. If one group contains the highest value and the other the lowest value, then we may choose (i) to count the number of values in the one group exceeding all numbers in the other, (ii) to count the number of values in the other group falling below all those in the one, and (iii) to sum these

---

[5]For a derivation of this formula and for a rationalization of the Friedman test, see M. Friedman, "The Use of Ranks to Avoid the Assumption of Normality Implicit in the Analysis of Variance," *Journal of American Statistical Association*, vol. 32, pp. 675–701; or M. Friedman, "A Comparison of Alternative Tests of Significance for the Problem of m Rankings," *Annuals of Mathematical Statistics,* vol. 11, pp. 86–92; or M. G. Kendall, "A New Measure of Rank Correlation," *Biometrika,* vol. 30, pp. 81–93.

[6] See Siegel, *Nonparametric Statistics,* pp. 166–73 and 280–81.

[7] See John Tukey, *Technometrics,* vol. 1, no. 1, p. 32.

[8] See ibid., p. 32.

counts (we *require* that neither count be zero). If the two groups are roughly the same size, then the critical values of the total count are, *roughly* 7, 10 and 13, i.e., 7 for a two-sided 5% level, 10 for a two-sided 1% level, and 13 for a two-sided 0.1% level.

Tukey's "complete description" of this test certainly has conciseness in its favor. Only three sentences were used to (*a*) provide the intuitive justification of the test, (*b*) outline the computations necessary, and (*c*) give the critical values for making a probability interpretation of the test. In reality, the Tukey quick test does make the following assumptions: (1) that the two samples are randomly selected, (2) that the two samples are mutually independent, (3) that the scale of measurement is at least an ordinal, (4) that the random variables are continuous (although some provision is made for the handling of ties), and (5) that either the two populations from which the samples were drawn have identical distributions or one population tends to yield a higher mean. Persons interested in more detail about this test are referred to W. J. Conover, *Practical Nonparametric Statistics,* John Wiley & Sons, 1971, pages 327–30.

SUMMARY (see Figure 17.5)

There are many specialized nonparametric tests available to the researcher that have not been discussed here. Students interested in this subject should refer to one or more of the sources given in the text or in the Related Readings at the end of the chapter. The reader is cautioned that where it is practical and valid to do so, the parametric test is preferred. Nonparametric tests are built on a relatively unsophisticated mathematical base of elementary probability theory. Usually, the derivation of the formulas can be handled and understood with no more mathematical training than a good high school algebra course. This weaker mathematical foundation is not mentioned as a distraction from the valid use of nonparametric tests, but the student should recognize that to a degree this does limit their applications. In spite of their recent upsurge in popularity, nonparametric tests can be traced back at least to 1710.[9] The mathematics upon which these tests are based date much earlier. It should be remembered that both parametric and nonparametric tests require that the sampling theory assumptions of random and independent selection be met. Nonparametric tests were *not* designed for use when nonprobability sampling plans preclude the valid use of parametric methods. Rather, their main source of strength comes

---

[9] See John Arbuthnott, "An Argument for Divine Providence, Taken from the Constant Regularity Observed in the Births of Both Sexes," *Philosophical Transactions,* vol. 27 (1710), pp. 186–90.

FIGURE 17.5
Summary of chi-square and nonparametric tests

Name of Test	Test Characteristics	Comments and/or Questions to Be Answered
Chi-square:		
Goodness of fit	Specifies some theoretical distribution as being descriptive of the population.	Test measures how closely the observed data approximate the theoretical or expected frequencies.
Independence of principles of classification	Used to determine if a set of classification criteria is meaningful.	Answers the question of whether or not the principles of classification are independent. H_0 states there is no relationship between the criteria.
McNemar test	Can be used with nominal scale of measurement for testing the significance of changes "before" and "after."	Extension of the chi-square test. Deals with two related samples; therefore parametric tests are not appropriate.
Kolmogorov-Smirnov test	Concerned with the degree of agreement between the set of sample observations and some specified theoretical distribution. Nominal scale adequate.	Related to the chi-square test in that the KS is a test of goodness of fit. Unlike the chi-square test the KS test deals with specifying the theoretical cumulative frequency distribution (CFD_e) and comparing it with the observed cumulative frequency distribution (CFD_o). KS test is more powerful than chi-square. Also applicable to very small samples which may not be appropriate for chi-square test.
Sign test	Used to determine if two controlled conditions have different effects. Nominal scale adequate.	Based upon the differences between two pairs of observations and ignores magnitudes. Variable must have a continuous distribution. Because the sign test deals with two related samples the parametric tests are not appropriate.
Wilcoxon matched-pairs signed-ranks test	Used to determine if two controlled conditions have different effects. At least ordinal scale required.	Basically the same test as the sign test except that this one takes into consideration the magnitudes of the differences. More powerful statistically than the sign test. Samples are related so parametric tests inappropriate.

FIGURE 17.5
(continued)

Name of Test	Test Characteristics	Comments and/or Questions to Be Answered
Mann-Whitney *U* test	Can be used to determine whether or not two independent samples are drawn from the same population or from two populations having the same mean. At least ordinal level of measurement required.	Good substitute for the parametric test of the significance of the difference between two means. Samples must be independent. When n_1 and n_2 are large this test may be a good approximation to the randomization test.
Randomization test	Used to test the significance of the difference between two means.	Samples must be independent. Good substitute for the parametric *t*-test of the significance of the difference between two means when n_1 and n_2 are less than 30.
One-sample runs test for randomness	Designed to test the hypothesis that the sample is random.	Test is based upon the order or sequence in which the observations are made.
Friedman two-way analysis of variance by ranks	Tests the H_0 whether or not two related samples have been drawn from the same population.	Measurements to determine the ranks must be done in ordinal scale or higher. Because the test deals with matched samples $n_1 = n_2$.
Tukey's quick test	Tests whether or not two random variables are identically distributed against the hypothesis that the two means are equal.	Its use stems from the fact that it is easier to apply than the Mann-Whitney but it is less powerful than the latter. Assumes that the two samples are randomly selected, that the two samples are mutually independent, that the scale of measurement is at least ordinal and the random variables are continuous.
Rank correlation	Gives measure of association between two variables based upon the ranks. Rankings should be done on data collected at ordinal scale or higher.	Rankings give weight to items in terms of their position not magnitude; therefore, useful where extreme items may distort Personian correlation results. Items may be ranked from high to low or vice versa.

from the fact that nonparametric tests can be used with data of a lower level of measurement. This characteristic has made them popular in the social sciences where oftentimes precise higher order measurements are impossible or impractical to obtain. Chapter 18 deals with another approach to data analysis that differs from the classical statistics. Here we look at the application of some subjective probabilities.

SYMBOLS

χ_c^2, computed value of chi-square
χ_e^2, table value of chi-square, "expected" value
f_e, expected or theoretical frequencies
f_o, observed or sample frequencies
df, degrees of freedom
n, sample size
T, sum of the positive or negative ranks
R_1, sum of the ranks assigned to items in sample n_1
R_2, sum of the ranks assigned to items in sample n_2
N^*, equals sums of items in sample n_1 + sample n_2
t, equals the number of sample items tied for a given rank and Student's t distribution
U, Mann-Whitney U test statistic
R, number of runs
r_r, Spearman coefficient of rank correlation
d, difference between correlative ranks of $X - Y$

PROBLEMS

17.1. If 200 random drawings from a bridge deck of 52 cards (items replaced after each draw) produced 60 spades, 40 hearts, 44 clubs, and 56 diamonds, would you consider these results unusual at the .05 level of significance? At the 0.1 level?

17.2. If 50 tosses of a 6-sided die yields the following results:

Number on Face of Die	Frequency of Occurrence
1	8
2	6
3	12
4	11
5	9
6	4
	50

Would you consider the die to be honest and the tosses random at the .10 level of significance?

17.3. Assume 1,000 tosses of 4 coins yields the following distribution:

Number of Heads	Frequency
0	180
1	190
2	220
3	210
4	200
	1,000

Has some factor other than chance caused the distribution to differ from what might be expected? Use the .05 level of significance.

17.4. A firm selling a product issues a money-back guarantee that 99 percent of any given lot will meet some stated specifications. A random sample of 500 items from this supplier revealed that 460 were acceptable and 40 were not. On the basis of this evidence, is the firm's claim valid at the .01 level of significance? Would you return the goods?

17.5. Ggollek's, a large breakfast cereal producer, had decided to introduce a new product. The management requests the market research unit to determine which of two packages might have the greater customer appeal. Using a nationwide panel of 2,000 families, each is given two entirely different packages with identical contents but identified as brand X and brand Y. The following results are reported to the statistician:

Favor brand X	825
Favor brand Y	750
No preference	425
	2,000

Is there a significant difference between customer preference for X or Y? Use .01 level of significance. *Note:* This problem also appears in Chapter 11. Reach a decision using chi-square analysis and compare the results with those of Chapter 11.

17.6. Seniors in the public and private high schools of Cuyahoga County, Ohio, were surveyed in the spring of 1966 to determine their post high school plans. (See *Economic Review,* April 1967, Federal Reserve Bank of Cleveland.) The following information was part of the data gathered:

	Plan to Attend College					
	Boys		Girls		Total	
Family Income	No.	%	No.	%	No.	%
Under $5,000	244	5.5	254	7.2	498	6.2
$5,000–$9,999	2,352	52.5	1,913	54.6	4,265	53.4
$10,000 and over	1,881	42.0	1,339	38.2	3,220	40.4
Total	4,477	100.0	3,506	100.0	7,983	100.0

Comparable data for the United States as a whole are:

Family Income	Plan to Attend College (percentage)		
	Boys	Girls	Total
Under $5,000	2.5	4.8	3.6
$5,000–$9,999	50.6	52.1	51.3
$10,000 and over	46.9	43.1	45.1
Total	100.0	100.0	100.0

Do the high school seniors in Cuyahoga County who plan to attend college come from different economic backgrounds than one might expect based upon U.S. data? Use the .05 level of significance.

17.7. What distinguishes the nonparametric from the parametric statistical tests? What is meant by the term "distribution-free" test?

17.8. What are the four basic scales of measurement? What are the distinguishing features of each?

17.9. Which scales of measurement are appropriate for the valid application of (*a*) parametric statistical tests and (*b*) nonparametric?

17.10. What are some of the advantages of the nonparametric tests over parametric tests?

17.11. Distinguish clearly between tests of goodness of fit and contingency tables.

17.12. What are the basic assumptions necessary for the proper application of a chi-square test?

RELATED READINGS

CHERNOFF, H., and SAVAGE, I. R. "Asymptotic Normality and Efficiency of Certain Nonparametric Test Statistics." *Annuals of Mathematical Statistics,* vol. 29 (1958), pp. 972–94.

COCHRAN, W. G. "The Chi-Square Test of Goodness of Fit." *Annuals of Mathematical Statistics,* vol. 23 (1952), pp. 315–45.

COOMBS, C. H. *A Theory of Psychological Scaling.* University of Michigan, Engineering Research Institute, Bulletin No. 34 (1952).

DIXON, W. J. "Power under Normality of Several Nonparametric Tests." *Annuals of Mathematical Statistics,* vol. 25 (1954), pp. 610–14.

FRASER, D. A. S. "Most Powerful Rank-Type Tests." *Annuals of Mathematical Statistics,* vol. 28 (1957), pp. 1040–43.

GAUMNITZ, JACK E. "Nonparametric Tests for Randomness and Independence of Data." *Fortran Applications in Business Administration,* vol. II, pp. 133–50. Ann Arbor: Graduate School of Business Administration, University of Michigan, 1971. A computer program written in Fortran IV language.

HALPERIN, M. "Extension of the Wilcoxon-Mann-Whitney Test to Samples Censored at the Same Fixed Point." *Journal of the American Statistical Association,* vol. 55 (1960), pp. 125–38.

HOTELLING, H. "The Behavior of Some Standard Statistical Tests under Nonstandard Conditions." *Proceedings of the Fourth Berkeley Symposium on Mathematical Statistics and Probability,* vol. I, pp. 319–59. Edited by Jerzy Neyman. Berkeley: University of California Press, 1961.

———, and PABST, MARGARET R. "Rank Correlation and Tests of Significance Involving No Assumption of Normality." *Annuals of Mathematical Statistics,* vol. 7 (1936), pp. 29–43.

KENDALL, M. G., and SUNDRUM, R. M. "Distribution-Free Methods and Order Properties." *Review of International Statistical Institute,* vol. 3 (1953), pp. 124–34.

MANN, H. B. "Nonparametric Tests against Trend." *Econometrica,* vol. 13 (1945), pp. 245–59.

———, and WHITNEY, D. R. "On a Test of Whether One of Two Random Variables Is Stochastically Larger than the Other." *Annuals of Mathematical Statistics,* vol. 18 (1947), pp. 50–60.

MILLER, RUPERT G., JR. "Sequential Signed-Rank Test." *Journal of the American Statistical Association,* vol. 65, no. 332 (December 1970), pp. 1554–61.

MOOD, A. M. "The Distribution Theory of Runs." *Annuals of Mathematical Statistics,* vol. 11 (1940), pp. 367–92.

MOORE, G. H., and WALLIS, W. A. "Time Series Significance Tests Based on Signs of Differences." *Journal of the American Statistical Association,* vol. 38 (1943), pp. 153–64.

PAGE, ELLIS BATTEN. "Ordered Hypotheses for Multiple Treatments: A Significance Test for Linear Ranks." *Journal of the American Statistical Association,* vol. 58 (1963), pp. 216–30. This article illustrates a significance test of a monotonic relationship among an expected ordering among treatment effects in a two-way analysis of variance. The L-test assumes data of ordinal measurement or higher.

PRATT, J. W. "Remarks on Zeros and Ties in the Wilcoxon Signed-Rank Procedures." *Journal of the American Statistical Association,* vol. 54 (1959), pp. 655–67.

SAVAGE, I. R. "Bibliography of Nonparametric Statistics and Related Topics." *Journal of American Statistical Association,* vol. 48 (1953), pp. 844–906.

SCHEFFE, H., and TUKEY, J. W. "Nonparametric Estimation: I. Validation of Order Statistics." *Annuals of Mathematical Statistics,* vol. 16 (1945), pp. 187–92.

SMITH, K. "Distribution-Free Statistical Methods and the Concept of Power Efficiency." In *Research Methods in the Behavioral Sciences,* edited by L. Festinger and D. Katz. New York: Dryden Press, 1953, pp. 536–77.

STEVENS, S. S. "On the Theory of Scales of Measurement." *Science,* vol. 103 (1946), pp. 677–80.

TUKEY, J. W. "A Survey of Sampling from Contaminated Distributions." In *Contributions to Probability and Statistics: Essays in Honor of Harold*

Hotelling, edited by Ingram Olkin et al. Stanford, Calif.: Stanford University Press, 1960, pp. 448–85.

WILCOXON, F. *Some Rapid Approximate Statistical Procedures.* New York: American Cyanamid Co., 1949 (pamphlet).

———; KATTI, S. K.; and WILCOX, ROBERTA A. *Critical Values and Probability Levels for the Wilcoxon Rank Sum Test and the Wilcoxon Signed Rank Test.* Lederle Laboratories Division, American Cyanamid Co., Pearl River, New York, and Department of Statistics, The Florida State University, Tallahassee, Florida, August 1963.

ZAR, JERROLD H. "Significance Testing of the Spearman Rank Correlation Coefficient." *Journal of the American Statistical Association,* September 1972, pp. 578–83.

Part VIII

Elementary decision making using prior information

To me, there is something superbly symbolic in the fact that an astronaut, sent up as an assistant to a series of computers, found that he worked more accurately and more intelligently than they. Inside the capsule man is still in charge.

Adlai E. Stevenson

18

Introduction to Bayesian decision theory

When possible make the decision now, even if action is in the future.
A reviewed decision usually is better than one reached at the last moment.

William B. Given, Jr.

THE MATERIAL presented in Chapters 1 through 11 and 15 and 16 provides the necessary background for understanding the basic theoretical considerations of the *classical school* in terms of statistical inference and interpretations of a probability. For example, the tests of hypotheses procedures presented in Chapter 10 were designed to test a statistical statement about a population (the null hypothesis), given a level of significance (α). We had two alternative courses of action, and our main task was to establish some criteria (decision rule) for choosing between these two acts. We reached a decision based upon an event (sample evidence) evaluated in the light of our criteria or decision rule. Our conclusions and course of action were based upon the so-called *objective* interpretation of a probability stated in terms of a relative frequency distribution. *Our interest was focused upon the course of action while the test related to the null hypothesis.*

As indicated in Chapter 1, a supplementary analysis of these problems is provided by the subjective approach to the application of probability theory.[1] This school of thought emphasizes the *selection* of the courses of action rather than the hypotheses themselves. The reader is cautioned that this brief section is intended only to acquaint the nonstatistician

[1]This basic introduction attempts to summarize briefly some of the fundamental concepts of the "Bayesian School" as represented by Robert Schlaifer given in *Probability and Statistics for Business Decisions*, 1959, and *Introduction to Statistics for Business Decisions*, 1961. Both pioneering texts were published by McGraw-Hill Book Co., New York.

with the Bayesian objectives. Students interested in a more vigorous pursuit of the subject should become familiar with the references given at the end of the chapter. Many universities have specific courses on this material and they are excellent complementary cognates for the more traditional statistical inference of the classical school. For example, Schlaifer's second volume cited above ". . . aims at a *unified* treatment of classical and Bayesian statistics, intended to bring out from the very first the essential agreements between the two schools. . . ."[2] An illustration of the subjectivist's solution to statistical decision-making problems is shown below.

Payoff table

Schlaifer assumes that when the consequences of various alternative courses of action are based upon conditions of uncertainty, the reasonable method of choosing the "best" solution is to assign (1) values to consequences and (2) probabilities to events. Then the rational choice is to select the act with the highest expected value. This process appears diagrammatically in Figure 18.1.

Assume a drug manufacturer is considering marketing a new compound that will require a prescription for various size dosages. Unfortunately, costs also will vary depending upon the size of the pill. What size pill should the manufacturer produce in order to maximize profits? In order to keep the problem relatively simple and not overshadow the points to be made, let us make the following assumptions:

1. The pill can be produced in almost any size; however, only four different sizes will be considered. Size S_1 requires the patient take one pill each dose; size S_2 requires that the patient take two pills each dose;

FIGURE 18.1

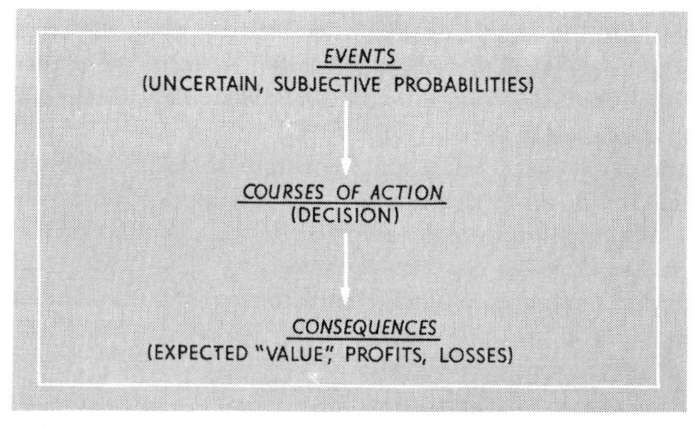

[2] Schlaifer, *Introduction to Statistics for Business Decisions,* p. v.

size S_3 requires that the patient take three pills each dose; and Size S_0 requires the patient take one-half pill each dose. The latter size would require the patient to cut the pill on a scored line, but it can be produced and marketed at the lowest cost of all four products if the company can sell at least five million of them. *These would be the alternative courses of action.*

2. *The events might be* (*a*) the product is successful, that is, at least five million pills can be sold; and (*b*) the product is unsuccessful, that is, only 100,000 can be sold.

3. The cost of producing and distributing each size pill in amounts of 100 or more is as shown in Table 18.1.

TABLE 18.1

Pill Size	Cost, 5 Million Sales	Cost, 100,000 Sales
S_1	$1.50	$1.00
S_2	1.60	1.25
S_3	1.75	1.50
S_0	1.00	1.75

4. The profit per 100 pills is as shown in Table 18.2.

TABLE 18.2

Pill Size	Profit, 5 Million Sales*	Profit or Loss, 100,000 Sales*
S_1	$2.60	$ .10
S_2	2.75	.05
S_3	3.00	−.25
S_0	3.50	−.50

* Number of bottles of 100 pills each.

Given the *event,* the relationship between the courses of action and the consequences are shown in Table 18.3 in terms of profits.

We note that the sequence of reasoning is from events → course of action → consequences. However, the event is uncertain in the real world, namely, whether or not the product will be successful. The consequences (profits) are the result of a course of action, given an event. If the product is successful, the "best" course of action would be to produce and distribute pill S_0 since the profits would be $17.5 million. However, if the product is unsuccessful, then S_1 would be preferred. We

TABLE 18.3
Payoff table: Profits

Event	Course of Action (pill size)			
	S_1	S_2	S_3	S_0
Product is successful (E_1)	$13.00 mil.*	$13.75 mil.	$15.00 mil.	$17.50 mil.
Product is unsuccessful (E_2)	$10 thou.†	$5 thou.	−$25 thou.	−$50 thou.

* ($2.60)(5 mil.) = $13.00 mil.
† ($0.10)(100,000) = $10,000.

may also evaluate the consequences in terms of costs or opportunity loss. Tables 18.4 and 18.5 show these values.

Table 18.4 reflects the fact that costs differ depending upon sales whereas Table 18.3 indicated that profits varied with sales. Table 18.5 is in terms of opportunity losses. Generally we refer to such a summary as a *loss* table. Schlaifer[3] defines opportunity loss of a decision as: ". . . the *difference* between the cost or profit *actually* realized under that decision and the cost or profit which *would have been* realized if the decision had been the best one possible for the event which actually occurred."

We note that even though we select the best possible course of action in the light of the information *before* the fact, this decision often turns out incorrect *after* full knowledge is available. For example, in terms of Table 18.3 the "best" decision would be to produce and distribute pill size S_0 *given* event E_1. If "after the fact" we discover the product is unsuccessful, our "best" decision would have been to produce and market

TABLE 18.4
Payoff table: Costs

Event	Course of Action (pill size)			
	S_1	S_2	S_3	S_0
Product is successful (E_1)	$7.5 mil.*	$8.0 mil.	$8.75 mil.	$5.0 mil.
Product is unsuccessful (E_2)	$100 thou.†	$125 thou.	$150 thou.	$175 thou.

* ($1.50)(5 mil.) = $7.5 mil.
† ($1.00)(100,000) = $100,000.

[3] Ibid., p. 70.

TABLE 18.5
Payoff table: Opportunity loss

Event	Course of Action (pill size)			
	S_1	S_2	S_3	S_0
Product is successful (E_1)	$4.50 mil.*	$3.75 mil.	$2.50 mil.	$.0
Product is unsuccessful (E_2)	$0.0	$5.0 thou.†	$35.0 thou.	$60.0 thou.

* $17.50 mil − $13.00 mil. = $4.50 mil.
† $10.0 thou. − $5.0 thou. = $5.0 thou.

S_1. Where the event is uncertain and we must reach a decision on the basis of limited information, such results are bound to happen (remember the Edsel?). However, we should be aware of the losses which are the result of imperfect information. This is the major point of this introduction to Bayesian inference, and it will be more developed later.

Practical business executives could not care less about the actual computation of the opportunity loss. They may regret it, but the figures of Table 18.5 are probably only of interest to his curiosity. Business executives are more interested in the risk of loss *before* they make their final decision as to which course of action they take. That is, they are concerned with the *expected loss* of the course of action they are contemplating. If this expected loss is too large business executives will want to reduce it by (1) postponing their decision until more information is available or by (2) hedging the risk. Therefore, we shall now direct our attention to the determination of the *expected opportunity loss* (Tables 18.6 and 18.7).

Expected opportunity loss

As a means of illustrating the computations, let us examine the expected opportunity loss if management decides to produce and distribute pill S_2. Table 18.6 is based upon the conditional losses of Table 18.5, and

TABLE 18.6
Expected loss: Pill size S_2

Event	$P(E)$ Prior	Conditional Loss	Expected Loss
Product is successful (E_1)	.60	$3.75 mil.	$2.25 mil.
Product is unsuccessful (E_2)	.40	$5.0 thou.	.002 mil.
	1.00		$2.252 mil.

TABLE 18.7
Expected losses and profits: All possible decisions

Course of Action	Expected Losses	Expected Profits
S_1	$2.700 mil.	$ 7.804 mil.
S_2	2.252 mil.	8.252 mil.
S_3	1.514 mil.	8.990 mil.
S_0	.024 mil.	10.480 mil.

the probabilities of E_1 and E_2 are based upon management's subjective judgment of the probabilities of success. Table 18.7 shows the expected losses and profits of every possible decision in our example computed by the same method. The reader should check his or her understanding of the principles by determining the values for S_1, S_0, and S_3.

Comparison of courses of action in terms of expected losses

On an intuitive basis, we would expect management to select the course of action with the *highest expected profits* which might be achieved by producing and distributing the pill with the *lowest expected loss.* As Schlaifer points out, this is a fact that may be demonstrated. In any decision problem: "The difference between the expected profits of any two acts is equal in magnitude but opposite in sign to the difference between their expected loss."[4]

Table 18.7 indicates that the expected profits for S_2 is higher than the expected profits of S_1 by $.448 million ($8.252 − $7.804 = .448). The expected *loss* of producing and distributing S_2 is $.448 million lower than S_1, $2.700 − $2.252 = $.448. The reader may wish to verify that the values for the other courses of action are as shown in Table 18.7.

Cost of risk or uncertainty

Given the fact that in the real world events E_1 and E_2 are uncertain, the manufacturer must assume some risk which is inherent in situations where the information available for decision making is incomplete or if the manufacturer acts irrationally. Table 18.7 shows that the best possible course of action for the drug manufacturer is to produce and distribute pill size S_0. *In this case the expected opportunity loss is $24,000 and might be viewed as the cost of the risk involved.* We might view this cost as the difference between the best course of action that management can expect to do with the available information and what the decision makers might do if they had *perfect* knowledge.

[4] Ibid., p. 75.

TABLE 18.8

Course of Action	Expected Losses (millions)	Expected Profits (millions)
S_3 vs. S_2	$ .738 lower	$ .738 higher
S_3 vs. S_1	1.186 lower	1.186 higher
S_0 vs. S_1	2.676 lower	2.676 higher
S_0 vs. S_2	2.228 lower	2.228 higher
S_0 vs. S_3	1.490 lower	1.490 higher

Schlaifer makes a distinction between the *cost of uncertainty* (the $24,000 in our example) and the *cost of irrationality.* He defines these terms as follows:

> *Cost of uncertainty:* The expected opportunity loss of the *best possible* decision under a given probability distribution.
>
> *Cost of irrationality:* The amount by which the expected opportunity loss of the chosen decision exceeds the cost of uncertainty under a given probability distribution.[5]

For example, we have stated that the cost of uncertainty in this problem is $24,000. The cost of irrationality for course of action S_1 is $2.700 million − $.024 million = $2.676 million. It pays to be rational! (See Table 18.8.)

Subjective probability: Bayes theorem and decision problems

As indicated in Chapter 1, there are two schools of thought in terms of a probability interpretation. The classical school maintains that probability is applicable only to events that can be repeated under the same conditions. Disciples of this school view probability as a *long-run relative frequency* and prefer to be known as *objectivists.* Given this position, of course, there are many problems that are classified by the objectivists as being inappropriate for a probability interpretation simply because there is no long-run relative frequency in sight.

The so-called *Bayesian* school, labeled as *subjectivists,* regard probability as a measure of the degree of personal belief in a proposition. That is, these people assign probabilities to events on the basis of someone's subjective evaluation based on experience, past performance, judgment, or whatever else is deemed useful. It is an interpretation of their "degree of confidence" in assessing a specified outcome. It should be noted that using subjective probability in this way it is related to the same concept that the objectivists use in determining the *proportion of successes* in the long-run. Some critics argue that the Bayesians assume that there is a

[5] Ibid., p. 76.

straight-line relationship between subjective probabilities and dollars in the utility theory concept. Their point is that this may not be so because utilities vary among decision makers even for the same probabilities.

One major advantage to viewing probabilities as the Bayesian school does is that the decision-making model developed above places the selection of a course of action on a rational basis. *If we fail to utilize our procedures simply because the objective probabilities are unavailable, then we are back to decision making based on intuition, hunch, or really subjective judgment. On the other hand, if we use this information to form subjective probabilities for the various events, then we can act rationally from that point on.* It is here where Bayes theorem (see Chapter 7) is useful in producing more objective (and probably better) decisions using our very subjective probabilities!

Also in the "real world" there is no objective probability. The concept of probability itself is a theoretical one. When we assign probabilities in the real world based on random samples or tosses of an honest coin, they are actually subjective in nature based on knowledge, intuition, experience, or judgment. One can easily defend the argument that when one deals with *real world* situations you are using subjective probabilities. The assignment of such probabilities to a unique event (unheard of for the classical school) might be viewed as the logical extension of the theory rather than as a radical departure. Perhaps, one may argue that the acceptance or rejection of subjective probabilities is subjective itself.

Some writers contend that the distinction between the two schools of thought has been exaggerated because both formulate problems in terms of events or states of nature, courses of action, and consequences of such action. The typical Bayesian sometimes is charged with overemphasizing *decision-making* problems at the expense of minimizing the *statistical inference.* These critics claim that the subjectivists are more concerned with decisions than conclusions. Bayesians counter that the classical statistician cannot objectively evaluate the seriousness of Type I and II errors in their usual power curve approach to such problems. The basis for this disagreement stems from the contention that losses usually vary with the alternative values that the population parameter might assume. The Bayesian claims that the classicist does not have a *formal* structure for including these losses in the decision-making process. The classicist rejects this notion and insists that the Bayesian misunderstands and overstates the value of a "formal structure" for recognizing these losses. Bayesians argue that by providing a loss function (their formal structure) which specifies the seriousness of these errors and by assigning prior probabilities to events, their approach supplements and extends the classical analysis. The classical statistician will retort that the Bayesian tends to grossly understate the problem of the errors of measurement associated with the assignment of the subjective probabilities and the establishment of the loss function. For example, it is not clear that under conditions of no information that all events should be considered equally likely as Bayesians frequently do. Also, there is a lively controversy going on as to how to phrase questions to a decision maker in order to determine the prior probability distribution or to even quantify the measurements.

Additionally, some classicists will dismiss the Bayesian criticisms of their hypothesis testing procedures on the grounds that it is a statistical inference problem not a decision situation. This, of course, is a powerful point to refute from a theoretical point of view. In any event, it seems that this academic self-flagellation has been helpful in stimulating intellectual discussions and the end result being a strengthening of the dynamic field of statistics.

Bayes theorem. As given in Chapter 7, Bayes theorem is as follows:

$$P(A_i|B) = \frac{P(A_i)P(B|A_i)}{P(A_1)P(B|A_1) + \cdots + P(A_n)P(B|A_n)} \qquad (18.1)$$

The formula is, of course, a mathematical statement of conditional probability to simplify calculations. Bayes theorem is used to revise probabilities based on sample evidence. These revised probabilities are termed *posterior probabilities* (see Chapter 7). Having accepted the usefulness of the concept of subjective probabilities, let us now recast the drug problem to utilize Bayes theorem. In the process we should examine the concepts of the expected value of perfect information (EVPI) and the value of imperfect information (VII).

Expected value of perfect information (EVPI)

Most firms that conduct market research prior to taking any management action on a marketing problem associated with a new product usually do not know *precisely* the value of any additional information that might be made available to the decision makers. That is, in many cases it is assumed that more information is "better" without really costing it out. Quite possibly in many situations more information is better; however, in some cases the impact on expected profits may not warrant the time and expense of gathering new information.

In terms of our drug problem, one might ask (1) should additional information be obtained prior to making the marketing decision? and (2) how valuable is this new information going to be? Naturally, in most instances it is difficult, if not impossible, to place an exact value on any specific new information in terms of its contribution to expected profits. However, it generally is possible to place an upper bound on the value of any additional data. Indeed, one may calculate the value of perfect information, that is, the precise knowledge of what event will happen. The expected value of perfect information (EVPI) may be defined as the expected profits (or savings) from knowing the exact event that will occur. Therefore, the EVPI is exactly equal to the expected opportunity loss of the best action. The reader will recall that the expected opportunity loss is the *additional* profit associated with making the best decision. Consequently, if the decision makers did have perfect information they could always pick the best course of action and will save the amount of the expected opportunity loss. If one multiplies the opportunity loss by the probabilities of each event occurring we obtain (1) the *expected* opportunity loss and (2) the expected value of perfect information.

The values of the expected profits, costs, and opportunity loss given in Tables 18.3, 18.4, and 18.5 are based on the *prior* information that the probability of the product being successful (E_1) is .60. However, let us assume that pill size (S_0) is test marketed and the results are not so encouraging. The test market survey indicates that the probability of the product being successful is only .30. The test also indicates a probability of .50 that the product S_0 might be unsuccessful. Using Bayes theorem we may now compute the *posterior probability* that it will turn out to be successful and then reexamine our decision to market S_0 in view of the *expected* opportunity loss.

The *prior probabilities* are:

$$P(E_1) = .60; P(E_2) = .40$$

The *conditional probabilities* of S_0 based on the test market are:

$$P(E_1) = .30; P(E_2) = .50$$

TABLE 18.9
Payoff table: Expected opportunity profits

Event	$P(E)$ Posterior	Course of Action (pill size)			
		S_1	S_2	S_3	S_0
Product is successful (E_1).........	.47	\$6.1100 mil.	\$6.4625 mil.	\$7.0500 mil.	\$8.2250 mil.
Product is unsuccessful (E_2)......	.53	5.3 thou.	2.65 thou.	−13.25 thou.	−26.5 thou.
Expected profits......		\$6.115 mil.*	\$6.465 mil.	\$7.037 mil.	\$8.199 mil.

* (\$2.60) (5 mil.) (.47) + (\$.10) (100,000) (.53) = \$6.115 mil.

TABLE 18.10
Payoff table: Expected opportunity losses

Event	P(E) Posterior	Course of Action (pill size)			
		S_1	S_2	S_3	S_0
Product is successful (E_1)........	.47	$2.1150 mil.	$1.7625 mil.	$ 1.1750 mil.	$ .0
Product is unsuccessful (E_2).....	.53	.0	2.65 thou.	18.55 thou.	31.8 thou.
Expected losses........		$2.115 mil.*	$1.765 mil.	$1.194 mil.	$ .032 mil.

* ($4.50 mil.)(.47) + ($.0)(.53) = $2.1150 mil.

Then from Bayes theorem the posterior probability is:

$$P(E_1S_0) = \frac{(.60)(.30)}{(.60)(.30) + (.40)(.50)}$$

$$= .47$$

The test market results then change management's estimate of the probability that we might be successful. Perhaps, we should reconsider our decision in the light of this additional evidence. Table 18.9 computes the expected profits given this new information; Table 18.10 shows the expected losses; and Table 18.11 indicates both the expected losses and profits for all possible decisions.

On the assumption that the test market results would be comparable for other pill sizes (which might be challenged), the best decision still would be to produce and market S_0.

Comparing our courses of action in terms of expected losses we find the figures shown in Table 18.12 on the following page.

TABLE 18.11
Expected opportunity losses and profits: All possible decisions

Course of Action	Expected Opportunity Losses	Expected Opportunity Profits
S_1............	$2.115 mil.	$6.115 mil.
S_2............	1.765 mil.	6.465 mil.
S_3............	1.194 mil.	7.037 mil.
S_0............	.032 mil.	8.199 mil.

TABLE 18.12

Course of Action	Expected Opportunity Losses	Expected Opportunity Profits
S_3 vs. S_2.......	$ .572 mil. lower	$ .572 mil. higher
S_3 vs. S_1.......	.921 mil. lower	.922* mil. higher
S_0 vs. S_1.......	2.083 mil. lower	2.083 mil. higher
S_0 vs. S_2.......	1.733 mil. lower	1.733 mil. higher
S_0 vs. S_3.......	1.163 mil. lower	1.162 mil. higher

* Rounding error.

While it is not so decisive that S_0 is the preferred product to market based on the new information, Tables 18.10 and 18.11 both show that pill S_0 appears to be the best risk. The cost of this risk is now $32,000 (compared to $24,000 previously) given the modification of our subjective probabilities by the test market results. Of course, to this figure one must add the cost of obtaining the new information. If the survey test market results cost $50,000 then the cost of the risk really is $82,000. Again this cost of uncertainty involved actually is the difference between the best course of action that management can expect to take with the available information, and what the decision makers might do if they had *perfect* knowledge about the market. The cost of irrationality–that is, marketing S_1 instead of S_0–would be $2.115 million minus $0.032 million minus $0.050 million equals $2.033 million.

Value of imperfect information

In reality, the above calculations help us to understand the value of imperfect information. The EVPI determines the *upper bound* of the value of additional data obtained in our test market. Given our knowledge of sampling theory and the behavior of individuals we know that the information obtained through the test market survey probably is *imperfect*. Its imperfection lies in the fact that the test market results are unlikely to predict *precisely* what will happen on a national basis. Possibly, additional test markets might give a different picture. That is, we do not know for certain if the new drug will be successful; however, the additional information can be helpful to management if the data improve the chances of making a correct decision. And, normally making the correct decision will have favorable profit implications and might enrich the so-called "bottom line."

Generally, we can evaluate the value of any additional information only if we have some estimate of reliability attached to the new data. In terms of this particular problem the results of the test market changed the probabilities that were originally selected on an intuitive basis. Possibly the decision makers may have some reservations about the probable success of marketing S_0 even after receiving this new test market information. Should they gather additional information from another test market? To help them make this decision we might first calculate the EVPI by referring to Table 18.3. If we knew for certain that the product would be successful (that is, event E_1 has a probability of 1.0 of occurring) then we might calculate the value of perfect information. From Table 18.3 we note that the profits from marketing S_0 would be $17.5 million if the product is successful. If we assume that the probability of product success is 1.0 it is abundantly clear that under this ideal situation that considerable potential value may be obtained through the collection of new information. In this case the EVPI is equal to the expected opportunity loss of the best action, that is, marketing S_0. However, Tables 18.9, 18.10, 18.11, and 18.12 reflect the results of the new information provided by the test market. Given the fact that the probability of product success is less than 1.0, the question remains: do we go ahead on the basis of this imperfect information or do we buy still more data? Without *any* information about the probability of the success or failure of the drug the company stands to lose only $50,000. If that is true then

FIGURE 18.2
Decision tree for drug problem: Market S_0

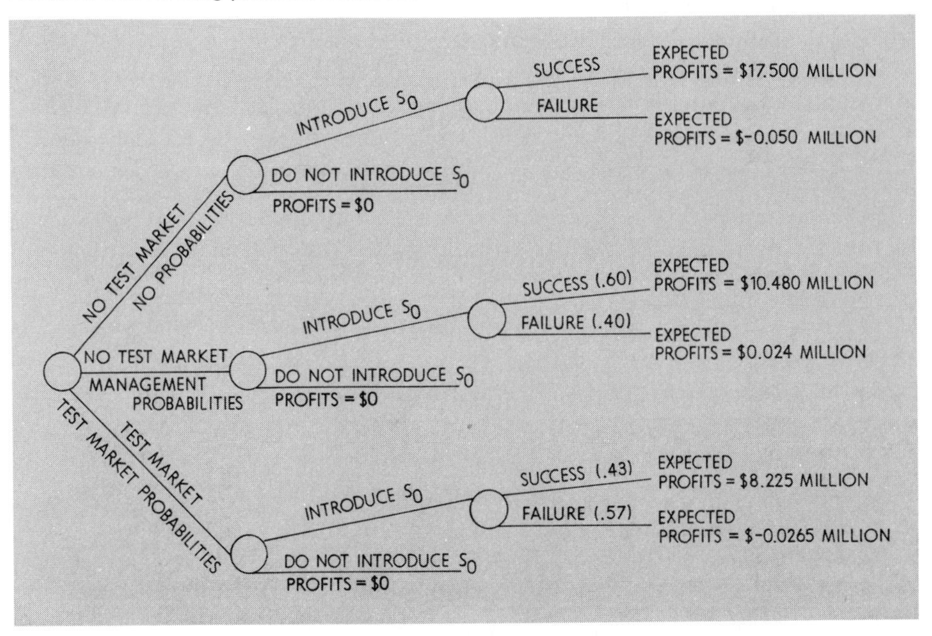

the cost of additional information might be a questionable expense. The reader is cautioned here that our unrealistic hypothetical example does not include the cost of developmental efforts and promotional costs in our loss figures. Also, Figure 18.2 makes it clear that if the probabilities of product success change then the expected profits may vary significantly. Conceivably, management may still wish to collect information about potential product success even though the cost of this data might exceed or equal any possible losses. In addition, the expected profits under the test market hypothesis would need to be adjusted to reflect the survey costs of $50,000.

Figure 18.2 indicates the drug problem's essential features as we have outlined it so far from: (1) the point where no probabilities were assigned to product success or failure, (2) the point of view of the management's intuitive estimates of the probabilities, where $P(S) = .60$ and $P(F) = .40$, and (3) the point of view reflected by the test market results where $P(S) = .47$ and $P(F) = .53$.

New information and the revision of the probabilities

Let us now extend this problem a little further to make the application of Bayes theorem a bit more realistic. To do this we will assume that management decides that it would like more information prior to making a final decision. Table 18.13 shows the management's original estimates of probable product success or failure *plus* some new information and assumptions about expected losses.

If the product is successful (sales are high enough to market 5,000,000 bottles of 100 pills over a reasonable period of time) then the net profit would be $17.5 million. On the other hand, if the sales are only 100,000 units over this same time span then losses will be $10 million. The probable action indicated by Table 18.13 would be to market S_0; how-

TABLE 18.13
Payoff table for decision on the introduction of drug S_0 (dollars are in mililons)

Event	Probability	Course of Action	
		Introduce S_0	Do Not Introduce S_0
Product is successful, (5,000,000 sales over a period of time), E_1	.40	$17.5 million	$0
Product is unsuccessful, (100,000 sales over a period of time), E_2	.60	−10.0 million	0
Expected Values*		$ 1.0 million	$0

* (.40)($17.5m) + (.60)(−$10.0m) = $1 million.

ever, management may have questions about their estimates of the probability of potential success and may wish to gather a little more refined estimates. (For this problem let us assume that we are starting anew with respect to gathering test market information.) Initially, the company statistician computes the EVPI. He or she refers to the opportunity loss associated with not introducing S_0. That is, (\$17.5 million) (.40) + (\$0) (.60) = \$7 million. Therefore, it appears that the new information potentially has a high value in the decision-making process.

Let us now assume that three test markets are surveyed nationally. This information is combined with the estimates of the company statistician about the probabilities of the potential success or failure of drug S_0. These probabilities not only reflect the estimated sales based on the three test markets but also on past experience with similar market analysis and are shown in Table 18.14.

Application of Bayes theorem. In order to extend this analysis the decision makers need to know the *conditional probabilities* of the sales of drug S_0 5,000,000 or 100,000. (Table 18.14 merely indicates the probabilities of success or failure of the test markets *given* a certain level of sales.) What is needed are the conditional probabilities of each level of sales given success, mixed or failure results in the test markets. This requires the computation of *joint probabilities* as indicated in Table 18.15. Note that it is coincidental that the total probabilities of .40 and .60 are the same as management's original estimates. The computations for Table 18.15 were done as follows:

1. To determine the joint probability of both a high sales level (H) and a successful test market prediction (S) we complete the equation

$$\begin{aligned} P(H, S) &= P(H)P(S|H) \\ &= (.40)(.35) \\ &= .14 \end{aligned} \tag{18.2}$$

TABLE 18.14
Conditional probabilities of test markets

Test Market Results	Given Level of Sales: 5,000,000 (H)	Given Level of Sales: 100,000 (L)
Test market results suggest success (S)	.35	.15
Test market results are mixed (M)	.40	.45
Test market results suggest failure (F)	.25	.40
Total	1.00	1.00

TABLE 18.15
Joint probability table, drug S_0

	Test Market Results			
Sales	Success (S)	Mixed (M)	Failure (F)	Total
5,000,000 (H)	.14	.16	.10	.40
100,000 (L)	.09	.27	.24	.60
Total	.23	.43	.34	1.00
	$P(S)$	$P(M)$	$P(F)$	

That is, the probability of a high level of sales (.40 from Table 18.13) is multiplied by the conditional probability of a successful test market prediction (.35 from Table 18.14) which gives the joint probability of .14.

2. To determine the joint probability of both a low level of sales (L) and a successful test market prediction (S) we complete the equation

$$P(L, S) = P(L)P(S|L) \tag{18.3}$$
$$P(L, S) = (.60)\ (.15)$$
$$= .09$$

3. To determine the joint probability of both a high level of sales (H) and a mixed (M) test market prediction we complete the equation

$$P(H, M) = P(H)P(M|H) \tag{18.4}$$
$$= (.40)\ (.45)$$
$$= .16$$

4. To determine the joint probability of both a low level of sales (L) and a mixed (M) test market prediction we complete the equation

$$P(L, M) = P(L)P(M|L) \tag{18.5}$$
$$= (.60)\ (.45)$$
$$= .27$$

5. To determine the joint probability of both a high (H) level of sales and a test market prediction of failure (F) we complete the equation

$$P(H, F) = P(H)P(F|H) \tag{18.6}$$
$$= (.40)\ (.25)$$
$$= .10$$

6. To determine the joint probability of both a low level of sales (L) and a test market prediction of failure (F) we complete the equation

$$
\begin{aligned}
P(L, F) &= P(L)P(F|L) \\
&= (.60)\,(.40) \\
&= .24
\end{aligned} \tag{18.7}
$$

The final row in Table 18.15 (.23, .43, .34) indicate the *marginal probabilities* for success, mixed, and failure test market predictions. These values will be used later. In order to complete the data analysis to reach a decision we still need to combine some more probabilities. That is, we need conditional probabilities for the high and low levels of sales *given* the test market prediction of success, mixed, or failure. For example, the probability of a high level of sales (H), given a test market prediction of success (S) is

$$
\begin{aligned}
P(H|S) &= \frac{P(H, S)}{P(S)} = \frac{(.14)}{(.23)} \\
&= .61
\end{aligned} \tag{18.8}
$$

and the probability of a low level of sales (L) given a test market prediction of success (S) is

$$
\begin{aligned}
P(L|S) &= \frac{P(L, S)}{P(S)} = \frac{(.09)}{(.23)} \\
&= .39
\end{aligned} \tag{18.9}
$$

Also,

$$
\begin{aligned}
P(H|M) &= \frac{P(H, M)}{P(M)} = \frac{(.16)}{(.43)} \\
&= .37
\end{aligned} \tag{18.10}
$$

$$
\begin{aligned}
P(L|M) &= \frac{P(L, M)}{P(M)} = \frac{(.27)}{(.43)} \\
&= .63
\end{aligned} \tag{18.11}
$$

And,

$$
\begin{aligned}
P(H|F) &= \frac{P(H, F)}{P(F)} = \frac{(.10)}{(.34)} \\
&= .29
\end{aligned} \tag{18.12}
$$

$$
\begin{aligned}
P(L|F) &= \frac{P(L, F)}{P(F)} = \frac{(.24)}{(.34)} \\
&= .71
\end{aligned} \tag{18.13}
$$

Readers familiar with the handling of probabilities will recognize the above calculations as exercises in the use of the concept of conditional probabilities. The importance of these calculations stem from Bayes theorem who first emphasized the concept in the 17th century. Modifying Formula (18.1) using Formula (18.2) and Formula (18.3) we obtain

$$P(H|S) = \frac{P(H|S)}{P(S)} = \frac{P(H)P(S|H)}{[P(H)P(S|H) + P(L)P(S|L)]} \quad (18.14)$$

In this form Bayes theorem reflects our hoped for state of the world (high sales of drug S_0) *given* the test market results predicting success. The probabilities are expressed in terms of the conditional probabilities of a successful market *given* the two states of our world (high or low sales) and the simple probabilities of these two events.

Figure 18.3 summarizes the results of these calculations of the probabilities and applies them to our drug problem expressing the results in expected profits. The expected profits shown in the circles were calculated as follows:

No test market: (.40) ($17.5m) + (.60) (−$10.0m) = $1.0 million
Test markets predict:
  Success: (.61) ($17.5m) + (.39) (−$10.0m) = $6.8 million
  Mixed: (.37) ($17.5m) + (.63) (−$10.0m) = $0.2 million
  Failure: (.29) ($17.5m) + (.71) (−$10.0m) = −$2.0 million

Decision. The expected profits based on the three test market results were −$4.7 million *below* the best course of action without the surveys.[6]

FIGURE 18.3
Decision tree for drug problem to market S_0 (with revised probabilities)

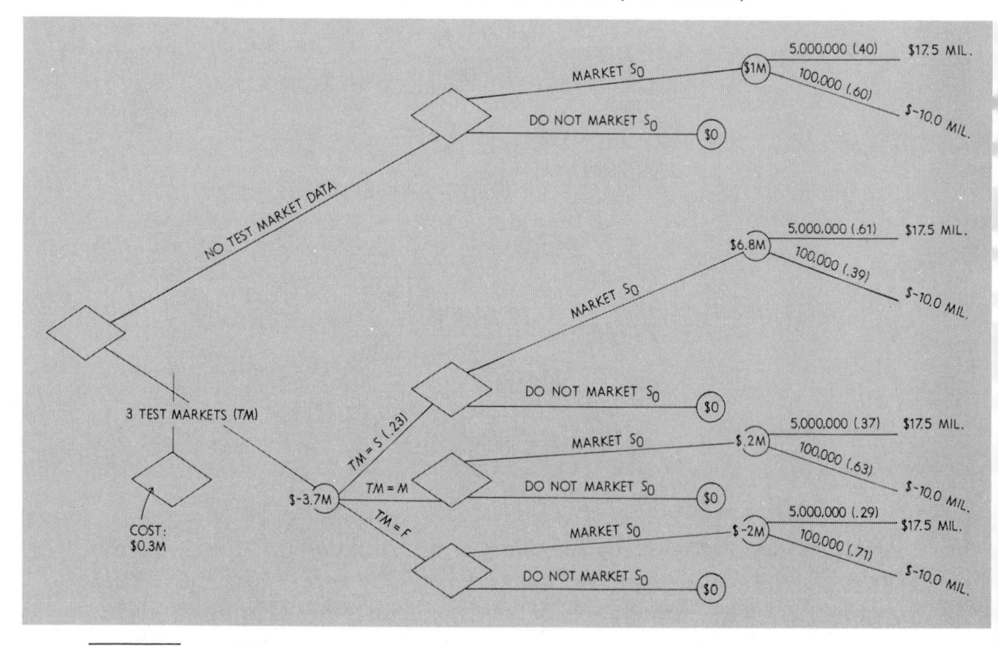

[6] (.23)($17.5m) + (.43)(−$10.0m) = −$3.675m; the expected profits with no test market data is $1.0m.

It was assumed that expected profits for introducing drug S_0 occur only if the test market data predict success (sales of 5,000,000). If the test market results are mixed or predict failure then S_0 should not be introduced. Therefore, the value of imperfect information was $4.7 million. Because this exceeds the cost of the test marketing ($3.0 million), the new information is worth obtaining. The reader should recognize: (1) that Table 18.14 reflects the probabilities associated with the reliability of the data from the three test markets as well as previous evidence of this type, and (2) from our knowledge of sampling theory we know that that the test market information will not be perfect (i.e., predict the results with 100 percent accuracy). Nonetheless, the test marketing seems to be merited given the risks involved.

SUMMARY

The purpose of this example is to illustrate: (1) the use of Bayes theorem and posterior probabilities, (2) the value of new information, and (3) the uncertainty that remains even with additional information based on sample evidence. In the real world this new data, normally collected by some sampling design, usually is not perfect. Nonetheless, the value of the information can be measured if some reliability can be given to the survey data. Then Bayes theorem can be applied to calculate the probabilities used in evaluating this new information.

PROBLEMS

18.1 Contrast the application of probability theory by the *classical school* with that of the Bayesian statistician.

18.2. Clearly define the following:
a. Expected loss.
b. Expected opportunity loss.
c. Cost of uncertainty.
d. Cost of irrationality.

18.3. Distinguish between conditional probabilities and posterior probabilities.

18.4. Verify the expected profits of Table 18.9.

18.5. Assume that the sample survey results discussed on page 630 indicated that the pill size (S_1) where the patient must take only one pill each dose has a probability of .40 of being successful. Determine the expected losses and the expected profits for this size pill and compare the results with those in the text for pill size (S_0) where the sample results indicated product success at .30. What should management do?

18.6. A firm is considering introducing a new consumer nondurable product in a highly competitive market requiring large expenditures for promotion and product development. From the firm's standpoint the product will either be a success (sales of 10,000,000 over a 3-year period) or a failure (sales of 200,000 over a 3-year period). The payoff table for the marketing decision is shown below: (The probabilities are the estimate of management.)

Payoff table

		Course of Action	
Event	Probability	Market	Do Not Market
10,000,000 sales (S_h)	.20	\$ 25.0 million	\$0
200,000 sales (S_1)	.80	−7.0 million	0
Expected values		\$ −0.6 million	\$0

a. What course of action is indicated by the above table?

b. Given the fact that the decision maker stands a chance to make a large profit, the decision maker may wish to obtain some additional information. What is the EVPI in this case?

c. The marketing research vice president suggests that a survey of consumers be taken to determine the probable success or failure of the product. Some new information may be collected on this basis for a cost of \$250,000. The vice president's probabilities associated with the possible reliability of the survey are given below:

Conditional probabilities of survey information reliability given actual sales

	Given Level of Sales	
Survey Prediction	10,000,000	200,000
Success (S)	.30	.05
Mixed (M)	.30	.50
Failure (F)	.40	.45
	1.00	1.00

Note: Assume that the product will not be marketed unless the survey predicts success.

Given this new information, what do you recommend with respect to the collection of some additional data?

d. What is the value of imperfect information in this case?

e. Draw a decision tree to summarize your calculations.

18.7. Assume that the reliability of the survey information in Problem 18.6 is given as below instead of as previously assumed.

Conditional probabilities of survey information reliability given actual sales

	Given Level of Sales	
Survey Prediction	10,000,000	200,000
Success (S)	.45	.10
Mixed (M)	.45	.50
Failure (F)	.10	.40
	1.00	1.00

a. Now what is the expected value of this new information?
b. Should the new information be obtained?

RELATED READINGS

BOISSEAU, HENRY J. "A Program for Solving Elementary Bayesian Decision Problems." *Fortran Applications in Business Administration,* vol. II, pp. 65–78. Ann Arbor: University of Michigan Graduate School of Business Administration, 1971.

BORSCH, KARL. *The Economics of Uncertainty.* Working Paper No. 70. Los Angeles: University of California Press, Western Management Sciences Institute, March 1965.

BOWMAN, E. H. "Consistency and Optimality in Managerial Decision Making." *Management Science,* vol. 9 (January 1963), pp. 310–21.

CORNFIELD, JEROME. "A Bayesian Test of Some Classical Hypotheses—with Applications to Sequential Clinical Trials." *Journal of the American Statistical Association,* vol. 61, no. 315 (September 1966), pp. 577–94.

DYCKMAN, T. S.; SMIDT, S.; and MCADAMS, A. K. *Management Decision Making under Uncertainty,* chaps. 10–12. New York: Macmillan Co., 1969.

EDWARDS, WARD, and PHILLIPS, L. D. "Man as Transducer for Probabilities in Bayesian Command and Control Systems." In *Human Judgments and Optimality,* edited by M. W. Shelly and G. L. Bryan. New York: John Wiley & Sons, Inc., 1964, pp. 360–401.

ELLSBERG, DANIEL. "Risk, Ambiguity, and the Savage Axioms: Comment." *Quarterly Journal of Economics,* November 1961, p. 690 ff.

ERICSON, W. A. "Subjective Bayesian Models in Sampling Finite Populations: Stratification." In *New Developments in Survey Sampling.* New York: John Wiley & Sons, Inc., 1969, pp. 326–57.

GREEN, PAUL; PETERS, WILLIAM S.; and ROBINSON, P. J. "A Behavioral Experiment in Decision-Making under Uncertainty." *Journal of Purchasing,* August 1965, pp. 18–31.

HAMBURG, MORRIS. "Bayesian Decision Theory and Statistical Quality Control." *Industrial Quality Control,* vol. 19, no. 6 (December 1962), pp. 10–14.

HERTZ, DAVID B. "Risk Analysis in Capital Expenditure Decisions." *Harvard Business Review,* January–February 1964, p. 95 ff.

HOUGH, LOUIS. "Mathematical Decision Methods for the Purchasing Agent." *Journal of Purchasing,* August 1965, pp. 41–53.

PARSONS, ROBERT A. "Bayesian Decision Theory Determination of Optimal Decision Rules." *Fortran Applications in Business Administration,* vol. II, pp. 289–300. Ann Arbor: University of Michigan Graduate School of Business Administration, 1971.

RAIFFA, HOWARD. "Bayesian Decision Theory." In *Recent Developments in Information and Decision Processes,* edited by Robert E. Machol and Paul Gray. New York: Macmillan Co., 1962, pp. 92–101.

———. *Decision Analysis, Introductory Lectures on Choices under Uncertainty.* Reading, Mass.: Addison-Wesley Publishing Co., Inc., 1968.

———, and SCHLAIFER, ROBERT. *Applied Statistical Decision Theory.* Boston: Harvard University, Graduate School of Business Administration, Division of Research, 1961.

ROBERTS, HARRY V. "Bayesian Statistics in Marketing." *Journal of Marketing,* vol. 27, no. 1 (January 1963), pp. 1–4.

SAVAGE, LEONARD J. "Bayesian Statistics." In *Recent Developments in Information and Decision Processes,* edited by Robert E. Machol and Paul Gray. New York: Macmillan Co., 1962, pp. 161–94.

SCHLAIFER, ROBERT. *Introduction to Statistics for Business Decisions.* New York: McGraw-Hill Book Co., 1961.

———. *Probability and Statistics for Business Decisions.* New York: McGraw-Hill Book Co., 1959.

Acquire new knowledge whilst thinking over the old,
and you may become a teacher of others.

Confucius

Appendixes

Appendix A

Table of squares and square roots

N	N^2	$\sqrt{N}$	$\sqrt{10N}$	$1/N$
1	1	1.000 000	3.162 278	1.0000000
2	4	1.414 214	4.472 136	.5000000
3	9	1.732 051	5.477 226	.3333333
4	16	2.000 000	6.324 555	.2500000
5	25	2.236 068	7.071 068	.2000000
6	36	2.449 490	7.745 967	.1666667
7	49	2.645 751	8.366 600	.1428571
8	64	2.828 427	8.944 272	.1250000
9	81	3.000 000	9.486 833	.1111111
10	100	3.162 278	10.00000	.1000000
11	121	3.316 625	10.48809	.09090909
12	144	3.464 102	10.95445	.08333333
13	169	3.605 551	11.40175	.07692308
14	196	3.741 657	11.83216	.07142857
15	225	3.872 983	12.24745	.06666667
16	256	4.000 000	12.64911	.06250000
17	289	4.123 106	13.03840	.05882353
18	324	4.242 641	13.41641	.05555556
19	361	4.358 899	13.78405	.05263158
20	400	4.472 136	14.14214	.05000000
21	441	4.582 576	14.49138	.04761905
22	484	4.690 416	14.83240	.04545455
23	529	4.795 832	15.16575	.04347826
24	576	4.898 979	15.49193	.04166667
25	625	5.000 000	15.81139	.04000000
26	676	5.099 020	16.12452	.03846154
27	729	5.196 152	16.43168	.03703704
28	784	5.291 503	16.73320	.03571429
29	841	5.385 165	17.02939	.03448276
30	900	5.477 226	17.32051	.03333333
31	961	5.567 764	17.60682	.03225806
32	1 024	5.656 854	17.88854	.03125000
33	1 089	5.744 563	18.16590	.03030303
34	1 156	5.830 952	18.43909	.02941176
35	1 225	5.916 080	18.70829	.02857143
36	1 296	6.000 000	18.97367	.02777778
37	1 369	6.082 763	19.23538	.02702703
38	1 444	6.164 414	19.49359	.02631579
39	1 521	6.244 998	19.74842	.02564103
40	1 600	6.324 555	20.00000	.02500000
41	1 681	6.403 124	20.24846	.02439024
42	1 764	6.480 741	20.49390	.02380952
43	1 849	6.557 439	20.73644	.02325581
44	1 936	6.633 250	20.97618	.02272727
45	2 025	6.708 204	21.21320	.02222222
46	2 116	6.782 330	21.44761	.02173913
47	2 209	6.855 655	21.67948	.02127660
48	2 304	6.928 203	21.90890	.02083333
49	2 401	7.000 000	22.13594	.02040816
50	2 500	7.071 068	22.36068	.02000000

N	N^2	$\sqrt{N}$	$\sqrt{10N}$	$1/N$ .0
50	2 500	7.071 068	22.36068	2000000
51	2 601	7.141 428	22.58318	1960784
52	2 704	7.211 103	22.80351	1923077
53	2 809	7.280 110	23.02173	1886792
54	2 916	7.348 469	23.23790	1851852
55	3 025	7.416 198	23.45208	1818182
56	3 136	7.483 315	23.66432	1785714
57	3 249	7.549 834	23.87467	1754386
58	3 364	7.615 773	24.08319	1724138
59	3 481	7.681 146	24.28992	1694915
60	3 600	7.745 967	24.49490	1666667
61	3 721	7.810 250	24.69818	1639344
62	3 844	7.874 008	24.89980	1612903
63	3 969	7.937 254	25.09980	1587302
64	4 096	8.000 000	25.29822	1562500
65	4 225	8.062 258	25.49510	1538462
66	4 356	8.124 038	25.69047	1515152
67	4 489	8.185 353	25.88436	1492537
68	4 624	8.246 211	26.07681	1470588
69	4 761	8.306 624	26.26785	1449275
70	4 900	8.366 600	26.45751	1428571
71	5 041	8.426 150	26.64583	1408451
72	5 184	8.485 281	26.83282	1388889
73	5 329	8.544 004	27.01851	1369863
74	5 476	8.602 325	27.20294	1351351
75	5 625	8.660 254	27.38613	1333333
76	5 776	8.717 798	27.56810	1315789
77	5 929	8.774 964	27.74887	1298701
78	6 084	8.831 761	27.92848	1282051
79	6 241	8.888 194	28.10694	1265823
80	6 400	8.944 272	28.28427	1250000
81	6 561	9.000 000	28.46050	1234568
82	6 724	9.055 385	28.63564	1219512
83	6 889	9.110 434	28.80972	1204819
84	7 056	9.165 151	28.98275	1190476
85	7 225	9.219 544	29.15476	1176471
86	7 396	9.273 618	29.32576	1162791
87	7 569	9.327 379	29.49576	1149425
88	7 744	9.380 832	29.66479	1136364
89	7 921	9.433 981	29.83287	1123596
90	8 100	9.486 833	30.00000	1111111
91	8 281	9.539 392	30.16621	1098901
92	8 464	9.591 663	30.33150	1086957
93	8 649	9.643 651	30.49590	1075269
94	8 836	9.695 360	30.65942	1063830
95	9 025	9.746 794	30.82207	1052632
96	9 216	9.797 959	30.98387	1041667
97	9 409	9.848 858	31.14482	1030928
98	9 604	9.899 495	31.30495	1020408
99	9 801	9.949 874	31.46427	1010101
100	10 000	10.00000	31.62278	1000000

Squares and square roots (*continued*)

N	N^2	$\sqrt{N}$	$\sqrt{10N}$	$1/N$ .0
100	10 000	10.00000	31.62278	10000000
101	10 201	10.04988	31.78050	09900990
102	10 404	10.09950	31.93744	09803922
103	10 609	10.14889	32.09361	09708738
104	10 816	10.19804	32.24903	09615385
105	11 025	10.24695	32.40370	09523810
106	11 236	10.29563	32.55764	09433962
107	11 449	10.34408	32.71085	09345794
108	11 664	10.39230	32.86335	09259259
109	11 881	10.44031	33.01515	09174312
110	12 100	10.48809	33.16625	09090909
111	12 321	10.53565	33.31666	09009009
112	12 544	10.58301	33.46640	08928571
113	12 769	10.63015	33.61547	08849558
114	12 996	10.67708	33.76389	08771930
115	13 225	10.72381	33.91165	08695652
116	13 456	10.77033	34.05877	08620690
117	13 689	10.81665	34.20526	08547009
118	13 924	10.86278	34.35113	08474576
119	14 161	10.90871	34.49638	08403361
120	14 400	10.95445	34.64102	08333333
121	14 641	11.00000	34.78505	08264463
122	14 884	11.04536	34.92850	08196721
123	15 129	11.09054	35.07136	08130081
124	15 376	11.13553	35.21363	08064516
125	15 625	11.18034	35.35534	08000000
126	15 876	11.22497	35.49648	07936508
127	16 129	11.26943	35.63706	07874016
128	16 384	11.31371	35.77709	07812500
129	16 641	11.35782	35.91657	07751938
130	16 900	11.40175	36.05551	07692308
131	17 161	11.44552	36.19392	07633588
132	17 424	11.48913	36.33180	07575758
133	17 689	11.53256	36.46917	07518797
134	17 956	11.57584	36.60601	07462687
135	18 225	11.61895	36.74235	07407407
136	18 496	11.66190	36.87818	07352941
137	18 769	11.70470	37.01351	07299270
138	19 044	11.74734	37.14835	07246377
139	19 321	11.78983	37.28270	07194245
140	19 600	11.83216	37.41657	07142857
141	19 881	11.87434	37.54997	07092199
142	20 164	11.91638	37.68289	07042254
143	20 449	11.95826	37.81534	06993007
144	20 736	12.00000	37.94733	06944444
145	21 025	12.04159	38.07887	06896552
146	21 316	12.08305	38.20995	06849315
147	21 609	12.12436	38.34058	06802721
148	21 904	12.16553	38.47077	06756757
149	22 201	12.20656	38.60052	06711409
150	22 500	12.24745	38.72983	06666667

N	N^2	$\sqrt{N}$	$\sqrt{10N}$	$1/N$ .00
150	22 500	12.24745	38.72983	6666667
151	22 801	12.28821	38.85872	6622517
152	23 104	12.32883	38.98718	6578947
153	23 409	12.36932	39.11521	6535948
154	23 716	12.40967	39.24283	6493506
155	24 025	12.44990	39.37004	6451613
156	24 336	12.49000	39.49684	6410256
157	24 649	12.52996	39.62323	6369427
158	24 964	12.56981	39.74921	6329114
159	25 281	12.60952	39.87480	6289308
160	25 600	12.64911	40.00000	6250000
161	25 921	12.68858	40.12481	6211180
162	26 244	12.72792	40.24922	6172840
163	26 569	12.76715	40.37326	6134969
164	26 896	12.80625	40.49691	6097561
165	27 225	12.84523	40.62019	6060606
166	27 556	12.88410	40.74310	6024096
167	27 889	12.92285	40.86563	5988024
168	28 224	12.96148	40.98780	5952381
169	28 561	13.00000	41.10961	5917160
170	28 900	13.03840	41.23106	5882353
171	29 241	13.07670	41.35215	5847953
172	29 584	13.11488	41.47288	5813953
173	29 929	13.15295	41.59327	5780347
174	30 276	13.19091	41.71331	5747126
175	30 625	13.22876	41.83300	5714286
176	30 976	13.26650	41.95235	5681818
177	31 329	13.30413	42.07137	5649718
178	31 684	13.34166	42.19005	5617978
179	32 041	13.37909	42.30839	5586592
180	32 400	13.41641	42.42641	5555556
181	32 761	13.45362	42.54409	5524862
182	33 124	13.49074	42.66146	5494505
183	33 489	13.52775	42.77850	5464481
184	33 856	13.56466	42.89522	5434783
185	34 225	13.60147	43.01163	5405405
186	34 596	13.63818	43.12772	5376344
187	34 969	13.67479	43.24350	5347594
188	35 344	13.71131	43.35897	5319149
189	35 721	13.74773	43.47413	5291005
190	36 100	13.78405	43.58899	5263158
191	36 481	13.82027	43.70355	5235602
192	36 864	13.85641	43.81780	5208333
193	37 249	13.89244	43.93177	5181347
194	37 636	13.92839	44.04543	5154639
195	38 025	13.96424	44.15880	5128205
196	38 416	14.00000	44.27189	5102041
197	38 809	14.03567	44.38468	5076142
198	39 204	14.07125	44.49719	5050505
199	39 601	14.10674	44.60942	5025126
200	40 000	14.14214	44.72136	5000000

Squares and square roots (*continued*)

N	N^2	$\sqrt{N}$	$\sqrt{10N}$	$1/N$ .00
200	40 000	14.14214	44.72136	5000000
201	40 401	14.17745	44.83302	4975124
202	40 804	14.21267	44.94441	4950495
203	41 209	14.24781	45.05552	4926108
204	41 616	14.28286	45.16636	4901961
205	42 025	14.31782	45.27693	4878049
206	42 436	14.35270	45.38722	4854369
207	42 849	14.38749	45.49725	4830918
208	43 264	14.42221	45.60702	4807692
209	43 681	14.45683	45.71652	4784689
210	44 100	14.49138	45.82576	4761905
211	44 521	14.52584	45.93474	4739336
212	44 944	14.56022	46.04346	4716981
213	45 369	14.59452	46.15192	4694836
214	45 796	14.62874	46.26013	4672897
215	46 225	14.66288	46.36809	4651163
216	46 656	14.69694	46.47580	4629630
217	47 089	14.73092	46.58326	4608295
218	47 524	14.76482	46.69047	4587156
219	47 961	14.79865	46.79744	4566210
220	48 400	14.83240	46.90416	4545455
221	48 841	14.86607	47.01064	4524887
222	49 284	14.89966	47.11688	4504505
223	49 729	14.93318	47.22288	4484305
224	50 176	14.96663	47.32864	4464286
225	50 625	15.00000	47.43416	4444444
226	51 076	15.03330	47.53946	4424779
227	51 529	15.06652	47.64452	4405286
228	51 984	15.09967	47.74935	4385965
229	52 441	15.13275	47.85394	4366812
230	52 900	15.16575	47.95832	4347826
231	53 361	15.19868	48.06246	4329004
232	53 824	15.23155	48.16638	4310345
233	54 289	15.26434	48.27007	4291845
234	54 756	15.29706	48.37355	4273504
235	55 225	15.32971	48.47680	4255319
236	55 696	15.36229	48.57983	4237288
237	56 169	15.39480	48.68265	4219409
238	56 644	15.42725	48.78524	4201681
239	57 121	15.45962	48.88763	4184100
240	57 600	15.49193	48.98979	4166667
241	58 081	15.52417	49.09175	4149378
242	58 564	15.55635	49.19350	4132231
243	59 049	15.58846	49.29503	4115226
244	59 536	15.62050	49.39636	4098361
245	60 025	15.65248	49.49747	4081633
246	60 516	15.68439	49.59839	4065041
247	61 009	15.71623	49.69909	4048583
248	61 504	15.74802	49.79960	4032258
249	62 001	15.77973	49.89990	4016064
250	62 500	15.81139	50.00000	4000000
250	62 500	15.81139	50.00000	4000000
251	63 001	15.84298	50.09990	3984064
252	63 504	15.87451	50.19960	3968254
253	64 009	15.90597	50.29911	3952569
254	64 516	15.93738	50.39841	3937008
255	65 025	15.96872	50.49752	3921569
256	65 536	16.00000	50.59644	3906250
257	66 049	16.03122	50.69517	3891051
258	66 564	16.06238	50.79370	3875969
259	67 081	16.09348	50.89204	3861004
260	67 600	16.12452	50.99020	3846154
261	68 121	16.15549	51.08816	3831418
262	68 644	16.18641	51.18594	3816794
263	69 169	16.21727	51.28353	3802281
264	69 696	16.24808	51.38093	3787879
265	70 225	16.27882	51.47815	3773585
266	70 756	16.30951	51.57519	3759398
267	71 289	16.34013	51.67204	3745318
268	71 824	16.37071	51.76872	3731343
269	72 361	16.40122	51.86521	3717472
270	72 900	16.43168	51.96152	3703704
271	73 441	16.46208	52.05766	3690037
272	73 984	16.49242	52.15362	3676471
273	74 529	16.52271	52.24940	3663004
274	75 076	16.55295	52.34501	3649635
275	75 625	16.58312	52.44044	3636364
276	76 176	16.61325	52.53570	3623188
277	76 729	16.64332	52.63079	3610108
278	77 284	16.67333	52.72571	3597122
279	77 841	16.70329	52.82045	3584229
280	78 400	16.73320	52.91503	3571429
281	78 961	16.76305	53.00943	3558719
282	79 524	16.79286	53.10367	3546099
283	80 089	16.82260	53.19774	3533569
284	80 656	16.85230	53.29165	3521127
285	81 225	16.88194	53.38539	3508772
286	81 796	16.91153	53.47897	3496503
287	82 369	16.94107	53.57238	3484321
288	82 944	16.97056	53.66563	3472222
289	83 521	17.00000	53.75872	3460208
290	84 100	17.02939	53.85165	3448276
291	84 681	17.05872	53.94442	3436426
292	85 264	17.08801	54.03702	3424658
293	85 849	17.11724	54.12947	3412969
294	86 436	17.14643	54.22177	3401361
295	87 025	17.17556	54.31390	3389831
296	87 616	17.20465	54.40588	3378378
297	88 209	17.23369	54.49771	3367003
298	88 804	17.26268	54.58938	3355705
299	89 401	17.29162	54.68089	3344482
300	90 000	17.32051	54.77226	3333333

Squares and square roots (*continued*)

N	N^2	$\sqrt{N}$	$\sqrt{10N}$	$1/N$ .00
300	90 000	17.32051	54.77226	3333333
301	90 601	17.34935	54.86347	3322259
302	91 204	17.37815	54.95453	3311258
303	91 809	17.40690	55.04544	3300330
304	92 416	17.43560	55.13620	3289474
305	93 025	17.46425	55.22681	3278689
306	93 636	17.49286	55.31727	3267974
307	94 249	17.52142	55.40758	3257329
308	94 864	17.54993	55.49775	3246753
309	95 481	17.57840	55.58777	3236246
310	96 100	17.60682	55.67764	3225806
311	96 721	17.63519	55.76737	3215434
312	97 344	17.66352	55.85696	3205128
313	97 969	17.69181	55.94640	3194888
314	98 596	17.72005	56.03570	3184713
315	99 225	17.74824	56.12486	3174603
316	99 856	17.77639	56.21388	3164557
317	100 489	17.80449	56.30275	3154574
318	101 124	17.83255	56.39149	3144654
319	101 761	17.86057	56.48008	3134796
320	102 400	17.88854	56.56854	3125000
321	103 041	17.91647	56.65686	3115265
322	103 684	17.94436	56.74504	3105590
323	104 329	17.97220	56.83309	3095975
324	104 976	18.00000	56.92100	3086420
325	105 625	18.02776	57.00877	3076923
326	106 276	18.05547	57.09641	3067485
327	106 929	18.08314	57.18391	3058104
328	107 584	18.11077	57.27128	3048780
329	108 241	18.13836	57.35852	3039514
330	108 900	18.16590	57.44563	3030303
331	109 561	18.19341	57.53260	3021148
332	110 224	18 22087	57.61944	3012048
333	110 889	18.24829	57.70615	3003003
334	111 556	18.27567	57.79273	2994012
335	112 225	18.30301	57.87918	2985075
336	112 896	18.33030	57.96551	2976190
337	113 569	18.35756	58.05170	2967359
338	114 244	18.38478	58.13777	2958580
339	114 921	18.41195	58.22371	2949853
340	115 600	18.43909	58.30952	2941176
341	116 281	18.46619	58.39521	2932551
342	116 964	18.49324	58.48077	2923977
343	117 649	18.52026	58.56620	2915452
344	118 336	18.54724	58.65151	2906977
345	119 025	18.57418	58.73670	2898551
346	119 716	18.60108	58.82176	2890173
347	120 409	18.62794	58.90671	2881844
348	121 104	18.65476	58.99152	2873563
349	121 801	18.68154	59.07622	2865330
350	122 500	18.70829	59.16080	2857143
350	122 500	18.70829	59.16080	2857143
351	123 201	18.73499	59.24525	2849003
352	123 904	18.76166	59.32959	2840909
353	124 609	18.78829	59.41380	2832861
354	125 316	18.81489	59.49790	2824859
355	126 025	18.84144	59.58188	2816901
356	126 736	18.86796	59.66574	2808989
357	127 449	18.89444	59.74948	2801120
358	128 164	18.92089	59.83310	2793296
359	128 881	18.94730	59.91661	2785515
360	129 600	18.97367	60.00000	2777778
361	130 321	19.00000	60.08328	2770083
362	131 044	19.02630	60.16644	2762431
363	131 769	19.05256	60.24948	2754821
364	132 496	19.07878	60.33241	2747253
365	133 225	19.10497	60.41523	2739726
366	133 956	19.13113	60.49793	2732240
367	134 689	19.15724	60.58052	2724796
368	135 424	19.18333	60.66300	2717391
369	136 161	19.20937	60.74537	2710027
370	136 900	19.23538	60.82763	2702703
371	137 641	19.26136	60.90977	2695418
372	138 384	19.28730	60.99180	2688172
373	139 129	19.31321	61.07373	2680965
374	139 876	19.33908	61.15554	2673797
375	140 625	19.36492	61.23724	2666667
376	141 376	19.39072	61.31884	2659574
377	142 129	19.41649	61.40033	2652520
378	142 884	19.44222	61.48170	2645503
379	143 641	19.46792	61.56298	2638522
380	144 400	19.49359	61.64414	2631579
381	145 161	19.51922	61.72520	2624672
382	145 924	19.54483	61.80615	2617801
383	146 689	19.57039	61.88699	2610966
384	147 456	19.59592	61.96773	2604167
385	148 225	19.62142	62.04837	2597403
386	148 996	19.64688	62.12890	2590674
387	149 769	19.67232	62.20932	2583979
388	150 544	19.69772	62.28965	2577320
389	151 321	19.72308	62.36986	2570694
390	152 100	19.74842	62.44998	2564103
391	152 881	19.77372	62.52999	2557545
392	153 664	19.79899	62.60990	2551020
393	154 449	19.82423	62.68971	2544529
394	155 236	19.84943	62.76942	2538071
395	156 025	19.87461	62.84903	2531646
396	156 816	19.89975	62.92853	2525253
397	157 609	19.92486	63.00794	2518892
398	158 404	19.94994	63.08724	2512563
399	159 201	19.97498	63.16645	2506266
400	160 000	20.00000	63.24555	2500000

Squares and square roots (*continued*)

N	N^2	$\sqrt{N}$	$\sqrt{10N}$	$1/N$ .00
400	160 000	20.00000	63.24555	2500000
401	160 801	20.02498	63.32456	2493766
402	161 604	20.04994	63.40347	2487562
403	162 409	20.07486	63.48228	2481390
404	163 216	20.09975	63.56099	2475248
405	164 025	20.12461	63.63961	2469136
406	164 836	20.14944	63.71813	2463054
407	165 649	20.17424	63.79655	2457002
408	166 464	20.19901	63.87488	2450980
409	167 281	20.22375	63.95311	2444988
410	168 100	20.24846	64.03124	2439024
411	168 921	20.27313	64.10928	2433090
412	169 744	20.29778	64.18723	2427184
413	170 569	20.32240	64.26508	2421308
414	171 396	20.34699	64.34283	2415459
415	172 225	20.37155	64.42049	2409639
416	173 056	20.39608	64.49806	2403846
417	173 889	20.42058	64.57554	2398082
418	174 724	20.44505	64.65292	2392344
419	175 561	20.46949	64.73021	2386635
420	176 400	20.49390	64.80741	2380952
421	177 241	20.51828	64.88451	2375297
422	178 084	20.54264	64.96153	2369668
423	178 929	20.56696	65.03845	2364066
424	179 776	20.59126	65.11528	2358491
425	180 625	20.61553	65.19202	2352941
426	181 476	20.63977	65.26868	2347418
427	182 329	20.66398	65.34524	2341920
428	183 184	20.68816	65.42171	2336449
429	184 041	20.71232	65.49809	2331002
430	184 900	20.73644	65.57439	2325581
431	185 761	20.76054	65.65059	2320186
432	186 624	20.78461	65.72671	[illegible]14815
433	187 489	20.80865	65.80274	2309469
434	188 356	20.83267	65.87868	2304147
435	189 225	20.85665	65.95453	2298851
436	190 096	20.88061	66.03030	2293578
437	190 969	20.90454	66.10598	2288330
438	191 844	20.92845	66.18157	2283105
439	192 721	20.95233	66.25708	2277904
440	193 600	20.97618	66.33250	2272727
441	194 481	21.00000	66.40783	2267574
442	195 364	21.02380	66.48308	2262443
443	196 249	21.04757	66.55825	2257336
444	197 136	21.07131	66.63332	2252252
445	198 025	21.09502	66.70832	2247191
446	198 916	21.11871	66.78323	2242152
447	199 809	21.14237	66.85806	2237136
448	200 704	21.16601	66.93280	2232143
449	201 601	21.18962	67.00746	2227171
450	202 500	21.21320	67.08204	2222222

N	N^2	$\sqrt{N}$	$\sqrt{10N}$	$1/N$ .00
450	202 500	21.21320	67.08204	2222222
451	203 401	21.23676	67.15653	2217295
452	204 304	21.26029	67.23095	2212389
453	205 209	21.28380	67.30527	2207506
454	206 116	21.30728	67.37952	2202643
455	207 025	21.33073	67.45369	2197802
456	207 936	21.35416	67.52777	2192982
457	208 849	21.37756	67.60178	2188184
458	209 764	21.40093	67.67570	2183406
459	210 681	21.42429	67.74954	2178649
460	211 600	21.44761	67.82330	2173913
461	212 521	21.47091	67.89698	2169197
462	213 444	21.49419	67.97058	2164502
463	214 369	21.51743	68.04410	2159827
464	215 296	21.54066	68.11755	2155172
465	216 225	21.56386	68.19091	2150538
466	217 156	21.58703	68.26419	2145923
467	218 089	21.61018	68.33740	2141328
468	219 024	21.63331	68.41053	2136752
469	219 961	21.65641	68.48357	2132196
470	220 900	21.67948	68.55655	2127660
471	221 841	21.70253	68.62944	2123142
472	222 784	21.72556	68.70226	2118644
473	223 729	21.74856	68.77500	2114165
474	224 676	21.77154	68.84766	2109705
475	225 625	21.79449	68.92024	2105263
476	226 576	21.81742	68.99275	2100840
477	227 529	21.84033	69.06519	2096436
478	228 484	21.86321	69.13754	2092050
479	229 441	21.88607	69.20983	2087683
480	230 400	21.90890	69.28203	2083333
481	231 361	21.93171	69.35416	2079002
482	232 324	21.95450	69.42622	2074689
483	233 289	21.97726	69.49820	2070393
484	234 256	22.00000	69.57011	2066116
485	235 225	22.02272	69.64194	2061856
486	236 196	22.04541	69.71370	2057613
487	237 169	22.06808	69.78539	2053388
488	238 144	22.09072	69.85700	2049180
489	239 121	22.11334	69.92853	2044990
490	240 100	22.13594	70.00000	2040816
491	241 081	22.15852	70.07139	2036660
492	242 064	22.18107	70.14271	2032520
493	243 049	22.20360	70.21396	2028398
494	244 036	22.22611	70.28513	2024291
495	245 025	22.24860	70.35624	2020202
496	246 016	22.27106	70.42727	2016129
497	247 009	22.29350	70.49823	2012072
498	248 004	22.31591	70.56912	2008032
499	249 001	22.33831	70.63993	2004008
500	250 000	22.36068	70.71068	2000000

Squares and square roots (*continued*)

N	N^2	$\sqrt{N}$	$\sqrt{10N}$	$1/N$ .00
500	250 000	22.36068	70.71068	2000000
501	251 001	22.38303	70.78135	1996008
502	252 004	22.40536	70.85196	1992032
503	253 009	22.42766	70.92249	1988072
504	254 016	22.44994	70.99296	1984127
505	255 025	22.47221	71.06335	1980198
506	256 036	22.49444	71.13368	1976285
507	257 049	22.51666	71.20393	1972387
508	258 064	22.53886	71.27412	1968504
509	259 081	22.56103	71.34424	1964637
510	260 100	22.58318	71.41428	1960784
511	261 121	22.60531	71.48426	1956947
512	262 144	22.62742	71.55418	1953125
513	263 169	22.64950	71.62402	1949318
514	264 196	22.67157	71.69379	1945525
515	265 225	22.69361	71.76350	1941748
516	266 256	22.71563	71.83314	1937984
517	267 289	22.73763	71.90271	1934236
518	268 324	22.75961	71.97222	1930502
519	269 361	22.78157	72.04165	1926782
520	270 400	22.80351	72.11103	1923077
521	271 441	22.82542	72.18033	1919386
522	272 484	22.84732	72.24957	1915709
523	273 529	22.86919	72.31874	1912046
524	274 576	22.89105	72.38784	1908397
525	275 625	22.91288	72.45688	1904762
526	276 676	22.93469	72.52586	1901141
527	277 729	22.95648	72.59477	1897533
528	278 784	22.97825	72.66361	1893939
529	279 841	23.00000	72.73239	1890359
530	280 900	23.02173	72.80110	1886792
531	281 961	23.04344	72.86975	1883239
532	283 024	23.06513	72.93833	1879699
533	284 089	23.08679	73.00685	1876173
534	285 156	23.10844	73.07530	1872659
535	286 225	23.13007	73.14369	1869159
536	287 296	23.15167	73.21202	1865672
537	288 369	23.17326	73.28028	1862197
538	289 444	23.19483	73.34848	1858736
539	290 521	23.21637	73.41662	1855288
540	291 600	23.23790	73.48469	1851852
541	292 681	23.25941	73.55270	1848429
542	293 764	23.28089	73.62065	1845018
543	294 849	23.30236	73.68853	1841621
544	295 936	23.32381	73.75636	1838235
545	297 025	23.34524	73.82412	1834862
546	298 116	23.36664	73.89181	1831502
547	299 209	23.38803	73.95945	1828154
548	300 304	23.40940	74.02702	1824818
549	301 401	23.43075	74.09453	1821494
550	302 500	23.45208	74.16198	1818182

N	N^2	$\sqrt{N}$	$\sqrt{10N}$	$1/N$ .00
550	302 500	23.45208	74.16198	1818182
551	303 601	23.47339	74.22937	1814882
552	304 704	23.49468	74.29670	1811594
553	305 809	23.51595	74.36397	1808318
554	306 916	23.53720	74.43118	1805054
555	308 025	23.55844	74.49832	1801802
556	309 136	23.57965	74.56541	1798561
557	310 249	23.60085	74.63243	1795332
558	311 364	23.62202	74.69940	1792115
559	312 481	23.64318	74.76630	1788909
560	313 600	23.66432	74.83315	1785714
561	314 721	23.68544	74.89993	1782531
562	315 844	23.70654	74.96666	1779359
563	316 969	23.72762	75.03333	1776199
564	318 096	23.74868	75.09993	1773050
565	319 225	23.76973	75.16648	1769912
566	320 356	23.79075	75.23297	1766784
567	321 489	23.81176	75.29940	1763668
568	322 624	23.83275	75.36577	1760563
569	323 761	23.85372	75.43209	1757469
570	324 900	23.87467	75.49834	1754386
571	326 041	23.89561	75.56454	1751313
572	327 184	23.91652	75.63068	1748252
573	328 329	23.93742	75.69676	1745201
574	329 476	23.95830	75.76279	1742160
575	330 625	23.97916	75.82875	1739130
576	331 776	24.00000	75.89466	1736111
577	332 929	24.02082	75.96052	1733102
578	334 084	24.04163	76.02631	1730104
579	335 241	24.06242	76.09205	1727116
580	336 400	24.08319	76.15773	1724138
581	337 561	24.10394	76.22336	1721170
582	338 724	24.12468	76.28892	1718213
583	339 889	24.14539	76.35444	1715266
584	341 056	24.16609	76.41989	1712329
585	342 225	24.18677	76.48529	1709402
586	343 396	24.20744	76.55064	1706485
587	344 569	24.22808	76.61593	1703578
588	345 744	24.24871	76.68116	1700680
589	346 921	24.26932	76.74634	1697793
590	348 100	24.28992	76.81146	1694915
591	349 281	24.31049	76.87652	1692047
592	350 464	24.33105	76.94154	1689189
593	351 649	24.35159	77.00649	1686341
594	352 836	24.37212	77.07140	1683502
595	354 025	24.39262	77.13624	1680672
596	355 216	24.41311	77.20104	1677852
597	356 409	24.43358	77.26578	1675042
598	357 604	24.45404	77.33046	1672241
599	358 801	24.47448	77.39509	1669449
600	360 000	24.49490	77.45967	1666667

Squares and square roots (*continued*)

N	N^2	$\sqrt{N}$	$\sqrt{10N}$	$1/N$ .00
600	360 000	24.49490	77.45967	1666667
601	361 201	24.51530	77.52419	1663894
602	362 404	24.53569	77.58866	1661130
603	363 609	24.55606	77.65307	1658375
604	364 816	24.57641	77.71744	1655629
605	366 025	24.59675	77.78175	1652893
606	367 236	24.61707	77.84600	1650165
607	368 449	24.63737	77.91020	1647446
608	369 664	24.65766	77.97435	1644737
609	370 881	24.67793	78.03845	1642036
610	372 100	24.69818	78.10250	1639344
611	373 321	24.71841	78.16649	1636661
612	374 544	24.73863	78.23043	1633987
613	375 769	24.75884	78.29432	1631321
614	376 996	24.77902	78.35815	1628664
615	378 225	24.79919	78.42194	1626016
616	379 456	24.81935	78.48567	1623377
617	380 689	24.83948	78.54935	1620746
618	381 924	24.85961	78.61298	1618123
619	383 161	24.87971	78.67655	1615509
620	384 400	24.89980	78.74008	1612903
621	385 641	24.91987	78.80355	1610306
622	386 884	24.93993	78.86698	1607717
623	388 129	24.95997	78.93035	1605136
624	389 376	24.97999	78.99367	1602564
625	390 625	25.00000	79.05694	1600000
626	391 876	25.01999	79.12016	1597444
627	393 129	25.03997	79.18333	1594896
628	394 384	25.05993	79.24645	1592357
629	395 641	25.07987	79.30952	1589825
630	396 900	25.09980	79.37254	1587302
631	398 161	25.11971	79.43551	1584786
632	399 424	25.13961	79.49843	1582278
633	400 689	25.15949	79.56130	1579779
634	401 956	25.17936	79.62412	1577287
635	403 225	25.19921	79.68689	1574803
636	404 496	25.21904	79.74961	1572327
637	405 769	25.23886	79.81228	1569859
638	407 044	25.25866	79.87490	1567398
639	408 321	25.27845	79.93748	1564945
640	409 600	25.29822	80.00000	1562500
641	410 881	25.31798	80.06248	1560062
642	412 164	25.33772	80.12490	1557632
643	413 449	25.35744	80.18728	1555210
644	414 736	25.37716	80.24961	1552795
645	416 025	25.39685	80.31189	1550388
646	417 316	25.41653	80.37413	1547988
647	418 609	25.43619	80.43631	1545595
648	419 904	25.45584	80.49845	1543210
649	421 201	25.47548	80.56054	1540832
650	422 500	25.49510	80.62258	1538462

N	N^2	$\sqrt{N}$	$\sqrt{10N}$	$1/N$ .00
650	422 500	25.49510	80.62258	1538462
651	423 801	25.51470	80.68457	1536098
652	425 104	25.53429	80.74652	1533742
653	426 409	25.55386	80.80842	1531394
654	427 716	25.57342	80.87027	1529052
655	429 025	25.59297	80.93207	1526718
656	430 336	25.61250	80.99383	1524390
657	431 649	25.63201	81.05554	1522070
658	432 964	25.65151	81.11720	1519757
659	434 281	25.67100	81.17881	1517451
660	435 600	25.69047	81.24038	1515152
661	436 921	25.70992	81.30191	1512859
662	438 244	25.72936	81.36338	1510574
663	439 569	25.74879	81.42481	1508296
664	440 896	25.76820	81.48620	1506024
665	442 225	25.78759	81.54753	1503759
666	443 556	25.80698	81.60882	1501502
667	444 889	25.82634	81.67007	1499250
668	446 224	25.84570	81.73127	1497006
669	447 561	25.86503	81.79242	1494768
670	448 900	25.88436	81.85353	1492537
671	450 241	25.90367	81.91459	1490313
672	451 584	25.92296	81.97561	1488095
673	452 929	25.94224	82.03658	1485884
674	454 276	25.96151	82.09750	1483680
675	455 625	25.98076	82.15838	1481481
676	456 976	26.00000	82.21922	1479290
677	458 329	26.01922	82.28001	1477105
678	459 684	26.03843	82.34076	1474926
679	461 041	26.05763	82.40146	1472754
680	462 400	26.07681	82.46211	1470588
681	463 761	26.09598	82.42272	1468429
682	465 124	26.11513	82.58329	1466276
683	466 489	26.13427	82.64381	1464129
684	467 856	26.15339	82.70429	1461988
685	469 225	26.17250	82.76473	1459854
686	470 596	26.19160	82.82512	1457726
687	471 969	26.21068	82.88546	1455604
688	473 344	26.22975	82.94577	1453488
689	474 721	26.24881	83.00602	1451379
690	476 100	26.26785	83.06624	1449275
691	477 481	26.28688	83.12641	1447178
692	478 864	26.30589	83.18654	1445087
693	480 249	26.32489	83.24662	1443001
694	481 636	26.34388	83.30666	1440922
695	483 025	26.36285	83.36666	1438849
696	484 416	26.38181	83.42661	1436782
697	485 809	26.40076	83.48653	1434720
698	487 204	26.41969	83.54639	1432665
699	488 601	26.43861	83.60622	1430615
700	490 000	26.45751	83.66600	1428571

Squares and square roots (*continued*)

N	N^2	$\sqrt{N}$	$\sqrt{10N}$	$1/N$ .00
700	490 000	26.45751	83.66600	1428571
701	491 401	26.47640	83.72574	1426534
702	492 804	26.49528	83.78544	1424501
703	494 209	26.51415	83.84510	1422475
704	495 616	26.53300	83.90471	1420455
705	497 025	26.55184	83.96428	1418440
706	498 436	26.57066	84.02381	1416431
707	499 849	26.58947	84.08329	1414427
708	501 264	26.60827	84.14274	1412429
709	502 681	26.62705	84.20214	1410437
710	504 100	26.64583	84.26150	1408451
711	505 521	26.66458	84.32082	1406470
712	506 944	26.68333	84.38009	1404494
713	508 369	26.70206	84.43933	1402525
714	509 796	26.72078	84.49852	1400560
715	511 225	26.73948	84.55767	1398601
716	512 656	26.75818	84.61678	1396648
717	514 089	26.77686	84.67585	1394700
718	515 524	26.79552	84.73488	1392758
719	516 961	26.81418	84.79387	1390821
720	518 400	26.83282	84.85281	1388889
721	519 841	26.85144	84.91172	1386963
722	521 284	26.87006	84.97058	1385042
723	522 729	26.88866	85.02941	1383126
724	524 176	26.90725	85.08819	1381215
725	525 625	26.92582	85.14693	1379310
726	527 076	26.94439	85.20563	1377410
727	528 529	26.96294	85.26429	1375516
728	529 984	26.98148	85.32292	1373626
729	531 441	27.00000	85.38150	1371742
730	532 900	27.01851	85.44004	1369863
731	534 361	27.03701	85.49854	1367989
732	535 824	27.05550	85.55700	1366120
733	537 289	27 07397	85.61542	1364256
734	538 756	27.09243	85.67380	1362398
735	540 225	27.11088	85.73214	1360544
736	541 696	27.12932	85.79044	1358696
737	543 169	27.14774	85.84870	1356852
738	544 644	27.16616	85.90693	1355014
739	546 121	27.18455	85.96511	1353180
740	547 600	27.20294	86.02325	1351351
741	549 081	27.22132	86.08136	1349528
742	550 564	27.23968	86.13942	1347709
743	552 049	27.25803	86.19745	1345895
744	553 536	27.27636	86.25543	1344086
745	555 025	27.29469	86.31338	1342282
746	556 516	27.31300	86.37129	1340483
747	558 009	27.33130	86.42916	1338688
748	559 504	27.34959	86.48699	1336898
749	561 001	27.36786	86.54479	1335113
750	562 500	27.38613	86.60254	1333333

N	N^2	$\sqrt{N}$	$\sqrt{10N}$	$1/N$ .00
750	562 500	27.38613	86.60254	1333333
751	564 001	27.40438	86.66026	1331558
752	565 504	27.42262	86.71793	1329787
753	567 009	27.44085	86.77557	1328021
754	568 516	27.45906	86.83317	1326260
755	570 025	27.47726	86.89074	1324503
756	571 536	27.49545	86.94826	1322751
757	573 049	27.51363	87.00575	1321004
758	574 564	27.53180	87.06320	1319261
759	576 081	27.54995	87.12061	1317523
760	577 600	27.56810	87.17798	1315789
761	579 121	27.58623	87.23531	1314060
762	580 644	27.60435	87.29261	1312336
763	582 169	27.62245	87.34987	1310616
764	583 696	27.64055	87.40709	1308901
765	585 225	27.65863	87.46428	1307190
766	586 756	27.67671	87.52143	1305483
767	588 289	27.69476	87.57854	1303781
768	589 824	27.71281	87.63561	1302083
769	591 361	27.73085	87.69265	1300390
770	592 900	27.74887	87.74964	1298701
771	594 441	27.76689	87.80661	1297017
772	595 984	27.78489	87.86353	1295337
773	597 529	27.80288	87.92042	1293661
774	599 076	27.82086	87.97727	1291990
775	600 625	27.83882	88.03408	1290323
776	602 176	27.85678	88.09086	1288660
777	603 729	27.87472	88.14760	1287001
778	605 284	27.89265	88.20431	1285347
779	606 841	27.91057	88.26098	1283697
780	608 400	27.92848	88.31761	1282051
781	609 961	27.94638	88.37420	1280410
782	611 524	27.96426	88.43076	1278772
783	613 089	27.98214	88.48729	1277139
734	614 656	28.00000	88.54377	1275510
785	616 225	28.01785	88.60023	1273885
786	617 796	28.03569	88.65664	1272265
787	619 369	28.05352	88.71302	1270648
788	620 944	28.07134	88.76936	1269036
789	622 521	28.08914	88.82567	1267427
790	624 100	28.10694	88.88194	1265823
791	625 681	28.12472	88.93818	1264223
792	627 264	28.14249	88.99438	1262626
793	628 849	28.16026	89.05055	1261034
794	630 436	28.17801	89.10668	1259446
795	632 025	28.19574	89.16277	1257862
796	633 616	28.21347	89.21883	1256281
797	635 209	28.23119	89.27486	1254705
798	636 804	28.24889	89.33085	1253133
799	638 401	28.26659	89.38680	1251564
800	640 000	28.28427	89.44272	1250000

Squares and square roots (*continued*)

N	N^2	$\sqrt{N}$	$\sqrt{10N}$	$1/N$ .00
800	640 000	28.28427	89.44272	1250000
801	641 601	28.30194	89.49860	1248439
802	643 204	28.31960	89.55445	1246883
803	644 809	28.33725	89.61027	1245330
804	646 416	28.35489	89.66605	1243781
805	648 025	28.37252	89.72179	1242236
806	649 636	28.39014	89.77750	1240695
807	651 249	28.40775	89.83318	1239157
808	652 864	28.42534	89.88882	1237624
809	654 481	28.44293	89.94443	1236094
810	656 100	28.46050	90.00000	1234568
811	657 721	28.47806	90.05554	1233046
812	659 344	28.49561	90.11104	1231527
813	660 969	28.51315	90.16651	1230012
814	662 596	28.53069	90.22195	1228501
815	664 225	28.54820	90.27735	1226994
816	665 856	28.56571	90.33272	1225490
817	667 489	28.58321	90.38805	1223990
818	669 124	28.60070	90.44335	1222494
819	670 761	28.61818	90.49862	1221001
820	672 400	28.63564	90.55385	1219512
821	674 041	28.65310	90.60905	1218027
822	675 684	28.67054	90.66422	1216545
823	677 329	28.68798	90.71935	1215067
824	678 976	28.70540	90.77445	1213592
825	680 625	28.72281	90.82951	1212121
826	682 276	28.74022	90.88454	1210654
827	683 929	28.75761	90.93954	1209190
828	685 584	28.77499	90.99451	1207729
829	687 241	28.79236	91.04944	1206273
830	688 900	28.80972	91.10434	1204819
831	690 561	28.82707	91.15920	1203369
832	692 224	28.84441	91.21403	1201923
833	693 889	28.86174	91.26883	1200480
834	695 556	28.87906	91.32360	1199041
835	697 225	28.89637	91.37833	1197605
836	698 896	28.91366	91.43304	1196172
837	700 569	28.93095	91.48770	1194743
838	702 244	28.94823	91.54234	1193317
839	703 921	28.96550	91.59694	1191895
840	705 600	28.98275	91.65151	1190476
841	707 281	29.00000	91.70605	1189061
842	708 964	29.01724	91.76056	1187648
843	710 649	29.03446	91.81503	1186240
844	712 336	29.05168	91.86947	1184834
845	714 025	29.06888	91.92388	1183432
846	715 716	29.08608	91.97826	1182033
847	717 409	29.10326	92.03260	1180638
848	719 104	29.12044	92.08692	1179245
849	720 801	29.13760	92.14120	1177856
850	722 500	29.15476	92.19544	1176471

N	N^2	$\sqrt{N}$	$\sqrt{10N}$	$1/N$ .00
850	722 500	29.15476	92.19544	1176471
851	724 201	29.17190	92.24966	1175088
852	725 904	29.18904	92.30385	1173709
853	727 609	29.20616	92.35800	1172333
854	729 316	29.22328	92.41212	1170960
855	731 025	29.24038	92.46621	1169591
856	732 736	29.25748	92.52027	1168224
857	734 449	29.27456	92.57429	1166861
858	736 164	29.29164	92.62829	1165501
859	737 881	29.30870	92.68225	1164144
860	739 600	29.32576	92.73618	1162791
861	741 321	29.34280	92.79009	1161440
862	743 044	29.35984	92.84396	1160093
863	744 769	29.37686	92.89779	1158749
864	746 496	29.39388	92.95160	1157407
865	748 225	29.41088	93.00538	1156069
866	749 956	29.42788	93.05912	1154734
867	751 689	29.44486	93.11283	1153403
868	753 424	29.46184	93.16652	1152074
869	755 161	29.47881	93.22017	1150748
870	756 900	29.49576	93.27379	1149425
871	758 641	29.51271	93.32738	1148106
872	760 384	29.52965	93.38094	1146789
873	762 129	29.54657	93.43447	1145475
874	763 876	29.56349	93.48797	1144165
875	765 625	29.58040	93.54143	1142857
876	767 376	29.59730	93.59487	1141553
877	769 129	29.61419	93.64828	1140251
878	770 884	29.63106	93.70165	1138952
879	772 641	29.64793	93.75500	1137656
880	774 400	29.66479	93.80832	1136364
881	776 161	29.68164	93.86160	1135074
882	777 924	29.69848	93.91486	1133787
883	779 689	29.71532	93.96808	1132503
884	781 456	29.73214	94.02127	1131222
885	783 225	29.74895	94.07444	1129944
886	784 996	29.76575	94.12757	1128668
887	786 769	29.78255	94.18068	1127396
888	788 544	29.79933	94.23375	1126126
889	790 321	29.81610	94.28680	1124859
890	792 100	29.83287	94.33981	1123596
891	793 881	29.84962	94.39280	1122334
892	795 664	29.86637	94.44575	1121076
893	797 449	29.88311	94.49868	1119821
894	799 236	29.89983	94.55157	1118568
895	801 025	29.91655	94.60444	1117318
896	802 816	29.93326	94.65728	1116071
897	804 609	29.94996	94.71008	1114827
898	806 404	29.96665	94.76286	1113586
899	808 201	29.98333	94.81561	1112347
900	810 000	30.00000	94.86833	1111111

Squares and square roots (*concluded*)

N	N^2	$\sqrt{N}$	$\sqrt{10N}$	$1/N$ .00
900	810 000	30.00000	94.86833	1111111
901	811 801	30.01666	94.92102	1109878
902	813 604	30.03331	94.97368	1108647
903	815 409	30.04996	95.02631	1107420
904	817 216	30.06659	95.07891	1106195
905	819 025	30.08322	95.13149	1104972
906	820 836	30.09983	95.18403	1103753
907	822 649	30.11644	95.23655	1102536
908	824 464	30.13304	95.28903	1101322
909	826 281	30.14963	95.34149	1100110
910	828 100	30.16621	95.39392	1098901
911	829 921	30.18278	95.44632	1097695
912	831 744	30.19934	95.49869	1096491
913	833 569	30.21589	95.55103	1095290
914	835 396	30.23243	95.60335	1094092
915	837 225	30.24897	95.65563	1092896
916	839 056	30.26549	95.70789	1091703
917	840 889	30.28201	95.76012	1090513
918	842 724	30.29851	95.81232	1089325
919	844 561	30.31501	95.86449	1088139
920	846 400	30.33150	95.91663	1086957
921	848 241	30.34798	95.96874	1085776
922	850 084	30.36445	96.02083	1084599
923	851 929	30.38092	96.07289	1083424
924	853 776	30.39737	96.12492	1082251
925	855 625	30.41381	96.17692	1081081
926	857 476	30.43025	96.22889	1079914
927	859 329	30.44667	96.28084	1078749
928	861 184	30.46309	96.33276	1077586
929	863 041	30.47950	96.38465	1076426
930	864 900	30.49590	96.43651	1075269
931	866 761	30.51229	96.48834	1074114
932	868 624	30.52868	96.54015	1072961
933	870 489	30.54505	96.59193	1071811
934	872 356	30.56141	96.64368	1070664
935	874 225	30.57777	96.69540	1069519
936	876 096	30.59412	96.74709	1068376
937	877 969	30.61046	96.79876	1067236
938	879 844	30.62679	96.85040	1066098
939	881 721	30.64311	96.90201	1064963
940	883 600	30.65942	96.95360	1063830
941	885 481	30.67572	97.00515	1062699
942	887 364	30.69202	97.05668	1061571
943	889 249	30.70831	97.10819	1060445
944	891 136	30.72458	97.15966	1059322
945	893 025	30.74085	97.21111	1058201
946	894 916	30.75711	97.26253	1057082
947	896 809	30.77337	97.31393	1055966
948	898 704	30.78961	97.36529	1054852
949	900 601	30.80584	97.41663	1053741
950	902 500	30.82207	97.46794	1052632

N	N^2	$\sqrt{N}$	$\sqrt{10N}$	$1/N$ .00
950	902 500	30.82207	97.46794	1052632
951	904 401	30.83829	97.51923	1051525
952	906 304	30.85450	97.57049	1050420
953	908 209	30.87070	97.62172	1049318
954	910.116	30.88689	97.67292	1048218
955	912 025	30.90307	97.72410	1047120
956	913 936	30.91925	97.77525	1046025
957	915 849	30.93542	97.82638	1044932
958	917 764	30.95158	97.87747	1043841
959	919 681	30.96773	97.92855	1042753
960	921 600	30.98387	97.97959	1041667
961	923 521	31.00000	98.03061	1040583
962	925 444	31.01612	98.08160	1039501
963	927 369	31.03224	98.13256	1038422
964	929 296	31.04835	98.18350	1037344
965	931 225	31.06445	98.23441	1036269
966	933 156	31.08054	98.28530	1035197
967	935 089	31.09662	98.33616	1034126
968	937 024	31.11270	98.38699	1033058
969	938 961	31.12876	98.43780	1031992
970	940 900	31.14482	98.48858	1030928
971	942 841	31.16087	98.53933	1029866
972	944 784	31.17691	98.59006	1028807
973	946 729	31.19295	98.64076	1027749
974	948 676	31.20897	98.69144	1026694
975	950 625	31.22499	98.74209	1025641
976	952 576	31.24100	98.79271	1024590
977	954 529	31.25700	98.84331	1023541
978	956 484	31.27299	98.89388	1022495
979	958 441	31.28898	98.94443	1021450
980	960 400	31.30495	98.99495	1020408
981	962 361	31.32092	99.04544	1019368
982	964 324	31.33688	99.09591	1018330
983	966 289	31.35283	99.14636	1017294
984	968 256	31.36877	99.19677	1016260
985	970 225	31.38471	99.24717	1015228
986	972 196	31.40064	99.29753	1014199
987	974 169	31.41656	99.34787	1013171
988	976 144	31.43247	99.39819	1012146
989	978 121	31.44837	99.44848	1011122
990	980 100	31.46427	99.49874	1010101
991	982 081	31.48015	99.54898	1009082
992	984 064	31.49603	99.59920	1008065
993	986 049	31.51190	99.64939	1007049
994	988 036	31.52777	99.69955	1006036
995	990 025	31.54362	99.74969	1005025
996	992 016	31.55947	99.79980	1004016
997	994 009	31.57531	99.84989	1003009
998	996 004	31.59114	99.89995	1002004
999	998 001	31.60696	99.94999	1001001
1000	1 000 000	31.62278	100.00000	1000000

Source: From Frederick E. Croxton and Dudley J. Cowden, *Practical Business Statistics,* 2d ed. (© 1934, 1948 by Prentice-Hall, Inc., Englewood Cliffs, N.J.), pp. 524–33. Reprinted by permission of the publisher.

Appendix B

Four-place logarithms

N	0	1	2	3	4	5	6	7	8	9
10	0000	0043	0086	0128	0170	0212	0253	0294	0334	0374
11	0414	0453	0492	0531	0569	0607	0645	0682	0719	0755
12	0792	0828	0864	0899	0934	0969	1004	1038	1072	1106
13	1139	1173	1206	1239	1271	1303	1335	1367	1399	1430
14	1461	1492	1523	1553	1584	1614	1644	1673	1703	1732
15	1761	1790	1818	1847	1875	1903	1931	1959	1987	2014
16	2041	2068	2095	2122	2148	2175	2201	2227	2253	2279
17	2304	2330	2355	2380	2405	2430	2455	2480	2504	2529
18	2553	2577	2601	2625	2648	2672	2695	2718	2742	2765
19	2788	2810	2833	2856	2878	2900	2923	2945	2967	2989
20	3010	3032	3054	3075	3096	3118	3139	3160	3181	3201
21	3222	3243	3263	3284	3304	3324	3345	3365	3385	3404
22	3424	3444	3464	3483	3502	3522	3541	3560	3579	3598
23	3617	3636	3655	3674	3692	3711	3729	3747	3766	3784
24	3802	3820	3838	3856	3874	3892	3909	3927	3945	3962
25	3979	3997	4014	4031	4048	4065	4082	4099	4116	4133
26	4150	4166	4183	4200	4216	4232	4249	4265	4281	4298
27	4314	4330	4346	4362	4378	4393	4409	4425	4440	4456
28	4472	4487	4502	4518	4533	4548	4564	4579	4594	4609
29	4624	4639	4654	4669	4683	4698	4713	4728	4742	4757
30	4771	4786	4800	4814	4829	4843	4857	4871	4886	4900
31	4914	4928	4942	4955	4969	4983	4997	5011	5024	5038
32	5051	5065	5079	5092	5105	5119	5132	5145	5159	5172
33	5185	5198	5211	5224	5237	5250	5263	5276	5289	5302
34	5315	5328	5340	5353	5366	5378	5391	5403	5416	5428
35	5441	5453	5465	5478	5490	5502	5514	5527	5539	5551
36	5563	5575	5587	5599	5611	5623	5635	5647	5658	5670
37	5682	5694	5705	5717	5729	5740	5752	5763	5775	5786
38	5798	5809	5821	5832	5843	5855	5866	5877	5888	5899
39	5911	5922	5933	5944	5955	5966	5977	5988	5999	6010
40	6021	6031	6042	6053	6064	6075	6085	6096	6107	6117
41	6128	6138	6149	6160	6170	6180	6191	6201	6212	6222
42	6232	6243	6253	6263	6274	6284	6294	6304	6314	6325
43	6336	6345	6355	6365	6375	6385	6395	6405	6415	6425
44	6435	6444	6454	6464	6474	6484	6493	6503	6513	6522
45	6532	6542	6551	6561	6571	6580	6590	6599	6609	6618
46	6628	6637	6646	6656	6665	6675	6684	6693	6702	6712
47	6721	6730	6739	6749	6758	6767	6776	6785	6794	6803
48	6812	6821	6830	6839	6848	6857	6866	6875	6884	6893
49	6902	6911	6920	6928	6937	6946	6955	6964	6972	6981
50	6990	6998	7007	7016	7024	7033	7042	7050	7059	7067
51	7076	7084	7093	7101	7110	7118	7126	7135	7143	7152
52	7160	7168	7177	7185	7193	7202	7210	7218	7226	7235
53	7243	7251	7259	7267	7275	7284	7292	7300	7308	7316
54	7324	7332	7340	7348	7356	7364	7372	7380	7388	7396

Four-place logarithms (*continued*)

N	0	1	2	3	4	5	6	7	8	9
55	7404	7412	7419	7427	7435	7443	7451	7459	7466	7474
56	7482	7490	7497	7505	7513	7520	7528	7536	7543	7551
57	7559	7566	7574	7582	7589	7597	7604	7612	7619	7627
58	7634	7642	7649	7657	7664	7672	7679	7686	7694	7701
59	7709	7716	7723	7731	7738	7745	7752	7760	7767	7774
60	7782	7789	7796	7803	7810	7818	7825	7832	7839	7846
61	7853	7860	7868	7875	7882	7889	7896	7903	7910	7917
62	7924	7931	7938	7945	7952	7959	7966	7973	7980	7987
63	7993	8000	8007	8014	8021	8028	8035	8041	8048	8055
64	8062	8069	8075	8082	8089	8096	8102	8109	8116	8122
65	8129	8136	8142	8149	8156	8162	8169	8176	8182	8189
66	8195	8202	8209	8215	8222	8228	8235	8241	8248	8254
67	8261	8267	8274	8280	8287	8293	8299	8306	8312	8319
68	8325	8331	8338	8344	8351	8357	8363	8370	8376	8382
69	8388	8395	8401	8407	8414	8420	8426	8432	8439	8445
70	8451	8457	8463	8470	8476	8482	8488	8494	8500	8506
71	8513	8519	8525	8531	8537	8543	8549	8555	8561	8567
72	8573	8579	8585	8591	8597	8603	8609	8615	8621	8627
73	8633	8639	8645	8651	8657	8663	8669	8675	8681	8686
74	8692	8698	8704	8710	8716	8722	8727	8733	8739	8745
75	8751	8756	8762	8768	8774	8779	8785	8791	8797	8802
76	8808	8814	8820	8825	8831	8837	8842	8848	8854	8859
77	8865	8871	8876	8882	8887	8893	8899	8904	8910	8915
78	8921	8927	8932	8938	8943	8949	8954	8960	8965	8971
79	8976	8982	8987	8993	8998	9004	9009	9015	9020	9025
80	9031	9036	9042	9047	9053	9058	9063	9069	9074	9079
81	9085	9090	9096	9101	9106	9112	9117	9122	9128	9133
82	9138	9143	9149	9154	9159	9165	9170	9175	9180	9186
83	9191	9196	9201	9206	9212	9217	9222	9227	9232	9238
84	9243	9248	9253	9258	9263	9269	9274	9279	9284	9289
85	9294	9299	9304	9309	9315	9320	9325	9330	9335	9340
86	9345	9350	9355	9360	9365	9370	9375	9380	9385	9390
87	9395	9400	9405	9410	9415	9420	9425	9430	9435	9440
88	9445	9450	9455	9460	9465	9469	9474	9479	9484	9489
89	9494	9499	9504	9509	9513	9518	9523	9528	9533	9538
90	9542	9547	9552	9557	9562	9566	9571	9576	9581	9586
91	9590	9595	9600	9605	9609	9614	9619	9624	9628	9633
92	9638	9643	9647	9652	9657	9661	9666	9671	9675	9680
93	9685	9689	9694	9699	9703	9708	9713	9717	9722	9727
94	9731	9736	9741	9745	9750	9754	9759	9763	9768	9773
95	9777	9782	9786	9791	9795	9800	9805	9809	9814	9818
96	9823	9827	9832	9836	9841	9845	9850	9854	9859	9863
97	9868	9872	9877	9881	9886	9890	9894	9899	9903	9908
98	9912	9917	9921	9926	9930	9934	9939	9943	9948	9952
99	9956	9961	9965	9969	9974	9978	9983	9987	9991	9996

Appendix C

The normal distribution

A VARIATE normally distributed takes all values from $-\infty$ to $+\infty$, and frequency at each size of the variate is given by a mathematical law:

$$f = \frac{1}{\sigma\sqrt{2\pi}} e^{-x^2/2\sigma^2}$$

This formula says that the logarithm of the frequency at any distance, x, from the center of the distribution is less than the logarithm of the frequency at the center of the distribution by a quantity proportional to x^2. Therefore the normal curve is symmetrical. The curve is shown in Figure C.1.

Table C.1 clearly shows that if $x = 1.96$, the area included between zero and x is .4750, or 47.5 percent of the total curve. Any other value of $(0 - X)/\sigma = x/\sigma = z$ can be read from the table.

Thus, if a variate is normally distributed with mean $\mu = 100$ and standard deviation $\sigma = 10$, there are only $2\frac{1}{2}$ chances in 100 that an item will exceed $\mu + 1.96\sigma = 100 + 19.60 = 119.60$. Also, there are only $2\frac{1}{2}$ chances in 100 that an item will be less than $\mu - 1.96\sigma = 100 - 19.60 = 80.40$.

The normal curve table shown here is designed for simplicity of use.

FIGURE C.1
Normal distribution

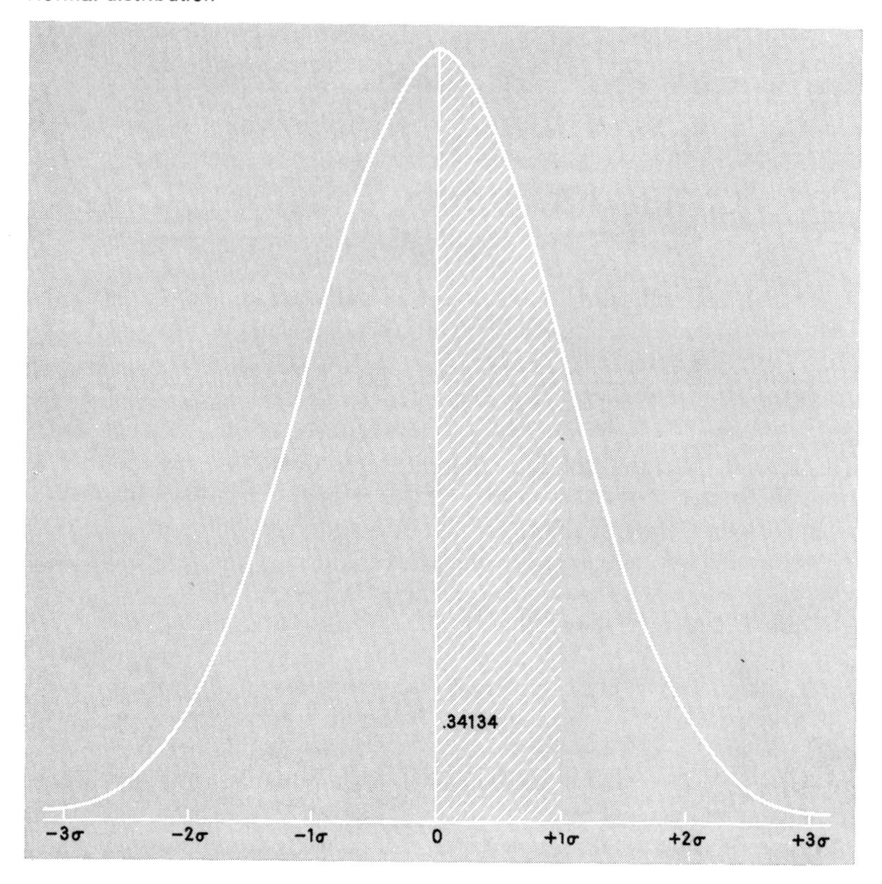

TABLE C.1
Table of areas from mean to distances Z from mean for normal probability distribution

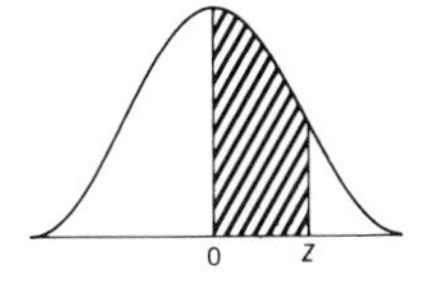

Z	.00	.01	.02	.03	.04	.05	.06	.07	.08	.09
.0	.0000	.0040	.0080	.0120	.0160	.0199	.0239	.0279	.0319	.0359
.1	.0398	.0438	.0478	.0517	.0557	.0596	.0636	.0675	.0714	.0753
.2	.0793	.0832	.0871	.0910	.0948	.0987	.1026	.1064	.1103	.1141
.3	.1179	.1217	.1255	.1293	.1331	.1368	.1406	.1443	.1480	.1517
.4	.1554	.1591	.1628	.1664	.1700	.1736	.1772	.1808	.1844	.1879
.5	.1915	.1950	.1985	.2019	.2054	.2088	.2123	.2157	.2190	.2224
.6	.2257	.2291	.2324	.2357	.2389	.2422	.2454	.2486	.2518	.2549
.7	.2580	.2612	.2642	.2673	.2704	.2734	.2764	.2794	.2823	.2852
.8	.2881	.2910	.2939	.2967	.2995	.3023	.3051	.3078	.3106	.3133
.9	.3159	.3186	.3212	.3238	.3264	.3289	.3315	.3340	.3365	.3389
1.0	.3413	.3438	.3461	.3485	.3508	.3531	.3554	.3577	.3599	.3621
1.1	.3643	.3665	.3686	.3708	.3729	.3749	.3770	.3790	.3810	.3830
1.2	.3849	.3869	.3888	.3907	.3925	.3944	.3962	.3980	.3997	.4015
1.3	.4032	.4049	.4066	.4082	.4099	.4115	.4131	.4147	.4162	.4177
1.4	.4192	.4207	.4222	.4236	.4251	.4265	.4279	.4292	.4306	.4319
1.5	.4332	.4345	.4357	.4370	.4382	.4394	.4406	.4418	.4429	.4441
1.6	.4452	.4463	.4474	.4484	.4495	.4505	.4515	.4525	.4535	.4545
1.7	.4554	.4564	.4573	.4582	.4591	.4599	.4608	.4616	.4625	.4633
1.8	.4641	.4649	.4656	.4664	.4671	.4678	.4686	.4693	.4699	.4706
1.9	.4713	.4719	.4726	.4732	.4738	.4744	.4750	.4756	.4761	.4767
2.0	.4772	.4778	.4783	.4788	.4793	.4798	.4803	.4808	.4812	.4817
2.1	.4821	.4826	.4830	.4834	.4838	.4842	.4846	.4850	.4854	.4857
2.2	.4861	.4864	.4868	.4871	.4875	.4878	.4881	.4884	.4887	.4890
2.3	.4893	.4896	.4898	.4901	.4904	.4906	.4909	.4911	.4913	.4916
2.4	.4918	.4920	.4922	.4925	.4927	.4931	.4931	.4932	.4934	.4936
2.5	.4938	.4940	.4941	.4943	.4945	.4946	.4948	.4949	.4951	.4952
2.6	.4953	.4955	.4956	.4957	.4959	.4960	.4961	.4962	.4963	.4964
2.7	.4965	.4966	.4967	.4968	.4969	.4970	.4971	.4972	.4973	.4974
2.8	.4974	.4975	.4976	.4977	.4977	.4978	.4979	.4979	.4980	.4981
2.9	.4981	.4982	.4982	.4983	.4984	.4984	.4985	.4985	.4986	.4986
3.0	.49865	.4987	.4987	.4988	.4988	.4989	.4989	.4989	.4990	.4990
4.0	.4999683									

Illustration: For $Z = 1.96$, shaded area is .4750 out of total area of 1.

TABLE C.2
Ordinates of the normal curve

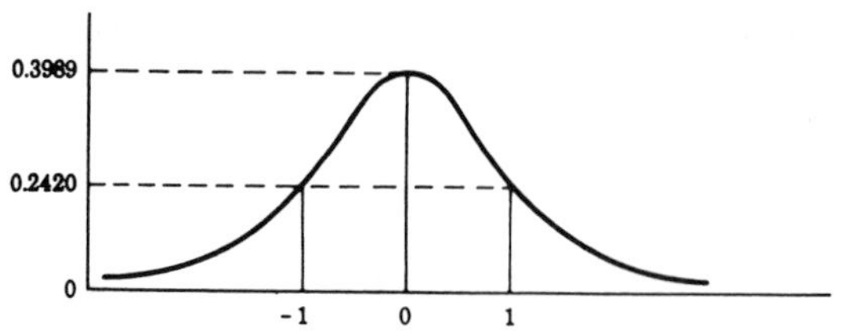

Example

$$Z = \frac{x - \mu}{\sigma}$$

$Z = 0$: ordinate $= 0.3939$

$Z = 1$: ordinate $= 0.2420$

z	.00	.01	.02	.03	.04	.05	.06	.07	.08	.09
.0	.3989	.3989	.3989	.3988	.3986	.3984	.3982	.3980	.3977	.3973
.1	.3970	.3965	.3961	.3956	.3951	.3945	.3939	.3932	.3925	.3918
.2	.3910	.3902	.3894	.3885	.3876	.3867	.3857	.3847	.3836	.3825
.3	.3814	.3802	.3790	.3778	.3765	.3752	.3739	.3725	.3712	.3697
.4	.3683	.3668	.3653	.3637	.3621	.3605	.3589	.3572	.3555	.3538
.5	.3521	.3503	.3485	.3467	.3448	.3429	.3410	.3391	.3372	.3352
.6	.3332	.3312	.3292	.3271	.3251	.3230	.3209	.3187	.3166	.3144
.7	.3123	.3101	.3079	.3056	.3034	.3011	.2989	.2966	.2943	.2920
.8	.2897	.2874	.2850	.2827	.2803	.2780	.2756	.2732	.2709	.2685
.9	.2661	.2637	.2613	.2589	.2565	.2541	.2516	.2492	.2468	.2444
1.0	.2420	.2396	.2371	.2347	.2323	.2299	.2275	.2251	.2227	.2203
1.1	.2179	.2155	.2131	.2107	.2083	.2059	.2036	.2012	.1989	.1965
1.2	.1942	.1919	.1895	.1872	.1849	.1826	.1804	.1781	.1758	.1736
1.3	.1714	.1691	.1669	.1647	.1626	.1604	.1582	.1561	.1539	.1518
1.4	.1497	.1476	.1456	.1435	.1415	.1394	.1374	.1354	.1334	.1315
1.5	.1295	.1276	.1257	.1238	.1219	.1200	.1182	.1163	.1145	.1127
1.6	.1109	.1092	.1074	.1057	.1040	.1023	.1006	.0989	.0973	.0957
1.7	.0940	.0925	.0909	.0893	.0878	.0863	.0848	.0833	.0818	.0804
1.8	.0790	.0775	.0761	.0748	.0734	.0721	.0707	.0694	.0681	.0669
1.9	.0656	.0644	.0632	.0620	.0608	.0596	.0584	.0573	.0562	.0551
2.0	.0540	.0529	.0519	.0508	.0498	.0488	.0478	.0468	.0459	.0449
2.1	.0440	.0431	.0422	.0413	.0404	.0396	.0387	.0379	.0371	.0363
2.2	.0355	.0347	.0339	.0332	.0325	.0317	.0310	.0303	.0297	.0290
2.3	.0283	.0277	.0270	.0264	.0258	.0252	.0246	.0241	.0235	.0229
2.4	.0224	.0219	.0213	.0208	.0203	.0198	.0194	.0189	.0184	.0180
2.5	.0175	.0171	.0167	.0163	.0158	.0154	.0151	.0147	.0143	.0139
2.6	.0136	.0132	.0129	.0126	.0122	.0119	.0116	.0113	.0110	.0107
2.7	.0104	.0101	.0099	.0096	.0093	.0091	.0088	.0086	.0084	.0081
2.8	.0079	.0077	.0075	.0073	.0071	.0069	.0067	.0065	.0063	.0061
2.9	.0060	.0058	.0056	.0055	.0053	.0051	.0050	.0048	.0047	.0046
3.0	.0044	.0043	.0042	.0040	.0039	.0038	.0037	.0036	.0035	.0034
3.1	.0033	.0032	.0031	.0030	.0029	.0028	.0027	.0026	.0025	.0025
3.2	.0024	.0023	.0022	.0022	.0021	.0020	.0020	.0019	.0018	.0018
3.3	.0017	.0017	.0016	.0016	.0015	.0015	.0014	.0014	.0013	.0013
3.4	.0012	.0012	.0012	.0011	.0011	.0010	.0010	.0010	.0009	.0009
3.5	.0009	.0008	.0008	.0008	.0008	.0007	.0007	.0007	.0007	.0006
3.6	.0006	.0006	.0006	.0005	.0005	.0005	.0005	.0005	.0005	.0004
3.7	.0004	.0004	.0004	.0004	.0004	.0004	.0003	.0003	.0003	.0003
3.8	.0003	.0003	.0003	.0003	.0003	.0002	.0002	.0002	.0002	.0002
3.9	.0002	.0002	.0002	.0002	.0002	.0002	.0002	.0002	.0001	.0001

Source: This table is reproduced from Taro Yamane, *Statistics, an Introductory Analysis*, published by Harper & Row, Publishers, New York, and by permission of the author and publishers.

TABLE C.3
Areas under the normal curve

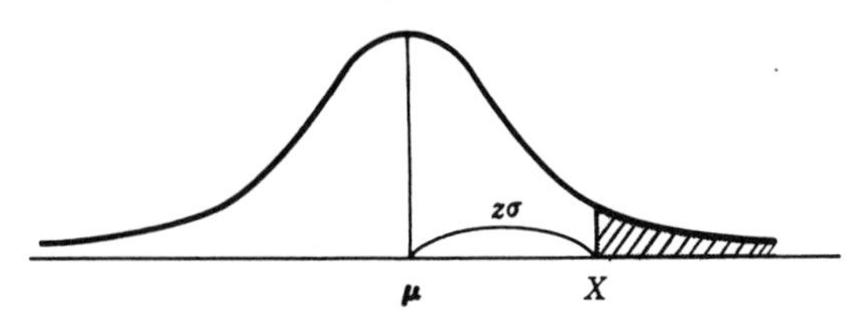

Example

$$Z = \frac{X - \mu}{\sigma}$$

$$P[Z > 1] = .1587$$

$$P[Z > 1.96] = .0250$$

Normal Deviate z	.00	.01	.02	.03	.04	.05	.06	.07	.08	.09
0.0	.5000	.4960	.4920	.4880	.4840	.4801	.4761	.4721	.4681	.4641
0.1	.4602	.4562	.4522	.4483	.4443	.4404	.4364	.4325	.4286	.4247
0.2	.4207	.4168	.4129	.4090	.4052	.4013	.3974	.3936	.3897	.3859
0.3	.3821	.3783	.3745	.3707	.3669	.3632	.3594	.3557	.3520	.3483
0.4	.3446	.3409	.3372	.3336	.3300	.3264	.3228	.3192	.3156	.3121
0.5	.3085	.3050	.3015	.2981	.2946	.2912	.2877	.2843	.2810	.2776
0.6	.2743	.2709	.2676	.2643	.2611	.2578	.2546	.2514	.2483	.2451
0.7	.2420	.2389	.2358	.2327	.2296	.2266	.2236	.2206	.2177	.2148
0.8	.2119	.2090	.2061	.2033	.2005	.1977	.1949	.1922	.1894	.1867
0.9	.1841	.1814	.1788	.1762	.1736	.1711	.1685	.1660	.1635	.1611
1.0	.1587	.1562	.1539	.1515	.1492	.1469	.1446	.1423	.1401	.1379
1.1	.1357	.1335	.1314	.1292	.1271	.1251	.1230	.1210	.1190	.1170
1.2	.1151	.1131	.1112	.1093	.1075	.1056	.1038	.1020	.1003	.0985
1.3	.0968	.0951	.0934	.0918	.0901	.0885	.0869	.0853	.0838	.0823
1.4	.0808	.0793	.0778	.0764	.0749	.0735	.0721	.0708	.0694	.0681
1.5	.0668	.0655	.0643	.0630	.0618	.0606	.0594	.0582	.0571	.0559
1.6	.0548	.0537	.0526	.0516	.0505	.0495	.0485	.0475	.0465	.0455
1.7	.0446	.0436	.0427	.0418	.0409	.0401	.0392	.0384	.0375	.0367
1.8	.0359	.0351	.0344	.0336	.0329	.0322	.0314	.0307	.0301	.0294
1.9	.0287	.0281	.0274	.0268	.0262	.0256	.0250	.0244	.0239	.0233
2.0	.0228	.0222	.0217	.0212	.0207	.0202	.0197	.0192	.0188	.0183
2.1	.0179	.0174	.0170	.0166	.0162	.0158	.0154	.0150	.0146	.0143
2.2	.0139	.0136	.0132	.0129	.0125	.0122	.0119	.0116	.0113	.0110
2.3	.0107	.0104	.0102	.0099	.0096	.0094	.0091	.0089	.0087	.0084
2.4	.0082	.0080	.0078	.0075	.0073	.0071	.0069	.0068	.0066	.0064
2.5	.0062	.0060	.0059	.0057	.0055	.0054	.0052	.0051	.0049	.0048
2.6	.0047	.0045	.0044	.0043	.0041	.0040	.0039	.0038	.0037	.0036
2.7	.0035	.0034	.0033	.0032	.0031	.0030	.0029	.0028	.0027	.0026
2.8	.0026	.0025	.0024	.0023	.0023	.0022	.0021	.0021	.0020	.0019
2.9	.0019	.0018	.0018	.0017	.0016	.0016	.0015	.0015	.0014	.0014
3.0	.0013	.0013	.0013	.0012	.0012	.0011	.0011	.0011	.0010	.0010

Source: This table is reproduced from Taro Yamane, *Statistics, an Introductory Analysis,* published by Harper & Row, Publishers, New York, and by permission of the author and publishers.

Appendix D

The *t* distribution

IF A VARIABLE X is normally distributed, then the statistic has a

$$x/\sigma = \frac{X - \mu}{\sigma}$$

normal distribution. But the statistic

$$x/\hat{\sigma} = \frac{X - \mu}{\hat{\sigma}}$$

has a t distribution, $\hat{\sigma}$ being the square root of an unbiased estimate of σ^2. That is,

$$\hat{\sigma} = \sqrt{\hat{\sigma}^2} \qquad \text{and} \qquad \hat{\sigma}^2 = \frac{\Sigma(X - \bar{X})^2}{n - 1}$$

However, even though the variable X is not normally distributed, if the sample means $(\bar{X})$ are normally distributed, then, of course, the statistic

$$x/\sigma = \frac{\bar{X} - \mu}{\sigma_{\bar{x}}}$$

has a normal distribution. But the statistic

$$x/\hat{\sigma} = \frac{\bar{X} - \mu}{\hat{\sigma}_{\bar{x}}}$$

has a t distribution.

To test an hypothesis concerning a sample mean, $\bar{X}$, using either the normal distribution or the t distribution assumes that the sampling distribution of the $\bar{X}$ is normal. Therefore the only distinction to be made between use of the normal curve and the t distribution is that the normal distribution is used with σ and the t distribution is used with $\hat{\sigma}$.

However, when $n = 30$, $\hat{\sigma}_x \doteq \sigma_x$; and therefore, $\hat{\sigma}_{\bar{x}} \doteq \sigma_{\bar{x}}$. With $\hat{\sigma}_{\bar{x}} \doteq \sigma_{\bar{x}}$ the distribution approximates a normal curve. This can be seen in Figures D.2 and D.3.

Thus, as n gets larger and larger, $\hat{\sigma}_{\bar{x}}$ approaches $\sigma_{\bar{x}}$, and the distribution of t approaches the normal curve.

FIGURE D.1
Normal distribution

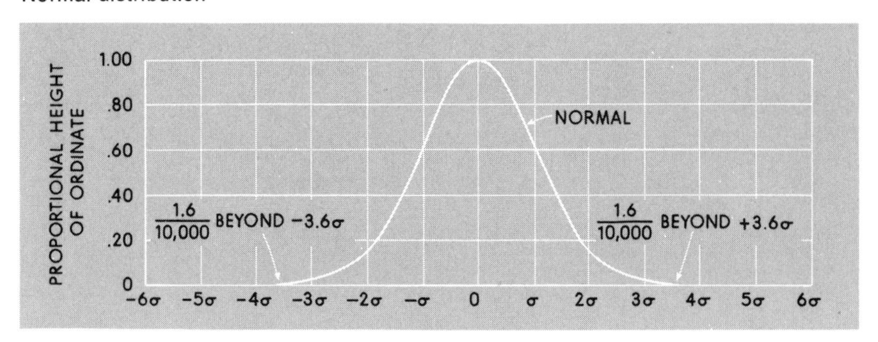

FIGURE D.2
Comparison of the normal distribution and the t distribution when $n = 20$

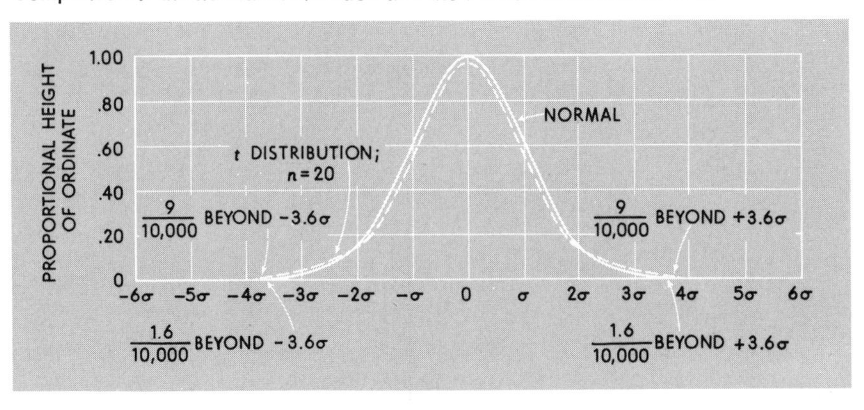

FIGURE D.3
Comparison of the normal distribtuion and the t distribution when $n = 5$

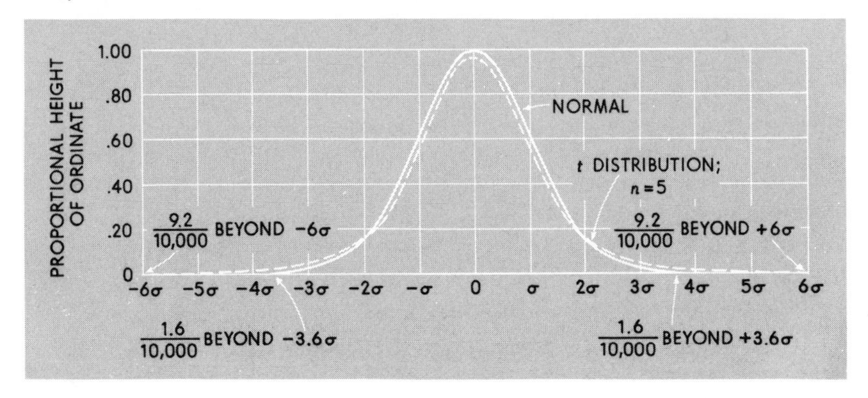

TABLE D.1

Values of *t* (probabilities refer to the sum of the two-tail areas; for a single tail divide the probability by 2)

	Probability (*P*).												
n	·9	·8	·7	·6	·5	·4	·3	·2	·1	·05	·02	·01	·001
1	·158	·325	·510	·727	1·000	1·376	1·963	3·078	6·314	12·706	31·821	63·657	636·619
2	·142	·289	·445	·617	·816	1·061	1·386	1·886	2·920	4·303	6·965	9·925	31·598
3	·137	·277	·424	·584	·765	·978	1·250	1·638	2·353	3·182	4·541	5·841	12·941
4	·134	·271	·414	·569	·741	·941	1·190	1·533	2·132	2·776	3·747	4·604	8·610
5	·132	·267	·408	·559	·727	·920	1·156	1·476	2·015	2·571	3·365	4·032	6·859
6	·131	·265	·404	·553	·718	·906	1·134	1·440	1·943	2·447	3·143	3·707	5·959
7	·130	·263	·402	·549	·711	·896	1·119	1·415	1·895	2·365	2·998	3·499	5·405
8	·130	·262	·399	·546	·706	·889	1·108	1·397	1·860	2·306	2·896	3·355	5·041
9	·129	·261	·398	·543	·703	·883	1·100	1·383	1·833	2·262	2·821	3·250	4·781
10	·129	·260	·397	·542	·700	·879	1·093	1·372	1·812	2·228	2·764	3·169	4·587
11	·129	·260	·396	·540	·697	·876	1·088	1·363	1·796	2·201	2·718	3·106	4·437
12	·128	·259	·395	·539	·695	·873	1·083	1·356	1·782	2·179	2·681	3·055	4·318
13	·128	·259	·394	·538	·694	·870	1·079	1·350	1·771	2·160	2·650	3·012	4·221
14	·128	·258	·393	·537	·692	·868	1·076	1·345	1·761	2·145	2·624	2·977	4·140
15	·128	·258	·393	·536	·691	·866	1·074	1·341	1·753	2·131	2·602	2·947	4·073
16	·128	·258	·392	·535	·690	·865	1·071	1·337	1·746	2·120	2·583	2·921	4·015
17	·128	·257	·392	·534	·689	·863	1·069	1·333	1·740	2·110	2·567	2·898	3·965
18	·127	·257	·392	·534	·688	·862	1·067	1·330	1·734	2·101	2·552	2·878	3·922
19	·127	·257	·391	·533	·688	·861	1·066	1·328	1·729	2·093	2·539	2·861	3·883
20	·127	·257	·391	·533	·687	·860	1·064	1·325	1·725	2·086	2·528	2·845	3·850
21	·127	·257	·391	·532	·686	·859	1·063	1·323	1·721	2·080	2·518	2·831	3·819
22	·127	·256	·390	·532	·686	·858	1·061	1·321	1·717	2·074	2·508	2·819	3·792
23	·127	·256	·390	·532	·685	·858	1·060	1·319	1·714	2·069	2·500	2·807	3·767
24	·127	·256	·390	·531	·685	·857	1·059	1·318	1·711	2·064	2·492	2·797	3·745
25	·127	·256	·390	·531	·684	·856	1·058	1·316	1·708	2·060	2·485	2·787	3·725
26	·127	·256	·390	·531	·684	·856	1·058	1·315	1·706	2·056	2·479	2·779	3·707
27	·127	·256	·389	·531	·684	·855	1·057	1·314	1·703	2·052	2·473	2·771	3·690
28	·127	·256	·389	·530	·683	·855	1·056	1·313	1·701	2·048	2·467	2·763	3·674
29	·127	·256	·389	·530	·683	·854	1·055	1·311	1·699	2·045	2·462	2·756	3·659
30	·127	·256	·389	·530	·683	·854	1·055	1·310	1·697	2·042	2·457	2·750	3·646
40	·126	·255	·388	·529	·681	·851	1·050	1·303	1·684	2·021	2·423	2·704	3·551
60	·126	·254	·387	·527	·679	·848	1·046	1·296	1·671	2·000	2·390	2·660	3·460
120	·126	·254	·386	·526	·677	·845	1·041	1·289	1·658	1·980	2·358	2·617	3·373
∞	·126	·253	·385	·524	·674	·842	1·036	1·282	1·645	1·960	2·326	2·576	3·291

Source: This table reprinted from Table III of R. A. Fisher and F. Yates, *Statistical Tables for Biological Agricultural and Medical Research,* published by Oliver and Boyd, Ltd., Edinburgh, by permission of the authors and publishers.

Appendix E

Binomial distribution function

$$P\{X\} = \frac{n!}{X!(n-X)!}\,p^X q^{n-X}$$

n	X	.05	.10	.15	.20	.25	.30	.35	.40	.45	.50
						p					
1	0	.9500	.9000	.8500	.8000	.7500	.7000	.6500	.6000	.5500	.5000
	1	.0500	.1000	.1500	.2000	.2500	.3000	.3500	.4000	.4500	.5000
2	0	.9025	.8100	.7225	.6400	.5625	.4900	.4225	.3600	.3025	.2500
	1	.0950	.1800	.2550	.3200	.3750	.4200	.4550	.4800	.4950	.5000
	2	.0025	.0100	.0225	.0400	.0625	.0900	.1225	.1600	.2025	.2500
3	0	.8574	.7290	.6141	.5120	.4219	.3430	.2746	.2160	.1664	.1250
	1	.1354	.2430	.3251	.3840	.4219	.4410	.4436	.4320	.4084	.3750
	2	.0071	.0270	.0574	.0960	.1406	.1890	.2389	.2880	.3341	.3750
	3	.0001	.0010	.0034	.0080	.0156	.0270	.0429	.0640	.0911	.1250
4	0	.8145	.6561	.5220	.4096	.3164	.2401	.1785	.1296	.0915	.0625
	1	.1715	.2916	.3685	.4096	.4219	.4116	.3845	.3456	.2995	.2500
	2	.0135	.0486	.0975	.1536	.2109	.2646	.3105	.3456	.3675	.3750
	3	.0005	.0036	.0115	.0256	.0469	.0756	.1115	.1536	.2005	.2500
	4	.0000	.0001	.0005	.0016	.0039	.0081	.0150	.0256	.0410	.0625
5	0	.7738	.5905	.4437	.3277	.2373	.1681	.1160	.0778	.0503	.0312
	1	.2036	.3280	.3915	.4096	.3955	.3602	.3124	.2592	.2059	.1562
	2	.0214	.0729	.1382	.2048	.2637	.3087	.3364	.3456	.3369	.3125
	3	.0011	.0081	.0244	.0512	.0879	.1323	.1811	.2304	.2757	.3125
	4	.0000	.0004	.0022	.0064	.0146	.0284	.0488	.0768	.1128	.1562
	5	.0000	.0000	.0001	.0003	.0010	.0024	.0053	.0102	.0185	.0312
6	0	.7351	.5314	.3771	.2621	.1780	.1176	.0754	.0467	.0277	.0156
	1	.2321	.3543	.3993	.3932	.3560	.3025	.2437	.1866	.1359	.0938
	2	.0305	.0984	.1762	.2458	.2966	.3241	.3280	.3110	.2780	.2344
	3	.0021	.0146	.0415	.0819	.1318	.1852	.2355	.2765	.3032	.3125
	4	.0001	.0012	.0055	.0154	.0330	.0595	.0951	.1382	.1861	.2344
	5	.0000	.0001	.0004	.0015	.0044	.0102	.0205	.0369	.0609	.0938
	6	.0000	.0000	.0000	.0001	.0002	.0007	.0018	.0041	.0083	.0156
7	0	.6983	.4783	.3206	.2097	.1335	.0824	.0490	.0280	.0152	.0078
	1	.2573	.3720	.3960	.3670	.3115	.2471	.1848	.1306	.0872	.0547
	2	.0406	.1240	.2097	.2753	.3115	.3177	.2985	.2613	.2140	.1641
	3	.0036	.0230	.0617	.1147	.1730	.2269	.2679	.2903	.2918	.2734
	4	.0002	.0026	.0109	.0287	.0577	.0972	.1442	.1935	.2388	.2734
	5	.0000	.0002	.0012	.0043	.0115	.0250	.0466	.0774	.1172	.1641
	6	.0000	.0000	.0001	.0004	.0013	.0036	.0084	.0172	.0320	.0547
	7	.0000	.0000	.0000	.0000	.0001	.0002	.0006	.0016	.0037	.0078
8	0	.6634	.4305	.2725	.1678	.1002	.0576	.0319	.0168	.0084	.0039
	1	.2793	.3826	.3847	.3355	.2670	.1977	.1373	.0896	.0548	.0312
	2	.0515	.1488	.2376	.2936	.3115	.2065	.2587	.2090	.1569	.1094

Binomial distribution function (*continued*)

n	X	.05	.10	.15	.20	.25	p .30	.35	.40	.45	.50
8	3	.0054	.0331	.0839	.1468	.2076	.2541	.2786	.2787	.2568	.2188
	4	.0004	.0046	.0185	.0459	.0865	.1361	.1875	.2322	.2627	.2734
	5	.0000	.0004	.0026	.0092	.0231	.0467	.0808	.1239	.1719	.2188
	6	.0000	.0000	.0002	.0011	.0038	.0100	.0217	.0413	.0403	.1094
	7	.0000	.0000	.0000	.0001	.0004	.0012	.0033	.0079	.0164	.0312
	8	.0000	.0000	.0000	.0000	.0000	.0001	.0002	.0007	.0017	.0039
9	0	.6302	.3874	.2316	.1342	.0751	.0404	.0207	.0101	.0046	.0020
	1	.2985	.3874	.3679	.3020	.2253	.1556	.1004	.0605	.0339	.0176
	2	.0629	.1722	.2597	.3020	.3003	.2668	.2162	.1612	.1110	.0703
	3	.0077	.0446	.1069	.1762	.2336	.2668	.2716	.2508	.2119	.1641
	4	.0006	.0074	.0283	.0661	.1168	.1715	.2194	.2508	.2600	.2461
	5	.0000	.0008	.0050	.0165	.0389	.0735	.1181	.1672	.2128	.2461
	6	.0000	.0001	.0006	.0028	.0087	.0210	.0424	.0743	.1160	.1641
	7	.0000	.0000	.0000	.0003	.0012	.0039	.0098	.0212	.0407	.0703
	8	.0000	.0000	.0000	.0000	.0001	.0004	.0013	.0035	.0083	.0176
	9	.0000	.0000	.0000	.0000	.0000	.0000	.0001	.0003	.0008	.0020
10	0	.5987	.3487	.1969	.1074	.0563	.0282	.0135	.0060	.0025	.0010
	1	.3151	.3874	.3474	.2684	.1877	.1211	.0725	.0403	.0207	.0098
	2	.0746	.1937	.2759	.3020	.2816	.2335	.1757	.1209	.0763	.0439
	3	.0105	.0574	.1298	.2013	.2503	.2668	.2522	.2150	.1665	.1172
	4	.0010	.0112	.0401	.0881	.1460	.2001	.2377	.2508	.2384	.2051
	5	.0001	.0015	.0085	.0264	.0584	.1029	.1536	.2007	.2340	.2461
	6	.0000	.0001	.0012	.0055	.0162	.0368	.0689	.1115	.1596	.2051
	7	.0000	.0000	.0001	.0008	.0031	.0090	.0212	.0425	.0746	.1172
	8	.0000	.0000	.0000	.0001	.0004	.0014	.0043	.0106	.0229	.0439
	9	.0000	.0000	.0000	.0000	.0000	.0001	.0005	.0016	.0042	.0098
	10	.0000	.0000	.0000	.0000	.0000	.0000	.0000	.0001	.0003	.0010
11	0	.5688	.3138	.1673	.0859	.0422	.0198	.0088	.0036	.0014	.0005
	1	.3293	.3835	.3248	.2362	.1549	.0932	.0518	.0266	.0125	.0054
	2	.0867	.2131	.2866	.2953	.2581	.1998	.1395	.0887	.0513	.0269
	3	.0137	.0710	.1517	.2215	.2581	.2568	.2254	.1774	.1259	.0806
	4	.0014	.0158	.0536	.1107	.1721	.2201	.2428	.2365	.2060	.1611
	5	.0001	.0025	.0132	.0388	.0803	.1321	.1830	.2207	.2360	.2256
	6	.0000	.0003	.0023	.0097	.0268	.0566	.0985	.1471	.1931	.2256
	7	.0000	.0000	.0003	.0017	.0064	.0173	.0379	.0701	.1128	.1611
	8	.0000	.0000	.0000	.0002	.0011	.0037	.0102	.0234	.0462	.0806
	9	.0000	.0000	.0000	.0000	.0001	.0005	.0018	.0052	.0126	.0269
	10	.0000	.0000	.0000	.0000	.0000	.0000	.0002	.0007	.0021	.0054
	11	.0000	.0000	.0000	.0000	.0000	.0000	.0000	.0000	.0002	.0005
12	0	.5404	.2824	.1422	.0687	.0317	.0138	.0057	.0022	.0008	.0002
	1	.3413	.3766	.3012	.2062	.1267	.0712	.0368	.0174	.0075	.0029

Binomial distribution function (*continued*)

n	x	.05	.10	.15	.20	.25	.30	.35	.40	.45	.50
12	2	.0988	.2301	.2924	.2835	.2323	.1678	.1088	.0639	.0339	.0161
	3	.0173	.0852	.1720	.2362	.2581	.2397	.1954	.1419	.0923	.0537
	4	.0021	.0213	.0683	.1329	.1936	.2311	.2367	.2128	.1700	.1208
	5	.0002	.0038	.0193	.0532	.1032	.1585	.2039	.2270	.2225	.1934
	6	.0000	.0005	.0040	.0155	.0401	.0792	.1281	.1766	.2124	.2256
	7	.0000	.0000	.0006	.0033	.0115	.0291	.0591	.1009	.1489	.1934
	8	.0000	.0000	.0001	.0005	.0024	.0078	.0199	.0420	.0762	.1208
	9	.0000	.0000	.0000	.0001	.0004	.0015	.0048	.0125	.0277	.0537
	10	.0000	.0000	.0000	.0000	.0000	.0002	.0008	.0025	.0068	.0161
	11	.0000	.0000	.0000	.0000	.0000	.0000	.0001	.0003	.0010	.0029
	12	.0000	.0000	.0000	.0000	.0000	.0000	.0000	.0000	.0001	.0002
13	0	.5133	.2542	.1209	.0550	.0238	.0097	.0037	.0013	.0004	.0001
	1	.3512	.3672	.2774	.1787	.1029	.0540	.0259	.0113	.0045	.0016
	2	.1109	.2448	.2937	.2680	.2059	.1388	.0836	.0453	.0220	.0095
	3	.0214	.0997	.1900	.2457	.2517	.2181	.1651	.1107	.0660	.0349
	4	.0028	.0277	.0838	.1535	.2097	.2337	.2222	.1845	.1350	.0873
	5	.0003	.0055	.0266	.0691	.1258	.1803	.2154	.2214	.1989	.1571
	6	.0000	.0008	.0063	.0230	.0559	.1030	.1546	.1968	.2169	.2095
	7	.0000	.0001	.0011	.0058	.0186	.0442	.0833	.1312	.1775	.2095
	8	.0000	.0001	.0001	.0011	.0047	.0142	.0336	.0656	.1089	.1571
	9	.0000	.0000	.0000	.0001	.0009	.0034	.0101	.0243	.0495	.0873
	10	.0000	.0000	.0000	.0000	.0001	.0006	.0022	.0065	.0162	.0349
	11	.0000	.0000	.0000	.0000	.0000	.0001	.0003	.0012	.0036	.0095
	12	.0000	.0000	.0000	.0000	.0000	.0000	.0000	.0001	.0005	.0016
	13	.0000	.0000	.0000	.0000	.0000	.0000	.0000	.0000	.0000	.0001
14	0	.4877	.2288	.1028	.0440	.0178	.0068	.0024	.0008	.0002	.0001
	1	.3593	.3559	.2539	.1539	.0832	.0407	.0181	.0073	.0027	.0009
	2	.1229	.2570	.2912	.2501	.1802	.1134	.0634	.0317	.0141	.0056
	3	.0259	.1142	.2056	.2501	.2402	.1943	.1366	.0845	.0462	.0222
	4	.0037	.0349	.0998	.1720	.2202	.2290	.2022	.1549	.1040	.0611
	5	.0004	.0078	.0352	.0860	.1468	.1963	.2178	.2066	.1701	.1222
	6	.0000	.0013	.0093	.0322	.0734	.1262	.1759	.2066	.2088	.1833
	7	.0000	.0002	.0019	.0092	.0280	.0618	.1082	.1574	.1952	.2095
	8	.0000	.0000	.0003	.0020	.0082	.0232	.0510	.0918	.1398	.1833
	9	.0000	.0000	.0000	.0003	.0018	.0066	.0183	.0408	.0762	.1222
	10	.0000	.0000	.0000	.0000	.0003	.0014	.0049	.0136	.0312	.0611
	11	.0000	.0000	.0000	.0000	.0000	.0002	.0010	.0033	.0093	.0222
	12	.0000	.0000	.0000	.0000	.0000	.0000	.0001	.0005	.0019	.0056
	13	.0000	.0000	.0000	.0000	.0000	.0000	.0000	.0001	.0002	.0009
	14	.0000	.0000	.0000	.0000	.0000	.0000	.0000	.0000	.0000	.0001
15	0	.4633	.2059	.0874	.0352	.0134	.0047	.0016	.0005	.0001	.0000
	1	.3658	.3432	.2312	.1319	.0668	.0305	.0126	.0047	.0016	.0005
	2	.1348	.2669	.2856	.2309	.1559	.0916	.0476	.0219	.0090	.0032

Binomial distribution function (*continued*)

n	X	.05	.10	.15	.20	.25	p .30	.35	.40	.45	.50
15	3	.0307	.1285	.2184	.2501	.2252	.1700	.1110	.0634	.0318	.0139
	4	.0049	.0428	.1156	.1876	.2252	.2186	.1792	.1268	.0780	.0417
	5	.0006	.0105	.0449	.1032	.1651	.2061	.2123	.1859	.1404	.0916
	6	.0000	.0019	.0132	.0430	.0917	.1472	.1906	.2066	.1914	.1527
	7	.0000	.0003	.0030	.0138	.0393	.0811	.1319	.1771	.2013	.1964
	8	.0000	.0000	.0005	.0035	.0131	.0348	.0710	.1181	.1647	.1964
	9	.0000	.0000	.0001	.0007	.0034	.0116	.0298	.0612	.1048	.1527
	10	.0000	.0000	.0000	.0001	.0007	.0030	.0096	.0245	.0515	.0916
	11	.0000	.0000	.0000	.0000	.0001	.0006	.0024	.0074	.0191	.0417
	12	.0000	.0000	.0000	.0000	.0000	.0001	.0004	.0016	.0052	.0139
	13	.0000	.0000	.0000	.0000	.0000	.0000	.0001	.0003	.0010	.0032
	14	.0000	.0000	.0000	.0000	.0000	.0000	.0000	.0000	.0001	.0005
	15	.0000	.0000	.0000	.0000	.0000	.0000	.0000	.0000	.0000	.0000
16	0	.4401	.1853	.0743	.0281	.0100	.0033	.0010	.0003	.0001	.0000
	1	.3706	.3294	.2097	.1126	.0535	.0228	.0087	.0030	.0009	.0002
	2	.1463	.2745	.2775	.2111	.1336	.0732	.0353	.0150	.0056	.0018
	3	.0359	.1423	.2285	.2463	.2079	.1465	.0888	.0468	.0215	.0085
	4	.0061	.0514	.1311	.2001	.2252	.2040	.1553	.1014	.0572	.0278
	5	.0008	.0137	.0555	.1201	.1802	.2099	.2008	.1623	.1123	.0667
	6	.0001	.0028	.0180	.0550	.1101	.1649	.1982	.1983	.1684	.1222
	7	.0000	.0004	.0045	.0197	.0524	.1010	.1524	.1889	.1969	.1746
	8	.0000	.0001	.0009	.0055	.0197	.0487	.0923	.1417	.1812	.1964
	9	.0000	.0000	.0001	.0012	.0058	.0185	.0442	.0840	.1318	.1746
	10	.0000	.0000	.0000	.0002	.0014	.0056	.0167	.0392	.0755	.1222
	11	.0000	.0000	.0000	.0000	.0002	.0013	.0049	.0142	.0337	.0667
	12	.0000	.0000	.0000	.0000	.0000	.0002	.0011	.0040	.0115	.0278
	13	.0000	.0000	.0000	.0000	.0000	.0000	.0002	.0008	.0029	.0085
	14	.0000	.0000	.0000	.0000	.0000	.0000	.0000	.0001	.0005	.0018
	15	.0000	.000C	.0000	.0000	.0000	.0000	.0000	.0000	.0001	.0002
	16	.0000	.000C	.0000	.0000	.0000	.0000	.0000	.0000	.0000	.0000
17	0	.4181	.1668	.0631	.0225	.0075	.0023	.0007	.0002	.0000	.0000
	1	.3741	.3150	.1893	.0957	.0426	.0169	.0060	.0019	.0005	.0001
	2	.1575	.2800	.2673	.1914	.1136	.0581	.0260	.0102	.0035	.0010
	3	.0415	.1556	.2359	.2393	.1893	.1245	.0701	.0341	.0144	.0052
	4	.0076	.0605	.1457	.2093	.2209	.1868	.1320	.0796	.0411	.0182
	5	.0010	.0175	.0668	.1361	.1914	.2081	.1849	.1379	.0875	.0472
	6	.0001	.0039	.0236	.0680	.1276	.1784	.1991	.1839	.1432	.0944
	7	.0000	.0007	.0065	.0267	.0668	.1201	.1685	.1927	.1841	.1484
	8	.0000	.0001	.0014	.0084	.0279	.0644	.1134	.1606	.1883	.1855
	9	.0000	.0000	.0003	.0021	.0093	.0276	.0611	.1070	.1540	.1855

Binomial distribution function (*continued*)

n	X	.05	.10	.15	.20	.25	p .30	.35	.40	.45	.50
17	10	.0000	.0000	.0000	.0004	.0025	.0095	.0263	.0571	.1008	.1484
	11	.0000	.0000	.0000	.0001	.0005	.0026	.0090	.0242	.0525	.0944
	12	.0000	.0000	.0000	.0000	.0001	.0006	.0024	.0081	.0215	.0472
	13	.0000	.0000	.0000	.0000	.0000	.0001	.0005	.0021	.0068	.0182
	14	.0000	.0000	.0000	.0000	.0000	.0000	.0001	.0004	.0016	.0052
	15	.0000	.0000	.0000	.0000	.0000	.0000	.0000	.0001	.0003	.0010
	16	.0000	.0000	.0000	.0000	.0000	.0000	.0000	.0000	.0000	.0001
	17	.0000	.0000	.0000	.0000	.0000	.0000	.0000	.0000	.0000	.0000
18	0	.3972	.1501	.0536	.0180	.0056	.0016	.0004	.0001	.0000	.0000
	1	.3763	.3002	.1704	.0811	.0338	.0126	.0042	.0012	.0003	.0001
	2	.1683	.2835	.2556	.1723	.0958	.0458	.0190	.0069	.0022	.0006
	3	.0473	.1680	.2406	.2297	.1704	.1046	.0547	.0246	.0095	.0031
	4	.0093	.0700	.1592	.2153	.2130	.1681	.1104	.0614	.0291	.0117
	5	.0014	.0218	.0787	.1507	.1988	.2017	.1664	.1146	.0666	.0327
	6	.0002	.0052	.0301	.0816	.1436	.1873	.1941	.1655	.1181	.0708
	7	.0000	.0010	.0091	.0350	.0820	.1376	.1792	.1892	.1657	.1214
	8	.0000	.0002	.0022	.0120	.0376	.0811	.1327	.1734	.1864	.1669
	9	.0000	.0000	.0004	.0033	.0139	.0386	.0794	.1284	.1694	.1855
	10	.0000	.0000	.0001	.0008	.0042	.0149	.0385	.0771	.1248	.1669
	11	.0000	.0000	.0000	.0001	.0010	.0046	.0151	.0374	.0742	.1214
	12	.0000	.0000	.0000	.0000	.0002	.0012	.0047	.0145	.0354	.0708
	13	.0000	.0000	.0000	.0000	.0000	.0002	.0012	.0045	.0134	.0327
	14	.0000	.0000	.0000	.0000	.0000	.0000	.0002	.0011	.0039	.0117
	15	.0000	.0000	.0000	.0000	.0000	.0000	.0000	.0002	.0009	.0031
	16	.0000	.0000	.0000	.0000	.0000	.0000	.0000	.0000	.0001	.0006
	17	.0000	.0000	.0000	.0000	.0000	.0000	.0000	.0000	.0000	.0001
	18	.0000	.0000	.0000	.0000	.0000	.0000	.0000	.0000	.0000	.0000
19	0	.3774	.1351	.0456	.0144	.0042	.0011	.0003	.0001	.0000	.0000
	1	.3774	.2852	.1529	.0685	.0268	.0093	.0029	.0008	.0002	.0000
	2	.1787	.2852	.2428	.1540	.0803	.0358	.0138	.0046	.0013	.0003
	3	.0533	.1796	.2428	.2182	.1517	.0869	.0422	.0175	.0062	.0018
	4	.0112	.0798	.1714	.2182	.2023	.1491	.0909	.0467	.0203	.0074
	5	.0018	.0266	.0907	.1636	.2023	.1916	.1468	.0933	.0497	.0222
	6	.0002	.0069	.0374	.0955	.1574	.1916	.1844	.1451	.0949	.0518
	7	.0000	.0014	.0122	.0443	.0974	.1525	.1844	.1797	.1443	.0961
	8	.0000	.0002	.0032	.0166	.0487	.0981	.1489	.1797	.1771	.1442
	9	.0000	.0000	.0007	.0051	.0198	.0514	.0980	.1464	.1771	.1762
	10	.0000	.0000	.0001	.0013	.0066	.0220	.0528	.0976	.1449	.1762
	11	.0000	.0000	.0000	.0003	.0018	.0077	.0233	.0532	.0970	.1442
	12	.0000	.0000	.0000	.0000	.0004	.0022	.0083	.0237	.0529	.0961
	13	.0000	.0000	.0000	.0000	.0001	.0005	.0024	.0085	.0233	.0518
	14	.0000	.0000	.0000	.0000	.0000	.0001	.0006	.0024	.0082	.0222

Binomial distribution function (*concluded*)

n	X	.05	.10	.15	.20	.25	p .30	.35	.40	.45	.50
19	15	.0000	.0000	.0000	.0000	.0000	.0000	.0001	.0005	.0022	.0074
	16	.0000	.0000	.0000	.0000	.0000	.0000	.0000	.0001	.0005	.0018
	17	.0000	.0000	.0000	.0000	.0000	.0000	.0000	.0000	.0001	.0003
	18	.0000	.0000	.0000	.0000	.0000	.0000	.0000	.0000	.0000	.0000
	19	.0000	.0000	.0000	.0000	.0000	.0000	.0000	.0000	.0000	.0000
20	0	.3585	.1216	.0388	.0115	.0032	.0008	.0002	.0000	.0000	.0000
	1	.3774	.2702	.1368	.0576	.0211	.0068	.0020	.0005	.0001	.0000
	2	.1887	.2852	.2293	.1369	.0669	.0278	.0100	.0031	.0008	.0002
	3	.0596	.1901	.2428	.2054	.1339	.0718	.0323	.0123	.0040	.0011
	4	.0133	.0898	.1821	.2182	.1897	.1304	.0738	.0350	.0139	.0046
	5	.0022	.0319	.1028	.1746	.2023	.1789	.1272	.0746	.0365	.0148
	6	.0003	.0089	.0454	.1091	.1686	.1916	.1712	.1244	.0746	.0370
	7	.0000	.0020	.0160	.0545	.1124	.1643	.1844	.1659	.1221	.0739
	8	.0000	.0004	.0046	.0222	.0609	.1144	.1614	.1797	.1623	.1201
	9	.0000	.0001	.0011	.0074	.0271	.0654	.1158	.1597	.1771	.1602
	10	.0000	.0000	.0002	.0020	.0099	.0308	.0686	.1171	.1593	.1762
	11	.0000	.0000	.0000	.0005	.0030	.0120	.0336	.0710	.1185	.1602
	12	.0000	.0000	.0000	.0001	.0008	.0039	.0136	.0355	.0727	.1201
	13	.0000	.0000	.0000	.0000	.0002	.0010	.0045	.0146	.0366	.0739
	14	.0000	.0000	.0000	.0000	.0000	.0002	.0012	.0049	.0150	.0370
	15	.0000	.0000	.0000	.0000	.0000	.0000	.0003	.0013	.0049	.0148
	16	.0000	.0000	.0000	.0000	.0000	.0000	.0000	.0003	.0013	.0046
	17	.0000	.0000	.0000	.0000	.0000	.0000	.0000	.0000	.0002	.0011
	18	.0000	.0000	.0000	.0000	.0000	.0000	.0000	.0000	.0000	.0002
	19	.0000	.0000	.0000	.0000	.0000	.0000	.0000	.0000	.0000	.0000
	20	.0000	.0000	.0000	.0000	.0000	.0000	.0000	.0000	.0000	.0000

Source: This table is reprinted from Table I of Burington and May: *Handbook of Probability and Statistics,* Copyright, 1953, by McGraw-Hill Book Co., Inc., by permission of the authors and the publishers.

Linear interpolation with respect to p will generally not be accurate to more than two decimal places, and sometimes less.

For extensive tables of $Cx^n p^x (1 - p)^{n-x}$, see *Tables of the Binomial Probability Distribution,* National Bureau of Standards, Applied Mathematics Series 6, Washington, D.C., 1950.

Appendix F

The F distribution

WHENEVER THERE ARE two estimates of variances, such as explained $\hat{\sigma}^2_{yc}$ and unexplained $\hat{\sigma}^2_{yx}$ in correlation analysis the ratio of one estimate to the other

$$F = \frac{\hat{\sigma}^2_{y\hat{c}}}{\hat{\sigma}^2_{yx}} = \frac{r^2}{1 - r^2} \cdot \frac{n - m}{m - 1}$$

has a sampling distribution called the F *distribution* or the *ratio of variances distribution*. Here,

$$\hat{\sigma}^2_{y_c} = \frac{bxy}{m - 1} \quad \text{and} \quad \hat{\sigma}^2_{yx} = \frac{\Sigma y^2 - bxy}{n - m}$$

In simple linear correlation, $m = 2$. There is a different distribution for every sample size and method. Table F.1 gives the F value for the .05 level of confidence and just beside it the F value for the .01 and the .001 levels of confidence.

An example of how the table can be used is appropriate. Suppose a simple linear correlation of $n = 10$ items results in $r = .70$.

$$F_c = \frac{r^2}{1 - r^2} \cdot \frac{n - m}{m - 1} = \frac{.49}{.51} \cdot 8 = 7.69$$

In the F table, we read $n - m = n_2 = 10 - 2 = 8$, and $m - 1 = n_1 = 2 - 1 = 1$. The .05 level shows $F = 5.318$. $F_c = 7.69 > F = 5.32$; thus, we assume that the correlation probably did not occur by chance.

The chart following the F table, Figure F.1, shows $n_2 = n - m$ and $n_1 = m - 1$.

TABLE F.1

Table of 5 percent and 1 percent F for designated degrees of freedom in greater and smaller mean square (each entry is the percent point of F which is exceeded by the proportion of values of F listed at the head of its column for degrees of freedom n_1 listed in the major caption and n listed in the stub)

n_2	$n_1 = 1$			$n_1 = 2$			$n_1 = 3$			$n_1 = 4$			$n_1 = 5$		
	.05	.01	.001	.05	.01	.001	.05	.01	.001	.05	.01	.001	.05	.01	.001
1	161.45	4,052.2	405,284	199.50	4,999.5	500,000	215.71	5,403.3	540,379	224.58	5,624.6	562,500	230.16	5,763.7	576,405
2	18.513	98.503	998.5	19.000	99.000	999.0	19.164	99.166	999.2	19.247	99.249	999.2	19.296	99.299	999.3
3	10.128	34.116	167.5	9.552	30.817	148.5	9.277	29.457	141.1	9.117	28.710	137.1	9.014	23.237	134.6
4	7.709	21.198	74.14	6.944	18.000	61.25	6.591	16.694	56.18	6.388	15.977	53.44	6.256	15.522	51.71
5	6.608	16.258	47.04	5.786	13.274	36.61	5.410	12.060	33.20	5.192	11.392	31.09	5.050	10.967	29.75
6	5.987	13.745	35.51	5.143	10.925	27.00	4.757	9.779	23.70	4.534	9.148	21.90	4.387	8.746	20.81
7	5.591	12.246	29.22	4.737	9.547	21.69	4.347	8.451	18.77	4.120	7.847	17.19	3.972	7.460	16.21
8	5.318	11.259	25.42	4.459	8.649	18.49	4.066	7.591	15.83	3.838	7.006	14.39	3.688	6.632	13.49
9	5.117	10.561	22.86	4.256	8.022	16.39	3.863	6.992	13.90	3.633	6.422	12.56	3.482	6.057	11.71
10	4.965	10.044	21.04	4.103	7.559	14.91	3.708	6.552	12.55	3.478	5.994	11.28	3.326	5.636	10.48
11	4.844	9.646	19.69	3.982	7.206	13.81	3.587	6.217	11.56	3.357	5.668	10.35	3.204	5.316	9.58
12	4.747	9.330	18.64	3.885	6.927	12.97	3.490	5.953	10.80	3.259	5.412	9.63	3.106	5.064	8.89
13	4.667	9.074	17.81	3.806	6.701	12.31	3.410	5.739	10.21	3.179	5.205	9.07	3.025	4.862	8.35
14	4.600	8.862	17.14	3.739	6.515	11.78	3.344	5.564	9.73	3.112	5.035	8.62	2.958	4.695	7.92
15	4.543	8.683	16.59	3.682	6.359	11.34	3.287	5.417	9.34	3.056	4.893	8.25	2.901	4.556	7.57
16	4.494	8.531	16.12	3.634	6.226	10.97	3.239	5.292	9.00	3.007	4.773	7.94	2.852	4.437	7.27
17	4.451	8.400	15.72	3.592	6.112	10.66	3.197	5.185	8.73	2.965	4.669	7.68	2.810	4.336	7.02
18	4.414	8.285	15.38	3.555	6.013	10.39	3.160	5.092	8.49	2.928	4.579	7.46	2.773	4.248	6.81
19	4.381	8.185	15.08	3.522	5.926	10.16	3.127	5.010	8.28	2.895	4.500	7.26	2.740	4.171	6.61
20	4.351	8.096	14.82	3.493	5.849	9.95	3.098	4.938	8.10	2.866	4.431	7.10	2.711	4.103	6.46
21	4.325	8.017	14.59	3.467	5.780	9.77	3.072	4.874	7.94	2.840	4.369	6.95	2.685	4.042	6.32
22	4.301	7.945	14.38	3.443	5.719	9.61	3.049	4.817	7.80	2.817	4.313	6.81	2.661	3.988	6.19
23	4.279	7.881	14.19	3.422	5.664	9.47	3.028	4.765	7.67	2.795	4.264	6.69	2.640	3.939	6.08
24	4.260	7.823	14.03	3.403	5.614	9.34	3.009	4.718	7.55	2.776	4.218	6.59	2.621	3.895	5.98
25	4.242	7.770	13.88	3.385	5.568	9.22	2.991	4.676	7.45	2.759	4.177	6.49	2.603	3.855	5.88
26	4.225	7.721	13.74	3.369	5.526	9.12	2.975	4.637	7.36	2.743	4.140	6.41	2.587	3.818	5.80
27	4.210	7.677	13.61	3.354	5.488	9.02	2.960	4.601	7.27	2.728	4.106	6.33	2.572	3.785	5.73
28	4.196	7.636	13.50	3.340	5.453	8.93	2.947	4.568	7.19	2.714	4.074	6.25	2.558	3.754	5.66
29	4.183	7.598	13.39	3.328	5.421	8.85	2.934	4.538	7.12	2.701	4.045	6.19	2.545	3.725	5.59
30	4.171	7.563	13.29	3.316	5.390	8.77	2.922	4.510	7.05	2.690	4.018	6.12	2.534	3.699	5.53
40	4.085	7.314	12.61	3.232	5.178	8.25	2.839	4.313	6.60	2.606	3.828	5.70	2.450	3.514	5.13
60	4.001	7.077	11.97	3.150	4.977	7.76	2.758	4.126	6.17	2.525	3.649	5.31	2.368	3.339	4.76
120	3.920	6.851	11.38	3.072	4.786	7.31	2.680	3.949	5.79	2.447	3.480	4.95	2.290	3.174	4.42
∞	3.841	6.635	10.83	2.996	4.605	91	2.605	3.782	5.42	2.372	3.319	4.62	2.214	3.017	4.10

TABLE F.1
(continued)

n_2	$n_1 = 6$			$n_1 = 8$			$n_1 = 12$			$n_1 = 24$			$n_1 = \infty$		
	.05	.01	.001	.05	.01	.001	.05	.01	.001	.05	.01	.001	.05	.01	.001
1	233.99	5,859.0	585,937	238.88	5,981.6	598,144	243.91	6,106.3	610,667	249.05	6,234.6	623,497	254.32	6,366.0	636,619
2	19.330	99.332	999.3	19.371	99.374	999.4	19.413	99.416	999.4	19.454	99.458	999.5	19.496	99.501	999.5
3	8.941	27.911	132.8	8.845	27.489	130.6	8.745	27.052	128.3	8.638	26.598	125.9	8.527	26.125	123.5
4	6.163	15.207	50.53	6.041	14.799	49.00	5.912	14.374	47.41	5.774	13.929	45.77	5.628	13.463	44.05
5	4.950	10.672	28.84	4.818	10.289	27.64	4.678	9.838	26.42	4.527	9.467	25.14	4.365	9.020	23.78
6	4.284	8.466	20.03	4.147	8.102	19.03	4.000	7.718	17.99	3.841	7.313	16.89	3.669	6.880	15.75
7	3.866	7.191	15.52	3.726	6.840	14.63	3.575	6.469	13.71	3.410	6.074	12.73	3.230	5.650	11.69
8	3.581	6.371	12.86	3.438	6.029	12.04	3.284	5.667	11.19	3.115	5.279	10.30	2.928	4.859	9.34
9	3.374	5.802	11.13	3.230	5.467	10.37	3.073	5.111	9.57	2.900	4.729	8.72	2.707	4.311	7.81
10	3.217	5.386	9.92	3.072	5.057	9.20	2.913	4.706	8.45	2.737	4.327	7.64	2.538	3.909	6.76
11	3.095	5.069	9.05	2.948	4.745	8.35	2.788	4.397	7.63	2.609	4.021	6.85	2.405	3.602	6.00
12	2.996	4.821	8.38	2.849	4.499	7.71	2.687	4.155	7.00	2.505	3.780	6.25	2.296	3.361	5.42
13	2.915	4.620	7.86	2.767	4.302	7.21	2.604	3.960	6.52	2.420	3.587	5.78	2.206	3.165	4.97
14	2.848	4.456	7.43	2.699	4.140	6.80	2.534	3.800	6.13	2.349	3.427	5.41	2.131	3.004	4.60
15	2.790	4.318	7.09	2.641	4.004	6.47	2.475	3.666	5.81	2.288	3.294	5.10	2.066	2.868	4.31
16	2.741	4.202	6.81	2.591	3.890	6.19	2.425	3.553	5.55	2.235	3.181	4.85	2.010	2.753	4.06
17	2.699	4.102	6.56	2.548	3.791	5.96	2.381	3.455	5.32	2.190	3.083	4.63	1.960	2.653	3.85
18	2.661	4.015	6.35	2.510	3.705	5.76	2.342	3.371	5.13	2.150	2.999	4.45	1.917	2.566	3.67
19	2.628	3.939	6.18	2.477	3.631	5.59	2.308	3.296	4.97	2.114	2.925	4.29	1.878	2.489	3.52
20	2.599	3.871	6.02	2.447	3.564	5.44	2.278	3.231	4.82	2.083	2.859	4.15	1.843	2.421	3.38
21	2.573	3.812	5.88	2.421	3.506	5.31	2.250	3.173	4.70	2.054	2.801	4.03	1.812	2.360	3.26
22	2.549	3.758	5.76	2.397	3.453	5.19	2.226	3.121	4.58	2.028	2.749	3.92	1.783	2.305	3.15
23	2.528	3.710	5.65	2.375	3.406	5.09	2.204	3.074	4.48	2.005	2.702	3.82	1.757	2.256	3.05
24	2.508	3.667	5.55	2.355	3.363	4.99	2.183	3.032	4.39	1.984	2.659	3.74	1.733	2.211	2.97
25	2.490	3.627	5.46	2.337	3.324	4.91	2.165	2.993	4.31	1.964	2.620	3.66	1.711	2.169	2.89
26	2.474	3.591	5.38	2.321	3.288	4.83	2.148	2.958	4.24	1.946	2.585	3.59	1.691	2.132	2.82
27	2.459	3.558	5.31	2.305	3.256	4.76	2.132	2.926	4.17	1.930	2.552	3.52	1.672	2.096	2.75
28	2.445	3.528	5.24	2.291	3.226	4.69	2.118	2.896	4.11	1.915	2.522	3.46	1.654	2.064	2.70
29	2.432	3.499	5.18	2.278	3.198	4.64	2.104	2.869	4.05	1.901	2.495	3.41	1.638	2.034	2.64
30	2.421	3.474	5.12	2.266	3.173	4.58	2.092	2.843	4.00	1.887	2.469	3.36	1.622	2.006	2.59
40	2.336	3.291	4.73	2.180	2.993	4.21	2.004	2.665	3.64	1.793	2.288	3.01	1.509	1.805	2.23
60	2.254	3.119	4.37	2.097	2.823	3.87	1.917	2.496	3.31	1.700	2.115	2.69	1.389	1.601	1.90
120	2.175	2.956	4.04	2.016	2.663	3.55	1.834	2.336	3.02	1.608	1.950	2.40	1.254	1.380	1.56
∞	2.099	2.802	3.74	1.938	2.511	3.27	1.752	2.185	2.74	1.517	1.791	2.13	1.000	1.000	1.00

Source: Frederick E. Croxton and Dudley J. Cowden, *Practical Business Statistics,* 2d ed. (© 1934, 1948 by Prentice-Hall, Inc., Englewood Cliffs, N.J.), pp. 514 and 515. Reprinted by permission of the publisher.

Values of F at the .05 and .01 points were taken, by permission, from Maxine Merrington and Catherine M. Thompson, "New Tables of Statistical Variables," *Biometrika,* vol. 33, part 1, pp. 80, 81, 84, and 85. Values of F at the .001 point were taken from Table V of R. A. Fisher and F. Yates, *Statistical Tables for Biological, Agricultural and Medical Research,* Oliver and Boyd, Ltd., Edinburgh, 1938, by permission of the authors and publishers. The first reference gives F values to five digits at the .50, .25, .10, .05, .025, .01, and .005 points and for values of n_1 in addition to those shown in the above table. The second reference gives F values to three or more digits at the .20, .05, .01, and .001 points.

FIGURE F.1

Distribution of F for $n_1 = 1$, $n_2 = 5$; and $n_1 = 5$, $n_2 = 4$; horizontal and vertical scales extend to ∞

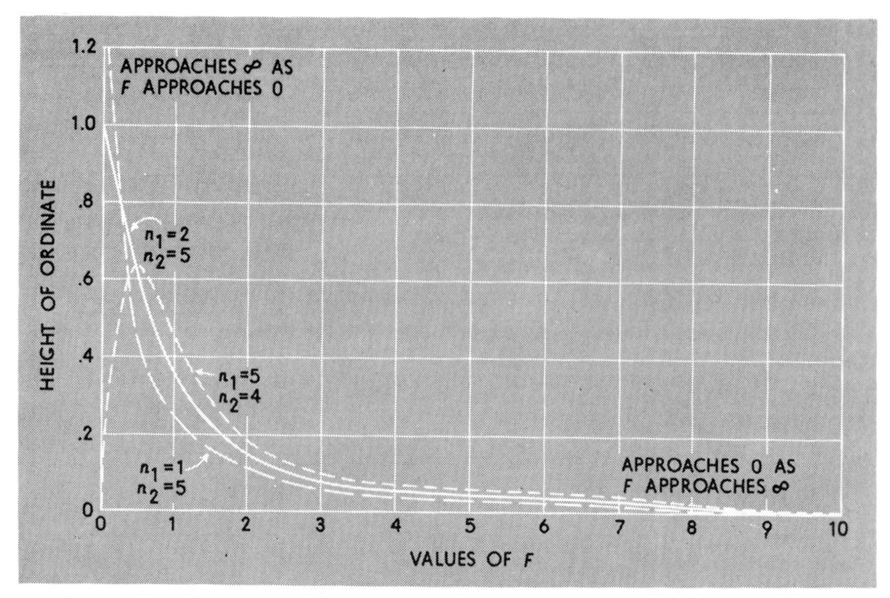

Source: Frederick E. Croxton and Dudley J. Cowden, *Applied General Statistics,* 2d ed. (New York: Prentice-Hall, Inc., 1955), p. 703.

Appendix G

Values of the correlation coefficient for different levels of significance

n	.05	.02	.01
1	.996917	.9995066	.9998766
2	.95000	.98000	.990000
3	.8783	.93433	.95873
4	.8114	.8822	.91720
5	.7545	.8329	.8745
6	.7067	.7887	.8343
7	.6664	.7498	.7977
8	.6319	.7155	.7646
9	.6021	.6851	.7348
10	.5760	.6581	.7079
11	.5529	.6339	.6835
12	.5324	.6120	.6614
13	.5139	.5923	.6411
14	.4973	.5742	.6226
15	.4821	.5577	.6055
16	.4683	.5425	.5897
17	.4555	.5285	.5751
18	.4438	.5155	.5614
19	.4329	.5034	.5487
20	.4227	.4921	.5368
25	.3809	.4451	.4869
30	.3494	.4093	.4487
35	.3246	.3810	.4182
40	.3044	.3578	.3932
45	.2875	.3384	.3721
50	.2732	.3218	.3541
60	.2500	.2948	.3248
70	.2319	.2737	.3017
80	.2172	.2565	.2830
90	.2050	.2422	.2673
100	.1946	.2301	.2540

Note: $n = N - m$

Source: This table is reprinted from Table V-A of R. H. Fisher, *Statistical Methods for Research Workers*, published by Oliver and Boyd, Ltd., Edinburgh, by permission of the author and publishers.

Appendix H

Poisson distribution—individual terms

THE TABLE presents individual Poisson probabilities for the number of occurrences X per unit of measurement for selected values of m, the mean number of occurrences per unit of measurement.

A blank space is left for values less than .0005.

$$f(x) = \frac{e^{-m}m^x}{x!}$$

x	.001	.002	.003	.004	.005	.006	.007	.008	.009	.01	.02	.03	.04	.05	.06	.07	.08	.09	.10	.15	x
0	999	998	997	996	995	994	993	992	991	990	980	970	961	951	942	932	923	914	905	861	0
1	001	002	003	004	005	006	007	008	009	010	020	030	038	048	057	065	074	082	090	129	1
2													001	001	002	002	003	004	005	010	2

x	.20	.25	.30	.40	.50	.60	.70	.80	.90	1.0	1.1	1.2	1.3	1.4	1.5	1.6	1.7	1.8	1.9	2.0	x
0	819	779	741	670	607	549	497	449	407	368	333	301	273	247	223	202	183	165	150	135	0
1	164	195	222	268	303	329	348	359	366	368	366	361	354	345	335	323	311	298	284	271	1
2	016	024	033	054	076	099	122	144	165	184	201	217	230	242	251	258	264	268	270	271	2
3	001	002	003	007	013	020	028	038	049	061	074	087	100	113	126	138	150	161	171	180	3
4				001	002	003	005	008	011	015	020	026	032	039	047	055	063	072	081	090	4
5							001	001	002	003	004	006	008	011	014	018	022	026	031	036	5
6										001	001	001	002	003	004	005	006	008	010	012	6
7														001	001	001	001	002	003	003	7
8																			001	001	8

x	2.1	2.2	2.3	2.4	2.5	2.6	2.7	2.8	2.9	3.0	3.1	3.2	3.3	3.4	3.5	3.6	3.7	3.8	3.9	4.0	x
0	122	111	100	091	082	074	067	061	055	050	045	041	037	033	030	027	025	022	020	018	0
1	257	244	231	218	205	193	181	170	160	149	140	130	122	113	106	098	091	085	079	073	1
2	270	268	265	261	257	251	245	238	231	224	216	209	201	193	185	177	169	162	154	147	2
3	189	197	203	209	214	218	220	222	224	224	224	223	221	219	216	212	209	205	200	195	3
4	099	108	117	125	134	141	149	156	162	168	173	178	182	186	189	191	193	194	195	195	4
5	042	048	054	060	067	074	080	087	094	101	107	114	120	126	132	138	143	148	152	156	5
6	015	017	021	024	028	032	036	041	045	050	056	061	066	072	077	083	088	094	099	104	6
7	004	005	007	008	010	012	014	016	019	022	025	028	031	035	039	042	047	051	055	060	7
8	001	002	002	002	003	004	005	006	007	008	010	011	013	015	017	019	022	024	027	030	8
9				001	001	001	001	002	002	003	003	004	005	006	007	008	009	010	012	013	9
10									001	001	001	001	002	002	002	003	003	004	005	005	10
11														001	001	001	001	001	002	002	11
12																			001	001	12

Poisson distribution—individual terms (*continued*)

$$f(x) = \frac{e^{-m}m^x}{x!}$$

x	4.1	4.2	4.3	4.4	4.5	4.6	4.7	4.8	4.9	5.0	5.1	5.2	5.3	5.4	5.5	5.6	5.7	5.8	5.9	6.0	x
0	017	015	014	012	011	010	009	008	007	007	006	006	005	005	004	004	003	003	003	002	0
1	068	063	058	054	050	046	043	040	036	034	031	029	026	024	022	021	019	018	016	015	1
2	139	132	125	119	112	106	100	095	089	084	079	075	070	066	062	058	054	051	048	045	2
3	190	185	180	174	169	163	157	152	146	140	135	129	124	119	113	108	103	098	094	089	3
4	195	194	193	192	190	188	185	182	179	175	172	168	164	160	156	152	147	143	138	134	4
5	160	163	166	169	171	173	174	175	175	175	175	175	174	173	171	170	168	166	163	161	5
6	109	114	119	124	128	132	136	140	143	146	149	151	154	156	157	158	159	160	160	161	6
7	064	069	073	078	082	087	091	096	100	104	109	113	116	120	123	127	130	133	135	138	7
8	033	036	039	043	046	050	054	058	061	065	069	073	077	081	085	089	092	096	100	103	8
9	015	017	019	021	023	026	028	031	033	036	039	042	045	049	052	055	059	062	065	069	9
10	006	007	008	009	010	012	013	015	016	018	020	022	024	026	029	031	033	036	039	041	10
11	002	003	003	004	004	005	006	006	007	008	009	010	012	013	014	016	017	019	021	023	11
12	001	001	001	001	002	002	002	003	003	003	004	005	005	006	007	007	008	009	010	011	12
13					001	001	001	001	001	001	002	002	002	002	003	003	004	004	005	005	13
14											001	001	001	001	001	001	001	002	002	002	14
15																	001	001	001	001	15

x	6.1	6.2	6.3	6.4	6.5	6.6	6.7	6.8	6.9	7.0	7.1	7.2	7.3	7.4	7.5	8.0	8.5	9.0	9.5	10.0	x
0	002	002	002	002	002	001	001	001	001	001	001	001	001	001	001						0
1	014	013	012	011	010	009	008	008	007	006	006	005	005	005	004	003	002	001	001		1
2	042	039	036	034	032	030	028	026	024	022	021	019	018	017	016	011	007	005	003	002	2
3	085	081	077	073	069	065	062	058	055	052	049	046	044	041	039	029	021	015	011	008	3
4	129	125	121	116	112	108	103	099	095	091	087	084	080	076	073	057	044	034	025	019	4
5	158	155	152	149	145	142	138	135	131	128	124	120	117	113	109	092	075	061	048	038	5
6	160	160	159	159	157	156	155	153	151	149	147	144	142	139	137	122	107	091	076	063	6
7	140	142	144	145	146	147	148	149	149	149	149	149	148	147	146	140	129	117	104	090	7
8	107	110	113	116	119	121	124	126	128	130	132	134	135	136	137	140	138	132	123	113	8
9	072	076	079	082	086	089	092	095	098	101	104	107	110	112	114	124	130	132	130	125	9
10	044	047	050	053	056	059	062	065	068	071	074	077	080	083	086	099	110	119	124	125	10
11	024	026	029	031	033	035	038	040	043	045	048	050	053	056	059	072	085	097	107	114	11
12	012	014	015	016	018	019	021	023	025	026	028	030	032	034	037	048	060	073	084	095	12
13	006	007	007	008	009	010	011	012	013	014	015	017	018	020	021	030	040	050	062	073	13
14	003	003	003	004	004	005	005	006	006	007	008	009	009	010	011	017	024	032	042	052	14
15	001	001	001	002	002	002	002	003	003	003	004	004	005	005	006	009	014	019	027	035	15
16			001	001	001	001	001	001	001	001	002	002	002	002	003	005	007	011	016	022	16
17									001	001	001	001	001	001	001	002	004	006	009	013	17
18																001	002	003	005	007	18
19																	001	001	002	004	19
20																		001	001	002	20
21																				001	21

Source: This table is reproduced from William A. Spurr and Charles P. Bonini, *Statistical Analysis for Business Decisions* (Homewood, Ill.: Richard D. Irwin, Inc., 1973), pp. 696–97, with the permission of the authors and the publishers.

Appendix I: Values of e^{-x}

THIS TABLE lists values of e^{-x} for values of X from 0 to 10. Intermediate values can be calculated by making use of the relationship $e^{-(a+b)} = e^{-a} \cdot e^{-b}$. For example, to find $e^{-1.21}$, use $e^{-1.0} = .368$ and $e^{-.21} = .811$; then $e^{-1.21} = (.386)(.811) = .298$.

X	e^{-x}	X	e^{-x}	X	e^{-x}	X	e^{-x}
.00	1.000	.40	.670	.80	.449	3.00	.04979
.01	.990	.41	.664	.81	.445	3.10	.04505
.02	.980	.42	.657	.82	.440	3.20	.04076
.03	.970	.43	.651	.83	.436	3.30	.03688
.04	.961	.44	.644	.84	.432	3.40	.03337
.05	.951	.45	.638	.85	.427	3.50	.03080
.06	.942	.46	.631	.86	.423	3.60	.02732
.07	.932	.47	.625	.87	.419	3.70	.02472
.08	.923	.48	.619	.88	.415	3.80	.02237
.09	.914	.49	.613	.89	.411	3.90	.02024
.10	.905	.50	.607	.90	.407	4.00	.01832
.11	.896	.51	.600	.91	.403	4.10	.01657
.12	.887	.52	.595	.92	.399	4.20	.01500
.13	.878	.53	.589	.93	.395	4.30	.01357
.14	.869	.54	.583	.94	.391	4.40	.01228
.15	.861	.55	.577	.95	.387	4.50	.01111
.16	.852	.56	.571	.96	.383	4.60	.01005
.17	.844	.57	.566	.97	.379	4.70	.00910
.18	.835	.58	.560	.98	.375	4.80	.00823
.19	.827	.59	.554	.99	.372	4.90	.00745
.20	.819	.60	.549	1.00	.368	5.00	.00674
.21	.811	.61	.543	1.10	.333	5.50	.00409
.22	.803	.62	.538	1.20	.301	6.00	.00248
.23	.795	.63	.533	1.30	.273	6.30	.00150
.24	.787	.64	.527	1.40	.247	7.00	.00091
.25	.779	.65	.522	1.50	.223	7.50	.00055
.26	.771	.66	.517	1.60	.202	8.00	.00034
.27	.763	.67	.512	1.70	.183	8.50	.00020
.28	.756	.68	.507	1.80	.165	9.00	.00012
.29	.748	.69	.502	1.90	.150	10.00	.00005
.30	.741	.70	.497	2.00	.135		
.31	.733	.71	.492	2.10	.122		
.32	.726	.72	.487	2.20	.111		
.33	.719	.73	.482	2.30	.100		
.34	.712	.74	.477	2.40	.091		
.35	.705	.75	.472	2.50	.082		
.36	.698	.76	.468	2.60	.074		
.37	.691	.77	.463	2.70	.067		
.38	.684	.78	.458	2.80	.061		
.39	.677	.79	.454	2.90	.055		

Source: This table is reproduced from William A. Spurr and Charles P. Bonini, *Statistical Analysis for Business Decisions* (Homewood, Ill.: Richard D. Irwin, Inc., 1973), p. 700, with the permission of the authors and the publishers.

Appendix J

Percentage points of the χ^2 distribution

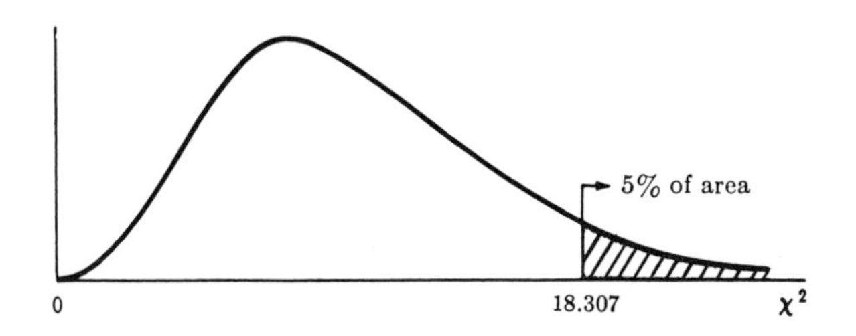

Example

For 10 degrees of freedom:

$P[\chi^2 > 18.307] = .05$

df	.99	.98	.95	.90	.80	.70	.50	.30	.20	.10	.05	.02	.01	.001
1	$.0^3157$	$.0^3628$	.00393	.0158	.0642	.148	.455	1.074	1.642	2.706	3.841	5.412	6.635	10.827
2	.0201	.0404	.103	.211	.446	.713	1.386	2.408	3.219	4.605	5.991	7.824	9.210	13.815
3	.115	.185	.352	.584	1.005	1.424	2.366	3.665	4.642	6.251	7.815	9.837	11.345	16.268
4	.297	.429	.711	1.064	1.649	2.195	3.357	4.878	5.989	7.779	9.488	11.668	13.277	18.465
5	.554	.752	1.145	1.610	2.343	3.000	4.351	6.064	7.289	9.236	11.070	13.388	15.086	20.517
6	.872	1.134	1.635	2.204	3.070	3.828	5.348	7.231	8.558	10.645	12.592	15.033	16.812	22.457
7	1.239	1.564	2.167	2.833	3.822	4.671	6.346	8.383	9.803	12.017	14.067	16.622	18.475	24.322
8	1.646	2.032	2.733	3.490	4.594	5.527	7.344	9.524	11.030	13.362	15.507	18.168	20.090	26.125
9	2.088	2.532	3.325	4.168	5.380	6.393	8.343	10.656	12.242	14.684	16.919	19.679	21.666	27.877
10	2.558	3.059	3.940	4.865	6.179	7.267	9.342	11.781	13.442	15.987	18.307	21.161	23.209	29.588
11	3.053	3.609	4.575	5.578	6.989	8.148	10.341	12.899	14.631	17.275	19.675	22.618	24.725	31.264
12	3.571	4.178	5.226	6.304	7.807	9.034	11.340	14.011	15.812	18.549	21.026	24.054	26.217	32.909
13	4.107	4.765	5.892	7.042	8.634	9.926	12.340	15.119	16.985	19.812	22.362	25.472	27.688	34.528
14	4.660	5.368	6.571	7.790	9.467	10.821	13.339	16.222	18.151	21.064	23.685	26.873	29.141	36.123
15	5.229	5.985	7.261	8.547	10.307	11.721	14.339	17.322	19.311	22.307	24.996	28.259	30.578	37.697
16	5.812	6.614	7.962	9.312	11.152	12.624	15.338	18.418	20.465	23.542	26.296	29.633	32.000	39.252
17	6.408	7.255	8.672	10.085	12.002	13.531	16.338	19.511	21.615	24.769	27.587	30.995	33.409	40.790
18	7.015	7.906	9.390	10.865	12.857	14.440	17.338	20.601	22.760	25.989	28.869	32.346	34.805	42.312
19	7.633	8.567	10.117	11.651	13.716	18.352	18.338	21.689	23.900	27.204	30.144	33.687	36.191	43.820
20	8.260	9.237	10.851	12.443	14.578	16.266	19.337	22.775	25.038	28.412	31.410	35.020	37.566	45.315
21	8.897	9.915	11.591	13.240	15.445	17.182	20.337	23.858	26.171	29.615	32.671	36.343	38.932	46.797
22	9.542	10.600	12.338	14.041	16.314	18.101	21.337	24.939	27.301	30.813	33.924	37.659	40.289	48.268
23	10.196	11.293	13.091	14.848	17.187	19.021	22.337	26.018	28.429	32.007	35.172	38.968	41.638	49.728
24	10.856	11.992	13.848	15.659	18.062	19.943	23.337	27.096	29.553	33.196	36.415	40.270	42.980	51.179
25	11.524	12.697	14.611	16.473	18.940	20.867	24.337	28.172	30.675	34.382	37.652	41.566	44.314	52.620
26	12.198	13.409	15.379	17.292	19.820	21.792	25.336	29.246	31.795	35.563	38.885	42.856	45.642	54.052
27	12.879	14.125	16.151	18.114	20.703	22.719	26.336	30.319	32.912	36.741	40.113	44.140	46.963	55.476
28	13.565	14.847	16.928	18.939	21.588	23.647	27.336	31.391	34.027	37.916	41.337	45.419	48.278	56.893
29	14.256	15.574	17.708	19.768	22.475	24.577	28.336	32.461	35.139	39.087	42.557	46.693	49.588	58.302
30	14.953	16.306	18.493	20.599	23.364	25.508	29.336	33.530	36.250	40.256	43.773	47.962	50.892	59.703

For larger values of n, the expression $\sqrt{2\chi^2} - \sqrt{2n-1}$ may be used as a normal deviate with unit variance remembering that the probability for χ^2 corresponds with that of a single tail of the normal curve.

Source: This table is abridged from Table III of Fisher & Yates: *Statistical Methods for Research Workers,* published by Oliver & Boyd Ltd., Edinburgh, and by permission of the authors and publishers.

Appendix K

Table of values of r for values of z

z	.00	.01	.02	.03	.04	.05	.06	.07	.08	.09
.0	.0000	.0100	.0200	.0300	.0400	.0500	.0599	.0699	.0798	.0898
.1	.0997	.1096	.1194	.1293	.1391	.1489	.1587	.1684	.1781	.1878
.2	.1974	.2070	.2165	.2260	.2355	.2449	.2543	.2636	.2729	.2821
.3	.2913	.3004	.3095	.3185	.3275	.3364	.3452	.3540	.3627	.3714
.4	.3800	.3885	.3969	.4053	.4136	.4219	.4301	.4382	.4462	.4542
.5	.4621	.4700	.4777	.4854	.4930	.5005	.5080	.5154	.5227	.5299
.6	.5370	.5441	.5511	.5581	.5649	.5717	.5784	.5850	.5915	.5980
.7	.6044	.6107	.6169	.6231	.6291	.6352	.6411	.6469	.6527	.6584
.8	.6640	.6696	.6751	.6805	.6858	.6911	.6963	.7014	.7064	.7114
.9	.7163	.7211	.7259	.7306	.7352	.7398	.7443	.7487	.7531	.7574
1.0	.7616	.7658	.7699	.7739	.7779	.7818	.7857	.7895	.7932	.7969
1.1	.8005	.8041	.8076	.8110	.8144	.8178	.8210	.8243	.8275	.8306
1.2	.8337	.8367	.8397	.8426	.8455	.8483	.8511	.8538	.8565	.8591
1.3	.8617	.8643	.8668	.8693	.8717	.8741	.8764	.8787	.8810	.8832
1.4	.8854	.8875	.8896	.8917	.8937	.8957	.8977	.8996	.9015	.9033
1.5	.9052	.9069	.9087	.9104	.9121	.9138	.9154	.9170	.9186	.9202
1.6	.9217	.9232	.9246	.9261	.9275	.9289	.9302	.9316	.9329	.9342
1.7	.9354	.9367	.9379	.9391	.9402	.9414	.9425	.9436	.9447	.9458
1.8	.9468	.9478	.9488	.9498	.9508	.9518	.9527	.9536	.9545	.9554
1.9	.9562	.9571	.9579	.9587	.9595	.9603	.9611	.9619	.9626	.9633
2.0	.9640	.9647	.9654	.9661	.9668	.9674	.9680	.9687	.9693	.9699
2.1	.9705	.9710	.9716	.9722	.9727	.9732	.9738	.9743	.9748	.9753
2.2	.9757	.9762	.9767	.9771	.9776	.9780	.9785	.9789	.9793	.9797
2.3	.9801	.9805	.9809	.9812	.9816	.9820	.9823	.9827	.9830	.9834
2.4	.9837	.9840	.9843	.9846	.9849	.9852	.9855	.9858	.9861	.9863
2.5	.9866	.9869	.9871	.9874	.9876	.9879	.9881	.9884	.9886	.9888
2.6	.9890	.9892	.9895	.9897	.9899	.9901	.9903	.9905	.9906	.9908
2.7	.9910	.9912	.9914	.9915	.9917	.9919	.9920	.9922	.9923	.9925
2.8	.9926	.9928	.9929	.9931	.9932	.9933	.9935	.9936	.9937	.9938
2.9	.9940	.9941	.9942	.9943	.9944	.9945	.9946	.9947	.9949	.9950
3.0	.9951									
4.0	.9993									
5.0	.9999									

Source: This table is reprinted from Table V-B of R. H. Fisher, *Statistical Methods for Research Workers* (11th ed.), published by Oliver and Boyd, Ltd., Edinburgh, 1950, by permission of the author and publishers.

Appendix L

TABLE L.1
Durbin-Watson statistic (d). Significance points of d_L and d_U: 5%

n	$k' = 1$		$k' = 2$		$k' = 3$		$k' = 4$		$k' = 5$	
	d_L	d_U	d_L	d_U	d_L	d_U	d_L	d_U	d_L	d_U
15	1.08	1.36	0.95	1.54	0.82	1.75	0.69	1.97	0.56	2.21
16	1.10	1.37	0.98	1.54	0.86	1.73	0.74	1.93	0.62	2.15
17	1.13	1.38	1.02	1.54	0.90	1.71	0.78	1.90	0.67	2.10
18	1.16	1.39	1.05	1.53	0.93	1.69	0.82	1.87	0.71	2.06
19	1.18	1.40	1.08	1.53	0.97	1.68	0.86	1.85	0.75	2.02
20	1.20	1.41	1.10	1.54	1.00	1.68	0.90	1.83	0.79	1.99
21	1.22	1.42	1.13	1.54	1.03	1.67	0.93	1.81	0.83	1.96
22	1.24	1.43	1.15	1.54	1.05	1.66	0.96	1.80	0.86	1.94
23	1.26	1.44	1.17	1.54	1.08	1.66	0.99	1.79	0.90	1.92
24	1.27	1.45	1.19	1.55	1.10	1.66	1.01	1.78	0.93	1.90
25	1.29	1.45	1.21	1.55	1.12	1.66	1.04	1.77	0.95	1.89
26	1.30	1.46	1.22	1.55	1.14	1.65	1.06	1.76	0.98	1.88
27	1.32	1.47	1.24	1.56	1.16	1.65	1.08	1.76	1.01	1.86
28	1.33	1.48	1.26	1.56	1.18	1.65	1.10	1.75	1.03	1.85
29	1.34	1.48	1.27	1.56	1.20	1.65	1.12	1.74	1.05	1.84
30	1.35	1.49	1.28	1.57	1.21	1.65	1.14	1.74	1.07	1.83
31	1.36	1.50	1.30	1.57	1.23	1.65	1.16	1.74	1.09	1.83
32	1.37	1.50	1.31	1.57	1.24	1.65	1.18	1.73	1.11	1.82
33	1.38	1.51	1.32	1.58	1.26	1.65	1.19	1.73	1.13	1.81
34	1.39	1.51	1.33	1.58	1.27	1.65	1.21	1.73	1.15	1.81
35	1.40	1.52	1.34	1.58	1.28	1.65	1.22	1.73	1.16	1.80
36	1.41	1.52	1.35	1.59	1.29	1.65	1.24	1.73	1.18	1.80
37	1.42	1.53	1.36	1.59	1.31	1.66	1.25	1.72	1.19	1.80
38	1.43	1.54	1.37	1.59	1.32	1.66	1.26	1.72	1.21	1.79
39	1.43	1.54	1.38	1.60	1.33	1.66	1.27	1.72	1.22	1.79
40	1.44	1.54	1.39	1.60	1.34	1.66	1.29	1.72	1.23	1.79
45	1.48	1.57	1.43	1.62	1.38	1.67	1.34	1.72	1.29	1.78
50	1.50	1.59	1.46	1.63	1.42	1.67	1.38	1.72	1.34	1.77
55	1.53	1.60	1.49	1.64	1.45	1.68	1.41	1.72	1.38	1.77
60	1.55	1.62	1.51	1.65	1.48	1.69	1.44	1.73	1.41	1.77
65	1.57	1.63	1.54	1.66	1.50	1.70	1.47	1.73	1.44	1.77
70	1.58	1.64	1.55	1.67	1.52	1.70	1.49	1.74	1.46	1.77
75	1.60	1.65	1.57	1.68	1.54	1.71	1.51	1.74	1.49	1.77
80	1.61	1.66	1.59	1.69	1.56	1.72	1.53	1.74	1.51	1.77
85	1.62	1.67	1.60	1.70	1.57	1.72	1.55	1.75	1.52	1.77
90	1.63	1.68	1.61	1.70	1.59	1.73	1.57	1.75	1.54	1.78
95	1.64	1.69	1.62	1.71	1.60	1.73	1.58	1.75	1.56	1.78
100	1.65	1.69	1.63	1.72	1.61	1.74	1.59	1.76	1.57	1.78

n = number of observations.
k' = number of independent variables.
Source: This table is reproduced from *Biometrika*, vol. 41, p. 173, 1951, with the permission of the Trustees.

TABLE L.2
Durbin-Watson statistic (d). Significance points of d_L and d_U: 1%

n	$k' = 1$		$k' = 2$		$k' = 3$		$k' = 4$		$k' = 5$	
	d_L	d_U	d_L	d_U	d_L	d_U	d_L	d_U	d_L	d_U
15	0.81	1.07	0.70	1.25	0.59	1.46	0.49	1.70	0.39	1.96
16	0.84	1.09	0.74	1.25	0.63	1.44	0.53	1.66	0.44	1.90
17	0.87	1.10	0.77	1.25	0.67	1.43	0.57	1.63	0.48	1.85
18	0.90	1.12	0.80	1.26	0.71	1.42	0.61	1.60	0.52	1.80
19	0.93	1.13	0.83	1.26	0.74	1.41	0.65	1.58	0.56	1.77
20	0.95	1.15	0.86	1.27	0.77	1.41	0.68	1.57	0.60	1.74
21	0.97	1.16	0.89	1.27	0.80	1.41	0.72	1.55	0.63	1.71
22	1.00	1.17	0.91	1.28	0.83	1.40	0.75	1.54	0.66	1.69
23	1.02	1.19	0.94	1.29	0.86	1.40	0.77	1.53	0.70	1.67
24	1.04	1.20	0.96	1.30	0.88	1.41	0.80	1.53	0.72	1.66
25	1.05	1.21	0.98	1.30	0.90	1.41	0.83	1.52	0.75	1.65
26	1.07	1.22	1.00	1.31	0.93	1.41	0.85	1.52	0.78	1.64
27	1.09	1.23	1.02	1.32	0.95	1.41	0.88	1.51	0.81	1.63
28	1.10	1.24	1.04	1.32	0.97	1.41	0.90	1.51	0.83	1.62
29	1.12	1.25	1.05	1.33	0.99	1.42	0.92	1.51	0.85	1.61
30	1.13	1.26	1.07	1.34	1.01	1.42	0.94	1.51	0.88	1.61
31	1.15	1.27	1.08	1.34	1.02	1.42	0.96	1.51	0.90	1.60
32	1.16	1.28	1.10	1.35	1.04	1.43	0.98	1.51	0.92	1.60
33	1.17	1.29	1.11	1.36	1.05	1.43	1.00	1.51	0.94	1.59
34	1.18	1.30	1.13	1.36	1.07	1.43	1.01	1.51	0.95	1.59
35	1.19	1.31	1.14	1.37	1.08	1.44	1.03	1.51	0.97	1.59
36	1.21	1.32	1.15	1.38	1.10	1.44	1.04	1.51	0.99	1.59
37	1.22	1.32	1.16	1.38	1.11	1.45	1.06	1.51	1.00	1.59
38	1.23	1.33	1.18	1.39	1.12	1.45	1.07	1.52	1.02	1.58
39	1.24	1.34	1.19	1.39	1.14	1.45	1.09	1.52	1.03	1.58
40	1.25	1.34	1.20	1.40	1.15	1.46	1.10	1.52	1.05	1.58
45	1.29	1.38	1.24	1.42	1.20	1.48	1.16	1.53	1.11	1.58
50	1.32	1.40	1.28	1.45	1.24	1.49	1.20	1.54	1.16	1.59
55	1.36	1.43	1.32	1.47	1.28	1.51	1.25	1.55	1.21	1.59
60	1.38	1.45	1.35	1.48	1.32	1.52	1.28	1.56	1.25	1.60
65	1.41	1.47	1.38	1.50	1.35	1.53	1.31	1.57	1.28	1.61
70	1.43	1.49	1.40	1.52	1.37	1.55	1.34	1.58	1.31	1.61
75	1.45	1.50	1.42	1.53	1.39	1.56	1.37	1.59	1.34	1.62
80	1.47	1.52	1.44	1.54	1.42	1.57	1.39	1.60	1.36	1.62
85	1.48	1.53	1.46	1.55	1.43	1.58	1.41	1.60	1.39	1.63
90	1.50	1.54	1.47	1.56	1.45	1.59	1.43	1.61	1.41	1.64
95	1.51	1.55	1.49	1.57	1.47	1.60	1.45	1.62	1.42	1.64
100	1.52	1.56	1.50	1.58	1.48	1.60	1.46	1.63	1.44	1.65

n = number of observations.
k' = number of independent variables.
Source: This table is reproduced from *Biometrika,* vol. 41, p. 175, 1951, with the permission of the Trustees.

Appendix M

Table of critical values of D

Table of critical values of D in the Kolmogorov-Smirnov one-sample test

Sample Size (N)	Level of Significance for D = Maximum $\lvert F_0(X) - S_N(X) \rvert$				
	.20	.15	.10	.05	.01
1	.900	.925	.950	.975	.995
2	.684	.726	.776	.842	.929
3	.565	.597	.642	.708	.828
4	.494	.525	.564	.624	.733
5	.446	.474	.510	.565	.669
6	.410	.436	.470	.521	.618
7	.381	.405	.438	.486	.577
8	.358	.381	.411	.457	.543
9	.339	.360	.388	.432	.514
10	.322	.342	.368	.410	.490
11	.307	.326	.352	.391	.468
12	.295	.313	.338	.375	.450
13	.284	.302	.325	.361	.433
14	.274	.292	.314	.349	.418
15	.266	.283	.304	.338	.404
16	.258	.274	.295	.328	.392
17	.250	.266	.286	.318	.381
18	.244	.259	.278	.309	.371
19	.237	.252	.272	.301	.363
20	.231	.246	.264	.294	.356
25	.21	.22	.24	.27	.32
30	.19	.20	.22	.24	.29
35	.18	.19	.21	.23	.27
Over 35	$\frac{1.07}{\sqrt{N}}$	$\frac{1.14}{\sqrt{N}}$	$\frac{1.22}{\sqrt{N}}$	$\frac{1.36}{\sqrt{N}}$	$\frac{1.63}{\sqrt{N}}$

Source: Adapted from Massey, F. J., Jr. 1951. The Kolmogorov-Smirnov test for goodness of fit. *J. Amer. Statist. Ass.*, 46, 70, with the kind permission of the author and publisher.

Indexes

Author index

E

F

G

H–I

J

K

Subject index

D

T

U–V

This book has been set in 10 and 9 point Caledonia, leaded 2 points. Part numbers are 24 pt. (large) Helvetica and part titles are 24 pt. (small) Helvetica. Chapter numbers are 30 pt. Helvetica Medium and chapter titles are 18 pt. Helvetica. The size of the type page is 27 x 45½ picas.